Lecture Notes in Computer Science 9840

More information about this series at http://www.springer.com/series/7407

Srečko Brlek · Christophe Reutenauer (Eds.)

Developments in Language Theory

20th International Conference, DLT 2016
Montréal, Canada, July 25–28, 2016
Proceedings

Springer

Editors
Srečko Brlek
Université du Québec à Montréal
Montreal, QC
Canada

Christophe Reutenauer
Département de mathématiques
Université du Québec à Montréal
Montreal, QC
Canada

ISSN 0302-9743 ISSN 1611-3349 (electronic)
Lecture Notes in Computer Science
ISBN 978-3-662-53131-0 ISBN 978-3-662-53132-7 (eBook)
DOI 10.1007/978-3-662-53132-7

Library of Congress Control Number: 2016946010

LNCS Sublibrary: SL1 – Theoretical Computer Science and General Issues

Printed on acid-free paper

This Springer imprint is published by Springer Nature
The registered company is Springer-Verlag GmbH Berlin Heidelberg

Preface

The DLT 2016 Conference was organized by the Laboratoire de Combinatoire et d'Informatique mathématique (LaCIM) during July 25–28, 2016. It was the 20th edition of a series initiated in 1993 by G. Rozenberg and A. Salomaa in Turku (Finland). These conferences took place every odd year in the first editions: Magdeburg, Germany (1995), Thessaloniki, Greece (1997), Aachen, Germany (1999), and Vienna, Austria (2001). Since then, the event was held in Europe on every odd year and outside Europe on every even year. The locations of DLT conferences since 2002 have been: Kyoto, Japan (2002), Szeged, Hungary (2003), Auckland, New Zealand (2004), Palermo, Italy (2005), Santa Barbara, California, USA (2006), Turku, Finland (2007), Kyoto, Japan (2008), Stuttgart, Germany (2009), London, Ontario, Canada (2010), Milan, Italy (2011), Taipei, Taiwan (2012), Marne-la-Vallée, France (2013), Ekaterinburg, Russia (2014), Liverpool (2015).

This series of International Conferences on Developments in Language Theory provides a forum for presenting current developments in formal languages and automata. Its scope is very general and includes, among others, the following topics and areas: combinatorial and algebraic properties of words and languages; grammars, acceptors and transducers for strings, trees, graphs, arrays; algebraic theories for automata and languages; codes; efficient text algorithms; symbolic dynamics; decision problems; relationships to complexity theory and logic; picture description and analysis; polyominoes and bidimensional patterns; cryptography; concurrency; cellular automata; bio-inspired computing; quantum computing.

This volume of *Lecture Notes in Computer Science* contains the papers that were presented at DLT 2016. There were 48 submissions and each of them was reviewed by at least three reviewers. The selection process was undertaken by the Program Committee with the help of generous reviewers who accepted to participate in the selection of 32 papers within a tight schedule. The present volume also includes the abstracts of the lectures given by four invited speakers

- Valérie Berthé: "Tree Sets: From Bifix Codes to Algebraic Word Combinatorics"
- Emilie Charlier: "Permutations and Shifts"
- Cédric Chauve: "Counting, Generating, and Sampling Tree Alignments"
- Janusz A. (John) Brzozowski: "Towards a Theory of Complexity for Regular Languages"

We warmly thank Valérie, Émilie, Cédric, and Janusz for delivering sound lectures intended for a large audience. We take this opportunity to thank all authors for their submissions and the anonymous reviewers who provided numerous and constructive reviews that led to the selection of high-standard contributions.

Special thanks are due to Alfred Hofmann and the *Lecture Notes in Computer Science* team at Springer for having granted us the opportunity to publish this special issue devoted to DLT 2016 and for their help during the final stages.

The organization of DLT 2016 benefited from the support of the Centre de Recherches Mathématiques (CRM) and the Canadian Research Chair in Algebra, Combinatorics and Computer Science. The reviewing process was facilitated by the EasyChair conference system created by Andrei Voronkov.

Finally, we were fortunate to have a number of collaborators who contributed to the success of the conference: the secretary Johanne Patoine, our postdoctoral fellows Mathieu Guay-Paquet and Nathan Williams, our students Mélodie Lapointe, Nadia Lafrenière, and Hugo Tremblay. Our warmest thanks for their invaluable assistance and contribution in the organization of the event.

June 2016

Srečko Brlek
Christophe Reutenauer

Organization

Steering Committee

Marie-Pierre Béal	Université Paris-Est-Marne-la-Vallée, France
Cristian S. Calude	University of Auckland, New Zealand
Volker Diekert	University of Stuttgart, Germany
Juraj Hromkovic	ETH Zürich, Switzerland
Oscar H. Ibarra	UCSB, Santa Barbara, USA
Masami Ito	Kyoto Sangyo University, Japan
Natasha Jonoska	University of South Florida, USA
Juhani Karhumäki (Chair)	Turku University, Finland
Martin Kutrib	University of Giessen, Germany
Michel Rigo	University of Liege, Belgium
Antonio Restivo	University of Palermo, Italy
Grzegorz Rozenberg	Leiden Institute of Advanced Computer Science, The Netherlands

Program Committee

Srečko Brlek (Chair)	Université du Québec à Montréal, Canada
Christophe Reutenauer (Co-chair)	Université du Québec à Montréal, Canada
Olivier Carton	Université Paris-Diderot, France
Manfred Droste	University of Leipzig, Germany
Vesa Halava	Turku University, Finland
Lila Kari	University of Western Ontario, Canada
Gregory Kucherov	Université Paris-Est, France
Edita Pelantová	Czech Technical University, Czech Republic
Jean-Éric Pin	Université Paris-Diderot, France
Igor Potapov	University of Liverpool, UK
Daniel Reidenbach	Loughborough University, UK
Michel Rigo	Université de Liège, Belgique
Marinella Sciortino	Università di Palermo, Italy
Jeffrey Shallit	University of Waterloo, Canada
Mikhail Volkov	Ural Federal University, Russia

Organizing Committee

Alexandre Blondin Massé (Chair)	LaCIM, Canada
Srečko Brlek (Co-chair)	LaCIM, Canada

Mathieu Guay-Paquet	LaCIM, Canada
Nadia Lafrenière	LaCIM, Canada
Mélodie Lapointe	LaCIM, Canada
Christophe Reutenauer	LaCIM, Canada
Hugo Tremblay	LaCIM, Canada
Nathan Williams	LaCIM, Canada

Additional Reviewers

A.V. Sreejith
Aleksi Saarela
Alessandra Cherubini
Alexander Meduna
Alexander Okhotin
Andreas Maletti
Antoine Durand-Gasselin
Antoine Meyer
Antonio Restivo
Arnaud Carayol
Arturo Carpi
Benjamin Monmege
Bernard Boigelot
Christof Löding
Christophe Reutenauer
Damien Jamet
Damián López
Daniel Reidenbach
Dominik D. Freydenberger
Edita Pelantova
Elena Pribavkina
Florin Manea
Frantisek Mraz
Gabriele Fici
Giovanni Pighizzini
Gregory Kucherov
Igor Potapov
Jacques Duparc
Jarkko Kari
Jarkko Peltomäki
Jean-Éric Pin
Joel Day
Julien Cassaigne
Karel Klouda
Laure Daviaud
Luca Breveglieri
Luigi Santocanale
Manfred Droste
Manfred Kufleitner
Marc Zeitoun
Marinella Sciortino
Markus Holzer
Markus Lohrey
Markus Whiteland
Mathieu Guay-Paquet
Maxime Crochemore
Michael Rao
Michal Kunc
Michel Rigo
Mika Hirvensalo
Narad Rampersad
Nathan Williams
Nils Jansen
Pascal Weil
Paul Bell
Paul Gastin
Pavel Semukhin
Petr Jancar
Philippe Schnoebelen
Pierre McKenzie
Reino Niskanen
Robert Mercas
Roman Kolpakov
Sabrina Mantaci
Sang-Ki Ko
Sebastian Maneth
Sergey Verlan
Srečko Brlek
Stepan Holub
Sylvain Lombardy
Tero Harju
Thomas Colcombet
Thomas Place
Valérie Berthé
Vesa Halava
Violetta Lonati
Vladimir Gusev
Volker Diekert
Vít Jelínek
Wojciech Plandowski
Zoltan Fülöp
Štepán Starosta

Abstracts of Invited Talks

Tree Sets: From Bifix Codes to Algebraic Word Combinatorics

Valérie Berthé

Université Paris-Diderot, Paris, France

Tree sets are languages defined with regard to a tree property: they are sets of factors of a family of infinite words that are defined in terms of the possible left and right extensions of their factors, with their extension graphs being trees. This class of words with linear factor complexity includes classical families such as Sturmian words, interval exchanges or else Arnoux-Rauzy words. We discuss here their combinatorial, ergodic and algebraic properties. This includes algebraic properties of their return words, and of maximal bifix codes defined with respect to their languages. This lecture is based on joint work with C. De Felice, V. Delecroix, F. Dolce, J. Leroy, D. Perrin, C. Reutenauer, G. Rindone.

Towards a Theory of Complexity for Regular Languages

Janusz A. (John) Brzozowski

University of Waterloo, Waterloo, Canada

The state complexity of a regular language is the number of states in a complete minimal deterministic finite automaton (DFA) recognizing the language. The state complexity of an operation on regular languages is the maximal state complexity of the result of the operation as a function of the state complexities of the operands. The state complexity of an operation gives a worst-case lower bound on the time and space complexity of the operation, and has been studied extensively for that reason. The first results on the state complexity of union, concatenation, Kleene star and four other less often used operations were stated without proof by Maslov in 1970, but this paper was unknown in the West for many years. In 1994, Yu, Zhuang and K. Salomaa studied the complexity of basic operations (union, intersection, concatenation, star and reversal) and provided complete proofs. Since then, many authors obtained numerous results for various subclasses of the class of regular languages, and for various operations. Moreover, other measures of complexity, including the size of the syntactic semi-group of a language, have been added. In this talk I will summarize the results obtained in the past few years in the area of complexity of regular languages and finite automata.

Permutations and Shifts

Émilie Charlier

Université de Liège, Liège, Belgique

The entropy of a symbolic dynamical system is usually defined in terms of the growth rate of the number of distinct allowed factors of length n. Bandt, Keller and Pompe showed that, for piecewise monotone interval maps, the entropy is also given by the number of permutations defined by consecutive elements in the trajectory of a point. This result was the starting point of several works of Elizalde where he investigates permutations in shift systems, notably in full shifts and in beta-shifts. The goal of this talk is to survey Elizalde's results. I will end by mentioning the case of negative beta-shifts, which has been simultaneously studied by Elizalde and Moore on the one hand, and by Steiner and myself on the other hand.

A full version is available at http://dlt2016.lacim.uqam.ca/en/files/charlier.pdf.

Counting, Generating and Sampling Tree Alignments

Cédric Chauve

Simon Fraser University, Burnaby, Canada

Pairwise alignment of ordered rooted trees is a natural extension of the classical pairwise sequence alignment, with applications in several fields, such as RNA secondary structure comparison for example. Motivated by this application, and the need to explore the space of possibly sub-optimal alignments, we introduce the notion of unambiguous tree alignment. We first take an enumerative combinatorics point of view and propose a decomposition scheme for unambiguous tree alignments, under the form of a context-free grammar, that leads to precise asymptotic enumerative results, by mean of basic analytic combinatorics. We then shift our focus to algorithmic questions, and show our grammar can be refined into a dynamic programming algorithm for sampling tree alignments under the Gibbs-Boltzmann probability distribution. We also provide some surprising average case complexity results on the tree alignment problem. This work, in collaboration with Yann Ponty and Julien Courtiel, illustrates the potential of considering algorithmic questions from the point of view of enumerating the solution space.

Contents

Context-Free Ambiguity Detection Using Multi-stack Pushdown Automata

H.J.S. Basten[1,2](✉)

[1] Basten Science & Software LLP, Zevenhuizen, The Netherlands
basten@bsns.nl

[2] Centrum Wiskunde & Informatica, Amsterdam, The Netherlands

Abstract. We propose a method for detecting ambiguity in context-free grammars using multi-stack pushdown automata. Since the ambiguity problem is undecidable in general, we use restricted MPDAs that have a limited configuration space. The analysis might thus not be complete, but it is able to detect both ambiguity and unambiguity. Our method is general in the type of automata used. We discuss the suitability of existing MPDAs in our setting and present a new class called bounded-balance MPDAs. These MPDAs allow for infinitely deep nesting/nesting intersection, as long as the nesting depth differences within each scope stay within the balance bound. We compare our contributions to various related MPDAs and ambiguity detection methods.

1 Introduction

Context-free ambiguity detection and related problems like intersection emptiness and inclusion are important in various fields like programming language development [5], program verification [13], model checking and bioinformatics [7]. For instance, context-free grammars are very suitable for specifying formal languages because they allow the definition of regular as well as nested language constructs. However, they have the often undesirable property that they can be ambiguous. Their combinatorial complexity makes ambiguities very hard to spot, which makes automated ambiguity detection essential.

Unfortunately, deciding the ambiguity of a grammar is undecidable in the general case. Still, various ambiguity detection methods exist that aim at being either sound or complete. They limit the possibly infinite search space to either a finite subset [3,6,9,12,13,22,25] or an infinite overapproximation that is checkable in finite time [4,7,21]. For practical purposes however, it is desirable for a method to be able to answer both 'ambiguous' and 'unambiguous'. In this paper we describe a novel way to search an infinite subset in finite time, without approximation. This allows us to detect both ambiguity and unambiguity.

We propose a framework for ambiguity detection using restricted multi-stack pushdown automata. These types of automata are often used in model checking [1,8,10,16,17,20] because they can represent concurrent recursive processes. In general they are Turing complete, but with certain restrictions their configuration space can be limited and searched in finite time. Our framework is general

S. Brlek and C. Reutenauer (Eds.): DLT 2016, LNCS 9840, pp. 1–12, 2016.
DOI: 10.1007/978-3-662-53132-7_1

in the type of MPDA used, allowing the reuse of results from the model checking literature.

In addition, we propose a new class of multi-pushdown automata called bounded-balance multi-stack pushdown automata. The balance of a scope—the part of a run between the matching push and pop of a symbol—is the number of symbols pushed but not popped on other stacks during this scope. In the context of language intersection, limiting the balances in a run has several advantages over limiting the number of contexts or phases. First, it enables the possibility to detect the unambiguity of a grammar. Second, it allows for full intersection of the regular structures within a grammar with other regular or nesting structures. Third, nesting-only intersection can reach unbounded depth, as long as the nesting depths within each scope do not differ more than the balance bound.

Outline. This paper is structured as follows. The next section starts by introducing some basic definitions and notational conventions. In Sect. 3 we propose our ambiguity detection method and discuss the use of different automata types. Section 4 presents bounded-balance multi-stack pushdown automata. In Sect. 5 we compare our ambiguity detection method and MPDA type to other approaches and MPDAs. Section 6 concludes.

2 Preliminaries and Notational Conventions

Throughout this paper we use the following definitions and notations.

2.1 Context-Free Grammars

A context-free grammar G is a 4-tuple (N, T, P, S) consisting of N, a finite set of *nonterminals*, T, a finite set of *terminals* (the alphabet), P, a finite subset of $N \times (N \cup T)^*$, called the *production rules*, and $S \in N$, the *start symbol.* The character ε represents the empty string. We use V to denote the set $N \cup T$ and T^ε for $T \cup \{\varepsilon\}$. The following characters are used to represent different symbols and strings: $a, b, \ldots$ are terminals, $A, B, \ldots$ are nonterminals, $\alpha, \beta, \ldots$ are strings in V^*, $u, v, \ldots$ are strings in T^*. A production (A, α) in P is written as $A \rightarrow \alpha$. We use the function $\mathsf{pid} : P \rightarrow \mathbb{N}$ to relate each production to a unique integer. Given the string $\alpha B \gamma$ and a production rule $B \rightarrow \beta$ from P, we can write $\alpha B \gamma \Longrightarrow \alpha\beta\gamma$—read $\alpha B \gamma$ directly derives $\alpha\beta\gamma$. The *language* of a grammar G is $\mathcal{L}(G) = \{u \mid S \Longrightarrow^+ u\}$. A nonterminal A is said to be self-embedding or *nesting* iff $A \Longrightarrow^+ uAv$, otherwise its language is regular.

The *parse tree* of a sentential form describes how it is derived from S, but disregards the order of the derivation steps. To represent parse trees we use bracketed strings, which are described by bracketed grammars [11]. From a grammar $G = (N, T, P, S)$ a *bracketed grammar* G_b can be constructed by adding unique terminals to the beginning and end of every production rule. The bracketed grammar G_b is defined as the 4-tuple (N, T_b, P_b, S), where $T_b = T \cup T_\langle \cup T_\rangle$,

$T_\langle = \{\langle_i \mid \exists p \in P : i = \mathsf{pid}(p)\}$, $T_\rangle = \{\rangle_i \mid \exists p \in P : i = \mathsf{pid}(p)\}$, and $P_b = \{A \to \langle_i\alpha\rangle_i \mid A \to \alpha \in P, i = \mathsf{pid}(A \to \alpha)\}$. V_b is defined as $T_b \cup N$. The homomorphism yield from V_b^* to V^* maps each string in T_b^* to T^*. It is defined by $\mathsf{yield}(\langle_i) = \varepsilon$, $\mathsf{yield}(\rangle_i) = \varepsilon$, and $\mathsf{yield}(a) = a$. $\mathcal{L}(G_b)$ describes exactly all parse trees of all strings in $\mathcal{L}(G)$. The set of ambiguous strings of G is $\mathcal{A}(G) = \{\mathsf{yield}(u) \mid u, v \in \mathcal{L}(G_b),\ u \neq v,\ \mathsf{yield}(u) = \mathsf{yield}(v)\}$. A grammar G is ambiguous iff $\mathcal{A}(G)$ is non-empty.

2.2 Pushdown Automata

A *pushdown automaton*, or PDA, M is a 6-tuple $(Q, T^\varepsilon, \Gamma, \Delta, q_0, F)$ consisting of: Q, a finite set of *states*, T^ε, a finite set of *input symbols*, Γ, a finite set of *stack symbols* containing a bottom-of-stack symbol $\bot$, $\Delta = \Delta_\to \cup \Delta_\downarrow \cup \Delta_\uparrow$ is the *transition relation*, $\Delta_\to$ over $Q \times T^\varepsilon \times Q$ are *shift transitions*, $\Delta_\downarrow$ over $Q \times \{\downarrow\} \times \Gamma \times Q$ are *push transitions*, $\Delta_\uparrow$ over $Q \times \{\uparrow\} \times \Gamma \times Q$ are *pop transitions*, $q_0 \in Q$, is the *start state*, $F \subseteq Q$, a finite set of *accepting states*. To distinguish between pushes and pops of stack symbols we define $\Gamma' = \{\downarrow, \uparrow\} \times \Gamma$. We use p to represent stack symbols in Γ, π for strings in Γ^* and φ for symbols in $T^\varepsilon \cup \Gamma'$. An element in $Q \times \Gamma^*$ is called a *configuration*, representing a state and stack contents. We assume every PDA to start with the initial configuration $c_0 = (q_0, \bot)$. The relation Δ defines state transitions. We write $q \xrightarrow{\alpha} q'$ for tuples in $\Delta_\to$, $q \xrightarrow{\downarrow p} q'$ for tuples in $\Delta_\downarrow$ and $q \xrightarrow{\uparrow p} q'$ for tuples in $\Delta_\uparrow$. Configuration transition is denoted with $\vdash$. We write $(q, \pi) \vdash^\alpha (q', \pi)$ if $q \xrightarrow{\alpha} q'$, $(q, \pi) \vdash^{\downarrow p} (q', \pi p)$ if $q \xrightarrow{\downarrow p} q'$ and $(q, \pi p) \vdash^{\uparrow p} (q', \pi)$ if $q \xrightarrow{\uparrow p} q'$. A *run* ρ is a sequence of $\vdash$ steps. We write $c_0 \vdash^\rho c_n$ if $\rho = p_1 \ldots p_n \in (T^\varepsilon \cup \Gamma')^+$ and for every $i \in [n]$ there exists c_i s.t. $c_{i-1} \vdash^{p_i} c_i$, where $[n]$ denotes the set $\{1 \ldots n\}$. The set of possible configurations of M is $\mathcal{C}(M) = \{c \mid c_0 \vdash^* c\}$. The set of accepting runs of M is $\mathcal{R}(M) = \{\rho \mid \exists q_f \in F,\ \pi \in \Gamma^* : c_0 \vdash^\rho (q_f, \pi)\}$.

A *multi-stack pushdown automaton*, or MPDA, M_n with n stacks is a tuple $(Q, T, \widetilde{\Gamma}_n, \Delta, q_0, F)$ where Q, T, Δ, q_0 and F are defined the same as for a PDA and $\widetilde{\Gamma}_n = \bigcup_{i=1}^n \Gamma_i$ are the n stack alphabets, each containing $\bot_i$. W.l.o.g. we assume all subsets $\Gamma_i \subset \widetilde{\Gamma}_n$ to be disjoint. We use $\{\pi_i\}_{i\in[n]}$ to denote a set of stacks. A configuration is a tuple over $Q \times \Gamma_1^* \times \cdots \times \Gamma_n^*$ and $c_0 = \{q_0, \{\bot_i\}_{i\in[n]}\}$. We write $(q, \{\pi_i\}_{i\in[n]}) \vdash^\alpha (q', \{\pi_i\}_{i\in[n]})$ if $q \xrightarrow{\alpha} q'$; $(q, \{\pi_i\}_{i\in[n]}) \vdash^{\downarrow p} (q', \{\pi'_i\}_{i\in[n]})$ if $q \xrightarrow{\downarrow p} q'$, $p \in \Gamma_j$, $\pi'_j = \pi_j p$ and $\pi'_i = \pi_i$ for $i \neq j$; and $(q, \{\pi_i\}_{i\in[n]}) \vdash^{\uparrow p} (q', \{\pi'_i\}_{i\in[n]})$ if $q \xrightarrow{\uparrow p} q'$, $p \in \Gamma_j$, $\pi_j = \pi'_j p$ and $\pi'_i = \pi_i$ for $i \neq j$.

3 Ambiguity Detection with MPDAs

We present a framework for ambiguity detection of context-free grammars using multi-stack automata.

3.1 Checking Ambiguity

Given a PDA M that defines the derivations of a context-free grammar G, we can express the ambiguity problem for the grammar using an MPDA. This can be done on the condition that there is a bijective relation between the runs of the PDA and parse trees of G, let us call it $\mathsf{tree} : \mathcal{R}(M) \to \mathcal{L}(G_b)$. Two different runs of the same input string then prove the ambiguity of G.

We build a 2-stack MPDA that simulates two runs of the PDA for the same input string. The states of this MPDA consist of pairs of states of the PDA. Both stacks can be modified independently of each other, but non-empty shifts are synchronized to ensure both runs parse the same input string. Different runs for the same input string both start from q_0 but eventually split up. There are two ways in which the runs can deviate from a common state: the runs can each take different transitions, or only one of the two continues independently until the next common shift. W.l.o.g. we distinguish three possible phases in this process:

1. the runs are not split up yet and alternately follow the same transitions;
2. the first run continues with independent transitions while the second run waits for the next shift;
3. both runs are in different states.

We add two additional fields to the state pairs to register these phases: an integer field denoting the current phase and a symbol from $T^\varepsilon \cup \widetilde{\Gamma}'_2$ to record the last action taken by the first run. The second field is used to synchronize shifts and internal transitions during phase 1, and to recognize phase transitions.

Definition 1. *Given a PDA* $M = (Q, T^\varepsilon, \Gamma, \Delta, q_0, F)$ *the ambiguity MPDA of* M *is a 2-stack MPDA* $M^a = (Q^a, T^\varepsilon, \widetilde{\Gamma}_2, \Delta^a, q_0^a, F^a)$, *where* $Q^a \subseteq Q \times Q \times (T^\varepsilon \cup \widetilde{\Gamma}'_2) \times [3]$, $q_0^a = (q_0, q_0, \bot, 1)$, $F^a = F \times F \times (T^\varepsilon \cup \widetilde{\Gamma}'_2) \times \{2,3\}$, $\Delta_{\downarrow\uparrow} = \Delta_\downarrow \cup \Delta_\uparrow$,

$$
\begin{aligned}
\Delta^a = \{ & (q,q,\bot,1) \xrightarrow{\varphi} (q',q,\varphi,1) \mid q \xrightarrow{\varphi} q' \in \Delta_{\downarrow\uparrow}\} \cup \\
\{ & (q',q,\varphi,1) \xrightarrow{\varphi} (q',q',\bot,1) \mid \} \cup \\
\{ & (q,q,\bot,1) \xrightarrow{\alpha} (q',q,\alpha,1) \mid q \xrightarrow{\alpha} q' \in \Delta_\to\} \cup \\
\{ & (q',q,\alpha,1) \xrightarrow{\varepsilon} (q',q',\bot,1) \mid \} \cup \\
\{ & (q,q,\bot,1) \xrightarrow{\varphi} (q',q,\bot,2) \mid q \xrightarrow{\varphi} q' \in \Delta_{\downarrow\uparrow}\} \cup \\
\{ & (q',q,\varphi,1) \xrightarrow{\varphi'} (q',q'',\bot,3) \mid q \xrightarrow{\varphi'} q'' \in \Delta_{\downarrow\uparrow},\ \varphi' \neq \varphi \vee q'' \neq q'\} \cup \\
\{ & (q',q,\alpha,1) \xrightarrow{\varepsilon} (q',q'',\bot,3) \mid q \xrightarrow{\alpha} q'' \in \Delta_\to,\ q'' \neq q'\} \cup \\
\{ & (q,q',\bot,y) \xrightarrow{\varphi} (q'',q',\bot,y) \mid q \xrightarrow{\varphi} q'' \in \Delta_{\downarrow\uparrow},\ y \in \{2,3\}\} \cup \\
\{ & (q,q',\bot,y) \xrightarrow{\varepsilon} (q'',q',\bot,y) \mid q \xrightarrow{\varepsilon} q'' \in \Delta_\to,\ y \in \{2,3\}\} \cup \\
\{ & (q,q',\bot,y) \xrightarrow{b} (q'',q',b,y) \mid q \xrightarrow{b} q'' \in \Delta_\to,\ y \in \{2,3\}\} \cup \\
\{ & (q,q',b,y) \xrightarrow{\varepsilon} (q,q'',\bot,3) \mid q' \xrightarrow{b} q'' \in \Delta_\to,\ y \in \{2,3\}\} \cup \\
\{ & (q,q',x,3) \xrightarrow{\varphi} (q,q'',x,3) \mid q' \xrightarrow{\varphi} q'' \in \Delta_{\downarrow\uparrow}\} \cup \\
\{ & (q,q',x,3) \xrightarrow{\varepsilon} (q,q'',x,3) \mid q' \xrightarrow{\varepsilon} q'' \in \Delta_\to\}.
\end{aligned}
$$

The input strings of runs leading to accepting states are the ambiguous strings of G. To test for ambiguity we choose a restricted MPDA class and compute the image of $\{q_0^a, \{\perp_i\}_{i\in[n]}\}$ under $\vdash^*$. If we can reach a state in F^a the MPDA's language is non-empty and G is ambiguous. On the other hand, if the chosen MPDA class allows the complete exploration of the configuration space of M^a and no accepting state can be reached then G is unambiguous. Otherwise, the problem remains unanswered. This is formalized by the following statements.

Lemma 2. *Given a grammar G, a PDA M and a bijective relation* tree : $\mathcal{R}(M) \rightarrow \mathcal{L}(G_b)$, *the language $\mathcal{L}(M^a)$ equals $\mathcal{A}(G)$.*

Definition 3. *Given an MPDA class C^m, a PDA M is MSA(C^m)-ambiguous iff M^a has at least one run that complies with the restrictions of C^m. The PDA is MSA(C^m)-unambiguous iff M^a is a member of C^m and $\mathcal{L}(M^a)$ is empty.*

Definition 4. *Given a PDA class C^p and an MPDA class C^m, a grammar G is MSA(C^p, C^m)-ambiguous iff a C^p-PDA of G is MSA(C^m)-ambiguous. Similarly, G is MSA(C^p, C^m)-unambiguous iff a C^p-PDA of G is MSA(C^m)-unambiguous.*

Definition 5. *Given a PDA class C^p and an MPDA class C^m, a grammar G is in MSA(C^p, C^m) iff it is MSA(C^p, C^m)-ambiguous or MSA(C^p, C^m)-unambiguous.*

Theorem 6. *Given a PDA class C^p, an MPDA class C^m and a grammar G, if G is MSA(C^p, C^m)-ambiguous then G is ambiguous.*

Theorem 7. *Given a PDA class C^p, an MPDA class C^m and a grammar G, if G is MSA(C^p, C^m)-unambiguous then G is unambiguous.*

3.2 Choice of Pushdown Automaton

Since our method is parametric in the type of PDA, it can apply different strategies for exploring parse trees. Furthermore, this allows for easy integration with existing parser implementations. The parse tree exploration depends on the way a PDA uses its stack. For instance, recursive descent parsers—like LL [15]—push on every entry of a production and pop on a reduce. This makes the stack depth correspond to parse tree height. The number of pushes in a run corresponds to the number of non-leaf parse tree nodes. Shift-reduce parsers—like LR [14]—push on every shift and pop when a production is reduced, followed by another push of the reduced nonterminal. In this case the number of pushes in a run corresponds to the total number of parse tree nodes.

In general, the less stack activity a PDA requires for a given language, the larger the set of parse trees that can be covered by the configuration space of the restricted MPDA. Reduce incorporated parsers [24] are aimed at reducing the stack activity of a parser. They use the PDA as a DFA for regular structures and only use the stack to record the derivation of nesting nonterminals. Parse trees of the regular structures are built using special ε-transitions that mark reductions. However, when a nesting nonterminal is also right or left recursive the stack is used to track these kinds of derivations as well. To reduce the stack activity even further—and use it purely for nesting—we can apply a similar strategy as Nederhof ([19] Sect. 4.2), which completely separates the regular structures in a grammar from the context-free ones. This way we can completely intersect the regular structures in a grammar with each other and with the nesting ones, and fully use the stack space for nesting/nesting intersection. We do not define such a PDA here, but only mention their possibility. We will call them Nesting Stack PDAs or NSPDAs.

3.3 Choice of Multi-stack Pushdown Automaton

Below we discuss various existing MPDA types and explore their suitability for detecting ambiguity and unambiguity. To detect the ambiguity of a grammar with a certain type of MPDA, it suffices to explore a single path in M^a to an accepting state. The more paths an MPDA can cover, the higher the chance of finding an ambiguous one. In advance we can state that this is possible with all MDPA types described below, to varying extents. However, to detect unambiguity we need to ensure no state in F^a can be reached at all. This requires the configuration space of the MPDA to cover all possible paths of M^a. In order to do so, an MPDA type should pose no restrictions on nesting depth, since every nesting nonterminal will create paths in (at least) phase 1 that push to infinite stack depths and pop out of these as well. We will see that no discussed MPDA is able to cover such paths.

Another criteria we will look at is to what extent an MPDA type is able to intersect the regular structures in a grammar with other regular structures as well as with nesting structures. Since the emptiness of both regular/regular intersection and regular/nesting intersection is decidable, making use of these results enlarges the class of grammars the MPDA can decide the ambiguity of. With NSPDAs all MPDAs allow full regular/regular intersection, since this requires no stack activity. For full regular/nesting intersection an MPDA should allow one stack to reach and return from infinite depths, while the other remains untouched.

Bounded nesting depth MPDAs [10] pose an intuitive restriction, which allows complete regular/regular intersection, but only limited regular/nesting and nesting/nesting intersection. They are thus suitable for ambiguity detection, but not for unambiguity detection.

Bounded-context switching MPDAs [20] use the concept of contexts—a part of a run in which only one stack can push and pop—and restrict runs to a limited number of contexts. This allows for complete regular/regular and regular/nesting intersection. However, the depth of nesting/nesting intersection is bounded since every alternate nesting requires a context-switch. Bounded-context MPDAs can be useful for finding ambiguity, but not for finding unambiguity.

Bounded-phase MPDAs [16] use the concept of phases, in which only one stack can pop, but others are free to push. These MPDAs cover a strictly larger search space than bounded-context MPDAs [17] and are thus better suitable for finding ambiguity. Stacks can nest simultaneously to unlimited depth, but only pop out together for a limited number of steps. Hence, they can still not completely explore all configurations of phase 1.

MPDAs with scope-bounded matching relations [17,18] require that every push is popped within a bounded number of rounds, or never at all. During a round all stacks are allowed one context each, in a predefined order. These MPDAs have a larger coverage than bounded-context MPDAs, but are incomparable with bounded-phase [17]. The fact that pushes do not have to be popped can be helpful for finding ambiguity, but not for detecting unambiguity if we require all pushes to be popped. In this case, the first push of any stack has to immediately start a scope and the MPDA reverts to a bounded-context exploration.

Budget bounded MPDAs [1] allow unlimited context switches for stacks whose depth is below a certain bound, and a limited number of contexts for as long as they are above this depth bound. In other words, once the depth limit is reached, a new scope is started which has to be closed within a bounded number of contexts. Budget bounded MPDAs are thus closely related to scope-bounded MPDAs, but because they also allow pops before the start of a scope they have a larger coverage. Nevertheless, there remains a bound on the nesting depth.

Ordered MPDAs [8], the earliest type of restricted MPDA, assume an ordering of the stacks and allow only the first non-empty stack to pop. All stacks can push freely at any time. At first sight this concept might not seem suitable for nesting/nesting intersection, because it does not allow simultaneous pops. However, ordered MDPAs can simulate bounded-phase MPDAs [2] and thus allow bounded nesting/nesting intersection. In fact, they are even more expressive than bounded-phase, which makes them at least equally suitable for detecting ambiguity and unambiguity.

Concluding, we can say that all discussed MPDA types are suitable to find ambiguities with our scheme, resulting in different exploration strategies of strings and prefixes. All MPDAs can either simultaneously push into bounded or unbounded nesting depths, and some can simultaneously pop a bounded number of steps as well. However, none of the MPDAs are able to let both stacks pop out of infinitely deep nestings together, making them unsuitable for detecting the unambiguity of context-free grammars in general. In the next section we describe a new type of restricted MPDA that does have this property.

4 Bounded-Balance Multi-Stack Pushdown Automata

We propose a new type of restricted MPDA called bounded-balance multi-stack pushdown automata, or BBMPDAs. They are MPDAs with an upper bound on the number of symbols that are pushed but not popped within each scope of matching push and pop transitions.

4.1 Definition

First we introduce the concepts of scope and balance. A *scope* is the part of a run between the push of a symbol and the pop of that symbol. We use $\mu \subseteq \mathbb{N} \times \mathbb{N}$ to hold matching transition indices that open and close a scope (as in [18]). Given a run $\rho = c_0 \vdash^{\varphi_1} c_1 \vdash^{\varphi_2} \dots \vdash^{\varphi_m} c_m$, a pair $(s,t) \in \mu_\rho$ iff $s < t$ and exists $p \in \Gamma_i$ for some $i \in [n]$ s.t. $\varphi_s = \downarrow p$, $\varphi_t = \uparrow p$ and

- for all $s < s' < t$, if $\varphi_{s'} = \downarrow p'$, $p' \in \Gamma_i$ then there exists $s' < t' < t$ such that $(s',t') \in \mu_\rho$, and
- for all $s < t' < t$, if $\varphi_{t'} = \uparrow p'$, $p' \in \Gamma_i$ then there exists $s < s' < t'$ such that $(s',t') \in \mu_\rho$.

The *balance* of a scope is the number of stack symbols that were pushed but not popped within the scope.

Definition 8. *The balance of a scope* $(s,t) \in \mu_\rho$ *is* $\mathsf{balance}(s,t) = |\{\, s' \mid (s',t') \in \mu_\rho,\ s < s' < t < t'\}|$.

Viewed differently, balance corresponds to the stack depth differences built up during a scope. By limiting the balances during runs, we acquire a new class of restricted MPDAs, which we call bounded-balance MPDAs.

Definition 9. *An n-MPDA* M_n *is a BB(k)MPDA iff for every run* $\rho \in \mathcal{R}(M_n)$ *and scope* $(s,t) \in \mu_\rho$ *it holds that* $\mathsf{balance}(s,t) \leq k$.

A finite balance bound allows for a finite representation of the possibly infinite configuration space of BBMPDAs. This makes testing for the BB(k) property decidable. In the following section we show how a BBMPDA can be simulated by a standard single stack PDA, which enables using existing techniques for configuration space exploration [23].

4.2 Configuration Exploration

BB(k)MPDAs can be simulated by a standard PDA that can pop from the topmost $k+1$ symbols of its stack, by temporarily remembering up to k stack symbols in its states. It has a single stack over $\widetilde{\Gamma}_n$, storing pushes of all stacks sequentially. When a certain stack needs to be popped, but its top symbol is not at the top of the simulating stack, intermediary symbols are popped and temporarily stored in the PDA states, until the required symbol is reached. This symbol is then popped as per usual, and temporarily stored symbols are pushed back to the stack again. The number of the stack to be popped is also stored in the states, so a series of borrows is always targeted at a single stack. To make sure that the order in which the symbols of the individual stacks are pushed and popped remains unchanged, a stack cannot be borrowed from once it has been targeted, i.e. only the top of the targeted stack can be popped. The number 0 is used to indicate no stack is currently targeted.

Definition 10. *Given an n-stack MPDA $M = (Q, T^\varepsilon, \widetilde{\Gamma}_n, \Delta, q_0, F)$ the k-borrowing PDA of M is $M_k^b = (Q^b, T^\varepsilon, \widetilde{\Gamma}_n, \Delta^b, (q_0, 0, \varepsilon), F \times \{0\} \times \{\varepsilon\})$, where $Q^b = Q \times \{0 \dots n\} \times (\widetilde{\Gamma}_n \cup \{\varepsilon\})^k$, $\Delta^b =$*

$$\begin{aligned}
&\{(q,0,\varepsilon) \xrightarrow{X} (q',0,\varepsilon) \mid q \xrightarrow{X} q' \in \Delta\} \cup && \textit{(copy of } \Delta\textit{)}\\
&\{(q,i,\pi) \xrightarrow{\uparrow p} (q,j,p\pi) \mid q \xrightarrow{\uparrow p'} q' \in \Delta_\uparrow,\, p' \in \Gamma_j,\, p \notin \Gamma_j,\, i \in \{0,j\}\} \cup && \textit{(borrows)}\\
&\{(q,i,\pi) \xrightarrow{\uparrow p} (q',0,\pi) \mid q \xrightarrow{\uparrow p} q' \in \Delta_\uparrow,\, p \in \Gamma_i\} \cup && \textit{(pops)}\\
&\{(q,0,p\pi) \xrightarrow{\downarrow p} (q,0,\pi) \mid \}. && \textit{(returns)}
\end{aligned}$$

The initial configuration of M_k^b is $((q_0, 0, \varepsilon), \bot_1 \dots \bot_n)$.

Theorem 11. *The k-borrowing PDA M_k^b of a BB(k)MPDA M simulates exactly all runs of M.*

Testing whether an MPDA is BB(k) for a fixed k comes down to constructing M_{k+1}^b and testing whether no states with borrowed stacks of size $k+1$ can be reached. Note that this scheme allows for incremental search with increasing k. The computational complexity depends on the chosen model checking algorithm, of which most are polynomial in the size of the PDA. The number of states and transitions of M_k^b is exponential in k, which puts our approach in EXPTIME.

4.3 Application to Ambiguity Detection

When applied in the ambiguity detection scheme of Sect. 3, BBMPDAs yield several desirable properties. First, they allow for full regular/regular intersection and regular/nesting intersection of the paths of M in M^a. In combination with NSPDAs, regular/regular intersection requires no stack activity and will not build up any balance. During regular/nesting intersection, which starts with the opening of a scope on one stack and ends when this scope is closed or the other stack becomes active, pushes on the active stack do add to the balance of the current scope of the other stack. However, this balance is only compared to the bound at the moment the regular/nesting intersection ends. During the intersection the nesting stack is allowed to grow and shrink indefinitely.

Second, full nesting/nesting intersection is also possible in case M^a meets the BB(k) condition. Both stacks are allowed to reach unbounded depths together and pop out of them as well, as long as the scope balances stay within the bound.

Third, BBMPDAs have the possibility of detecting the unambiguity of grammars with nesting structures, which is a consequence from the previous property. As the next theorem states, the scope balances in the paths of M^a in phase 1 will never be more than 1. Therefore, the configuration space of BB(k)MPDAs with $k \geq 1$ will at least cover all these paths. If in the continuations of these paths in phases 2 and 3, the scope balances stay within the balance bound as well and no end state is reached, the tested grammar is unambiguous.

Theorem 12. *Given a PDA M, for all partial runs $c_0 \dots \vdash^{\Uparrow p} ((q, q', x, 1), \{\pi_i\}_{i \in [n]})$ in phase 1 of M^a, the balance of the closed scope of p is at most 1.*

5 Comparisons and Related Work

In this section we compare our contributions to related MPDAs and ambiguity detection methods.

5.1 Multi-Stack Pushdown Automata

We show that BBMPDAs include bounded depth MPDAs but that they are incomparable with bounded-context switching MPDAs. This implies they are also incomparable with the larger MPDA classes mentioned in Sect. 3.3—bounded-phase, scope-bounded, budget-bounded and ordered MPDAs—since none of these can, in general, cover all paths in phase 1 of M^a. For detecting ambiguity however, these MPDAs and BBMPDAs are complementary.

Theorem 13. *If M is a n-stack MPDA with depth bound k, its scope balances are bounded by $k * (n-1)$.*

Theorem 14. *The class of BBMPDAs is incomparable with bounded-context switching MPDAs.*

Regarding simulation, any 1-stack MPDA with enough freedom to be a plain PDA can simulate BBMPDAs. With the exception of bounded depth MPDAs this is the case for all MPDA types mentioned above.

5.2 Ambiguity Detection Methods

In this section we will discuss related work in ambiguity detection and compare it with our approach if possible.

Bounded-Search Methods. There are several methods that enumerate strings in $\mathcal{L}(G)$ of bounded length and test them for ambiguity [3,6,9,12,13,22,25]. In general these approaches are only able to detect ambiguities, because they can never entirely cover $\mathcal{L}(G)$. In essence our approach also applies a bounded search, but with the difference that the search space can cover an infinitely large language. Depending on G and the types of PDA and MPDA we can cover $\mathcal{L}(G)$ entirely and detect unambiguity. However, with certain types of PDA and MPDA, our method can also be set up for bounded string exploration. For example, using a—nondeterministic—LR PDA with a bounded-context MPDA will result in the exploration of strings and prefixes of bounded length. An LR PDA pushes with every shift, requiring a context switch to allow both stacks to push. Every reduction requires a number of pops and a push, but each stack can perform all its reductions within one context, either after the last push or before the next push, requiring no additional context switches.

Conservative Approximation Methods. Contrary to bounded search, other methods apply conservative approximation to reduce the infinite $\mathcal{L}(G)$ to a limited space. This yields the possibility of detecting unambiguity, but prohibits detecting ambiguity in most cases. The ACLA-test [7] applies regular approximation to the languages of individual production rules and searches for the absence of horizontal and vertical ambiguities using intersection and overlap operations. The NU-test [21] approximates the set of parse trees of a grammar and searches for the absence of different trees for the same ambiguous string. An extension to the NU-test allows the detection of harmless productions, which are rules that do not contribute to any ambiguity [4]. The rules can be filtered from the grammar to incrementally improve the approximation.

A significant difference between both approximative methods and our approach is that they are less able to recognize the unambiguity of nesting structures. Due to the regular approximation they lose the ability to match the left and right contexts of nestings, i.e. count the nesting depth. The ACLA-test applies production unfolding to counter this disadvantage, but this is only possible up to a certain depth. As an example, both tests are not able the detect the unambiguity of the following grammar, which is MSA(LL(0), BB(3))-unambiguous.

$$S \to A \mid B,\ A \to aAb \mid ab,\ B \to aBb \mid a \tag{1}$$

6 Conclusion

We present a novel method for detecting ambiguity in context-free grammars using restricted multi-stack pushdown automata. It is able to find both ambiguity and unambiguity. We discuss the use of existing MPDA classes within our framework, as well as propose a new class called bounded-balance MPDAs. These MPDAs are particularly useful for language intersection since they allow for unbounded nesting/nesting intersection, as long as the nesting depth differences stay within the balance bound.

References

1. Abdulla, P., Atig, M., Rezine, O., Stenman, J.: Budget-bounded model-checking pushdown systems. Form. Methods Syst. Des. **45**(2), 273–301 (2014)
2. Atig, M.F., Bollig, B., Habermehl, P.: Emptiness of multi-pushdown automata is 2ETIME-complete. In: Ito, M., Toyama, M. (eds.) DLT 2008. LNCS, vol. 5257, pp. 121–133. Springer, Heidelberg (2008)
3. Axelsson, R., Heljanko, K., Lange, M.: Analyzing context-free grammars using an incremental SAT solver. In: Aceto, L., Damgård, I., Goldberg, L.A., Halldórsson, M.M., Ingólfsdóttir, A., Walukiewicz, I. (eds.) ICALP 2008, Part II. LNCS, vol. 5126, pp. 410–422. Springer, Heidelberg (2008)
4. Basten, H.J.S.: Tracking down the origins of ambiguity in context-free grammars. In: Cavalcanti, A., Deharbe, D., Gaudel, M.-C., Woodcock, J. (eds.) ICTAC 2010. LNCS, vol. 6255, pp. 76–90. Springer, Heidelberg (2010)
5. Basten, H.J.S.: Ambiguity detection for programming language grammars. Ph.D. thesis, Universiteit van Amsterdam (2011)
6. Basten, H.J.S., Vinju, J.J.: Faster ambiguity detection by grammar filtering. In: Proceedings of the Tenth Workshop on Language Descriptions, Tools and Applications (LDTA 2010), pp. 5:1–5:9. ACM (2010)
7. Brabrand, C., Giegerich, R., Møller, A.: Analyzing ambiguity of context-free grammars. Sci. Comput. Program. **75**(3), 176–191 (2010)

8. Breveglieri, L., Cherubini, A., Citrini, C., Reghizzi, S.C.: Multi-push-down languages and grammars. Int. J. Found. Comput. Sci. **07**(03), 253–291 (1996)
9. Cheung, B.S.N., Uzgalis, R.C.: Ambiguity in context-free grammars. In: Proceedings of the 1995 ACM Symposium on Applied Computing (SAC 1995), pp. 272–276. ACM, New York (1995)
10. Clarke, E., Kroning, D., Lerda, F.: A tool for checking ANSI-C programs. In: Jensen, K., Podelski, A. (eds.) TACAS 2004. LNCS, vol. 2988, pp. 168–176. Springer, Heidelberg (2004)
11. Ginsburg, S., Harrison, M.A.: Bracketed context-free languages. J. Comput. Syst. Sci. **1**(1), 1–23 (1967)
12. Gorn, S.: Detection of generative ambiguities in context-free mechanical languages. J. ACM **10**(2), 196–208 (1963)
13. Kieżun, A., Ganesh, V., Guo, P.J., Hooimeijer, P., Ernst, M.D.: HAMPI: a solver for string constraints. In: Proceedings of the 2009 International Symposium on Software Testing and Analysis (ISSTA 2009), pp. 105–116. ACM (2009)
14. Knuth, D.E.: On the translation of languages from left to right. Inf. Control **8**(6), 607–639 (1965)
15. Knuth, D.E.: Top-down syntax analysis. Acta Informatica **1**, 79–110 (1971)
16. La Torre, S., Madhusudan, P., Parlato, G.: A robust class of context-sensitive languages. In: 22nd Annual IEEE Symposium on Logic in Computer Science (LICS 2007), pp. 161–170. IEEE (2007)
17. La Torre, S., Napoli, M.: Reachability of multistack pushdown systems with scope-bounded matching relations. In: Katoen, J.-P., König, B. (eds.) CONCUR 2011. LNCS, vol. 6901, pp. 203–218. Springer, Heidelberg (2011)
18. La Torre, S., Parlato, G.: Scope-bounded multistack pushdown systems: fixed-point, sequentialization, and tree-width. In: IARCS Annual Conference on Foundations of Software Technology and Theoretical Computer Science (FSTTCS 2012), pp. 173–184 (2012)
19. Nederhof, M.: Practical experiments with regular approximation of context-free languages. Comput. Linguist. **26**(1), 17–44 (2000)
20. Qadeer, S., Rehof, J.: Context-bounded model checking of concurrent software. In: Halbwachs, N., Zuck, L.D. (eds.) TACAS 2005. LNCS, vol. 3440, pp. 93–107. Springer, Heidelberg (2005)
21. Schmitz, S.: Conservative ambiguity detection in context-free grammars. In: Arge, L., Cachin, C., Jurdziński, T., Tarlecki, A. (eds.) ICALP 2007. LNCS, vol. 4596, pp. 692–703. Springer, Heidelberg (2007)
22. Schröer, F.W.: AMBER, an ambiguity checker for context-free grammars. Technical report (2001). compilertools.net, http://accent.compilertools.net/Amber.html
23. Schwoon, S.: Model-checking pushdown systems. Ph.D. thesis, Technische Universität München, June 2002
24. Scott, E., Johnstone, A.: Generalized bottom up parsers with reduced stack activity. Comput. J. **48**(5), 565–587 (2005)
25. Vasudevan, N., Tratt, L.: Detecting ambiguity in programming language grammars. In: Erwig, M., Paige, R.F., Van Wyk, E. (eds.) SLE 2013. LNCS, vol. 8225, pp. 157–176. Springer, Heidelberg (2013)

Complementation of Branching Automata for Scattered and Countable Series-Parallel Posets

Nicolas Bedon(✉)

LITIS (EA 4108), Université de Rouen, Rouen, France
Nicolas.Bedon@univ-rouen.fr

Abstract. We prove the closure under complementation of the class of languages of scattered and countable N-free posets recognized by branching automata. The proof relies entirely on effective constructions.

Keywords: Transfinite N-free posets · Series-parallel posets · SP-rational languages · Automata · Commutative monoids

1 Introduction

Automata over finite words have been widely studied since their introduction by Kleene in the last fifties, because they are a natural model for sequential computation with bounded memory, and they are linked to many other areas, as for example formal logic, coding theory or formal series. The depth of those links and the richness of the results led the community to develop generalizations o f Kleene automata, as for example automata over trees [18], ω-words [6,17], ordinals [7], and more recently, over linear orderings [5].

Among those generalizations, Lodaya and Weil proposed a notion of *branching-automata* that are a natural model for parallel computation with the Fork/Join principle. The Fork/Join principle splits an execution flow f into n concurrent flows $f_1, \dots, f_n$ and joins $f_1, \dots, f_n$ before it continues. Divide-and-conquer concurrent programming naturally uses this Fork/Join principle. Traces of execution of programs are in this case finite N-free posets, or equivalently, finite series-parallel posets [19,22]. Lodaya and Weil extended some fundamental results of automata on words to branching-automata, as for example a Kleene-like Theorem or algebraic recognizability [13–16]. Unfortunately, and contrarily to the finite words case, the algebraic counterpart of branching automata may be infinite, leading to difficulties regarding the generalization of fundamental results over finite words to finite N-free posets. Kuske [11,12] extended branching-automata to recognition of ω-N-free posets, and established a connection with monadic second-order logic (MSO[<]) in the particular case of languages of N-free posets with bounded-size antichains. The logical characterization of languages of finite N-free posets recognized by branching automata of Lodaya and Weil is provided in [3] in the general case.

S. Brlek and C. Reutenauer (Eds.): DLT 2016, LNCS 9840, pp. 13–25, 2016.
DOI: 10.1007/978-3-662-53132-7_2

In [4], branching automata are generalized to N-free posets with finite antichains and countable and scattered chains, and a Kleene-like Theorem is provided. The connection with MSO[<] is established in [2] in the particular case of languages of N-free posets with bounded-size antichains. In this paper, we prove that the class of languages recognized by the generalization of branching automata of [4] is closed under complement. The (effective) proof relies on an algebraic approach of branching automata, on the use of Simon's factorization forests proposed by Colcombet in [9] for regular languages of linear orderings, and on the closure under complementation of the class of rational sets of finitely generated commutative monoids [10].

2 Notation and Basic Definitions

Let E be a set. We denote by $|E|$, $\mathcal{P}(E)$, $\mathcal{P}^+(E)$ and $\mathcal{M}^{>1}(E)$ respectively the cardinality of E, the set of subsets of E, the set of non-empty subsets of E and the set of multi-subsets of E with at least two elements. For any integer n, the set $\{1, \ldots, n\}$ is denoted by $[n]$ and the group of permutations of $[n]$ by S_n.

We start by some basic definitions on linear orderings. We refer to [20] for a survey on the subject. Let J be a set equipped with an order $<$. The ordering J is *linear* if all elements are comparable : for any distinct j and k in J, either $j < k$ or $k < j$. For any linear ordering J, we denote by $-J$ the backward linear ordering obtained from the set J with the reverse ordering. A linear ordering J is *dense* if for any $j, k \in J$ such that $j < k$, there exists an element i of J such that $j < i < k$. It is *scattered* if it contains no dense sub-ordering. The orderings $\omega = (\mathbb{N}, <)$ and $\zeta = (\mathbb{Z}, <)$ are scattered. Ordinals are also scattered orderings. We denote by $\mathcal{O}$ the class of countable ordinals and $\mathcal{S}$ the class of countable scattered linear orderings. An *interval* K of $J \in \mathcal{S}$ is a subset $K \subseteq J$ such that $\forall k_1, k_2 \in K, \forall j \in J$, if $k_1 < j < k_2$ then $j \in K$.

A *poset* $(P, <)$ is a set P partially ordered by $<$. In order to lighten the notation we often denote the poset $(P, <)$ by P. An *antichain* is a subset P' of P such that all elements of P' are incomparable (with $<$). The *width* of P is $\text{wd}(P) = \sup\{|E| : E \text{ is an antichain of } P\}$ where sup denotes the least upper bound of the set. If $x, y \in P$, we denote by $x^- = \{z \in P : z < x\}$, $x^+ = \{z \in P : x < z\}$ and $x \sim_< y$ if $x^- \cup x^+ \cup \{x\} = y^- \cup y^+ \cup \{y\}$. In this paper, we restrict to *countable scattered posets of finite width* which are thus partially ordered countable sets without any dense sub-ordering. Let $(P, <_P)$ and $(Q, <_Q)$ be two disjoint posets. The *parallel composition* of $(P, <_P)$ and $(Q, <_Q)$ is the poset $(P \cup Q, <)$ where $x < y$ if and only if ($x, y \in P$ and $x <_P y$) or ($x, y \in Q$ and $x <_Q y$). The *sum* (or *sequential composition*) $P + Q$ of P and Q is the poset $(P \cup Q, <)$ such that $x < y$ if and only if one of the following three conditions is true: (1): $x \in P$, $y \in P$ and $x <_P y$; (2): $x \in Q$, $y \in Q$ and $x <_Q y$; (3): $x \in P$ and $y \in Q$. The sum of two posets can be generalized to any linearly ordered sequence of pairwise disjoint posets: if J is a linear ordering and $((P_j, <_j))_{j \in J}$ is a sequence of posets, then $\sum_{j \in J} P_j = (\cup_{j \in J} P_j, <)$ such that $x < y$ if and only if ($x \in P_j$, $y \in P_j$ and $x <_j y$) or ($x \in P_j$ and

$y \in P_k$ and $j < k$). The sequence $((P_j, <_j))_{j \in J}$ is called a *J-factorization*, or *factorization* for short, of the poset $\sum_{j \in J} P_j$. A nonempty poset P is *sequential* if it admits a J-factorization where J contains at least two elements, or P is a singleton. It is a *parallel poset* otherwise. The only poset $(\emptyset, <)$ of width 0 is called *empty poset* and is denoted by ε. The class $SP^\diamond$ of *series-parallel* scattered and countable posets is the smallest class of posets containing ε, the singleton and closed under finite parallel composition and sum indexed by countable scattered linear orderings. It has a nice characterization in terms of graph properties: $SP^\diamond$ coincides with the class of scattered and countable N-free posets without infinite antichain (see [4]). We denote by $SP^{\diamond +} = SP^\diamond - \{\varepsilon\}$.

The sets of (Dedekind-MacNeille) cuts of a poset P is defined as a generalization of cuts of linear orderings. It is the set of all pairs (A, B), with $A, B \subseteq P$, such that B consists of all the elements of P greater than all the elements of A, and reciprocally, A consists of all the elements of P less than all the elements of B. The cuts are partially ordered with inclusion on the first component, and with the elements of P with $(A, B) < x$ if $x \in B$. The partially ordered set of all cuts of P is denoted $\hat{P}$, and we usually denote by $P \cup \hat{P}$ the partially ordered set consisting of the elements of P with its cuts. Note that an equivalence class of cuts of P for $\sim_<$ is totally ordered. The notation $\hat{P}^{\iota\iota'}$ with $\iota, \iota' \in \{[,]\}$ excludes or not the minimum and maximum elements from $\hat{P}$. We denote also by $\hat{P}^* = \hat{P} - \{(\emptyset, P), (P, \emptyset)\}$. We define the partial ordering $\preceq$ over the cuts of P by $(A, B) \preceq (A', B')$ if and only if $A \cup B = A' \cup B'$ and $A \subseteq A'$.

An *alphabet* is a nonempty set whose elements are called *letters*. In this paper, we use only finite alphabets, thus the term "finite" is omitted. A poset *labeled* by A is a poset $(P, <)$ equipped with a *labeling* map $P \to A$ which associates a letter to any element of P. The notion of a labeled poset corresponds to the notion of a *pomset* in the literature. Also, the finite labeled posets of width 1 correspond to the usual notion of words. In order to shorten the notation, we make no distinction between a poset and a labeled poset, except for operations. The *sequential product* (or *concatenation*, denoted by $P \cdot P'$ or PP' for short) and the *parallel product* (denoted by $P \parallel P'$) of labeled posets are respectively obtained by the sequential and parallel compositions of the corresponding (unlabeled) posets. The class of posets of $SP^\diamond$ labeled by A (or over A) is denoted by $SP^\diamond(A)$. We set $A^\diamond = \{P \in SP^\diamond(A) : \mathrm{wd}(P) \leq 1\}$. Observe that the elements of $A^\diamond$ are precisely the usual words on scattered and countable linear orderings, as defined in [5]. A *language* of $SP^\diamond(A)$ is a subset of $SP^\diamond(A)$. Let A and B be two alphabets and let $P \in SP^\diamond(A)$, $L \subseteq SP^\diamond(B)$ and $\xi \in A$. The labeled poset P in which each occurrence of the letter ξ is non-uniformly replaced by a labeled poset of the language L is denoted by $L \circ_\xi P$. The substitution, sequential and parallel products can be easily extended from labeled posets to languages of posets.

3 Rational Languages and Branching Automata

Let A be an alphabet and $\xi \in A$. Using the definition of substitution $\circ_\xi$, we define the iterated substitution on languages. By the way the usual rational operations on linear orderings are recalled. Let L and L' be languages of $SP^\diamond(A)$:

$$L \circ_\xi L' = \bigcup_{P \in L'} L \circ_\xi P, \qquad L^* = \{\prod_{j \in n} P_j | n \in \mathbb{N}, P_j \in L\}$$

$$L^{*\xi} = \bigcup_{i \in \mathbb{N}} L^{i\xi} \text{ with } L^{0\xi} = \{\xi\} \text{ and } L^{(i+1)\xi} = (\bigcup_{j \leq i} L^{j\xi}) \circ_\xi L$$

$$L^\omega = \{\prod_{j \in \omega} P_j | P_j \in L\} \qquad L^{-\omega} = \{\prod_{j \in -\omega} P_j | P_j \in L\}$$

$$L^\natural = \{\prod_{j \in \alpha} P_j | \alpha \in \mathcal{O}, P_j \in L\} \qquad L^{-\natural} = \{\prod_{j \in -\alpha} P_j | \alpha \in \mathcal{O}, P_j \in L\}$$

$$L \diamond L' = \{\prod_{j \in J \cup \hat{J}^*} P_j : J \in \mathcal{S} - \{0\} \text{ and } P_j \in L \text{ if } j \in J \text{ and } P_j \in L' \text{ if } j \in \hat{J}^*\}$$

A language $L \subseteq SP^\diamond(A)$ is *rational* if it is empty, or obtained from the letters of the alphabet A using usual rational operators : finite union $\cup$, finite concatenation $\cdot$, and finite iteration $*$, ω and $-\omega$ iterations, iteration and reverse iteration on ordinals $\natural$ and $-\natural$ as well as diamond operator $\diamond$, and using also the rational operators of finite parallel product $\|$, substitution $\circ_\xi$ and iterated substitution $^{*\xi}$, provided that the letter $\xi \in A$ appears only inside parallel factors. This latter condition excludes from the rational languages those of the form $(a\xi b)^{*\xi} = \{a^n \xi b^n : n \in \mathbb{N}\}$, for example, which are known to be not Kleene rational. Observe also that the usual Kleene rational languages are a particular case of the rational languages defined above, in which the operators $\|$, $\circ_\xi$ and $^{*\xi}$ are not allowed. Note also that the rational expressions are precisely those of Bruyère and Carton [5] over labeled posets on scattered and countable linear orderings, with additional operators $\|$, $\circ_\xi$ and $^{*\xi}$ for parallelism and substitution.

Example 1. Let $A = \{a, b, c\}$ and $L = c \circ_\xi (a \| (b\xi))^{*\xi}$. Then L is the smallest language containing c and such that if $x \in L$, then $a \| (bx) \in L$. Thus we have $L = \{c, a \| (bc), a \| (b(a \| (bc))), \dots\}$. □

Let L be a language where the letter ξ is not used. In order to lighten the notation we use the following abbreviation: $L^\circledast = \{\varepsilon\} \circ_\xi (L \| \xi)^{*\xi} = \{\|_{i<n} P_i : n \in \mathbb{N}, P_i \in L\}$ and $L^\oplus = L^\circledast - \{\varepsilon\}$. A subset L of $A^\circledast$ is *linear* if it has the form $L = a_1 \| \cdots \| a_k \| \left(\cup_{i \in I}(a_{i,1} \| \cdots \| a_{i,k_i})\right)^\circledast$ where the a_i and $a_{i,j}$ are elements of A and I is a finite set. It is *semi-linear* if it is a finite union of linear sets. The class of $\|$-*rational* languages of $A^\circledast$ is the smallest containing the empty set, $\{\varepsilon\}$, $\{a\}$ for all $a \in A$, and closed under finite union, parallel product $\|$, and finite parallel iteration $^\circledast$. The notions of rational, $\|$-rational, linear and semi-linear languages, which are defined over free algebras, also naturally apply to non-free algebras. It is known (see [10]) that the $\|$-rational sets of a commutative monoid M are precisely the semi-linear sets of M. Observe also that when L is a rational language of $SP^{\diamond+}(A)$, then $L \subseteq A^\circledast$ if and only if L is $\|$-rational.

We refer to [4] for a proof of the following Lemma:

Lemma 2 (Lemma 19 of [4]). *Let A be an alphabet and let ξ, X be two new symbols. Let $M \subseteq SP^{\diamond}(A)$ and let $L \subseteq SP^{\diamond}(A \cup \{X\}) \setminus SP^{\diamond}(A)$. Then $M \circ_{\xi} (\xi \circ_X L)^{*\xi}$ is the unique solution of the equation $X = M + L$.*

Automata on countable, scattered and series-parallel posets are a generalization of automata on finite series-parallel posets [13–16], series-parallel ω-posets [12] and automata on linear orderings [5]. A *branching automaton* over an alphabet A is a tuple $\mathcal{A} = (Q, A, E, I, F)$ where Q is a finite set of *states*, $I \subseteq Q$ is the set of *initial states*, $F \subseteq Q$ the set of *final states*, and E is the set of *transitions* of $\mathcal{A}$. The set of transitions E is partitioned into $E = (E_{\mathrm{seq}}, E_{\mathrm{join}}, E_{\mathrm{fork}})$, according to the different kinds of transitions. The set $E_{\mathrm{seq}} \subseteq (Q \times A \times Q) \cup (Q \times \mathcal{P}^+(Q)) \cup (\mathcal{P}^+(Q) \times Q)$ contains the *sequential* transitions, which are usual transitions (elements of $(Q \times A \times Q)$) or *limit* transitions (elements of $(Q \times \mathcal{P}^+(Q)) \cup (\mathcal{P}^+(Q) \times Q)$). The sets $E_{\mathrm{fork}} \subseteq Q \times \mathcal{M}^{>1}(Q)$ and $E_{\mathrm{join}} \subseteq \mathcal{M}^{>1}(Q) \times Q$ are respectively the sets of *fork* and *join* transitions. Transitions $(p, a, q) \in Q \times A \times Q$ and $(P, q) \in \mathcal{P}^+(Q) \times Q$ are sometimes respectively denoted by $p \xrightarrow{a} q$ and $P \rightarrow q$. A path γ from a state p to a state q is either the empty poset (in this case $p = q$), or a non-empty poset labeled by transitions, with a unique minimum and a unique maximum element. The states p and q are respectively called *source* (or *origin*) and *destination* of γ. Two paths γ and γ' are *consecutive* if the destination of γ is also the source of γ'. The paths γ labeled by $P \in SP^{\diamond}(A)$ and of content $C(\gamma) \in \mathcal{P}^+(Q)$ in $\mathcal{A}$ are defined as follows. For all $p \in Q$ there is an empty path from p to p labeled by ε and of content $\{p\}$. For all sequential transition $t = (p, a, q)$, $\gamma = t$ is a path from p to q labeled by a and of content $\{p, q\}$. For any finite sequence $(\gamma_j)_{j \leq k}$ of paths (with $k \geq 1$) respectively labeled by $P_0, \ldots, P_k$, from $p_0, \ldots, p_k$ to $q_0, \ldots, q_k$, if $t = (p, \{p_0, \ldots, p_k\})$ is a fork transition and $t' = (\{q_0, \ldots, q_k\}, q)$ a join transition, then $\gamma = t(\|_{j \leq k} \gamma_j)t'$ is a path from p to q, labeled by $\|_{j \leq k} P_j$ and of content $C(\gamma) = \{p, q\}$: observe that $C(\gamma)$ does not depend on the parallel parts $\gamma_0, \ldots, \gamma_k$ of γ. Furthermore, if the paths $(\gamma_j)_{j \leq k}$ are consecutive with respective contents $(C(\gamma_j))_{j \leq k}$, then $\prod_{j \leq k} \gamma_j$ is a path labeled by $\prod_{j \leq k} P_j$ from the source of γ_0 to the destination of γ_k, and of content $\cup_{j \leq k} C_j$. Finally, for any sequence $(\gamma_j)_{j \in \omega}$ of consecutive paths respectively labeled by $(P_j)_{j \in \omega}$ and of contents $(C(\gamma_j))_{j \in \omega}$, if $R = \{q \in Q : \forall i \in \omega\ \exists j > i\ q \in C(\gamma_j)\}$, then for any transition $t = (R, q)$, $(\prod_{j \in \omega} \gamma_j)t$ is a path from the source of γ_0 and to q, labeled by $\prod_{j \in \omega} P_j$ and of content $(\cup_{j \in \omega} C_j) \cup \{q\}$. The case $-\omega$ is symmetrical to ω. In $\mathcal{A}$, a path γ from p to q labeled by P of content C is denoted by $\gamma : p \underset{C,\mathcal{A}}{\overset{P}{\Longrightarrow}} q$. The label, content or automaton can be omitted in the notation of a path when they are implicit or of no interest. A labeled poset is *accepted* by an automaton if it is the label of a successful path leading from an initial state to a final state. The language $L(\mathcal{A})$ is the set of labeled posets accepted by the automaton $\mathcal{A}$.

Note that branching automata without fork and join transitions are precisely the automata on scattered and countable linear orderings defined by Bruyère and Carton [5]. The same way, if limit transitions are removed, we get branching automata for finite labeled posets of Lodaya and Weil [13–16]. As for finite words, rational languages and branching automata for scattered series-parallel posets are connected with a Kleene-like Theorem:

Theorem 3 [4]. *Let $L \subseteq SP^{\diamond}(A)$. Then L is the language of a branching automaton if and only if it is rational.*

Example 4. The automaton $\mathcal{A} = ([6], \{a, b, c\}, E, \{1\}, \{6\})$ defined by $E_{\text{seq}} = \{(2, a, 4), (3, b, 5), (6, c, 1), (\{1, 6\}, 6), (6, \{1, 6\})\}$, $E_{\text{fork}} = \{(1, \{2, 3\})\}$ and $E_{\text{join}} = \{(\{4, 5\}, 6)\}$ verifies $L(\mathcal{A}) = (a \parallel b) \diamond c$. □

An automaton is *sequentially separated* if, for all pairs (p, q) of states, all labels of paths from p to q are parallel posets, or all labels of paths from p to q are sequential posets. For every automaton $\mathcal{A}$ there is a sequentially separated automaton $\mathcal{B}$ such that $L(\mathcal{A}) = L(\mathcal{B})$. Also, for every pair of states (p, q) of an automaton, it is decidable whether there is a path from p to q or not.

The following Theorem states the main result of this paper:

Theorem 5. *Let A be an alphabet. The class of rational languages of $SP^{\diamond+}(A)$ is effectively closed under complement.*

Section 5 is devoted to a sketch of its proof, which essentially relies on the algebraic approach of automata.

4 Algebras

We now focus on the definitions of algebras for the recognition of languages of $SP^{\diamond}(A)$, with A an alphabet. Recall that an algebra is finite if it is composed of a finite number of elements. Even if in this paper we deal with infinite algebras, we use notions of universal algebras which are usually defined on finite algebras, and that can be easily generalized to our case. We refer to [1] for the basic algebraic definitions. A semigroup $(S, \cdot)$ is a set S equipped with an associative binary operation $\cdot$ called *product*. A $\parallel$*-semigroup* [13–16] $(S, \cdot, \parallel)$ is an algebra such that $(S, \cdot)$ is a semigroup and $(S, \parallel)$ is a commutative semigroup. In ambiguous contexts, the $\cdot$ and $\parallel$ products are respectively called *sequential* (or *series*) and *parallel*. The $\diamond$*-semigroups* are a generalization of semigroups for the recognition of words of $A^{\diamond}$ (see [8] for more details): a $\diamond$*-semigroup* $(S, \prod)$ is a set equipped with a map $\prod$ (also called *sequential product*) which associates an element of S to any countable and linearly ordered sequence $s = (s_j)_{j \in J}$ (with $J \in \mathcal{S}$) of elements of S, such that $\prod(t) = t$ for any $t \in S$ and $\prod$ is associative (i.e. for any factorization of the sequence s into a sequence of sequences $(t_j)_{j \in J'}$, $\prod(s) = \prod((\prod t_j)_{j \in J'})$. Finally, a $\parallel$-$\diamond$*-semigroup* $(S, \prod, \parallel)$ is an algebra such that $(S, \prod)$ and $(S, \parallel)$ are respectively a $\diamond$– and a commutative semigroup. In order to lighten the notation we often denote an algebra by its set of elements: for example, we denote the semigroup $(S, \cdot)$ by S. We denote by S^1 the algebra S if S has an identity 1 for all its operations, $S \cup \{1\}$ otherwise. We also denote by A^+, $SP(A)$ and $A^{\diamond}$ respectively the free semigroup, $\parallel$-semigroup, and $\diamond$-semigroup over the alphabet A. In this paper we particularly focus on $SP^{\diamond}(A)$ which is the free $\|$-$\diamond$-semigroup over A. Let S and T be two algebras of the same type. A morphism $\varphi : S \to T$ *recognizes* a subset X of S if $\varphi^{-1}\varphi(X) = X$. We say that T *recognizes* X if there exists a morphism from S into T recognizing X. A subset X of an algebra S is *recognizable* if there exist a *finite* algebra T with the same type as S and a morphism $\varphi : S \to T$ that recognizes X. Recognizable languages of $SP^+(A)$ are rational. However, in general, rational languages of $SP^+(A)$ are not recognizable. As an example, $(a \| b)^\oplus$ is not recognizable, since its syntactic $\|$-semigroup is isomorphic to $\mathbb{Z}$ (see [13]). Let $(S, \prod, \|)$ be a $\|$-$\diamond$-semigroup. Its sequential product $\prod$ is a *finite projection* if there exists $X \subseteq S$ such that $(X, \prod)$ is a finite $\diamond$-semigroup and $\prod$ maps every sequential product of at least two elements of S to an element of X. By extension of the work of Wilke [23] on ω-words, when $\prod$ is a finite projection, it can be equivalently replaced by an associative binary sequential product $\cdot$ and two maps $\omega : S \to S$ and $-\omega : S \to S$ such that, for all $s, t \in S$, $s \cdot (t \cdot s)^\omega = (s \cdot t)^\omega$, $(s \cdot t)^{-\omega} \cdot s = (t \cdot s)^{-\omega}$, $(s^n)^\omega = s^\omega$ and $(s^n)^{-\omega} = s^{-\omega}$ for all $n \in \mathbb{N}^*$. Observe that it suffices to define ω and $-\omega$ over finitely many elements: the idempotents (for the sequential product) of S.

Example 6. Let $A = \{a, b\}$ and $L \subseteq SP^{\diamond+}(A)$ be the language of non-empty posets P such that P has width at most 2 and each letter a that appears into a parallel part of P is incomparable with a b. Let $S = (X, \prod, \|)$ be the finite $\|$-$\diamond$-semigroup defined by $X = \{a, b, ab, p, 0\}$, the following $\|$ commutative product: $a \| a = ab \| a = 0$, $p \| x = 0$ for all $x \in S$, $a \| b = ab \| b = ab \| ab = b \| b = p$ and the sequential product $\prod$ such that, for any non-empty sequence $(s_j)_{j \in J}$ ($J \in \mathcal{S} - \{\emptyset\}$) of elements of S, $\prod((s_j)_{j \in J}) = a$ if $(s_j)_{j \in J}$ contains only as, $\prod((s_j)_{j \in J}) = ab$ if $(s_j)_{j \in J}$ contains at least one a and one b, $\prod((s_j)_{j \in J}) = b$ if $(s_j)_{j \in J}$ contains only bs, and $\prod((s_j)_{j \in J}) = p$ if $(s_j)_{j \in J}$ contains only p, a, b, ab, with at least one p. The element 0 is a zero for both $\prod$ and $\|$. Let $\varphi : SP^{\diamond+}(A) \to S$ be the morphism defined by $\varphi(a) = a$ and $\varphi(b) = b$. Then $L = \varphi^{-1}(\{a, b, ab, p\})$. Furthermore, the sequential product of S is a finite projection since S has a finite number of elements. Then S can be equivalently defined by $W = (X, \cdot, \omega, -\omega, \|)$ where $x \cdot x' = \prod(x, x')$ and $x^\omega = x^{-\omega} = x$ for all $x, x' \in X$. □

The following notions are adapted from [9]. Let P be a partially ordered set and S a semigroup. A mapping σ from ordered pairs $(x, y) \in P^2$ such that $x \sim_< y$, to S, is an *additive labeling from P to S* if $\sigma(x, y)\sigma(y, z) = \sigma(x, z)$ for all $x < y < z$ in P. From a morphism of semigroups $\varphi : (SP^\diamond(A), \cdot) \to S$ and $P \in SP^\diamond(A)$, one can build an additive labeling $\varphi_P : (\hat{P}, \preceq) \to S$ with $\varphi_P((A, B), (A', B')) = \varphi(B \cap A')$. A *split of height n* of P is a mapping $s : P \to [n]$ ($n = 0$ is possible; in this case $P = \emptyset$). Two elements x, y such that $x \sim_< y$ and $s(x) = s(y) = k$ are *k-neighbors* if $s(z) \geq k$ for all $z \in [x, y]$ with $z \sim_< x$. Note that k-neighborhoodness is an equivalence relation over the elements of P. Let σ be an additive labeling from P to a semigroup S. Then a split s of P is *Ramseyan* for σ if for every equivalence class C for k-neighborhoodness there exists an idempotent e such that $\sigma(x, y) = e$ for all $x < y$ in C.

The notion of a *finite projection* of a semigroup S is self-understanding from its definition on $\|$-$\diamond$-semigroups. Theorem 4 of [9] can be reformulated for posets as follows:

Theorem 7. *For every poset $P \in SP^\diamond$, every semigroup S with a finite projection $fp(S)$ and additive labeling σ from P to S, there exists a Ramseyan split of P for σ of height at most $2|fp(S)|+1$.*

5 Sketch of the Proof of Theorem 5

Let A be an alphabet, $\mathcal{A} = (Q, A, E, I, F)$ a branching automaton, and $L = L(\mathcal{A})$. When $X \subseteq SP^{\diamond+}(A)$, we denote by $Seq(X)$ the set of sequential posets of X. Denote also by $L_{p,q}$ (resp. $L_{p,q,C}$ with $C \in \mathcal{P}^+(Q)$) the set of non-empty labels of paths from state p to state q (resp. of content C) in $\mathcal{A}$.

The proof of Theorem 5 consists in constructing a rational expression e for $SP^{\diamond+}(A) - L$. When $\phi : SP^{\diamond+}(A) \to S$ is a morphism of $\|$-$\diamond$-semigroups and $D \in \mathcal{P}(Q^2 \times \mathcal{P}^+(Q))$, denote by $\Delta_D^\phi = \{\phi(P) : p \overset{P}{\underset{C,\mathcal{A}}{\Longrightarrow}} q \text{ iff } (p,q,C) \in D\}$. The first step is to construct a $\|$-$\diamond$-semigroup from $\mathcal{A}$, by a generalization of the usual technique used to construct a finite semigroup from a Kleene automaton on finite words. This consists in defining a congruence $\sim_\mathcal{A}$ of $\|$-$\diamond$-semigroups over the posets of $SP^{\diamond+}(A)$, by $P \sim_\mathcal{A} P'$ if and only if P can be substituted by P' in any part of any path $\gamma : p \overset{R}{\underset{C,\mathcal{A}}{\Longrightarrow}} q$ of $\mathcal{A}$ in order to build another path $\gamma' : p \overset{R'}{\underset{C,\mathcal{A}}{\Longrightarrow}} q$ of same source, destination and content and whose label R' is R in which some occurrences of P have been replaced by P'. The natural morphism $\varphi_{\sim_\mathcal{A}} : SP^{\diamond+}(A) \to SP^{\diamond+}(A)/\sim_\mathcal{A}$ which associates to each poset $P \in SP^{\diamond+}(A)$ its equivalence class in $SP^{\diamond+}(A)/\sim_\mathcal{A}$ recognizes $L_{p,q,C}$ for each $p, q, C \in Q^2 \times \mathcal{P}^+(Q)$, and L. Note that $SP^{\diamond+}(A)/\sim_\mathcal{A}$ may be infinite, as it is illustrated by the following example.

Example 8. Consider the automaton $\mathcal{A}$ of Fig. 1 of language $L(\mathcal{A}) = (a \parallel b)^\oplus \diamond c$. For all $k_1, k_2, k_3, k_4 \in \mathbb{N}$ such that $k_1 - k_2 = k_3 - k_4$ and $k_2, k_4 > 0$ we have $a^{\|k_1} \parallel b^{\|k_2} \sim_\mathcal{A} a^{\|k_3} \parallel b^{\|k_4}$. Also, $P \sim_\mathcal{A} P'$ for all $P, P' \in (a \parallel b)^\oplus \diamond c - (a \parallel b)^\oplus$.

Let $S = \mathbb{Z} \cup \{a, b, c, 0c, c0, 0c0, \perp\}$. Equip S with a commutative parallel product with $z \parallel z' = z +_\mathbb{Z} z'$, $a \parallel z' = 1 \parallel z'$, $b \parallel z' = -1 \parallel z'$ for all $z, z' \in \mathbb{Z}$, $a \parallel b = 0$, and all other parallel product are sent to $\perp$. Equip also S with a sequential product such that for all sequence $s = (s_i)_{i \in I}$ of elements of S, $I \in \mathcal{S} - \{0, 1\}$, $\prod_{i \in I} s_i = 0c0$ if $s \in 0 \diamond c$, $c0c = c$, $z^2 = zx = xz = c^2 = \perp^2 = \perp$ for all $z \in \mathbb{Z} \cup \{a, b\}$, $x \in S$. As the sequential product of S is a finite projection and the idempotents for the sequential product are $0c, c0, \perp$, it can equivalently be defined by the binary product as above and $(0c)^\omega = 0c0 = (c0)^{-\omega}$, $(0c)^{-\omega} = 0c$, $(c0)^\omega = c0$. Note that $(S, \|)$ is finitely generated by $\{-1, 1, a, b, c, 0c, c0, 0c0, \perp\}$. Let $\varphi : SP^{\diamond+}(A) \to S$ defined by $\varphi(x) = x$ for all $x \in A$. Then $L = \varphi^{-1}(\{0, 0c0\})$. Furthermore, $SP^{\diamond+}(A)/\sim_\mathcal{A}$ is isomorphic to S.

□

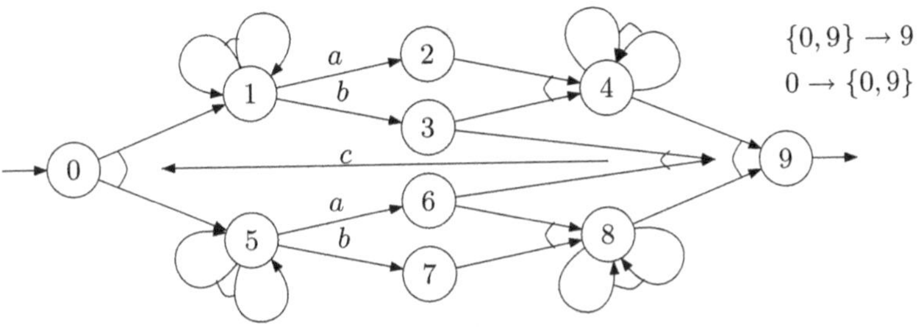

Fig. 1. An automaton $\mathcal{A}$ for $(a \parallel b)^{\oplus} \diamond c$. Fork transitions are $(0, \{1, 5\})$, $(5, \{5, 5\})$ and $(1, \{1, 1\})$, join transitions are $(\{2, 3\}, 4)$, $(\{6, 7\}, 8)$, $(\{4, 8\}, 9)$, $(\{4, 4\}, 4)$, $(\{8, 8\}, 8)$ and $(\{3, 6\}, 9)$

However, $(SP^{\diamond+}(A)/\sim_{\mathcal{A}}, \parallel)$ is finitely generated by $\varphi_{\sim_{\mathcal{A}}}(Seq(SP^{\diamond+}(A)))$, $\varphi_{\sim_{\mathcal{A}}}(L_{p,q})$ is a $\parallel$-rational set of $SP^{\diamond+}(A)/\sim_{\mathcal{A}}$ for all $p, q \in Q$, and thus, so is $\varphi_{\sim_{\mathcal{A}}}(L)$. We also have $\varphi_{\sim_{\mathcal{A}}}^{-1}(\Delta_D^{\varphi_{\sim_{\mathcal{A}}}}) = \Delta_D^{\mathrm{id}}$ for all $D \in \mathcal{P}(Q^2 \times \mathcal{P}^+(Q))$. Recall that the $\parallel$-rational sets of a commutative monoid M form a boolean algebra [10, Theorem 3], which is effective when M is finitely generated (as emphasized in [21]). As a consequence, $\Delta_D^{\varphi_{\sim_{\mathcal{A}}}}$ is a $\parallel$-rational set of $SP^{\diamond+}(A)/\sim_{\mathcal{A}}$ for all D.

As $SP^{\diamond+}(A) - L = \bigcup_{\substack{D\in\mathcal{P}(Q^2\times\mathcal{P}^+(Q))\\ D\cap I\times F\times\mathcal{P}^+(Q)=\emptyset}} \Delta_D^{\mathrm{id}}$, it suffices to show that $\varphi_{\sim_{\mathcal{A}}}^{-1}(\Delta_D^{\varphi_{\sim_{\mathcal{A}}}})$ is a rational set of $SP^{\diamond+}(A)$ for all D. We translate the problem into a $\parallel$-$\diamond$-semigroup $\mathbb{N}^{k*}$ with more properties than $SP^{\diamond+}(A)/\sim_{\mathcal{A}}$. Very informally speaking, denote by $G = \{g_1, \ldots, g_k\}$ the finite generator of $(SP^{\diamond+}(A)/\sim_{\mathcal{A}}, \parallel)$. We may suppose that $\mathcal{A}$ is sequentially separated. Thus that the elements of G are indecomposable with respect to the parallel product, that is to say, each $g_i \in G$ can not be written $g_i = s \parallel s'$ with $s, s' \in SP^{\diamond+}(A)/\sim_{\mathcal{A}}$. We are going to define a morphism $\mu : SP^{\diamond+}(A) \to \mathbb{N}^{k*}$ that enables, for every $P \in SP^{\diamond+}(A)$ whose maximal parallel factorization is $P = P_1 \parallel \cdots \parallel P_n$, the count of all i, $i \in [n]$, such that $\varphi_{\sim_{\mathcal{A}}}(P_i) = g_j$, for every $j \in [k]$. Also, every language recognized by $SP^{\diamond+}(A)/\sim_{\mathcal{A}}$ is recognized by $\mathbb{N}^{k*}$.

Denote by $(\mathbb{N}^{k*}, +)$ the commutative semigroup whose elements are k-tuples of non-negative integers, without $(0, \ldots, 0)$. It is generated by the k-tuples with all components set to 0, except one which is set to 1. For short we denote by 1^i the element of the generator of $\mathbb{N}^{k*}$ with the ith component set to 1. The (parallel) product $+$ of $(\mathbb{N}^{k*}, +)$ is the sum componentwise. Define a surjective morphism of commutative semigroups $\psi : (\mathbb{N}^{k*}, +) \to (SP^{\diamond+}(A)/\sim_{\mathcal{A}}, \parallel)$ by $\psi(1^i) = g_i$ for all $i \in [k]$. As $\psi^{-1}(g_i) = \{1^i\}$ for all $i \in [k]$, $\psi^{-1}(ss')$ is a singleton for all $s, s' \in SP^{\diamond+}(A)/\sim_{\mathcal{A}}$. Now we equip $(\mathbb{N}^{k*}, +)$ with a structure of $\parallel$-$\diamond$-semigroup by setting, for all $n, n_1, n_2 \in \mathbb{N}^{k*}$, $n_1 n_2 = \psi^{-1}(\psi(n_1)\psi(n_2))$, $n^{\omega} = \psi^{-1}((\psi(n))^{\omega})$ and $n^{-\omega} = \psi^{-1}((\psi(n))^{-\omega})$. This sequential product is a finite projection. We define a surjective morphism of $\parallel$-$\diamond$-semigroups $\mu : SP^{\diamond+}(A) \to \mathbb{N}^{k*}$ by $\mu(a) = \psi^{-1}\varphi_{\sim_{\mathcal{A}}}(a)$ for all $a \in A$. The diagram of Fig. 2 sums up the situation. For all $s \in SP^{\diamond+}(A)/\sim_{\mathcal{A}}$, $\varphi_{\sim_{\mathcal{A}}}^{-1}(s) = \mu^{-1}\psi^{-1}(s)$. We also have $\psi^{-1}(\Delta_D^{\varphi_{\sim_{\mathcal{A}}}}) = \Delta_D^{\mu}$ and $\mu^{-1}(\Delta_D^{\mu}) = \Delta_D^{\mathrm{id}}$ for all $D \in \mathcal{P}(Q^2 \times \mathcal{P}^+(Q))$. According to [10, Corollary III.2] Δ_D^{μ} is a $\parallel$-rational set of $\mathbb{N}^{k*}$, and thus semi-linear: it has the form $\Delta_D^{\mu} =$

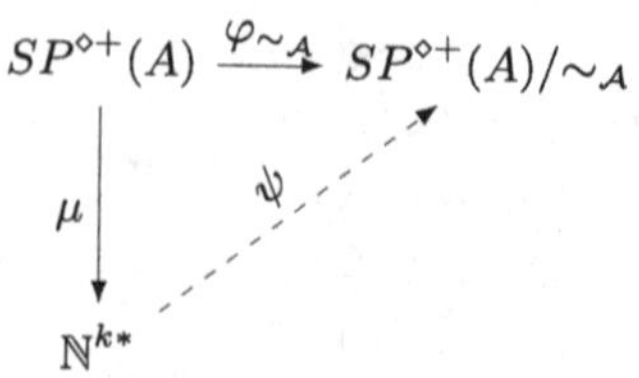

Fig. 2. The morphisms between the ||-◇-semigroups. Full arrows represent morphisms of ||-◇-semigroups, and dashed arrows morphisms of commutative semigroups

$\cup_{i\in I_D}(a_{D,i} + B_{D,i}^{\circledast})$ for some finite set I_D, $a_{D,i} \in \mathbb{N}^{k*}$, $B_{D,i}$ some finite part of $\mathbb{N}^{k*}$. For all $i \in I_D$ set $\Delta^{\mu}_{D,i} = a_{D,i} + B^{\circledast}_{D,i}$. We may assume that all the $\Delta^{\mu}_{D,i}$ are pairwise disjoint [10, Theorem IV]. Setting $B_{D,i} = \{b_{D,i,1}, \ldots, b_{D,i,l_{D,i}}\}$ it holds $\mu^{-1}(\Delta^{\mu}_{D,i}) = \mu^{-1}(a_{D,i}) \parallel \{\mu^{-1}(b_{D,i,1}) \cup \cdots \cup \mu^{-1}(b_{D,i,l_{D,i}})\}^{\circledast}$, so it just remain to show that

Lemma 9. *For all $n \in \mathbb{N}^{k*}$, $\mu^{-1}(n)$ is a rational set of $SP^{\diamond+}(A)$.*

Proof. (Sketch of) Let $\varphi : SP^{\diamond+}(A) \to S$ be a morphism of ||-◇-semigroups. For each j, M non-negative integers, $x \in S$, and $\iota, \iota' \in \{[,]\}$, define $S^{\iota\iota'}_{j,M}(x)$ (or $S^{\iota\iota'}_{j}(x)$ for short) as the posets P of $\varphi^{-1}(x)$ such that $\hat{P}^{\iota\iota'}$ admits a Ramseyan split s, for φ_P, of height M; and s is also a Ramseyan split of $\{(A, B) \in \hat{P}^{\iota\iota'} : A \cup B = P\}$, for φ_P, of height j.

Considering linear orderings only, Colcombet [9] expressed $S^{\iota\iota'}_{j+1}(x)$ with an equality that depends only of the $S^{\iota''\iota'''}_{j}(s)$, $s \in S$, $\iota'', \iota''' \in \{[,]\}$ and that uses only the rational operators for linear orderings:

$$S^{\iota\iota'}_{j+1}(x) = S^{\iota\iota'}_{j}(x) + \sum_{yz=x} S^{\iota[}_{j}(y)S^{]\iota'}_{j}(z) + \sum_{\substack{yez=x\\ e^2=e}} S^{\iota[}_{j}(y)C_{e,j+1}S^{]\iota'}_{j}(z)$$
$$+ \sum_{\substack{ye^{\omega}z=x\\ e^2=e}} S^{\iota[}_{j}(y)C^{\omega}_{e,j+1}S^{[\iota'}_{j}(z) + \sum_{\substack{ye^{-\omega}z=x\\ e^2=e}} S^{\iota]}_{j}(y)C^{-\omega}_{e,j+1}S^{]\iota'}_{j}(z) + \sum_{\substack{ye^{\zeta}z=x\\ e^2=e}} S^{\iota]}_{j}(y)C^{\zeta}_{e,j+1}S^{[\iota'}_{j}(z)$$

with $\varphi^{-1}(x) = S^{[]}_{2|S|}(x)$, $S^{][}_{0}(x) = \varphi^{-1}(x) \cap A$, $S^{[[}_{0}(x) = S^{]]}_{0}(x) = \varphi^{-1}(x) \cap \{\varepsilon\}$, $S^{[]}_{0}(x) = \emptyset$, and $C_{e,j+1}$, $C^{\omega}_{e,j+1}$, $C^{-\omega}_{e,j+1}$, $C^{\zeta}_{e,j+1}$ rational sets that depend only of the languages of the form $S^{\iota''\iota'''}_{j}(s)$, $s \in S$, $\iota'', \iota''' \in \{[,]\}$.

We adapt this to the case where $P \in SP^{\diamond+}(A)$, replacing φ by $\mu : SP^{\diamond+}(A) \to \mathbb{N}^{k*}$. There are several points to consider. First, technically the empty poset is not taken into consideration in the framework of posets. Second, $\hat{P}^{\iota\iota'}$ admits a Ramseyan split for φ_P and of height j if and only if $C = \{(A, B) \in \hat{P}^{\iota\iota'} : (A, B) = P\}$ admits a Ramseyan split for φ_P and of height j, and, for each P_i between two consecutive elements of C with $|P_i| > 1$ (thus $P_i = \parallel_{j\in J_i} P_j$ for some $|J_i| > 1$ and nonempty P_j), each $\hat{P}_j^{[]}$ admits itself a Ramseyan split for φ_P and of height $|2k + 1|$. And third, as $\mathbb{N}^{k*}$ is infinite but partitioned into finitely many $\Delta^{\mu}_{D,i}$ in which all elements are equivalent regarding to the sequential product, we need to replace any occurrence of some x involved in a sequential product in the right member of the equality above by some $\Delta^{\mu}_{D,i}$. The set $S^{\iota\iota'}_{j+1}(\Delta^{\mu}_{D,i})$ is composed of all the sequential posets of $S^{\iota\iota'}_{j+1}(g)$ for all $g \in G \cap \Delta^{\mu}_{D,i}$, and, if $\iota\iota' =][$, all the parallel posets and letters of $\mu^{-1}(\Delta^{\mu}_{D,i})$. For simplicity we write $\Delta^{\mu}_{D,i}x = y$ (resp. $x\Delta^{\mu}_{D,i} = y$) when $zx = y$ (resp. $xz = y$) for all $z \in \Delta^{\mu}_{D,i}$. With the help of Theorem 7, the equalities of Colcombet above can be rewritten in $\mathbb{N}^{k*}$ as

$$S^{\iota\iota'}_{j+1}(x) = S^{\iota\iota'}_{j}(x) + \sum_{\Delta^{\mu}_{D,i}\Delta^{\mu}_{D',i'}=x} S^{\iota[}_{j}(\Delta^{\mu}_{D,i})S^{]\iota'}_{j}(\Delta^{\mu}_{D',i'})$$
$$+ \sum_{\substack{\Delta^{\mu}_{D,i}e\Delta^{\mu}_{D',i'}=x\\ e^2=e}} S^{\iota[}_{j}(\Delta^{\mu}_{D,i})C_{e,j+1}S^{]\iota'}_{j}(\Delta^{\mu}_{D',i'}) + \sum_{\substack{\Delta^{\mu}_{D,i}e^{\omega}\Delta^{\mu}_{D',i'}=x\\ e^2=e}} S^{\iota[}_{j}(\Delta^{\mu}_{D,i})C^{\omega}_{e,j+1}S^{[\iota'}_{j}(\Delta^{\mu}_{D',i'})$$
$$+ \sum_{\substack{\Delta^{\mu}_{D,i}e^{-\omega}\Delta^{\mu}_{D',i'}=x\\ e^2=e}} S^{\iota]}_{j}(\Delta^{\mu}_{D,i})C^{-\omega}_{e,j+1}S^{]\iota'}_{j}(\Delta^{\mu}_{D',i'}) + \sum_{\substack{\Delta^{\mu}_{D,i}e^{\zeta}\Delta^{\mu}_{D',i'}=x\\ e^2=e}} S^{\iota]}_{j}(\Delta^{\mu}_{D,i})C^{\zeta}_{e,j+1}S^{[\iota'}_{j}(\Delta^{\mu}_{D',i'})$$

where $C_{e,j+1}$, $C^{\omega}_{e,j+1}$, $C^{-\omega}_{e,j+1}$, $C^{\zeta}_{e,j+1}$ are rational sets that depend only of the languages of the form $S^{\iota''\iota'''}_{j}(\Delta^{\mu}_{D,i})$ and can be obtained precisely as in the case of linear orderings (see [9, Proof of Theorem 6]), $D \in \mathcal{P}^{+}(Q)$, $i \in I_D$, $\iota'', \iota''' \in \{[,]\}$, and with

$$S^{\iota\iota'}_{j+1}(\Delta^{\mu}_{D,i}) = \begin{cases} S^{\iota\iota'}_{j+1}(a_{D,i}) + S^{\iota\iota'}_{j}(\Delta^{\mu}_{D,i}) & \text{if } \Delta^{\mu}_{D,i} = a_{D,i} + B^{\circledast}_{D,i} \\ (\sum_{b \in B_{D,i}} S^{\iota\iota'}_{j+1}(b)) + S^{\iota\iota'}_{j}(\Delta^{\mu}_{D,i}) & \text{if } \Delta^{\mu}_{D,i} = B^{\circledast}_{D,i} \end{cases} \tag{1}$$

$$S^{][}_{0}(\Delta^{\mu}_{D,i}) = \begin{cases} S^{][}_{0}(a_{D,i}) + S^{[]}_{2k+1}(a_{D,i}) \parallel (\sum_{b \in B_{D,i}} S^{[]}_{2k+1}(b))^{\oplus} \\ \qquad \text{if } \Delta^{\mu}_{D,i} = a_{D,i} + B^{\circledast}_{D,i}, \\ (\sum_{b \in B_{D,i}} S^{][}_{0}(b)) + (\sum_{b \in B_{D,i}} S^{[]}_{2k+1}(b)) \parallel (\sum_{b \in B_{D,i}} S^{[]}_{2k+1}(b))^{\oplus} \\ \qquad \text{if } \Delta^{\mu}_{D,i} = B^{\circledast}_{D,i} \end{cases} \tag{2}$$

$$S^{][}_{0}(x) = (\mu^{-1}(x) \cap A) \sum_{y+z=x} S^{[]}_{2k+1}(y) \parallel S^{[]}_{2k+1}(z) \tag{3}$$

$$S^{[[}_{0}(x) = S^{]]}_{0}(x) = S^{[]}_{0}(x) = S^{[[}_{0}(\Delta^{\mu}_{D,i}) = S^{]]}_{0}(\Delta^{\mu}_{D,i}) = S^{[]}_{0}(\Delta^{\mu}_{D,i}) = \emptyset \tag{4}$$

Note that the choices for y, z in (3) are finite since we are in $\mathbb{N}^{k*}$. This gives a finite system of equations, where recursion occurs only in parallel parts, and whose solution is rational with the help of Lemma 2. As $\mu^{-1}(n) = S^{[]}_{2k+1}(n)$ then $\mu^{-1}(n)$ is rational for all $n \in \mathbb{N}^{k*}$.

Immediately, $\mu^{-1}(\Delta_D^\mu)$ and $\varphi_{\sim_A}^{-1}(\Delta_D^{SP^{\diamond+}(A)/\sim_A})$ are rational sets of $SP^{\diamond+}(A)$ for all $D \in \mathcal{P}(Q^2 \times \mathcal{P}^+(Q))$. Note that all the constructions are effective.

References

1. Almeida, J.: Finite Semigroups and Universal Algebra. Series in Algebra, vol. 3. World Scientific, Singapore (1994)
2. Bedon, N.: Logic and bounded-width rational languages of posets over countable scattered linear orderings. In: Artemov, S., Nerode, A. (eds.) LFCS 2009. LNCS, vol. 5407, pp. 61–75. Springer, Heidelberg (2008)
3. Bedon, N.: Logic and branching automata. Log. Meth. Comput. Sci. **11**(4:2), 1–38 (2015)
4. Bedon, N., Rispal, C.: Series-parallel languages on scattered and countable posets. Theor. Comput. Sci. **412**(22), 2356–2369 (2011)
5. Bruyère, V., Carton, O.: Automata on linear orderings. J. Comput. Syst. Sci. **73**(1), 1–24 (2007)
6. Büchi, J.R.: On a decision method in the restricted second-order arithmetic. In: Proceedings of International Congress on Logic, Methodology and Philosophy of Science, Berkeley 1960 (1962)
7. Büchi, J.R.: Transfinite automata recursions and weak second order theory of ordinals. In: Proceedings of International Congress on Logic, Methodology, and Philosophy of Science 1964 (1965)
8. Carton, O., Rispal, C.: Complementation of rational sets on countable scattered linear orderings. Int. J. Found. Comput. Sci. **16**(4), 767 (2005)
9. Colcombet, T.: Factorisation forests for infinite words and applications to countable scattered linear orderings. Theor. Comput. Sci. **411**, 751–764 (2010)
10. Eilenberg, S., Schützenberger, M.P.: Rational sets in commutative monoids. J. Algebra **13**(2), 173–191 (1969)
11. Kuske, D.: Infinite series-parallel posets: logic and languages. In: Welzl, E., Montanari, U., Rolim, J.D.P. (eds.) ICALP 2000. LNCS, vol. 1853, pp. 648–662. Springer, Heidelberg (2000)
12. Kuske, D.: Towards a language theory for infinite N-free pomsets. Theor. Comput. Sci. **299**, 347–386 (2003)
13. Lodaya, K., Weil, P.: A Kleene iteration for parallelism. In: Arvind, V., Sarukkai, S. (eds.) FST TCS 1998. LNCS, vol. 1530, pp. 355–367. Springer, Heidelberg (1998)
14. Lodaya, K., Weil, P.: Series-parallel posets: algebra, automata and languages. In: Meinel, C., Morvan, M. (eds.) STACS 1998. LNCS, vol. 1373, pp. 555–565. Springer, Heidelberg (1998)
15. Lodaya, K., Weil, P.: Series-parallel languages and the bounded-width property. Theor. Comput. Sci. **237**(1–2), 347–380 (2000)
16. Lodaya, K., Weil, P.: Rationality in algebras with a series operation. Inform. Comput. **171**, 269–293 (2001)
17. Muller, D.E.: Infinite sequences and finite machines. In: Proceedings of Fourth Annual Symposium on Switching circuit theory and logical design. IEEE (1963)
18. Rabin, M.O.: Decidability of second-order theories and automata on infinite trees. Trans. Am. Math. Soc. **141**, 1–35 (1969)
19. Rival, I.: Optimal linear extension by interchanging chains. Proc. AMS **89**(3), 387–394 (1983)
20. Rosenstein, J.G.: Linear Orderings. Academic Press, New York (1982)

21. Sakarovitch, J.: Elements of Automata Theory. Cambridge University Press, Cambridge (2009)
22. Valdes, J., Tarjan, R.E., Lawler, E.L.: The recognition of series parallel digraphs. SIAM J. Comput. **11**, 298–313 (1982)
23. Wilke, T.: An algebraic theory for regular languages of finite and infinite words. Int. J. Algebra Comput. **3**(4), 44–489 (1993)

Cayley Automatic Groups and Numerical Characteristics of Turing Transducers

Dmitry Berdinsky(✉)

Department of Computer Science, The University of Auckland,
Private Bag 92019, Auckland 1142, New Zealand
berdinsky@gmail.com

Abstract. This paper is devoted to the problem of finding characterizations for Cayley automatic groups. The concept of Cayley automatic groups was recently introduced by Kharlampovich, Khoussainov and Miasnikov. We address this problem by introducing three numerical characteristics of Turing transducers: growth functions, Følner functions and average length growth functions. These three numerical characteristics are the analogs of growth functions, Følner functions and drifts of simple random walks for Cayley graphs of groups. We study these numerical characteristics for Turing transducers obtained from automatic presentations of labeled directed graphs.

Keywords: Cayley automatic groups · Turing transducers · Growth function · Følner function · Random walk

1 Introduction

This paper contributes to the field of automatic structures [11–13] with particular emphasis on Cayley automatic groups [10]. Recall that a finitely generated group G is called Cayley automatic if for some set of generators S the labeled directed Cayley graph $\Gamma(G, S)$ is an automatic structure (or, FA–presentable). All automatic groups in the sense of Thurston are Cayley automatic. However, the class of Cayley automatic groups is considerably wider than the class of automatic groups. For example, all finitely generated nilpotent groups of nilpotency class at most two and all fundamental groups of three–dimensional manifolds are Cayley automatic [10]. The Baumslag–Solitar groups are Cayley automatic [1]. Cayley automatic groups retain the key algorithmic properties which hold for automatic groups: the word problem for Cayley automatic groups is decidable in quadratic time, the conjugacy problem for Cayley biautomatic groups is decidable, and the first order theory for Cayley graphs of Cayley automatic groups is decidable.

Oliver and Thomas found a characterization of FA–presentable groups by showing that a finitely generated group is FA–presentable if and only if it is virtually abelian [16]. Their result is based partly on the celebrated Gromov's theorem on groups of polynomial growth. But, the problem of finding characterizations for Cayley automatic groups is more complicated, and it seems to

S. Brlek and C. Reutenauer (Eds.): DLT 2016, LNCS 9840, pp. 26–37, 2016.
DOI: 10.1007/978-3-662-53132-7_3

require new approaches. In this paper we address this problem by introducing some numerical characteristics for Turing transducers of the special class $\mathcal{T}$.

In Sect. 2 we define the class of Turing transducers $\mathcal{T}$. Then we show that automatic presentations of Cayley graphs of groups can be expressed in terms of Turing transducers of the class $\mathcal{T}$. This explains why study of admissible asymptotic behavior for some numerical characteristics of Turing transducers of the class $\mathcal{T}$ is relevant to the problem of finding characterizations for Cayley automatic groups. In Sect. 3 we introduce three numerical characteristics for Turing transducers of the class $\mathcal{T}$. In this paper, wreath products of groups are used as the source of examples of Cayley automatic groups. Therefore, in Sect. 4 we briefly recall basic definitions for wreath products of groups. In Sect. 5 we discuss asymptotic behavior of the numerical characteristics of Turing transducers of the class $\mathcal{T}$. Section 6 concludes the paper.

2 Turing Transducers of the Class $\mathcal{T}$ and Automatic Presentations of Labeled Directed Graphs

Recall that a $(k+1)$–tape Turing transducer T for $k \geqslant 1$ is a multi–tape Turing machine which has one input tape and k output tapes. See, e.g., [14, Sect. 10] for the definition of Turing transducers. The special class of Turing transducers $\mathcal{T}$ that we consider in this paper is described as follows. Let us be given a $(k+1)$–tape Turing transducer $T \in \mathcal{T}$ and an input word $x \in \Sigma^*$. Initially, the input word x appears on the input tape, the output tapes are completely blank and all heads are over the leftmost cells. First the heads of T move synchronously from the left to the right until the end of the input x. Then the heads make a finite number of steps (probably no steps) further to the right, where this number of steps is bounded from above by some constant which depends on T. After that, the heads of T move synchronously from the right to the left until it enters a final state with all heads over the leftmost cells.

We say that T accepts x if T enters an accepting state; otherwise, T rejects x. Let $L \subseteq \Sigma^*$ be the set of inputs accepted by T. We say that T translates $x \in L$ into the outputs $y_1, \dots, y_k$ if for the word x fed to T as an input, T returns the word y_i on the ith output tape of T for every $i = 1, \dots, k$. It is assumed that for every input $x \in L$, the output $y_i \in L$ for every $i = 1, \dots, k$. Let $L' \subseteq L^k$ be the set of all k–tuples of outputs $(y_1, \dots, y_k)$. We say that T translates L into L'.

For a given finite alphabet Σ put $\Sigma_\diamond = \Sigma \cup \{\diamond\}$, where $\diamond \notin \Sigma$. The convolution of n words $w_1, \dots, w_n \in \Sigma^*$ is the string $w_1 \otimes \cdots \otimes w_n$ of length $\max\{|w_1|, \dots, |w_n|\}$ over the alphabet $\Sigma_\diamond^n$ defined as follows. The kth symbol of the string is $(\sigma_1, \dots, \sigma_n)$, where σ_i, $i = 1, \dots, n$ is the kth symbol of w_i if $k \leqslant |w_i|$ and $\diamond$ otherwise. The convolution $\otimes R$ of a n–ary relation $R \subseteq \Sigma^{*n}$ is defined as $\otimes R = \{w_1 \otimes \cdots \otimes w_n | (w_1, \dots, w_n) \in R\}$. Recall that a n–tape synchronous finite automaton is a finite automaton over the alphabet $\Sigma_\diamond^n \setminus \{(\diamond, \dots, \diamond)\}$. Let $T \in \mathcal{T}$. Lemma 1 below shows connection between Turing transducers of the class $\mathcal{T}$ and multi–tape synchronous finite automata.

Lemma 1. *There exists a $(k+1)$–tape synchronous finite automaton $\mathcal{M}$ such that a convolution $x \otimes y_1 \otimes \cdots \otimes y_k \in \Sigma_\diamond^{(k+1)*}$ is accepted by $\mathcal{M}$ iff T translates the input x into the outputs $y_1, \dots, y_k$. In particular, the language L is regular.*

Proof. The lemma can be obtained straightforwardly from the following two well known facts. The first fact is that the class of regular languages is closed under reverse. The second fact is as follows. Let the convolutions $\otimes R_1$ and $\otimes R_2$ of two relations $R_1 = \{(x,y)|x,y \in \Sigma^*\}$ and $R_2 = \{(y,z)|y,z \in \Sigma^*\}$ be accepted by two–tape synchronous finite automata. Then the convolution $\otimes R$ of the relation $R = \{(x,z)|\exists y[(x,y) \in R_1 \wedge (y,z) \in R_2]\}$ is accepted by a two–tape synchronous finite automaton. □

In other words, one can say that multi–tape synchronous finite automata simulate Turing transducers of the class $\mathcal{T}$. In different context, the notion of simulation for finite automata appeared, e.g., in [3,4].

For a given k, put $\Sigma_k = \{1, \dots, k\}$. Let $T \in \mathcal{T}$ be a $(k+1)$–tape Turing transducer translating a language L into $L' \subseteq L^k$. We construct the labeled directed graph Γ_T with the labels from Σ_k as follows. The set of vertices $V(\Gamma_T)$ is identified with L. For given $u, v \in L$ there is an oriented edge (u,v) labeled by $j \in \Sigma_k$ if T translates u into some outputs $w_1, \dots, w_k$ such that $w_j = v$. It is easy to see that each vertex of the graph Γ_T has k outgoing edges labeled by $1, \dots, k$.

Let Γ be a labeled directed graph for which every vertex has k outgoing edges labeled by $1, \dots, k$. Recall that Γ is called automatic if there exists a bijection between a regular language and the set of vertices $V(\Gamma)$ such that for every $j \in \Sigma_k$ the set of oriented edges labeled by j is accepted by a synchronous two–tape finite automaton. From Lemma 1 we obtain that Γ_T is automatic. Suppose that Γ is automatic. Lemma 2 below shows that Γ can be obtained as Γ_T for some $(k+1)$–tape Turing transducer $T \in \mathcal{T}$.

Lemma 2. *There exists a $(k+1)$–tape Turing transducer $T \in \mathcal{T}$ for which $\Gamma_T \cong \Gamma$.*

Proof. The lemma can be obtained from the following fact. Let $R = \{(x,y)|x,y \in L\}$ be a binary relation such that $\otimes R$ is recognized by a two–tape synchronous finite automaton, where L is a regular language. Suppose that for every $x \in L$ there exists exactly one $y \in L$ such that $(x,y) \in R$. Then there exists a two–tape Turing transducer $T_R \in \mathcal{T}$ for which T_R translates x into y iff $(x,y) \in R$ and T_R rejects x iff $x \notin L$. The construction of the Turing transducer T_R can be found, e.g., in [6, Theorem 2.3.10]. The resulting $(k+1)$–tape Turing transducer $T \in \mathcal{T}$ is obtained as the combination of k two–tape Turing transducers $T_{R_1}, \dots, T_{R_k}$, where $R_1, \dots, R_k$ are the binary relations defined by the directed edges of Γ labeled by $1, \dots, k$, respectively. □

Lemmas 1 and 2 together imply the following theorem.

Theorem 3. *The labeled directed graph Γ is automatic iff there exists a Turing transducer $T \in \mathcal{T}$ for which $\Gamma \cong \Gamma_T$.* □

Let $\Gamma(G, S)$ be a Cayley graph for some set of generators $S = \{s_1, \ldots, s_k\}$. Let us fix an order of elements in S as $s_1, \ldots, s_k$. We say that the Cayley graph $\Gamma(G, S)$ is presented by $T \in \mathcal{T}$ if, after changing labels from j to s_j for every $j \in \Sigma_k$ in Γ_T, $\Gamma_T \cong \Gamma(G, S)$. The isomorphism $\Gamma_T \cong \Gamma(G, S)$ defines the bijection $\psi : L \to G$ up to the choice of the word of L corresponding to the identity $e \in G$. By Theorem 3 we obtain that if $\Gamma(G, S)$ is presented by $T \in \mathcal{T}$, then G is a Cayley automatic group and T provides an automatic presentation for the Cayley graph $\Gamma(G, S)$. Moreover, for each automatic presentation of $\Gamma(G, S)$ there is a corresponding Turing transducer $T \in \mathcal{T}$ for which $\Gamma(G, S)$ is presented by T.

3 Numerical Characteristics of Turing Transducers

We now introduce three numerical characteristics for Turing transducers of the class $\mathcal{T}$. Let $T \in \mathcal{T}$ be a $(k+1)$–tape Turing transducer translating a language L into $L' \subseteq L^k$. Given a word $w \in L$, feed w to T. Let $w_1, \ldots, w_k \in L$ be the outputs of T for w. We denote by $T(w)$ the set $T(w) = \{w_1, \ldots, w_k\}$. Given a set $W \subseteq L$, we denote by $T(W)$ the set $T(W) = \bigcup\limits_{w \in W} T(w)$. Let us choose a word $w_0 \in L$. Put $W_0 = \{w_0\}$, $W_1 = T(W_0)$ and, for $i > 1$, put $W_{i+1} = T(W_i)$. Let $V_n = \bigcup\limits_{i=0}^{n} W_i$, $n \geqslant 0$. Put $b_n = \#V_n$.

- We call the sequence $b_n, n = 0, \ldots, \infty$ the growth function of the pair (T, w_0).

For a given finite set $W \subseteq L$ put

$$\partial W = \{w \in W | T(w) \nsubseteq W\}.$$

In other words, ∂W is the set of words $w \in W$ for which at least one of the outputs of T for w is not in W. Define the function $F\o l(\varepsilon) : (0, 1) \to \mathbb{N}$ as

$$F\o l(\varepsilon) = \min\{\#W | \#\partial W < \varepsilon \#W\}.$$

It is assumed that the function $F\o l(\varepsilon)$ is defined on the whole interval $(0, 1)$, i.e., for every $\varepsilon \in (0, 1)$ the set $\{W | \#\partial W < \varepsilon \#W\}$ is not empty.

- We call the sequence $f_n = F\o l(\frac{1}{n}), n = 1, \ldots, \infty$ the Følner function of T.

Let M be a finite multiset consisting of some words of L. We denote by $T(M)$ the multiset obtained as follows. Initially, $T(M)$ is empty. Then, for every word w in M add the outputs of T for w to $T(M)$. If w has the multiplicity m in M, then this procedure must be repeated m times. Let M_0 be the multiset consisting of the word w_0 with the multiplicity one. Put $M_1 = T(M_0)$ and, for $i > 1$, put $M_{i+1} = T(M_i)$. The total number of elements (multiplicities are taken into account) in the multiset M_n is k^n. Put ℓ_n to be

$$\ell_n = \frac{\sum\limits_{w \in M_n} m_w |w|}{k^n}, \qquad (1)$$

where m_w is the multiplicity of a word w in M_n and $|w|$ is the length of w. In other words, ℓ_n is the average length of the words in the multiset M_n.

- We call the sequence $\ell_n, n = 1, \ldots, \infty$ the average length growth function of the pair (T, w_0).

4 Wreath Products of Groups: Basic Notation

Most of the labeled directed graphs in this paper are obtained as Cayley graphs of wreath products of groups. For the sake of convenience we describe basic notation for restricted wreath products $A \wr B$ in the present section. For more details on wreath products see, e.g., [9]. For given two groups A and B, we denote by $A^{(B)}$ the set of all functions $f : B \to A$ having finite supports. Recall that a function $f : B \to A$ has finite support if the set $\{x \in B | f(x) \neq e\}$ is finite, where e is the identity of A. Given $f \in A^{(B)}$ and $b \in B$, we define $f^b \in A^{(B)}$ as follows. Put $f^b(x) = f(bx)$ for all $x \in B$. The group $A \wr B$ is the set product $A^{(B)} \times B$ with the group multiplication given by $(f, b) \cdot (f', b') = (ff'^{b^{-1}}, bb')$.

We denote by $\mathfrak{i}_A$ the embedding $\mathfrak{i}_A : A \to A \wr B$ for which $\mathfrak{i}_A : a \mapsto (f_a, e)$, where e is the identity of the group B and $f_a \in A^{(B)}$ is the function $f_a : B \to A$ such that $f_a(e) = a$ and $f_a(x)$ is the identity of the group A for every $x \neq e$. We denote by $\mathfrak{i}_B$ the embedding $\mathfrak{i}_B : B \to A \wr B$ for which $\mathfrak{i}_B : b \mapsto (\mathbf{e}, b)$, where $\mathbf{e}$ is the identity of the group $A^{(B)}$; in other words, $\mathbf{e}$ is the function which maps all elements of B to the identity of the group A. For the sake of convenience we will identify A and B with the subgroups $\mathfrak{i}_A(A) \leqslant A \wr B$ and $\mathfrak{i}_B(B) \leqslant A \wr B$, respectively. Let $S_A = \{a_1, \ldots, a_n\} \subseteq A$ and $S_B = \{b_1, \ldots, b_m\} \subseteq B$ be some sets of generators of the groups A and B, respectively. Then the set $S = \mathfrak{i}_A(S_A) \cup \mathfrak{i}_B(S_B)$ is a set of generators of $A \wr B$. The Cayley graph $\Gamma(A \wr B, S)$ can be obtained as follows. The vertices of $\Gamma(A \wr B, S)$ are the elements of $A \wr B$, i.e., all pairs (f, b) such that $f \in A^{(B)}$ and $b \in B$. The right multiplication of an element (f, b) by $a_i, i = 1, \ldots, n$ is $(f, b)a_i = (\widehat{f}, b)$, where $\widehat{f}(s) = f(s)$ if $s \neq b$ and $\widehat{f}(b) = f(b)a_i$. The right multiplication of an element (f, b) by $b_j, j = 1, \ldots, m$ is $(f, b)b_j = (f, bb_j)$.

5 Asymptotic Behavior of the Numerical Characteristics

In this section we discuss asymptotic behavior of the numerical characteristics of Turing transducers of the class $\mathcal{T}$.

5.1 Growth Functions and Følner Functions

We first consider the behavior of growth function $b_n, n = 0, \ldots, \infty$ for Turing transducers of the class $\mathcal{T}$.

Let G be a group with a finite set of generators $Q \subseteq G$. Put $S = Q \cup Q^{-1}$. Recall that the growth function of the pair (G, Q) is the function $\#B_n, n = 0, \ldots, \infty$, where $\#B_n$ is the number of elements in the ball $B_n = \{g \in G | \ell_S(g) \leqslant n\}$. Let $T \in \mathcal{T}$ be a Turing transducer translating a language L into $L' \subseteq L^k$, where $k = \#S$. Choose any word $w_0 \in L$. The following claim is straightforward.

Claim. Suppose that the Cayley graph $\Gamma(G, S)$ is presented by T. Then the growth function b_n of the pair (T, w_0) coincides with the growth function of the pair (G, Q). □

One of the important questions in the group theory is whether or not for a given pair (G, Q) the growth series is rational. A similar question naturally arises for a pair (T, w_0). It is easy to show an example of a pair $(T, w_0), T \in \mathcal{T}$ for which the growth series is not rational.

Example 4. Stoll proved that the growth series of the Heisenberg group H_5 with respect to the standard set of generators is not rational [18]. The Cayley graph of H_5 is automatic [10, Example 6.7]. Therefore, we obtain that there exists a pair $(T, w_0), T \in \mathcal{T}$ for which the growth series $\sum b_n z^n$ is not rational. □

Moreover, a Turing transducer of the class $\mathcal{T}$ may have a function $b_n, n = 0, \dots, \infty$ of intermediate growth.

Example 5. Miasnikov and Savchuk constructed an example of a 4–regular automatic graph which has intermediate growth [15]. Therefore, we obtain that there exists a pair $(T, w_0), T \in \mathcal{T}$ for which the function $b_n, n = 0, \dots, \infty$ has intermediate growth. □

We now consider the behavior of Følner function $f_n, n = 1, \dots, \infty$ for Turing transducers of the class $\mathcal{T}$. Følner functions were first considered by A. Vershik for Cayley graphs of amenable groups [19]. Recall first some necessary definitions regarding Følner functions [7].

Let G be an amenable group with a finite set of generators $Q \subseteq G$. Put $S = Q \cup Q^{-1}$. Let E be the set of directed edges of $\Gamma(G, S)$. For a given finite set $U \subseteq G$ the boundary ∂U is defined as

$$\partial U = \{u \in U | \exists v \in G[(u, v) \in E \wedge v \notin U]\}.$$

The function $F\o l_{G,Q} : (0, 1) \to \mathbb{N}$ is defined as

$$F\o l_{G,Q}(\varepsilon) = \min\{\#U | \#\partial U < \varepsilon \#U\}.$$

The Følner function $\mathrm{F\o l}_{G,Q} : \mathbb{N} \to \mathbb{N}$ is defined as $\mathrm{F\o l}_{G,Q}(n) = F\o l_{G,Q}(\frac{1}{n})$. The following claim is straightforward.

Claim. Suppose that the Cayley graph $\Gamma(G, S)$ is presented by a Turing transducer $T \in \mathcal{T}$. Then for the Følner function f_n of T, $f_n = \mathrm{F\o l}_{G,Q}(n)$. □

In this subsection we say that $f_1(n) \sim f_2(n)$ if there exists $K \in \mathbb{N}$ such that $f_1(Kn) \geqslant \frac{1}{K} f_2(n)$ and $f_2(Kn) \geqslant \frac{1}{K} f_1(n)$, i.e., $f_1(n)$ and $f_2(n)$ are equivalent up to a quasi–isometry. Let $Q' \subseteq G$ be another set generating G. Then $\mathrm{F\o l}_{G,Q}(n) \sim \mathrm{F\o l}_{G,Q'}(n)$. In this subsection Følner functions are considered up to quasi–isometries. So, instead of $\mathrm{F\o l}_{G,Q}(n)$, we will write $\mathrm{F\o l}_{\mathrm{G}}(n)$.

Let $G_1 = \mathbb{Z} \wr \mathbb{Z}$. Put $G_{i+1} = G_i \wr \mathbb{Z}, i \geqslant 1$. It is shown [7, Example 3] that $\mathrm{F\o l}_{G_i}(n) \sim n^{(n^i)}$. It follows from [2, Theorem 3] that for every integer $i \geqslant 1$ there exists a Turing transducer $T_i \in \mathcal{T}$ for which a Cayley graph of G_i is presented by T_i. The following theorem shows that the logarithm of Følner functions for Turing transducers of the class $\mathcal{T}$ can grow faster than any given polynomial.

Theorem 6. *For every integer $i \geqslant 1$ there exists a Turing transducer of the class $\mathcal{T}$ for which $f_n \sim n^{(n^i)}$.* □

Remark 7. Consider the group $\mathbb{Z} \wr (\mathbb{Z} \wr \mathbb{Z})$. It is shown [7, Example 4] that $\text{Føl}_{\mathbb{Z}\wr(\mathbb{Z}\wr\mathbb{Z})}(n) \sim n^{(n^n)}$. In particular, $\text{Føl}_{\mathbb{Z}\wr(\mathbb{Z}\wr\mathbb{Z})}(n)$ grows faster than $\text{Føl}_{G_i}(n)$ for every $i \geqslant 1$. However, it is not known whether or not there exists a Turing transducer $T \in \mathcal{T}$ for which a Cayley graph of $\mathbb{Z} \wr (\mathbb{Z} \wr \mathbb{Z})$ is presented by T. □

5.2 Random Walk and Average Length Growth Functions

Recall first some necessary definitions [20]. Let G be an infinite group with a set of generators $Q = \{s_1, \dots, s_m\} \subseteq G$.

Put $S = Q \cup Q^{-1} = \{s_1, \dots, s_m, s_1^{-1}, \dots, s_m^{-1}\}$. For a given $g \in G$ we denote by $\ell_S(g)$ the minimal length of a word representing g in terms of S. We denote by B_n the ball of the radius n, $B_n = \{g \in G | \ell_S(g) \leqslant n\}$. Let μ be a symmetric measure defined on S, i.e., $\mu(s) = \mu(s^{-1})$ for all $s \in S$. The convolution $\mu^{*n}(g)$ on B_n is defined as

$$\mu^{*n}(g) = \sum_{g=g_1\cdots g_n} \prod_{i=1,\dots,n} \mu(g_i),$$

where $g_i \in S$, $i = 1, \dots, n$.

Let $c_n(g)$ be the number of words of length n over the alphabet S representing the element $g \in G$. If μ is the uniform measure on S, then $\mu^{*n}(g) = \frac{c_n(g)}{(2m)^n}$. Therefore, $\mu^{*n}(g)$ is the probability that a n–step simple symmetric random walk on the Cayley graph $\Gamma(G, S)$, which starts at the identity $e \in G$, ends up at the vertex $g \in G$. In this paper we consider only uniform measures μ. We denote by $E_{\mu^{*n}}[\ell_S]$ the average value of the functional ℓ_S on the ball B_n with respect to the measure μ^{*n}. For some Cayley graphs of wreath products of groups we will show asymptotic behavior of $E_{\mu^{*n}}[\ell_S]$ of the form $E_{\mu^{*n}}[\ell_S] \asymp f(n)$, where $g(n) \asymp f(n)$ means that $\delta_1 f(n) \leqslant g(n) \leqslant \delta_2 f(n)$ for some constants $\delta_2 \geqslant \delta_1 > 0$.

Let $T \in \mathcal{T}$ be a Turing transducer translating a language L into L'. Suppose that the Cayley graph $\Gamma(G, S)$ is presented by T. Let us choose any word $w_0 \in L$. The Turing transducer T provides the bijection $\psi : L \to G$ such that $\psi^{-1}(e) = w_0$. Therefore, we can consider the average of the functional $|w|$ on the ball B_n with respect to the measure μ^{*n}, where $|w|$ is the length of a word $w \in L$. The following claim is straightforward.

Claim. For a n–step symmetric simple random walk on the Cayley graph $\Gamma(G, S)$, $E_{\mu^{*n}}[|w|] = \ell_n$, where ℓ_n is the nth element of the average length growth function of the pair (T, w_0). □

The following proposition relates ℓ_n and $E_{\mu^{*n}}[\ell_S]$.

Proposition 8. *There exist constants C_1 and C_2 such that $\ell_n \leqslant C_1 E_{\mu^{*n}}[\ell_S] + C_2$ for all n.*

Proof. Recall that, by definition, there exists a constant c such that for every input $x \in L$ and an output $y_j \in L, j = 1, \dots, 2m$, $|y_j| \leqslant |x| + c$. Put $C_1 = c$ and $C_2 = |w_0|$. Therefore, we obtain that the inequality $\ell_n \leqslant C_1 E_{\mu^{*n}}[\ell_S] + C_2$ holds for all n. □

It is easy to give examples of Turing transducers of the class $\mathcal{T}$ for which $\ell_n \asymp \sqrt{n}$ and the growth function b_n is polynomial using a unary–like representation of integers. See Example 9 below.

Example 9. Let $Q = \{s_1, \dots, s_m\}$ be the standard set of generators of the group $\mathbb{Z}^m$, where $s_i = (\delta_i^1, \dots, \delta_i^m)$ and $\delta_i^j = 1$ if $i = j$, $\delta_i^j = 0$ if $i \neq j$. Put $S = Q \cup Q^{-1}$. It can be seen that there exists a $(2m+1)$–tape Turing transducer $T \in \mathcal{T}$ translating a language L into a language $L' \subseteq L^{2m}$ for which $\Gamma(\mathbb{Z}^m, S)$ is presented by T. It is easy to see that a language L and an isomorphism between Γ_T and $\Gamma(\mathbb{Z}^m, S)$ can be chosen in a way that $\ell_S(g) = |w|$, where $g \in \mathbb{Z}^m$ and $w \in L$ is the word corresponding to g. In particular, put the empty word ϵ to be the representative of the identity $(0, \dots, 0) \in \mathbb{Z}^m$. Therefore, for such a Turing transducer T, $\ell_n = E_{\mu^{*n}}[\ell_S]$. For a symmetric simple random walk on the m–dimensional grid, $E_{\mu^{*n}}[\ell_S] \asymp \sqrt{n}$. For the proof see, e.g., [17]. So, for the pair (T, ϵ), $\ell_n \asymp \sqrt{n}$. The growth function b_n of (T, ϵ) is polynomial. Thus, we obtain $(2m+1)$–tape Turing transducers $T_m, m = 1, \dots, \infty$ for which $\ell_n \asymp \sqrt{n}$ and the growth function b_n is polynomial. □

A more complicated technique is required in order to show an example of a Turing transducer of the class $\mathcal{T}$ for which $\ell_n \asymp \sqrt{n}$ and the growth function b_n is exponential. We will construct such a Turing transducer in Lemma 11.

Let H be a group with a set of generators $S_H = \{t_1, \dots, t_k\}$. Consider the group $\mathbb{Z}_2 \wr H$. Let $h \in \mathbb{Z}_2^{(H)}$ be the function $h : H \to \mathbb{Z}_2$ such that $h(g) = e$ if $g \neq e$ and $h(e) = a$, where a is the nontrivial element of $\mathbb{Z}_2$. Let $Q = \{t, th, ht, hth | t \in S_H\}$ be the set of generators of the group $\mathbb{Z}_2 \wr H$. Put $S = Q \cup Q^{-1}$. Consider a symmetric simple random walk on the Cayley graph $\Gamma(\mathbb{Z}_2 \wr H, S)$. It is easy to see that a n–step random walk on $\Gamma(\mathbb{Z}_2 \wr H, S)$ corresponds to a n–step random walk on H. Put $P = S_H \cup S_H^{-1}$. Let R_n be the number of different vertices visited after walking n steps on $\Gamma(H, P)$. We call R_n the range of a n–step random walk on $\Gamma(H, P)$. In the following proposition the asymptotic behavior of $E_{\mu^{*n}}[\ell_S]$ is expressed in terms of $E_{\mu^{*n}}[R_n]$ – the average range for a n–step random walk on $\Gamma(H, P)$.

Proposition 10. *Let H and S be as above. For a symmetric simple random walk on $\Gamma(\mathbb{Z}_2 \wr H, S)$, $E_{\mu^{*n}}[\ell_S] \asymp E_{\mu^{*n}}[R_n]$.*

Proof. For the proof see [5, Lemma 2]. □

Lemma 11. *There exists a set of generators S_1 of the lamplighter group $\mathbb{Z}_2 \wr \mathbb{Z}$ for which the following statements hold.*

(a) For a simple symmetric random walk on $\Gamma(\mathbb{Z}_2 \wr \mathbb{Z}, S_1)$, $E_{\mu^{*n}}[\ell_{S_1}] \asymp \sqrt{n}$.
(b) There exists a Turing transducer $T_1 \in \mathcal{T}$ such that $\Gamma(\mathbb{Z}_2 \wr \mathbb{Z}, S_1)$ is presented by T_1 and $\ell_n \asymp \sqrt{n}$.

Proof. Let us consider the lamplighter group $\mathbb{Z}_2 \wr \mathbb{Z}$. Let t be a generator of the subgroup $\mathbb{Z} \leqslant \mathbb{Z}_2 \wr \mathbb{Z}$ and $h \in \mathbb{Z}_2^{(\mathbb{Z})} \leqslant \mathbb{Z}_2 \wr \mathbb{Z}$ be the function $h : \mathbb{Z} \to \mathbb{Z}_2$ such that $h(z) = e$ if $z \neq 0$ and $h(0) = a$. Let $Q_1 = \{t, th, ht, hth\}$ be the set of generators of $\mathbb{Z}_2 \wr \mathbb{Z}$ and $S_1 = Q_1 \cup Q_1^{-1}$. For a simple symmetric random walk on $\Gamma(\mathbb{Z}, \{t, t^{-1}\})$, $E_{\mu^{*n}}[R_n] \sim \sqrt{n}$, where $\sim$ here means asymptotic equivalence. For the proof see, e.g., [17]. Therefore, from Proposition 10 we obtain that for a simple symmetric random walk on $\Gamma(\mathbb{Z}_2 \wr \mathbb{Z}, S_1)$, $E_{\mu^{*n}}[\ell_{S_1}] \asymp \sqrt{n}$.

Let $Q_1' = \{t, h\}$ be a set of generators of $\mathbb{Z}_2 \wr \mathbb{Z}$. Put $S_1' = Q_1' \cup Q_1'^{-1} = \{t, t^{-1}, h\}$. In [2, Theorem 2] we constructed an automatic presentation of the Cayley graph $\Gamma(\mathbb{Z}_2 \wr \mathbb{Z}, S_1')$, the bijection $\psi_1 : L_1 \to \mathbb{Z}_2 \wr \mathbb{Z}$, for which the inequalities $\frac{1}{3}\ell_{S_1'}(g) + \frac{2}{3} \leqslant |w| \leqslant \ell_{S_1'}(g) + 1$ hold for all $g \in \mathbb{Z}_2 \wr \mathbb{Z}$, where L_1 is a regular language, $w = \psi_1^{-1}(g) \in L_1$ is the word corresponding to g and $|w|$ is the length of w. It is easy to see that $\frac{1}{2}\ell_{S_1}(g) \leqslant \ell_{S_1'}(g) \leqslant 3\ell_{S_1}(g)$. Therefore, we obtain that $\frac{1}{6}\ell_{S_1}(g) + \frac{2}{3} \leqslant |w| \leqslant 3\ell_{S_1}(g) + 1$ for all $g \in \mathbb{Z}_2 \wr \mathbb{Z}$. This implies that $\frac{1}{6}E_{\mu^{*n}}[\ell_{S_1}] + \frac{2}{3} \leqslant E_{\mu^{*n}}[|w|] \leqslant 3E_{\mu^{*n}}[\ell_{S_1}] + 1$. The bijection $\psi_1 : L_1 \to \mathbb{Z}_2 \wr \mathbb{Z}$ provides an automatic presentation for the Cayley graph $\Gamma(\mathbb{Z}_2 \wr \mathbb{Z}, S_1)$. By Lemma 2, we obtain that there exists a 9–tape Turing transducer $T_1 \in \mathcal{T}$ translating the language L_1 into some language $L_1' \subseteq L_1^8$ for which $\Gamma(\mathbb{Z}_2 \wr \mathbb{Z}, S_1)$ is presented by T_1. Therefore, we obtain that for T_1, $\ell_n \asymp \sqrt{n}$. Since the growth function of the group $\mathbb{Z}_2 \wr \mathbb{Z}$ is exponentinal, the growth function b_n of T_1 is exponential. □

It is easy to give examples of Turing transducers of the class $\mathcal{T}$ for which $\ell_n \asymp n$ and the growth function b_n is exponential. See Example 12 below.

Example 12. Let F_m be the free group over m generators $s_1, \dots, s_m$. Put $Q = \{s_1, \dots, s_m\}$ and $S = Q \cup Q^{-1}$. There exists a natural automatic presentation of the Cayley graph $\Gamma(F_m, S)$, the bijection $\psi : L \to F_m$, for which L is the language of all reduced words over the alphabet S. In particular, the empty word ϵ represents the identity $e \in F_m$. The bijection ψ maps a word $w \in L$ into the corresponding group element of F_m. It is clear that $\ell_S(g) = |w|$, where $w = \psi^{-1}(g)$. For a symmetric simple random walk on $\Gamma(F_m, S)$, $E_{\mu^{*n}}[\ell_S] \asymp n$. Therefore, $E_{\mu^{*n}}[|w|] \asymp n$. Therefore, for each $m > 1$ we obtain the pair $(T, \epsilon), T \in \mathcal{T}$ for which $\ell_n \asymp n$. Since the growth function of the free group F_m is exponential, the growth function b_n of the pair (T, ϵ) is exponential. □

Is there a Turing transducer of the class $\mathcal{T}$ for which ℓ_n grows between $\sqrt{n}$ and n? We will answer on this question positively in Theorem 14 which follows from Proposition 13 below.

Let G be a group with a set of generators $S_G = \{g_1, \dots, g_m\}$. Put $P = S_G \cup S_G^{-1}$. Assume that for a symmetric simple random walk on $\Gamma(G, P)$, $\ell_n(\mu) \asymp n^\alpha$ for some $0 < \alpha \leqslant 1$. Consider the wreath product $G \wr \mathbb{Z}$. Let t be a generator of the subgroup $\mathbb{Z} \leqslant G \wr \mathbb{Z}$. Let $h_i \in G^{(\mathbb{Z})} \leqslant G \wr \mathbb{Z}$, $i = 1, \dots, m$ be the functions $h_i : \mathbb{Z} \to G$ such that $h_i(z) = e$ if $z \neq 0$ and $h_i(0) = g_i$. Put $Q = \{h_i^p t h_j^q \,|\, i, j = 1, \dots, m; p, q = -1, 0, 1\}$ to be the set of generators of the group $G \wr \mathbb{Z}$ and $S = Q \cup Q^{-1}$. Consider a n–step random walk on $\Gamma(G \wr \mathbb{Z}, S)$. The following proposition shows asymptotic behavior of $E_{\mu^{*n}}[\ell_S]$.

Proposition 13. *Let G, S and α be as above. For a symmetric simple random walk on $\Gamma(G \wr \mathbb{Z}, S)$, $E_{\mu^{*n}}[\ell_S] \asymp n^{\frac{1+\alpha}{2}}$.*

Proof. For the proof see [8, Lemma 3]. □

Theorem 14. *For every $\alpha < 1$ there exists a Turing transducer $T \in \mathcal{T}$ for which $\ell_n \asymp n^{\beta}$ for some β such that $\alpha < \beta < 1$ and the growth function b_n is exponential.*

Proof. Let us consider the sequence of wreath products $G_m, m = 1, \ldots, \infty$ such that $G_1 = \mathbb{Z}_2 \wr \mathbb{Z}$ and $G_{m+1} = G_m \wr \mathbb{Z}$, $m \geqslant 1$. From Lemma 11(a) and Proposition 13 we obtain that for every $m > 1$ there exists a proper set of generators $Q_m \subseteq G_m$ such that for a symmetric simple random walk on the Cayley graph $\Gamma(G_m, S_m)$, $E_{\mu^{*n}}[\ell_{S_m}] \asymp n^{1-\frac{1}{2^m}}$, where $S_m = Q_m \cup Q_m^{-1}$. It follows from [2, Theorem 3] that for every $m > 1$ there is an automatic presentation of the Cayley graph $\Gamma(G_m, S'_m)$, the bijection $\psi_m : L_m \to G_m$, for which the inequalities $\delta'_1 \ell_{S'_m}(g) + \lambda'_1 \leqslant |w| \leqslant \delta'_2 \ell_{S'_m}(g) + \lambda'_2$ hold for all $g \in G_m$ for some constants $\delta'_2 > \delta'_1 > 0, \lambda'_1, \lambda'_2$, where L_m is a regular language and $S'_m = Q'_m \cup {Q'_m}^{-1}$ for some proper set of generators $Q'_m \subseteq G_m$, and $w = \psi_m^{-1}(g)$ is the word representing g. Therefore, the inequalities $\delta_1 \ell_{S_m}(g) + \lambda_1 \leqslant |w| \leqslant \delta_2 \ell_{S_m}(g) + \lambda_2$ hold for all $g \in G_m$ for some constants $\delta_2 > \delta_1 > 0, \lambda_1, \lambda_2$. This implies that $\delta_1 E_{\mu^{*n}}[\ell_{S_m}] + \lambda_1 \leqslant E_{\mu^{*n}}[|w|] \leqslant \delta_2 E_{\mu^{*n}}[\ell_{S_m}] + \lambda_2$. Therefore, $E_{\mu^{*n}}[|w|] \asymp n^{1-\frac{1}{2^m}}$.

For every $m > 1$ the bijection $\psi_m : L_m \to G_m$ provides an automatic presentation of the Cayley graph $\Gamma(G_m, S_m)$. It follows from Lemma 2 that there is a $(k_m + 1)$–tape Turing transducer $T_m \in \mathcal{T}$ translating the language L_m into $L'_m \subseteq L_m^{k_m}$ for which, after proper relabeling, $\Gamma_{T_m} \cong \Gamma(G_m, S_m)$. The numbers $k_m, m = 1, \ldots, \infty$ can be obtained recurrently as follows. It is easy to see that $k_{m+1} = 2(k_m + 1)^2$ for $m \geqslant 1$ and $k_1 = 8$, which is simply the number of elements in S_1 (see Lemma 11). So, we obtain that for $T_m, m > 1$, $\ell_n \asymp n^{1-\frac{1}{2^m}}$. For every $m > 1$, since the growth function of the group G_m is exponential, the growth function b_n of T_m is exponential. □

6 Discussion

In this paper we addressed the problem of finding characterizations of Cayley automatic groups. Our approach was to define and then study three numerical characteristics of Turing transducers of the special class $\mathcal{T}$. This class of Turing transducers was obtained from automatic presentations of labeled directed graphs. The numerical characteristics that we defined are the analogs of growth functions, Følner functions and drifts of simple random walks for Cayley graphs of groups. We hope that further study of asymptotic behavior of these three numerical characteristics of Turing transducers of the class $\mathcal{T}$ will yield some characterizations for Cayley automatic groups.

Two open questions are apparent from the results of Sect. 5.

- Theorem 6 shows that for every integer $i \geqslant 1$ there exists a Turing transducer of the class $\mathcal{T}$ for which $f_n \sim n^{(n^i)}$. Is there a Turing transducer $T \in \mathcal{T}$ for which the Følner function grows faster than $n^{(n^i)}$ for all $i \geqslant 1$?

- Theorem 14 tells us that for every $\alpha < 1$ there exists a Turing transducer $T \in \mathcal{T}$ for which $\ell_n \asymp n^\beta$ for some β such that $\alpha < \beta < 1$. Is there a Turing transducer $T \in \mathcal{T}$ for which ℓ_n grows faster than n^α for every $\alpha < 1$ but slower than n?

Acknowledgments. The author thanks Bakhadyr Khoussainov and the anonymous reviewers for useful suggestions. The author thanks Sunny Daniels for proofreading a draft of this paper and making several changes to it.

References

1. Berdinsky, D., Khoussainov, B.: On automatic transitive graphs. In: Shur, A.M., Volkov, M.V. (eds.) DLT 2014. LNCS, vol. 8633, pp. 1–12. Springer, Heidelberg (2014)
2. Berdinsky, D., Khoussainov, B.: Cayley automatic representations of wreath products. Int. J. Found. Comput. Sci. **27**(2), 147–159 (2016)
3. Calude, C., Calude, E., Khoussainov, B.: Deterministic automata simulation, universality and minimality. Ann. Pure Appl. Logic **90**(1), 263–276 (1997)
4. Calude, C.S., Calude, E., Khoussainov, B.: Finite nondeterministic automata: simulation and minimality. Theor. Comput. Sci. **242**(1), 219–235 (2000)
5. Dyubina, A.: An example of the rate of growth for a random walk on a group. Russ. Math. Surv. **54**(5), 1023–1024 (1999)
6. Epstein, D.B.A., Cannon, J.W., Holt, D.F., Levy, S.V.F., Paterson, M.S., Thurston, W.P.: Word Processing in Groups. Jones and Barlett Publishers, Boston (1992)
7. Erschler, A.: On isoperimetric profiles of finitely generated groups. Geom. Dedicata **100**(1), 157–171 (2003)
8. Erschler, A.: On the asymptotics of drift. J. Math. Sci. **121**(3), 2437–2440 (2004)
9. Kargapolov, M.I., Merzljakov, J.I.: Fundamentals of the Theory of Groups. Springer, New York (1979)
10. Kharlampovich, O., Khoussainov, B., Miasnikov, A.: From automatic structures to automatic groups. Groups Geom. Dyn. **8**(1), 157–198 (2014)
11. Khoussainov, B., Minnes, M.: Three lectures on automatic structures. In: Proceedings of Logic Colloquium, pp. 132–176 (2007)
12. Khoussainov, B., Nerode, A.: Automatic presentations of structures. In: Leivant, D. (ed.) LCC 1994. LNCS, vol. 960, pp. 367–392. Springer, Berlin Heidelberg (1995)
13. Khoussainov, B., Nerode, A.: Open questions in the theory of automatic structures. Bull. EATCS **94**, 181–204 (2008)
14. Meduna, A.: Automata and Languages: Theory and Applications. Springer, London (2000)
15. Miasnikov, A., Savchuk, D.: An example of an automatic graph of intermediate growth. Ann. Pure Appl. Logic **166**(10), 1037–1048 (2015)
16. Oliver, G.P., Thomas, R.M.: Automatic presentations for finitely generated groups. In: Diekert, V., Durand, B. (eds.) STACS 2005. LNCS, vol. 3404, pp. 693–704. Springer, Heidelberg (2005)
17. Spitzer, F.: Principles of Random Walk. Van Nostrand, Princeton (1964)
18. Stoll, M.: Rational and transcendental growth series for the higher Heisenberg groups. Invent. Math. **126**, 85–109 (1996)

19. Vershik, A.: Countable groups that are close to finite ones. In: Greenleaf, F.P. (ed.) Invariant Means on Topological Groups and their Applications. Mir, Moscow (1973). (Appendix, in Russian). A revised English translation: Amenability and approximation of infinite groups. Selecta Math. **2**(4), 311–330 (1982)
20. Vershik, A.: Numerical characteristics of groups and corresponding relations. J. Math. Sci. **107**(5), 4147–4156 (2001)

A Perfect Class of Context-Sensitive Timed Languages

Devendra Bhave[1(✉)], Vrunda Dave[1], S.N. Krishna[1], Ramchandra Phawade[1], and Ashutosh Trivedi[1,2]

[1] IIT Bombay, Mumbai, India
devendra@cse.iitb.ac.in
[2] CU Boulder, Boulder, USA

Abstract. Perfect languages—a term coined by Esparza, Ganty, and Majumdar—are the classes of languages that are closed under Boolean operations and enjoy decidable emptiness problem. Perfect languages form the basis for decidable automata-theoretic model-checking for the respective class of models. Regular languages and visibly pushdown languages are paradigmatic examples of perfect languages. Alur and Dill initiated the language-theoretic study of timed languages and introduced timed automata capturing a timed analog of regular languages. However, unlike their untimed counterparts, timed regular languages are not perfect. Alur, Fix, and Henzinger later discovered a perfect subclass of timed languages recognized by event-clock automata. Since then, a number of perfect subclasses of timed context-free languages, such as event-clock visibly pushdown languages, have been proposed. There exist examples of perfect languages even beyond context-free languages:—La Torre, Madhusudan, and Parlato characterized first perfect class of context-sensitive languages via multistack visibly pushdown automata with an explicit bound on number of stages where in each stage at most one stack is used. In this paper we extend their work for timed languages by characterizing a perfect subclass of timed context-sensitive languages called dense-time multistack visibly pushdown languages and provide a logical characterization for this class of timed languages.

Keywords: Perfect languages · Context-sensitive languages · Multi-stack automata · Timed languages

1 Introduction

A class $\mathcal{C}$ of languages is called *perfect* [10] if it is closed under Boolean operations and permits algorithmic emptiness-checking. Perfect languages are the key ingredient for the Vardi-Wolper recipe for automata-theoretic model-checking:—given a system specification $\mathcal{S}$ and a system implementation $\mathcal{M}$ as languages in $\mathcal{C}$, the model-checking involves deciding the emptiness of the language $\mathcal{M} \cap \neg\mathcal{S} \in \mathcal{C}$. The class of ($\omega$-)regular languages is a well-known class of perfect languages,

S. Brlek and C. Reutenauer (Eds.): DLT 2016, LNCS 9840, pp. 38–50, 2016.
DOI: 10.1007/978-3-662-53132-7_4

while other classes of languages such as context-free languages (CFLs) or context-sensitive languages (CSLs) are, in general, not perfect. CFLs are not perfect since they are not closed under intersection and complementation, although emptiness is decidable.On the other hand, CSLs are closed under Boolean operations but emptiness, in general, is undecidable for CSLs [6].

Alur and Madhusudan [4] discovered a perfect subclass of CFLs, called visibly pushdown languages (VPLs), characterized by *visibly pushdown automata* (VPA) that operate over words that dictate the stack operations. This notion is formalized by giving an explicit partition of the alphabet into three disjoint sets of *call*, *return*, and *internal* symbols and the VPA must push one symbol to stack while reading a call symbol, and must pop one symbol (given stack is non-empty) while reading a return symbol, and must not touch the stack while reading an internal symbol. This visibility enables closure of these automata under all of the Boolean operations, while retaining the decidable emptiness property. Building upon this work, La Torre et al. [11] introduced a perfect class of CSLs, called multistack visibly pushdown languages (MVPLs), recognized by VPA with multiple stacks (and call-return symbols for each stack) where the number of switches between various stacks for popping-purposes is bounded.

Example 1. $L = \{a^n b^n : n \geq 0\}$ is a VPL with a as call and b as return symbol for the unique stack, whereas $L' = \{a_1^n a_2^m b_1^n b_2^m : n, m \geq 0\}$ is a MVPL considering a_i and b_i as call and return symbols, respectively, for stack-i where $i \in \{1, 2\}$. Finally, $L'' = \{a^n b^n c^n : n \geq 0\}$ is neither VPL nor MVPL for any partition of alphabets as call and respectively alphabets of various stacks.

In this paper we introduce a timed extension of this context-sensitive language and study language-theoretic properties of the class in [14]. We characterize a perfect subclass of timed context-sensitive languages and provide a logical characterization for this class of timed languages.

Quest for Perfect Timed Languages. Alur and Dill [2] initiated automata-theoretic study of timed languages and characterized the class of timed-regular languages as the languages defined by timed automata. Unlike untimed regular languages, Alur and Dill showed that timed regular languages are not perfect as they are not closed under complementation. However, the emptiness of timed automata is a decidable using a technique known as region-construction. To overcome the limitation of timed automata for model-checking, Alur et al. introduced a perfect class of timed languages called the event-clock automata [3] (ECA) that achieves the closure under Boolean operations by making clock resets visible—the reset of each clock variable is determined by a fixed class of events and hence *visible* just by looking at the input word. The decidability of the emptiness for ECA follows from the decidability of regular timed languages.

Two of the well-known models for context-free timed languages include recursive timed automata (RTAs) [15] and dense-time pushdown automata (dtPDAs) [1]. RTAs generalize recursive state machines with clock variables, while dtPDAs generalize pushdown automata with clocks and stack with variable ages. In general, the emptiness problem for the RTA in undecidable, however [15] characterizes classes of RTA with decidable emptiness problem. However, without any further restrictions, such as event-clock or visible stack, the languages captured by these classes are not perfect, since they strictly generalize both timed regular languages and CFLs. Tang and Ogawa in [16] proposed a first perfect timed context-free language class characterized by *event-clock visibly pushdown automata* (ECVPA) that generalized both ECA and VPA. For the proposed model they showed determinizability as well as closure under Boolean operations, and proved the decidability of the emptiness problem. However, ECVPAs, unlike dtPDAs, do not support pushing the clocks on the stack. We proposed [7] a generalization of ECVPA called dense-time visibly pushdown automata (dtVPA), that are strictly more expressive than ECVPA as they support stack with variable ages (like dtPDA) and showed that dtVPA characterize a perfect timed context-free language.

Contributions. We study a class of timed context-sensitive languages called dense-time multistack visibly pushdown languages (dtMVPLs), characterized by dense-time visibly pushdown multistack automata (dtMVPA), that generalize MVPLs with multiple stacks with ages as shown in the following example.

Example 2. Consider the timed language whose untimed component is of the form $\{a^y b^z c^y d^z \mid y, z \geq 1\}$ with the critical timing restrictions among various symbols in the following manner. The first c must appear after 1 time-unit of last a, the first d must appear within 3 time-unit after last b, and finally the last b must appear within 2 time units of the beginning and last d must appear precisely at 4 time unit. This language is accepted by a dtMVPA with two stacks shown in Fig. 1. Let a and c (b and d, resp.) be call and return symbols for the first (second, resp.) stack. Stack alphabets for first stack is $\Gamma^1 = \{\alpha, \$\}$ and for second stack is $\Gamma^2 = \{\beta, \$\}$. In the figure a clock x_a measures the time since the occurrence of last a, while constraints $pop(\gamma) \in I$ checks if the age of the popped symbol is in a given interval I. The correctness of the model is easy to verify.

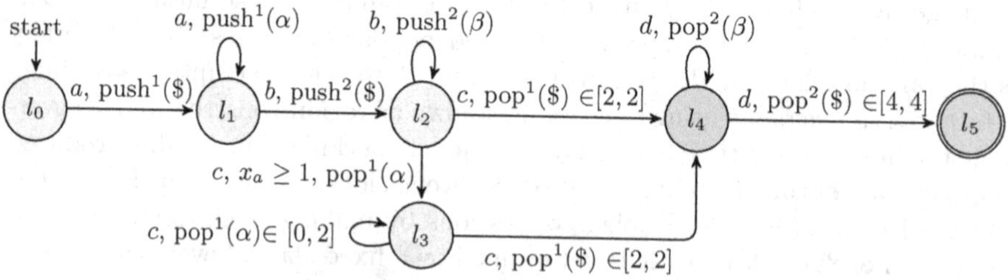

Fig. 1. Dense-time multistack visibly pushdown automata used in Example 2

In this paper we show dtMVPLs are closed under Boolean operations and enjoy decidable emptiness problem. Although, the emptiness problem for restrictions of context sensitive languages has been studied extensively [5, 9, 12–14], ours is the first attempt to formalize perfect dense-time context-sensitive languages. We will also give a logical characterization of this class of languages. We believe that dtMVPLs provide an expressive yet decidable model-checking framework for concurrent time-critical software systems (See [8] for an example).

The paper is organized as follows. We begin by introducing dense-time visibly pushdown multistack automata in the next section. In Sect. 3 we show closure under Boolean operations for this model, followed by logical characterization in Sect. 4. Details of the proof for decidability of emptiness can be found in [8].

2 Dense-Time Visibly Pushdown Multistack Automata

We assume that the reader is comfortable with standard concepts from automata theory (such as context-free languages, pushdown automata, MSO logic), concepts from timed automata (such as clocks, event clocks, clock constraints, and valuations), and visibly pushdown automata. Due to space limitation, we only give a very brief introduction of required concepts in this section, and for a detailed background on these concepts we refer the reader to [2–4].

A finite timed word over Σ is a sequence $(a_1, t_1), (a_2, t_2), \ldots, (a_n, t_n) \in (\Sigma \times \mathbb{R}_{\geq 0})^*$ such that $t_i \leq t_{i+1}$ for all $1 \leq i \leq n-1$. Alternatively, we can represent timed words as tuple $(\langle a_1, \ldots, a_n \rangle, \langle t_1, \ldots, t_n \rangle)$. We use both of these formats depending on technical convenience. We represent the set of finite timed words over Σ by $T\Sigma^*$. Before we introduce our model, we recall the definitions of event-clock automata and visibly pushdown automata.

2.1 Preliminaries

Event-clock automata (ECA) [3] are a determinizable subclass of timed automata [2] that for every action $a \in \Sigma$ implicitly associate two clocks x_a and y_a, where the "recorder" clock x_a records the time of the last occurrence of action a, and the "predictor" clock y_a predicts the time of the next occurrence of action a. Hence, event-clock automata do not permit explicit reset of clocks and it is implicitly governed by the input timed word. This property makes ECA determinizable and closed under all Boolean operations.

Notice that since clock resets are "visible" in input timed word, the clock valuations after reading a prefix of the word is also determined by the timed word. For example, for a timed word $w = (a_1, t_1), (a_2, t_2), \ldots, (a_n, t_n)$, the value of the event clock x_a at position j is $t_j - t_i$ where i is the largest position preceding j where an action a occurred. If no a has occurred before the jth position, then the value of x_a is undefined denoted by a special symbol $\vdash$. Similarly, the value of y_a at position j of w is undefined if symbol a does not occur in w after the jth position. Otherwise, it is $t_k - t_j$ where k is the first occurrence of a after j.

We write C for the set of all event clocks and we use $\mathbb{R}^{\vdash}_{>0}$ for the set $\mathbb{R}_{>0} \cup \{\vdash\}$. Formally, the clock valuation after reading j-th prefix of the input timed word w, $\nu_j^w : C \mapsto \mathbb{R}^{\vdash}_{>0}$, is defined in the following way: $\nu_j^w(x_q) = t_j - t_i$ if there exists an $0 \leq i < j$ such that $a_i = q$ and $a_k \neq q$ for all $i < k < j$, otherwise $\nu_j^w(x_q) = \vdash$

(undefined). Similarly, $\nu_j^w(y_q) = t_m - t_j$ if there is $j < m$ such that $a_m = q$ and $a_l \neq q$ for all $j < l < m$, otherwise $\nu_j^w(y_q) = \vdash$. A clock constraint over C is a boolean combination of constraints of the form $z \sim c$ where $z \in C$, $c \in \mathbb{N}$ and $\sim \in \{\leq, \geq\}$. Given a clock constraint $z \sim c$ over C, we write $\nu_i^w \models (z \sim c)$ to denote if $\nu_j^w(z) \sim c$. For any boolean combination φ, $\nu_i^w \models \varphi$ is defined in an obvious way: if $\varphi = \varphi_1 \wedge \varphi_2$, then $\nu_i^w \models \varphi$ iff $\nu_i^w \models \varphi_1$ and $\nu_i^w \models \varphi_2$. Likewise, the other boolean combinations are defined.

Definition 3. *An event clock automaton is a tuple $A = (L, \Sigma, L^0, F, E)$ where L is a set of finite locations, Σ is a finite alphabet, $L^0 \in L$ is the set of initial locations, $F \in L$ is the set of final locations, and E is a finite set of edges of the form $(\ell, \ell', a, \varphi)$ where ℓ, ℓ' are locations, $a \in \Sigma$, and φ is a clock constraint.*

The class of languages accepted by event-clock automata are closed under boolean operations with decidable emptiness property [3].

Visibly pushdown automata [4] are a determinizable subclass of pushdown automata that operate over words that dictate the stack operations. This notion is formalized by giving an explicit partition of the alphabet into three disjoint sets of *call*, *return*, and *internal* symbols and the visibly pushdown automata must push one symbol to stack while reading a call symbol, and must pop one symbol (given stack is non-empty) while reading a return symbol, and must not touch the stack while reading the internal symbol.

Definition 4. *A visibly pushdown alphabet is a tuple $\Sigma = \langle \Sigma_c, \Sigma_r, \Sigma_{int} \rangle$ where Σ is partitioned into a* call *alphabet Σ_c, a* return *alphabet Σ_r, and an* internal *alphabet Σ_{int}. A* visibly pushdown automata (VPA) *over $\Sigma = \langle \Sigma_c, \Sigma_r, \Sigma_{int} \rangle$ is a tuple $(L, \Sigma, \Gamma, L^0, \delta, F)$ where L is a finite set of locations including a set $L^0 \subseteq L$ of initial locations, Γ is a finite stack alphabet with special end-of-stack symbol $\perp$, $\Delta \subseteq (L \times \Sigma_c \times L \times (\Gamma \backslash \perp)) \cup (L \times \Sigma_r \times \Gamma \times L) \cup (L \times \Sigma_{int} \times L)$ is the transition relation, and $F \subseteq L$ is final locations.*

The class of languages accepted by visibly pushdown automata are closed under boolean operations with decidable emptiness property [4].

2.2 Dense-Time Visibly Pushdown Multistack Automata (dtMVPA)

We introduce the dense-time visibly pushdown automata as an event-clock automaton equipped with multiple (say $n \geq 1$) timed stacks along with a visibly pushdown alphabet $\Sigma = \langle \Sigma_c^h, \Sigma_r^h, \Sigma_{int}^h \rangle_{h=1}^n$ where $\Sigma_x^i \cap \Sigma_x^j = \emptyset$ for $i \neq j$, and $x \in \{c, r, int\}$. Due to space limitation and notational convenience, we assume that the partitioning function is one-to-one, i.e. each symbol $a \in \Sigma^h$ has unique recorder x_a and predictor y_a clocks assigned to it. Let Γ^h be the stack alphabet of the h-th stack. Let $\Gamma = \bigcup_{h=1}^n \Gamma^h$ and let $\Sigma^h = \langle \Sigma_c^h, \Sigma_r^h, \Sigma_{int}^h \rangle$. Let C_{Σ^h} (or C_h when Σ^h is clear) be a finite set of event clocks. Let $\Phi(C_h)$ be the set of *clock constraints* over C_h and $\mathcal{I}$ be the set of intervals.

Definition 5. *A dense-time visibly pushdown multistack automata (dtMVPAs) over $\langle \Sigma_c^h, \Sigma_r^h, \Sigma_{int}^h \rangle_{h=1}^n$ is a tuple $(L, \Sigma, \Gamma, L^0, F, \Delta{=}(\Delta_c^h \cup \Delta_r^h \cup \Delta_{int}^h)_{h=1}^n)$ where*

- *L is a finite set of locations including a set $L^0 \subseteq L$ of initial locations,*
- *Γ^h is the finite alphabet of the hth stack with special end-of-stack symbol $\bot_h$,*
- *$\Delta_c^h \subseteq (L \times \Sigma_c^h \times \Phi(C_h) \times L \times (\Gamma^h \backslash \bot_h))$ is the set of call transitions,*
- *$\Delta_r^h \subseteq (L \times \Sigma_r^h \times \mathcal{I} \times \Gamma^h \times \Phi(C_h) \times L)$ is set of return transitions,*
- *$\Delta_{int}^h \subseteq (L \times \Sigma_{int}^h \times \Phi(C_h) \times L)$ is set of internal transitions, and*
- *$F \subseteq L$ is the set of final locations.*

Let $w = (a_0, t_0), \ldots, (a_e, t_e)$ be a timed word. A configuration of the dtMVPA is a tuple $(\ell, \nu_i^w, (((\gamma^1\sigma^1, age(\gamma^1\sigma^1)), \ldots, (\gamma^n\sigma^n, age(\gamma^n\sigma^n)))$ where ℓ is the current location of the dtMVPA, ν_i^w gives the valuation of all the event clocks at position $i \leq |w|$, $\gamma^h\sigma^h \in \Gamma^h(\Gamma^h)^*$ is the content of stack h with γ^h being the topmost symbol and σ^h is the string representing the stack content below γ^h, while $age(\gamma^h\sigma^h)$ is a sequence of real numbers encoding the ages of all the stack symbols (the time elapsed since each of them was pushed on to the stack). We follow the assumption that $age(\bot^h) = \langle \vdash \rangle$ (undefined). If for some string $\sigma^h \in (\Gamma^h)^*$ we have that $age(\sigma^h) = \langle t_1, t_2, \ldots, t_g \rangle$ and for $\tau \in \mathbb{R}_{\geq 0}$ we write $age(\sigma^h) + \tau$ for the sequence $\langle t_1 + \tau, t_2 + \tau, \ldots, t_g + \tau \rangle$. For a sequence $\sigma^h = \langle \gamma_1^h, \ldots, \gamma_g^h \rangle$ and a member γ^h we write $\gamma^h :: \sigma^h$ for $\langle \gamma^h, \gamma_1^h, \ldots, \gamma_g^h \rangle$.

A run of a dtMVPA on $w = (a_0, t_0), \ldots, (a_e, t_e)$ is a sequence of configurations $(\ell_0, \nu_0^w, (\langle \bot^1 \rangle, \langle \vdash \rangle), \ldots, (\langle \bot^n \rangle, \langle \vdash \rangle))$, $(\ell_1, \nu_1^w, ((\sigma_1^1, age(\sigma_1^1)), \ldots, (\sigma_1^n, age(\sigma_1^n)))$, $\ldots, (\ell_{e+1}, \nu_{e+1}^w, (\sigma_{e+1}^1, age(\sigma_{e+1}^1)), \ldots, (\sigma_{e+1}^n, age(\sigma_{e+1}^n))))$ where $\ell_i \in L$, $\ell_0 \in L^0$, $\sigma_i^h \in (\Gamma^h \cup \{\bot^h\})^+$, and for each i, $0 \leq i \leq e$, we have:

- If $a_i \in \Sigma_c^h$, then there is $(\ell_i, a_i, \varphi, \ell_{i+1}, \gamma^h) \in \Delta_c^h$ such that $\nu_i^w \models \varphi$. The symbol $\gamma^h \in \Gamma^h \backslash \{\bot^h\}$ is then pushed onto the stack h, and its age is initialized to zero, i.e. $(\sigma_{i+1}^h, age(\sigma_{i+1}^h)) = (\gamma^h :: \sigma_i^h, 0 :: (age(\sigma_i^h) + (t_i - t_{i-1})))$. All symbols in all other stacks are unchanged, and age by $t_i - t_{i-1}$.
- If $a_i \in \Sigma_r^h$, then there is $(\ell_i, a_i, I, \gamma^h, \varphi, \ell_{i+1}) \in \Delta_r^h$ such that $\nu_i^w \models \varphi$. Also, $\sigma_i^h = \gamma^h :: \kappa \in \Gamma^h(\Gamma^h)^*$ and $age(\gamma^h) + (t_i - t_{i-1}) \in I$. The symbol γ^h is popped from stack h obtaining $\sigma_{i+1}^h = \kappa$ and $age(\sigma_{i+1}^h) = age(\sigma_i^h) + (t_i - t_{i-1})$. However, if $\gamma^h = \langle \bot^h \rangle$, then γ^h is not popped. The contents of all other stacks remains unchanged, and simply age by $(t_i - t_{i-1})$.
- If $a_i \in \Sigma_{int}^h$, then there is $(\ell_i, a_i, \varphi, \ell_{i+1}) \in \Delta_{int}^h$ such that $\nu_i^w \models \varphi$. In this case all stacks remain unchanged i.e. $\sigma_i^h = \sigma_{i+1}^h$, and $age(\sigma_{i+1}^h) = age(\sigma_i^h) + (t_i - t_{i-1})$ for all $1 \leq h \leq n$. All symbols in all stacks age by $t_i - t_{i-1}$.

A run ρ of a dtMVPA M is accepting if it terminates in a final location. A timed word w is an accepting word if there is an accepting run of M on w. The language $L(M)$ of a dtMVPA M, is the set of all timed words w accepted by M.

A dtMVPA $M = (L, \Sigma, \Gamma, L^0, F, \Delta)$ is said to be *deterministic* if it has exactly one start location, and for every configuration and input action exactly one transition is enabled. Formally, we have the following conditions: for every $(\ell, a, \phi_1, \ell', \gamma_1), (\ell, a, \phi_2, \ell'', \gamma_2) \in \Delta_c^h$, $\phi_1 \wedge \phi_2$ is unsatisfiable; for every $(\ell, a, I_1, \gamma, \phi_1, \ell'), (\ell, a, I_2, \gamma, \phi_2, \ell'') \in \Delta_r^h$, either $\phi_1 \wedge \phi_2$ is unsatisfiable or $I_1 \cap I_2 = \emptyset$; and for every $(\ell, a, \phi_1, \ell'), (\ell, a, \phi_2, \ell') \in \Delta_{int}^h$, $\phi_1 \wedge \phi_2$ is unsatisfiable. An ECMVPA is a dtMVPA where the stacks are untimed. A ECMVPA $(L, \Sigma, \Gamma, L^0, F, \Delta)$ is an dtMVPA if $I = [0, +\infty]$ for every $(\ell, a, I, \gamma, \phi, \ell') \in \Delta_r^h$.

Let $\Sigma = \langle \Sigma_c^h, \Sigma_r^h, \Sigma_{int}^h \rangle_{h=1}^n$ be a visibly pushdown alphabet. A *context* over $\Sigma^h = \langle \Sigma_c^h, \Sigma_r^h, \Sigma_{int}^h \rangle$ is a timed word in $(\Sigma^h)^*$. The empty word ε is a context. For ease, we assume in this paper that any context *has at least one symbol from* Σ. A *round* over Σ is a timed word w over Σ of the form $w_1 w_2 \dots w_n$ such that each w_h is a context over Σ^h. A k-round over Σ is a timed word w that can be obtained as a concatenation of k rounds over Σ. That is, $w = u_1 u_2 \dots u_k$, where each u_i is a round. Let $Round(\Sigma, k)$ denote the set of all k-round timed words over Σ. For any fixed k, a k-round dtMVPA over Σ is a tuple $A = (k, L, \Sigma, \Gamma, L^0, F, \Delta)$ where $M = (L, \Sigma, \Gamma, L^0, F, \Delta)$ is a dtMVPA over Σ. The language accepted by A is $L(A) = L(M) \cap Round(\Sigma, k)$ and is called k-round dense time multistack visibly push down language. The class of k-round dense time multistack visibly push down languages is denoted k-dtMVPL. The set $\bigcup_{k \geq 1} k$-dtMVPL is denoted *bd*-dtMVPL, and is the class of dense time multistack visibly push down languages with a bounded number of rounds. We define k-ECMVPL and *bd*-ECMVPL in a similar fashion. Also, we write k-dtMVPA and k-ECMVPA to denote k-round dtMVPA and k-round ECMVPA. The key result of the paper is the following.

Theorem 6 (A Perfect Timed Context-Sensitive Language). *The classes of languages accepted by k-dtMVPA and k-ECMVPA are perfect:— they are closed under Boolean operations with decidable emptiness problem.*

We sketch key lemmas towards this proof in the following section. As an application of this theorem we show Monadic second-order logic characterization of the languages accepted by k-dtMVPA in Sect. 4.

3 Proof of Theorem 6

The closure under union and intersection for both k-dtMVPA and k-ECMVPA is straightforward and is sketched in [8]. In order to show closure under complementation, the main hurdle is to show determinizability of these automata. We sketch the key ideas required to get determinizability for k-ECMVPA in Sect. 3.1 and for k-dtMVPA in Sect. 3.2. The decidability of the emptiness problem for k-ECMVPA follows as for every k-ECMVPA, via region construction [3], one can get an untimed-bisimilar k-MVPA, which has a decidable emptiness [14]. In Sect. 3.2 we show that for every k-dtMVPA we get an emptiness-preserving k-ECMVPA and hence this result in combination with previous remark yield decidability of emptiness for k-dtMVPA.

3.1 Determinizability of k-ECMVPA

For the determinizability proof the key observation is the since the words accepted by A is a catenation of k rounds, and the stacks (or contexts) do not interfere with each other, the k-ECMVPA A can be considered as a "composition" of n ECVPA $A_1, \dots, A_n$, with stack of each A_i corresponds to i-th stack of the k-ECMVPA. A has to simulate the n ECVPAs in a round robin fashion for k rounds.

If $w \in L(A)$, then $w = u_1u_2 \ldots u_k$, and $u_i = u_{i1}u_{i2} \ldots u_{in}$, where u_{ij} is the jth context in the ith round. Starting in an initial location ℓ_{11}, control is passed to A_1, which runs on u_{11} and enters location $\ell'_{11} = \ell_{12}$. Let $\nu'_{11} = \nu_{12}$ be the values of all clocks after processing u_{11}. At this point of time, A_2 runs on u_{12} starting in location ℓ_{12}, and so on, until A_n runs on u_{1n} starting in location ℓ_{1n}. Now first round is over, and u_1 is processed. A_n ends in some location $\ell'_{1n} = \ell_{21}$. Now A_1 starts again in ℓ_{21} and processes u_{21}. The values of all recorders and predictors change according to the time that elapsed during the simulation of $A_2, \ldots, A_n$. It must be noted that between two consecutive rounds i and $i+1$ of any A_j, none of the clocks pertaining to A_j get reset; they only reflect the time that has elapsed since the last round of A_j. This continues for k rounds, until u_{kn} is processed. A_j processes in order, $u_{1j}, u_{2j}, \ldots, u_{kj}$ over $(\Sigma^j)^*$ for $1 \leq j \leq n$. In round i, $1 \leq i \leq k$, each A_j, $1 \leq j \leq n-1$, starts in location ℓ_{ij}, runs on u_{ij} and "computes" a location ℓ_{ij+1}. Similarly, A_n moves from round i to round $i+1$, by starting in ℓ_{in}, runs on u_{in} and computes a location ℓ_{i+11}. The $(i+1)$th round begins in this location with A_1 running on u_{i+11}. Thus, by stitching together the locations needed to switch from A_j to A_{j+1}, we can obtain a simulation of A.

Let $u_{ij} = (a_j^1, t_{ij}^1) \ldots (a_j^{last}, t_{ij}^{last})$, where $t_{ij}^1, \ldots, t_{ij}^{last}$ are the time stamps on reading u_{ij}. Let $\kappa_j = u_{1j}(\#_1, t_{1j}^{last})u_{2j}(\#_2, t_{2j}^{last}) \ldots u_{kj}(\#_k, t_{kj}^{last})$. The new symbols $\#_i$ help disambiguate A_j processing $u_{1j}, \ldots, u_{kj}$ in k rounds. We first focus on each ECVPA A_j which processes $u_{1j}, u_{2j}, \ldots, u_{kj}$. Let $cmax$ be the maximum constant used in clock constraints of Σ^j in the ECMVPA A. Let $\mathcal{I} = \{[0,0], [0,1], \ldots, [cmax, cmax], [cmax, \infty)\}$ be a set of intervals. A *correct sequence of round switches* for A_j with respect to κ_j is a sequence of pairs $V_j = P_{1j}P_{2j} \ldots P_{kj}$, where $P_{hj} = ((\ell_{hj}, I_{hj}), \ell'_{hj})$, $2 \leq h \leq k$, $P_{1j} = ((\ell_{1j}, \nu_{1j}), \ell'_{1j})$ and $I_{hj} \in \mathcal{I}$ such that

1. Starting in ℓ_{1j}, with the jth stack containing $\bot_j$, and an initial valuation ν_{1j} of all recorders and predictors of Σ^j, the ECMVPA A processes u_{1j} and reaches some ℓ'_{1j} with stack content σ_{2j} and clock valuation ν'_{1j}. The processing of u_{2j} by A then starts at location ℓ_{2j}, and a time $t \in I_{2j}$ has elapsed between the processing of u_{1j} and u_{2j}. Thus, A starts processing u_{2j} in (ℓ_{2j}, ν_{2j}) where ν_{2j} is the valuation of all recorders and predictors updated from ν'_{1j} with respect to t. The stack content remains same as σ_{2j} when the processing of u_{2j} begins.
2. In general, starting in (ℓ_{hj}, ν_{hj}), $h > 1$ with the jth stack containing σ_{hj}, and ν_{hj} obtained from ν_{h-1j} by updating all recorders and predictors based on the time interval I_{hj} that records the time elapse between processing u_{hj-1} and u_{hj}, A processes u_{hj} and reaches (ℓ'_{hj}, ν'_{hj}) with stack content σ_{h+1j}. The processing of u_{h+1j} starts after time $t \in I_{h+1}$ has elapsed since processing u_{hj} in a location ℓ_{h+1j}, and stack content being σ_{h+1j}.

Lemma 7 *(Round Switching Lemma for A_j). Let $A = (k, L, \Sigma, \Gamma, L^0, F, \Delta)$ be a k-ECMVPA. Let $w = u_1 u_2 \dots u_k$ with $u_i = u_{i1} u_{i2} \dots u_{in}$, $1 \leq i \leq k$. Then we can construct a ECVPA A_j over $\Sigma^j \cup \{\#_1, \dots, \#_k\}$ which reaches a location V_j on reading κ_j iff V_j is a correct sequence of round switches for A_j.*

Proof. Recall that κ_j is defined by annotating $u_{1j} u_{2j} \dots u_{kj}$ with new symbols $\{\#_1, \dots, \#_k\}$ and appropriate time stamps. Let $V_j = P_{1j} \dots P_{kj}$ be a correct sequence of round switches for A_j. Given the k-ECMVPA $A = (k, L, \Sigma, \Gamma, L^0, F, \Delta)$ with w, the ECVPA A_j is constructed by simulating the transitions of A on Σ^j by guessing V_j in its initial location. The alphabet of A_j is $\Sigma^j \cup \{\#_1, \dots, \#_n\}$, and hence has event clocks $x_a, x_{\#_i}$, $a \in \Sigma^j$. Whenever A_j reads the $\#_i$, the control location as well as the valuation of all recorders and predictors are changed according to P_{i+1j}, $1 \leq i \leq k-1$. On reading $\#_k$, A_j enters the location V_j from its current location ℓ'_{kj}. The locations of A_j are $V_j \cup \{(i, \ell_{ij}, V_j), (i, \ell_{ij}, V_j, \#), (i, \ell_{ij}, V_j, a) \mid 1 \leq i \leq k, \ell \in L, a \in \Sigma^j, V_j \in ((L \times I) \times L)^k\}, \cup((L \times I) \times L)^k, I \in \mathcal{I}$. The set of initial locations are $\{(1, \ell_{1j}, V_j) \mid V_j \in ((L \times I) \times L)^k, I \in \mathcal{I}\}$. Starting in $(1, \ell_{1j}, V_j)$, A_j processes u_{1j}. When the last symbol a of u_{1j} is read, it enters a location $(1, \ell'_{1j}, V_j, a)$. From this location, only $\#_1$ transitions are enabled. On reading $\#_1$, we move from $(1, \ell'_{1j}, V_j, a)$ to a location $(2, \ell_{2j}, V_j, \#)$, where $P_2 = ((\ell_{2j}, I_{2j}), \ell'_{2j})$ and $P_1 = ((\ell_{1j}, \nu_{1j}), \ell'_{1j})$, after checking no time elapse since a (check $x_a = 0$). This ensures that no time is spent in processing $\#_1$ after u_{1j}. Now A_j starts processing u_{2j} starting in location $(2, \ell_{2j}, V_j, \#)$. From $(2, \ell_{2j}, V_j, \#)$, on reading a symbol $a \in \Sigma^j$, we check that the time elapse since $\#_1$ lies in the interval I_{2j} (check $x_{\#_1} \in I_{2j}$) as given by P_2 and so on. When round k is reached, A_j starts processing in some location $(k, \ell_{kj}, V_j, \#)$, and reaches (k, ℓ'_{kj}, V_j, a). When $\#_k$ is read, A_j enters location V_j. The details of transitions δ^j of A_j can be found in [8]. It is easy to see that V_j is reached by A_j only when the guessed V_j in the initial location is a correct sequence of round switches for A_j. □

While each V_j describes the correct sequence of round switches, $1 \leq j \leq n$, the sequence $V_1 V_2 \dots V_n$ is called a *globally correct sequence* iff we can stitch together the individual V_i's to obtain a complete simulation of A on w by moving across contexts and rounds. For instance, consider $V_j = P_{1j} P_{2j} \dots P_{kj}$ and $V_{j+1} = P_{1j+1} P_{2j+1} \dots P_{kj+1}$ for $1 \leq j \leq n-1$. Recall that $P_{ij} = ((\ell_{ij}, I_{ij}), \ell'_{ij})$ and $P_{ij+1} = ((\ell_{ij+1}, I_{ij+1}), \ell'_{ij+1})$ for $1 \leq i \leq k$. The sequence $V_1 V_2 \dots V_n$ is globally correct iff $\ell'_{ij} = \ell_{ij+1}$, $j \leq n-1$ and $\ell'_{in} = \ell_{i+11}$ for $1 \leq i \leq k$.

Lemma 8. *Let $w = u_1 u_2 \dots u_k$ be a timed word in Round(Σ, k), with $A = (k, L, \Sigma, \Gamma, L^0, F, \Delta)$ being a k-ECMVPA over Σ, and let $u_i = u_{i1} u_{i2} \dots u_{in}$ and κ_j be as defined above. Then $w \in L(A)$ iff for $1 \leq j \leq n$, there exists a correct switching sequence V_j of the ECVPA A_j for κ_j such that $V_1 V_2 \dots V_n$ is a globally correct sequence for A with $\ell_{11} \in L^0$ and $\ell'_{kn} \in F$.*

Proof. The proof essentially shows how one can simulate A by composing the A_j's using a globally correct sequence $V_1 V_2 \dots V_n$. The idea is to simulate each A_j one after the other, allowing A_{j+1} to begin on u_{ij+1} iff the location reached ℓ'_{ij}

at the end of u_{ij} by A_j matches with ℓ_{ij+1}, the proposed starting location of A_{j+1} on u_{ij+1}. Lets construct a composition of $A_1, \ldots, A_n$ which runs on w, and accepts w iff there exists a globally correct sequence $V_1 V_2 \ldots V_n$. The initial locations are of the form $(p_1, p_2, \ldots, p_n, 1, 1)$, where the last two entries denote the current round number and context number and p_j is an initial location of A_j. The transitions Δ of the composition are defined using the transitions δ^j of A_j.

In some chosen initial location, we first run A_1 updating only the first entry p_1 of the tuple until u_{11} is completely read. The first entry of the tuple then has the form $p'_1 = (1, \ell'_{11}, V_1, a)$ where a is the last symbol of u_{11}. When A_1 reads $\#_1$, the current location in the composition is $(p'_1, p_2, \ldots, p_n, 1, 1)$. In the composition of $A_1, \ldots, A_n$, since there are no #'s to be read, we start simulation of A_2 on u_{12} from $(p'_1, p_2, \ldots, p_n, 1, 1)$ iff p_2 is $(2, \ell_{12}, V_2)$ such that the ℓ'_{11} in p_1 is same as the ℓ_{12} in p_2. We then add the transition from $(p'_1, p_2, \ldots, p_n, 1, 1)$ to $(p''_1 = (2, \ell_{21}, V_1, a), q, \ldots, p_n, 1, 2)$ where q is obtained from p_2 by a transition in A_2 on the first symbol of u_{12}. The a in p''_1 is the last symbol of u_{11} taken from $p'_1 = (1, \ell'_{11}, V_1, a)$, and the ℓ_{21} in p''_1 is obtained from $P_{21} = ((\ell_{21}, I_{21}), \ell'_{21})$ of V_1. We continue like this till we reach u_{1n}, the last context in round 1, and reach some location $(s_1, s_2, \ldots, s_{n-1}, p'_n, 1, 1)$ with $s_1 = (2, \ell_{21}, V_1, a_1), s_2 = (2, \ell_{22}, V_2, a_2), \ldots, s_{n-1} = (2, \ell_{2\ n-1}, V_{n-1}, a_{n-1})$ and $p'_n = (1, \ell'_{1n}, V_n, a_n)$.

Now, to start the second round, that is on u_{21}, we allow the transition from the above location iff $\ell'_{1n} = \ell_{21}$ and if $x_{a_1} \in I_{21}$ and we start simulating A_1 again, after updating p'_n, the context and round number. That is, we have the transition $(s_1, \ldots, s_{n-1}, p'_n, 1, n)$ on the first symbol of u_{21} to $(r, \ldots, s_{n-1}, s_n, 2, 1)$ where $s_n = (2, \ell_{2\ n}, V_n, a_n)$ iff $\ell'_{1n} = \ell_{21}$ and $x_{a_1} \in I_{21}$. Also, r is obtained from s_1 by a transition of A_1 on the first symbol of u_{21}. The check $x_{a_1} \in I_{21}$ is consistent with the check of $x_{\#_1} \in I_{21}$ in A_1. From $(r, \ldots, s_{n-1}, s_n, 2, 1)$, the processing of u_{21} happens as in A_1, and we continue till we finish processing u_{2n}. The same checks are repeated at the start of each fresh round.

So we have a run on w in the composition only when we have a globally correct sequence. On completing u_{kn}, we reach location $(V_1, \ldots, V_{n-1}, V_n, k, n)$, each V_j obtained from the individual A_j. We define the accepting locations of the composition to be $\{(V_1, \ldots, V_n) \mid P_{kn} = (\ell'_{kn}, [0, \infty)), \ell'_{kn} \in F\}$. Clearly, whenever there is a run in A on w that ends up in $\ell'_{kn} \in F$, we have an accepting run on w in the composition. □

The key idea of the determinization of k-ECMVPA follows from Lemma 8 and the determinizability of ECVPA [16]. Details are given in [8].

Theorem 9. *k-ECMVPAs are determinizable.*

3.2 Determinizability of k-dtMVPA

Given a k-dtMVPA M, we first construct (untiming construction) a k-ECMVPA M' and a morphism h such that $L(M) = h(L(M'))$. We then use the determinizability of k-ECMVPA (Theorem 9) to obtain a deterministic k-ECMVPA M'' such that $L(M') = L(M'')$. We then show how to obtain a k-dtMVPA D from M'' preserving the determinism of M'' such that $L(D) = h(L(M'')) = h(L(M')) = L(M)$.

We give an intuition to the untiming construction. Each time a symbol is pushed on to a stack (say stack i), we guess its age (the time interval) at the time of popping. For instance, in the dtMVPA M, while pushing a symbol a if the guessed constraint is $< \kappa$ for $\kappa \in \mathbb{N}$, then in the ECMVPA M', we push the symbol $(a, < \kappa, first)$ in the stack i, if this is the first symbol for which the guessed age is $< \kappa$. If $< \kappa$ has already been guessed as the age for a symbol pushed earlier, then we push $(a, < \kappa)$ onto the stack i. The guess $<_i \kappa$ is remembered in the finite control of the M'. Thus, for each symbol a pushed in stack i of the M, we push in stack i of the M', either $(a, < \kappa, first)$ or $(a, < \kappa)$ and remember $<_i \kappa$ in the finite control as a set of obligations. This information $<_i \kappa$ is retained in the finite control until popping the symbol $(a, < \kappa, first)$ from stack i. New symbols $<_i \kappa$ are added as internal symbols to the M'. The number of these symbols is finite since we have finitely many stacks and there is a maximum constant used in age comparisons of the M. After pushing $(a, < \kappa, first)$ onto the stack i, we read the internal symbol $<_i \kappa$, ensuring no time elapse since the last input symbol. Thus the event clock $x_{<_i \kappa}$ is reset at the same time as pushing $(a, < \kappa, first)$ on the stack. While popping $(a, < \kappa, first)$, we check that the value of the event clock $x_{<_i \kappa}$ is less than κ. Constraints of the form $> \kappa$ are handled similarly. Since the n stacks do not interfere with each other, this construction (adding extra symbols $<_i \kappa$ one per stack, retaining these symbols in the finite control until popping $(a, < \kappa, first)$ from stack i) can be done for all stacks, mimicking the timed stack. Note that the language accepted by the M is $h(L(M'))$, where h is the morphism which erases symbols of the form $<_i \kappa$ and $>_i \kappa$ from $L(M')$. This gives an ECMVPA preserving emptiness of the dtMVPA. We can determinize the ECMVPA M' obtaining $det(M')$ using Theorem 9. It remains to eliminate the transitions on the new symbols $<_i \kappa$ and $>_i \kappa$ from $det(M')$ and argue that the resulting machine stays deterministic and accepts $L(M)$.

Theorem 10. *k-dtMVPAs have decidable emptiness and are determinizable.*

4 Logical Characterization of k-dtMVPA

We consider a timed word $w = (a_0, t_0), (a_1, t_1), \ldots, (a_m, t_m)$ over alphabet $\Sigma = \langle \Sigma_c^i, \Sigma_{int}^i, \Sigma_r^i \rangle_{i=1}^n$ as a *word structure* over the universe $U = \{1, 2, \ldots, |w|\}$ of positions in the timed word. The predicates in the structure are $Q_a(i)$ for $a \in \Sigma$ which evaluates to true at position i iff $w[i] = a$, where $w[i]$ denotes the ith position of w. Following [11], we use the matching relation $\mu_j(i, k)$ which evaluates to true iff the ith position is a call and the kth position is its matching return corresponding to the jth stack. We also introduce three predicates $\lhd_a$, $\rhd_a$, and θ_j capturing the following relations: For an interval I, the predicate $\lhd_a(i) \in I$ evaluates to true on the structure iff $\nu_i^w(x_a) \in I$ for recorder x_a; the predicate $\rhd_a(i) \in I$ evaluates to true iff $\nu_i^w(y_a) \in I$ for predictor y_a; the predicate $\theta_j(i) \in I$ evaluates to true iff $w[i] \in \Sigma_r^j$, and there is some $k < i$ such that $\mu_j(k, i)$ evaluates to true and $t_i - t_k \in I$. The predicate $\theta_j(i)$ measures the time elapse between position k, where a call was made on the stack j and its matching return, i. This time elapse is the age of the symbol pushed on to the stack during the call at position k. Since position i is the matching return, this symbol is popped at position i; if the age lies in the interval I, the predicate evaluates to true. We define MSO(Σ), the MSO logic over Σ, as:

$$\varphi := Q_a(x) \mid x \in X \mid \mu_j(x,y) \mid \lhd_a(x) \in I \mid \rhd_a(x) \in I \mid \theta_j(x) \in I \mid \neg\varphi \mid \varphi \vee \varphi \mid \exists\, x.\varphi \mid \exists\, X.\varphi$$

where $a \in \Sigma$, $x_a \in C_\Sigma$, x is a first order variable and X is a second order variable.

The models of a formula $\phi \in$ MSO(Σ) are timed words w over Σ. The semantics of this logic is standard where first order variables are interpreted over positions of w and second order variables over subsets of positions. We define the language $L(\varphi)$ of an MSO sentence φ as the set of all words satisfying φ. Words in $Round(\Sigma, k)$, for some k rounds, can be captured by an MSO formula $Bd_k(\psi)$. For instance if $k = 1$, and n stacks, the formula $\exists x_1.(Q_{a^1}(x_1) \wedge \forall y_1(y_1 \leq x_1 \rightarrow Q_{a^1}(y_1)) \wedge \exists x_2.(x_1 < x_2 \wedge Q_{a^2}(x_2) \wedge \forall y_2(x_1 < y_2 < x_2 \rightarrow Q_{a^2}(y_2)) \wedge \ldots \wedge \exists x_n(x_{n-1} < x_n \wedge Q_{a^n}(x_n) \wedge last(x_n) \wedge \forall y_n(x_{n-1} < y_n < x_n \rightarrow Q_{a^n}(y_n)))))$, where $a^i \in \Sigma^i$ and $last(x)$ denotes x is the last position, captures a round. This can be extended to capture k-round words. Conjuncting the formula obtained from a dtMVPA M with $Bd_k(\psi)$ accepts only those words which lie in $L(M) \cap Round(\Sigma, k)$. Likewise, if one considers any MSO formula $\zeta = \varphi \wedge Bd_k(\psi)$, it can be shown that the dtMVPA M constructed for ζ will be a k-dtMVPA. The two directions, dtMVPA to MSO, as well as MSO to dtMVPA can be handled using standard techniques, and can be found in [8].

Theorem 11. *A language L over Σ is accepted by an k-dtMVPA iff there is a MSO sentence φ over Σ such that $L(\varphi) \cap Round(\Sigma, k) = L$.*

References

1. Abdulla, P., Atig, M., Stenman, J.: Dense-timed pushdown automata. In: LICS, pp. 35–44 (2012)
2. Alur, R., Dill, D.: A theory of timed automata. TCS **126**, 183–235 (1994)
3. Alur, R., Fix, L., Henzinger, T.A.: Event-clock automata: a determinizable class of timed automata. TCS **211**(1–2), 253–273 (1999)
4. Alur, R., Madhusudan, P.: Visibly pushdown languages. In: Symposium on Theory of Computing, pp. 202–211 (2004)
5. Atig, M.F.: Model-checking of ordered multi-pushdown automata. Log. Methods Comput. Sci. **8**(3), 1–31 (2012)
6. Bar-Hillel, Y., Perles, M., Shamir, E.: On formal properties of simple phrase structure grammars. Zeitschrift für Phonetik, Sprachwissenschaft und Kommunikationsforschung **14**, 143–172 (1961)
7. Bhave, D., Dave, V., Krishna, S.N., Phawade, R., Trivedi, A.: A logical characterization for dense-time visibly pushdown automata. In: Dediu, A.-H., Janoušek, J., Martín-Vide, C., Truthe, B. (eds.) LATA 2016. LNCS, vol. 9618, pp. 89–101. Springer, Heidelberg (2016). doi:10.1007/978-3-319-30000-9_7

8. Bhave, D., Dave, V., Krishna, S.N., Phawade, R., Trivedi, A.: A perfect class of context-sensitive timed languages. Technical report, IIT Bombay (2016). www.cse.iitb.ac.in/internal/techreports/reports/TR-CSE-2016-80.pdf
9. Czerwiński, W., Hofman, P., Lasota, S.: Reachability problem for weak multi-pushdown automata. In: Koutny, M., Ulidowski, I. (eds.) CONCUR 2012. LNCS, vol. 7454, pp. 53–68. Springer, Heidelberg (2012)
10. Esparza, J., Ganty, P., Majumdar, R.: A perfect model for bounded verification. In: LICS, pp. 285–294. IEEE Computer Society (2012)
11. La Torre, S., Madhusudan, P., Parlato, G.: A robust class of context-sensitive languages. In: LICS, pp. 161–170 (2007)
12. La Torre, S., Napoli, M., Parlato, G.: Scope-bounded pushdown languages. In: Shur, A.M., Volkov, M.V. (eds.) DLT 2014. LNCS, vol. 8633, pp. 116–128. Springer, Heidelberg (2014)
13. La Torre, S., Napoli, M., Parlato, G.: A unifying approach for multistack pushdown automata. In: Csuhaj-Varjú, E., Dietzfelbinger, M., Ésik, Z. (eds.) MFCS 2014, Part I. LNCS, vol. 8634, pp. 377–389. Springer, Heidelberg (2014)
14. La Torre, S., Madhusudan, P., Parlato, G.: The language theory of bounded context-switching. In: López-Ortiz, A. (ed.) LATIN 2010. LNCS, vol. 6034, pp. 96–107. Springer, Heidelberg (2010)
15. Trivedi, A., Wojtczak, D.: Recursive timed automata. In: Bouajjani, A., Chin, W.-N. (eds.) ATVA 2010. LNCS, vol. 6252, pp. 306–324. Springer, Heidelberg (2010)
16. Van Tang, N., Ogawa, M.: Event-clock visibly pushdown automata. In: Nielsen, M., Kučera, A., Miltersen, P.B., Palamidessi, C., Tůma, P., Valencia, F. (eds.) SOFSEM 2009. LNCS, vol. 5404, pp. 558–569. Springer, Heidelberg (2009)

Position Automaton Construction for Regular Expressions with Intersection

Sabine Broda, António Machiavelo, Nelma Moreira, and Rogério Reis(✉)

CMUP, Faculdade de Ciências da Universidade do Porto, Porto, Portugal
{sbb,nam,rvr}@dcc.fc.up.pt, ajmachia@fc.up.pt

Abstract. Positions and derivatives are two essential notions in the conversion methods from regular expressions to equivalent finite automata. Partial derivative based methods have recently been extended to regular expressions with intersection. In this paper, we present a position automaton construction for those expressions. This construction generalizes the notion of position making it compatible with intersection. The resulting automaton is homogeneous and has the partial derivative automaton as its quotient.

1 Introduction

The position automaton, introduced by Glushkov [12], permits the conversion of a simple regular expression (involving only the sum, concatenation and star operations) into an equivalent nondeterministic finite automaton (NFA) without ε-transitions. The states in the position automaton ($\mathcal{A}_{\mathsf{pos}}$) correspond to the positions of letters in the corresponding regular expression plus an additional initial state. McNaughton and Yamada [15] also used the positions of a regular expression to define an automaton, however they directly computed a deterministic version of the position automaton. The position automaton has been well studied [3,8] and is considered the *standard* automaton simulation of a regular expression [16]. Some of its interesting properties are: homogeneity, i.e. for each state, all in-transitions have the same label (letter); whenever deterministic, these automata characterize certain families of unambiguous regular expressions, and can be computed in quadratic time [4]; other automata simulations of regular expressions are quotients of the $\mathcal{A}_{\mathsf{pos}}$, e.g. partial derivative automata ($\mathcal{A}_{\mathsf{pd}}$) [9] and follow automata [14].

Many authors observed that the position automaton construction could not directly be extended to regular expressions with intersection [3,6], as intersection (and also complementation) is not compatible with the notion of position. In fact, considering positions of letters in the expression $(ab^\star) \cap a$, whose language is $\{a\}$, we obtain the regular expression $(a_1 b_2^\star) \cap a_3$. Interpreting a_1 and a_3 as distinct alphabet symbols, the language described by this expression is empty and there is

This work was partially supported by CMUP (UID/MAT/00144/2013), which is funded by FCT (Portugal) with national (MEC) and European structural funds through the programs FEDER, under the partnership agreement PT2020.

S. Brlek and C. Reutenauer (Eds.): DLT 2016, LNCS 9840, pp. 51–63, 2016.
DOI: 10.1007/978-3-662-53132-7_5

no longer a correspondence between the languages of $(ab^\star)\cap a$ and $(a_1b_2^\star)\cap a_3$, as it is the case for expressions without intersection. However, the conversions from expressions to automata based on the notion of derivative or partial derivative can still be extended to regular expressions with intersection [2,5,7]. In this paper, we present a position automaton construction for regular expressions with intersection by generalizing the notion of position. Instead of positions, sets of positions are considered in such a way that marking a regular expression is made compatible with the intersection operation. We also show that the partial derivative automaton is a quotient of the position automaton.

2 Preliminaries

In this section we recall the basic definitions we use throughout this paper and the notation. For further details we refer to [13,17].

Let Σ be an *alphabet* (set of letters). A *word* over Σ is a finite sequence of letters, where ε is the empty word. The size of a word x, $|x|$, is the number of alphabet symbols in x. $\Sigma^\star$ denotes the set of all words over Σ, and a *language* over Σ is any subset of $\Sigma^\star$. The *concatenation* of two languages L_1 and L_2 is defined by $L_1 \cdot L_2 = \{ xy \mid x \in L_1, y \in L_2 \}$, and $L^\star$ denotes the set $\{ x_1x_2\cdots x_n \mid n \geq 0, x_i \in L \}$. The *left quotient* of a language $L \subseteq \Sigma^\star$ w.r.t. a word $x \in \Sigma^\star$ is the language $x^{-1}L = \{ y \mid xy \in L \}$.

The set $\mathsf{RE}_\cap$ of *regular expressions with intersection* over Σ is defined by the following grammar

$$\alpha, \beta := \emptyset \mid \varepsilon \mid a \in \Sigma \mid (\alpha + \beta) \mid (\alpha \cdot \beta) \mid (\alpha^\star) \mid (\alpha \cap \beta), \tag{1}$$

where the concatenation operator $\cdot$ is often omitted. We consider $\mathsf{RE}_\cap$ expressions modulo the standard equations for $\emptyset$ and ε, i.e. $\alpha+\emptyset = \emptyset+\alpha = \alpha\cdot\varepsilon = \varepsilon\cdot\alpha = \alpha$, $\alpha\cdot\emptyset = \emptyset\cdot\alpha = \alpha\cap\emptyset = \emptyset\cap\alpha = \emptyset$, and $\emptyset^\star = \varepsilon$. Throughout this paper we often refer to regular expressions with intersection just as regular expressions. The set of alphabet symbols with occurrences in α is denoted by Σ_α. Expressions containing no occurrence of the operator $\cap$ are called *simple regular expressions.* A *linear regular expression* is a regular expression in which every alphabet symbol occurs at most once. We let $|\alpha|$, $|\alpha|_\Sigma$ and $|\alpha|_\cap$ denote for $\alpha \in \mathsf{RE}_\cap$ the number of symbols, the number of occurrences of alphabet symbols and the number of occurrences of the binary operator $\cap$, respectively. The language $\mathcal{L}(\alpha)$ for $\alpha \in \mathsf{RE}_\cap$ is defined as usual, with $\mathcal{L}(\alpha\cap\beta) = \mathcal{L}(\alpha)\cap\mathcal{L}(\beta)$. The language of $S \subseteq \mathsf{RE}_\cap$ is $\mathcal{L}(S) = \cup_{\alpha\in S}\mathcal{L}(\alpha)$. Given an expression $\alpha \in \mathsf{RE}_\cap$, we define $\varepsilon(\alpha) = \varepsilon$ if $\varepsilon \in \mathcal{L}(\alpha)$, and $\varepsilon(\alpha) = \emptyset$ otherwise. A recursive definition of $\varepsilon : \mathsf{RE}_\cap \longrightarrow \{\emptyset, \varepsilon\}$ is given by the following: $\varepsilon(a) = \varepsilon(\emptyset) = \emptyset$, $\varepsilon(\varepsilon) = \varepsilon(\alpha^\star) = \varepsilon$, $\varepsilon(\alpha+\beta) = \varepsilon(\alpha) + \varepsilon(\beta)$, and $\varepsilon(\alpha\beta) = \varepsilon(\alpha\cap\beta) = \varepsilon(\alpha)\cdot\varepsilon(\beta)$.

A *nondeterministic finite automaton* (NFA) is a tuple $\mathcal{A} = \langle S, \Sigma, S_0, \delta, F\rangle$, where S is a finite set of states, Σ is a finite alphabet, $S_0 \subseteq S$ a set of initial states, $\delta : S \times \Sigma \longrightarrow \mathcal{P}(S)$ the transition function, and $F \subseteq S$ a set of final states. The extension of δ to sets of states and words is defined by $\delta(X, \varepsilon) = X$ and $\delta(X, ax) = \delta(\cup_{s\in X}\delta(s, a), x)$. A word $x \in \Sigma^\star$ is accepted by $\mathcal{A}$ if and only if $\delta(S_0, x) \cap F \neq \emptyset$. The *language of* $\mathcal{A}$, $\mathcal{L}(\mathcal{A})$, is the set of words accepted by $\mathcal{A}$. The *right language of a state* s, $\mathcal{L}_s$, is the language accepted by $\mathcal{A}$ if we take $S_0 = \{s\}$. An NFA is *initially connected* or *accessible* if each state is reachable from an initial state and it is *trimmed* if, moreover, the right language of each state is non-empty. Given $\mathcal{A}$, we denote by $\mathcal{A}^{\mathsf{ac}}$ and $\mathcal{A}^{\mathsf{t}}$ the result of removing unreachable states from $\mathcal{A}$ and trimming $\mathcal{A}$, respectively. It is clear that $\mathcal{L}(\mathcal{A}) = \mathcal{L}(\mathcal{A}^{\mathsf{ac}}) = \mathcal{L}(\mathcal{A}^{\mathsf{t}})$.

We say that an equivalence relation $\equiv$ over S is right invariant w.r.t. $\mathcal{A}$ iff

1. $\forall s, t \in S,\ s \equiv t \wedge s \in F \implies t \in F$
2. $\forall s, t \in S, \forall a \in \Sigma,\ s \equiv t \implies \forall s_1 \in \delta(s, a)\ \exists t_1 \in \delta(t, a), s_1 \equiv t_1$.

If $\equiv$ is right invariant, then we can define the quotient automaton $\mathcal{A}/_{\equiv}$ in the usual way, and $\mathcal{L}(\mathcal{A}/_{\equiv}) = \mathcal{L}(\mathcal{A})$.

The notions of partial derivatives and partial derivative automata were introduced by Antimirov [1] for simple regular expressions. Bastos et al. [2] presented an extension of the Antimirov construction from $\mathsf{RE}_\cap$ expressions.

Definition 1. *For $\alpha \in \mathsf{RE}_\cap$ and $a \in \Sigma$, the set $\partial_a(\alpha)$ of* partial derivatives *of α w.r.t. a is defined by:*

$$\partial_a(\emptyset) = \partial_a(\varepsilon) = \emptyset \qquad \partial_a(\alpha + \beta) = \partial_a(\alpha) \cup \partial_a(\beta)$$

$$\partial_a(b) = \begin{cases} \{\varepsilon\}, & \text{if } a = b \\ \emptyset & \text{otherwise} \end{cases} \qquad \partial_a(\alpha\beta) = \begin{cases} \partial_a(\alpha) \odot \beta \cup \partial_a(\beta), & \text{if } \varepsilon(\alpha) = \varepsilon \\ \partial_a(\alpha)\beta, & \text{otherwise} \end{cases}$$

$$\partial_a(\alpha^\star) = \partial_a(\alpha) \odot \alpha^\star \qquad \partial_a(\alpha \cap \beta) = \partial_a(\alpha) \mathbin{\dot\cap} \partial_a(\beta),$$

where for $S, T \subseteq \mathsf{RE}_\cap$ and $\beta \in \mathsf{RE}_\cap$, $S \odot \beta = \{\, \alpha\beta \mid \alpha \in S \,\}$, $\beta \odot S = \{\, \beta\alpha \mid \alpha \in S \,\}$, and $S \mathbin{\dot\cap} T = \{\, \alpha \cap \beta \mid \alpha \in S, \beta \in T \,\}$.

This definition is extended to any word w by $\partial_\varepsilon(\alpha) = \{\alpha\}$, $\partial_{wa}(\alpha) = \bigcup_{\alpha_i \in \partial_w(\alpha)} \partial_a(\alpha_i)$, and $\partial_w(R) = \bigcup_{\alpha_i \in R} \partial_w(\alpha_i)$, where $R \subseteq \mathsf{RE}_\cap$. The set of partial derivatives of an expression α is $\partial(\alpha) = \bigcup_{w \in \Sigma^\star} \partial_w(\alpha)$. As for simple regular expressions, the partial derivative automaton of an expression $\alpha \in \mathsf{RE}_\cap$ is defined by $\mathcal{A}_{\mathsf{pd}}(\alpha) = \langle \partial(\alpha), \Sigma, \{\alpha\}, \delta_{\mathsf{pd}}, F_{\mathsf{pd}} \rangle$, where $F_{\mathsf{pd}} = \{\, \gamma \in \partial(\alpha) \mid \varepsilon(\gamma) = \varepsilon \,\}$ and $\delta_{\mathsf{pd}}(\gamma, a) = \partial_a(\gamma)$. It follows that $\mathcal{L}(\mathcal{A}_{\mathsf{pd}}(\alpha))$ is exactly $\mathcal{L}(\alpha)$ and by construction $\mathcal{A}_{\mathsf{pd}}(\alpha)$ is accessible. Bastos et al. [2] showed also that $|\partial(\alpha)| \leq 2^{|\alpha|_\Sigma - |\alpha|_\cap - 1} + 1$ and asymptotically and on average an upper bound for the number of states is $(1.056 + o(1))^n$, where n is the size of the expression.

3 Indexed Expressions

Given an alphabet Σ and a nonempty set of indexes $J \subseteq \mathbb{N}$, let $\Sigma_J = \{\, a_j \mid a \in \Sigma, j \in J \,\}$. An *indexed regular expression* is a regular expression over the alphabet Σ_J such that for all $a_i, b_j \in \Sigma_J$ occurring in the expression, $a \neq b$ implies $i \neq j$. We let $\rho, \rho_1, \rho_2, \ldots$ denote indexed regular expressions. If ρ is an indexed expression, then $\overline{\rho}$ is the regular expression over the alphabet Σ obtained by removing the indexes. The set of all indexes occurring in ρ is denoted by $\mathsf{ind}(\rho) = \{\, i \mid a_i \in \Sigma_\rho \,\}$. Given an indexed expression ρ and $i \in \mathsf{ind}(\rho)$, $\ell_\rho(i)$ is the letter indexed by i in ρ. From now on, we will simply write $\ell(i)$ for $\ell_\rho(i)$ since it will always be clear that we are referring to a specific expression ρ. Given an indexed expression ρ, let

$$\mathcal{I}_\rho = \{\, \mathrm{I} \subseteq \mathsf{ind}(\rho) \mid \mathrm{I} \neq \emptyset \text{ and } \forall i_1, i_2 \in \mathrm{I}, \ell(i_1) = \ell(i_2) \,\}.$$

For $\mathrm{I} \in \mathcal{I}_\rho$ we extend the definition of ℓ by $\ell(\mathrm{I}) = \ell(i)$, $i \in \mathrm{I}$. Finally, we say that ρ is *well-indexed* if for all subterms of ρ of the form $\rho_1 \cap \rho_2$ one has $\mathsf{ind}(\rho_1) \cap \mathsf{ind}(\rho_2) = \emptyset$.

Example 2. For $\rho = a_1(a_4 b_5^\star \cap a_4)$ one has $\overline{\rho} = a(ab^\star \cap a)$, $\mathsf{ind}(\rho) = \{1,4,5\}$, $\ell(4) = \ell(\{1,4\}) = a$ and $\mathcal{I}_\rho = \{\{1\},\{4\},\{5\},\{1,4\}\}$. However, this expression is not well-indexed, since a_4 occurs on both sides of an intersection.

Definition 3. *Consider an indexed expression ρ. For $L \subseteq \mathcal{I}_\rho^\star$ and $x = \mathrm{I}_1 \cdots \mathrm{I}_n \in L$, we define $\ell(x) = \ell(\mathrm{I}_1)\cdots\ell(\mathrm{I}_n)$ and $\ell(L) = \{\, \ell(x) \mid x \in L \,\}$. The* indexed intersection *of two words $x = \mathrm{I}_1 \cdots \mathrm{I}_m, y = \mathrm{J}_1 \cdots \mathrm{J}_n \in \mathcal{I}_\rho^\star$ is defined by $x \cap_{\mathcal{I}} y = (\mathrm{I}_1 \cup \mathrm{J}_1)\cdots(\mathrm{I}_n \cup \mathrm{J}_n)$ if $\ell(x) = \ell(y)$*[1]*, and undefined otherwise. Then, the* indexed intersection *of two languages $L_1, L_2 \in \mathcal{I}_\rho^\star$ is defined as follows:*

$$L_1 \cap_{\mathcal{I}} L_2 = \{\, x \cap_{\mathcal{I}} y \mid x \in L_1, y \in L_2 \,\}.$$

We define the index-language *$\mathcal{L}_{\mathcal{I}}(\rho) \subseteq \mathcal{I}_\rho^\star$ associated to ρ as follows.*

$$\begin{array}{lll} \mathcal{L}_{\mathcal{I}}(\emptyset) = \emptyset, & \mathcal{L}_{\mathcal{I}}(a_i) = \{\{i\}\}, & \mathcal{L}_{\mathcal{I}}(\rho_1 + \rho_2) = \mathcal{L}_{\mathcal{I}}(\rho_1) \cup \mathcal{L}_{\mathcal{I}}(\rho_2), \\ \mathcal{L}_{\mathcal{I}}(\varepsilon) = \{\varepsilon\}, & \mathcal{L}_{\mathcal{I}}(\rho^\star) = \mathcal{L}_{\mathcal{I}}(\rho)^\star, & \mathcal{L}_{\mathcal{I}}(\rho_1 \cdot \rho_2) = \mathcal{L}_{\mathcal{I}}(\rho_1) \cdot \mathcal{L}_{\mathcal{I}}(\rho_2), \\ & & \mathcal{L}_{\mathcal{I}}(\rho_1 \cap \rho_2) = \mathcal{L}_{\mathcal{I}}(\rho_1) \cap_{\mathcal{I}} \mathcal{L}_{\mathcal{I}}(\rho_2). \end{array}$$

Example 4. For $\rho = (a_1a_2 + b_3 + a_4)^\star \cap (a_5 + b_6)^\star$, we have $\mathcal{L}_{\mathcal{I}}(\rho) = \{\{4,5\},\{3,6\}, \{1,5\}\{2,5\},\{4,5\}\{4,5\},\{4,5\}\{3,6\},\ldots\}$, and $\ell(\mathcal{L}_{\mathcal{I}}(\rho)) = \{a, b, aa, ab, \ldots\}$ (since $\ell(\{1,5\}\{2,5\}) = \ell(\{4,5\}\{4,5\}) = aa$).

Proposition 5. *Given an indexed expression ρ, one has $\ell(\mathcal{L}_{\mathcal{I}}(\rho)) = \mathcal{L}(\overline{\rho})$.*

4 A Position Automaton for $\mathsf{RE}_\cap$ Expressions

Let $\alpha \in \mathsf{RE}_\cap$. We define the set of positions in α by $\mathsf{pos}(\alpha) = \{1, \ldots, |\alpha|_\Sigma\}$. As usual, we let $\overline{\alpha}$ denote the expression obtained from α by indexing each letter with its position in α. The same notation is used to remove the indexes, as already stated, thus, $\overline{\overline{\alpha}} = \alpha$. Note that for $\alpha \in \mathsf{RE}_\cap$, the indexed expression $\overline{\alpha}$ is always linear (thus well-indexed), and also $\mathsf{pos}(\alpha) = \mathsf{ind}(\overline{\alpha})$.

[1] Note that $\ell(x) = \ell(y)$ implies that $m = n$ and that $\ell(x \cap_{\mathcal{I}} y) = \ell(x) = \ell(y)$.

Given an indexed linear expression ρ we define the following sets:

$$\begin{aligned}\mathsf{First}'(\rho) &= \{\, \mathrm{I} \mid \mathrm{I}x \in \mathcal{L}_{\mathcal{I}}(\rho) \,\},\\ \mathsf{Last}'(\rho) &= \{\, \mathrm{I} \mid x\mathrm{I} \in \mathcal{L}_{\mathcal{I}}(\rho) \,\},\\ \mathsf{Follow}'(\rho) &= \{\, (\mathrm{I},\mathrm{J}) \mid x\mathrm{I}\mathrm{J}y \in \mathcal{L}_{\mathcal{I}}(\rho) \,\}.\end{aligned}$$

Then, given $\alpha \in \mathsf{RE}_\cap$, we define $\mathsf{First}(\alpha) = \mathsf{First}'(\overline{\alpha}), \mathsf{Last}(\alpha) = \mathsf{Last}'(\overline{\alpha})$, and $\mathsf{Follow}(\alpha) = \mathsf{Follow}'(\overline{\alpha})$.

Definition 6. *The* position automaton *of an expression* $\alpha \in \mathsf{RE}_\cap$ *is*

$$\mathcal{A}_{\mathsf{pos}}(\alpha) = \langle S_{\mathsf{pos}}, \Sigma, \{\{0\}\}, \delta_{\mathsf{pos}}, F_{\mathsf{pos}}\rangle,$$

where $S_{\mathsf{pos}} = \{\{0\}\} \cup \{\mathrm{I} \in \mathcal{I}_{\overline{\alpha}} \mid x\mathrm{I}y \in \mathcal{L}_{\mathcal{I}}(\overline{\alpha})$ *for some* $x, y \in \mathcal{I}_{\overline{\alpha}}^\star \}$,

$$\delta_{\mathsf{pos}} = \{\, (\mathrm{I}, \ell(\mathrm{J}), \mathrm{J}) \mid (\mathrm{I},\mathrm{J}) \in \mathsf{Follow}(\alpha) \,\} \cup \{\, (\{0\}, \ell(\mathrm{I}), \mathrm{I}) \mid \mathrm{I} \in \mathsf{First}(\alpha) \,\},$$

$$F_{\mathsf{pos}} = \begin{cases} \mathsf{Last}(\alpha) \cup \{\{0\}\}, & \textit{if } \varepsilon(\alpha) = \varepsilon;\\ F_{\mathsf{pos}} = \mathsf{Last}(\alpha), & \textit{otherwise.}\end{cases}$$

Proposition 7. *Given an expression* $\alpha \in \mathsf{RE}_\cap$, *one has* $\mathcal{L}(\mathcal{A}_{\mathsf{pos}}(\alpha)) = \mathcal{L}(\alpha)$.

Note that for regular expressions without intersection (simple regular expressions) the automaton is, by the definition of $\mathcal{L}_{\mathcal{I}}$, isomorphic to the classic position automaton, with the difference that now states are labelled with singletons $\{i\}$ instead of $i \in \mathsf{pos}(\alpha) \cup \{0\}$. We now give definitions for recursively computing sets corresponding to First, Last and Follow. These definitions lead to supersets of the corresponding sets but we will proof that extra elements can be discarded and if we trim the resulting NFA we obtain $\mathcal{A}_{\mathsf{pos}}$.

Definition 8. *Given a indexed well-indexed expression* ρ, *let* $\mathsf{Fst}(\rho) \subseteq \mathcal{I}_\rho$ *be inductively defined as follows,*

$$\begin{aligned}\mathsf{Fst}(\emptyset) &= \mathsf{Fst}(\varepsilon) = \emptyset\\ \mathsf{Fst}(a_i) &= \{\{i\}\}\\ \mathsf{Fst}(\rho^\star) &= \mathsf{Fst}(\rho)\end{aligned}
\qquad
\begin{aligned}\mathsf{Fst}(\rho_1 + \rho_2) &= \mathsf{Fst}(\rho_1) \cup \mathsf{Fst}(\rho_2)\\ \mathsf{Fst}(\rho_1 \cdot \rho_2) &= \begin{cases}\mathsf{Fst}(\rho_1) \cup \mathsf{Fst}(\rho_2), & \textit{if } \varepsilon(\rho_1) = \varepsilon\\ \mathsf{Fst}(\rho_1), & \textit{otherwise}\end{cases}\\ \mathsf{Fst}(\rho_1 \cap \rho_2) &= \mathsf{Fst}(\rho_1) \otimes \mathsf{Fst}(\rho_2).\end{aligned}$$

where for $F_1, F_2 \subseteq \mathcal{I}_\rho$, $F_1 \otimes F_2 = \{\, \mathrm{I}_1 \cup \mathrm{I}_2 \mid \ell(\mathrm{I}_1) = \ell(\mathrm{I}_2)$ *and* $\mathrm{I}_1 \in F_1, \mathrm{I}_2 \in F_2 \,\}$.

By construction, all elements $\mathrm{I} \in \mathsf{Fst}(\rho)$ are non-empty and such that $\ell(i_1) = \ell(i_2)$ for all $i_1, i_2 \in \mathrm{I}$, guaranting that $\otimes$ is well defined and $\mathsf{Fst}(\rho) \subseteq \mathcal{I}_\rho$.

Example 9. We have $\mathsf{Fst}(a_1^\star b_2^\star \cap a_3) = \mathsf{Fst}(a_1^\star b_2^\star) \otimes \mathsf{Fst}(a_3) = \{\{1\},\{2\}\} \otimes \{\{3\}\} = \{\{1,3\}\}$.

Definition 10. *Given a well-indexed expression* ρ, *the set* $\mathsf{Lst}(\rho) \subseteq \mathcal{I}_\rho$ *is defined as* $\mathsf{Fst}(\rho)$, *with the difference that for concatenation we have:*

$$\mathsf{Lst}(\rho_1 \cdot \rho_2) = \begin{cases} \mathsf{Lst}(\rho_1) \cup \mathsf{Lst}(\rho_2), & \text{if } \varepsilon(\rho_2) = \varepsilon \\ \mathsf{Lst}(\rho_2), & \text{otherwise.} \end{cases}$$

The set $\mathsf{Fol}(\rho) \subseteq \mathcal{I}_\rho \times \mathcal{I}_\rho$ *is inductively defined as follows,*

$$\begin{array}{ll} \mathsf{Fol}(\emptyset) = \mathsf{Fol}(\varepsilon) = \mathsf{Fol}(a_i) = \emptyset & \mathsf{Fol}(\rho_1 + \rho_2) = \mathsf{Fol}(\rho_1) \cup \mathsf{Fol}(\rho_2) \\ \mathsf{Fol}(\rho^\star) = \mathsf{Fol}(\rho) \cup \mathsf{Lst}(\rho) \times \mathsf{Fst}(\rho) & \mathsf{Fol}(\rho_1 \cap \rho_2) = \mathsf{Fol}(\rho_1) \otimes \mathsf{Fol}(\rho_2) \end{array}$$
$$\mathsf{Fol}(\rho_1 \cdot \rho_2) = \mathsf{Fol}(\rho_1) \cup \mathsf{Fol}(\rho_2) \cup \mathsf{Lst}(\rho_1) \times \mathsf{Fst}(\rho_2).$$

where, for $S_1, S_2 \subseteq \mathcal{I}_\rho \times \mathcal{I}_\rho$,

$$S_1 \otimes S_2 = \{ (\mathrm{I}_1 \cup \mathrm{I}_2, \mathrm{J}_1 \cup \mathrm{J}_2) \mid (\mathrm{I}_1, \mathrm{J}_1) \in S_1, (\mathrm{I}_2, \mathrm{J}_2) \in S_2 \text{ and } \ell(\mathrm{I}_1) = \ell(\mathrm{I}_2), \ell(\mathrm{J}_1) = \ell(\mathrm{J}_2) \}.$$

In the next definition we will use the projection functions on the first and second coordinates, π_1 and π_2, respectively.

Definition 11. *Given* $\alpha \in \mathsf{RE}_\cap$, *let* $\mathcal{A}_{\mathsf{posi}}(\alpha) = \langle S_{\mathsf{posi}}, \Sigma, \{\{0\}\}, \delta_{\mathsf{posi}}, F_{\mathsf{posi}} \rangle$ *be the NFA where* $S_{\mathsf{posi}} = \{\{0\}\} \cup \mathsf{Fst}(\overline{\alpha}) \cup \mathsf{Lst}(\overline{\alpha}) \cup \pi_1(\mathsf{Fol}(\overline{\alpha})) \cup \pi_2(\mathsf{Fol}(\overline{\alpha}))$, *and* δ_{posi} *and* F_{posi} *are defined as* δ_{pos} *and* F_{pos}, *substituting the functions* First, Last *and* Follow, *by* Fst, Lst *and* Fol, *respectively.*

We will now show that $\mathcal{L}(\mathcal{A}_{\mathsf{pos}}(\alpha)) = \mathcal{L}(\mathcal{A}_{\mathsf{posi}}(\alpha))$, and that $\mathcal{A}_{\mathsf{pos}}(\alpha)$ is obtained by trimming $\mathcal{A}_{\mathsf{posi}}(\alpha)$, as the result of the two following lemmas. An example is presented at the end of this section.

Lemma 12. *Given an indexed linear expression* ρ, *one has:* 1) $\mathsf{First}'(\rho) \subseteq \mathsf{Fst}(\rho)$; 2) $\mathsf{Last}'(\rho) \subseteq \mathsf{Lst}(\rho)$; 3) $\mathsf{Follow}'(\rho) \subseteq \mathsf{Fol}(\rho)$.

Example 13. For $\rho = (a_1 \cap b_2)c_3d_4$, we have $(\{3\}, \{4\}) \in \mathsf{Fol}(\rho)$, but $(\{3\}, \{4\}) \notin \mathsf{Follow}(\rho)$. Thus, $\mathsf{Fol}(\rho) \nsubseteq \mathsf{Follow}(\rho)$.

The previous Lemma shows that for any $\alpha \in \mathsf{RE}_\cap$, $\mathcal{A}_{\mathsf{pos}}(\alpha)$ is a subautomaton of $\mathcal{A}_{\mathsf{posi}}(\alpha)$, and thus $\mathcal{L}(\mathcal{A}_{\mathsf{pos}}(\alpha)) \subseteq \mathcal{L}(\mathcal{A}_{\mathsf{posi}}(\alpha))$. We now show that both recognize the same language and can be made isomorphic by trimming $\mathcal{A}_{\mathsf{posi}}$.

Lemma 14. *Given an indexed linear expression* ρ *and some* $n \geq 1$, *if* $\mathrm{I}_n \in \mathsf{Lst}(\rho)$ *and there exist* $\mathrm{I}_1, \ldots, \mathrm{I}_n \in \mathcal{I}_\rho$ *such that*

$$(\{0\}, \ell(\mathrm{I}_1), \mathrm{I}_1), (\mathrm{I}_1, \ell(\mathrm{I}_2), \mathrm{I}_2), \ldots, (\mathrm{I}_{n-1}, \ell(\mathrm{I}_n), \mathrm{I}_n) \in \delta_{\mathsf{posi}},$$

then $\mathrm{I}_1 \cdots \mathrm{I}_n \in \mathcal{L}_{\mathcal{I}}(\rho)$.

Theorem 15. *For any* $\alpha \in \mathsf{RE}_\cap$, $\mathcal{L}(\mathcal{A}_{\mathsf{pos}}(\alpha)) = \mathcal{L}(\mathcal{A}_{\mathsf{posi}}(\alpha))$.

From these results, it follows that if we trim the automaton $\mathcal{A}_{\mathsf{posi}}$ we obtain exactly $\mathcal{A}_{\mathsf{pos}}$.

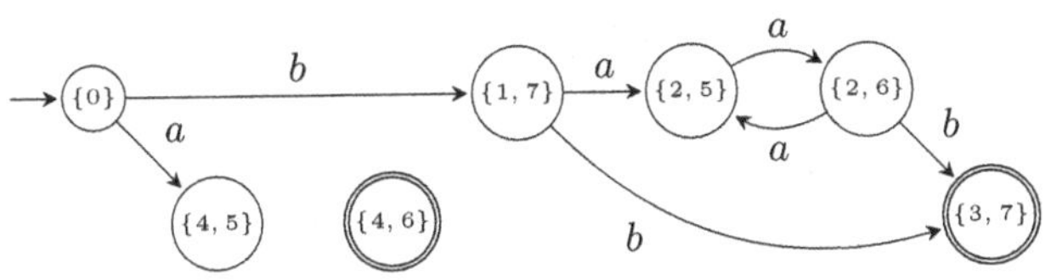

Fig. 1. $\mathcal{A}_{\mathsf{posi}}((ba^{\star}b + a) \cap (aa + b)^{\star})$

Example 16. Consider $\alpha = (ba^{\star}b + a) \cap (aa + b)^{\star}$. Then $\overline{\alpha} = (b_1a_2^{\star}b_3 + a_4) \cap (a_5a_6 + b_7)^{\star}$, $\mathsf{Fst}(\overline{\alpha}) = \{\{1,7\},\{4,5\}\}$, $\mathsf{Lst}(\overline{\alpha}) = \{\{3,7\},\{4,6\}\}$, and $\mathsf{Fol}(\overline{\alpha}) = \{(\{2,5\},\{2,6\}),(\{2,6\},\{2,5\}),(\{2,6\},\{3,7\}),(\{1,7\},\{2,5\}),(\{1,7\},\{3,7\})\}$.

The automaton $\mathcal{A}_{\mathsf{posi}}(\alpha)$ is represented in Fig. 1. The trimmed automaton, $\mathcal{A}_{\mathsf{posi}}(\alpha)^{\mathsf{t}}$, is obtained removing the states labeled by $\{4,5\}$ and $\{4,6\}$, and the correspondent transitions.

5 A c-Continuation Automaton for $\mathsf{RE}_{\cap}$ Expressions

In the case of simple regular expressions, Champarnaud and Ziadi [9] defined a nondeterministic automaton isomorphic to the position automaton, called the c-continuation automaton, in order to show that the partial derivative automaton can be seen as a quotient of the position automaton. With the same purpose, in this section, we present a c-continuation automaton for expressions with intersection. Moreover, instead of considering derivatives of regular expressions [5], we use partial derivatives to restate some known results for simple regular expressions.

The notion of continuation was defined by Berry and Sethy [3], and developed by Champarnaud and Ziadi [9], by Ilie and Yu [14], and by Chen and Yu [10]. Given $a \in \Sigma$ and a linear simple expression α, the set of partial derivatives $\partial_{xa}(\alpha)$, for any word $x \in \Sigma^{\star}$, is either $\emptyset$ or has a unique element γ called the *continuation* of a in α. Note that using partial derivatives, *continuations* and non-null c-*continuations* coincide. Furthermore, the continuation can be obtained by some refinement of the inductive definition of partial derivatives, exploring the linearity of α. In order to establish similar results for linear well-indexed expressions, we introduce the notion of *partial index-derivative* of a well-indexed expression ρ w.r.t. an index $\mathrm{I} \in \mathcal{I}_{\rho}$.

Given a well-indexed expression ρ, a subexpression τ of ρ, and a set of indexes $\mathrm{I} \in \mathcal{I}_{\rho}$, let $\mathrm{I}\big|_{\tau}$ denote the set of indexes in I that occur in τ. This definition is naturally extended to words $x = \mathrm{I}_1 \cdots \mathrm{I}_n \in \mathcal{I}_{\rho}^{\star}$ by $x\big|_{\tau} = \mathrm{I}_1\big|_{\tau} \cdots \mathrm{I}_n\big|_{\tau}$, for $n \geq 0$.

Definition 17. *The set of partial index-derivatives of a well-indexed expression ρ by $\mathrm{I} \in \mathcal{I}_\rho \cup \{\emptyset\}$, $\partial_{\mathrm{I}}(\rho)$, is defined by*

$$\partial_{\mathrm{I}}(\emptyset) = \partial_{\mathrm{I}}(\varepsilon) = \emptyset$$

$$\partial_{\mathrm{I}}(a_i) = \begin{cases} \{\varepsilon\}, & \text{if } \mathrm{I} = \{i\} \\ \emptyset, & \text{otherwise} \end{cases} \qquad \begin{aligned} \partial_{\mathrm{I}}(\rho^\star) &= \partial_{\mathrm{I}}(\rho) \odot \rho^\star \\ \partial_{\mathrm{I}}(\rho_1 + \rho_2) &= \partial_{\mathrm{I}}(\rho_1) \cup \partial_{\mathrm{I}}(\rho_2) \end{aligned}$$

$$\partial_{\mathrm{I}}(\rho_1 \cdot \rho_2) = \begin{cases} \partial_{\mathrm{I}}(\rho_1) \odot \rho_2 \cup \partial_{\mathrm{I}}(\rho_2), & \text{if } \varepsilon(\rho_1) = \varepsilon \\ \partial_{\mathrm{I}}(\rho_1) \odot \rho_2, & \text{otherwise} \end{cases}$$

$$\partial_{\mathrm{I}}(\rho_1 \cap \rho_2) = \begin{cases} \partial_{\mathrm{I}|_{\rho_1}}(\rho_1) \Cap \partial_{\mathrm{I}|_{\rho_2}}(\rho_2), & \text{if } \mathrm{I} = \mathrm{I}\big|_{\rho_1} \cup \mathrm{I}\big|_{\rho_2} \\ \emptyset, & \text{otherwise.} \end{cases}$$

The set of partial index-derivatives of ρ by a word $x \in \mathcal{I}_\rho^\star$ is then inductively defined by $\partial_\varepsilon(\rho) = \{\rho\}$ and $\partial_{x\mathrm{I}}(\rho) = \bigcup_{\rho' \in \partial_x(\rho)} \partial_{\mathrm{I}}(\rho')$. If S is a set of well-indexed expressions, $\partial_x(S) = \bigcup_{\rho \in S} \partial_x(\rho)$.

It is straightforward to see that $\partial_\emptyset(\rho) = \emptyset$ for all ρ. Although $\emptyset \notin \mathcal{I}_\rho$, the notion of partial index-derivative includes the derivative by an empty set of indexes, in order to guarantee that the derivative of an intersection is well-defined. Also note that the partial index-derivative of a well-indexed expression is still well-indexed. Finally, the set of partial index-derivatives of ρ by all $\mathrm{I} \in \mathcal{I}_\rho$ can be calculated simultaneously using an extension of the linear form defined by Antimirov [1], i.e. considering pairs (I, ρ') where $\rho' \in \partial_{\mathrm{I}}(\rho)$. The following lemma is proved by induction on n.

Lemma 18. *If $x = \mathrm{I}_1 \cdots \mathrm{I}_n$ and $\partial_x(\rho) \neq \emptyset$, then $x = x\big|_\rho$.*

Example 19. We have $\partial_{\{1,3\}}(a_1^\star b_2^\star \cap a_3) = \partial_{\{1\}}(a_1^\star b_2^\star) \Cap \partial_{\{3\}}(a_3) = \{a_1^\star b_2^\star \cap \varepsilon\}$.

Proposition 20. *Consider a well-indexed expression ρ and $\mathrm{I} \in \mathcal{I}_\rho$. Then, $\mathrm{I}^{-1}\mathcal{L}_{\mathcal{I}}(\rho) = \mathcal{L}_{\mathcal{I}}(\partial_{\mathrm{I}}(\rho))$ and $\mathcal{L}_{\mathcal{I}}(\rho) = \mathcal{L}_{\mathcal{I}}\left(\bigcup_{\mathrm{I} \in \mathcal{I}_\rho} (\mathrm{I} \odot \partial_{\mathrm{I}}(\rho)) \cup \varepsilon(\rho)\right)$.*

Corollary 21. *For every well-indexed expression $\rho \in \mathsf{RE}_\cap$ and word $x \in \mathcal{I}_\rho^\star$, one has $x^{-1}\mathcal{L}_{\mathcal{I}}(\rho) = \mathcal{L}_{\mathcal{I}}(\partial_x(\rho))$ and $\mathcal{L}_{\mathcal{I}}(\rho) = \mathcal{L}_{\mathcal{I}}(\bigcup_{x \in \mathcal{I}_\rho^\star} (x \odot \partial_x(\rho)) \cup \varepsilon(\rho))$.*

The following is an adaptation, for partial index-derivatives and intersection, of a result due to Berry and Sethi [3].

Proposition 22. *Consider a linear indexed expression $\rho \in \mathsf{RE}_\cap$ and $x\mathrm{I} \in \mathcal{I}_\rho^\star$, and let $\mathsf{suff}(x)$ denote the set of all suffixes of x. The partial index-derivative $\partial_{x\mathrm{I}}(\rho)$ of ρ satisfies:*

$$\partial_{x\mathrm{I}}(\emptyset) = \partial_{x\mathrm{I}}(\varepsilon) = \emptyset,$$

$$\partial_{x\mathrm{I}}(a_i) = \begin{cases} \{\varepsilon\}, & \text{if } x\mathrm{I} = \{i\}, \\ \emptyset, & \text{otherwise}, \end{cases}$$

$$\partial_{x\mathrm{I}}(\rho_1 + \rho_2) = \begin{cases} \partial_{x\mathrm{I}}(\rho_1), & \text{if } x\mathrm{I} = (x\mathrm{I})\big|_{\rho_1}, \\ \partial_{x\mathrm{I}}(\rho_2), & \text{if } x\mathrm{I} = (x\mathrm{I})\big|_{\rho_2}, \\ \emptyset & \text{otherwise} \end{cases}$$

$$\partial_{x\mathrm{I}}(\rho_1 \cdot \rho_2) = \begin{cases} \partial_{x\mathrm{I}}(\rho_1) \odot \rho_2, & \text{if } x\mathrm{I} = (x\mathrm{I})\big|_{\rho_1}, \\ \partial_{z\mathrm{I}}(\rho_2), & \text{if } x = yz, \varepsilon(\partial_y(\rho_1)) = \varepsilon, z\mathrm{I} = (z\mathrm{I})\big|_{\rho_2}, \\ \emptyset, & \text{otherwise}, \end{cases}$$

$$\partial_{x\mathrm{I}}(\rho^\star) \subseteq \bigcup_{v \in \mathsf{suff}(x)} \partial_{v\mathrm{I}}(\rho) \odot \rho^\star,$$

$$\partial_{x\mathrm{I}}(\rho_1 \cap \rho_2) = \begin{cases} \partial_{(x\mathrm{I})|_{\rho_1}}(\rho_1) \Cap \partial_{(x\mathrm{I})|_{\rho_2}}(\rho_2), & \text{if } x\mathrm{I} = (x\mathrm{I})\big|_{\rho_1} \cap_{\mathcal{I}} (x\mathrm{I})\big|_{\rho_2}, \\ \emptyset, & \text{otherwise.} \end{cases}$$

The previous proposition implies that if $\partial_{x\mathrm{I}}(\rho) \neq \emptyset$, then it has only one element for every $x \in \mathcal{I}_\rho^\star$. This fact is proved in Proposition 24 and the unique element (if exists) is defined below.

Definition 23. *Given a linear indexed expression ρ and a set of indexes* I, *the c-continuation $\mathsf{c}_\mathrm{I}(\rho)$ of ρ w.r.t.* I *is defined by the following rules.*

$$\mathsf{c}_\mathrm{I}(\emptyset) = \mathsf{c}_\mathrm{I}(\varepsilon) = \emptyset \qquad \mathsf{c}_\mathrm{I}(\rho^\star) = \mathsf{c}_\mathrm{I}(\rho)\rho^\star$$

$$\mathsf{c}_\mathrm{I}(a_i) = \begin{cases} \varepsilon, & \text{if } \mathrm{I} = \{i\} \\ \emptyset, & \text{otherwise} \end{cases} \qquad \mathsf{c}_\mathrm{I}(\rho_1 + \rho_2) = \begin{cases} \mathsf{c}_\mathrm{I}(\rho_1), & \text{if } \mathsf{c}_\mathrm{I}(\rho_1) \neq \emptyset \\ \mathsf{c}_\mathrm{I}(\rho_2), & \text{otherwise} \end{cases}$$

$$\mathsf{c}_\mathrm{I}(\rho_1 \cdot \rho_2) = \begin{cases} \mathsf{c}_\mathrm{I}(\rho_1) \cdot \rho_2, & \text{if } \mathsf{c}_\mathrm{I}(\rho_1) \neq \emptyset \\ \mathsf{c}_\mathrm{I}(\rho_2), & \text{otherwise} \end{cases}$$

$$\mathsf{c}_\mathrm{I}(\rho_1 \cap \rho_2) = \begin{cases} \mathsf{c}_{\mathrm{I}|_{\rho_1}}(\rho_1) \cap \mathsf{c}_{\mathrm{I}|_{\rho_2}}(\rho_2), & \text{if } \mathrm{I} = \mathrm{I}\big|_{\rho_1} \cup \mathrm{I}\big|_{\rho_1} \\ \emptyset, & \text{otherwise.} \end{cases}$$

It is easy to verify that $\mathsf{c}_\mathrm{I}(\rho) \neq \emptyset$ implies $\mathrm{I} \subseteq \mathsf{ind}(\rho)$, i.e. $\mathrm{I}\big|_\rho = \mathrm{I}$.

Proposition 24. *Consider a linear indexed expression ρ and $\mathrm{I} \in \mathcal{I}_\rho$. Then, for every $x \in \mathcal{I}_\rho^\star$ such that $\partial_{x\mathrm{I}}(\rho) \neq \emptyset$, one has $\partial_{x\mathrm{I}}(\rho) = \{\mathsf{c}_\mathrm{I}(\rho)\}$ and $\mathsf{c}_\mathrm{I}(\rho) \neq \emptyset$.*

Proof. We proceed by induction on the structure of ρ. For $\emptyset$ and ε the set of partial index-derivatives is $\emptyset$. Let ρ be a_i. We need to prove that $\forall \mathrm{I} \in \mathcal{I}_{a_i} \forall x \in \mathcal{I}_{a_i}^\star\ (\partial_{x\mathrm{I}}(a_i) \neq \emptyset \implies \partial_{x\mathrm{I}}(a_i) = \{\mathsf{c}_\mathrm{I}(a_i)\} \neq \{\emptyset\})$. Let $\partial_{x\mathrm{I}}(a_i) \neq \emptyset$, then by Proposition 22, $\partial_{x\mathrm{I}}(a_i) = \{\varepsilon\}$ and $x\mathrm{I} = \{i\}$. Then $\mathrm{I} = \{i\}$ and $\mathsf{c}_\mathrm{I}(a_i) = \varepsilon$. Thus, we conclude that $\partial_{x\mathrm{I}}(a_i) = \{\mathsf{c}_\mathrm{I}(a_i)\} \neq \{\emptyset\}$. Let us suppose that for ρ_i, $i = 1, 2$ we have $\forall \mathrm{I} \in \mathcal{I}_{\rho_i} \forall x \in \mathcal{I}_{\rho_i}^\star\ (\partial_{x\mathrm{I}}(\rho_i) \neq \emptyset \implies \partial_{x\mathrm{I}}(\rho_i) = \{\mathsf{c}_\mathrm{I}(\rho_i)\} \neq \{\emptyset\})$. Let $\rho = \rho_1 + \rho_2$ be such that $\partial_{x\mathrm{I}}(\rho_1 + \rho_2) \neq \emptyset$. Then, $\partial_{x\mathrm{I}}(\rho_1 + \rho_2) = \partial_{x\mathrm{I}}(\rho_i)$

with $x\mathrm{I} = (x\mathrm{I})\big|_{\rho_i}$, for some $i \in \{1,2\}$. By the induction hypothesis, $\partial_{x\mathrm{I}}(\rho_i) = \{\mathsf{c}_\mathrm{I}(\rho_i)\} \neq \{\emptyset\}$. Thus, $\mathsf{c}_\mathrm{I}(\rho_i) \neq \emptyset$ and $\mathsf{c}_\mathrm{I}(\rho_1+\rho_2) = \mathsf{c}_\mathrm{I}(\rho_i)$. Let $\rho = \rho_1\rho_2$. If $\partial_{x\mathrm{I}}(\rho_1\rho_2) \neq \emptyset$ then we have to consider two cases. Let $\partial_{x\mathrm{I}}(\rho_1\rho_2) = \partial_{x\mathrm{I}}(\rho_1) \odot \rho_2$ and $x\mathrm{I} = (x\mathrm{I})\big|_{\rho_1}$. Then, $\partial_{x\mathrm{I}}(\rho_1) \neq \emptyset$ and $\partial_{x\mathrm{I}}(\rho_1) = \{\mathsf{c}_\mathrm{I}(\rho_1)\}$. We conclude that $\mathsf{c}_\mathrm{I}(\rho_1) \neq \emptyset$ and $\mathsf{c}_\mathrm{I}(\rho_1\rho_2) = \mathsf{c}_\mathrm{I}(\rho_1)$. In the second case, $\partial_{x\mathrm{I}}(\rho_1\rho_2) = \partial_{z\mathrm{I}}(\rho_2) \neq \emptyset$, $x = yz$, $\varepsilon(\partial_y(\rho_1)) = \varepsilon$ and $z\mathrm{I} = (z\mathrm{I})\big|_{\rho_2}$. We conclude that $y = y\big|_{\rho_1}$ and $\mathrm{I} = \mathrm{I}\big|_{\rho_2}$. Then, $\mathsf{c}_\mathrm{I}(\rho_1) = \emptyset$ and $\mathsf{c}_\mathrm{I}(\rho_1\rho_2) = \mathsf{c}_\mathrm{I}(\rho_2)$. By the induction hypothesis, $\partial_{z\mathrm{I}}(\rho_2) = \{\mathsf{c}_\mathrm{I}(\rho_2)\}$ and the result follows. Let $\rho = \rho_1^\star$. If $\partial_{x\mathrm{I}}(\rho_1^\star) \neq \emptyset$, we can write $\partial_{x\mathrm{I}}(\rho_1^\star) = \partial_{v_1\mathrm{I}}(\rho_1) \odot \rho_1^\star \cup \cdots \cup \partial_{v_n\mathrm{I}}(\rho_1) \odot \rho_1^\star$, with $n \geq 1$, such that for all $1 \leq i \leq n$, $x = u_iv_i$ and $\partial_{v_i\mathrm{I}}(\rho_1) \odot \rho_1^\star \neq \emptyset$. By the induction hypothesis, each nonempty set of partial index-derivatives $\partial_{v_i\mathrm{I}}(\rho_1)$ is equal to $\{\mathsf{c}_\mathrm{I}(\rho_1)\} \neq \{\emptyset\}$. Thus, $\partial_{x\mathrm{I}}(\rho_1^\star) = \{\mathsf{c}_\mathrm{I}(\rho_1)\rho_1^\star\}$. Finally, let $\rho = \rho_1 \cap \rho_2$ be such that $\partial_{x\mathrm{I}}(\rho_1 \cap \rho_2) \neq \emptyset$. Then $\partial_{x\mathrm{I}}(\rho_1 \cap \rho_2) = \partial_{(x\mathrm{I})|_{\rho_1}}(\rho_1) \cap \partial_{(x\mathrm{I})|_{\rho_2}}(\rho_2)$, $x\mathrm{I} = (x\mathrm{I})\big|_{\rho_1} \cap_{\mathcal{I}} (x\mathrm{I})\big|_{\rho_2}$ and $\partial_{(x\mathrm{I})|_{\rho_i}}(\rho_i) \neq \emptyset$, for $i = 1,2$. Moreover, $\partial_{(x\mathrm{I})|_{\rho_i}}(\rho_i) = \{\mathsf{c}_{\mathrm{I}|_{\rho_i}}(\rho_i)\}$. The result follows by the induction hypothesis and from the definition of $\mathsf{c}_\mathrm{I}(\rho_1 \cap \rho_2)$. □

This result guarantees that, given a linear indexed expression ρ and $\mathrm{I} \in \mathcal{I}_\rho$, all sets of partial index-derivatives $\partial_{x\mathrm{I}}(\rho)$ different from $\emptyset$ are singletons with an unique c-continuation $\mathsf{c}_\mathrm{I}(\rho)$ of ρ w.r.t. I.

Lemma 25. *Consider a linear indexed expression ρ. Then, $\mathrm{I} \in \mathsf{Lst}(\rho)$ if and only if $\varepsilon(\mathsf{c}_\mathrm{I}(\rho)) = \varepsilon$.*

Lemma 26. *Consider a linear indexed expression ρ and sets of indexes $\mathrm{I}, \mathrm{J} \in \mathcal{I}_\rho^\star$. Then, $(\mathrm{I},\mathrm{J}) \in \mathsf{Fol}(\rho)$ if and only if $\mathrm{J} \in \mathsf{Fst}(\mathsf{c}_\mathrm{I}(\rho))$.*

Definition 27. *The c-continuation automaton of an expression $\alpha \in \mathsf{RE}_\cap$ is*

$$\mathcal{A}_\mathsf{c}(\alpha) = \langle S_\mathsf{c}, \Sigma, \{(\{0\}, \mathsf{c}_{\{0\}}(\overline{\alpha}))\}, \delta_\mathsf{c}, F_\mathsf{c}\rangle,$$

where $S_\mathsf{c} = \{\ (\mathrm{I}, \mathsf{c}_\mathrm{I}(\overline{\alpha})) \mid \mathrm{I} \in S_\mathsf{posi}\ \}$, $F_\mathsf{c} = \{\ (\mathrm{I}, \mathsf{c}_\mathrm{I}(\overline{\alpha})) \mid \varepsilon(\mathsf{c}_\mathrm{I}(\overline{\alpha})) = \varepsilon\ \}$, $\mathsf{c}_{\{0\}}(\overline{\alpha}) = \overline{\alpha}$, $\delta_\mathsf{c} = \{\ ((\mathrm{I}, \mathsf{c}_\mathrm{I}(\overline{\alpha})), \ell(\mathrm{J}), (\mathrm{J}, \mathsf{c}_\mathrm{J}(\overline{\alpha}))) \mid \mathrm{J} \in \mathsf{Fst}(\mathsf{c}_\mathrm{I}(\overline{\alpha}))\ \}$.

By Lemmas 25 and 26, and considering $\varphi : S_\mathsf{c} \to S_\mathsf{posi}$ such that $\varphi((\mathrm{I}, \mathsf{c}_\mathrm{I}(\overline{\alpha}))) = \mathrm{I}$, the following holds.

Theorem 28. *For $\alpha \in \mathsf{RE}_\cap$, we have $\mathcal{A}_\mathsf{posi}(\alpha) \simeq \mathcal{A}_\mathsf{c}(\alpha)$.*

Example 29. Consider the expression $\overline{\alpha} = (b_1a_2^\star b_3 + a_4) \cap (a_5a_6 + b_7)^\star$, from Example 16, and let $\rho_2 = (a_5a_6 + b_7)^\star$. We have the following c-continuations: $\mathsf{c}_{\{1,7\}}(\overline{\alpha}) = a_2^\star b_3 \cap \rho_2$, $\mathsf{c}_{\{4,5\}}(\overline{\alpha}) = \varepsilon \cap a_6\rho_2$, $\mathsf{c}_{\{4,6\}}(\overline{\alpha}) = \varepsilon \cap \rho_2$, $\mathsf{c}_{\{2,5\}}(\overline{\alpha}) = a_2^\star b_3 \cap a_6\rho_2$, $\mathsf{c}_{\{2,6\}}(\overline{\alpha}) = a_2^\star b_3 \cap \rho_2$, and $\mathsf{c}_{\{3,7\}}(\overline{\alpha}) = \varepsilon \cap \rho_2$.

6 The $\mathcal{A}_\mathsf{pd}$ as a Quotient of $\mathcal{A}_\mathsf{pos}$

Using $\mathcal{A}_\mathsf{c}$ we show that the partial derivative automaton $\mathcal{A}_\mathsf{pd}$ is a quotient of $\mathcal{A}_\mathsf{pos}$. This extends the corresponding result for simple regular expressions, although the proof cannot use the same technique. Recall that, for a simple regular expression α, one builds $\mathcal{A}_{\mathsf{pd}}(\overline{\alpha})$, and then shows that when its transitions are unmarked, the result $\overline{\mathcal{A}_{\mathsf{pd}}(\overline{\alpha})}$ is isomorphic to a quotient of $\mathcal{A}_{\mathsf{c}}(\alpha)$. However, with $\alpha \in \mathsf{RE}_\cap$, this method cannot be used because, as mentioned in the introduction, intersection does not commute with marking. For $\alpha \in \mathsf{RE}_\cap$, we will present a direct isomorphism between $\mathcal{A}_{\mathsf{pd}}(\alpha)$ and a quotient of $\mathcal{A}_{\mathsf{c}}(\alpha)$. The next lemmas will be needed to build that isomorphism.

Lemma 30. *Consider a linear indexed expression ρ and $\mathrm{I} \in \mathcal{I}_\rho$. If $\mathrm{I} \in \mathsf{Fst}(\rho)$, then $\mathsf{c}_\mathrm{I}(\rho) \neq \emptyset$ and $\mathsf{c}_\mathrm{I}(\rho) \in \partial_\mathrm{I}(\rho)$.*

Lemma 31. *Consider a linear indexed expression ρ and $\mathrm{I}, \mathrm{J} \in \mathcal{I}_\rho$, such that $\mathrm{J} \in \mathsf{Fst}(\mathsf{c}_\mathrm{I}(\rho))$. Then, $\mathsf{c}_\mathrm{J}(\rho) \in \partial_\mathrm{J}(\mathsf{c}_\mathrm{I}(\rho))$.*

Lemma 32. *Consider well-indexed expressions ρ', ρ and $\mathrm{I} \in \mathcal{I}_\rho$, such that $\rho' \in \partial_\mathrm{I}(\rho)$. Then, $\overline{\rho'} \in \partial_{\ell(\mathrm{I})}(\overline{\rho})$.*

Lemma 33. *Consider a well-indexed expression ρ, $a \in \Sigma$ and $\beta \in \partial_a(\overline{\rho})$. Then, there exist $\mathrm{I} \in \mathcal{I}_\rho$ and $\rho' \in \partial_\mathrm{I}(\rho)$ with $\ell(\mathrm{I}) = a$ and $\overline{\rho'} = \beta$. Furthermore, for $x = a_1 \cdots a_n \in \Sigma^\star$, if $\beta \in \partial_x(\overline{\rho})$, there exist $\mathrm{I}_1 \cdots \mathrm{I}_n \in \mathcal{I}_\rho^\star$ and $\rho' \in \partial_{\mathrm{I}_1 \cdots \mathrm{I}_n}(\rho)$ with $\ell(\mathrm{I}_1 \cdots \mathrm{I}_n) = x$ and $\overline{\rho'} = \beta$.*

Given $\alpha \in \mathsf{RE}_\cap$, consider $\mathcal{A}_{\mathsf{c}}(\alpha)$ and the equivalence relation $\equiv_\ell$ on S_{c} given by $(\mathrm{I}, \mathsf{c}_\mathrm{I}(\overline{\alpha})) \equiv_\ell (\mathrm{J}, \mathsf{c}_\mathrm{J}(\overline{\alpha}))$ if and only if $\overline{\mathsf{c}_\mathrm{I}(\overline{\alpha})} = \overline{\mathsf{c}_\mathrm{J}(\overline{\alpha})}$, for $\mathrm{I}, \mathrm{J} \in \mathcal{I}_{\overline{\alpha}} \cup \{\{0\}\}$.

Lemma 34. *The relation $\equiv_\ell$ is right invariant w.r.t. $\mathcal{A}_{\mathsf{c}}$.*

Theorem 35. *For $\alpha \in \mathsf{RE}_\cap$, $\mathcal{A}_{\mathsf{pd}}(\alpha) \simeq \mathcal{A}_{\mathsf{c}}(\alpha)^{\mathsf{ac}}/{\equiv_\ell}$.*

Proof. Let $\mathcal{A}_{\mathsf{c}}(\alpha)^{\mathsf{ac}}/{\equiv_\ell} = (S_\ell, \Sigma, \delta_\ell, [(\{0\}, \overline{\alpha})], F_\ell)$. We define the map $\varphi : S_\ell \rightarrow \partial(\alpha)$, by $\varphi([(\mathrm{I}, \mathsf{c}_\mathrm{I}(\overline{\alpha}))]) = \overline{\mathsf{c}_\mathrm{I}(\overline{\alpha})}$. We have to show that: 1) φ is well-defined; 2) φ is bijective; 3) $\varphi(\delta_\ell(s, a)) = \delta_{\mathsf{pd}}(\varphi(s), a)$ for every $s \in S_\ell, a \in \Sigma$; 4) $\varphi(F_\ell) = F_{\mathsf{pd}}$; 5) $\varphi([(\{0\}, \mathsf{c}_{\{0\}}(\overline{\alpha}))]) = \alpha$.

Claim 1 follows from Lemmas 30 and 31. The last two are obvious. That φ is injective follows from the definition of $\equiv_\ell$. Furthermore, if $\beta \in \partial(\alpha)$, then there are terms $\beta_0 = \alpha, \beta_1, \ldots, \beta_n = \beta$ and letters $a_1, \ldots, a_n \in \Sigma$, with $n \geq 0$, such that $\beta_{i+1} \in \partial_{a_{i+1}}(\beta_i)$ for $0 \leq i \leq n-1$. It follows from Lemma 33 that there exist $\mathrm{I}_1 \cdots \mathrm{I}_n \in \mathcal{I}_\rho^\star$ and $\rho' \in \partial_{\mathrm{I}_1 \cdots \mathrm{I}_n}(\overline{\alpha})$ with $\ell(\mathrm{I}_1 \cdots \mathrm{I}_n) = a_1 \cdots a_n$ and $\overline{\rho'} = \beta$. Furthermore, by Proposition 24, we know that $\partial_{\mathrm{I}_1 \cdots \mathrm{I}_n}(\overline{\alpha}) = \{\mathsf{c}_{\mathrm{I}_n}(\overline{\alpha})\}$, with $\mathsf{c}_{\mathrm{I}_n}(\overline{\alpha}) \neq \emptyset$. Thus, $[(\mathrm{I}_n, \mathsf{c}_{\mathrm{I}_n}(\overline{\alpha}))] \in S_\ell$ and we conclude that φ is surjective. For 3) we consider both inclusions. Consider $\beta \in \varphi(\delta_\ell(s, a))$, for $s \in S_\ell$ and $a \in \Sigma$. Then, there exist $\mathrm{I}, \mathrm{J} \in \mathcal{I}_{\overline{\alpha}}$ such that $[(\mathrm{I}, \mathsf{c}_\mathrm{I}(\overline{\alpha}))] = s$, $\overline{\mathsf{c}_\mathrm{J}(\overline{\alpha})} = \beta$, $(\mathrm{J}, \mathsf{c}_\mathrm{J}(\overline{\alpha})) \in \delta_{\mathsf{c}}((\mathrm{I}, \mathsf{c}_\mathrm{I}(\overline{\alpha})), \ell(\mathrm{J}))$ and $\ell(\mathrm{J}) = a$, i.e. $\mathrm{J} \in \mathsf{Fst}(\mathsf{c}_\mathrm{I}(\overline{\alpha}))$. By Lemma 31, we have $\mathsf{c}_\mathrm{J}(\overline{\alpha}) \in \partial_\mathrm{J}(\mathsf{c}_\mathrm{I}(\overline{\alpha}))$ and by Lemma 32, $\overline{\mathsf{c}_\mathrm{J}(\overline{\alpha})} \in \partial_a(\overline{\mathsf{c}_\mathrm{I}(\overline{\alpha})})$. Thus, $\overline{\mathsf{c}_\mathrm{J}(\overline{\alpha})} \in \delta_{\mathsf{pd}}(\overline{\mathsf{c}_\mathrm{I}(\overline{\alpha})}, a)$. Now, let $\beta \in \delta_{\mathsf{pd}}(\tau, a)$, where $\tau = \overline{\mathsf{c}_\mathrm{I}(\overline{\alpha})}$, for some $\mathrm{I} \in \mathcal{I}_{\overline{\alpha}}$ and $a \in \Sigma$. Then, there is a sequence of terms $\tau_0 = \alpha, \tau_1, \ldots, \tau_n = \tau$ and a sequence of letters $a_1, \ldots, a_n \in \Sigma$ such that $\tau_{i+1} \in \partial_{a_{i+1}}(\tau_i)$, for $0 \leq i \leq n-1$, and $\beta \in \partial_a(\tau)$, i.e. $\beta \in \partial_{a_1 \cdots a_n a}(\alpha)$. By Lemma 33, there exist $\mathrm{J}_1, \ldots, \mathrm{J}_n, \mathrm{J} \in \mathcal{I}_{\overline{\alpha}}$, with $\ell(\mathrm{J}_1 \cdots \mathrm{J}_n \mathrm{J}) = a_1 \cdots a_n a$, and $\rho' \in$

$\partial_{J_1 \cdots J_n J}(\overline{\alpha})$ such that $\overline{\rho'} = \beta$. By Proposition 24, $\rho' = \mathsf{c}_J(\overline{\alpha})$. On the other hand, it is straightforward to show by structural induction on a well-indexed expression ρ, that $\partial_J(\rho) \neq \emptyset$ implies $J \in \mathsf{Fst}(\rho)$. Thus, $[(J, \mathsf{c}_J(\overline{\alpha}))] \in \delta_\ell([(I, \mathsf{c}_I(\overline{\alpha}))], \ell(J))$ and consequently $\beta = \overline{\mathsf{c}_J(\overline{\alpha})} \in \varphi(\delta_\ell([(I, \mathsf{c}_I(\overline{\alpha}))], a))$. □

Example 36. Consider $\alpha = (ba^\star b + a) \cap (aa + b)^\star$ from Examples 16 and 29. Set $\beta = (aa + b)^\star$. For the positions present in $\mathcal{A}_{\mathsf{c}}(\alpha)^{\mathsf{ac}}$, we have $\overline{\mathsf{c}_{\{4,5\}}(\overline{\alpha})} = \varepsilon \cap a\beta$, $\overline{\mathsf{c}_{\{3,7\}}(\overline{\alpha})} = \varepsilon \cap \beta$, $\overline{\mathsf{c}_{\{2,5\}}(\overline{\alpha})} = a^\star b \cap a\beta$, and $\overline{\mathsf{c}_{\{1,7\}}(\overline{\alpha})} = \overline{\mathsf{c}_{\{2,6\}}(\overline{\alpha})} = a^\star b \cap \beta$. Merging states $(\{1,7\}, \mathsf{c}_{\{1,7\}}(\overline{\alpha}))$ and $(\{2,6\}, \mathsf{c}_{\{2,6\}}(\overline{\alpha}))$ in $\mathcal{A}_{\mathsf{c}}(\alpha)^{\mathsf{ac}}$, one obtains an NFA isomorphic to $\mathcal{A}_{\mathsf{pd}}(\alpha)$, which is represented in Fig. 2.

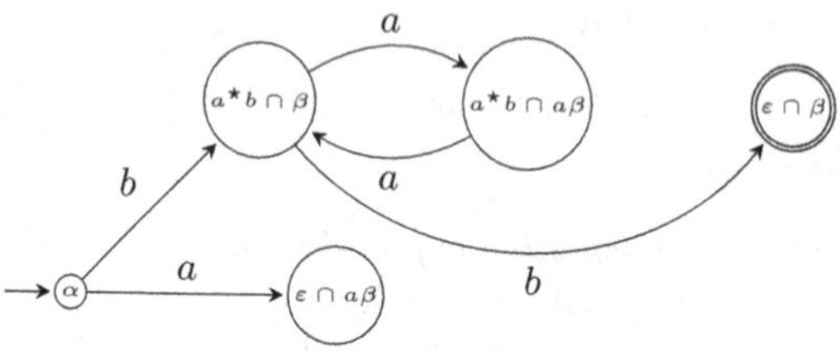

Fig. 2. $\mathcal{A}_{\mathsf{pd}}((ba^\star b + a) \cap (aa + b)^\star)$

7 Final Remarks

For simple regular expressions of size n, the size of $\mathcal{A}_{\mathsf{pos}}(\alpha)$ is $O(n^2)$, and using $\mathcal{A}_{\mathsf{c}}(\alpha)$ it is possible to efficiently compute $\mathcal{A}_{\mathsf{pd}}(\alpha)$ [9]. For regular expressions with intersection the conversion to NFA's has exponential computational complexity [11] and both the size of $\mathcal{A}_{\mathsf{pos}}$ and $\mathcal{A}_{\mathsf{pd}}$ can be exponential in the size of the regular expression. On the average case, however, the size of these automata seem to be much smaller [2], and thus feasible for practical applications. In this scenario, algorithms for building $\mathcal{A}_{\mathsf{pd}}$ using $\mathcal{A}_{\mathsf{pos}}$ seem worthwhile to develop.

References

1. Antimirov, V.: Partial derivatives of regular expressions and finite automaton constructions. Theoret. Comput. Sci. **155**(2), 291–319 (1996)
2. Bastos, R., Broda, S., Machiavelo, A., Moreira, N., Reis, R.: On the state complexity of partial derivative automata for regular expressions with intersection. In: Câmpeanu, C., Manea, F., Shallit, J. (eds.) DCFS 2016. LNCS, vol. 9777, pp. 45–59. Springer, Heidelberg (2016). doi:10.1007/978-3-319-41114-9_4
3. Berry, G., Sethi, R.: From regular expressions to deterministic automata. Theoret. Comput. Sci. **48**, 117–126 (1986)
4. Brüggemann-Klein, A.: Regular expressions into finite automata. Theoret. Comput. Sci. **48**, 197–213 (1993)

5. Brzozowski, J.: Derivatives of regular expressions. JACM **11**(4), 481–494 (1964)
6. Caron, P., Champarnaud, J.-M., Mignot, L.: Partial derivatives of an extended regular expression. In: Dediu, A.-H., Inenaga, S., Martín-Vide, C. (eds.) LATA 2011. LNCS, vol. 6638, pp. 179–191. Springer, Heidelberg (2011)
7. Caron, P., Champarnaud, J., Mignot, L.: A general framework for the derivation of regular expressions. RAIRO - Theor. Inf. Appl. **48**(3), 281–305 (2014)
8. Caron, P., Ziadi, D.: Characterization of Glushkov automata. Theoret. Comput. Sci. **233**(1–2), 75–90 (2000)
9. Champarnaud, J.M., Ziadi, D.: Canonical derivatives, partial derivatives and finite automaton constructions. Theoret. Comput. Sci. **289**, 137–163 (2002)
10. Chen, H., Yu, S.: Derivatives of regular expressions and an application. In: Dinneen, M.J., Khoussainov, B., Nies, A. (eds.) Computation, Physics and Beyond. LNCS, vol. 7160, pp. 343–356. Springer, Heidelberg (2012)
11. Gelade, W.: Succinctness of regular expressions with interleaving, intersection and counting. Theor. Comput. Sci. **411**(31–33), 2987–2998 (2010)
12. Glushkov, V.M.: The abstract theory of automata. Russ. Math. Surv. **16**, 1–53 (1961)
13. Hopcroft, J.E., Ullman, J.D.: Introduction to Automata Theory, Languages and Computation. Addison Wesley, Reading (1979)
14. Ilie, L., Yu, S.: Follow automata. Inf. Comput. **186**(1), 140–162 (2003)
15. McNaughton, R., Yamada, H.: Regular expressions and state graphs for automata. IEEE Trans. Elect. Comput. **9**, 39–47 (1960)
16. Sakarovitch, J.: Elements of Automata Theory. Cambridge University Press, Cambridge (2009)
17. Yu, S.: Regular languages. In: Rozenberg, G., Salomaa, A. (eds.) Handbook of Formal Languages, vol. 1, pp. 41–110. Springer, Heidelberg (1997)

A Language-Theoretical Approach to Descriptive Complexity

Michaël Cadilhac(✉), Andreas Krebs, and Klaus-Jörn Lange

Wilhelm-Schickard-Institut, Universität Tübingen, Sand 13, Tübingen, Germany
michael@cadilhac.name, mail@krebs-net.de, lange@informatik.uni-tuebingen.de

Abstract. Logical formulas are naturally decomposed into their subformulas and circuits into their layers. How are these decompositions expressed in a purely language-theoretical setting? We address that question, and in doing so, introduce a product directly on languages that parallels formula composition. This framework makes an essential use of languages of higher-dimensional words, called *hyperwords*, of arbitrary dimensions. It is shown here that the product thus introduced is associative over classes of languages closed under the product itself; this translates back to extra freedom in the way formulas and circuits can be decomposed.

Keywords: Logic · Languages · Descriptive complexity · Hyperwords · Circuits

1 Introduction

The theory of constant-depth polysize unbounded-fan-in circuits (hereafter simply *circuits*) abounds in fine classes of languages and open problems about their relationships. Some of the main classes of focus in the literature are:

- AC^0, the class of languages recognized by circuits with Boolean gates;
- TC^0, based on AC^0 circuits with additional *threshold* gates, which output 1 if the majority of their input bits is 1;
- NC^1, which, while being usually defined with log-depth, polysize, bounded-fan-in Boolean circuits, is also characterized by AC^0 circuits with additional *regular* oracle gates, which output 1 if their input is in a prescribed regular language.

Strikingly, all these classes admit characterizations that rely on language recognition by first-order logic formulas—these are the classical results of [1,5], that we recall in Proposition 16 (see [10] for a lovely account). In this framework, the variables of a logical formula range over the positions in an input word, and the language described by the formula is the set of words satisfying it. Similarly, algebraic characterizations of AC^0 and NC^1 relying on *programs over finite monoids* [1] and of TC^0 relying on recognition by *typed monoids* [6] are known. This however is not a mere coincidence, and tokens of the pervasiveness of this

S. Brlek and C. Reutenauer (Eds.): DLT 2016, LNCS 9840, pp. 64–76, 2016.
DOI: 10.1007/978-3-662-53132-7_6

interplay between logic, circuits, and algebra were unveiled in more general settings [2,7,10,12], including in restrictions of these classes to a linear number of gates [3]. Each time, these results are shown inductively by identifying building blocks (simple formulas, simple circuits, etc.) and an appropriate composition operation (substitution, stacking of circuits, etc.).

There is, however, a missing link in this picture: a purely language-theoretical construct that would unify these frameworks. As they all are used *in fine* as language specifications, this calls for a better understanding of their building blocks and compositions, without appeal to a specific model of computation. This is what we aim for in this article.

Higher Dimensions. A prominent feature of our study is its reliance on words of higher dimension, that we call *hyperwords.* Contrary to previous works where pictures are 2-dimensional, i.e., mappings from $\{1,2,\ldots,m\} \times \{1,2,\ldots,n\}$ to some alphabet [9], our hyperwords are labeled squares, cubes, etc., and more generally, mappings from $\{1,2,\ldots,n\}^d$ to an alphabet. Going to higher dimensions constitutes a severe change that is prompted by multiple considerations: **1.** In the logical framework, composition of formulas (the so-called "substitution") is a process that replaces letter predicates $\mathfrak{c}_a(x)$, asserting that there there is an a at position x in the input, by a formula with one distinguished variable. Generalizing this substitution to a greater number of variables naturally leads to consider letter predicates of the form $\mathfrak{c}_a(x_1, x_2, \ldots, x_d)$, hence formulas recognizing d-dimensional hyperwords. **2.** In the circuit framework, one can speak of the language accepted by a circuit with n inputs. However, a *layer* of the circuit may have a polynomial number of input gates, say n^d, and thus accepts a hyperword of dimension d. **3.** Since the early stages of descriptive complexity, there has been a great interest in quantifiers that bind more than one variable. For instance, the *majority of pairs* quantifier, $(\mathrm{MAJ}_2\ x, y)[\varphi]$, asserts that there is a majority of positions (i,j) of the input word making $\varphi(x := i, y := j)$ true. Barrington, Immerman, and Straubing conjectured in the seminal paper [1] that MAJ_2 is more powerful than the majority quantifier over a single variable, and this was proven in [8]. A quantifier of that type, a so-called *Lindström quantifier,* is entirely described by a set of hyperwords; for instance, the truth value of $(\mathrm{MAJ}_2\ x, y)[\varphi]$ depends solely on whether the 2-dimensional hyperword mapping (i,j) to the truth value of $\varphi(x := i, y := j)$ contains a majority of "true." Thus again, quantifiers are determined by hyperword languages.

Our contributions are the following:

1. We adapt the traditional logic framework to the description of hyperword languages, and define a notion of substitution that extends the one for single variable formulas (see Sect. 4);
2. We introduce a purely language-theoretical framework, relying on hyperword languages, and a product over languages ("block product") that allows to express logic-defined languages (and thus ultimately languages of circuit families) independently of a model (see Sects. 3 and 6);

3. We show that the product thus defined verifies a certain associativity property: there is a trade-off between the possible bracketings of an expression and the dimensions of the languages therein (see Theorem 21).

2 Preliminaries

For an integer n, we write $[n]$ for the set $\{1, 2, \ldots, n\}$. For a function $f : X \to Y$, and for a set X', we write $f\lceil_{X'}$ for the function from $X \cap X'$ to Y that agrees with f on its domain. If $x_1, x_2, \ldots, x_e$ are some variables, we write $\boldsymbol{x}$ for the vector $(x_1, x_2, \ldots, x_e)$, and if $\boldsymbol{i}$ is a vector of same length, then $\boldsymbol{x} = \boldsymbol{i}$ is to be understood component-wise.

In the following, A and B will be alphabets, i.e., finite sets of symbols, and $\mathcal{V}$ will be a finite set of *variable symbols* included in[1] $\{\ldots, \mathfrak{v}_{-2}, \mathfrak{v}_{-1}, \mathfrak{v}_0, \mathfrak{v}_1, \mathfrak{v}_2, \ldots\}$, and we will use $x, y, x_1, x_2, \ldots$ to refer to these variables. Such sets $\mathcal{V}$ are naturally ordered, and we will often speak of the i first variables of $\mathcal{V}$.

A *stripped hyperword* over A of *dimension* $d \geq 0$ and *length* $n \geq 0$ is a map from $[n]^d$ to A; the set of stripped hyperwords of dimension d for any length is written $\mathcal{H}_d(A)$, and in particular, we have that $A^* = \mathcal{H}_1(A)$. We will also naturally identify A with $\mathcal{H}_0(A)$.

Hyperwords will *always* be paired with valuations of a (possibly empty) finite set of variables: we let $\mathcal{H}_d(A) \otimes \mathcal{V}$ be the set of pairs $W = (\mathsf{str}_W, \mathsf{val}_W)$ such that $\mathsf{str}_W \in \mathcal{H}_d(A)$ and $\mathsf{val}_W : \mathcal{V} \to \{1, \ldots, n\}$, with n the length of str_W. These objects will be called simply *hyperwords*, and we define the *length* of W, written $|W|$, to be that of str_W, its *strip* to be str_W, and its *valuation* to be val_W. A language of dimension d is then a set of hyperwords of this dimension, and we identify subsets of $\mathcal{H}_d(A)$ with languages in $\mathcal{H}_d(A) \otimes \emptyset$. Further, for a hyperword $W \in \mathcal{H}_d(A) \otimes \mathcal{V}$ and $\boldsymbol{i} \in [|W|]^d$, we write $W(\boldsymbol{i})$ for the letter $\mathsf{str}_W(\boldsymbol{i})$, and if $x \in \mathcal{V}$, then $W(\ldots, x, \ldots)$ denotes $W(\ldots, \mathsf{val}_W(x), \ldots)$. For a variable x that may or may not be in $\mathcal{V}$ and $i \in [|W|]$, we write $W_{x=i}$ for the hyperword with strip str_W and valuation val_W modified so that x is mapped to i (hence x is added to the domain of val_W if $x \notin \mathcal{V}$). Hyperwords of dimension 1 will usually be called *words*. For a language $L \subseteq \mathcal{H}_d(A) \otimes \mathcal{V}$, we denote its characteristic function by $\chi_L : \mathcal{H}_d(A) \otimes \mathcal{V} \to \{0, 1\}$.

3 Composing Languages

We begin with an intuitive presentation. Suppose we are given a language $L \subseteq \mathcal{H}_d(A) \otimes \{\mathfrak{v}_1\}$, and we wish to extract from it the language $L' \subseteq \mathcal{H}_d(A)$ of hyperwords in L that have an even valuation of $\mathfrak{v}_1$. In symbols, we want to define $L' = \{W \mid (\exists i \in 2\mathbb{N})[W_{\mathfrak{v}_1=i} \in L]\}$. When checking whether $W \in L'$, we are thus interested in the different values of $\chi_L(W_{\mathfrak{v}_1=i})$ for i ranging from 1

[1] We only make scarce use of the variables with nonpositive indexes explicitly, with the notable exception of the first part of the proof of Theorem 21.

to $|W|$; indeed, if K is the set of words over $\{0,1\}$ having at least one 1 in an even position, then $W \in L'$ if and only if:

$$\chi_L(W_{\mathfrak{v}_1=1}) \cdot \chi_L(W_{\mathfrak{v}_1=2}) \cdots \chi_L(W_{\mathfrak{v}_1=|W|}) \in K.$$

This construction is a particular example of the *block product*[2] of two languages, and we shall later write $L' = K \,\square\, L$. Our definition of the block product follows the definition of Lindström quantifiers (e.g., [1]) by making the following three generalizations:

1. We extend the valuation of $\mathfrak{v}_1$ to a *set* of variables; for instance, for two variables $\mathfrak{v}_1$ and $\mathfrak{v}_2$, rather than checking whether the *word* whose i-th letter is $\chi_L(W_{\mathfrak{v}_1=i})$ belongs to K, it should be checked whether the *hyperword* whose letter at position (i, j) is $\chi_L(W_{\mathfrak{v}_1=i,\mathfrak{v}_2=j})$ belongs to K;
2. The membership tests $\chi_L(W_{\mathfrak{v}_1=i})$ are allowed to range over a finite number of different languages L; this implies that K in our example is not simply a language over $\{0,1\}$, but over $\{0,1\}^k$ for some $k > 0$;
3. We introduce mappings from the truth values of these membership tests to different alphabets; in other words, we implement a mechanism to let K be over any alphabet.

Definition 1 (Simple join). *Let $(L_i)_{i\in[k]}$ be languages. When (and only when) all the L_i's share the same alphabet A, dimension d, and variable set $\mathcal{V}$, we write $\mathcal{L} = [L_1, L_2, \ldots, L_k]$ to denote the vector whose i-th component is L_i.*

This vector $\mathcal{L}$ is called a simple join *of length k over $\mathcal{H}_d(A) \otimes \mathcal{V}$, and we naturally extend the characteristic functions to such objects by letting, for any $W \in \mathcal{H}_d(A) \otimes \mathcal{V}$, $\chi_{\mathcal{L}}(W) = (\chi_{L_1}(W), \chi_{L_2}(W), \ldots, \chi_{L_k}(W)) \in \{0,1\}^k$.*

Definition 2 (Block product). *Let K be a language in $\mathcal{H}_e(B) \otimes \mathcal{V}$ and $\mathcal{L}$ be a simple join of length k over $\mathcal{H}_d(A) \otimes (\mathcal{X} \cup \mathcal{V})$ with $\mathcal{X} = \{x_1, x_2, \ldots, x_e\}$ the first e variables of $\mathcal{X} \cup \mathcal{V}$, in order. Further, let $g\colon \{0,1\}^k \to B$.*

Let $W \in \mathcal{H}_d(A) \otimes \mathcal{V}$. The transcript *$\tau(W) \in \mathcal{H}_e(B) \otimes \mathcal{V}$ of W is the hyperword with strip:*

$$\begin{aligned} [|W|]^e &\to B \\ (i_1, i_2, \ldots, i_e) &\mapsto g(\chi_{\mathcal{L}}(W_{\boldsymbol{x}=\boldsymbol{i}})), \end{aligned}$$

and valuation val_W. *The* block product *of K and $\mathcal{L}$ (with* alphabet replacement *g) is then $K \,\square_g\, \mathcal{L} = \{W \in \mathcal{H}_d(A) \otimes \mathcal{V} \mid \tau(W) \in K\}$.*

Notation 3. *We will often use alphabet replacements from $\{0,1\}^k$ to $\{0,1\}$. In this case, we see $0, 1$ as Boolean values, and use the notations $\wedge, \vee, \leftrightarrow, \ldots$ directly in the list $\mathcal{L}$. For instance, $L = K \,\square\, [L_1 \vee (L_2 \leftrightarrow L_3)]$ defines $g\colon \{0,1\}^3 \to \{0,1\}$*

[2] This nomenclature stems from the algebraic operation bearing the same name. There is a precise relationship between block products of monoids and block products of languages of words (Definition 2) that will be made explicit in an extended version of this article.

by $g(i, j, k) = i \vee (j \leftrightarrow k)$*, and then* $L = K \Box_g [L_1, L_2, L_3]$*. Further, we omit the alphabet replacement when it is the identity, and if* L *is a language, we write* $K \Box L$ *for* $K \Box [L]$*.*

The following operators do not directly relate to the block product. However, they will be part of our elementary set of tools to define more complex languages.

Definition 4 (Variable operators). *Let* $L \subseteq \mathcal{H}_d(A) \otimes \mathcal{V}$*. The following two operators respectively decrease and increase the number of variables used.*

The variable renaming *identifies and renames variables of* $\mathcal{V}$*. Let* $\sigma\colon \mathcal{V} \to \mathcal{V}'$ *be a given partial map, for a set* $\mathcal{V}'$ *of variables. First, extend* σ *to all of* $\mathcal{V}$ *by letting* $\sigma(x) = x$ *if* σ *was undefined on* x*. Then for a* valuation val *of* $\sigma(\mathcal{V})$*, write* $\sigma^{-1}(\mathsf{val})$ *for the valuation of* $\mathcal{V}$ *mapping* x *to* $\mathsf{val}(\sigma(x))$*. The* variable renaming *of* L *by* σ *is* $\mathsf{ren}(L, \sigma) = \{W \in \mathcal{H}_d(A) \otimes \sigma(\mathcal{V}) \mid (\mathsf{str}_W, \sigma^{-1}(\mathsf{val}_W)) \in L\}$*.*

The variable extension *augments the set of variables* $\mathcal{V}$ *with untested variables. Let* $\mathcal{V}'$ *be a finite set of variables, the* variable extension *of* L *by* $\mathcal{V}'$ *is* $\mathsf{var\text{-}ext}(L, \mathcal{V}') = \{W \in \mathcal{H}_d(A) \otimes (\mathcal{V} \cup \mathcal{V}') \mid (\mathsf{str}_W, \mathsf{val}_W\restriction_{\mathcal{V}}) \in L\}$*.*

4 The Descriptive Complexity Framework

We present a generalized version of the classical framework of descriptive complexity for expressing languages (e.g., [1,10]). The generalization lies essentially in the ability for a formula to recognize a language of hyperwords. A logic will be given by the set of allowed *quantifiers* and *numerical predicates*, which will have a preset semantics. As an example, we want to be able to write formulas such as $(\mathrm{MAJ}_{2,1}\, \mathfrak{v}_1, \mathfrak{v}_2)[\mathfrak{c}_a(\mathfrak{v}_1, \mathfrak{v}_2)]$, expressing that there is a majority of pairs of positions (i, j) such that the 2-dimensional input hyperword has an a in position (i, j). As usual (e.g., [1,8]), we also allow multiple formulas under the scope of a quantifier.

Definition 5 (Quantifier, numerical predicate). *An* (e, k)-ary quantifier *is a pair* (L, g) *where* $L \subseteq \mathcal{H}_e(A) \otimes \mathcal{V}$*, for some alphabet* A *and variable set* $\mathcal{V}$*, and* $g\colon \{0,1\}^k \to A$*. Intuitively,* e *will be the number of variables quantified and* k *the number of formulas over which the quantifier ranges.*

An e-ary numerical predicate *is a subset of* $\mathcal{H}_1(\{a\}) \otimes \{\mathfrak{v}_1, \mathfrak{v}_2, \ldots, \mathfrak{v}_e\}$*.*

Definition 6 (Logic). *Given a set of quantifiers* $\mathcal{Q}$ *and a set of numerical predicates* $\mathcal{N}$*, we define the logic* $\mathcal{Q}[\mathcal{N}]$ *as the set of following formulas with the provided semantics:*

- Syntax. *A formula of dimension* d *over the alphabet* A *is built from the following syntax, where the* x_i*'s are variables that are not necessarily distinct, except in Case 3:*

$$\begin{aligned}
\varphi ::= \quad & \mathfrak{c}_a(x_1, x_2, \ldots, x_d) \text{ where } a \in A && (1)\\
\mid\ & N(x_1, x_2, \ldots, x_e) \text{ for any } N \in \mathcal{N} \text{ of arity } e && (2)\\
\mid\ & (Q\, x_1, x_2, \ldots, x_e)[\varphi_1, \varphi_2, \ldots, \varphi_k] \text{ for any } Q \in \mathcal{Q} \text{ of arity } (e, k) && (3)\\
\mid\ & \varphi_1 \wedge \varphi_2 \mid \varphi_1 \vee \varphi_2 \mid \neg\varphi_1 && (4)
\end{aligned}$$

We rely on the usual vocabulary concerning variables: a variable used in a formula is bounded *if it always appears after being quantified, otherwise it is* free. *This includes the variables that may appear within a quantifier, e.g., if $Q = (L, g)$ is a quantifier where $L \subseteq \mathcal{H}_1(A) \otimes \mathcal{V}$, then all variables of $\mathcal{V}$ are free in $(Q\ x)[\varphi]$. If $\{\mathfrak{v}_{i_1}, \mathfrak{v}_{i_2}, \ldots, \mathfrak{v}_{i_n}\}$ are the free variables of φ, with $i_1 < i_2 < \cdots < i_n$, we write $\varphi(x_1, x_2, \ldots, x_k)$, with $k \le n$, for the formula φ with $\mathfrak{v}_{i_j}$ replaced by x_j, for all $j \in [k]$; we let the formulas obtained in this fashion also belong to $\mathcal{Q}[\mathcal{N}]$.*

- Semantics. *Let φ be a formula of dimension d over the alphabet A, and $\mathcal{V}$ a set containing all its free variables. A hyperword $W \in \mathcal{H}_d(A) \otimes \mathcal{V}$ is said to be* a model *of φ, written $W \models \varphi$, when (the cases refer to the above syntax):*
  - *(Case 1). $W(x_1, x_2, \ldots, x_d) = a$, recalling our use of $W(\ldots, x_i, \ldots)$ as short for $W(\ldots, \mathsf{val}_W(x_i), \ldots)$.*
  - *(Case 2). The word $(a^{|W|}, \{\mathfrak{v}_i \mapsto \mathsf{val}_W(x_i)\}_{i \in [e]})$ is in N.*
  - *(Case 3). $W' \in L$, where $Q = (L, g)$ with $L \subseteq \mathcal{H}_e(B) \otimes \mathcal{V}'$, and W' is defined as the hyperword with strip:*

$$\begin{aligned} [|W|]^e &\to B \\ (i_1, i_2, \ldots, i_e) &\mapsto g\big(\, (W_{\boldsymbol{x}=\boldsymbol{i}} \models \varphi_1) \cdot (W_{\boldsymbol{x}=\boldsymbol{i}} \models \varphi_2) \cdots (W_{\boldsymbol{x}=\boldsymbol{i}} \models \varphi_e) \,\big), \end{aligned}$$

    (where "$W_{\boldsymbol{x}=\boldsymbol{i}} \models \varphi_j$" is 1 if true, and 0 otherwise) and valuation $\mathsf{val}_W\lceil_{\mathcal{V}'}$.
  - *(Case 4). For $\wedge$, when $W \models \varphi_1$ and $W \models \varphi_2$, and likewise for $\vee$ and $\neg$.*

Finally, we let $L(\varphi)$, the language of φ, be $\{W \in \mathcal{H}_d(A) \otimes \mathcal{V} \mid W \models \varphi\}$, with $\mathcal{V}$ the set of free variables of φ, and also identify $\mathcal{Q}[\mathcal{N}]$ with the class of languages of its formulas.

Example 7 (Some standard quantifiers). The *first-order quantifiers* FO $= \{\exists, \forall\}$ are defined as follows. The $(1,1)$-ary quantifier $\exists$ consists of the pair (L, g) where g is the identity over $\{0,1\}$, and L the set of words $\{0,1\}^* \cdot 1 \cdot \{0,1\}^*$. The quantifier $\forall$ is defined similarly with $L = 1^*$.

The (e,k)-ary *majority quantifier* $\mathrm{MAJ}_{e,k}$ is the pair (L, g) where $g\colon \{0,1\}^k \to \{-k, \ldots, k\}$ computes the difference of the number of 1's and 0's and L consists of hyperwords of $\mathcal{H}_e(\{-k, \ldots, k\})$ such that the sum of all letters appearing is greater than 0. The *counting quantifier* $\exists^{=\mathfrak{v}_1}$ can be expressed correspondingly.

Example 8 (Some standard numerical predicates). The 2-ary numerical predicate $=$ is the set of words w such that $\mathsf{val}_w(\mathfrak{v}_1) = \mathsf{val}_w(\mathfrak{v}_2)$; we always assume that this predicate belongs to $\mathcal{N}$ when defining a logic. The 2-ary numerical predicate $<$ is defined similarly. Next, $+$ is a 3-ary numerical predicate for which $\mathsf{val}_w(\mathfrak{v}_1) + \mathsf{val}_w(\mathfrak{v}_2) = \mathsf{val}_w(\mathfrak{v}_3)$. The 2-ary numerical predicate $+1$ is the one for which the words verify $\mathsf{val}_w(\mathfrak{v}_1) + 1 = \mathsf{val}_w(\mathfrak{v}_2)$. The 1-ary numerical predicate max is the set of words w for which $\mathsf{val}_w(\mathfrak{v}_1) = |w|$.

Definition 9 (Substitution). *Let $\varphi \in \mathcal{Q}[\mathcal{N}]$ be a formula of dimension e over the alphabet B, and $\varphi_1, \varphi_2, \ldots, \varphi_k \in \mathcal{Q}[\mathcal{N}]$ be formulas of dimension d over the* *alphabet A. Further let* $g\colon \{0,1\}^k \to B$*. The formula* $\varphi \circ_g [\varphi_1, \varphi_2, \ldots, \varphi_k]$ *is obtained from* φ *by replacing its atomic formulas* $\mathfrak{c}_a(x_1, x_2, \ldots, x_e)$, $a \in B$, *by:*

$$\bigvee_{v \in g^{-1}(a)} \Big(\bigwedge_{i : v_i = 1} \varphi_i(x_1, x_2, \ldots, x_e) \wedge \bigwedge_{i : v_i = 0} \neg \varphi_i(x_1, x_2, \ldots, x_e) \Big).$$

This results in a formula of $\mathcal{Q}[\mathcal{N}]$ *of dimension d over the alphabet A called a* substitution *of* φ.

5 Examples

Example 10 (Existential quantification, logical and). Let $L_i \subseteq \mathcal{H}_d(A) \otimes \{\mathfrak{v}_1\}$, for $i = 1, 2$, be defined as $\{W \mid W \models \varphi_i\}$ for some formulas φ_i of dimension d with free variable $\mathfrak{v}_1$. We wish to express L defined by the formula $(\exists \mathfrak{v}_1)[\varphi_1 \wedge \varphi_2]$ using the block product. To this end, let $E = \{0,1\}^* \cdot 1 \cdot \{0,1\}^*$, we claim that:

$$L = E \,\Box\, [L_1 \wedge L_2].$$

Indeed, the transcript of a hyperword W has a 1 in position i iff, by definition, $\chi_{[L_1, L_2]}(W_{\mathfrak{v}_1 = i}) = (1, 1)$, that is, iff $W_{\mathfrak{v}_1 = i} \in L_1 \cap L_2$. The language E then checks that *there exists* one position of the transcript that contains a 1.

Example 11 (Identities). Example 10 seems to indicate that Boolean operations on languages ought to be expressed under the scope of a quantifier (existential in the example). This is correct, but does *not* come at the expense of introducing new variables, since we may speak about *0-dimensional* hyperwords, that is, letters. Thus any language L is equal to $\{1\} \,\Box\, L$, where the left-hand side is of dimension 0.

Now let $L \subseteq \mathcal{H}_d(A) \otimes \mathcal{V}$, we wish to express L by using it as the *left-hand side* of a block product. Let $\mathcal{V}' = \{\mathfrak{v}_1, \mathfrak{v}_2, \ldots, \mathfrak{v}_d\}$, and define for all $a \in A$ the language $C_a \subseteq \mathcal{H}_d(A) \otimes \mathcal{V}'$ to be the set of hyperwords W with $W(\mathfrak{v}) = a$. Finally, with $A = \{a_1, a_2, \ldots, a_\ell\}$, let $g\colon \{0,1\}^\ell \to A$ map $(b_1, b_2, \ldots, b_\ell)$ to a_i if $b_i = 1$ for some unique i—the other values of g being irrelevant. It then holds that:

$$L = L \,\Box_g\, [\textsf{var-ext}(C_{a_1}, \mathcal{V}), \textsf{var-ext}(C_{a_2}, \mathcal{V}), \ldots, \textsf{var-ext}(C_{a_\ell}, \mathcal{V})].$$

Example 12 (Boolean operations). Now given a Boolean expression on k variables, that is, a function $g\colon \{0,1\}^k \to \{0,1\}$, and a simple join $[L_1, L_2, \ldots, L_k]$, the language obtained by combining the languages using the expression is:

$$\{1\} \,\Box_g\, [L_1, L_2, \ldots, L_k].$$

In particular, we have:

$$L_1 \cup L_2 = \{1\} \,\Box\, [L_1 \vee L_2], \quad L_1 \cap L_2 = \{1\} \,\Box\, [L_1 \wedge L_2].$$

6 Logics and Their Language Classes

In this section, we show that, given a logic, the class of languages recognized by its formulas is the closure, under mainly block product, of a set of languages associated with its quantifiers and numerical predicates.

Definition 13 (Block closure). *A class of languages $\mathcal{C}$ is* block-closed *if it is closed under block products, variable extension, and variable renaming. Further, for a class of languages $\mathcal{C}$, we let $\boxast(\mathcal{C})$ be the smallest block-closed class that contains $\mathcal{C}$ and the languages $C_a^{A,d}$, defined for any alphabet A, $a \in A$, and $d \geq 0$, as:*

$$C_a^{A,d} = \{W \in \mathcal{H}_d(A) \otimes \{\mathfrak{v}_1, \mathfrak{v}_2, \ldots, \mathfrak{v}_d\} \mid W(\mathfrak{v}_1, \mathfrak{v}_2, \ldots, \mathfrak{v}_d) = a\}.$$

For a map $g: A \to B$ and a hyperword $W \in \mathcal{H}_d(A) \otimes \mathcal{V}$, write $g(W)$ for the hyperword W where each letter $a \in A$ of str_W is replaced by $g(a)$.

Theorem 14. *Let $\mathcal{Q}$ be a set of quantifiers and $\mathcal{N}$ be a set of numerical predicates. Let $\mathcal{Q}' = \{g^{-1}(L) \mid (L, g) \in \mathcal{Q}\}$. Then $\mathcal{Q}[\mathcal{N}] = \boxast(\mathcal{Q}' \cup \mathcal{N})$.*

Proof. ($\mathcal{Q}[\mathcal{N}] \subseteq \boxast(\mathcal{Q}' \cup \mathcal{N})$). This is proved by induction; let $\varphi \in \mathcal{Q}[\mathcal{N}]$ over A with free variables in $\mathcal{V}$, then:

- If $\varphi \equiv \mathfrak{c}_a(x_1, x_2, \ldots, x_d)$, then $L(\varphi) = \mathsf{ren}(C_a^{A,d}, \sigma)$, with $\sigma = \{\mathfrak{v}_i \mapsto x_i\}_{i \in [e]}$;
- If $\varphi \equiv N(x_1, x_2, \ldots, x_e)$ for $N \in \mathcal{N}$ of arity e, then $L(\varphi) = \mathsf{ren}(N, \sigma)$ with $\sigma = \{\mathfrak{v}_i \mapsto x_i\}_{i \in [e]}$;
- If $\varphi \equiv (Q\ x_1, x_2, \ldots, x_e)[\varphi_1, \varphi_2, \ldots, \varphi_k]$, with $Q = (L, g) \in \mathcal{Q}$ of arity (e, k), then let by induction $L_i = L(\varphi_i) \in \boxast(\mathcal{Q}' \cup \mathcal{N})$, for $i \in [k]$. Further, rename the variables of all the L_i's and $K = g^{-1}(L)$ so that $x_1, x_2, \ldots, x_e$ appear first among all the variables used, and extend these languages to a common set of variables. Then $L(\varphi) = K \,\square\, [L_1, L_2, \ldots, L_k]$;
- If $\varphi \equiv \varphi_1 \wedge \varphi_2$, then, noting that $\{1\}$, as 0-dimensional, is $C_1^{\{0,1\},0}$, and by Example 12, $L(\varphi) = C_1^{\{0,1\},0} \,\square\, [\mathsf{var\text{-}ext}(L(\varphi_1), \mathcal{V}) \wedge \mathsf{var\text{-}ext}(L(\varphi_2), \mathcal{V})]$;
- The cases $\varphi \equiv \varphi_1 \vee \varphi_2$ and $\varphi \equiv \neg\varphi_1$ are similar to the previous one.

Additionally, renaming of variables is achieved through ren. In each case, we inductively have that $L(\varphi) \in \boxast(\mathcal{Q}' \cup \mathcal{N})$.

($\boxast(\mathcal{Q}' \cup \mathcal{N}) \subseteq \mathcal{Q}[\mathcal{N}]$). Again, this is done by induction; let $L \in \boxast(\mathcal{Q}' \cup \mathcal{N})$, with $L \subseteq \mathcal{H}_d(A) \otimes \mathcal{V}$, then:

- If $L = N$ for $N \in \mathcal{N}$ of arity e, then $L = L(\varphi)$ for $\varphi \equiv N(\mathfrak{v}_1, \mathfrak{v}_2, \ldots, \mathfrak{v}_e)$ seen as a formula of dimension 1 over $\{a\}$;
- If $L = g^{-1}(L')$ for $Q = (L', g) \in \mathcal{Q}$, then $A = \{0, 1\}^k$ for some k. We then have that $L = L(\varphi)$ with:

$$\varphi \equiv (Q\ \mathfrak{v}_1, \mathfrak{v}_2, \ldots, \mathfrak{v}_d)\Big[\bigvee_{u \in \{0,1\}^k : u_1 = 1} \mathfrak{c}_u(\mathfrak{v}_1, \mathfrak{v}_2, \ldots, \mathfrak{v}_d), \\ \vdots \\ \bigvee_{u \in \{0,1\}^k : u_k = 1} \mathfrak{c}_u(\mathfrak{v}_1, \mathfrak{v}_2, \ldots, \mathfrak{v}_d)\Big];$$

- If $L = C_a^{A,d}$, then $L = L(\varphi)$ with $\varphi \equiv \mathfrak{c}_a(\mathfrak{v}_1, \mathfrak{v}_2, \ldots, \mathfrak{v}_d)$ seen as a formula over A;
- If $L = \mathsf{var\text{-}ext}(L', \mathcal{V}')$, then with φ' such that $L' = L(\varphi')$, define φ as the formula $\varphi' \wedge \bigwedge_{x \in \mathcal{V}'} x = x$. We thus have that $L(\varphi)$ is $L(\varphi')$ over the variables $\mathcal{V} \cup \mathcal{V}'$, hence $L = L(\varphi)$;
- If $L = \mathsf{ren}(L', \sigma)$, then we simply apply the renaming σ to the formula defining L';
- Finally, if $L = K \,\square_g\, [L_1, L_2, \ldots, L_k]$, let φ_i such that $L(\varphi_i) = L_i$ for all i, and φ_K such that $L(\varphi_K) = K$, then $L = L(\varphi_K \circ_g [\varphi_1, \varphi_2, \ldots, \varphi_k])$.z □

A salient property of this characterization is that there is no syntactic difference made between the languages coming from quantifiers, and those coming from numerical predicates. From this, we naturally derive the following restatement of Theorem 14 starting from languages:

Theorem 15. *Let $\mathcal{C}$ be a class of languages containing the numerical predicate $=$. Let $\mathcal{Q}$ be the set of quantifiers (L, g) such that $L \in \mathcal{C}$. It holds that $\mathcal{Q}[=] = \boxast(\mathcal{C})$.*

We note that Theorem 14 immediately implies that some complexity classes can be expressed as the block-closure of simple languages, namely:

Proposition 16. *The following equalities hold:*

- *DLOGTIME-uniform* $\mathsf{TC}^0 = \boxast(\{\mathrm{MAJ}_{2,1}, <\})$*;*
- *DLOGTIME-uniform* $\mathsf{NC}^1 = \boxast(\{\mathrm{MAJ}_{2,1}, \mathrm{S}_5, <\})$*, with S_5 the symmetric group on 5 elements, seen as the language of words $\sigma_1\sigma_2\cdots\sigma_n$, with each $\sigma_i \in S_5$, that evaluate to the identity permutation;*
- $\mathsf{P} = \boxast(\{\mathrm{MAJ}_{2,1}, \mathit{CVP}, <\})$*, where CVP is the* circuit valuation problem, *that is, encoding of Boolean circuits that evaluate to one.*

7 Associativity of the Block Product

In the context of the block product of algebraic structures,[3] it is well known that parenthesizing plays a crucial role. Indeed, the composition $(M \,\square\, N) \,\square\, K$ is sometimes called the *weak* product [3,11], by opposition to the *strong* one $M \,\square\, (N \,\square\, K)$, and it can be proved that the former recognizes, in general, less languages than the latter. Similarly—equivalently in fact [11,12]—the classical notion of formula substitution (akin to our definition but with formulas of dimension one) depends intrinsically on the parenthesizing: $\varphi_1 \circ (\varphi_2 \circ (\varphi_3 \circ \cdots))$ can express all formulas starting from formulas of depth 1 (i.e., formulas with one quantifier), while $((\cdots(\varphi_1 \circ \varphi_2) \circ \varphi_3) \circ \cdots)$ can only express formulas with two variables (that may be reused). Here, we show that we can get more freedom in the parenthesizing, provided that we allow products of languages of higher dimensions. We place this result in a purely language-theoretical framework

[3] The reader not versed in that topic can think of block products of monoids as block products of the languages of dimension 1 recognized by them.

(i.e., with languages and block products), and by Theorem 14 and its proof, it would carry over to the logical setting (i.e., with logical formulas and substitutions).

Naturally, as products of one-dimensional languages are nonassociative, we cannot hope for $K \Box (L_1 \Box L_2)$ to be equal to $(K \Box L_1) \Box L_2$ in general. We will however see in the proof of the forthcoming Theorem 21, that it is enough to provide a *dimensional jump* of L_1:

Definition 17 (Dimensional jump). *Let $L \subseteq \mathcal{H}_d(A) \otimes \mathcal{V}$. For $0 < c \leq |\mathcal{V}|$, we let $[\![L]\!]^c$, the c-dimensional jump of L, be the language of hyperwords W in $\mathcal{H}_{c+d}(A) \otimes \mathcal{V}$ defined as copies of L in the following sense. Let $\{x_1, x_2, \dots, x_c\}$ be the c first variables of $\mathcal{V}$. For $\boldsymbol{v} \in [|W|]^c$, define $W(\boldsymbol{v}, \bullet)$ as the d-dimensional hyperword of strip mapping $\boldsymbol{u} \in [|W|]^d$ to $W(\boldsymbol{v}, \boldsymbol{u})$, and of valuation val_W. Then:*

$$W \in [\![L]\!]^c \quad \Leftrightarrow \quad W(x_1, x_2, \dots, x_c, \bullet) \in L.$$

If $[L_1, \dots, L_k]$ is a join, we let $[\![L_1, \dots, L_k]\!]^c = [[\![L_1]\!]^c, \dots, [\![L_k]\!]^c]$.

Further, to treat simple lists, we will need the following symmetric operators that increase the dimension of hyperwords by a constant, the original hyperwords appearing in the first or the last components. With the notations of Definition 17:

Definition 18 (Dimensional extensions). *The* right dimensional extension *of $L \subseteq \mathcal{H}_d(A) \otimes \mathcal{V}$ for any $e > 0$, written $\mathsf{dim\text{-}ext}(L, e)$, is defined as the set $\{W \in \mathcal{H}_{d+e}(A) \otimes \mathcal{V} \mid (\forall \boldsymbol{v} \in [|W|]^e)[W(\boldsymbol{v}, \bullet) \in L]\}$.*

Similarly, its left dimensional extension *$\mathsf{dim\text{-}ext}(e, L)$ is the set of hyperwords $\{W \in \mathcal{H}_{e+d}(A) \otimes \mathcal{V} \mid (\forall \boldsymbol{v} \in [|W|]^e)[W(\bullet, \boldsymbol{v}) \in L]\}$.*

Finally, we will need to be able to "enlarge" the alphabets at hand:

Definition 19 (Alphabet product extension). *Let $L \subseteq \mathcal{H}_d(A) \otimes \mathcal{V}$ and B be an alphabet. The* right alphabet product extension *of L by B, written $\mathsf{alph\text{-}prod}(L, B)$, is the set of hyperwords in $\mathcal{H}_d(A \times B) \otimes \mathcal{V}$ such that dropping the second component of each letter gives a hyperword in L. The* left alphabet product extension *$\mathsf{alph\text{-}prod}(B, L)$ is defined symmetrically, resulting in hyperwords in $\mathcal{H}_d(B \times A) \otimes \mathcal{V}$.*

Lemma 20. *Any block-closed class $\boxast(\mathcal{C})$ containing the language 1^* is closed under dimensional jump, dimensional extensions, and alphabet product extensions.*

The aforementioned associativity property of the block product is then:

Theorem 21. *Every language of a block-closed class $\boxast(\mathcal{C})$ can be written from the languages of $\mathcal{C}$ and the languages $C_a^{A,d}$ using block products, variable extensions, variable renaming, dimensional jump and extensions, and alphabet product extensions, in such a way that no right-hand side of a block product contains a block product.*

Proof. Any language of $\circledast(\mathcal{C})$ can be written, by definition, from the languages of $\mathcal{C}$ and the languages $C_a^{A,d}$ using block products, variable extension, and variable renaming. It is not hard to show that the variable related operators and the dimensional jump can be pushed to the language level, so that a block product is never under the scope of such operators.

To show the main claim, we proceed inductively on the structure of the expression defining a language L of $\circledast(\mathcal{C})$, assuming that the variable operators are at the language level. The claim is true for languages of $\mathcal{C}$, their jumps, and their variable extensions and renaming.

We consider first a simplified situation. Let K, L_1, L_2 be languages of dimensions i, j, and k respectively. We claim that $K \Box (L_1 \Box L_2) = (K \Box [\![L_1]\!]^i) \Box L_2$, assuming that the left-hand side is well-defined.

Indeed, let W be a hyperword; we show that the transcript of W at the outermost product of the left-hand side is the same as the transcript of W at the innermost product of the right-hand side. This proves the equality, as the membership of W to either side depends only on this transcript.

The transcript W' of W at the outermost product of the left-hand side is the i-dimensional hyperword whose strip maps $\boldsymbol{v}$ to 1 iff $W'' = W_{\boldsymbol{x}=\boldsymbol{v}} \in L_1 \Box L_2$, where $\boldsymbol{x}$ denotes the i first variables of L_1. In turn, this holds iff the transcript of W'' at the innermost product of the left-hand side is in L_1; define U as the $(i+j)$-dimensional hyperword such that $U(\boldsymbol{v}, \bullet)$ is that transcript, for any $\boldsymbol{v}$ of dimension i, and valuation val_W. We have that $W'(\boldsymbol{v}) = 1$ iff $U(\boldsymbol{v}, \bullet)_{\boldsymbol{x}=\boldsymbol{v}} \in L_1$, that is, iff $U_{\boldsymbol{x}=\boldsymbol{v}} \in [\![L_2]\!]^i$. Now U is precisely the transcript of W at the outermost product of the *right*-hand side. Thus the transcript of U at the innermost product of the right-hand side is an i-dimensional hyperword whose strip maps $\boldsymbol{v}$ to 1 iff $U_{\boldsymbol{x}=\boldsymbol{v}} \in [\![L_2]\!]^i$, and this transcript is W'. This shows the equality.

We now introduce simple lists in two steps. Writing $[L_1, L_2] \Box_g \mathcal{L}$ for the simple list $[L_1 \Box_g \mathcal{L}, L_2 \Box_g \mathcal{L}]$, first note that:

$$K \Box_f \left([L_1, L_2] \Box_g \mathcal{L}\right) = (K \Box_f [\![L_1, L_2]\!]^i) \Box_g \mathcal{L}.$$

Now to treat the general case and conclude this proof, consider the expression $K \Box_f [L_1 \Box_g \mathcal{L}, L_2 \Box_{g'} \mathcal{L}']$. Clearly, for it to be well-defined, L_1 and L_2 must have the same set of variables, thus write $L_i \subseteq \mathcal{H}_{d_i}(A_i) \otimes \mathcal{V}$, $i = 1, 2$. Further, define $L_1' = \mathsf{alph\text{-}prod}(L_1, A_2)$ and $L_2' = \mathsf{alph\text{-}prod}(A_1, L_2)$. Using techniques similar to the above, we may assume that all the languages in $\mathcal{L}$ and $\mathcal{L}'$ are over the variables $\mathcal{X} \uplus \mathcal{V}$ and $\mathcal{X}' \uplus \mathcal{V}$, respectively, so that: 1. $|\mathcal{X}| = d_1$, $|\mathcal{X}'| = d_2$; 2. All the variables in $\mathcal{X}$ are smaller than those in $\mathcal{X}'$; and 3. All the variables in $\mathcal{X}'$ are smaller than those in $\mathcal{V}$. Finally, write $g''(\boldsymbol{u}, \boldsymbol{v}) = (g(\boldsymbol{u}), g(\boldsymbol{v}))$. It then readily holds that the above expression is equal to:

$$K \Box_f \Big(\big[\mathsf{dim\text{-}ext}(L_1, d_2), \mathsf{dim\text{-}ext}(d_1, L_2)\big] \Box_{g''} \big[\mathsf{var\text{-}ext}(\mathcal{L}, \mathcal{X}'), \mathsf{var\text{-}ext}(\mathcal{L}', \mathcal{X})\big]\Big),$$

where $\mathsf{var\text{-}ext}$ is applied component-wise to all languages of $\mathcal{L}$ and $\mathcal{L}'$. This concludes the proof, as this is of the simpler above form. □

Example 22 (Majorities). As alluded to, the majority of pairs quantifier $\mathrm{MAJ}_{2,1}$ is more powerful than the simple majority quantifier $\mathrm{MAJ}_{1,1}$, even when nested. It is thus interesting to see which quantifiers arise from Theorem 21.

Consider the language M of words over $\{0, 1\}$ containing more 1's than 0's. Let $L' = M \square (\text{var-ext}(M, \{\mathfrak{v}_1\}) \square L)$ be a well-defined language, where $\mathfrak{v}_1$ is the first variable of L. Then $L' = (M \square [\![\text{var-ext}(M, \{\mathfrak{v}_1\})]\!]^1) \square L$, by the proof of Theorem 21. Let $Z = (M \square [\![\text{var-ext}(M, \{\mathfrak{v}_1\})]\!]^1)$, which is a subset of $\mathcal{H}_2(\{0, 1\})$; we describe Z. A hyperword $W \in \mathcal{H}_2(\{0, 1\})$ is in Z iff its transcript is in M, by definition. This transcript has a 1 in position $i \in [|W|]$ iff $W(i, \bullet) \in M$. Thus, seeing two-dimensional hyperwords as arrays, a hyperword W is in Z iff there is a majority of rows of W that contain a majority of 1. There lies the intrinsic difference with $\mathrm{MAJ}_{2,1}$, a quantifier that would translate to a language of two-dimensional hyperwords having more 1's than 0's.

For two language classes $\mathcal{C}$ and $\mathcal{D}$, write $\mathcal{C} \square \mathcal{D}$ for the block closure of the set of languages $L \square \mathcal{L}'$ for all $L \in \mathcal{C}$ and $\mathcal{L}'$ a simple join of languages in $\mathcal{D}$.

Corollary 23. *For any classes $\mathcal{C}$, $\mathcal{D}$, $\mathcal{E}$ obtained as block closures, we have:*

$$\mathcal{C} \square (\mathcal{D} \square \mathcal{E}) = (\mathcal{C} \square \mathcal{D}) \square \mathcal{E}.$$

8 Conclusion

We presented a novel purely language-theoretical framework to express classes of languages described by logics. This addresses two shortcomings of the similar *algebraic* theory of typed monoids [6,7]. First, quantifiers on tuples can be expressed, providing for instance a shorter, arguably more compelling characterization of TC^0, and thus overcoming the limitation of "linear fan-in." Second, by allowing words of higher dimensions, we obtain a product mimicking the classical *block product* of algebraic structures that exhibits a property reminiscent of associativity—this may allow to translate techniques than only applied to weak parenthesizing (e.g., [4]) to a more general setting.

We believe that the results herein advocate for the use of hyperwords, leading to a unified framework in which the freedom of speaking of partial formulas (and hence partial circuits) is balanced by the dimensions used in expressing their composition.

Acknowledgments. We thank Charles Paperman for stimulating discussions.

References

1. Barrington, D.A.M., Immerman, N., Straubing, H.: On uniformity within NC^1. J. Comput. Syst. Sci **41**(3), 274–306 (1990)
2. Behle, C., Lange, K.J.: FO[<]-uniformity. In: Proceedings of the 21st Annual IEEE Conference on Computational Complexity (CCC 2006), pp. 183–189 (2006)

3. Behle, C., Krebs, A., Mercer, M.: Linear circuits, two-variable logic and weakly blocked monoids. In: Kučera, L., Kučera, A. (eds.) MFCS 2007. LNCS, vol. 4708, pp. 147–158. Springer, Heidelberg (2007)
4. Behle, C., Krebs, A., Reifferscheid, S.: Regular languages definable by majority quantifiers with two variables. In: Diekert, V., Nowotka, D. (eds.) DLT 2009. LNCS, vol. 5583, pp. 91–102. Springer, Heidelberg (2009)
5. Immerman, N.: Expressibility and parallel complexity. SIAM J. Comput. **18**(3), 625–638 (1989)
6. Krebs, A., Lange, K.J., Reifferscheid, S.: Characterizing TC^0 in terms of infinite groups. Theor. Comput. Syst. **40**(4), 303–325 (2007)
7. Krebs, A.: Typed semigroups, majority logic, and threshold circuits. Ph.D. thesis, Eberhard Karls University of Tübingen (2008)
8. Lautemann, C., McKenzie, P., Schwentick, T., Vollmer, H.: The descriptive complexity approach to LOGCFL. J. Comput. Syst. Sci. **62**(4), 629–652 (2001)
9. Rozenberg, G., Salomaa, A. (eds.): Handbook of Formal Languages: Volume 3 Beyond Words. Springer, Heidelberg (1997)
10. Straubing, H.: Finite Automata, Formal Logic, and Circuit Complexity. Birkhäuser, Boston (1994)
11. Straubing, H., Thérien, D.: Weakly iterated block products of finite monoids. In: Rajsbaum, S. (ed.) LATIN 2002. LNCS, vol. 2286, pp. 91–104. Springer, Heidelberg (2002)
12. Thérien, D., Wilke, T.: Nesting until and since in linear temporal logic. Theor. Comput. Syst. **37**(1), 111–131 (2004)

k-Abelian Equivalence and Rationality

Julien Cassaigne[1], Juhani Karhumäki[2], Svetlana Puzynina[3,4], and Markus A. Whiteland[2(✉)]

[1] Institut de Mathématiques de Marseille, Marseille, France
julien.cassaigne@math.cnrs.fr
[2] Department of Mathematics and Statistics, University of Turku, Turku, Finland
{karhumak,mawhit}@utu.fi
[3] LIP, ENS de Lyon, Université de Lyon, Lyon, France
[4] Sobolev Institute of Mathematics, Novosibirsk, Russia
s.puzynina@gmail.com

Abstract. Two words u and v are said to be k-abelian equivalent if, for each word x of length at most k, the number of occurrences of x as a factor of u is the same as for v. We study some combinatorial properties of k-abelian equivalence classes. Our starting point is a characterization of k-abelian equivalence by rewriting, so-called k-switching. We show that the set of lexicographically least representatives of equivalence classes is a regular language. From this we infer that the sequence of the numbers of equivalence classes is $\mathbb{N}$-rational. We also show that the set of words defining k-abelian singleton classes is regular.

Keywords: k-abelian equivalence · Regular languages · Rational sequences

1 Introduction

k-abelian equivalence has attracted quite a lot of interest recently, see, e.g., [1,2,8,10,12,15]. It is an equivalence relation extending abelian equivalence and allowing an infinitary approximation of the equality of words defined as follows: for an integer k, two words u and v are *k-abelian equivalent*, denoted by $u \sim_k v$, if, for each word w of length at most k, w occurs in u and v equally often.

k-abelian equivalence, originally introduced in [7], has been studied, e.g., in the following directions: avoiding *k-abelian powers* [6,15], estimating the number of k-abelian equivalence classes, that is, *k-abelian complexity* [11], analyzing the growth and the fluctuation of the *k-abelian complexity of infinite words* [1], analyzing *k-abelian palindromicity* [8], and studying *k-abelian singletons* [9].

J. Karhumäki—Supported by the Academy of Finland, grant 257857.
S. Puzynina—Supported by the LABEX MILYON (ANR-10-LABX-0070) of Université de Lyon, within the program "Investissements d'Avenir" (ANR-11-IDEX-0007) operated by the French National Research Agency (ANR).
M.A. Whiteland—Supported by the Academy of Finland, grant 257857.

S. Brlek and C. Reutenauer (Eds.): DLT 2016, LNCS 9840, pp. 77–88, 2016.
DOI: 10.1007/978-3-662-53132-7_7

We continue the approach of analyzing the structure of k-abelian equivalence classes. We also study some numerical properties of the equivalence classes.

Our starting point is a *k-switching* lemma, proved in [9], which allows a characterization of k-abelian equivalence in terms of rewriting. This is quite different from the other existing characterizations, so it is no surprise that it opens new perspectives of k-abelian equivalence. This is what we intend to explore here.

A fundamental observation from the characterization of k-abelian equivalence using k-switching is that certain languages related to k-abelian equivalence classes are *regular* (or *rational*). More precisely, the union of all singleton classes forms a regular language, for any parameter k, and any size m of the alphabet. Similarly, the set of lexicographically least (or greatest) representatives of k-abelian equivalence classes forms a regular language. Summing up all minimal elements of a fixed length we obtain the number of equivalence classes of words of this length. As a consequence, we conclude that the complexity function of k-abelian equivalence, that is, the function computing the number of the equivalence classes of all lengths, is a rational function.

Everything above is algorithmic. So, given the parameter k and the size m of the alphabet, we can algorithmically compute a rational generating function giving the numbers of all equivalence classes of words of length n. However, the automata involved are – due to the non-determinism and the complementation – so huge that in practice this can be done only for very small values of the parameters. We illustrate these in a few examples.

Inspired by the connection to automata theory, we study k-switching in connection with regular languages. We show that regular languages are closed under the *k-switching operation.* On the other hand, we show that regular languages are not closed under the transitive closure of this operation. Using the former result, we conclude that the union of k-abelian equivalence classes of size two is regular. On the other hand, it remains open whether this extends, instead of classes of size two, to larger classes. Another open problem is to determine the asymptotic behavior of the complexity function of equivalence classes.

2 Preliminaries and Notation

We recall some notation and basic terminology from the literature of combinatorics on words. We refer the reader to [13] for more on the subject.

The set of finite words over an *alphabet* Σ is denoted by Σ^* and the set of non-empty words is denoted by Σ^+. The empty word is denoted by ε. A set $L \subseteq \Sigma^*$ is called a *language.* We let $|w|$ denote the length of a word $w \in \Sigma^*$. By convention, we set $|\varepsilon| = 0$. The language of words of length n over the alphabet Σ is denoted by Σ^n.

For a word $w = a_1 a_2 \cdots a_n \in \Sigma^*$ and indices $1 \leq i \leq j \leq n$, we let $w[i,j]$ denote the factor $a_i \cdots a_j$. For $i > j$ we set $w[i,j] = \varepsilon$. Similarly, for $i < j$ we let $w[i,j)$ denote the factor $a_i \cdots a_{j-1}$, and we set $w[i,j) = \varepsilon$ when $i \geq j$. We say that a word $x \in \Sigma^*$ *has position i in w* if the word $w[i,|w|]$ has x as a prefix. For $u \in \Sigma^+$ we let $|w|_u$ denote the number of occurrences of u as a factor of w.

Two words $u, v \in \Sigma^*$ are *k-abelian equivalent*, denoted by $u \sim_k v$, if $|u|_x = |v|_x$ for all $x \in \Sigma^+$ with $|x| \leq k$. The relation $\sim_k$ is clearly an equivalence relation; we let $[u]_k$ denote the k-abelian equivalence class defined by u. A word u is called a *k-abelian singleton* if $|[u]_k| = 1$.

In [9], k-abelian equivalence is characterized in terms of rewriting, namely by *k-switching.* For this we define the following. Let $k \geq 1$ and let $u \in \Sigma^*$. Suppose that there exist $x, y \in \Sigma^{k-1}$, not necessarily distinct, and indices i, j, l and m, with $i < j \leq l < m$, such that x has positions i and l in u and y has positions j and m in u. In other words, we have

$$u = u[1, i) \cdot u[i, j) \cdot u[j, l) \cdot u[l, m) \cdot u[m, |u|],$$

where both $u[i, |u|]$ and $u[l, |u|]$ begin with x and both $u[j, |u|]$ and $u[m, |u|]$ begin with y. Furthermore, $u[i, j)$, $u[l, m) \neq \varepsilon$ but we allow $l = j$, in which case $y = x$ and $u[j, l) = \varepsilon$. We define a *k-switching* on u, denoted by $S_{u,k}(i, j, l, m)$, as

$$S_{u,k}(i, j, l, m) = u[1, i) \cdot u[l, m) \cdot u[j, l) \cdot u[i, j) \cdot u[m, |u|]. \tag{1}$$

A k-switching operation is illustrated in Fig. 1.

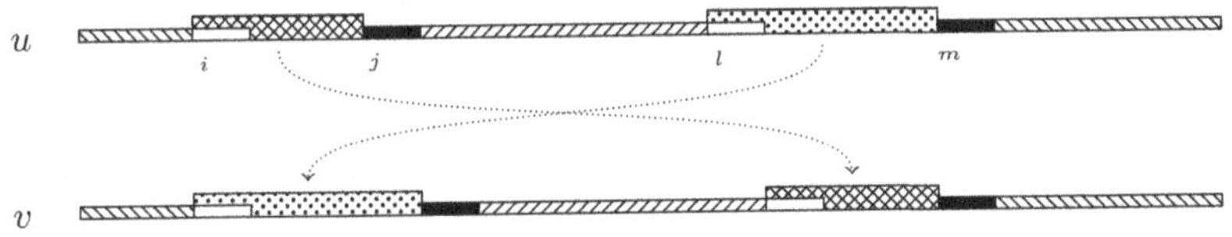

Fig. 1. Illustration of a k-switching. Here $v = S_{k,u}(i, j, l, m)$; the white rectangles symbolize x and the black rectangles symbolize y.

Example 1. Let $u = aababababaaabab$ and $k = 4$. Let then $x = aba$, $y = bab$, $i = 2$, $j = 3$, $l = 4$ and $m = 11$. We then have

$$u = a \cdot a \cdot b \cdot abababaaa \cdot bab$$
$$S_{u,4}(i, j, l, m) = a \cdot abababaaa \cdot b \cdot a \cdot bab.$$

Note here that the occurrences of x are overlapping. With $i = 2$, $j = l = 4$, and $m = 10$ we obtain the same word as above:

$$u = a \cdot ab \cdot abababaa \cdot abab$$
$$S_{u,4}(i, j, j, m) = a \cdot abababaa \cdot ab \cdot abab.$$

In this example we have $j = l$, whence $x = y = aba$ and $u[j, l) = \varepsilon$.

Let us define a relation R_k of Σ^* by uR_kv if and only if v is obtained from u by a k-switching. Now R_k is clearly symmetric, so that the reflexive and transitive closure R_k^* of R_k is an equivalence relation on Σ^*. In [9], k-abelian equivalence is characterized using R_k^*:

Lemma 2. *For $u, v \in \Sigma^*$, we have $u \sim_k v$ if and only if $uR_k^* v$.*

We need a few basic properties of *regular* (or *rational*) languages, such as equivalent definitions of regular languages with various models of finite automata, e.g., nondeterministic finite automata which can read the empty word (ε-NFA), and some basic closure properties of regular languages. We refer to [3] for this knowledge. In addition to classical language theoretical properties, we use the theory of *languages with multiplicities.* This counts how many times a word occurs in a language. This leads to the theory of $\mathbb{N}$-*rational sets.* Using the terminology of [16], a multiset over Σ^* is called $\mathbb{N}$-*rational* if it is obtained from finite multisets by applying finitely many times the rational operations *product, union,* and taking *quasi-inverses*, i.e., *iteration* restricted to ε-free languages. Further, a unary $\mathbb{N}$-rational subset is referred to as an $\mathbb{N}$-*rational sequence.* We refer to [16] for more on this topic. The basic result we need is (see [16]):

Proposition 3. *Let $\mathcal{A}$ be a nondeterministic finite automaton over the alphabet Σ. The function $f_{\mathcal{A}} : \Sigma^* \to \mathbb{N}$ defined as*

$$f_{\mathcal{A}}(w) = \#\ \textit{of accepting paths of } w \textit{ in } \mathcal{A}$$

is $\mathbb{N}$-rational. In particular, the function $\ell_{\mathcal{A}} : \mathbb{N} \to \mathbb{N}$,

$$\ell_{\mathcal{A}}(n) = \#\ \textit{of accepting paths of length } n \textit{ in } \mathcal{A} \tag{2}$$

is an $\mathbb{N}$-rational sequence. Consequently, the generating function *for $\ell_{\mathcal{A}}$ is a rational function.*

3 Properties of k-Switchings

Our starting point for the study of structural properties of k-abelian equivalence classes is the characterization of k-abelian equivalence in terms of k-switchings. We proceed to describe a k-switching operation on languages. We show that this operation preserves regularity. That is, given a regular language L, the language obtained by this operation is also regular. This result will be used later on.

We now describe k-switchings on languages. For a language $L \subset \Sigma^*$, we define the k-switching of L, denoted by $R_k(L)$, as the language

$$R_k(L) = \{w \in \Sigma^* \mid wR_k v \text{ for some } v \in L\}.$$

Similarly, we define $R_k^*(L) = \bigcup_{n \in \mathbb{N}} R_k^n(L) = \bigcup_{w \in L} [w]_k$.

Note that, from a regular language L, it is straightforward to identify all words that admit a k-switching (i.e., the words on the top row of Fig. 1). It is not at all clear that, by performing all possible k-switchings on all words of L (i.e., taking the union of all words on the bottom row of Fig. 1), the obtained language is also regular. We give a direct automata theoretic construction to show this.

Theorem 4. *Let L be a regular language. Then $R_k(L)$ is also regular.*

Proof. For a language L and fixed words $x, y \in \Sigma^{k-1}$, consider the language

$$R_{x,y}(L) = \{w \in \Sigma^* \mid w = S_{k,u}(i,j,l,m) \text{ for some } i < j \leq l < m, u \in L, \text{ with } u[i,i+k-1) = u[l,l+k-1) = x \text{ and } u[j,j+k-1) = u[m,m+k-1) = y\}.$$

We will construct, for a regular language L recognized by a deterministic finite automaton $\mathcal{A} = (Q, \Sigma, \delta, p_{\text{init}}, F)$, an ε-NFA $\hat{\mathcal{A}}$ which recognizes $R_{x,y}(L)$. The claim then follows for $R_k(L)$, as $R_k(L) = \bigcup_{x,y\in\Sigma^{k-1}} R_{x,y}(L)$ is a finite union of regular languages.

In essence, $\hat{\mathcal{A}}$ is a cartesian product of form $\hat{\mathcal{A}} = \mathcal{A}_1 \times \mathcal{A}_x \times \mathcal{A}_y \times \mathcal{A}_x \times \mathcal{A}_y$. The first component automaton $\mathcal{A}_1$ consists of $5|Q|^4$ copies of $\mathcal{A}$, some of which are connected by ε-transitions. The second and fourth components are copies of an automaton $\mathcal{A}_x$ recognizing the language $x\Sigma^*$ and the third and fifth components are copies of an automaton $\mathcal{A}_y$ recognizing the language $y\Sigma^*$. The components 2, 3, 4, and 5 are initiated according to the computations performed in $\mathcal{A}_1$. We shall now make this construction more formal.

We first construct $\mathcal{A}_1 = (Q_1, \Sigma, \delta_1, \tilde{p}_{\text{init}}, F_1)$ as follows. For each state $p \in Q$, we have $p^{(c,(p_1,p_2),(p_3,p_4))} \in Q_1$ for all $c = 1, \dots, 5$ and $p_r \in Q$, $r = 1, \dots, 4$. We also add the initial state $\tilde{p}_{\text{init}}$, from which we have ε-transitions to all the states of form $p_{\text{init}}^{(1,(p_1,p_2),(p_3,p_4))}$, $p_1, p_2, p_3, p_4 \in Q$. Thus the computation of $\mathcal{A}_1$ begins with an ε-transition. We then add the following ε-transitions for all $p_1, p_2, p_3, p_4 \in Q$:

$$p_1^{(1,(p_1,p_2),(p_3,p_4))} \xrightarrow{\varepsilon} p_2^{(2,(p_1,p_2),(p_3,p_4))}, \quad p_3^{(2,(p_1,p_2),(p_3,p_4))} \xrightarrow{\varepsilon} p_4^{(3,(p_1,p_2),(p_3,p_4))},$$
$$p_2^{(3,(p_1,p_2),(p_3,p_4))} \xrightarrow{\varepsilon} p_1^{(4,(p_1,p_2),(p_3,p_4))}, \quad p_4^{(4,(p_1,p_2),(p_3,p_4))} \xrightarrow{\varepsilon} p_3^{(5,(p_1,p_2),(p_3,p_4))}.$$

Otherwise the computation of $\mathcal{A}_1$ respects the original automaton, that is,

$$\delta_1(p^{(i,(p_1,p_2),(p_3,p_4))}, a) = q^{(i,(p_1,p_2),(p_3,p_4))}$$

if and only if there is a transition $\delta(p, a) = q$ in $\mathcal{A}$. Finally, F_1 consists of all states of form $f^{(5,(p_1,p_2),(p_3,p_4))}$, where $f \in F$ and $p_1, p_2, p_3, p_4 \in Q$.

We remark the following about $\mathcal{A}_1$. Firstly, once the first ε-transition is taken, the states p_1, p_2, p_3, and p_4 are fixed for the remainder of the computation. Secondly, the states p_r, $r = 1, \dots, 4$, determine between which states an ε-transition can be performed. Furthermore, the parameter c counts the number of ε-transitions performed. The parameters c, p_1, p_2, p_3, and p_4 together determine at which time and between which states an ε-transition can be performed.

We now describe the behavior of the rest of the component automata of $\hat{\mathcal{A}}$. For $s \in \{2, \dots, 5\}$, the sth component automaton of $\hat{\mathcal{A}}$ is initiated during the sth ε-transition performed in $\mathcal{A}_1$ (the first ε-transition being the first computation step of $\mathcal{A}_1$). We also require from $\hat{\mathcal{A}}$ that, after the second and fourth ε-transition performed in $\mathcal{A}_1$, at least one letter is read before performing the next ε-transition. This is not required after the third ε-transition. Note that these requirements can be encoded, e.g., into the parameter c of the states in $\mathcal{A}_1$. Finally, $\hat{\mathcal{A}}$ accepts if and only if all its components are in accepting states.

We first show that $R_{x,y}(L) \subseteq L(\hat{\mathcal{A}})$. In order to see this, let $u \in L$ and let $v = S_{k,u}(i,j,l,m) \in R_{x,y}(L)$. Let q_t, $t = 1, \ldots, |u|$, denote the state $\delta(p_{\text{init}}, u[1,t))$ (note that some of the states q_t can be the same). We then find an accepting computation of $\mathcal{A}_1$ for v as follows. We first take the ε-transition from $\tilde{p}_{\text{init}}$ to the state $p_{\text{init}}^{(1,(q_i,q_l),(q_j,q_m))}$. After this, the computation is as in Fig. 2 by following the dashed lines. The computation of $\mathcal{A}$ on u follows the continuous lines. Note that the other components of $\hat{\mathcal{A}}$ also end up in accepting states, since by the definition of the k-switching $S_{k,u}(i,j,l,m)$, x and y have positions in v corresponding to the initiations of the copies of the automata $\mathcal{A}_x$ and $\mathcal{A}_y$. Thus $R_{x,y}(L) \subseteq L(\hat{\mathcal{A}})$.

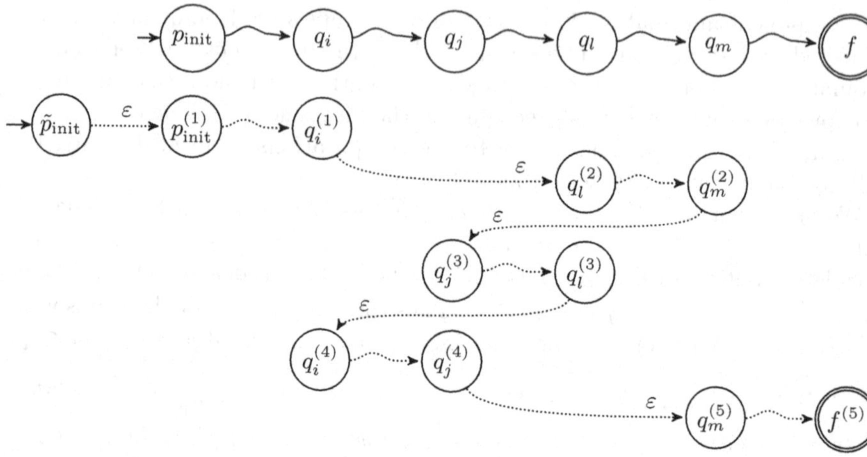

Fig. 2. The computation of automaton $\mathcal{A}$ on an accepted word u (in continuous lines) and a computation of $\mathcal{A}_1$ on $S_{k,u}(i,j,l,m)$ (in dotted lines). We have abbreviated the states $q_r^{(c,(q_i,q_l),(q_j,q_m))}$ by $q_r^{(c)}$ (for $c \in \{1, \ldots, 5\}$, $r \in \{\text{init}, i, j, l, m\}$).

We now show the converse. For this, let $v \in L(\hat{\mathcal{A}})$ and consider an accepting path of $\hat{\mathcal{A}}$ on v. By construction, the automaton $\mathcal{A}_1$ starts with an ε-transition to a state $p_{\text{init}}^{(1,(p_1,p_2),(p_3,p_4))}$. After this, the computation contains four more ε-transitions, suppose they occur just before reading the ith, jth, lth and mth letter, with $i < j \leq l < m$, respectively. (Here we use the requirement for not allowing an ε-transition immediately after the second and fourth ε-transitions.) Furthermore, by the acceptance of the other component automata of $\hat{\mathcal{A}}$, x has positions i and l, and y has positions j and m in v. We claim that $u = S_{k,v}(i,j,l,m) \in L$. It then follows, by the symmetry of the k-switching relation, that $v \in R_{x,y}(L)$. Indeed, turning back to the computation of $\mathcal{A}_1$ on v, we obtain the following paths in $\mathcal{A}$:

1. a path from p_{init} to p_1 labeled by $v[1,i)$,
2. a path from p_2 to p_3 labeled by $v[i,j)$,
3. a path from p_4 to p_2 labeled by $v[j,l)$,

4. a path from p_1 to p_4 labeled by $v[l, m)$, and
5. a path from p_3 to an accepting state of $\mathcal{A}$ labeled by $v[m, |v|]$.

Thus $u = v[1, i)v[l, m)v[j, l)v[i, j)v[m, |v|] \in L$, as was claimed. □

Remark 5. This result may also be proved using MSO logic for words, as suggested by one of the anonymous referees.

The following example shows that the family of regular languages is not closed under the language operation R_k^*.

Example 6. Fix $k \geq 1$ and let $L = (ab^k)^+$. It is straightforward to verify by, e.g., comparing the number of occurrences of factors of length k, that

$$R_k^*(L) = \left\{ ab^{r_1}ab^{r_2}\cdots ab^{r_n} \mid n \geq 1, r_i \geq k-1, \sum_{i=1}^{n} r_i = nk \right\}.$$

Let now h be a morphism defined by $h(a) = ab^{k-1}$ and $h(b) = b$. It is again straightforward to show that $h^{-1}(R_k^*(L)) = \{w \in a\{a,b\}^* \mid |w|_a = |w|_b\}$, which is clearly not regular. It follows that $R_k^*(L)$ is not regular.

4 On the Number of k-Abelian Equivalence Classes

In this section we focus on the number $\mathcal{P}_{k,m}(n)$ of k-abelian equivalence classes of words of length n over Σ, $|\Sigma| = m$, where k and an m are fixed. We first recall a result from [11]:

Theorem 7. *We have, for k and m fixed, $\mathcal{P}_{k,m}(n) = \Theta(n^{m^{k-1}(m-1)})$, where the constants in Θ depend on k and m.*

We are also interested in the number $\mathcal{S}_{k,m}(n)$ of k-abelian singletons of length n over Σ, $|\Sigma| = m$, where k and an m are fixed. We recall a result proved in [9].

Theorem 8. *For k and m fixed, we have $\mathcal{S}_{k,m}(n) = \mathcal{O}(n^{N_m(k-1)-1})$, where the constants in $\mathcal{O}$ depend on k and m. Here $N_m(l) = \frac{1}{l}\sum_{d|l}\varphi(d)m^{l/d}$ is the number of conjugacy classes (or necklaces) of words in Σ^l, where $|\Sigma| = m$.*

The main result of this section is the following:

Theorem 9. *The sequences $\mathcal{P}_{k,m}(n)$ and $\mathcal{S}_{k,m}(n)$ are $\mathbb{N}$-rational.*

In order to prove this, we define the following languages. Here $\leq$ denotes a lexicographic ordering of Σ^*.

$$L_{\text{min}} = \{w \in \Sigma^* \mid w \leq u \text{ for all } w \sim_k u\},$$
$$L_{\text{max}} = \{w \in \Sigma^* \mid w \geq u \text{ for all } w \sim_k u\}, \text{ and}$$
$$L_{\text{sing}} = \{w \in \Sigma^* \mid |[w]_k| = 1\}.$$

In other words, L_{min} (resp., L_{max}) is the language of lexicographically minimal (resp., maximal) representatives of k-abelian equivalence classes, while L_{sing} is the language of k-abelian singletons. We also recall a technical lemma from [9], a refinement of Lemma 2.

Lemma 10. *Let $u \sim_k v$ with $u \neq v$. Let p be the longest common prefix of u and v. Then there exists $z \in \Sigma^*$ such that $zR_k u$ and the longest common prefix of z and v has length at least $|p| + 1$.*

Lemma 11. *The languages $L_{\min}$, $L_{\max}$, and L_{sing} are regular languages.*

Proof. Let u be the minimal element in $[u]_k$. If there exists a k-switching on u which yields a new element, it has to be lexicographically greater than u. In particular, u does not contain factors from the language

$$((xb\Sigma^* \cap \Sigma^* y)\, \Sigma^* \cap \Sigma^* x)\, a\Sigma^* \cap \Sigma^* y,$$

where $x, y \in \Sigma^{k-1}$, $a, b \in \Sigma$, $a < b$. On the other hand, by the above lemma, any word u avoiding such factors is lexicographically least in $[u]_k$. We thus have

$$L_{\min} = \bigcap_{\substack{x,y \in \Sigma^{k-1} \\ a,b \in \Sigma,\ a<b}} \overline{\Sigma^* \left(((xb\Sigma^* \cap \Sigma^* y)\, \Sigma^* \cap \Sigma^* x)\, a\Sigma^* \cap \Sigma^* y\right) \Sigma^*}, \tag{3}$$

where, for a regular expression R, $\overline{R}$ denotes the *complement* language $\Sigma^* \backslash R$.

Similarly, for $L_{\max}$, by reversing $a < b$ to $a > b$ in (3), we obtain the claim.

Finally, $L_{\text{sing}} = L_{\min} \cap L_{\max}$ so that L_{sing} is regular. Another, perhaps more informative, way to see this is as follows: for k-abelian singletons, we are avoiding all possible k-switchings that give a different word. By requiring $a \neq b$, as opposed to $a < b$, in (3), we obtain the expression for L_{sing}. □

Proof (of Theorem 9). Consider first the language $L_{\min}$ and a DFA $\mathcal{A}$ recognizing it. We transform the automaton to a unary NFA $\mathcal{A}'$ by identifying all input letters. Since $\mathcal{A}$ is deterministic, the transformation is *faithful*, that is, for each word w accepted by $\mathcal{A}$, there exists a unique corresponding accepting path in $\mathcal{A}'$, and vice versa. By the construction of $\mathcal{A}'$, $\ell_{\mathcal{A}'}(n) = \mathcal{P}_{k,m}(n)$ for all $n \in \mathbb{N}$, from which the claim follows for $\mathcal{P}_{k,m}$. The case for $\mathcal{S}_{k,m}$ is similar. □

Remark 12. Let A be the adjacency matrix of the unary automaton $\mathcal{A}'$ described above. It is known that, for all large enough n,

$$\ell_{\mathcal{A}'}(n) = \sum_{\lambda \in Eig(A)} p_\lambda(n) \lambda^n \tag{4}$$

where the summation is taken over all distinct eigenvalues of A, and p_λ is a complex polynomial of degree at most $\mu_\lambda - 1$. Here μ_λ is the multiplicity of λ as a root of the *minimal polynomial* of A (see for instance [3,17]).

4.1 Complexities for Small Values of k and m

We now give some examples illustrating the results obtained above for small values of k and m. We also compute closed formulas for $\mathcal{P}_{k,m}$ and $\mathcal{S}_{k,m}$ for some small values of k and m.

Example 13. In Fig. 3, we have two minimal DFAs, one recognizing the minimal representatives of 2-abelian equivalence classes and the other recognizing 2-abelian singletons over $\Sigma = \{a, b\}$. The sink states are not included in the figures. We also note that all other states are accepting, since the languages are defined by avoiding certain patterns.

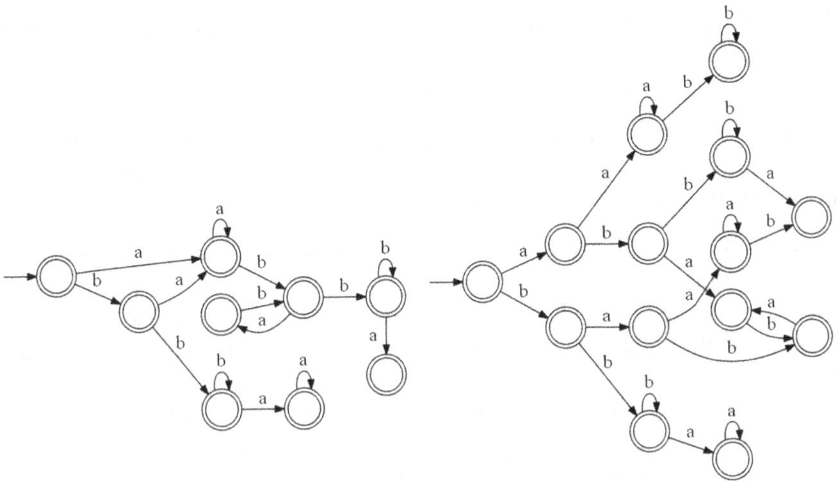

Fig. 3. DFAs recognizing the minimal representatives of 2-abelian equivalence classes (left) and 2-abelian singletons (right) over the alphabet $\{a, b\}$.

Using the idea of the proof of Theorem 9, we first construct deterministic automata for $L_{\min}$ and L_{sing} for small k and m. We then use the automata to compute the function ℓ as in Remark 12. We state these conclusions without proofs:

Proposition 14.

For all $n \geq 1$, $\mathcal{P}_{2,2}(n) = n^2 - n + 2$,

for all $n \geq 2$, $\mathcal{P}_{2,3}(n) = \frac{1}{18}n^4 - \frac{5}{18}n^3 + \frac{65}{36}n^2 - \frac{23}{6}n - \frac{1}{8}(-1)^n + \frac{2}{27}e^{-\frac{\pi i}{3}}(e^{\frac{2\pi i}{3}})^n + \frac{2}{27}e^{\frac{\pi i}{3}}(e^{-\frac{2\pi i}{3}})^n + \frac{1307}{216}$, *and*

for all $n \geq 4$, $\mathcal{P}_{3,2}(n) = \frac{1}{960}n^6 + \frac{7}{320}n^5 + \frac{67}{384}n^4 - \frac{19}{32}n^3 + \frac{1457}{480}n^2 - (\frac{1569}{640} + \frac{3}{128}(-1)^n)n + \frac{741}{256} + \frac{27}{256}(-1)^n$.

Proposition 15.

For all $n \geq 4$, $\mathcal{S}_{2,2}(n) = 2n + 4$,

for all $n \geq 6$, $\mathcal{S}_{2,3}(n) = 3n^2 + 27n - 63$, *and*

for all $n \geq 9$, $\mathcal{S}_{3,2}(n) = \frac{1}{2}n^2 + 16n + \frac{2}{3}(e^{\frac{2\pi i}{3}n} + (e^{-\frac{2\pi i}{3}})^n) - \frac{535}{12} - \frac{3}{4}(-1)^n$.

The formulae for $\mathcal{P}_{2,2}$ and $\mathcal{S}_{2,2}$ have previously been proved, using different methods, in [5,9], respectively. We note that Eero Harmaala (private communication) has previously computed the values for $\mathcal{P}_{2,3}$ and $\mathcal{P}_{3,2}$ ($n = 2, \ldots, 18$ and $n = 4, \ldots, 21$, respectively). We also note that computing the first few values of $S_{2,3}(n)$ and $S_{3,2}(n)$ is an easy task. The *On-Line Encyclopedia of Integer Sequences* (http://oeis.org, accessed June 10, 2016) does not contain any of the above sequences.

The methods used here are far from being practical for computing closed formulae for larger values of k and m, as is illustrated by the following example.

Example 16. For the binary alphabet, the number of states in the minimal DFA recognizing $L_{\min}$ for $k = 2, 3, 4$ is $10, 49$, and 936, respectively. This makes computing a closed formula for $\mathcal{P}_{4,2}$ already a computationally challenging problem.

Remark 17. The exponential blow-up of the computation time is due to complementation and non-determinism of the automata obtained from the regular expressions (3). Also, by Theorem 7, the automaton obtained from (3) has to grow necessarily exponentially with respect to k when the alphabet is fixed; some of the polynomials p_λ in (4) have degree $m^{k-1}(m-1)$.

For the case of k-abelian singletons, Theorem 8 does not give a large blow-up immediately, though in [9] it is conjectured that $\mathcal{S}_{k,m}(n) = \Theta(n^{N_m(k-1)-1})$, which would also yield a large blow-up in the number of states.

5 Towards a Structure of Fixed Sized Equivalence Classes

The regularity of the languages $L_{\min}$ and L_{sing} raises questions for the structure of larger equivalence classes. We are thus interested in the k-abelian equivalence classes of fixed cardinality. We employ the result of Theorem 4 to obtain a first step in this direction.

Proposition 18. *The language* $L_2 = \{w \in \Sigma^* \mid |[w]_k| = 2\}$ *is a regular language.*

Proof. Consider the regular language $L = \Sigma^* \backslash (L_{\min} \cup L_{\max})$: we have

$$L = \{w \in \Sigma^* \mid |[w]_k| \geq 3 \text{ and } w \text{ is not minimal or maximal}\},$$

since all classes containing at most two elements are removed. By Lemma 2, $R_k(R_k(L)) \cup R_k(L) \cup L$ then gives exactly the language

$$L' = \{w \in \Sigma^* \mid |[w]_k| \geq 3\},$$

and by Lemma 2, L' is regular. Finally, the complement of L' is the language $\{w \in \Sigma^* \mid |[w]_k| \leq 2\}$. We thus have that $L_2 = \overline{L'} \backslash L_{\text{sing}}$ is a regular language. □

Larger classes were not considered here, but we have no reason to suspect that the corresponding languages would not be regular. In fact, we suspect that modifications of Theorem 4 could yield methods, similar to the ones used in the above, to obtain some structure of larger classes.

6 Open Problems and Future Research

The topic of this paper opens up new aspects of k-abelian equivalence, and presents a series of questions. Though explicit formulas for the functions $\mathcal{P}_{k,m}$ and $\mathcal{S}_{k,m}$ were obtained, it remains to compute the corresponding generating functions (which, by our results, are rational functions).

To conclude, we suggest the following open problems.

- What are the generating functions for $\mathcal{P}_{k,m}$ and $\mathcal{S}_{k,m}$?
- When is $\mathcal{P}_{k,m}(n) \sim Cn^{m^{k-1}(m-1)}$ for some constant C? This is the case for small values of k and m.
- Is the language of words w having $|[w]_k| = l$, where l is a fixed constant, a regular language? For $l = 2$, this is settled in the positive by Proposition 18.

Acknowledgments. The automata used to calculate the functions in Propositions 14 and 15 were constructed using the java package `dk.brics.automaton` [14]. The automata in Fig. 3 were created using the software Graphviz [4]. We would like to thank the anonymous referees for valuable comments which helped to improve the presentation.

References

1. Cassaigne, J., Karhumäki, J., Saarela, A.: On growth and fluctuation of k-Abelian complexity. In: 10th International Computer Science Symposium Computer Science - Theory and Applications, CSR 2015, Proceedings, Listvyanka, Russia, 13–17 July 2015, pp. 109–122 (2015). http://dx.doi.org/10.1007/978-3-319-20297-6_8
2. Ehlers, T., Manea, F., Mercas, R., Nowotka, D.: k-Abelian pattern matching. J. Discrete Algorithms **34**, 37–48 (2015). http://dx.doi.org/10.1016/j.jda.2015.05.004
3. Eilenberg, S.: Automata, Languages, and Machines, vol. A. Academic Press Inc., New York (1974)
4. Gansner, E.R., North, S.C.: An open graph visualization system and its applications to software engineering. Softw. Prac. Experience **30**(11), 1203–1233 (2000). http://www.graphviz.org
5. Huova, M., Karhumäki, J., Saarela, A., Saari, K.: Local squares, periodicity and finite automata. In: Rainbow of Computer Science - Dedicated to Hermann Maurer on the Occasion of His 70th Birthday, pp. 90–101 (2011). http://dx.doi.org/10.1007/978-3-642-19391-0_7
6. Huova, M., Saarela, A.: Strongly k-Abelian repetitions. In: 9th International Conference on Combinatoricson Words, WORDS 2013, Turku, Finland, Proceedings, pp. 161–168, 19–20 September 2013. http://dx.doi.org/10.1007/978-3-642-40579-2_18
7. Karhumäki, J.: Generalized Parikh mappings and homomorphisms. Inf. control **47**(3), 155–165 (1980). http://dx.doi.org/10.1016/S0019-9958(80)90493-3
8. Karhumäki, J., Puzynina, S.: On k-Abelian palindromic rich and poor words. In: 18th International Conference on Developments in Language Theory, DLT 2014, Proceedings, Ekaterinburg, Russia, 26–29 August 2014, pp. 191–202 (2014). http://dx.doi.org/10.1007/978-3-319-09698-8_17

9. Karhumäki, J., Puzynina, S., Rao, M., Whiteland, M.A.: On cardinalities of k-Abelian equivalence classes. Theor. Comput. Sci. (2016). doi:10.1016/j.tcs.2016.06.010
10. Karhumäki, J., Puzynina, S., Saarela, A.: Fine and Wilf's theorem for k-Abelian periods. Int. J. Found. Comput. Sci. **24**(7), 1135–1152 (2013). http://dx.doi.org/10.1142/S0129054113400352
11. Karhumäki, J., Saarela, A., Zamboni, L.Q.: On a generalization of Abelian equivalence and complexity of infinite words. J. Comb. Theor. Ser. A **120**(8), 2189–2206 (2013). http://dx.doi.org/10.1016/j.jcta.2013.08.008
12. Karhumäki, J., Saarela, A., Zamboni, L.Q.: Variations of the Morse-Hedlund theorem for k-Abelian equivalence. In: 18th International Conference on Developments in Language Theory, DLT 2014, Proceedings, Ekaterinburg, Russia, 26–29 August 2014, pp. 203–214 (2014). http://dx.doi.org/10.1007/978-3-319-09698-8_18
13. Lothaire, M. (ed.): Combinatorics on Words, 2nd edn. Cambridge University Press, Cambridge (1997). http://dx.doi.org/10.1017/CBO9780511566097, Cambridge Books Online
14. Møller, A.: dk.brics.automaton - finite-state automata and regular expressions for Java (2010). http://www.brics.dk/automaton/
15. Rao, M., Rosenfeld, M.: Avoidability of long k-abelian repetitions. Mathematics of Computation (published electronically, 18 February 2016). http://dx.doi.org/10.1090/mcom/3085
16. Salomaa, A., Soittola, M.: Automata-Theoretic Aspects of Formal Power Series. Texts and Monographs in Computer Science. Springer, New York (1978). http://dx.doi.org/10.1007/978-1-4612-6264-0
17. Weintraub, S.H.: Jordan canonical form: theory and practice. In: Synthesis Lectures on Mathematics and Statistics, Morgan & Claypool Publishers (2009). http://dx.doi.org/10.2200/S00218ED1V01Y200908MAS006

Schützenberger Products in a Category

Liang-Ting Chen[1] and Henning Urbat[2(✉)]

[1] Department of Information and Computer Sciences, University of Hawaii at Manoa, Honolulu, HI, USA
ltchen@hawaii.edu

[2] Institut Für Theoretische Informatik, Technische Universität Braunschweig, Braunschweig, Germany
urbat@iti.cs.tu-bs.de

Abstract. The Schützenberger product of monoids is a key tool for the algebraic treatment of language concatenation. In this paper we generalize the Schützenberger product to the level of monoids in an algebraic category $\mathscr{D}$, leading to a uniform view of the corresponding constructions for monoids (Schützenberger), ordered monoids (Pin), idempotent semirings (Klíma and Polák), and algebras over a field (Reutenauer). In addition, assuming that $\mathscr{D}$ is part of a Stone-type duality, we derive a characterization of the languages recognized by Schützenberger products.

1 Introduction

Since the early days of automata theory, it has been known that regular languages are precisely the languages recognized by finite monoids. This observation is the origin of algebraic language theory. One of the classical and ongoing challenges of this theory is the algebraic treatment of the concatenation of languages. The most important tool for this purpose is the *Schützenberger product* $M \diamond N$ of two monoids M and N, introduced in [23]. Its key property is that it recognizes all marked products of languages recognized by M and N. Later, Reutenauer [22] showed that $M \diamond N$ is a "smallest" monoid with this property: any language recognized by $M \diamond N$ is a boolean combination of such marked products.

In the past decades, the original notion of language recognition by finite monoids has been refined to other algebraic structures, namely to ordered monoids by Pin [16], to idempotent semirings by Polák [19], and to associative algebras over a field by Reutenauer [21]. For all these structures, a Schützenberger product was introduced separately [15,17,21]. Moreover, Reutenauer's characterization of the languages recognized by Schützenberger products has been adapted to ordered monoids and idempotent semirings, replacing boolean combinations by positive boolean combinations [17] and finite unions [15], respectively.

L.-T. Chen—Acknowledges support from AFOSR.
H. Urbat—Acknowledges support from DFG under project AD 187/2-1.

S. Brlek and C. Reutenauer (Eds.): DLT 2016, LNCS 9840, pp. 89–101, 2016.
DOI: 10.1007/978-3-662-53132-7_8

This paper presents a unifying approach to Schützenberger products, covering the aforementioned constructions and results as special cases. Our starting point is the observation that all the algebraic structures appearing above (monoids, ordered monoids, idempotent semirings, and algebras over a field $\mathbb{K}$) are *monoids* interpreted in some variety $\mathscr{D}$ of algebras or ordered algebras, viz. $\mathscr{D}$ = sets, posets, semilattices, and $\mathbb{K}$-vector spaces, respectively. Next, we note that these categories $\mathscr{D}$ are related to the category $\mathbb{S}$-**Mod** of modules over some semiring $\mathbb{S}$. Indeed, semilattices and vector spaces are precisely modules over the two-element idempotent semiring $\mathbb{S} = \{0, 1\}$ and the field $\mathbb{S} = \mathbb{K}$, respectively. And every set or poset freely *generates* a semilattice (i.e. a module over $\{0, 1\}$), viz. the semilattice of finite subsets or finitely generated down-sets. Precisely speaking, each of the above categories $\mathscr{D}$ admits a *monoidal adjunction*

$$\mathbb{S}\text{-}\mathbf{Mod} \underset{F}{\overset{U}{\rightleftarrows}} \mathscr{D} \qquad (1.1)$$

for some semiring $\mathbb{S}$, where U is a forgetful functor and F is a free construction.

In this paper we introduce the Schützenberger product at the level of an abstract monoidal adjunction (1.1): for any two $\mathscr{D}$-monoids M and N, we construct a $\mathscr{D}$-monoid $M \diamond N$ that recognizes all marked products of languages recognized by M and N (Theorem 32), and prove that $M \diamond N$ is a "smallest" $\mathscr{D}$-monoid with this property (Theorem 37). Further, we derive a characterization of the languages recognized by $M \diamond N$ in the spirit of Reutenauer's theorem [22]. To this end, we consider another variety $\mathscr{C}$ that is *dual* to $\mathscr{D}$ on the level of finite algebras. For example, for $\mathscr{D}$ = sets we choose $\mathscr{C}$ = boolean algebras, since Stone's representation theorem gives a dual equivalence between finite boolean algebras and finite sets. We then prove that every language recognized by $M \diamond N$ is a "$\mathscr{C}$-algebraic combination" of languages recognized by M and N and their marked products (Theorem 40). The explicit use of duality makes our proof conceptually different from the original ones.

By instantiating (1.1) to the proper adjunctions, we recover the Schützenberger product for monoids, ordered monoids, idempotent semirings and algebras over a field, and obtain a new Schützenberger product for algebras over a commutative semiring. Moreover, our Theorems 32 and 40 specialize to the corresponding results [15,17,22] for (ordered) monoids and idempotent semirings. In the case of $\mathbb{K}$-algebras, Theorem 40 appears to be a new result. Apart from that, we believe that the main contribution of our paper is the identification of a categorical setting for language concatenation. We hope that the generality and the conceptual nature of our approach can contribute to an improved understanding of the various ad hoc constructions and separate results appearing in the literature.

Related work. In recent years, categorical approaches to algebraic language theory have been a growing research topic. The present paper is a natural continuation of [2], where we showed that the construction of syntactic monoids works at the level of closed monoidal categories (see also [13]), allowing for a uniform treatment of syntactic (ordered) monoids, idempotent semirings and algebras over a field. The systematic use of duality in algebraic language theory originates in the work of Gehrke et al. [11], who interpreted Eilenberg's variety theorem in terms of Stone duality. In [1,3,9] we extended their approach to an abstract Stone-type duality, leading to a uniform view of Eilenberg-type theorems for regular languages. See also [4,24]. Recently, Bojańczyk [6] proposed to use *monads* instead of monoids to get a categorical grasp on languages beyond finite words. By combining this idea with our duality framework, we established in [8,25] a variety theorem that covers most Eilenberg-type correspondences known in the literature, e.g. for ∞-languages, tree languages, and cost functions.

2 Preliminaries

A variety $\mathscr{D}$ of algebras or ordered algebras is *commutative* [10] if, for any $A, B \in \mathscr{D}$, the set $[A, B]$ of morphisms from A to B is an algebra of $\mathscr{D}$ with operations (and order) taken pointwise in B. Examples include **Set** (sets), **Pos** (posets) and $\mathbb{S}$-**Mod** (modules over a commutative semiring $\mathbb{S}$ with $0, 1$). Recall that an $\mathbb{S}$-module is a commutative monoid $(M, +, 0)$ with a scalar product $\cdot : \mathbb{S} \times M \to M$ satisfying $(r + s)x = rx + sx$, $r(x + y) = rx + ry$, $(rs)x = r(sx)$, $0x = 0$, $1x = 1$ and $r0 = 0$. Two special cases are the category **JSL** of join-semilattices with 0 (choose $\mathbb{S} = \{0, 1\}$, the two-element semiring with $1 + 1 = 1$), and the category $\mathbb{K}$-**Vec** of vector spaces over a field $\mathbb{K}$ (choose $\mathbb{S} = \mathbb{K}$).

Notation 1. Let $\mathscr{A}$, $\mathscr{B}$, $\mathscr{C}$, $\mathscr{D}$ always denote commutative varieties of algebras or ordered algebras. We write $\Psi = \Psi_{\mathscr{D}}$: **Set** $\to \mathscr{D}$ for the left adjoint to the forgetful functor $|-|: \mathscr{D} \to$ **Set**; thus ΨX is the free algebra of $\mathscr{D}$ over a set X. For simplicity, we assume that X is a subset of $|\Psi X|$ and the universal map $X \rightarrowtail |\Psi X|$ is the inclusion. Denote by $\mathbf{1}_{\mathscr{D}} = \Psi 1$ the free one-generated algebra.

Example 2. (1) For $\mathscr{D}$ = **Set** or **Pos** we have $\Psi X = X$ (discretely ordered).
(2) For $\mathscr{D}$ = **JSL** we get $\Psi X = (\mathcal{P}_f X, \cup)$, the semilattice of finite subsets of X.
(3) For $\mathscr{D} = \mathbb{S}$-**Mod** we have $\Psi X = \mathbb{S}^{(X)}$, the $\mathbb{S}$-module of all finite-support functions $X \to \mathbb{S}$ with sum and scalar product defined pointwise.

Definition 3. Let A, B, $C \in \mathscr{D}$. By a *bimorphism* from A, B to C is meant a function $f: |A| \times |B| \to |C|$ such that the maps $f(a, -): |B| \to |C|$ and $f(-, b): |A| \to |C|$ carry morphisms of $\mathscr{D}$ for every $a \in |A|$ and $b \in |B|$. A *tensor product* of A and B is a universal bimorphism $t_{A,B}: |A| \times |B| \to |A \otimes B|$, in the sense that for any bimorphism $f: |A| \times |B| \to |C|$ there is a unique $f': A \otimes B \to C$ in $\mathscr{D}$ with $f' \circ t_{A,B} = f$. We denote by $a \otimes b$ the element $t_{A,B}(a, b) \in |A \otimes B|$.

Example 4. In **Set** and **Pos** we have $A \otimes B = A \times B$. In $\mathbb{S}$-**Mod**, $A \otimes B$ is the usual tensor product of $\mathbb{S}$-modules, and $t_{A,B}$ is the universal $\mathbb{S}$-bilinear map.

Remark 5. (1) Tensor products exist in any commutative variety $\mathscr{D}$, see [10].
(2) $\otimes$ is associative and commutative and has unit $\mathbf{1}_{\mathscr{D}}$, i.e. there are natural isomorphisms $\alpha_{A,B,C}: (A \otimes B) \otimes C \cong A \otimes (B \otimes C)$, $\sigma_{A,B}: A \otimes B \cong B \otimes A$, $\rho_A: A \otimes \mathbf{1}_{\mathscr{D}}$ and $\lambda_A: \mathbf{1}_{\mathscr{D}} \otimes A \cong A$.

(3) Given $f\colon A \to C$ and $g\colon B \to D$ in $\mathscr{D}$, denote by $f \otimes g\colon A \otimes B \to C \otimes D$ the morphism induced by the bimorphism $|A| \times |B| \xrightarrow{f \times g} |C| \times |D| \xrightarrow{t_{C,D}} |C \otimes D|$.

Definition 6. A $\mathscr{D}$-monoid $(M, 1, \bullet)$ consists of an object M of $\mathscr{D}$ and a monoid $(|M|, 1, \bullet)$ whose multiplication $|M| \times |M| \xrightarrow{\bullet} |M|$ is a bimorphism of $\mathscr{D}$. A morphism $h\colon (M, 1, \bullet) \to (N, 1, \bullet)$ of $\mathscr{D}$-monoids is a morphism of $\mathscr{D}$ preserving the unit and multiplication. We denote the category of $\mathscr{D}$-monoids by $\mathbf{Mon}(\mathscr{D})$.

Example 7. Monoids in $\mathscr{D} = \mathbf{Set}, \mathbf{Pos}, \mathbf{JSL}$ and $\mathbb{S}$-$\mathbf{Mod}$ are precisely monoids, ordered monoids, idempotent semirings, and associative algebras over $\mathbb{S}$.

Proposition 8 (see[1]). *The free $\mathscr{D}$-monoid on a set Σ is carried by $\Psi\Sigma^* \in \mathscr{D}$, the free algebra in $\mathscr{D}$ on the set Σ^* of finite words over Σ. Its multiplication extends the concatenation of words in Σ^*, and its unit is the empty word ε.*

Example 9. (1) In $\mathscr{D} = \mathbf{Set}$ or $\mathbf{Pos}$ we have $\Psi\Sigma^* = \Sigma^*$ (discretely ordered).
(2) In $\mathscr{D} = \mathbf{JSL}$ we have $\Psi\Sigma^* = \mathcal{P}_f \Sigma^*$, the idempotent semiring of all finite languages over Σ w.r.t. union and concatenation of languages.
(3) In $\mathscr{D} = \mathbf{Mod}(\mathbb{S})$ we get $\Psi\Sigma^* = \mathbb{S}[\Sigma]$, the $\mathbb{S}$-algebra of all polynomials $\Sigma_{i=1}^n c(w_i) w_i$ (equivalently, finite-support functions $c\colon \Sigma^* \to \mathbb{S}$) w.r.t. the usual sum, scalar product and multiplication of polynomials.

Remark 10. Since the multiplication $\bullet\colon |M| \times |M| \to |M|$ of a $\mathscr{D}$-monoid $(M, 1, \bullet)$ forms a bimorphism, it corresponds to a morphism $\mu_M\colon M \otimes M \to M$ in $\mathscr{D}$, mapping $m \otimes m' \in |M \otimes M|$ to $m \bullet m' \in |M|$. Likewise, the unit $1 \in |M|$ corresponds to the morphism $\iota_M\colon 1_{\mathscr{D}} \to M$ sending the generator of $1_{\mathscr{D}}$ to 1. We can thus represent a $\mathscr{D}$-monoid $(M, 1, \bullet)$ as the triple (M, ι_M, μ_M).

Remark 11. For any two $\mathscr{D}$-monoids M and N, the tensor product $M \otimes N$ in $\mathscr{D}$ carries a $\mathscr{D}$-monoid structure with unit $1_{\mathscr{D}} \xrightarrow{\cong} 1_{\mathscr{D}} \otimes 1_{\mathscr{D}} \xrightarrow{\iota_M \otimes \iota_N} M \otimes N$ and multiplication $(M \otimes N) \otimes (M \otimes N) \xrightarrow{\cong} (M \otimes M) \otimes (N \otimes N) \xrightarrow{\mu_M \otimes \mu_N} M \otimes N$, see e.g. [20]. Equivalently, the unit of $M \otimes N$ is the element $1_M \otimes 1_N$, and the multiplication is determined by $(m \otimes n) \bullet (m' \otimes n') = (m \bullet_M m') \otimes (n \bullet_N n')$.

Definition 12. A *monoidal functor* $(G, \theta)\colon \mathscr{C} \to \mathscr{D}$ is a functor $G\colon \mathscr{C} \to \mathscr{D}$ with a morphism $\theta_1\colon 1_{\mathscr{D}} \to G1_{\mathscr{C}}$ and morphisms $\theta_{A,B}\colon GA \otimes GB \to G(A \otimes B)$ natural in $A, B \in \mathscr{C}$ such that the following squares commute (omitting indices):

$$
\begin{array}{ccc}
(GA \otimes GB) \otimes GC & \xrightarrow{\alpha} & GA \otimes (GB \otimes GC) \\
{\scriptstyle \theta \otimes GC} \downarrow & & \downarrow {\scriptstyle GA \otimes \theta} \\
G(A \otimes B) \otimes GC & & GA \otimes G(B \otimes C) \\
{\scriptstyle \theta} \downarrow & & \downarrow {\scriptstyle \theta} \\
G((A \otimes B) \otimes C) & \xrightarrow{G\alpha} & G(A \otimes (B \otimes C))
\end{array}
$$

$$
\begin{array}{ccc}
GA \otimes 1_{\mathscr{D}} & \xrightarrow{GA \otimes \theta} & GA \otimes G1_{\mathscr{C}} \\
{\scriptstyle \rho} \downarrow & & \downarrow {\scriptstyle \theta} \\
GA & \xleftarrow{G\rho} & G(A \otimes 1_{\mathscr{C}})
\end{array}
$$

$$
\begin{array}{ccc}
1_{\mathscr{D}} \otimes GA & \xrightarrow{\theta \otimes GA} & G1_{\mathscr{C}} \otimes GA \\
{\scriptstyle \lambda} \downarrow & & \downarrow {\scriptstyle \theta} \\
GA & \xleftarrow{G\lambda} & G(1_{\mathscr{C}} \otimes A)
\end{array}
$$

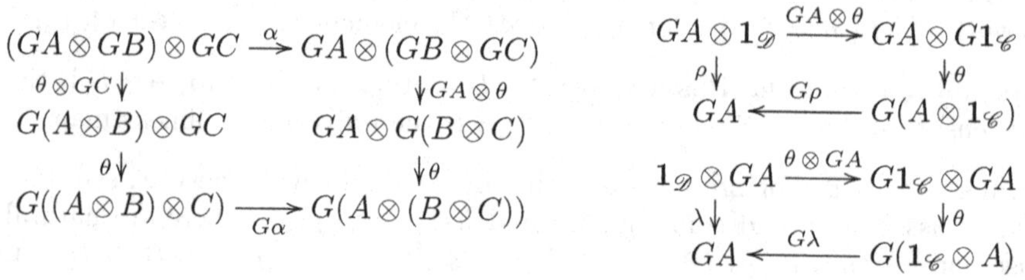

Given another monoidal functor $(G', \theta')\colon \mathscr{C} \to \mathscr{D}$, a natural transformation $\varphi\colon G \to G'$ is called *monoidal* if the following diagrams commute:

$$\begin{array}{ccc} GA \otimes GB & \xrightarrow{\varphi_A \otimes \varphi_B} & G'A \otimes G'B \\ \theta \downarrow & & \downarrow \theta' \\ G(A \otimes B) & \xrightarrow[\varphi_{A \otimes B}]{} & G'(A \otimes B) \end{array} \qquad \begin{array}{ccc} & \mathbf{1}_{\mathscr{D}} & \\ \theta \swarrow & & \searrow \theta' \\ G\mathbf{1}_{\mathscr{C}} & \xrightarrow[\varphi_{\mathbf{1}_{\mathscr{C}}}]{} & G'\mathbf{1}_{\mathscr{C}} \end{array}$$

Example 13. (1) The functor $|-|\colon \mathscr{D} \to \mathbf{Set}$ is monoidal w.r.t. the universal map $1 \rightarrowtail |\mathbf{1}_{\mathscr{D}}| = |\Psi 1|$ and the bimorphisms $t_{A,B}\colon |A| \times |B| \to |A \otimes B|$. Its left adjoint $\Psi\colon \mathbf{Set} \to \mathscr{D}$ is also monoidal: there is a natural *isomorphism* $\theta_{X,Y}\colon \Psi X \otimes \Psi Y \cong \Psi(X \times Y)$ with $\theta_{X,Y}^{-1}(x,y) = x \otimes y$ for $(x,y) \in X \times Y$. Together with $\theta_1 = id\colon \mathbf{1}_{\mathscr{D}} \to \Psi 1$, this makes Ψ a monoidal functor.

(2) In particular, the functors $|-|\colon \mathbf{JSL} \to \mathbf{Set}$ and $\mathcal{P}_f\colon \mathbf{Set} \to \mathbf{JSL}$ (see Example 2(2)) are monoidal w.r.t. the morphisms chosen as in (1).

(3) The forgetful functor $U\colon \mathbf{JSL} \to \mathbf{Pos}$ has a left adjoint $\mathcal{D}_f\colon \mathbf{Pos} \to \mathbf{JSL}$ constructed as follows. For any poset A and $X_0 \subseteq A$ denote by $\downarrow X_0 := \{\, a \in A : a \leq x \text{ for some } x \in X_0 \,\}$ the down-set generated by X_0. Then $\mathcal{D}_f$ maps a poset A to $\mathcal{D}_f(A) := \{\, X \subseteq A : X = \downarrow X_0 \text{ for some finite } X_0 \subseteq A \,\}$, the semilattice (w.r.t. union) of finitely generated down-sets of A, and a monotone map $h\colon A \to B$ to the semilattice morphism $\mathcal{D}_f(h)\colon \mathcal{D}_f(A) \to \mathcal{D}_f(B)$ with $\mathcal{D}_f(h)(X) = \downarrow h[X]$. Both U and $\mathcal{D}_f$ carry monoidal functors; the required morphisms, see Definition 12, are chosen in analogy to $|-|$ and $\mathcal{P}_f$ in (2).

(4) As a trivial example, the identity functor $\mathsf{Id}\colon \mathscr{D} \to \mathscr{D}$ is monoidal w.r.t. the identity morphisms $id\colon \mathbf{1}_{\mathcal{D}} \to \mathsf{Id}(\mathbf{1}_{\mathcal{D}})$ and $id\colon \mathsf{Id}(A) \otimes \mathsf{Id}(B) \to \mathsf{Id}(A \otimes B)$.

The importance of monoidal functors is that they preserve monoid structures:

Lemma 14. *Let $(G, \theta)\colon \mathscr{C} \to \mathscr{D}$ be a monoidal functor. Then G lifts to the functor $\overline{G}\colon \mathbf{Mon}(\mathscr{C}) \to \mathbf{Mon}(\mathscr{D})$ mapping a $\mathscr{C}$-monoid (M, ι, μ) to the $\mathscr{D}$-monoid $(GM,\ \mathbf{1}_{\mathscr{D}} \xrightarrow{\theta} G\mathbf{1}_{\mathscr{C}} \xrightarrow{G\iota} GM,\ GM \otimes GM \xrightarrow{\theta} G(M \otimes M) \xrightarrow{G\mu} GM)$, and a $\mathscr{C}$-monoid morphism h to Gh.*

Example 15. (1) $\mathcal{P}_f\colon \mathbf{Set} \to \mathbf{JSL}$ lifts to $\overline{\mathcal{P}}_f\colon \mathbf{Mon}(\mathbf{Set}) \to \mathbf{Mon}(\mathbf{JSL})$, mapping a monoid M to the semiring $\overline{\mathcal{P}}_f M$ of finite subsets of M, with union as addition, and multiplication $XY = \{\, xy\colon x \in X,\ y \in Y \,\}$.

(2) $\mathcal{D}_f\colon \mathbf{Pos} \to \mathbf{JSL}$ lifts to $\overline{\mathcal{D}}_f\colon \mathbf{Mon}(\mathbf{Pos}) \to \mathbf{Mon}(\mathbf{JSL})$, mapping an ordered monoid M to the semiring $\overline{\mathcal{D}}_f(M)$ of finitely generated down-sets of M, with union as addition, and multiplication $XY = \downarrow\{\, xy : x \in X,\ y \in Y \,\}$.

Lemma 16. *Let $(G, \theta)\colon \mathscr{A} \to \mathscr{B}$ and $(H, \sigma)\colon \mathscr{B} \to \mathscr{C}$ be monoidal functors. Then the composite $HG\colon \mathscr{A} \to \mathscr{C}$ is a monoidal functor w.r.t. to $H(\theta_1) \circ \sigma_1\colon \mathbf{1}_{\mathscr{C}} \to HG(\mathbf{1}_{\mathscr{A}})$ and $H(\theta_{A,B}) \circ \sigma_{GA,GB}\colon HGA \otimes HGB \to HG(A \otimes B)$.*

Definition 17. A *monoidal adjunction* between $\mathscr{C}$ and $\mathscr{D}$ is an adjunction $F \dashv U\colon \mathscr{C} \to \mathscr{D}$ such that U and F are monoidal functors and the unit $\eta\colon \mathsf{Id}_{\mathscr{D}} \to UF$ and counit $\varepsilon\colon FU \to \mathsf{Id}_{\mathscr{C}}$ are monoidal natural transformations.

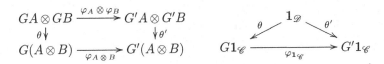

Example 18. $\mathsf{Id} \dashv \mathsf{Id}\colon \mathscr{D} \to \mathscr{D}$, $\mathcal{D}_f \dashv U\colon \mathbf{JSL} \to \mathbf{Pos}$ and $\Psi \dashv |{-}|\colon \mathscr{D} \to \mathbf{Set}$ are monoidal adjunctions. We call the latter the *monoidal adjunction of* $\mathscr{D}$.

Remark 19. If $(H \dashv V\colon \mathscr{C} \to \mathscr{B}, \eta', \varepsilon')$ and $(G \dashv U\colon \mathscr{B} \to \mathscr{A}, \eta, \varepsilon)$ are monoidal adjunctions, so is the composite adjunction $(HG \dashv UV\colon \mathscr{C} \to \mathscr{A},\ U\eta' G \circ \eta,\ \varepsilon' \circ H\varepsilon V)$. Here HG and UV are the composites of Lemma 16.

Definition 20. A monoidal adjunction $F \dashv U\colon \mathscr{C} \to \mathscr{D}$ is called a *concrete monoidal adjunction* if its composite with the monoidal adjunction of $\mathscr{D}$ is the monoidal adjunction of $\mathscr{C}$.

3 Languages and Algebraic Recognition

In this section we set the scene for our approach to Schützenberger products. Fix a commutative variety $\mathscr{D}$ of algebras or ordered algebras, a commutative semiring $\mathbb{S} = (S, +, \cdot, 0, 1)$, and a concrete monoidal adjunction $F \dashv U\colon \mathbb{S}\text{-}\mathbf{Mod} \to \mathscr{D}$ (i.e. apply Definition 20 to $\mathscr{C} = \mathbb{S}\text{-}\mathbf{Mod}$). We denote the unit by $\eta\colon \mathsf{Id} \to UF$. This gives the diagram of functors below. Here $\mathbb{S}\text{-}\mathbf{Alg} = \mathbf{Mon}(\mathbb{S}\text{-}\mathbf{Mod}))$ is the category of $\mathbb{S}$-algebras (see Example 7), $\overline{U}$ and $\overline{F}$ are the liftings of U and F (see Lemma 14), the vertical functors are the forgetful functors, and Ψ and $\mathbb{S}^{(-)}$ are the left adjoints to the forgetful functors of $\mathscr{D}$ and $\mathbb{S}\text{-}\mathbf{Mod}$, see Example 2.

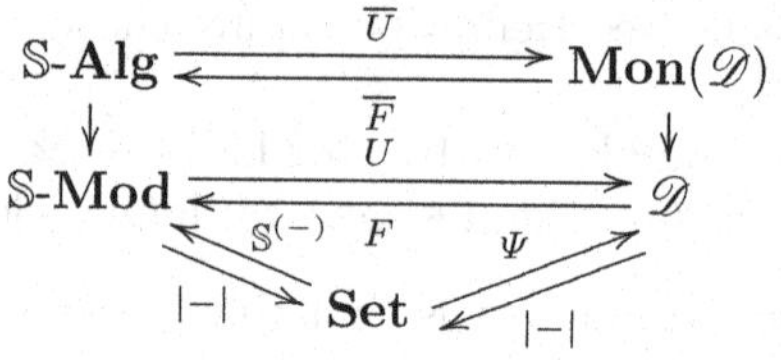

Example 21. In our applications we will choose the concrete monoidal adjunctions listed below. (The third and last column will be explained later.)

Notation 22. We can view the semiring $\mathbb{S}$ as (i) an $\mathbb{S}$-algebra $\mathbb{S}_{\mathbf{Alg}} \in \mathbb{S}\text{-}\mathbf{Alg}$ with scalar product given by the multiplication of $\mathbb{S}$, (ii) a $\mathscr{D}$-monoid $\mathbb{S}_{\mathbf{Mon}} \in \mathbf{Mon}(\mathscr{D})$ (by applying $\overline{U}$ to $\mathbb{S}_{\mathbf{Alg}}$), (iii) an $\mathbb{S}$-module $\mathbb{S}_{\mathbf{Mod}} \in \mathbb{S}\text{-}\mathbf{Mod}$ (by applying the forgetful functor to $\mathbb{S}_{\mathbf{Alg}}$) and (iv) an object $\mathbb{S}_{\mathscr{D}}$ of $\mathscr{D}$ (by applying U to $\mathbb{S}_{\mathbf{Mod}}$). The $\mathscr{D}$-monoid $\mathbb{S}_{\mathbf{Mon}}$ is carried by the object $\mathbb{S}_{\mathscr{D}}$, and its multiplication is a morphism of $\mathscr{D}$ that we denote by $\sigma\colon \mathbb{S}_{\mathscr{D}} \otimes \mathbb{S}_{\mathscr{D}} \to \mathbb{S}_{\mathscr{D}}$. For ease of notation we will usually drop the indices and simply write $\mathbb{S}$ for $\mathbb{S}_{\mathscr{D}}$, $\mathbb{S}_{\mathbf{Mod}}$, etc.

Definition 23. (1) A *language* (a.k.a. a *formal power series*) over a finite alphabet Σ is a map $L\colon \Sigma^* \to S$. Denote by $L_{\mathscr{D}}\colon \Psi\Sigma^* \to \mathbb{S}$ the adjoint transpose of L w.r.t. the adjunction $\Psi \dashv |{-}|\colon \mathscr{D} \to \mathbf{Set}$. A $\mathscr{D}$-monoid morphism $f\colon \Psi\Sigma^* \to M$ *recognizes* L if there is a morphism $p\colon M \to \mathbb{S}$ in $\mathscr{D}$ with $L_{\mathscr{D}} = p \circ f$. In this case, we also say that M *recognizes* L (via f and p).

	$\mathbb{S}$	$\mathscr{C}$	$\mathscr{D}$	$\mathbb{S}$**-Mod** $\underset{F}{\overset{U}{\rightleftarrows}}$ $\mathscr{D}$	$\mathscr{D}$-monoids	$M \diamond N$ carried by
1	$\{0,1\}$	**BA**	**Set**	**JSL** $\underset{\mathcal{P}_f}{\overset{\lvert - \rvert}{\rightleftarrows}}$ **Set**	monoids	$M \times \mathcal{P}_f(M \times N) \times N$
2	$\{0,1\}$	**DL**	**Pos**	**JSL** $\underset{\mathcal{D}_f}{\overset{U}{\rightleftarrows}}$ **Pos**	ord. monoids	$M \times \mathcal{D}_f(M \times N) \times N$
3	$\{0,1\}$	**JSL**	**JSL**	**JSL** $\underset{\mathsf{Id}}{\overset{\mathsf{Id}}{\rightleftarrows}}$ **JSL**	id. semirings	$M \times (M * N) \times N$
4	$\mathbb{K}$	$\mathbb{K}$**-Vec**	$\mathbb{K}$**-Vec**	$\mathbb{K}$**-Vec** $\underset{\mathsf{Id}}{\overset{\mathsf{Id}}{\rightleftarrows}}$ $\mathbb{K}$**-Vec**	$\mathbb{K}$-algebras	$M \times (M \otimes N) \times N$
5	$\mathbb{S}$	?	$\mathbb{S}$**-Mod**	$\mathbb{S}$**-Mod** $\underset{\mathsf{Id}}{\overset{\mathsf{Id}}{\rightleftarrows}}$ $\mathbb{S}$**-Mod**	$\mathbb{S}$-algebras	$M \times (M * N) \times N$

(2) The *marked Cauchy product* of two languages K, L: $\Sigma^* \to S$ w.r.t. a letter $a \in \Sigma$ is the language KaL: $\Sigma^* \to S$ with $(KaL)(u) = \sum_{u=vaw} K(v) \cdot L(w)$.

For $\mathbb{S} = \{0,1\}$, a language L: $\Sigma^* \to \{0,1\}$ corresponds to a classical language $L \subseteq \Sigma^*$ by taking the preimage of 1. Under this identification, we have $KaL = \{vaw\colon v \in K, w \in L\}$. Our concept of language recognition by $\mathscr{D}$-monoids originates in [2] and specializes to several related notions from the literature:

Example 24. (1) $\mathscr{D}$ = **Set** with $\mathbb{S} = \{0,1\}$: a map p: $M \to \{0,1\}$ corresponds to a subset $p^{-1}[1] \subseteq M$. Thus a monoid morphism f: $\Sigma^* \to M$ recognizes the language $L \subseteq \Sigma^*$ iff L is the preimage under f of some subset of M. This is the classical notion of language recognition by a monoid, see e.g. [18].

(2) $\mathscr{D}$ = **Pos** with $\mathbb{S} = \{0,1\}$: given an ordered monoid M, a monotone map p: $M \to \{0,1\}$ defines an upper set $p^{-1}[1] \subseteq M$. Hence a monoid morphism f: $\Sigma^* \to M$ recognizes $L \subseteq \Sigma^*$ iff L is the preimage under f of some upper set of M. This notion of recognition is due to Pin [16].

(3) $\mathscr{D}$ = **JSL** with $\mathbb{S} = \{0,1\}$: for any idempotent semiring M, a semilattice morphism p: $M \to \{0,1\}$ defines an ideal $I = p^{-1}[0]$, i.e. a nonempty down-set closed under joins. Hence a language $L \subseteq \Sigma$ is recognized by a semiring morphism f: $\mathcal{P}_f\Sigma^* \to M$ via p iff $L^{\mathsf{C}} = \Sigma^* \cap f^{-1}[I]$. Here we identify Σ^* with the set of all singleton languages $\{w\}$, $w \in \Sigma^*$. This is the concept of language recognition by idempotent semirings introduced by Polák [19].

(4) $\mathscr{D}$ = $\mathbb{S}$**-Mod**: given an $\mathbb{S}$-algebra M, a formal power series L: $\Sigma^* \to S$ is recognized by f: $\mathbb{S}[\Sigma] \to M$ via p: $M \to \mathbb{S}$ iff $L_{\mathbb{S}\textbf{-Mod}} = p \circ f$. This notion of recognition is due to Reutenauer [21]. If $\mathbb{S}$ is a commutative ring, the power series recognizable by $\mathbb{S}$-algebras of finite type (i.e. $\mathbb{S}$-algebras whose underlying $\mathbb{S}$-module is finitely generated) are precisely rational power series.

4 The Schützenberger Product

We are ready to introduce the Schützenberger product for $\mathscr{D}$-monoids. Fix two $\mathscr{D}$-monoids $(M, 1, \bullet)$ and $(N, 1, \bullet)$, and write xy for $x \bullet y$. Our goal is to construct a $\mathscr{D}$-monoid $M \diamond N$ that recognizes all marked products of languages recognized by M and N, and is a "smallest" such $\mathscr{D}$-monoid (Theorems 32, 37, 40).

Construction 25. As a preliminary step, we define a $\mathscr{D}$-monoid $M * N$ as follows. Call a family $\{ f_i \colon A \to B_i \}_{i \in I}$ in $\mathscr{D}$ *separating* if the morphism $f\colon A \to \prod_i B_i$ with $f(a) = (f_i(a))_{i\in I}$ is injective (resp. order-reflecting when $\mathscr{D}$ is a variety of ordered algebras). Any family $\{f_i\}$ yields a separating family $\{ f_i'\colon A' \to B_i \}_{i\in I}$ by factorizing $f = m \circ \pi$ with π surjective and m injective (resp. order-reflecting), and setting $f_i' := p_i \circ m$, where p_i is the projection. Now consider the family of all morphisms $\sigma \circ (p \otimes q)\colon M \otimes N \to \mathbb{S}$, where $p\colon M \to \mathbb{S}$ and $q\colon N \to \mathbb{S}$ are arbitrary morphisms in $\mathscr{D}$. Applying the above construction to this family $\{\sigma \circ (p \otimes q)\}_{p,q}$ gives an algebra $M * N$ in $\mathscr{D}$, a surjective morphism $\pi\colon M \otimes N \twoheadrightarrow M * N$, and a separating family $\{ p * q\colon M * N \to \mathbb{S} \}_{p,q}$, making the following diagram commute for all p and q:

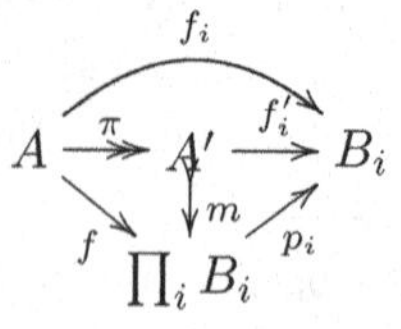

$$M \otimes N \xrightarrow[\pi]{} M * N \xrightarrow[p*q]{} \mathbb{S}, \qquad M \otimes N \xrightarrow{p \otimes q} \mathbb{S} \otimes \mathbb{S} \xrightarrow{\sigma} \mathbb{S} \tag{4.1}$$

Notation 26. For any $m \in |M|$ and $n \in |N|$, we write $m * n$ for the element $\pi(m \otimes n) \in |M * N|$.

Lemma 27. *There exists a (unique) $\mathscr{D}$-monoid structure on $M * N$ such that $\pi : M \otimes N \twoheadrightarrow M * N$ is a $\mathscr{D}$-monoid morphism. The multiplication is determined by $(m * n) \bullet (m' * n') = (mm') * (nn')$, and the unit is $1 * 1$.*

Example 28. For $\mathscr{D} =$ **Set**, **Pos** or $\mathbb{K}$-**Vec**, the family $\{ \sigma \circ (p \otimes q) \}_{p,q}$ is already separating, and therefore $M * N = M \otimes N$ and $p * q = \sigma \circ (p \otimes q)$. For $\mathscr{D} =$ **JSL** and in case M and N are *finite* idempotent semirings, we can describe the idempotent semiring $M * N$ as follows. For any subset $X \subseteq M \times N$, let $[X] \subseteq M \times N$ consist of those elements $(m, n) \in M \times N$ such that, for all ideals $I \subseteq M$ and $J \subseteq N$ with $m \notin I$ and $n \notin J$, there exists some $(x, y) \in X$ with $x \notin I$ and $y \notin J$. This gives us the closure operator $X \mapsto [X]$ on the power set of $M \times N$ in [15]. One can show that $M * N$ is isomorphic to the idempotent semiring of all closed subsets of $M \times N$, with sum and product defined by $[X] \vee [Y] = [X \cup Y]$ and $[X][Y] = [XY]$, where $XY = \{ xy : x \in X, y \in Y \}$.

Definition 29. The *Schützenberger product* of M and N is the $\mathscr{D}$-monoid $M \diamond N$ carried by the product $M \times UF(M * N) \times N$ in $\mathscr{D}$ and equipped with the following monoid structure: representing elements $(m, a, n) \in |M| \times |F(M * N)| \times |N|$ as upper triangular matrices $\begin{pmatrix} m & a \\ 0 & n \end{pmatrix}$, the multiplication and unit are given by

$$\begin{pmatrix} m & a \\ 0 & n \end{pmatrix} \begin{pmatrix} m' & a' \\ 0 & n' \end{pmatrix} = \begin{pmatrix} mm' & \eta(m * 1) \cdot a' + a \cdot \eta(1 * n') \\ 0 & nn' \end{pmatrix} \quad \text{and} \quad \begin{pmatrix} 1 & 0 \\ 0 & 1 \end{pmatrix}.$$

Here $\eta\colon M * N \to UF(M * N)$ is the universal map, and the sum, product and 0 in the upper right components are taken in the $\mathbb{S}$-algebra $\overline{F}(M * N)$.

Lemma 30. *$M \diamond N$ is a well-defined $\mathscr{D}$-monoid, and the product projections $\pi_M\colon M \diamond N \to M$ and $\pi_N\colon M \diamond N \to N$ are $\mathscr{D}$-monoid morphisms.*

Example 31. For the categories and adjunctions of Example 21, we recover four notions of Schützenberger products known in the literature, and obtain a new Schützenberger product for $\mathbb{S}$-algebras:

(1) $\mathscr{D} = \mathbf{Set}$: given monoids M and N, the Schützenberger product $M \diamond N$ is carried by the set $M \times \mathcal{P}_f(M \times N) \times N$, with multiplication and unit

$$\begin{pmatrix} m & X \\ 0 & n \end{pmatrix}\begin{pmatrix} m' & X' \\ 0 & n' \end{pmatrix} = \begin{pmatrix} mm' & mX' \cup Xn' \\ 0 & nn' \end{pmatrix} \quad \text{and} \quad \begin{pmatrix} 1 & \emptyset \\ 0 & 1 \end{pmatrix},$$

where $mX' = \{\,(my, z) : (y, z) \in X'\,\}$ and $Xn' = \{\,(y, zn')\colon (y, z) \in X\,\}$. This is the original construction of Schützenberger [23].

(2) $\mathscr{D} = \mathbf{Pos}$: for ordered monoids M and N, the Schützenberger product $M \diamond N$ is carried by the poset $M \times \mathcal{D}_f(M \times N) \times N$ with multiplication and unit

$$\begin{pmatrix} m & X \\ 0 & n \end{pmatrix}\begin{pmatrix} m' & X' \\ 0 & n' \end{pmatrix} = \begin{pmatrix} mm' & \downarrow(mX' \cup Xn') \\ 0 & nn' \end{pmatrix} \quad \text{and} \quad \begin{pmatrix} 1 & \emptyset \\ 0 & 1 \end{pmatrix}.$$

This construction is due to Pin [17].

(3) $\mathscr{D} = \mathbf{JSL}$: given idempotent semirings M and N, the Schützenberger product $M \diamond N$ is carried by the semilattice $M \times (M * N) \times N$. If M and N are finite, Example 28 shows that $M * N$ is the idempotent semiring of closed subsets of $M \times N$, and the multiplication and unit of $M \diamond N$ are given by

$$\begin{pmatrix} m & X \\ 0 & n \end{pmatrix}\begin{pmatrix} m' & X' \\ 0 & n' \end{pmatrix} = \begin{pmatrix} mm' & [mX' \cup Xn'] \\ 0 & nn' \end{pmatrix} \quad \text{and} \quad \begin{pmatrix} 1 & \emptyset \\ 0 & 1 \end{pmatrix}.$$

For the finite case, this construction is due to Klíma and Polák [15].

(4) $\mathscr{D} = \mathbb{K}\text{-}\mathbf{Vec}$: for $\mathbb{K}$-algebras M and N, the Schützenberger product $M \diamond N$ is carried by the vector space $M \times (M \otimes N) \times N$ with multiplication and unit

$$\begin{pmatrix} m & z \\ 0 & n \end{pmatrix}\begin{pmatrix} m' & z' \\ 0 & n' \end{pmatrix} = \begin{pmatrix} mm' & mz' + zn' \\ 0 & nn' \end{pmatrix} \quad \text{and} \quad \begin{pmatrix} 1 & 0 \otimes 0 \\ 0 & 1 \end{pmatrix},$$

where $mz' = (mm_0) \otimes n_0$ for $z' = m_0 \otimes n_0$, and extending via bilinearity for arbitrary z; similarly for zn'. This construction is due to Reutenauer [21].

(5) $\mathscr{D} = \mathbb{S}\text{-}\mathbf{Mod}$: given $\mathbb{S}$-algebras M and N, the Schützenberger product $M \diamond N$ is carried by the $\mathbb{S}$-module $M \times (M * N) \times N$ with multiplication and unit

$$\begin{pmatrix} m & z \\ 0 & n \end{pmatrix}\begin{pmatrix} m' & z' \\ 0 & n' \end{pmatrix} = \begin{pmatrix} mm' & mz' + zn' \\ 0 & nn' \end{pmatrix} \quad \text{and} \quad \begin{pmatrix} 1 & 0 * 0 \\ 0 & 1 \end{pmatrix},$$

where $mz' = (mm_0) * n_0$ for $z' = m_0 * n_0$, and similarly for zn'. This example specializes to (3) and (4) by taking $\mathbb{S} = \{0, 1\}$ and $\mathbb{S} = \mathbb{K}$, respectively, but appears to be new construction for other semirings $\mathbb{S}$.

The following theorem gives the key property of $M \diamond N$.

Theorem 32. *Let $K, L\colon \Sigma^* \to S$ be languages recognized by M and N, respectively. Then $M \diamond N$ recognizes the languages K, L and KaL for all $a \in \Sigma$.*

Next, we aim to show that $M \diamond N$ is a "smallest" $\mathscr{D}$-monoid satisfying the statement of the above theorem. This requires further assumptions on our setting.

Notation 33. Recall from (4.1) the morphism $p * q\colon M * N \to \mathbb{S}$. We denote its adjoint transpose w.r.t. the adjunction $F \dashv U$ by $\overline{p * q}\colon F(M * N) \to \mathbb{S}$.

Assumption 34. From now on, suppose that:

(i) $\mathscr{D}$ is locally finite, i.e. every finitely generated algebra of $\mathscr{D}$ is finite.
(ii) Epimorphisms in $\mathscr{D}$ and $\mathbb{S}$-**Mod** are surjective.
(iii) $\mathscr{D}(M, \mathbb{S})$, $\mathscr{D}(N, \mathbb{S})$, and $\{U(\overline{p * q})\colon UF(M * N) \to \mathbb{S}\}_{p\colon M \to \mathbb{S},\, q\colon N \to \mathbb{S}}$ are separating families of morphisms in $\mathscr{D}$.
(iv) There is a locally finite variety $\mathscr{C}$ of algebras such that the full subcategories $\mathscr{C}_f$ and $\mathscr{D}_f$ on *finite* algebras are dually equivalent. We denote the equivalence functor by $E\colon \mathscr{D}_f^{op} \simeq \mathscr{C}_f$.
(v) The semiring $\mathbb{S}$ is finite, and $E(\mathbb{S}) \cong \mathbf{1}_{\mathscr{C}}$.

Let us indicate the intuition behind our assumptions. First, (i) and (ii) imply that $M \diamond N$ is finite if M and N are. This is important, as one is usually interested in language recognition by *finite* $\mathscr{D}$-monoids. (iii) expresses that the semiring $\mathbb{S}$ has enough structure to separate elements of M, N and $UF(M * N)$, the three components of the Schützenberger product $M \diamond N$, by suitable morphisms into $\mathbb{S}$. This technical condition on $\mathbb{S}$ is the crucial ingredient for proving the "smallness" of $M \diamond N$ (Theorem 37). Finally, the variety $\mathscr{C}$ in (iv) and (v) will be used to determine, via duality, the algebraic operations to express languages recognized by $M \diamond N$ in terms of languages recognized by M and N (Theorem 40).

Example 35. The categories and adjunctions of Example 21(1)–(4) satisfy our assumptions. Here we briefly sketch the dualities; see [1,3] for details.

(1) For $\mathscr{D} = \mathbf{Set}$, choose $\mathscr{C} = \mathbf{BA}$ (boolean algebras). Stone duality [14] gives a dual equivalence $E\colon \mathbf{Set}_f^{op} \simeq \mathbf{BA}_f$ mapping a finite set to the boolean algebra of all subsets.
(2) For $\mathscr{D} = \mathbf{Pos}$, choose $\mathscr{C} = \mathbf{DL}$ (distributive lattices with 0, 1). Birkhoff duality [5] gives a dual equivalence $E\colon \mathbf{Pos}_f^{op} \simeq \mathbf{DL}_f$ mapping a finite poset to the lattice of all down-sets.
(3) For $\mathscr{D} = \mathbf{JSL}$, choose $\mathscr{C} = \mathbf{JSL}$. The dual equivalence $E\colon \mathbf{JSL}_f^{op} \simeq \mathbf{JSL}_f$ maps a finite semilattice $(X, \vee)$ to its opposite semilattice $(X, \wedge)$, see [14].
(4) For $\mathscr{D} = \mathbb{K}\text{-}\mathbf{Vec}$, $\mathbb{K}$ a *finite* field, choose $\mathscr{C} = \mathbb{K}\text{-}\mathbf{Vec}$. The dual equivalence $E\colon \mathbb{K}\text{-}\mathbf{Vec}_f \simeq \mathbb{K}\text{-}\mathbf{Vec}_f^{op}$ maps a space X to its dual space $X^* = \hom(X, \mathbb{K})$.

Notation 36. For any $\mathscr{D}$-monoid morphism $f\colon \Psi\Sigma^* \to M \diamond N$, put

$$\mathbb{L}_{M,N}(f) := \{\, K, L, KaL \mid a \in \Sigma, \pi_M \circ f \text{ recognizes } K, \pi_N \circ f \text{ recognizes } L \,\}$$

Theorem 37. *Let* $f\colon \Psi\Sigma^* \to M \diamond N$ *and* $e\colon \Psi\Sigma^* \to P$ *be two* $\mathscr{D}$*-monoid morphisms. If* e *is surjective and recognizes all languages in* $\mathbb{L}_{M,N}(f)$*, then there exists a (necessarily unique)* $\mathscr{D}$*-monoid morphism* $h\colon P \to M \diamond N$ *with* $h \circ e = f$.

Using our duality framework, this theorem can be rephrased in terms of language operations. Recall that $E(\mathbb{S}) \cong \mathbf{1}_{\mathscr{C}}$ by Assumption 34(v). Putting $O_{\mathscr{C}} := E(\mathbf{1}_{\mathscr{D}})$, we obtain a bijection $i\colon S \cong \mathscr{D}(\mathbf{1}_{\mathscr{D}}, \mathbb{S}) \cong \mathscr{C}(E(\mathbb{S}), E(\mathbf{1}_{\mathscr{D}})) \cong \mathscr{C}(\mathbf{1}_{\mathscr{C}}, O_{\mathscr{C}}) \cong |O_{\mathscr{C}}|$.

Definition 38. For any n-ary operation symbol γ in the signature of $\mathscr{C}$ and languages $L_1, \ldots, L_n\colon \Sigma^* \to S$, the language $\overline{\gamma}(L_1, \ldots, L_n)\colon \Sigma^* \to S$ is given by $\overline{\gamma}(L_1, \ldots, L_n)(u) := i^{-1}(\gamma^{O_{\mathscr{C}}}(i(L_1 u), \ldots, i(L_n u)))$. The operations $\overline{\gamma}$ are called the *$\mathscr{C}$-algebraic operations* on the set of languages over Σ.

Example 39. $O_{\mathbf{BA}} \cong \{0, 1\}$ is the two-element boolean algebra, and the **BA**-algebraic operations are precisely the boolean operations (union, intersection, complement, $\emptyset$, Σ^*) on languages. For example, the operation symbol $\vee$ induces the language operation $(K \overline{\vee} L)(u) = K(u) \vee L(u)$ corresponding to the union of languages. Similarly, for $\mathscr{C} = \mathbf{DL}$ we get union, intersection, $\emptyset$, Σ^*, for $\mathscr{C} = \mathbf{JSL}$ we get union and $\emptyset$, and for $\mathscr{C} = \mathbb{K}$-**Vec** we get sum, scalar product and $\emptyset$.

All our constructions and results so far apply to arbitrary $\mathscr{D}$-monoids. However, in the following theorem we need to restrict to *finite* $\mathscr{D}$-monoids. Recall that the *derivatives* of a language $L\colon \Sigma^* \to S$ are the languages $a^{-1}L$, $La^{-1}\colon \Sigma^* \to S$ (where $a \in \Sigma$) defined by $(a^{-1}L)(u) = L(au)$ and $(La^{-1})(u) = L(ua)$.

Theorem 40. *Let* M *and* N *be finite* $\mathscr{D}$*-monoids and* $f\colon \Psi\Sigma^* \to M \diamond N$ *be a* $\mathscr{D}$*-monoid morphism. Then every language recognized by* f *lies in the closure of* $\mathbb{L}_{M,N}(f)$ *under the* $\mathscr{C}$*-algebraic operations and derivatives.*

Our proof uses the *Local Variety Theorem* of [1]: for any finite set $\mathcal{V}$ of recognizable languages closed under $\mathscr{C}$-algebraic operations and derivatives, there is a finite $\mathscr{D}$-monoid recognizing precisely the languages of $\mathcal{V}$. Coincidentally, for each of our categories of Example 21(1)–(4) it suffices to take the closure of $\mathbb{L}_{M,N}(f)$ under $\mathscr{C}$-algebraic operations, as this set is already derivative-closed. For example, for $\mathscr{C} = \mathbb{K}$-**Vec** we have $a^{-1}(KaL) = (a^{-1}K)aL + K(\varepsilon)L$, i.e. $a^{-1}(KaL)$ is a linear combination of languages in $\mathbb{L}_{M,N}(f)$ and thus lies in the closure of $\mathbb{L}_{M,N}(f)$ under $\mathbb{K}$-**Vec**-operations. For $\mathscr{D}$ = **Set**, **Pos** and **JSL**, Theorem 40 then gives

Corollary 41. (Reutenauer [22], Pin [17], Klíma and Polák [15]). *Let* M *and* N *be finite monoids [ordered monoids, idempotent semirings]. Then any language recognized by the Schützenberger product* $M \diamond N$ *is a boolean combination [positive boolean combination, finite union] of languages of the form* K, L *and* KaL*, where* K *is recognized by* M*,* L *is recognized by* N*, and* $a \in \Sigma$.

For $\mathscr{D} = \mathbb{K}$-**Vec**, we obtain a new result for formal power series:

Corollary 42. *Let* M *and* N *be finite algebras over a finite field* $\mathbb{K}$*. Then any language recognized by* $M \diamond N$ *is a linear combination of power series of the form* K, L *and* KaL*, where* K *is recognized by* M*,* L *is recognized by* N*, and* $a \in \Sigma$.

5 Conclusions and Future Work

We presented a categorical framework that encompasses all known instances of Schützenberger products in the setting of regular languages. Two related constructions are the Schützenberger products for *ω-semigroups* [7], and for *boolean spaces with internal monoids* [12]. Neither of these structures are monoids in the categorical sense, and thus are not covered by our present setting. The use of monads as in [6,8,25] might pave the way to extending the scope of our work.

References

1. Adámek, J., Milius, S., Myers, R.S.R., Urbat, H.: Generalized Eilenberg theorem I: local varieties of languages. In: Muscholl, A. (ed.) FOSSACS 2014 (ETAPS). LNCS, vol. 8412, pp. 366–380. Springer, Heidelberg (2014). http://arxiv.org/pdf/1501.02834v1.pdf
2. Adámek, J., Milius, S., Urbat, H.: Syntactic monoids in a category. In: Proceedings of CALCO 2015. LIPIcs, Schloss Dagstuhl-Leibniz-Zentrum für Informatik (2015)
3. Adámek, J., Myers, R., Milius, S., Urbat, H.: Varieties of languages in a category. In: LICS 2015. IEEE (2015)
4. Ballester-Bolinches, A., Cosme-Llopez, E., Rutten, J.: The dual equivalence of equations and coequations for automata. Inform. Comput. **244**, 49–75 (2015)
5. Birkhoff, G.: Rings of sets. Duke Math. J. **3**(3), 443–454 (1937)
6. Bojańczyk, M.: Recognisable languages over monads. In: Potapov, I. (ed.) DLT 2015. LNCS, vol. 9168, pp. 1–13. Springer, Heidelberg (2015)
7. Carton, O.: Mots infinis, ω-semigroupes et topologie. Technical report, Université Paris 7 , report LITP-TH 93–08 (1993)
8. Chen, L.T., Adámek, J., Milius, S., Urbat, H.: Profinite monads, profinite equations and Reiterman's theorem. In: Jacobs, B., Löding, C. (eds.) Proceedings of FoSSaCS 2016. LNCS, vol. 9634, pp. 531–547. Springer, Heidelberg (2016). http://arxiv.org/abs/1511.02147
9. Chen, L.T., Urbat, H.: A fibrational approach to automata theory. In: Proceedings of CALCO 2015. LIPIcs, Schloss Dagstuhl-Leibniz-Zentrum für Informatik (2015)
10. Davey, B.A., Davis, G.: Tensor products and entropic varieties. Algebra Univers. **21**(1), 68–88 (1985)
11. Gehrke, M., Grigorieff, S., Pin, J.É.: Duality and equational theory of regular languages. In: Aceto, L., Damgård, I., Goldberg, L.A., Halldórsson, M.M., Ingólfsdóttir, A., Walukiewicz, I. (eds.) ICALP 2008, Part II. LNCS, vol. 5126, pp. 246–257. Springer, Heidelberg (2008)
12. Gehrke, M., Petrisan, D., Reggio, L.: The schützenberger product for syntactic spaces. In: Proceedings of ICALP 2016 (2016, to appear). Preprint: http://arxiv.org/abs/1603.08264
13. Goguen, J.A.: Discrete-time machines in closed monoidal categories. Int. J. Comput. Syst. Sci. **10**(1), 1–43 (1975)
14. Johnstone, P.T.: Stone Spaces. Cambridge University Press, Cambridge (1982)
15. Klíma, O., Polák, L.: On schützenberger products of semirings. In: Gao, Y., Lu, H., Seki, S., Yu, S. (eds.) DLT 2010. LNCS, vol. 6224, pp. 279–290. Springer, Heidelberg (2010)
16. Pin, J.É.: A variety theorem without complementation. Russ. Math. **39**, 80–90 (1995)

17. Pin, J.É.: Algebraic tools for the concatenation product. Theor. Comput. Sci. **292**, 317–342 (2003)
18. Pin, J.É.: Mathematical foundations of automata theory (October 2015). http://www.liafa.jussieu.fr/~jep/PDF/MPRI/MPRI.pdf
19. Polák, L.: Syntactic semiring of a language. In: Sgall, J., Pultr, A., Kolman, P. (eds.) MFCS 2001. LNCS, vol. 2136, pp. 611–620. Springer, Heidelberg (2001)
20. Porst, H.E.: On categories of monoids, comonoids, and bimonoids. Quaestiones Math. **31**, 127–139 (2008)
21. Reutenauer, C.: Séries formelles et algèbres syntactiques. J. Algebra **66**, 448–483 (1980)
22. Reutenauer, C.: Sur les variétés de langages et de monoídes. In: 4th GI-Conference on Theoretical Computer Science. LNCS, vol. 67, pp. 260–265. Springer (1979)
23. Schützenberger, M.P.: On finite monoids having only trivial subgroups. Inform. Control **8**, 190–194 (1965)
24. Uramoto, T.: Semi-galois categories: the classical Eilenberg variety theory. In: Proceeding of LICS 2016 (2016, to appear). Preprint: http://arxiv.org/abs/1512.04389
25. Urbat, H., Adámek, J., Chen, L.T., Milius, S.: One Eilenberg theorem to rule them all (2016). Preprint: http://arxiv.org/abs/1602.05831

Outfix-Guided Insertion
(Extended Abstract)

Da-Jung Cho[1], Yo-Sub Han[1], Timothy Ng[2], and Kai Salomaa[2(✉)]

[1] Department of Computer Science, Yonsei University,
50, Yonsei-Ro, Seodaemun-Gu, Seoul 120-749, Republic of Korea
{dajungcho,emmous}@yonsei.ac.kr
[2] School of Computing, Queen's University, Kingston, ON K7L 2N8, Canada
{ng,ksalomaa}@cs.queensu.ca

Abstract. Motivated by work on bio-operations on DNA strings, we consider an outfix-guided insertion operation that can be viewed as a generalization of the overlap assembly operation on strings studied previously. As the main result we construct a finite language L such that the outfix-guided insertion closure of L is nonregular. We consider also the closure properties of regular and (deterministic) context-free languages under the outfix-guided insertion operation and decision problems related to outfix-guided insertion. Deciding whether a language recognized by a deterministic finite automaton is closed under outfix-guided insertion can be done in polynomial time.

Keywords: Language operations · Closure properties · Regular languages

1 Introduction

Gene insertion and deletion are basic operations occurring in DNA recombination in molecular biology. Recombination creates a new DNA strand by cutting, substituting, inserting, deleting or combining other strands. Possible errors in this process directly affect DNA strands and impair the function of genes. Errors in DNA recombination cause mutation that plays a part in normal and abnormal biological processes such as cancer, the immune system, protein synthesis and evolution [1]. Since mutational damage may or may not be easily identifiable, researchers deliberately generate mutations so that the structure and biological activity of genes can be examined in detail. *Site-directed mutagenesis* is one of the most important techniques in laboratory for generating mutations on specific sites of DNA using PCR (polymerase chain reaction) based methods [7,11]. For a site-directed insertion mutagenesis by PCR, the mutagenic primers are typically designed to include the desired change, which could be base addition [15,16]. This enzymatic reaction occurs in the test tube with a DNA strand and predesigned primers in which the DNA strand includes a target region, and a predesigned

S. Brlek and C. Reutenauer (Eds.): DLT 2016, LNCS 9840, pp. 102–113, 2016.
DOI: 10.1007/978-3-662-53132-7_9

primer includes a complementary region of the target region. The complementary region of primers leads it to hybridize the target DNA region and generate a desired insertion on a specific site as a mutation. Figure 1 illustrates the procedure of site-directed insertion mutagenesis by PCR.

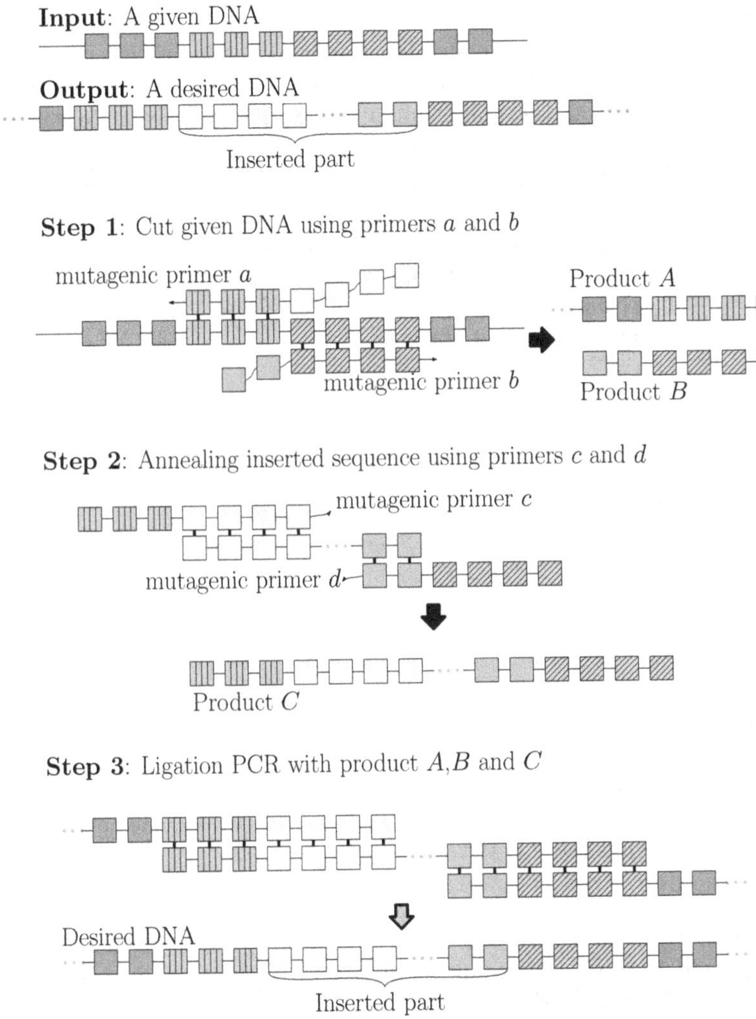

Fig. 1. An example of site-directed insertion mutagenesis by PCR. Given a DNA sequence and four predesigned primers a, b, c and d, two primers a and b lead the DNA sequence to break and extend into two products A and B under enzymatic reaction (Step 1). Two primers c and d complementarily bind to desired insertion region according to the overlapping region and extend into product C (Step 2). Then, the products A, B and C join together to create recombinant DNA that include the desired insertion (Step 3).

In formal language theory, the insertion of a string means adding a substring to a given string and deletion of a string means removing a substring. The insertions occurring in DNA strands are in some sense context-sensitive and Kari and Thierrin [13] modeled such bio-operations using *contextual insertions and deletions* [8,19]. A finite set of insertion-deletion rules, together with a finite set of axioms, can be viewed as a language generating device. Contextual insertion-deletion systems in the study of molecular computing have been used e.g. by Daley et al. [3], Krassovitskiy et al. [14] and Takahara and Yokomori [21]. Further theoretical studies on the computational power of insertion-deletion systems were done e.g. by Margenstern et al. [17] and Păun et al. [18]. Enaganti et al. [6] have studied related operations to model the action of DNA polymerase enzymes.

We formalize site-directed insertion mutagenesis by PCR and define a new operation *outfix-guided insertion* that *partially* inserts a string y into a string x when two non-empty substrings of x match with an outfix of y, see Fig. 2(b). The outfix-guided insertion is an overlapping variant of the ordinary insertion operation, analogously as the overlap assembly [2,4,5], cf. Fig. 2(a), is a variant of the ordinary string concatenation operation.

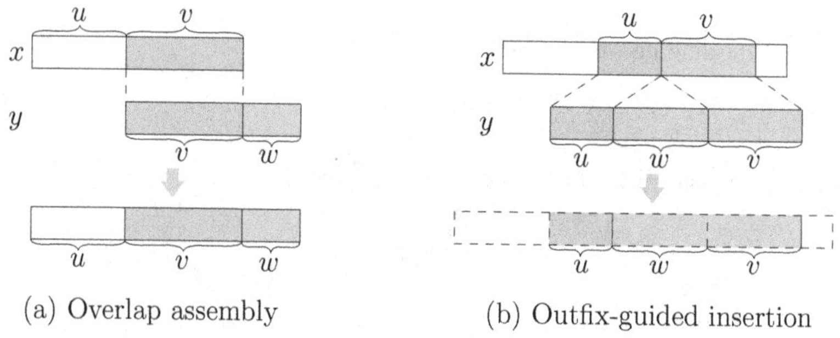
u
v
x
y
v
w
u
v
w
(a) Overlap assembly
u
v
x
y
u
w
v
u
w
v
(b) Outfix-guided insertion

Fig. 2. (a) If suffix v of x overlaps with the prefix v of y, then the overlap assembly operation partially catenates x and y appending suffix w of y to x. (b) If the outfix of y consisting of u and v matches the substring uv of x, then the outfix-guided insertion operation inserts w between u and v in the string x.

This paper investigates the language theoretic closure properties of outfix-guided insertion and iterated outfix-guided insertion. Note that since outfix-guided insertion, similarly as overlap assembly, is not associative, there are more than one way to define the iteration of the operation. We consider a general outfix-guided insertion closure of a language which is defined analogously as the iterated overlap assembly by Enaganti et al. [4]. Iterated (overlap) assembly is defined by Csuhaj-Varju et al. [2] in a different way, which we call right one-sided iteration of an operation.

It is fairly easy to see that regular languages are closed under outfix-guided insertion. Closure of regular languages under outfix-guided insertion closure turns out to be less obvious. It is well known that regular languages are not closed under the iteration of the ordinary (non-overlapping) insertion operation [12]. However, the known counter-examples, nor their variants, do not work for iterated outfix-guided insertion. Here using a more involved construction we show that there exists even a finite language L such that the outfix-guided insertion closure of L is nonregular. On the other hand, we show that the outfix-guided insertion closure of a unary regular language is always regular.

It is well known that context-free languages are closed under ordinary (non-iterated) insertion. We show that context-free languages are not closed under outfix-guided insertion. The outfix-guided insertion of a regular language into a context-free language (or vice versa) is always context-free. Also we establish that a similar closure property does not hold for the deterministic context-free and the regular languages. Finally in the last section we consider decision problems on whether a language is closed under outfix-guided insertion (or og-closed). We give a polynomial time algorithm to decide whether a language recognized by a deterministic finite automaton (DFA) is og-closed. We show that for a given context-free language L the question of deciding whether or not L is og-closed is undecidable. Most proofs are omitted in this extended abstract.

2 Definition of (Iterated) Outfix-Guided Insertion

We assume the reader to be familiar with the basics of formal languages, in particular, with the classes of regular languages and (deterministic) context-free languages [20,22]. More details on variants of the insertion operation and iterated insertion can be found in [12].

The symbol Σ stands always for a finite alphabet, Σ^* is the set of strings over Σ, $|w|$ is the length of a string $w \in \Sigma^*$, w^R is the reversal of w and ε is the empty string. For $i \in \mathbb{N}$, $\Sigma^{\geq i}$ is the set of strings of length at least i.

If $w = xy$, $x, y \in \Sigma^*$, we say that x is a prefix of w and y is a suffix of w. If $w = xyz$, $x, y, z \in \Sigma^*$, we say that (x, z) is an *outfix* of w. If additionally $x \neq \varepsilon$ and $z \neq \varepsilon$, (x, z) is a *non-trivial outfix* of w. Sometimes (in particular, when talking about the outfix-guided insertion operation) we refer to an outfix (x, z) simply as a string xz (when it is known from the context what are the components x and z). For example, with $\Sigma = \{a, b, c\}$ and $w = abca$ the non-trivial outfixes of w are aa, aba, aca and $abca$.

We begin by recalling some notions associated with the non-overlapping insertion operation.[1] The non-overlapping insertion of a string y into a string x is defined as the set of strings $x \overset{\text{nol}}{\leftarrow} y = \{x_1 y x_2 \mid x = x_1 x_2\}$. The insertion operation is extended in the natural way for languages by setting $L_1 \overset{\text{nol}}{\leftarrow} L_2 = \bigcup_{x \in L_1, y \in L_2} x \overset{\text{nol}}{\leftarrow} y$. Following Kari [12] we define the *left-iterated insertion* of L_2 into L_1 inductively by setting

$$\mathbb{LI}^{(0)}(L_1, L_2) = L_1 \text{ and } \mathbb{LI}^{(i+1)}(L_1, L_2) = \mathbb{LI}^{(i)}(L_1, L_2) \overset{\text{nol}}{\leftarrow} L_2,\ i \geq 0.$$

[1] We use the term "non-overlapping" to make the distinction clear to outfix-guided insertion which will be the main topic of this paper.

The *left-iterated insertion closure of* L_2 *into* L_1 is $\mathbb{LI}^*(L_1, L_2) = \bigcup_{i=0}^{\infty} \mathbb{LI}^{(i)}(L_1, L_2)$. It is well known that the iterated non-overlapping insertion operation does not preserve regularity [10,12]. The left-iterated insertion closure of the string ab into itself is nonregular because $\mathbb{LI}^*(ab, ab) \cap a^*b^* = \{a^ib^i \mid i \geq 0\}$.

Next we define the main notion of this paper which can be viewed as a generalization of the overlap assembly operation [2,4]. The "inside part" of a string y can be outfix-guided inserted into a string x if a non-trivial outfix of y overlaps with a substring of x in a position where the insertion occurs. This differs from contextual insertion (as defined in [13]) in the sense that y must actually contain the outfix that is matched with a substring of x (and additionally [13] specifies a set of contexts where an insertion can occur).

Definition 1. *The* outfix-guided insertion *of a string* y *into a string* x *is defined as*

$$x \overset{\text{ogi}}{\leftarrow} y = \{x_1uzvx_2 \mid x = x_1uvx_2,\ y = uzv,\ u \neq \varepsilon, v \neq \varepsilon\}.$$

Using the above notations, when $x_1uzvx_2 \in x \overset{\text{ogi}}{\leftarrow} y$ we say that the nonempty substrings u and v are the *matched parts.* Note that the matched parts form a non-trivial outfix of the inserted string y.

Since we are almost exclusively dealing with outfix-guided insertion, in the following for notational simplicity we write just $\leftarrow$ in place of $\overset{\text{ogi}}{\leftarrow}$. Outfix-guided insertion is extended in the usual way for languages by setting $L_1 \leftarrow L_2 = \bigcup_{w_i \in L_i, i=1,2} w_1 \leftarrow w_2$.

Example 2. Outfix-guided insertion is not associative. Let $\Sigma = \{a, b, c, d\}$. Now $abcd \in (acd \leftarrow abc) \leftarrow abcd$ but $abc \leftarrow abcd = \emptyset$.

Since outfix-guided insertion is non-associative we define the $(i+1)$th iterated operation (analogously as was done with iterated overlap assembly [4]) by inserting to a string of the ith iteration another string of the ith iteration.

Definition 3. *For a language* L *define inductively*

$$\mathbb{OGI}^{(0)}(L) = L, \quad \textit{and } \mathbb{OGI}^{(i+1)}(L) = \mathbb{OGI}^{(i)}(L) \leftarrow \mathbb{OGI}^{(i)}(L),\ \ i \geq 0.$$

The outfix-guided insertion closure *of* L *is* $\mathbb{OGI}^*(L) = \bigcup_{i=0}^{\infty} \mathbb{OGI}^{(i)}(L)$.

For talking about specific iterated outfix-guided insertions, we use the notation $x \overset{[y]}{\Rightarrow} z$ to indicate that string z is in $x \leftarrow y$, $x, y, z \in \Sigma^*$. A sequence of steps

$$x \overset{[y_1]}{\Rightarrow} z_1 \overset{[y_2]}{\Rightarrow} z_2 \overset{[y_3]}{\Rightarrow} \cdots \overset{[y_m]}{\Rightarrow} z_m,\ \ m \geq 1,$$

is called a derivation of z_m from x.

When we want to specify the matched substrings, they are indicated by underlining. If $x = x_1uvx_2$ derives z by inserting uzv (where u and v are the matched prefix and suffix, respectively,) this denoted

$$x_1\underline{uv}x_2 \overset{\underline{u}y\underline{v}}{\Rightarrow} z.$$

Also, sometimes underlining is done only in the inserted string if this makes it clear what must be the matched substrings in the original string.

By a *trivial derivation step* we mean a derivation $x \overset{[x]}{\Rightarrow} x$ where x is obtained from itself by selecting the outfix to consist of the entire string x. Every string of length at least two can be obtained from itself using a trivial derivation step. This means, in particular, that for any language L, $L - (\Sigma \cup \{\varepsilon\}) \subseteq \mathbb{OGI}^{(1)}(L)$. The sets $\mathbb{OGI}^{(i)}(L)$, $i \geq 1$, cannot contain strings of length less than two and, consequently $\mathbb{OGI}^{(i)}(L) \subseteq \mathbb{OGI}^{(i+1)}(L)$, for all $i \geq 1$.

Definition 3 iterates the outfix-guided insertion by inserting a string from the ith iteration of the operation into another string in the ith iteration. Since the operation is non-associative we can define iterated insertion in more than one way. The right one-sided insertion of L_2 into L_1 outfix-guided inserts a string of L_2 into L_1 and the iteration of the operation inserts a string obtained in the process into L_1. The iterated left one-sided outfix-guided insertion is defined symmetrically. In fact, when considering iterated ordinary insertion, Kari [12] uses a definition that we call left one-sided iterated insertion (and the operation was defined as $\mathbb{LI}^*(L_1, L_2)$ above). Csuhaj-Varju et al. [2] define iterated overlap assembly using right one-sided iteration of the operation.

Definition 4. Let L_1 and L_2 be languages. The *right one-sided iterated insertion of L_2 into L_1* is defined inductively by setting $\mathbb{ROGI}^{(0)}(L_1, L_2) = L_2$ and $\mathbb{ROGI}^{(i+1)}(L_1, L_2) = L_1 \leftarrow \mathbb{ROGI}^{(i)}(L_1, L_2)$, $i \geq 0$. The *right one-sided insertion closure* of L_2 into L_1 is $\mathbb{ROGI}^*(L_1, L_2) = \bigcup_{i=0}^{\infty} \mathbb{ROGI}^{(i)}(L_1, L_2)$.

The *left one-sided iterated insertion of L_2 into L_1* is defined inductively by setting $\mathbb{LOGI}^{(0)}(L_1, L_2) = L_1$ and $\mathbb{LOGI}^{(i+1)}(L_1, L_2) = \mathbb{LOGI}^{(i)}(L_1, L_2) \leftarrow L_2$, $i \geq 0$. The *left one-sided insertion closure* of L_2 into L_1 is $\mathbb{LOGI}^*(L_1, L_2) = \bigcup_{i=0}^{\infty} \mathbb{LOGI}^{(i)}(L_1, L_2)$.

The one-sided iterated insertion closures are defined for two argument languages. Naturally it would be possible to extend also the definition of unrestricted iterated outfix-guided insertion for two arguments. Note that for any language L, $\mathbb{OGI}^{(1)}(L) = \mathbb{LOGI}^{(1)}(L, L) = \mathbb{ROGI}^{(1)}(L, L) = L \leftarrow L$. On the other hand, the iterated version of unrestricted outfix-guided insertion is considerably more general than the one-sided variants. For any language L, $\mathbb{ROGI}^*(L, L)$ and $\mathbb{LOGI}^*(L, L)$ are always included in $\mathbb{OGI}^*(L)$ and, in general, the inclusions can be strict.

Example 5. Let $\Sigma = \{a, b, c\}$ and $L_1 = \{aacc\}$, $L_2 = \{abc\}$. Now $\mathbb{ROGI}^*(L_1, L_2) = a^+bc^+$. For example, by inserting abc into $aacc$ derives $aabcc$:

$$a\underline{acc} \overset{\underline{a}b\underline{c}}{\Rightarrow} aabcc. \tag{1}$$

A right one-sided iterated insertion of L_2 into L_1 could then be continued, for example, as $a\underline{acc} \overset{\underline{a}ab\underline{cc}}{\Rightarrow} aaabcc$. In this way right one-sided derivations can generate all strings of a^+bc^+. Since all inserted strings must contain the symbol b, the first matched part must always belong to a^+ and the second matched part must belong to c^+. This means that $\mathbb{ROGI}^*(L_1, L_2) \subseteq a^+bc^+$.

On the other hand, $\mathbb{LOGI}^*(L_1, L_2) = \{aabcc, aacc\}$. In a left one-sided iterated insertion of L_2 into L_1, the only non-trivial derivation step is (1).

By denoting $L_3 = L_1 \cup L_2$, it can be verified that

$$\mathbb{OGI}^*(L_3) = \mathbb{ROGI}^*(L_3, L_3) = \mathbb{LOGI}^*(L_3, L_3) = a^+bc^+ \cup a^2a^*c^2c^*.$$

The next example illustrates that unrestricted outfix-guided insertion closure of a language L' can be larger than $\mathbb{LOGI}^*(L', L')$. The language L used in the proof of Theorem 9 in the next section gives an example where the unrestricted insertion closure is larger than $\mathbb{ROGI}^*(L, L)$ (as explained before Proposition 15).

Example 6. Let $\Sigma = \{a, b, c, d, e, f\}$ and $L' = \{abce, bcde, acdef\}$. We note that $a\underline{bce} \overset{\underline{bcde}}{\Rightarrow} abcde$. Furthermore, it is easy to verify that by outfix-guided inserting strings of L' into $L' \cup \{abcde\}$ one cannot produce more strings and, thus, $\mathbb{LOGI}^*(L', L') = L' \cup \{abcde\}$. On the other hand, we have

$$\underline{acdef} \overset{a\underline{bcde}}{\Rightarrow} abcdef \in \mathbb{OGI}^{(2)}(L').$$

3 Outfix-Guided Insertion and Regular Languages

As can be expected, the family of regular languages is closed under outfix-guided insertion. On the other hand, the answer to the question whether regular languages are closed under outfix-guided insertion closure seems less clear. From Kari [12] we recall that it is easy to construct examples that establish the non-closure of regular languages under iterated non-overlapping insertion. However, such straightforward counter-examples do not work for the unrestricted outfix-guided insertion closure. Using a more involved construction we establish that even the outfix-guided insertion closure of a finite language need not be regular. The nonclosure of regular languages under right one-sided insertion closure is established by a more straightforward construction.

Proposition 7. *If L_1 and L_2 are regular, then so is $L_1 \leftarrow L_2$.*

It seems difficult to extend the proof of Proposition 7 for outfix-guided insertion closure because on strings with iterated insertions, the computations on corresponding prefix-suffix pairs can, in general, depend on each other and when processing a part inserted in between, an NFA would need to keep track of such pairs, as opposed to simply keep track of a set of states. On the other hand, it is not equally easy as in the case of non-overlapping iterated insertion to construct a counter-example, i.e., a regular language whose outfix-guided insertion closure is nonregular.

Next we show that regular languages, indeed, are not closed under iterated outfix-guided insertion. For the construction we use the following technical lemma.

Lemma 8. *Let* $\Sigma = \{a_1, a_2, a_3, b_1, b_2, b_3\}$ *and define*

$$L_1 = \{a_3a_1a_2b_1,\ a_2b_2b_1b_3,\ a_1a_2a_3b_2,\ a_3b_3b_2b_1,\ a_2a_3a_1b_3,\ a_1b_1b_3b_2\}.$$

Then $L_1 \leftarrow L_1 = L_1$.

Theorem 9. *There exists a finite language* L *such that* $\mathbb{OGI}^*(L)$ *is nonregular.*

Proof *(Sketch).* Let $\Sigma = \{a_1, a_2, a_3, b_1, b_2, b_3\}$ and define

$$L = \{\$a_3a_1b_1b_3\$,\ a_3a_1a_2b_1,\ a_2b_2b_1b_3,\ a_1a_2a_3b_2,\ a_3b_3b_2b_1,\ a_2a_3a_1b_3,\ a_1b_1b_3b_2\}.$$

Note that $L - \{\$a_3a_1b_1b_3\$\}$ is equal to the language L_1 from Lemma 8. For ease of discussion we introduce names for the strings of L_1: $y_1 = a_3a_1a_2b_1$, $y_2 = a_2b_2b_1b_3$, $y_3 = a_1a_2a_3b_2$, $y_4 = a_3b_3b_2b_1$, $y_5 = a_2a_3a_1b_3$, $y_6 = a_1b_1b_3b_2$. and define the finite set

$$S_{\text{middle}} = \{a_1b_1,\ a_1a_2b_1, a_1a_2b_2b_1,\ a_1a_2a_3b_2b_1,\ a_1a_2a_3b_3b_2b_1,\ a_1a_2a_3a_1b_3b_2b_1\}.$$

We claim that

$$\mathbb{OGI}^*(L) = \{\$a_3(a_1a_2a_3)^i z(b_3b_2b_1)^i b_3\$ \mid i \geq 0,\ z \in S_{\text{middle}}\}. \tag{2}$$

To establish the inclusion from right to left, we note that

$$\$a_3a_1b_1b_3\$ \overset{[y_1]}{\Rightarrow} \$a_3a_1a_2b_1b_3\$ \overset{[y_2]}{\Rightarrow} \$a_3a_1a_2b_2b_1b_3\$ \overset{[y_3]}{\Rightarrow} \$a_3a_1a_2a_3b_2b_1b_3\$ \overset{[y_4]}{\Rightarrow}$$
$$\$a_3a_1a_2a_3b_3b_2b_1b_3\$ \overset{[y_5]}{\Rightarrow} \$a_3a_1a_2a_3a_1b_3b_2b_1b_3\$ \overset{[y_6]}{\Rightarrow} \$a_3a_1a_2a_3a_1b_1b_3b_2b_1b_3\$ = w_1.$$

The first five insertions generate the strings $\$a_3zb_3\$$, $z \in S_{\text{middle}}$, and the last string w_1 again has "middle part" $a_3a_1b_1b_3$. By cyclically outfix-guided inserting the strings $y_1, \ldots y_6$ into w_1 we get all strings $\$a_3(a_1a_2a_3)z(b_3b_2b_1)b_3\$$, $z \in S_{\text{middle}}$, and the string $\$a_3(a_1a_2a_3)^2a_1b_1(b_3b_2b_1)^2b_3\$$. By simple induction it follows that $\mathbb{OGI}^*(L)$ contains the right side of (2).

To establish the converse inclusion, we verify using Lemma 8 that all strings obtained by iterated outfix-guided insertion from strings of L must be obtained as above, that is, all non-trivial derivations producing new strings must be as above. ∎

We conjecture that the iterated outfix-guided insertion closure of a regular language need not be even context-free. However, a construction of such a language would seem to be considerably more complicated than the construction used in the proof of Theorem 9.

Contrasting the result of Theorem 9 we show that unary regular languages are closed under iterated outfix-guided insertion. The construction is based on a technical lemma which shows that, for unary languages, outfix-guided insertion closure can be represented as a variant of the iterated overlap assembly [2,4].

Definition 10. Let $x, y \in \Sigma^*$. The *2-overlap catenation* of x and y, $x \overline{\odot}^2 y$, is defined as

$$x \overline{\odot}^2 y = \{z \in \Sigma^+ \mid (\exists u, w \in \Sigma^*)(\exists v \in \Sigma^{\geq 2})\ x = uv, y = vw, z = uvw\}.$$

For $L \subseteq \Sigma^*$, we define inductively $2\mathbb{OC}^{(0)}(L) = L$ and $2\mathbb{OC}^{(i+1)}(L) = 2\mathbb{OC}^{(i)}(L) \overline{\odot}^2 2\mathbb{OC}^{(i)}(L)$, $i \geq 0$. The *2-overlap catenation closure* of L is $2\mathbb{OC}^*(L) = \bigcup_{i=0}^{\infty} 2\mathbb{OC}^{(i)}(L)$.

Due to commutativity of unary languages we get the following property which will be crucial for establishing closure of unary regular languages under outfix-guided insertion closure.

Lemma 11. *If $x, y \in a^*$ are unary strings, then $x \leftarrow y = x \overline{\odot}^2 y$.*

Corollary 12. *If L is a unary language then $\mathbb{OGI}^*(L) = 2\mathbb{OC}^*(L)$.*

The 2-overlap closure of a regular language is always regular. The construction does not depend on a language being unary, so we state the result for regular languages over an arbitrary alphabet. Csuhaj-Varju et al. [2] have shown that iterated overlap assembly preserves regularity. The proof of Lemma 13 is inspired by Theorem 4 of [2] but does not follow from it because [2] defines iteration of operations as right one-sided iteration and, furthermore, 2-overlap catenation has an additional length restriction on the overlapping strings.

Lemma 13. *The 2-overlap catenation closure of a regular language is regular.*

By Corollary 12 and Lemma 13 we have shown that unary regular languages are closed under outfix-guided insertion closure, contrasting the result of Theorem 9 for general regular languages.

Theorem 14. *The outfix-guided insertion closure of a unary regular language is always regular.*

The left and right one-sided insertion closures are restricted variants of the general outfix-guided insertion closure, so Theorem 9 does not, at least not directly, imply the existence of regular languages L_1 and L_2 such that $\mathbb{LOGI}^*(L_1, L_2)$ or $\mathbb{ROGI}^*(L_1, L_2)$ are non-regular. Here we show that the one-sided outfix-guided insertion closures are not, in general, regularity preserving. For left-one one-sided outfix-guided insertion closure the construction is similar to that used in the proof of Theorem 9. However, this construction does not work for right one-sided closure because if L is the language used in the proof of Theorem 9, then $\mathbb{ROGI}^*(L, L)$ is the finite language $L \cup \{\$a_3a_1a_2b_1b_3\$\}$.

Proposition 15. *There exist finite languages L_1, L_2, L_3 and L_4 such that $\mathbb{ROGI}^*(L_1, L_2)$ and $\mathbb{LOGI}^*(L_3, L_4)$ are non-regular.*

4 Outfix-Guided Insertion and Context-Free Languages

It is well known that the family of context-free languages is closed under ordinary insertion. Contrasting the result of Proposition 7 we show that context-free languages are not closed under (non-iterated) outfix-guided insertion.

Theorem 16. *There exists a context-free language L such that $L \leftarrow L$ is not context-free.*

It follows that context-free languages are not closed under one-sided outfix-guided iteration because, for any language L, $\mathbb{OGI}^{(1)}(L) = \mathbb{ROGI}^{(1)}(L, L) = \mathbb{LOGI}^{(1)}(L, L) = L \leftarrow L$. On the other hand, the outfix-guided insertion of a regular (respectively, context-free) language into a context-free (respectively, regular) language is always regular.

Theorem 17. *If L_1 is context-free and L_2 is regular, then $L_1 \leftarrow L_2$ and $L_2 \leftarrow L_1$ are context-free.*

The analogy of Theorem 17 does not hold for deterministic context-free languages. Techniques for proving that a language is not deterministic context-free are known already from [9].

Theorem 18. *If L_1 is deterministic context-free and L_2 is regular, the languages $L_1 \leftarrow L_2$ or $L_2 \leftarrow L_1$ need not be deterministic context-free.*

Theorem 16 raises the question how complex languages can be obtained from context-free languages using iterated outfix-guided insertion. Note that if L_1 and L_2 are context-free, it is easy to verify that $L_1 \leftarrow L_2$ is at least deterministic context-sensitive.

Proposition 19. *If L_1 and L_2 are context-free then $\mathbb{ROGI}^*(L_1, L_2)$ and $\mathbb{LOGI}^*(L_1, L_2)$ are context-sensitive.*

In the proof of Proposition 19 it is sufficient to know that the languages L_1 and L_2 are context-sensitive, and as a consequence it follows that context-sensitive languages are closed under one-sided outfix-guided insertion closure.

Corollary 20. *If L_1 and L_2 are context-sensitive then so are $\mathbb{ROGI}^*(L_1, L_2)$ and $\mathbb{LOGI}^*(L_1, L_2)$.*

We conjecture that, for any context-free language L, $\mathbb{OGI}^*(L)$ must be context-sensitive. Constructing a linear bounded automaton for $\mathbb{OGI}^*(L)$ is more difficult than in the case of the right or left one-sided insertion closures, because a direct simulation of a derivation of $w \in \mathbb{OGI}^*(L)$ (i.e., simulation of the iterated outfix-guided insertion steps producing w) would need to remember, at a given time, an unbounded number of substrings of the input.

Also we do not know how to make the procedure in the proof of Proposition 19 deterministic and it remains open whether the one-sided outfix-guided insertion closures of context-free languages are always deterministic context-sensitive.

5 Deciding Closure Under Outfix-Guided Insertion

We say that a language L is *closed* under outfix-guided insertion, or *og-closed* for short, if outfix-guided inserting strings of L into L does not produce strings outside of L, that is, $(L \leftarrow L) \subseteq L$.

A natural algorithmic problem is then to decide for a given language L whether or not L is og-closed. If L is regular, by Proposition 7, we can decide whether or not L is og-closed. For a given DFA A, Proposition 7 yields only an NFA for the language $L(A) \leftarrow L(A)$. In general, the NFA equivalence or inclusion problem is PSPACE complete [22], however, inclusion of an NFA language in the language $L(A)$ can be tested efficiently when A is deterministic.

Proposition 21. *There is a polynomial time algorithm to decide whether for a given DFA A the language $L(A)$ is og-closed.*

The method used in Proposition 21 does not yield an efficient algorithm if the regular language L is specified by an NFA. The complexity of deciding og-closure of a language accepted by an NFA remains open. On the other hand, using a reduction from the Post Correspondence Problem it follows that the question whether or not a context-free language is og-closed in undecidable.

Theorem 22. *For a given context-free language L, the question whether or not L is og-closed is undecidable.*

6 Conclusion

Analogously with the recent overlap assembly operation [2,4], we have introduced an overlapping insertion operation on strings and have studied closure and decision properties of the outfix-guided insertion operation. While closure properties of non-iterated outfix-guided insertion are straightforward to establish, the questions become more involved for the outfix-guided insertion closure. As the main result we have shown that the outfix-guided insertion closure of a finite language need not be regular.

Much work remains to be done on outfix-guided insertion. One of the main open questions is to determine upper bounds for the complexity of the outfix-guided insertion closures of regular languages. Does there exist regular languages L such that the outfix-guided insertion closure of L is non-context-free?

Acknowledgments. Cho and Han were supported by the Basic Science Research Program through NRF funded by MEST (2015R1D1A1A01060097), the Yonsei University Future-leading Research Initiative of 2015 and the International Cooperation Program managed by NRF of Korea (2014K2A1A2048512). Ng and Salomaa were supported by Natural Sciences and Engineering Research Council of Canada Grant OGP0147224.

References

1. Bertram, J.S.: The molecular biology of cancer. Mol. Asp. Med. **21**(6), 167–223 (2000)
2. Csuhaj-Varju, E., Petre, I., Vaszil, G.: Self-assembly of strings and languages. Theoret. Comput. Sci. **374**, 74–81 (2007)
3. Daley, M., Kari, L., Gloor, G., Siromoney, R.: Circular contextual insertions/deletions with applications to biomolecular computation. In: String Processing and Information Retrieval Symposium, pp. 47–54 (1999)
4. Enaganti, S., Ibarra, O., Kari, L., Kopecki, S.: On the overlap assembly of strings and languages. Nat. Comput. (2016). dx.doi.org/10.1007/s11047-015-9538-x
5. Enaganti, S.K., Ibarra, O.H., Kari, L., Kopecki, S.: Further remarks on DNA overlap assembly, manuscript (2016)
6. Enaganti, S.K., Kari, L., Kopecki, S.: A formal language model of dna polymerase enzymatic activity. Fundam. Inform. **138**, 179–192 (2015)
7. Flavell, R., Sabo, D., Bandle, E., Weissmann, C.: Site-directed mutagenesis: effect of an extracistronic mutation on the in vitro propagation of bacteriophage qbeta RNA. Proc. Natl. Acad. Sci. **72**(1), 367–371 (1975)
8. Galiukschov, B.: Semicontextual grammars (in Russian). Mat. Log. Mat. Lingvistika 38–50 (1981)
9. Ginsburg, S., Greibach, S.: Deterministic context free languages. Inf. Control **9**, 620–648 (1966)
10. Haussler, D.: Insertion languages. Inf. Sci. **31**, 77–89 (1983)
11. Hemsley, A., Arnheim, N., Toney, M.D., Cortopassi, G., Galas, D.J.: A simple method for site-directed mutagenesis using the polymerase chain reaction. Nucleic Acids Res. **17**(16), 6545–6551 (1989)
12. Kari, L.: On insertion and deletion in formal languages. Ph.D. thesis, University of Turku (1991)
13. Kari, L., Thierrin, G.: Contextual insertions/deletions and computability. Inf. Comput. **131**(1), 47–61 (1996)
14. Krassovitskiy, A., Rogozhin, Y., Verlan, S.: Computational power of insertion-deletion (P) systems with rules of size two. Nat. Comput. **10**, 835–852 (2011)
15. Lee, J., Shin, M.K., Ryu, D.K., Kim, S., Ryu, W.S.: Insertion and deletion mutagenesis by overlap extension PCR. In: Braman, J. (ed.) In Vitro Mutagenesis Protocols, 3rd edn, pp. 137–146. Humana Press, New York (2010)
16. Liu, H., Naismith, J.H.: An efficient one-step site-directed deletion, insertion, single and multiple-site plasmid mutagenesis protocol. BMC Biotechnol. **8**(1), 91–101 (2008)
17. Margenstern, M., Păun, G., Rogozhin, Y., Verlan, S.: Context-free insertion-deletion systems. Theoret. Comput. Sci. **330**(2), 339–348 (2005)
18. Păun, G., Pérez-Jiménez, M.J., Yokomori, T.: Representations and characterizations of languages in Chomsky hierarchy by means of insertion-deletion systems. Int. J. Found. Comput. Sci. **19**(4), 859–871 (2008)
19. Păun, G.: On semicontextual grammars. Bull. Math. Soc. Sci. Math. Rouman. **28**, 63–68 (1984)
20. Shallit, J.: A Second Course in Formal Languages and Automata Theory. Cambridge University Press, Cambridge (2009)
21. Takahara, A., Yokomori, T.: On the computational power of insertion-deletion systems. Nat. Comput. **2**, 321–336 (2003)
22. Yu, S.: Regular languages. In: Salomaa, A., Rozenberg, G. (eds.) Handbook of Formal Languages, vol. I, pp. 41–110. Springer, Heidelberg (1997)

Both Ways Rational Functions

Christian Choffrut and Bruno Guillon(✉)

IRIF, CNRS and Université Paris 7 Denis Diderot, Paris, France
guillonb@liafa.unif-paris-diderot.fr

Abstract. We consider binary relations on words which can be recognized by finite two-tape devices in two different ways: the traditional way where the two tapes are scanned in the same direction and a new one where they are scanned in different directions. The devices of the former type define the family of rational relations, while those of the latter define an a priori really different family. We characterize the partial functions that are in the intersection of the two families. We state a conjecture for the intersection for general, nonfunctional, relations.

Keywords: Rational relations · Finite automata · Two-way transducers · Two-tape automata · Word relations

1 Introduction

A binary word relation is a set of pairs of words, i.e., a subset of the direct product of two free monoids. Rabin and Scott introduced in 1959 the notion of finite two-tape (actually multitape) automata as a natural extension of finite (one-tape) automata and used them as recognition device for pairs of words, [8]. The two words are stored on two tapes and are scanned at different speeds but in the same direction, from left to right. The machine has no write capability and its memory is finite. In 1965 Elgot and Mezei proved a Kleene-like theorem showing that the set of relations thus defined is precisely the set of rational subsets of the direct product of two free monoids, [4]. Decision issues were investigated in [6].

The question we tackle is the following. Modify the Rabin-Scott model so that scanning the two-tapes is done in opposite directions, the remaining features being otherwise kept. Under which condition can the same relation be recognized in these two different models? We somehow improperly call *both ways rational* the family BwRat of such binary relations which is the subject of this contribution.

We now explain how we came across the problem. Three years ago we started investigating the expressive power of two-way transducers which are nothing more than finite two-way automata provided with a one-way output tape. They can be viewed as accepting devices with two tapes which, contrary to Rabin-Scott model, play asymmetric roles since one is two-way (traditionally viewed as an input tape) and the other is one-way (the output tape). The two-way transducers still remain ill-understood except for the deterministic case, [5], which is no wonder because

S. Brlek and C. Reutenauer (Eds.): DLT 2016, LNCS 9840, pp. 114–124, 2016.
DOI: 10.1007/978-3-662-53132-7_10

two-way automata obtained from two-way transducers by eliminating the one-way output tape pose challenging longstanding open problems such as that of the cost of simulating a nondeterministic machine (a 2NFA) by a deterministic one (a 2DFA), [10]. What is lacking for transducers is operations on relations that would mimic the behavior of the tapes. In [2] we used the notions of Hadamard product and Hadamard star of a relation which capture the idea that the input tape can be scanned repeatedly from left to right. This happens to be sufficient in the case where the two tapes contain words on unary alphabets but it is clearly too weak to solve the general case, [7].

The literature on one-tape two-way automata considers weaker versions of the model such as sweeping [11] or rotating [10] automata. We proceed similarly by imposing the most possible drastic restrictions on the move of the input head of two-way transducers: the head makes a unique traversal from left to right or it makes a unique traversal from right to left. Of course the first constraint is equivalent to the initial model of Rabin and Scott but not the second. The both ways rational relations are the relations that are recognized by both of these two restricted transducers.

Now in order to present our main result we need to introduce a couple of definitions. The *left-reverse* of a relation in $\Sigma^* \times \Delta^*$ is the relation obtained by taking the mirror image of the first component of its elements. A relation is *factorizable* if it is a composition[1] of two rational relations through a unary alphabet, i.e., if it is equal to $R \circ S$ for some rational $R \subseteq \Sigma^* \times \Gamma^*$ and $S \subseteq \Gamma^* \times \Delta^*$ with $|\Gamma| = 1$.

In this paper we settle the case where R is the graph of a partial function (abbreviated as "function"), i.e., for all $u \in \Sigma^*$, $v, w \in \Delta^*$ the condition $(u, v), (u, w) \in R$ implies $v = w$. For this particular case, we give the characterization below.

Theorem 1. *Given a function $f : \Sigma^* \to \Delta^*$ the following conditions are equivalent:*

(1) f is both ways rational
(2) f is factorizable
(3) f is rational and its image[2] is a finite union of subsets of the form xy^*z for some $x, y, z \in \Delta^*$.

This characterization yields a procedure for testing membership of a rational function to BwRat. We conjecture that the equivalence of the first two points extends to the whole of BwRat. The arguments in favor of this claim is that all factorizable relations are both ways rational and that these two families satisfy the same closure properties under the usual operations. It is not difficult to work out an example of nonfunctional relation in BwRat that does not satisfy Point 3.

---

[1] We compose the relations from left to right: $R \circ S$ denotes the relation $\{(u, v) \mid \exists w,\ (u, w) \in R \text{ and } (w, v) \in S\}$.

[2] The image of a relation R is the subset $\{v \in \Delta^* \mid \exists u \in \Sigma^*,\ (u, v) \in R\}$.

We now turn to the discussion of the material of this manuscript. In Sect. 2 we collect all the basic definitions along with the different operators on binary relations such as inverse and the three types of reversals. In Sect. 3 we concentrate on the two families BwRat and Fact and show that the second one is included in the first one. We investigate their closure and non-closure properties and observe that these two families behave alike. The main result, which characterizes the relations in BwRat that are functions, is proved in Sect. 4. We evaluate the complexity of determining whether or not a given rational relation is in BwRat.

Due to space constraints, we are obliged to omit some proofs.

2 Preliminaries

In this section, we recall the two main families of subsets of a given monoid M, and their well-known properties and characterizations. Then we introduce some additional operations on relations, namely the inverse and three kinds of reversals.

2.1 Rational and Recognizable Subsets

The family of *rational* subsets, denoted $\text{Rat}(M)$, is the smallest family $\mathcal{F}$ of subsets of M which contains the finite subsets and which is closed under set union, *set concatenation* ($X, Y \in \mathcal{F} \Rightarrow X \cdot Y = \{xy \mid x \in X,\ y \in Y\} \in \mathcal{F}$) and *Kleene star* ($X \in \mathcal{F} \Rightarrow X^* = \{x_1 \cdots x_n \mid n \geq 0,\ x_i \in X\} \in \mathcal{F}$). We use the convention that the product $x_1 \cdots x_n$ reduces to the identity element of the monoid if $n = 0$.

A subset $X \subseteq M$ is *recognizable* if it is the inverse image of a morphism of M onto a finite monoid. The family of such relations is denoted $\text{Rec}(M)$. When M is a free monoid, Kleene Theorem asserts that $\text{Rat}(M) = \text{Rec}(M)$ holds.

2.2 Free Monoids and Direct Products Thereof

We denote by Σ^* the free monoid generated by the set Σ. Its elements are *words* and its identity element is the *empty word* denoted 1. For all words $u \in \Sigma^*$, we denote by $|u|$ its *length*, by $|u|_c$ the number of occurrences of the letter c in u and by $\overline{u}$ the *reverse* of u, i.e., if $u = a_1 \cdots a_n$ we have $\overline{u} = a_n \cdots a_1$ and we set $\overline{1} = 1$.

We are mainly interested in direct products of free monoids, say $\Sigma^* \times \Delta^*$, where the operation is the componentwise concatenation: $(u_1, v_1)(u_2, v_2) = (u_1u_2, v_1v_2)$. We use the term "relations" for their subsets. In the case of direct products of free monoids, the previous two families of subsets possess nice characterizations which facilitate the study of their properties.

We start by considering $\text{Rat}(\Sigma^* \times \Delta^*)$. We take for granted that the reader is acquainted with the classical notion of finite one tape deterministic and nondeterministic finite automaton, [9, Chapter IV]. A *two-tape nondeterministic automaton* introduced by Rabin and Scott in 1965, [8], is a structure

$\mathcal{A} = (Q, \Sigma, \Delta, I, E, F)$ where Q is a finite set of *states*, $I \subseteq Q$ and $F \subseteq Q$ are respectively the subsets of *initial* and *final* states and where

$$E \subseteq (Q \times \Sigma \times \{1\} \times Q) \cup (Q \times \{1\} \times \Delta \times Q) \tag{1}$$

is a set of *transitions*. The notions of *successful* paths, *labels* and subsets *accepted* by $\mathcal{A}$ are natural extensions of those of ordinary finite automata. In particular the label of a path is the componentwise concatenation of the labels of the transitions in the path. The fundamental result is that the family of relations accepted by two-tape finite automata is precisely the family $\mathrm{RAT}(\Sigma^* \times \Delta^*)$ of rational subsets of the monoid $\Sigma^* \times \Delta^*$ [4].

The second family of relations is characterized as follows.

Theorem 2 (Elgot-Mezei, [4]). *A subset $R \subseteq \Sigma^* \times \Delta^*$ is recognizable if and only if it is a finite union of subsets of the form $X \times Y$ where $X \in \mathrm{REC}(\Sigma^*)$ and $Y \in \mathrm{REC}(\Delta^*)$.*

We assume the reader is familiar with the main closure properties of rational relations, such as the composition of relations, the intersection with recognizable relations, and the strict inclusion of $\mathrm{REC}(\Sigma^* \times \Delta^*)$ in $\mathrm{RAT}(\Sigma^* \times \Delta^*)$ when Σ and Δ are nonempty, cf. [1,3,9].

2.3 Elementary Operators on Binary Relations

Our main result is on a decomposition of relations. Since we are led to manipulate compositions of (binary) relations, we find it more appropriate to compose them left to right. Thus, for two binary relations R and S, the *composition* $R \circ S$ is the relation $\{(x, y) \mid \exists z,\ (x, z) \in R \text{ and } (z, y) \in S\}$. We identify partial functions, simply *functions* in the sequel, with binary relations R such that $(x, y),\ (x, z) \in R$ implies $y = z$. Given an alphabet Θ, we define the *identity* I_Θ and its *reverse* J_Θ as follows:

$$\mathrm{I}_\Theta = \{(u, u) \mid u \in \Theta^*\} \qquad \text{and} \qquad \mathrm{J}_\Theta = \{(\overline{u}, u) \mid u \in \Theta^*\}.$$

Observe that I_Θ is rational but J_Θ is not [1, p. 65] and that $\mathrm{J}_\Theta \circ \mathrm{J}_\Theta = \mathrm{I}_\Theta$.

Definition 1. *Given $R \subseteq \Sigma^* \times \Delta^*$, we set*

- inverse *of R:* $R^{-1} = \{(v, u) \mid (u, v) \in R\}$,
- reversal *of R:* $\mathrm{J}_\Sigma \circ R \circ \mathrm{J}_\Delta = \{(\overline{u}, \overline{v}) \mid (u, v) \in R\}$,
- left-reversal *of R:* $\mathrm{J}_\Sigma \circ R = \{(\overline{u}, v) \mid (u, v) \in R\}$,
- right-reversalof *R:* $R \circ \mathrm{J}_\Delta = \{(u, \overline{v}) \mid (u, v) \in R\}$.

The following proposition is not difficult to check.

Proposition 1. *Given a relation $R \subseteq \Sigma^* \times \Delta^*$, the next claims are equivalent: (1) R is rational, (2) R^{-1} is rational, (3) $\mathrm{J}_\Sigma \circ R \circ \mathrm{J}_\Delta$ is rational.*

As an immediate consequence, we get:

Corollary 1. *Let $R \subseteq \Sigma^* \times \Delta^*$. Then $\mathrm{J}_\Sigma \circ R$ is rational if and only if $R \circ \mathrm{J}_\Delta$ is rational.*

3 Both Ways Rational Relations

3.1 Formal Definitions

We recall the definition of the families of relations under study.

Definition 2. *A relation* $R \subseteq \Sigma^* \times \Delta^*$ *is* both ways rational *if both* R *and* $\mathrm{J}_\Sigma \circ R$ *are rational. The family of such relations is denoted* BwRat$(\Sigma^* \times \Delta^*)$ *or simply* BwRat *when it is clear from the context.*

Definition 3. *A rational relation* $R \subseteq \Sigma^* \times \Delta^*$ *is* factorizable *if there exists a unary alphabet* $\Gamma = \{a\}$ *and two rational relations* $S \subseteq \Sigma^* \times \Gamma^*$ *and* $T \subseteq \Gamma^* \times \Delta^*$ *such that* $R = S \circ T$. *The family of such relations is denoted* Fact$(\Sigma^* \times \Delta^*)$ *or simply* Fact *when it is clear from the context.*

The following statement gives two interesting families of examples of relations in BwRat. Since Proposition 3 asserts that the family BwRat is closed under composition, it leads to a sufficient condition for a relation to be in BwRat, namely being in Fact, (cf. Corollary 2). We suspect it is also necessary. The main result of this paper, Theorem 1, is to show that it is indeed necessary when the relation is a function, see Sect. 4.

Proposition 2. *If* $|\Sigma| = 1$ *or* $|\Delta| = 1$ *then any rational relation in* $\Sigma^* \times \Delta^*$ *is both ways rational, i.e.,* BwRat$(\Sigma^* \times \Delta^*) =$ Rat$(\Sigma^* \times \Delta^*)$.

3.2 Closure Properties

We investigate the closure properties under natural operators of the two families Fact of factorizable relations and BwRat of both ways rational relations. The main objective of this section is to support the conjecture that the two families coincide. For each operation considered, either both families are closed or both are not.

Proposition 3. *The families* BwRat *and* Fact *are closed under inverse, the three types of reversals, composition and union.*

The first consequence of the previous properties is the inclusion of Fact in BwRat.

Corollary 2. *All factorizable relations are both ways rational, i.e.,* Fact $\subseteq$ BwRat.

As a second consequence we get an interesting, though not surprising, family of examples of relations in Fact.

Corollary 3. *Let* $R \subseteq \Sigma^* \times \Delta^*$ *be rational and assume it satisfies the condition* $(u, v) \in R$ *implies* $(u, v') \in R$ *for all* $|v'| = |v|$. *Then* $R \in$ Fact.

Now we verify that recognizable relations are in Fact and thus, by Corollary 2, in BwRat. This simple result but it serves, in conjunction with Proposition 5, as a means to prove that certain relations are not in Fact or BwRat.

Proposition 4. *If R is a recognizable relation, it belongs to* FACT *and thus to* BWRAT.

Proof. By Theorem 2 and since by Proposition 3 the family FACT is closed under union, it suffices to consider the case $R = X \times Y$ with $X \in \text{REC}(\Sigma^*)$ and $Y \in \text{REC}(\Delta^*)$. Let $\Gamma = \{a\}$ for some new symbol a. Then $X \times \{a\}$ and $\{a\} \times Y$ are rational and therefore $X \times Y = X \times \{a\} \circ \{a\} \times Y$ is in FACT by definition. □

Summarizing the situation whenever $|\Sigma| > 1$ and $|\Delta| > 1$, the following inclusions hold: $\text{REC} \subsetneq \text{FACT} \subseteq \text{BWRAT} \subsetneq \text{RAT}$.

This provides us with an extra closure property because the situation is similar to that of rational relations. Indeed, the family of recognizable relations is a "small" subfamily of rational relations. It is very well-known that though the rational relations are not closed under intersection, the intersection of a rational relation and a recognizable relation is rational, e.g., [9, Proposition IV.1.8]. A similar property holds for the two families as proved in the next proposition.

Proposition 5. *Consider two relations R, $S \subseteq \Sigma^* \times \Delta^*$ where S is recognizable. If R is in* FACT *(resp. in* BWRAT*), then $R \cap S$ is in* FACT, *(resp. in* BWRAT*).*

Proof. First of all, observe that, since the intersection distributes over the union and because both the families BWRAT and FACT are closed under union, it is sufficient to prove that the results hold when $S = X \times Y$ for some $X \in \text{REC}(\Sigma^*)$ and $Y \in \text{REC}(\Delta^*)$.

R ∈ Fact: We observe the following general equality where $T_0 \subseteq \Sigma^* \times \Gamma^*$, $T_1 \subseteq \Gamma^* \times \Delta^*$, $A \subseteq \Sigma^*$ and $B \subseteq \Delta^*$.

$$(T_0 \circ T_1) \cap (A \times B) = (T_0 \cap (A \times \Gamma^*)) \circ (T_1 \cap (\Gamma^* \times B))$$

We apply this equality to the case $R \in \text{FACT}$, by specifying that $R = T_0 \circ T_1$ with T_0 and T_1 rational, $|\Gamma| = 1$, $A = X$ and $B = Y$. Then:

$$R \cap (X \times Y) = (T_0 \cap (X \times \Gamma^*)) \circ (T_1 \cap (\Gamma^* \times Y))$$

The relations $T_0 \cap (X \times \Gamma^*)$ and $T_1 \cap (\Gamma^* \times Y)$ are rational and their composition is in FACT by definition.

R ∈ BwRat: By using the equality $\text{J}_\Sigma \circ (R \cap S) = (\text{J}_\Sigma \circ R) \cap (\text{J}_\Sigma \circ S)$ and the notation $\overline{X} = \{x \mid x \in X\}$, we have:

$$\text{J}_\Sigma \circ (R \cap S) = (\text{J}_\Sigma \circ R) \cap (\text{J}_\Sigma \circ (X \times Y)) = (\text{J}_\Sigma \circ R) \cap (\overline{X} \times Y)$$

Since $\overline{X}$ is recognizable and $\text{J}_\Sigma \circ R$ is rational by hypothesis, its intersection with the recognizable relation $\overline{X} \times Y$ is also rational. □

The next result is yet another property shared by the two families.

Proposition 6. *Let $R, S \subseteq \Sigma^* \times \Delta^*$. If R is in* FACT *(resp.* BWRAT*) and if S is recognizable, then $S \cdot R$ and $R \cdot S$ are in* FACT *(resp.* BWRAT*).*

3.3 Non-closure Properties

We call *Hadamard product* of two binary relations $R, S \subseteq \Sigma^* \times \Delta^*$ the relation (cf. [2])

$$R \odot S = \{(u, vw) \in \Sigma^* \times \Delta^* \mid (u,v) \in R,\ (u,w) \in S\}.$$

The families FACT and BWRAT are not closed under concatenation, Hadamard product and Kleene star.

Proposition 7. *With $\Sigma = \{a, b, \#\}$, the relations*

$$R_1 = \{(w, a^p) \mid w \in \{a,b\}^*,\ p = |w|_a\}, \qquad R_2 = \{(\#, 1)\},$$
$$R_3 = \{(w, b^r) \mid w \in \{a,b\}^*,\ r = |w|_b\}$$

are in FACT. *However the relation $R_1 \cdot R_2 \cdot R_3$ is not in* BWRAT *and the relation $R_1 \odot R_3$ is not rational. The relation $R_1 \cdot R_2 \cup R_2 \cdot R_3$ is in* FACT *but the relation $(R_1 \cdot R_2 \cup R_2 \cdot R_3)^*$ is not in* BWRAT.

4 The Case of Functions

4.1 A Sufficient Condition Concerning the Image

The following general result is the key argument for the both ways rational functions. We recall that the *image* of a binary relation $R \subseteq \Sigma^* \times \Delta^*$ is the subset IMAGE$(R) = \{v \in \Delta^* \mid \exists u \in \Sigma^*,\ (u,v) \in R\}$.

Proposition 8. *Let $R \subseteq \Sigma^* \times \Delta^*$ be a rational relation whose image is a finite union of subsets of the form xy^*z. Then R is factorizable, cf. Definition 3.*

Proof. Indeed, assume that the image of R is IMAGE$(R) = \bigcup_{i=0}^n x_i y_i^* z_i$. Each relation $R_i = R \cap (\Sigma^* \times x_i y_i^* z_i)$ is rational because it is the intersection of a rational and a recognizable relation. Define, for each i, the rational relation $S_i = \{(x_i y_i^k z_i, a^p) \mid p = k(n+1) + i\}$ which is one-to-one, i.e., $S_i \circ S_i^{-1}$ is the identity on $x_i y_i^* z_i$. Furthermore, $(\bigcup_{i=0}^n S_i) \circ (\bigcup_{i=0}^n S_i^{-1})$ is the identity on IMAGE(R). Indeed, this relation contains all pairs $(x_i y_i^k z_i, x_j y_j^\ell z_j)$ such that $k(n+1) + i = \ell(n+1) + j$. This implies $i = j$ and $k = \ell$. Then, observing that $R \circ (\bigcup_{i=0}^n S_i) \subseteq \Sigma^* \times a^*$ and $\bigcup_{i=0}^n S_i^{-1} \subseteq a^* \times \Sigma^*$, the relation $R = (R \circ (\bigcup_{i=0}^n S_i)) \circ (\bigcup_{i=0}^n S_i^{-1})$ belongs to FACT. □

4.2 Regular Subsets of Words Invariant Under Reversal-Like Operations

As will be explained in the proof of Proposition 10, we need a result concerning the regular subsets of a free monoid consisting of words invariant under operations akin to word reversal. More precisely, let α be an involution of Γ which we extend to an antiisomorphism of the free monoid: $\alpha(uv) = \alpha(v)\alpha(u)$. We investigate the regular sets consisting of words which are invariant under α. E.g., with α equal to the identity, the operation consists of taking the reversal, in which case the subsets in question consist of palindromes.

Proposition 9. *Let Γ be an alphabet and let α be the extension to Γ^* of an antiisomorphism acting as an involution on Γ. A subset X of Σ^* is a rational subset all elements of which are invariant under α, if and only if X is a finite union of subsets of the form*

$$x(yz)^*y\alpha(x) \qquad \text{with } y = \alpha(y) \text{ and } z = \alpha(z).$$

Proof. The condition is clearly sufficient, so we prove that it is necessary. We start with a couple of lemmas concerning a trim[3] deterministic automaton recognizing X. We let q_- be its initial state and Q be its set of states. A *maximal strongly connected component*, abbreviated *SCC*, is a maximum subset of states P such that for any pair $p, q \in P$ there exists a word taking p to q. It is *trivial* if it reduces to a unique state without loop on it.

Lemma 1. *Let u, v, with $|u| \leq |v|$, taking the initial state q_- to the same nontrivial SCC. Then u is a prefix of v.*

Proof. Assume first that u and v take q_- to the same state p, i.e., $q_- \cdot u = q_- \cdot v = p$. Since p belongs to a nontrivial SCC, we may choose a word w with $|w| > |v|$, taking p to a final state. Then we have

$$uw = \alpha(uw) = \alpha(w)\alpha(u) \qquad \text{and} \qquad vw = \alpha(vw) = \alpha(w)\alpha(v),$$

and thus u and v are prefixes of $\alpha(w)$.

Now if $q_- \cdot u = p$ and $q_- \cdot v = p'$, let v' take p' to p. By the above observation u and vv' are prefixes one another, thus so are u and v. □

Lemma 2. *All nontrivial* SCC *consist of a unique cycle.*

Proof. Let q be a state in a SCC. Let w be a word taking the initial state to q and assume there exist two different simple cycles around q, say labeled by u and v. By the previous Lemma wu and wv are prefix of one another and thus so are u and v contradicting the simplicity of the cycles. □

Lemma 3. *Two nontrivial SCCs are not accessible from one another.*

Proof. Assume by contradiction that a nontrivial SCC C' is accessible from a different SCC C. Let v be a word labeling a path starting in C in state q and arriving in C' in state p. Without loss of generality, we may assume that the only state of C visited by this path is q, or equivalently that if a is the first letter of v, i.e., $v = av'$, then the state $q \cdot a$ does not belong to C. Let w label a path from the initial state q_- to q, x label a nontrivial loop around q and y label a nontrivial loop around p.

Then for some integer n the word wvy^n is longer than the word wxv. By Lemma 1 wxv is a prefix of wvy^n which implies that x and v are prefixes one another. This contradicts the fact that x and v start with two different letters. □

[3] An ϵ-free automaton is *trim* if each of its state q is *accessible* (i.e., there exists a path from the initial state to q) and *co-accessible* (i.e., there exists a path from q to some accepting state). Every finite automaton is equivalent to some trim automaton.

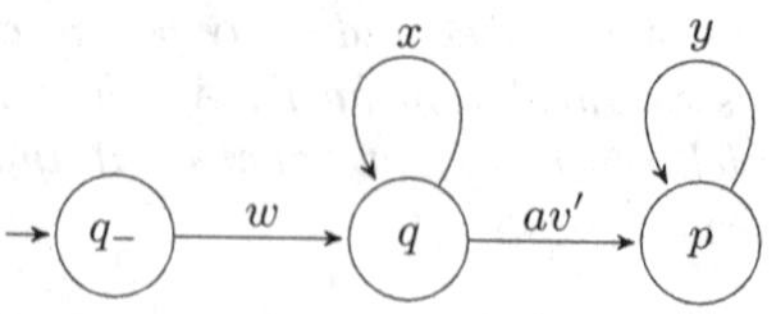

Fig. 1. Two hypothetical different connected components accessible from one another.

We now turn to the proof of the Proposition 9 which involves classical elements of combinatorics on words. We assume X is not finite otherwise there is nothing to prove. The previous lemmas show that X decomposes into a finite set along with finitely many subsets of the form xy^*z where x labels a path from the initial state to a state q belonging to a nontrivial SCC, y labels the unique cycle around q and z takes q to a final state (Fig. 1).

The elements of xy^*z are invariant under the action of α. Without loss of generality, assume that $|x| \le |z|$. Consider an integer n such that $|x|+(n-1)\,|y| \ge |z|$. In the equation $xy^nz = \alpha(z)\alpha(y)^n\alpha(x)$, by comparing the prefixes of length $|z|+|y|$ in both handsides, see Fig. 2, for some $p > 0$ we obtain $xy^py_1 = \alpha(z)\alpha(y)$ for some $y_1y_2 = y$. Since $|y| = |\alpha(y)|$ this yields $\alpha(y) = y_2y_1$ and $\alpha(z) = xy^{p-1}y_1$. Now, $y = y_1y_2$ implies $\alpha(y) = \alpha(y_2)\alpha(y_1) = y_2y_1$ and thus $y_1 = \alpha(y_1)$ and $y_2 = \alpha(y_2)$.

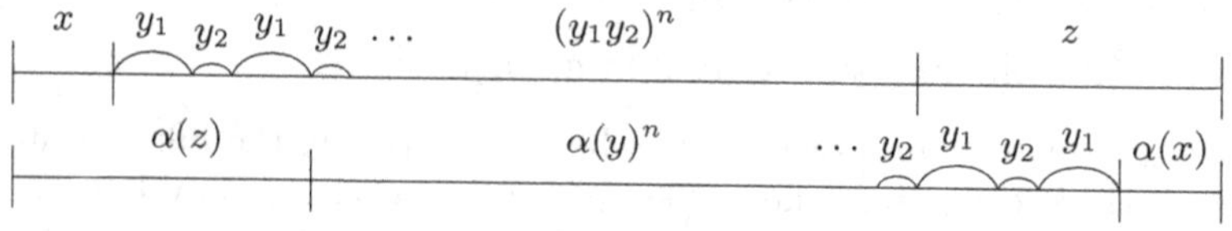

Fig. 2. Invariance of xy^nz.

Because of $z = \alpha(y_1)\alpha(y)^{p-1}\alpha(x)$ we get

$$xy^nz = xy^n\alpha(y_1)\alpha(y)^{p-1}\alpha(x) = xy^{n+p-1}y_1\alpha(x) = x(y_1y_2)^{n+p-1}y_1\alpha(x)$$

as claimed in Proposition 9.

4.3 Back to Both Ways Rational Functions

Proposition 10. *Let R be a function belonging to* BwRat. *Then its image is a finite union of subsets of the form xy^*z.*

Proof. Compute $S = R^{-1} \circ (R \circ \mathrm{J}_\Delta) = \{(v, \overline{v}) \mid \exists u \in \Sigma^*,\ (u, v) \in R\}$. If R is in BwRat then by Corollary 1 the above relation is rational and length-preserving, thus by [3, Thm IX 6.1.], S is a rational subset of the free monoid generated by the pairs in $\Theta = \Delta \times \Delta$. Let α be the involution of Θ defined by

$\alpha(a,b) = (b,a)$ and extend it to Θ^* as an antiisomorphism: $\alpha(uv) = \alpha(v)\alpha(u)$. Then all elements in S are invariant under α. By virtue of Proposition 9, S is a finite union of subsets of the form $x(yz)^*y\alpha(x)$ with $y = \alpha(y)$ and $z = \alpha(z)$. Consequently, denoting $x = (x_1, x_2)$, $y = (y_1, y_2)$, $z = (z_1, z_2)$, $\alpha(x) = (t_1, t_2) \in \Theta^*$, S is a finite union of subsets of the form

$$\{(x_1(y_1z_1)^n y_1 t_1, x_2(y_2z_2)^n y_2 t_2) \mid n \in \mathbb{N}\} \tag{2}$$

with the extra condition that $(\overline{x_2}, \overline{x_1}) = \alpha(x_1, x_2) = (t_1, t_2)$. Then by projecting Expression 2 on the first component we get the subset $x_1(y_1z_1)^*y_1\overline{x_2}$ which completes the proof. □

Proof (of Theorem 1). The implication 2 ⇒ 1 is due to Corollary 2. The implication 1 ⇒ 3 is proved in Proposition 10. The implication 3 ⇒ 2 is proved in Proposition 8. □

4.4 Complexity Considerations

Here we show briefly how the membership problem is decidable, i.e., we are given a two-tape automaton $\mathcal{A} = (Q, \Sigma, \Delta, I, E, F)$ recognizing a partial function in $\Sigma^* \to \Delta^*$ whose transition set is as in (1) and we ask whether or not it is both ways rational. By Theorem 1 it suffices to investigate the image of the function. A finite automaton recognizing the image is obtained by ignoring the second component of the quadruples in E, i.e., by considering the finite automaton with the same set of states, the same subsets of initial and final states and with the transition set

$$E' = \{(q, y, p) \mid \exists x \in \Sigma \cup \{1\}, (q, x, y, p) \in E\}.$$

Then we compute an equivalent deterministic automaton by eliminating the transitions labeled by 1 (the so-called ϵ-transitions), followed by the subset construction and finally by eliminating the states that are not accessible or coaccessible. This sequence of constructions can be achieved in exponential time relative to the number of states of $\mathcal{A}$. Then, because of Lemmas 2 and 3, we must verify that the nontrivial maximal strongly connected components of this automaton (SCC for short) consist of simple cycles which are not accessible one another. To that purpose, we run Tarjan's algorithm which computes the SCCs in linear time. Verifying that each SCC is reduced to a cycle is also done in linear time because a SCC is a simple cycle if and only if each state is the origin of a unique transition. It then remains to verify that the SCCs are not accessible one from another and this is achieved by a depth-first search.

Consequently, given a two-tape automaton $\mathcal{A}$ defining a function, testing whether or not the function is both ways rational can be determined in time exponential relative to the size of $\mathcal{A}$.

References

1. Berstel, J.: Transductions and Context-Free Languages. B. G. Teubner, Stuttgart (1979)
2. Choffrut, C., Guillon, B.: An algebraic characterization of unary two-way transducers. In: Csuhaj-Varjú, E., Dietzfelbinger, M., Ésik, Z. (eds.) MFCS 2014, Part I. LNCS, vol. 8634, pp. 196–207. Springer, Heidelberg (2014)
3. Eilenberg, S.: Automata, Languages and Machines, vol. A. Academic Press, New York (1974)
4. Elgot, C.C., Mezei, J.E.: On relations defined by generalized finite automata. IBM J. **10**, 47–68 (1965)
5. Engelfriet, J., Hoogeboom, H.: MSO definable string transductions and two-way finite-state transducers. ACM Trans. Comput. Log. **2**(2), 216–254 (2001)
6. Fischer, P.C., Rosenberg, A.L.: Multitape one-way nonwriting automata. J. Comput. Syst. Sci. **2**(1), 88–101 (1968)
7. Guillon, B.: Sweeping weakens two-way transducers even with a unary output alphabet. In: Proceedings of Seventh Workshop on NCMA 2015, Porto, Portugal, August 31 – September 1, 2015, pp. 91–108 (2015)
8. Rabin, M., Scott, D.: Finite automata and their decision problems. IBM J. Res. Dev. **3**(2), 125–144 (1959)
9. Sakarovitch, J.: Elements of Automata Theory. Cambridge University Press, New York (2009)
10. Sakoda, W.J., Sipser, M.: Nondeterminism and the size of two way finite automata. In: Proceedings of the 10th Annual ACM Symposium on Theory of Computing, 1–3 May 1978, San Diego, California, USA, pp. 275–286 (1978)
11. Sipser, M.: Lower bounds on the size of sweeping automata. In: Proceedings of the 11th Annual ACM Symposium on Theory of Computing, April 30 – May 2, 1979, Atlanta, Georgia, USA, pp. 360–364 (1979)

Aperiodic String Transducers

Luc Dartois[1(✉)], Ismaël Jecker[1], and Pierre-Alain Reynier[2]

[1] Université Libre de Bruxelles, Brussel, Belgium
{ldartois,ijecker}@ulb.ac.be
[2] Aix-Marseille Université, CNRS, LIF UMR 7279, Marseille, France
pierre-alain.reynier@lif.univ-mrs.fr

Abstract. Regular string-to-string functions enjoy a nice triple characterization through deterministic two-way transducers (2DFT), streaming string transducers (SST) and MSO definable functions. This result has recently been lifted to FO definable functions, with equivalent representations by means of *aperiodic* 2DFT and *aperiodic* 1-bounded SST, extending a well-known result on regular languages. In this paper, we give three direct transformations: (*i*) from 1-bounded SST to 2DFT, (*ii*) from 2DFT to copyless SST, and (*iii*) from k-bounded to 1-bounded SST. We give the complexity of each construction and also prove that they preserve the aperiodicity of transducers. As corollaries, we obtain that FO definable string-to-string functions are equivalent to SST whose transition monoid is finite and aperiodic, and to aperiodic copyless SST.

Keywords: Transducer · Streaming · Two-way · Monoid · Aperiodic

1 Introduction

The theory of regular languages constitutes a cornerstone in theoretical computer science. Initially studied on languages of finite words, it has since been extended in numerous directions, including finite and infinite trees. Another natural extension is moving from languages to transductions. We are interested in this work in string-to-string transductions, and more precisely in string-to-string functions. One of the strengths of the class of regular languages is their equivalent presentation by means of automata, logic, algebra and regular expressions. The class of so-called *regular string functions* enjoys a similar multiple presentation. It can indeed be alternatively defined using deterministic two-way finite state transducers (2DFT), using Monadic Second-Order graph transductions interpreted on strings (MSOT) [8], and using the model of streaming string transducers (SST) [1]. More precisely, regular string functions are equivalent to different classes of SST, namely copyless SST [1] and k-bounded SST, for every positive integer k [3]. Different papers [1–3, 8] have proposed transformations between 2DFT, MSOT and SST, summarized on Fig. 1.

This work is supported by the ARC project Transform (French speaking community of Belgium), the Belgian FNRS PDR project Flare, and the PHC project VAST (35961QJ) funded by Campus France and WBI.

S. Brlek and C. Reutenauer (Eds.): DLT 2016, LNCS 9840, pp. 125–137, 2016.
DOI: 10.1007/978-3-662-53132-7_11

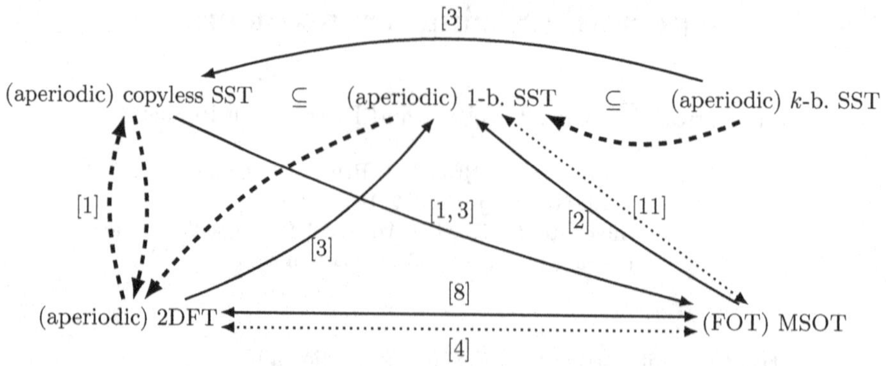

Fig. 1. Summary of transformations between equivalent models. k-b. stands for k-bounded. Plain (resp. dotted) arrows concern regular models (resp. bracketed models). Original constructions presented in this paper are depicted by thick dashed arrows and are valid for both regular and aperiodic versions of the models.

The connection between automata and logic, which has been very fruitful for model-checking for instance, also needs to be investigated in the framework of transductions. As it has been done for regular languages, an important objective is then to provide similar logic-automata connections for subclasses of regular functions, providing decidability results for these subclasses. As an illustration, the class of rational functions (accepted by one-way finite state transducers) owns a simple characterization in terms of logic, as shown in [9]. The corresponding logical fragment is called order-preserving MSOT. The decidability of the one-way definability of a two-way transducer proved in [10] thus yields the decidability of this fragment inside the class of MSOT.

The first-order logic considered with order predicate constitutes an important fragment of the monadic second order logic. It is well known that languages definable using this logic are equivalent to those recognized by finite state automata whose transition monoid is aperiodic (as well as other models such as star-free regular expressions). These positive results have motivated the study of similar connections between first-order definable string transformations (FOT) and restrictions of state-based transducers models. Two recent works provide such characterizations for 1-bounded SST and 2DFT respectively [4,11]. To this end, the authors study a notion of transition monoid for these transducers, and prove that FOT is expressively equivalent to transducers whose transition monoid is aperiodic by providing back and forth transformations between FOT and 1-bounded aperiodic SST (resp. aperiodic 2DFT). In particular, [11] lets as an open problem whether FOT is also equivalent to aperiodic copyless SST and to aperiodic k-bounded SST, for every positive integer k. It is also worth noticing that these characterizations of FOT, unlike the case of languages, do not allow to decide the class FOT inside the class MSOT. Indeed, while decidability for languages relies on the syntactic congruence of the language, no such canonical object exists for the class of regular string transductions.

In this work, we aim at improving our understanding of the relationships between 2DFT and SST. We first provide an original transformation from 1-bounded (or copyless) SST to 2DFT, and study its complexity. While the existing construction used MSO transformations as an intermediate formalism, resulting in a non-elementary complexity, our construction is in double exponential time, and in single exponential time if the input SST is copyless. Conversely, we describe a direct construction from 2DFT to copyless SST, which is similar to that of [1], but avoids the use of an intermediate model. These constructions also allow to establish links between the crossing degree of a 2DFT, and the number of variables of an equivalent copyless (resp. 1-bounded) SST, and conversely. Last, we provide a direct construction from k-bounded SST to 1-bounded SST, while the existing one was using copyless SST as a target model and not 1-bounded SST [3]. These constructions are represented by thick dashed arrows on Fig. 1.

In order to lift these constructions to aperiodic transducers, we introduce a new transition monoid for SST, which is intuitively more precise than the existing one. We use this new monoid to prove that the three constructions we have considered above preserve the aperiodicity of the transducer. As a corollary, this implies that FOT is equivalent to both aperiodic copyless and k-bounded SST, for every integer k, two results that were stated as conjectures in [11]. Omitted proofs can be found in [7].

2 Definitions

2.1 Words, Languages and Transducers

Given a finite alphabet A, we denote by A^* the set of finite words over A, and by ϵ the empty word. The length of a word $u \in A^*$ is its number of symbols, denoted by $|u|$. For all $i \in \{1, \dots, |u|\}$, we denote by $u[i]$ the i-th letter of u.

A *language* over A is a set $L \subseteq A^*$. Given two alphabets A and B, a *transduction* from A to B is a relation $R \subseteq A^* \times B^*$. A transduction R is *functional* if it is a function. The transducers we will introduce will define transductions. We will say that two transducers T, T' are equivalent whenever they define the same transduction.

Automata. A *deterministic two-way finite state automaton* (2DFA) over a finite alphabet A is a tuple $\mathcal{A} = (Q, q_0, F, \delta)$ where Q is a finite set of states, $q_0 \in Q$ is the initial state, $F \subseteq Q$ is a set of final states, and δ is the transition function, of type $\delta : Q \times (A \uplus \{\vdash, \dashv\}) \to Q \times \{+1, 0, -1\}$. The new symbols $\vdash$ and $\dashv$ are called *endmarkers*.

An input word u is given enriched by the endmarkers, meaning that $\mathcal{A}$ reads the input $\vdash u \dashv$. We set $u[0] = \vdash$ and $u[|u|+1] = \dashv$. Initially the head of $\mathcal{A}$ is on the first cell $\vdash$ in state q_0 (the cell at position 0). When $\mathcal{A}$ reads an input symbol, depending on the transitions in Δ, its head moves to the left (-1), or stays at the same position (0), or moves to the right ($+1$). To ensure the fact that the reading of $\mathcal{A}$ does not go out of bounds, we assume that there is no transition moving to the left (resp. to the right) on input symbol $\vdash$ (resp. $\dashv$). $\mathcal{A}$ stops as soon as it reaches the endmarker $\dashv$ in a final state.

A *configuration* of $\mathcal{A}$ is a pair $(q, i) \in Q \times \mathbb{N}$ where q is a state and i is a position on the input tape. A *run* ρ of $\mathcal{A}$ is a finite sequence of configurations. The run $\rho = (p_1, i_1) \dots (p_m, i_m)$ is a run on an input word $u \in A^*$ of length n if $i_m \leqslant n{+}1$, and for all $k \in \{1, \dots, m{-}1\}$, $0 \leqslant i_k \leqslant n{+}1$ and $(p_k, u[i_k], p_{k+1}, i_{k+1}{-}i_k) \in \Delta$. It is *accepting* if $p_1 = q_0$, $i_1 = 0$, and m is the only index where both $i_m = n{+}1$ and $p_m \in F$. The language of a 2DFA $\mathcal{A}$, denoted by $L(\mathcal{A})$, is the set of words u such that there exists an accepting run of $\mathcal{A}$ on u.

Transducers. *Deterministic two-way finite state transducers* (2DFT) over A extend 2DFA with a one-way left-to-right output tape. They are defined as 2DFA except that the transition relation δ is extended with outputs: $\delta : Q \times (A \uplus \{\vdash, \dashv\}) \to B^* \times Q \times \{-1, 0, +1\}$. When a transition (q, a, v, q', m) is fired, the word v is appended to the right of the output tape.

A run of a 2DFT is a run of its underlying automaton, i.e. the 2DFA obtained by ignoring the output (called its *underlying input automaton*). A run ρ may be simultaneously a run on a word u and on a word $u' \neq u$. However, when the input word is given, there is a unique sequence of transitions associated with ρ. Given a 2DFT T, an input word $u \in A^*$ and a run $\rho = (p_1, i_1) \dots (p_m, i_m)$ of T on u, the output of ρ on u is the word obtained by concatenating the outputs of the transitions followed by ρ. If ρ contains a single configuration, this output is simply ϵ. The transduction defined by T is the relation $R(T)$ defined as the set of pairs $(u, v) \in A^* \times B^*$ such that v is the output of an accepting run ρ on the word u. As T is deterministic, such a run is unique, thus $R(T)$ is a function.

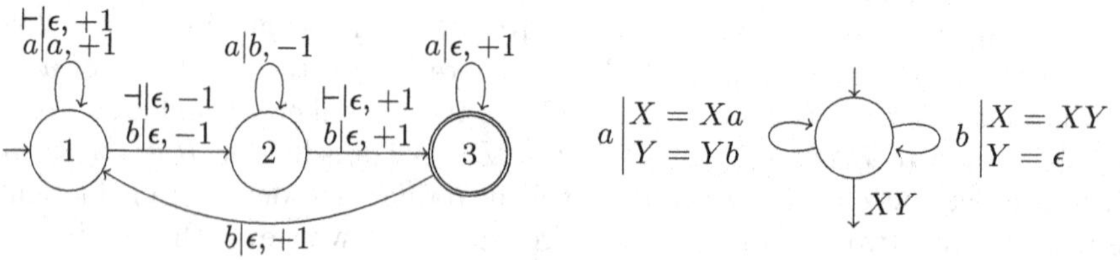

Fig. 2. Aperiodic 2DFT (left) and SST (right) realizing the function f.

Streaming String Transducers. Let $\mathcal{X}$ be a finite set of variables denoted by $X, Y, \dots$ and B be a finite alphabet. A substitution σ is defined as a mapping $\sigma : \mathcal{X} \to (B \cup \mathcal{X})^*$. Let $\mathcal{S}_{\mathcal{X},B}$ be the set of all substitutions. Any substitution σ can be extended to $\hat{\sigma} : (B \cup \mathcal{X})^* \to (B \cup \mathcal{X})^*$ in a straightforward manner. The composition $\sigma_1\sigma_2$ of two substitutions σ_1 and σ_2 is defined as the standard function composition $\hat{\sigma_1}\sigma_2$, i.e. $\hat{\sigma_1}\sigma_2(X) = \hat{\sigma_1}(\sigma_2(X))$ for all $X \in \mathcal{X}$. We say that a string $u \in (B \cup \mathcal{X})^*$ is *k-linear* if each $X \in \mathcal{X}$ occurs at most k times in u. A substitution σ is k-linear if $\sigma(X)$ is *k-linear* for all X. It is *copyless* if for any variable X, there exists at most one variable Y such that X occurs in $\sigma(Y)$, and X occurs at most once in $\sigma(Y)$.

A *streaming string transducer* (SST) is a tuple $T = (A, B, Q, q_0, Q_f, \delta, \mathcal{X}, \rho, F)$ where (Q, q_0, Q_f, δ) is a one-way automaton, A and B are finite sets of input and output alphabets respectively, $\mathcal{X}$ is a finite set of variables, $\rho : \delta \to \mathcal{S}_{\mathcal{X},B}$ is a variable update and $F : Q_f \rightharpoonup (\mathcal{X} \cup B)^*$ is the output function.

Example 1. As an example, let $f : \{a,b\}^* \to \{a,b\}^*$ be the function mapping any word $u = a^{k_0}ba^{k_1}\cdots ba^{k_n}$ to the word $f(u) = a^{k_0}b^{k_0}a^{k_1}b^{k_1}\cdots a^{k_n}b^{k_n}$ obtained by adding after each block of consecutive a a block of consecutive b of the same length. Since each word u over A can be uniquely written $u = a^{k_0}ba^{k_1}\cdots ba^{k_n}$ with some k_i being possibly equal to 0, the function f is well defined. We give in Fig. 2a 2DFT and an SST that realize f.

The concept of a run of an SST is defined in an analogous manner to that of a finite state automaton. The sequence $\langle\sigma_{r,i}\rangle_{0 \leqslant i \leqslant |r|}$ of substitutions induced by a run $r = q_0 \xrightarrow{a_1} q_1 \xrightarrow{a_2} q_2 \ldots q_{n-1} \xrightarrow{a_n} q_n$ is defined inductively as the following: $\sigma_{r,i} = \sigma_{r,i-1}\rho(q_{i-1}, a_i)$ for $1 < i \leqslant |r|$ and $\sigma_{r,1} = \rho(q_0, a_1)$. We denote $\sigma_{r,|r|}$ by σ_r and say that σ_r is induced by r.

If r is accepting, i.e. $q_n \in Q_f$, we can extend the output function F to r by $F(r) = \sigma_\epsilon \sigma_r F(q_n)$, where σ_ϵ substitutes all variables by their initial value ϵ. For all words $u \in A^*$, the output of u by T is defined only if there exists an accepting run r of T on u, and in that case the output is denoted by $T(u) = F(r)$. The transformation $R(T)$ is then defined as the set of pairs $(u, T(u)) \in A^* \times B^*$.

An SST T is copyless if for every transition $t \in \delta$, the variable update $\rho(t)$ is copyless. Given an integer $k \in \mathbb{N}_{>0}$, we say that T is *k-bounded* if all its runs induce k-linear substitutions. It is *bounded* if it is k-bounded for some k.

The following theorem gives the expressiveness equivalence of the models we consider. We do not give the definitions of MSO graph transductions as our results will only involve state-based transducers (see [9] for more details).

Theorem 2 [1,3,8]. *Let $f : A^* \to B^*$ be a function over words. Then the following conditions are equivalent:*

- *f is realized by an* MSO *graph transduction,*
- *f is realized by a* 2DFT,
- *f is realized by a copyless* SST,
- *f is realized by a bounded* SST.

2.2 Transition Monoid of Transducers

A (finite) monoid M is a (finite) set equipped with an associative internal law $\cdot_M$ having a neutral element for this law. A morphism $\eta : M \to N$ between monoids is an application from M to N that preserves the internal laws, meaning that for all x and y in M, $\eta(x \cdot_M y) = \eta(x) \cdot_N \eta(y)$. When the context is clear, we will write xy instead of $x \cdot_M y$. A monoid M divides a monoid N if there exists an onto morphism from a submonoid of N to M. A monoid M is said to be *aperiodic* if there exists a least integer n, called the *aperiodicity index* of M, such that for all elements x of M, we have $x^n = x^{n+1}$.

Given an alphabet A, the set of words A^* is a monoid equipped with the concatenation law, having the empty word as neutral element. It is called the *free monoid* on A. A finite monoid M *recognizes* a language L of A^* if there exists an onto morphism $\eta : A^* \rightarrow M$ such that $L = \eta^{-1}(\eta(L))$. It is well-known that the languages recognized by finite monoids are exactly the regular languages.

The monoid we construct from a machine is called its *transition monoid.* We are interested here in aperiodic machines, in the sense that a machine is aperiodic if its transition monoid is aperiodic. We now give the definition of the transition monoid for a 2DFT and an SST.

Deterministic Two-Way Finite State Transducers. As in the case of automata, the transition monoid of a 2DFT T is the set of all possible behaviors of T on a word. The following definition comes from [4], using ideas from [14] amongst others.

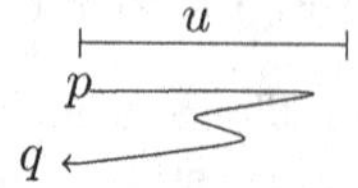

As a word can be read in both ways, the possible runs are split into four relations over the set of states Q of T. Given an input word u, we define the left-to-left behavior $\text{bh}_{\ell\ell}(u)$ as the set of pairs (p, q) of states of T such that there exists a run over u starting on the first letter of u in state p and exiting u on the left in state q (see Figure on the right). We define in an analogous fashion the left-to-right, right-to-left and right-to-right behaviors denoted respectively $\text{bh}_{\ell r}(u)$, $\text{bh}_{r\ell}(u)$ and $\text{bh}_{rr}(u)$. Then the transition monoid of a 2DFT is defined as follows:

Let $T = (Q, A, \delta, q_0, F)$ be a 2DFT. The *transition monoid* of T is $A^*/{\sim_T}$ where $\sim_T$ is the conjunction of the four relations $\sim_{ll}$, $\sim_{lr}$, $\sim_{rl}$ and $\sim_{rr}$ defined for any words u, u' of A^* as follows: $u \sim_{xy} u'$ iff $\text{bh}_{xy}(u) = \text{bh}_{xy}(u')$, for $x, y \in \{\ell, r\}$. The neutral element of this monoid is the class of the empty word ϵ, whose behaviors $bh_{xy}(\epsilon)$ is the identity function if $x \neq y$, and is the empty relation otherwise.

Note that since the set of states of T is finite, each behavior relation is of finite index and consequently the transition monoid of T is also finite. Let us also remark that the transition monoid of T does not depend on the output and is in fact the transition monoid of the underlying 2DFA.

Streaming String Transducers. A notion of transition monoid for SST was defined in [11]. We give here its formal definition and refer to [11] for advanced considerations. In order to describe the behaviors of an SST, this monoid describes the possible flows of variables along a run. Since we give later an alternative definition of transition monoid for SST, we will call it the *flow transition monoid* (FTM).

Let T be an SST with states Q and variables $\mathcal{X}$. The *flow transition monoid* M_T of T is a set of square matrices over the integers enriched with a new absorbent element $\bot$. The matrices are indexed by elements of $Q \times \mathcal{X}$. Given an input word u, the image of u in M_T is the matrix m such that for all states p, q and all variables X, Y, $m[p, X][q, Y] = n \in \mathbb{N}$ (resp. $m[p, X][q, Y] = \bot$) if, and only if, there exists a run r of T over u from state p to state q, and X occurs n times in $\sigma_r(Y)$ (resp. iff there is no run of T over u from state p to state q).

Note that if T is k-bounded, then for all word w, all the coefficients of its image in M_T are bounded by k. The converse also holds. Then M_T is finite if, and only if, T is k-bounded, for some k.

It can be checked that the machines given in Example 1 are aperiodic. Theorem 2 extends to aperiodic subclasses and to first-order logic, as in the case of regular languages [12,13]. These results as well as our contributions to these models are summed up in Fig. 1.

Theorem 3 [4,11]. *Let $f : A^* \rightarrow B^*$ be a function over words. Then the following conditions are equivalent:*

- *f is realized by a* FO *graph transduction,*
- *f is realized by an aperiodic* 2DFT,
- *f is realized by an aperiodic* 1*-bounded* SST.

3 Substitution Transition Monoid

In this section, we give an alternative take on the definition of the transition monoid of an SST, and show that both notions coincide on aperiodicity and boundedness. The intuition for this monoid, that we call the *substitution transition monoid*, is for the elements to take into account not only the multiplicity of the output of each variable in a given run, but also the order in which they appear in the output. It can be seen as an enrichment of the classic view of transition monoids as the set of functions over states equipped with the law of composition. Given a substitution $\sigma \in \mathcal{S}_{\mathcal{X},B}$, let us denote $\tilde{\sigma}$ the projection of σ on the set $\mathcal{X}$, i.e. we forget the parts from B. The substitutions $\tilde{\sigma}$ are homomorphisms of $\mathcal{X}^*$ which form an (infinite) monoid. Note that in the case of a 1-bounded SST, each variable occurs at most once in $\tilde{\sigma}(Y)$.

Substitution Transition Monoid of an SST. Let T be an SST with states Q and variables $\mathcal{X}$. The *substitution transition monoid* (STM) of T, denoted M_T^σ, is a set of partial functions $f : Q \rightharpoonup Q \times \mathcal{S}_{\mathcal{X},\emptyset}$. Given an input word u, the image of u in M_T^σ is the function f_u such that for all states p, $f_u(p) = (q, \tilde{\sigma}_r)$ if, and only if, there exists a run r of T over u from state p to state q that induces the substitution $\tilde{\sigma}_r$. This set forms a monoid when equipped with the following composition law: Given two functions $f_u, f_v \in M_T^\sigma$, the function f_{uv} is defined by $f_{uv}(q) = (q'', \tilde{\sigma} \circ \tilde{\sigma}')$ whenever $f_u(q) = (q', \tilde{\sigma})$ and $f_v(q') = (q'', \tilde{\sigma}')$.

We now make a few remarks about this monoid. Let us first observe that the FTM of T can be recovered from its STM. Indeed, the matrix m associated with a word u in M_T is easily deduced from the function f_u in M_T^σ. This observation induces an onto morphism from M_T^σ to M_T, and consequently the FTM of an SST divides its STM. This proves that if the STM is aperiodic, then so is the FTM since aperiodicity is preserved by division of monoids. Similarly, copyless and k-bounded SST (given $k \in \mathbb{N}_{>0}$) are characterized by means of their

STM. This transition monoid can be separated into two main components: the first one being the transition monoid of the underlying deterministic one-way automaton, which can be seen as a set of functions $Q \to Q$, while the second one is the monoid $\mathcal{S}_{\mathcal{X}}$ of homomorphisms on $\mathcal{X}$, equipped with the composition. The aware reader could notice that the STM can be written as the wreath product of the transformation semigroup $(\mathcal{X}^*, \mathcal{S}_{\mathcal{X}})$ by (Q, Q^Q). However, as the monoid of substitution is obtained through the closure under composition of the homomorphisms of a given SST, it may be infinite.

The next theorem proves that aperiodicity for both notions coincide, since the converse comes from the division of STM by FTM.

Theorem 4. *Let T be a k-bounded SST with ℓ variables. If its* FTM *is aperiodic with aperiodicity index n then its* STM *is aperiodic with aperiodicity index at most $n + (k+1)\ell$.*

4 From 1-Bounded SST to 2DFT

The existing transformation of a 1-bounded (or even copyless) SST into an equivalent 2DFT goes through MSO transductions, yielding a non-elementary complexity. We present here an original construction whose complexity is elementary.

Theorem 5. *Let T be a 1-bounded SST with n states and m variables. Then we can effectively construct a deterministic 2-way transducer that realizes the same function. If T is 1-bounded (resp. copyless), then the* 2DFT *has $O(m2^{m2^m}n^n)$ states (resp. $O(mn^n)$).*

Proof. We define the 2DFT as the composition of a left-to-right sequential transducer, a right-to-left sequential transducer and a 2-way transducer. Remark that this proves the result as two-way transducers are closed under composition with sequential ones [5]. The left-to-right sequential transducer does a single pass on the input word and outputs the same word enriched with the transition used by the SST in the previous step. The right-to-left transducer uses this information to enrich each position of the input word with the set of useful variables, i.e. the variables that flow to an output variable according to the partial run on the suffix read. The two sequential transducers are quite standard. They realize length-preserving functions that simply enrich the input word with new information. The last transducer is more interesting: it uses the enriched information to follow the output structure of T. The *output structure* of a run is a labeled and directed graph such that, for each variable X useful at a position j, we have two nodes X_i^j and X_o^j linked by a path whose concatenated labels form the value stored in X at position j of the run (see [11] and Fig. 3).

The transition function of the two-way transducer is described in Fig. 4. It first reaches the end of the word and picks the first variable to output. It then rewinds the run using the information stored by the first sequential transducer, producing the said variable using the local update function. When it has finished to compute and produce a variable X, it switches to the following one

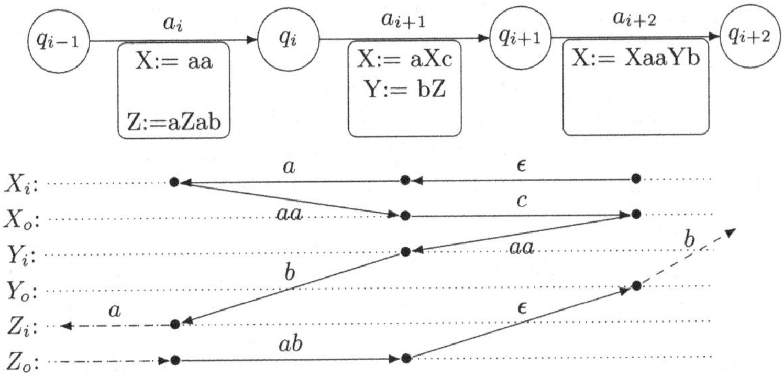

Fig. 3. The output structure of a partial run of an SST used in the proof of Theorem 5.

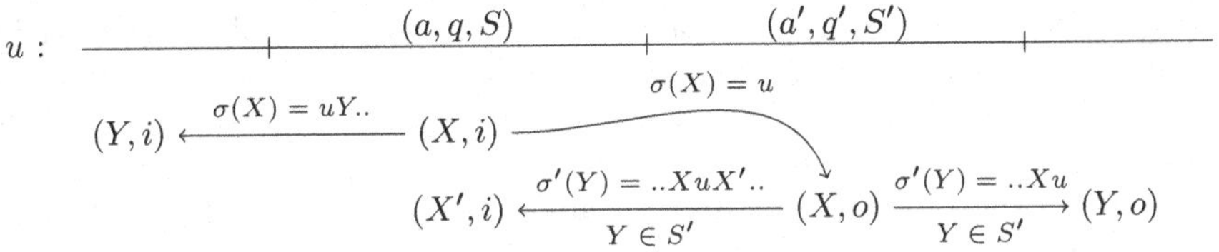

Fig. 4. The third transducer follows the output structure. States indexed by i correspond to the beginning of a variable, while states indexed by o correspond the end. σ (resp. σ') stand for the substitution at position a (resp. a').

using the information of the second transducer to know which variable Y X is flowing to, and starts producing it. Note that such a Y is unique thanks to the 1-boundedness property. If T is copyless, then this information is local and the second transducer can be bypassed. From [4,6], we can infer that the composition of a one-way transducer of size n with a two-way transducer of size m can be done by a two-way transducer of size $O(mn^n)$. Then given a 1-bounded SST with n states and m variables, we can construct a deterministic two-way transducer of size $O(m2^{m2^m}n^n)$. If T is copyless, the second sequential transducer is omitted, resulting in a size of $O(mn^n)$.

Theorem 6. *Let T be an aperiodic 1-bounded SST. Then the equivalent* 2DFT *constructed using Theorem 5 is also aperiodic.*

Proof. The aperiodicity of the three transducers gives the result as aperiodicity is preserved by composition of a one-way by a two-way [4]. The aperiodicity of the two sequential transducers is straightforward since their runs depend respectively on the underlying automaton and the update function. The aperiodicity of the 2DFT comes from the fact that since it follows the output structure of the SST, its partial runs are induced by the flow of variables and their order in the substitutions, which is an information contained in the FTM and thus aperiodic thanks to Theorem 4.

5 From 2DFT to Copyless SST

In [1], the authors give a procedure to construct a copyless SST from a 2DFT, using the intermediate model of heap based transducers. We give here a direct construction with similar complexity. This simplified presentation allows us to prove that the construction preserves the aperiodicity.

Theorem 7. *Let T be a* 2DFT *with n states. Then we can effectively construct a copyless SST with $O((2n)^{2n})$ states and $2n-1$ variables that computes the same function.*

Sketch of Proof. The main idea is for the constructed SST to keep track of the right-to-right behavior of the prefix read until the current position, similarly to the construction of Shepherdson [14]. This information can be updated upon reading a new letter, constructing a one-way machine recognizing the same input language. The idea from [3] is to have one variable per possible right-to-right run, which is bounded by the number of states. However, since two right-to-right runs from different starting states can merge, this construction results in a 1-bounded SST. To obtain copylessness, we keep track of these merges and the order in which they appear. Different variables are used to store the production of each run before the merge, and one more variable stores the production after.

The states of the copyless SST are represented by sets of labeled trees having the states of the input 2DFT as leaves. Each inner vertex represents one merging, and two leaves have a common ancestor if the right-to-right runs from the corresponding states merge at some point. Each tree then models a set of right-to-right runs that all end in a same state. Note that it is necessary to also store the end state of these runs. For each vertex, we use one variable to store the production of the partial run corresponding to the outgoing edge.

Given such a state and an input letter, the transition function can be defined by adding to the set of trees the local transitions at the given letter, and then reducing the resulting graph in a proper way (see Fig. 5).

Finally, as merges occur upon two disjoint sets of states of the 2DFT (initially singletons), the number of merges, and consequently the number of inner vertices of our states, is bounded by $n-1$. Therefore, an input 2DFT with n states can be realized by an SST having $2n-1$ variables. Finally, as states are labeled graphs, Cayley's formula yields an exponential bound on the number of states.

Theorem 8. *Let T be an aperiodic* 2DFT. *Then the equivalent* SST *constructed using Theorem 7 is also aperiodic.*

Proof. If the input 2DFT is aperiodic of index n, then for any word w, w^n and w^{n+1} merge the same partial runs for the four kinds of behaviors, by definition, and in fact the merges appear in the same order. As explained earlier, the state q_1 (resp. q_2) reached by the constructed SST over the inputs uw^n (resp. uw^{n+1}) represents the merges of the right-to-right runs of T over uw^n (resp. uw^{n+1}). Since these runs can be decomposed in right-to right runs over u and partial runs

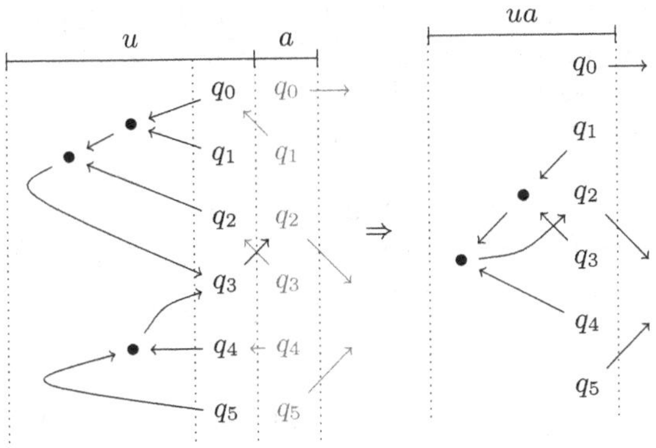

Fig. 5. Left: The state of the SST is represented in black. The red part corresponds to the local transitions of the 2DFT. Right: After reading a, we reduce the new forest by eliminating the useless branches and shortening the unlabeled linear paths. (Color figure online)

over w^n and w^{n+1}, the merge equivalence between w^n and w^{n+1} implies that $q_1 = q_2$. Moreover, since variables are linked to these merges, the aperiodicity of the merge equivalence implies the aperiodicity of both the underlying automaton and the substitution function of the SST, concluding the proof.

Corollary 9. *Let $f : A^* \to B^*$ be a function over words. Then f is realized by a* FO *graph transduction iff it is realized by an aperiodic copyless* SST.

6 From k-Bounded to 1-Bounded SST

The existing construction from k-bounded to 1-bounded, presented in [3], builds a copyless SST. We present an alternative construction that, given a k-bounded SST, directly builds an equivalent 1-bounded SST which preserves aperiodicity.

Theorem 10. *Given a k-bounded* SST *T with n states and m variables, we can effectively construct an equivalent 1-bounded* SST. *This new* SST *has $n2^N$ states and mkN variables, where $N = O(n^n(k+1)^{nm^2})$ is the size of the flow transition monoid M_T.*

Proof. In order to move from a k-bounded SST to a 1-bounded SST, the natural idea is to use copies of each variable. However, we cannot maintain k copies of each variable all the time: suppose that X flows into Y and Z, which both occur in the final output. If we have k copies of X, we cannot produce in a 1-bounded way (and we do not need to) k copies of Y and k copies of Z.

Now, if we have access to a look-ahead information, we can guess how many copies of each variable are needed, and we can easily construct a copyless SST by using exactly the right number of copies for each variable and at each step. The construction relies on this observation. We simulate a look-ahead through a subset construction, having copies of each variable for each possible behavior of the suffix. Then given a variable and the behavior of a suffix, we can maintain the exact number of variables needed and perform a copyless substitution to a potential suffix for the next step. However, since the SST is not necessarily co-deterministic, a given suffix can have multiple successors, and the result is that its variables flow to variables of different suffixes. As variables of different suffixes are never recombined, we obtain a 1-bounded SST.

Theorem 11. *Let T be an aperiodic k-bounded* SST*. Then the equivalent* 1*-bounded* SST *constructed using Theorem 10 is also aperiodic.*

As a corollary, we obtain that for the class of aperiodic bounded SST is expressively equivalent to first-order definable string-to-string transductions.

Corollary 12. *Let $f : A^* \to B^*$ be a function over words. Then f is realized by a* FO *graph transduction iff it is realized by an aperiodic bounded* SST *($k \in \mathbb{N}_{>0}$).*

7 Perspectives

There is still one model equivalent to the generic machines whose aperiodic subclass elude our scope yet, namely the *functional two-way transducers*, which correspond to non-deterministic two-way transducers realizing a function. To complete the picture, a natural approach would then be to consider the constructions from [15] and prove that aperiodicity is preserved. One could also think of applying this approach to other varieties of monoids, such as the $\mathcal{J}$-trivial monoids, equivalent to the boolean closure of existential first-order formulas $\mathcal{B}\Sigma_1[<]$. Unfortunately, the closure of such transducers under composition requires some strong properties on varieties (at least closure under semidirect product) which are not satisfied by varieties less expressive than the aperiodic. Consequently the construction from SST to 2DFT cannot be applied. On the other hand, the other construction could apply, providing one inclusion. Then an interesting question would be to know where the corresponding fragment of logic would position.

References

1. Alur, R., Černý, P.: Expressiveness of streaming string transducers. In: FSTTCS. LIPIcs, vol. 8, pp. 1–12. Schloss Dagstuhl, Leibniz-Zentrum für Informatik (2010)
2. Alur, R., Durand-Gasselin, A., Trivedi, A.: From monadic second-order definable string transformations to transducers. In: LICS, pp. 458–467 (2013)
3. Alur, R., Filiot, E., Trivedi, A.: Regular transformations of infinite strings. In: LICS, pp. 65–74 (2012)
4. Carton, O., Dartois, L.: Aperiodic two-way transducers and fo-transductions. In: CSL. LIPIcs, vol. 41, pp. 160–174. Schloss Dagstuhl, Leibniz-Zentrum für Informatik (2015)

5. Chytil, M.P., Jákl, V.: Serial composition of 2-way finite-state transducers and simple programs on strings. In: Salomaa, A., Steinby, M. (eds.) ICALP 1977. LNCS, vol. 52, pp. 135–137. Springer, Heidelberg (1977)
6. Dartois, L.: Méthodes algébriques pour la théorie des automates. Ph.D. thesis, LIAFA-Université Paris Diderot, Paris (2014)
7. Dartois, L., Jecker, I., Reynier, P.A.: Aperiodic string transducers. CoRR abs/1506.04059 (2016). http://arXiv.org/abs/1506.04059
8. Engelfriet, J., Hoogeboom, H.J.: MSO definable string transductions and two-way finite-state transducers. ACM Trans. Comput. Log. **2**(2), 216–254 (2001)
9. Filiot, E.: Logic-automata connections for transformations. In: Banerjee, M., Krishna, S.N. (eds.) ICLA. LNCS, vol. 8923, pp. 30–57. Springer, Heidelberg (2015)
10. Filiot, E., Gauwin, O., Reynier, P.A., Servais, F.: From two-way to one-way finite state transducers. In: LICS, pp. 468–477. IEEE Computer Society (2013). lics13.pdf
11. Filiot, E., Krishna, S.N., Trivedi, A.: First-order definable string transformations. In: FSTTCS 2014. LIPIcs, vol. 29, pp. 147–159. Schloss Dagstuhl, Leibniz-Zentrum für Informatik
12. McNaughton, R., Papert, S.: Counter-Free Automata. The M.I.T. Press, Cambridge, London (1971)
13. Schützenberger, M.P.: On finite monoids having only trivial subgroups. Inf. Control **8**, 190–194 (1965)
14. Shepherdson, J.C.: The reduction of two-way automata to one-way automata. IBM J. Res. Dev. **3**(2), 198–200 (1959)
15. de Souza, R.: Uniformisation of two-way transducers. In: Dediu, A.-H., Martín-Vide, C., Truthe, B. (eds.) LATA 2013. LNCS, vol. 7810, pp. 547–558. Springer, Heidelberg (2013)

An Automata Characterisation for Multiple Context-Free Languages

Tobias Denkinger(✉)

Faculty of Computer Science, Technische Universität Dresden,
Nöthnitzer Str. 46, 01062 Dresden, Germany
tobias.denkinger@tu-dresden.de

Abstract. We introduce tree stack automata as a new class of automata with storage and identify a restricted form of tree stack automata that recognises exactly the multiple context-free languages.

1 Introduction

Prominent classes of languages are often defined with the help of their generating mechanism, e.g. context-free languages are defined via context-free grammars, tree-adjoining languages via tree-adjoining grammars, and indexed languages via indexed grammars. To achieve a better understanding of how languages from a specific language class can be recognised, it is natural to ask for an automaton model. For context-free languages, this question is answered with pushdown automata [2,11], yield languages of tree-adjoining grammars are recognised by embedded pushdown automata [14, Sect. 3], and indexed languages are recognised by nested stack automata [1].

Mildly context-sensitive grammars are currently prominent in natural language processing as they are able to express the non-projective constituents and dependencies that occur in natural languages [9,10]. Multiple context-free grammars [13] describe many mildly context-sensitive grammars. Yet, to the author's knowledge, there is no corresponding automaton model. Thread automata [3,4], introduced by Villemonte de la Clergerie to describe parsing strategies for mildly context-sensitive grammar formalisms, already come close to such an automaton model. A construction of thread automata from ordered simple range concatenation grammars (which are equivalent to multiple context-free languages) was given [4, Sect. 4]. A construction for the converse direction as well as proofs of correctness, however, were not provided.

Based on the idea of thread automata, we introduce a new automaton model, tree stack automata, and formalise it using automata with storage [5,12] in the notation of Herrmann and Vogler [7], see Sect. 3. Tree stack automata possess, in addition to the usual finite state control, the ability to manipulate a tree-shaped stack that has the tree's root at its bottom. We find a restriction of tree stack automata that makes them equivalent to multiple context-free grammars and we give a constructive proof for this equivalence, see Sect. 4. An extended version of this paper can be found at http://arxiv.org/abs/1606.02975.

S. Brlek and C. Reutenauer (Eds.): DLT 2016, LNCS 9840, pp. 138–150, 2016.
DOI: 10.1007/978-3-662-53132-7_12

2 Preliminaries

In this section we fix some notation and briefly recall formalisms used throughout this paper. We denote the set of natural numbers (including 0) by $\mathbb{N}$, $\mathbb{N}\backslash\{0\}$ by $\mathbb{N}_+$, and $\{1,\ldots,n\}$ by $[n]$ for every $n \in \mathbb{N}$. The reflexive, transitive closure of some endorelation r is denoted as r^*. For two sets A and B, we denote the set of partial functions from A to B by $A \rightharpoonup B$. The operator $\rightharpoonup$ shall be right associative. Let $f\colon A \rightharpoonup B$, $a \in A$, and $b \in B$. The *domain of* f, denoted by $\mathrm{dom}(f)$, is the subset of A for which f is defined. If $\mathrm{dom}(f) = A$ we call f *total.* We define $f[a \mapsto b]$ as the partial function from A to B such that $f[a \mapsto b](a) = b$ and $f[a \mapsto b](a') = f(a')$ for every $a' \in \mathrm{dom}(f)\backslash\{a\}$. We sometimes construe partial functions as relations in the usual manner. Let S be a countable set (of *sorts*) and $s \in S$. An *S-sorted set* is a tuple (B, sort) where B is a set and $\mathrm{sort}\colon B \rightarrow S$ is total. We denote the preimage of s under sort by B_s and abbreviate (B, sort) by B; sort will always be clear from the context. Let A be a set and $L \subseteq A^*$. We call L *prefix-closed* if for every $w \in A^*$ and $a \in A$ we have that $wa \in L$ implies $w \in L$. An *alphabet* is a finite set (of *symbols*). Let Γ be an alphabet. The set of *trees over* Γ, denoted by T_Γ, is the set of partial functions from $\mathbb{N}_+^*$ to Γ with finite and prefix-closed domain. The usual definition of trees [6, Sect. 2] additionally requires that for every $\rho \in \mathbb{N}_+^*$ and $n \geq 2$: if ρn is in the domain of a tree then $\rho(n-1)$ is as well; we drop this restriction here.

Parallel multiple context-free grammars. We fix a set $X = \{x_i^j \mid i, j \in \mathbb{N}_+\}$ of *variables*. Let Σ be an alphabet. The set of *composition representations over* Σ is the $(\mathbb{N}_+^* \times \mathbb{N}_+)$-sorted set RF_Σ where for every $s_1, \ldots, s_\ell, s \in \mathbb{N}_+$ we define $X_{(s_1 \cdots s_\ell, s)} = \{x_i^j \mid i \in [\ell], j \in [s_i]\} \subseteq X$ and $(\mathrm{RF}_\Sigma)_{(s_1 \cdots s_\ell, s)} = \{[u_1, \ldots, u_s]_{(s_1 \cdots s_\ell, s)} \mid u_1, \ldots, u_s \in (\Sigma \cup X_{(s_1 \cdots s_\ell, s)})^*\}$ as a set of strings in which parentheses, brackets, commas, and the elements of $\mathbb{N}_+$, Σ, and $X_{(s_1 \cdots s_\ell, s)}$ are used as symbols. Let $f = [u_1, \ldots, u_s]_{(s_1 \cdots s_\ell, s)} \in \mathrm{RF}_\Sigma$. The *composition function of* f, also denoted by f, is the function from $(\Sigma^*)^{s_1} \times \cdots \times (\Sigma^*)^{s_\ell}$ to $(\Sigma^*)^s$ such that $f((w_1^1, \ldots, w_1^{s_1}), \ldots, (w_\ell^1, \ldots, w_\ell^{s_\ell})) = (u_1', \ldots, u_s')$ where $(u_1', \ldots, u_s')$ is obtained from $(u_1, \ldots, u_s)$ by replacing each occurrence of x_i^j by w_i^j for every $i \in [\ell]$ and $j \in [s_\ell]$. The set of all composition functions for some composition representation over Σ is denoted by F_Σ. From here on we no longer distinguish between composition representations and composition functions. We define the *fan-out of* f as s. We call f *linear* (*non-deleting*) if in $u_1 \cdots u_s$ every element of X occurs at most once (at least once, respectively). The subscript is dropped from f if its sort is clear from the context.

A *parallel multiple context-free grammar (short: PMCFG)* is a tuple $G = (N, \Sigma, I, R)$ where N is a finite $\mathbb{N}_+$-sorted set (of *non-terminals*), Σ is an alphabet (of *terminals*), $I \subseteq N_1$ (*initial non-terminals*), and $R \subseteq \bigcup_{k,s,s_1,\ldots,s_k \in \mathbb{N}} N_s \times (\mathrm{F}_\Sigma)_{(s_1 \cdots s_k, s)} \times (N_{s_1} \times \cdots \times N_{s_k})$ is finite (*rules*). A rule $(A, f, A_1 \cdots A_k) \in R$ is usually written as $A \rightarrow f(A_1, \ldots, A_k)$; it inherits its sort from f. A PMCFG that only contains rules with linear composition functions is called a *multiple context-free grammar (short: MCFG)*. An MCFG that contains only rules of fan-out at most k is called a k-MCFG.

For every $A \in N$, we recursively define the *set of derivations in* G *from* A as $\mathrm{D}_G(A) = \{r(d_1, \ldots, d_k) \mid r = A \to f(A_1, \ldots, A_k) \in R, \forall i \in [k]\colon d_i \in \mathrm{D}_G(A_i)\}$. The elements of $\mathrm{D}_G(A)$ can be construed as trees over R. Let $d \in \mathrm{D}_G(A)$. By projecting each rule in d on its second component, we obtain a term over F_Σ; the *tuple generated by* d, denoted by $[\![d]\!]$, is obtained by evaluating this term. We identify 1-tuples of strings with strings. The *set of (complete) derivations in* G is $\mathrm{D}_G = \bigcup_{A \in N} \mathrm{D}_G(A)$ ($\mathrm{D}_G^{\mathrm{c}} = \bigcup_{S \in I} \mathrm{D}_G(S)$, respectively). The *language of* G is $L(G) = \{[\![d]\!] \mid d \in \mathrm{D}_G^{\mathrm{c}}\}$.

Automata with storage. A *storage type* is a tuple $S = (C, P, F, C_\mathrm{i})$ where C is a set (of *storage configurations*), $P \subseteq \mathcal{P}(C)$ (*predicates*), $F \subseteq C \to C$ (*instructions*), and $C_\mathrm{i} \subseteq C$. An *automaton with storage* is a tuple $\mathcal{M} = (Q, S, \Sigma, q_\mathrm{i}, c_\mathrm{i}, \delta, Q_\mathrm{f})$ where Q is a finite set (of *states*), $S = (C, P, F, C_\mathrm{i})$ is a storage type, Σ is an alphabet (of *terminals*), $q_\mathrm{i} \in Q$ (*initial state*), $c_\mathrm{i} \in C_\mathrm{i}$ (*initial storage configuration*), $\delta \subseteq Q \times (\Sigma \cup \{\varepsilon\}) \times P \times F \times Q$ is finite (*transitions*), and $Q_\mathrm{f} \subseteq Q$ (*final states*).

Let $\tau = (q, \omega, p, f, q') \in \delta$ be a transition. We call q the *source state of* τ, p the *predicate of* τ, f the *instruction of* τ, and q' the *target state of* τ. A *configuration of* $\mathcal{M}$ is an tuple (q, c, w) where $q \in Q$, $c \in C$, and $w \in \Sigma^*$. We define the *run relation with respect to* τ as the binary relation $\vdash_\tau$ on the set of configurations of $\mathcal{M}$ such that $(q, c, w) \vdash_\tau (q', c', w')$ iff $(w = \omega w') \wedge (c \in p) \wedge (f(c) = c')$. The *set of runs in* $\mathcal{M}$ is the smallest set $\mathrm{R}_\mathcal{M} \subseteq \delta^*$ where for every $k \in \mathbb{N}$ and $\tau_1, \ldots, \tau_k \in \delta$, the string $\theta = \tau_1 \cdots \tau_k$ is in $\mathrm{R}_\mathcal{M}$ if there are $q_0, \ldots, q_k \in Q$, $c_0, \ldots, c_k \in C$, and $\omega_1, \ldots, \omega_k \in \Sigma \cup \{\varepsilon\}$ such that $(q_0, c_0, \omega_1 \cdots \omega_k) \vdash_{\tau_1} (q_1, c_1, \omega_2 \cdots \omega_k) \vdash_{\tau_2} \ldots \vdash_{\tau_k} (q_k, c_k, \varepsilon)$; we may then write $(q_0, c_0, \omega_1 \cdots \omega_k) \vdash_\theta (q_k, c_k, \varepsilon)$ or $(q_0, c_0) \vdash_\theta (q_k, c_k)$ or $[\![\theta]\!] = \omega_1 \cdots \omega_k$. The *set of valid runs in* $\mathcal{M}$, denoted by $\mathrm{R}_\mathcal{M}^\mathrm{v}$, contains exactly the runs $\theta \in \mathrm{R}_\mathcal{M}$ where $(q_\mathrm{i}, c_\mathrm{i}) \vdash_\theta (q, c)$ for some $q \in Q_\mathrm{f}$ and $c \in C$. For $\theta \in \mathrm{R}_\mathcal{M}^\mathrm{v}$ we say that $\mathcal{M}$ *recognises* $[\![\theta]\!]$. The *language of* $\mathcal{M}$ is $L(\mathcal{M}) = \{[\![\theta]\!] \mid \theta \in \mathrm{R}_\mathcal{M}^\mathrm{v}\}$.

3 Tree Stack Automata

Informally, a tree stack is a tree with a designated position in it. The root of the tree serves as *bottom-most symbol* and the leaves are *top-most symbols*. We allow the stack pointer to move *downward* (i.e. to the parent) and *upward* (i.e. to any child). We may *write* at any position except for the root. We may also *push* a symbol to any vacant child position of the current node. Formally, for an alphabet Γ, a *tree stack over* Γ is a tuple $(\xi[\varepsilon \mapsto @], \rho)$ where $\xi \in \mathrm{T}_\Gamma$, $@ \notin \Gamma$, and $\rho \in \mathrm{dom}(\xi) \cup \{\varepsilon\}$. The set of all tree stacks over Γ is denoted by $\mathrm{TS}(\Gamma)$. We define the following subsets (or predicates) of and partial functions on $\mathrm{TS}(\Gamma)$:

- $\mathrm{equals}(\gamma) = \{(\xi, \rho) \in \mathrm{TS}(\Gamma) \mid \xi(\rho) = \gamma\}$ for every $\gamma \in \Gamma$ and
- $\mathrm{bottom} = \{(\xi, \rho) \in \mathrm{TS}(\Gamma) \mid \rho = \varepsilon\}$.
- $\mathrm{id}\colon \mathrm{TS}(\Gamma) \to \mathrm{TS}(\Gamma)$ where $\mathrm{id}(\xi, \rho) = (\xi, \rho)$ for every $(\xi, \rho) \in \mathrm{TS}(\Gamma)$,

- push: $\mathbb{N} \to \Gamma \to \mathrm{TS}(\Gamma) \to \mathrm{TS}(\Gamma)$ where $\mathrm{push}_n(\gamma)(\xi, \rho) = (\xi[\rho n \mapsto \gamma], \rho n)$ for every $(\xi, \rho) \in \mathrm{TS}(\Gamma)$, $n \in \mathbb{N}$ with $\rho n \notin \mathrm{dom}(\xi)$, and $\gamma \in \Gamma$,
- up: $\mathbb{N} \to \mathrm{TS}(\Gamma) \to \mathrm{TS}(\Gamma)$ where $\mathrm{up}_n(\xi, \rho) = (\xi, \rho n)$ for every $(\xi, \rho) \in \mathrm{TS}(\Gamma)$ and $n \in \mathbb{N}$ with $\rho n \in \mathrm{dom}(\xi)$,
- down: $\mathrm{TS}(\Gamma) \to \mathrm{TS}(\Gamma)$ where $\mathrm{down}(\xi, \rho n) = (\xi, \rho)$ for every $(\xi, \rho n) \in \mathrm{TS}(\Gamma)$ with $n \in \mathbb{N}$, and
- set: $\Gamma \to \mathrm{TS}(\Gamma) \to \mathrm{TS}(\Gamma)$ where $\mathrm{set}(\gamma)(\xi, \rho) = (\xi[\rho \mapsto \gamma], \rho)$ for every $\gamma \in \Gamma$ and $(\xi, \rho) \in \mathrm{TS}(\Gamma)$ with $\rho \neq \varepsilon$.

We may denote a tree stack $(\xi, \rho) \in \mathrm{TS}(\Gamma)$ by writing ξ as a set and underlining the unique tuple of the form (ρ, γ) in this set. Consider for example a tree $\xi \in \mathrm{T}_{\{@,*,\#\}}$ with domain $\{\varepsilon, 2, 23\}$ such that $\xi\colon \varepsilon \mapsto @,\ 2 \mapsto *,\ 23 \mapsto \#$. We would then denote the tree stack $(\xi, 2) \in \mathrm{TS}(\{*, \#\})$ by $\{(\varepsilon, @), \underline{(2, *)}, (23, \#)\}$.

Definition 1. Let Γ be an alphabet. The *tree stack storage with respect to* Γ is the storage type $(\mathrm{TS}(\Gamma), P, F, \{\{\underline{(\varepsilon, @)}\}\})$, abbreviated by $\mathrm{TS}(\Gamma)$, where $P = \{\mathrm{bottom}, \mathrm{equals}(\gamma), \mathrm{TS}(\Gamma) \mid \gamma \in \Gamma\}$ and $F = \{\mathrm{id}, \mathrm{push}_n(\gamma), \mathrm{up}_n, \mathrm{down}, \mathrm{set}(\gamma) \mid \gamma \in \Gamma, n \in \mathbb{N}\}$. □

We call automata with tree stack storage *tree stack automata (short: TSA)*. In a storage configuration (ξ, ρ) of a TSA $\mathcal{M}$ we call ξ the *stack (of* $\mathcal{M}$*)* and ρ the *stack pointer (of* $\mathcal{M}$*)*.

Example 2. Let $\Sigma = \{\mathrm{a}, \mathrm{b}, \mathrm{c}, \mathrm{d}\}$ and $\Gamma = \{*, \#\}$. Consider the TSA $\mathcal{M} = ([5], \mathrm{TS}(\Gamma), \Sigma, 1, \{\underline{(\varepsilon, @)}\}, \delta, \{5\})$ where δ is shown in Fig. 2 (p. X). Figure 2 also shows the valid run $\tau_1\tau_2\tau_3\tau_4\tau_5\tau_6\tau_7\tau_8\tau_9$ in $\mathcal{M}$ recognising *abcd*. The language of $\mathcal{M}$ is $L(\mathcal{M}) = \{\mathrm{a}^n\mathrm{b}^n\mathrm{c}^n\mathrm{d}^n \mid n \in \mathbb{N}\}$ and thus not context-free.

While $\mathcal{M}$ only uses a monadic stack, branching is useful to recognise $L' = \{\mathrm{a}^i\mathrm{b}^j\mathrm{c}^i\mathrm{d}^j \mid i, j \in \mathbb{N}\}$; Fig. 1 shows a valid run of a TSA (that recognises L') on the word bd. □

Restricted TSA. Similar to Villemonte de la Clergerie [4], we are interested in how often any specific position in the stack is reached from below. For every TSA $\mathcal{M}$ we define $(c_{\mathcal{M}}(\theta)\colon \mathbb{N}_+^* \to \mathbb{N}_+ \mid \theta \in R_{\mathcal{M}}^{\mathrm{v}})$ as the family of total functions where $c_{\mathcal{M}}(\varepsilon)(\rho) = 0$ for every $\rho \in \mathbb{N}_+^*$, and for every $\theta\tau \in R_{\mathcal{M}}^{\mathrm{v}}$ with $\tau \in \delta$ we have $c_{\mathcal{M}}(\theta\tau) = c_{\mathcal{M}}(\theta)$ if τ has neither a push- nor up-instruction, and we have $c_{\mathcal{M}}(\theta\tau) = c_{\mathcal{M}}(\theta)[\rho \mapsto c_{\mathcal{M}}(\theta)(\rho) + 1]$ if τ has a push- or up-instruction and $\{\underline{(\varepsilon, @)}\} \vdash_{\theta\tau} (\xi, \rho)$ for some tree ξ. We call $\mathcal{M}$ *k-restricted* if $c_{\mathcal{M}}(\theta)(\rho) \leq k$ holds for every $\theta \in R_{\mathcal{M}}^{\mathrm{v}}$ and $\rho \in \mathbb{N}_+^*$. Note that $\mathcal{M}$ from Example 2 is 2-restricted.

Since (unrestricted) TSA can write at any position (except for ε) arbitrarily often, they can simulate Turing machines. It is apparent that 1-restricted TSA are exactly as powerful as pushdown automata. The power of k-restricted TSA for $k \geq 2$ is thus between the context-free and recursively enumerable languages.

Normal forms. We will see that loops that do not move the stack pointer as well as acceptance with non-ε stack pointers can be removed.

Let $\mathcal{M} = (Q, \mathrm{TS}(\Gamma), \Sigma, q_{\mathrm{i}}, \{(\varepsilon, @)\}, \delta, Q_{\mathrm{f}})$ be a TSA. For each $q, q' \in Q$ and $\gamma, \gamma' \in \Gamma \cup \{@\}$ we define $R_{\mathcal{M}}(q, q')|_{\mathrm{stay}}^{\gamma \to \gamma'}$ as the set of runs θ in $\mathcal{M}$ such that θ only uses set- or id-instructions and there are tree stacks $(\xi, \rho), (\zeta, \rho) \in \mathrm{TS}(\Gamma)$ with $\xi(\rho) = \gamma$, $\zeta(\rho) = \gamma'$, and $(q, (\xi, \rho)) \vdash_\theta (q', (\zeta, \rho))$.

Definition 3. We call a TSA $\mathcal{M} = (Q, \mathrm{TS}(\Gamma), \Sigma, q_{\mathrm{i}}, \{(\varepsilon, @)\}, \delta, Q_{\mathrm{f}})$ *cycle-free* if $R_{\mathcal{M}}(q, q)|_{\mathrm{stay}}^{\gamma \to \gamma} = \{\varepsilon\}$ for every $q \in Q$ and $\gamma \in \Gamma \cup \{@\}$. □

Lemma 4. *For every (k-restricted) TSA $\mathcal{M}$, there is a (k-restricted) cycle-free TSA $\mathcal{M}'$ such that $L(\mathcal{M}) = L(\mathcal{M}')$.*

Proof idea. Instead of performing all iterations of some non-empty loop $\theta \in R_{\mathcal{M}}(q, q)|_{\mathrm{stay}}^{\gamma \to \gamma} \setminus \{\varepsilon\}$ at the same position ρ in the stack, we insert additional push-instructions before each iteration of the loop. In order to find position ρ again after the desired number of iterations, we write symbols $*$ or $\#$ before every push, where a $*$ signifies that we have to perform at least two further down-instructions to reach ρ and $\#$ signifies that we will be at ρ after one more down-instruction. After returning to ρ, we enter a state $\tilde{q}$ that is equivalent to q except that it prevents us from entering the loop again. □

Definition 5. We say that a TSA $\mathcal{M}$ is in *stack normal form* if the stack pointer of $\mathcal{M}$ is ε whenever we reach a final state. □

Lemma 6. *For every (k-restricted) TSA $\mathcal{M}$, there is a (k-restricted) TSA $\mathcal{M}'$ in stack normal form such that $L(\mathcal{M}) = L(\mathcal{M}')$.*

Proof idea. We introduce a new state q_{f} as the only final state and add transitions such that, beginning from any original final state, we may perform down-instructions until the predicate bottom is satisfied and then enter state q_{f}. □

Note that $\mathcal{M}$ from Example 2 is cycle-free and in stack normal form.

4 The Equivalence of MCFG and Restricted TSA

4.1 Every MCFG Has an Equivalent Restricted TSA

The following construction applies the idea of Villemonte de la Clergerie [4, Sect. 4] to the case of parallel multiple context-free grammars where, additionally, we have to deal with copying, deletion, and permutation of argument components. The overall idea is to incrementally guess for an input word w a derivation d of G (that accepts w) on the stack while traversing the relevant components of the composition functions on the right-hand sides of already guessed rules (in d) left-to-right. This specific traversal of the derivation tree is ensured using states and stack symbols that encode positions in the rules of G.[1]

[1] The control flow of our constructed automaton is similar to that of the treewalk evaluator for attribute grammars [8, Sect. 3]. The two major differences are that the treewalk evaluator also treats inherited attributes (which are not present in PMCFGs) and that our constructed automaton generates the tree on the fly (while the treewalk evaluator is already provided with the tree).

Construction 7. Let $G = (N, \Sigma, I, R)$ be a PMCFG, $\Gamma = \{\Box\} \cup R \cup \bar{R}$, and $\bar{R} = \big\{\langle r, i, j\rangle \mid r = A \to [u_1, \ldots, u_s](A_1, \ldots, A_\ell) \in R, i \in [s], j \in \{0, \ldots, |u_i|\}\big\}$. Intuitively, an element $\langle r, i, j\rangle \in \bar{R}$ stands for the position in r right after the j-th symbol of the i-th component. The *automaton with respect to* G is $\mathcal{M}(G) = (Q, \mathrm{TS}(\Gamma), \Sigma, \Box, \{(\varepsilon, @)\}, \{\Box\}, \delta)$ where $Q = \{q, q_+, q_- \mid q \in \bar{R} \cup \{\Box\}\}$ and δ is the smallest set such that for every $r = S \to [u](A_1, \ldots, A_\ell) \in R$ with $S \in I$, we have the transitions

$$\begin{aligned}\mathrm{init}(r) &= \big(\Box, \varepsilon, \mathrm{TS}(\Gamma), \mathrm{push}_1(\Box), \langle r, 1, 0\rangle\big)\,,\\ \mathrm{suspend}_1(r, 1, \Box) &= \big(\langle r, 1, |u|\rangle, \varepsilon, \mathrm{equals}(\Box), \mathrm{set}(r), \Box_-\big)\,, \text{ and}\\ \mathrm{suspend}_2(\Box) &= \big(\Box_-, \varepsilon, \mathrm{TS}(\Gamma), \mathrm{down}, \Box\big) \text{ in } \delta;\end{aligned}$$

for every $r = A \to [u_1, \ldots, u_s](A_1, \ldots, A_\ell) \in R$, $i \in [s]$, $j \in [|u_i|]$ where $\sigma \in \Sigma$ is the j-th symbol in u_i, we have the transition

$$\mathrm{read}(r, i, j) = \big(\langle r, i, j-1\rangle, \sigma, \mathrm{TS}(\Gamma), \mathrm{id}, \langle r, i, j\rangle\big) \text{ in } \delta,$$

and for every $r = A \to [u_1, \ldots, u_s](A_1, \ldots, A_\ell) \in R$, $i \in [s]$, $j \in [|u_i|]$, $\kappa \in [\ell]$, $r' = A_\kappa \to [v_1, \ldots, v_{s'}](B_1, \ldots, B_{\ell'}) \in R$, $m \in [s']$ where $x_\kappa^m \in X$ is the j-th symbol in u_i, we have the transitions (abbreviating $\langle r, i, j\rangle$ by q)

$$\begin{aligned}\mathrm{call}(r, i, j, r') &= \big(\langle r, i, j-1\rangle, \varepsilon, \mathrm{TS}(\Gamma), \mathrm{push}_\kappa(q), \langle r', m, 0\rangle\big)\,,\\ \mathrm{resume}_1(r, i, j) &= \big(\langle r, i, j-1\rangle, \varepsilon, \mathrm{TS}(\Gamma), \mathrm{up}_\kappa, q_+\big)\,,\\ \mathrm{resume}_2(r, i, j, r') &= \big(\langle r, i, j-1\rangle_+, \varepsilon, \mathrm{equals}(r'), \mathrm{set}(q), \langle r', m, 0\rangle\big)\,,\\ \mathrm{suspend}_1(r', m, q) &= \big(\langle r', m, |v_m|\rangle, \varepsilon, \mathrm{equals}(q), \mathrm{set}(r'), q_-\big)\,, \text{ and}\\ \mathrm{suspend}_2(q) &= \big(q_-, \varepsilon, \mathrm{TS}(\Gamma), \mathrm{down}, q\big) \text{ in } \delta.\end{aligned}$$

□

Let us abbreviate a run $\mathrm{suspend}_1(r', m, q)\, \mathrm{suspend}_2(q)$ by $\mathrm{suspend}(r', m, q)$ and a run $\mathrm{resume}_1(r, i, j)\, \mathrm{resume}_2(r, i, j, r')$ by $\mathrm{resume}(r, i, j, r')$.

Example 8. Consider the MCFG $G = (\{S, A, B\}, \{a, b, c, d\}, \{S\}, R)$ where

$$\begin{aligned}R\colon\ r_1 = S \to [x_1^1 x_2^1 x_1^2 x_2^2](A, B) \quad r_2 &= A \to [\mathrm{a}x_1^1, \mathrm{c}x_1^2](A) \quad r_3 = A \to [\varepsilon, \varepsilon]()\\ r_4 &= B \to [\mathrm{b}x_1^1, \mathrm{d}x_1^2](B) \quad r_5 = B \to [\varepsilon, \varepsilon]()\,.\end{aligned}$$

Then $L(G) = \{\mathrm{a}^i \mathrm{b}^j \mathrm{c}^i \mathrm{d}^j \mid i, j \in \mathbb{N}\}$. Figure 1 shows that $\mathcal{M}(G)$ recognises bd. □

For the rest of Sect. 4.1 let $G = (N, \Sigma, I, R)$ *and* $\bar{R}$ *be defined as in Construction* 7.

Observation 9. The TSA $\mathcal{M}(G)$ is k-restricted if G is a k-MCFG. □

Lemma 10. $L(G) \subseteq L(\mathcal{M}(G))$.

Proof. For every $A \in N$ and every derivation $d = r(d_1, \ldots, d_m) \in \mathrm{D}_G(A)$ where $\mathrm{sort}(r) = (s_1 \cdots s_m, s)$ and $r = A \to [u_1, \ldots, u_s](B_1, \ldots, B_m)$, we recursively construct a tuple $(\theta^1, \ldots, \theta^s)$ of runs in $\mathcal{M}(G)$. For $d_1, \ldots, d_m$ we already have the tuples $(\theta_1^1, \ldots, \theta_1^{s_1}), \ldots, (\theta_m^1, \ldots, \theta_m^{s_m})$, respectively. For every $\kappa \in [s]$, let

$$
\begin{array}{ll}
 & (\Box, \{\underline{(\varepsilon, @)}\}) \\
\vdash_{\mathrm{init}(r_1)} & (\langle r_1,1,0\rangle, \{(\varepsilon, @), \underline{(1,\Box)}\}) \\
\vdash_{\mathrm{call}(r_1,1,1,r_3)} & (\langle r_3,1,0\rangle, \{(\varepsilon, @), (1,\Box), \underline{(11,\langle r_1,1,1\rangle)}\}) \\
\vdash_{\mathrm{suspend}(r_3,1,\langle r_1,1,1\rangle)} & (\langle r_1,1,1\rangle, \{(\varepsilon, @), \underline{(1,\Box)}, (11,r_3)\}) \\
\vdash_{\substack{\mathrm{call}(r_1,1,2,r_4)\\ \mathrm{read}(r_4,1,1)}} & (\langle r_4,1,1\rangle, \{(\varepsilon, @), (1,\Box), (11,r_3), \underline{(12,\langle r_1,1,2\rangle)}\}) \\
\vdash_{\mathrm{call}(r_4,1,2,r_5)} & (\langle r_5,1,0\rangle, \{(\varepsilon, @), (1,\Box), (11,r_3), (12,\langle r_1,1,2\rangle), \underline{(121,\langle r_4,1,2\rangle)}\}) \\
\vdash_{\mathrm{suspend}(r_5,1,\langle r_4,1,2\rangle)} & (\langle r_4,1,2\rangle, \{(\varepsilon, @), (1,\Box), (11,r_3), \underline{(12,\langle r_1,1,2\rangle)}, (121,r_5)\}) \\
\vdash_{\mathrm{suspend}(r_4,1,\langle r_1,1,2\rangle)} & (\langle r_1,1,2\rangle, \{(\varepsilon, @), \underline{(1,\Box)}, (11,r_3), (12,r_4), (121,r_5)\}) \\
\vdash_{\mathrm{resume}(r_1,1,3,r_3)} & (\langle r_3,2,0\rangle, \{(\varepsilon, @), (1,\Box), \underline{(11,\langle r_1,1,3\rangle)}, (12,r_4), (121,r_5)\}) \\
\vdash_{\mathrm{suspend}(r_3,2,\langle r_1,1,3\rangle)} & (\langle r_1,1,3\rangle, \{(\varepsilon, @), \underline{(1,\Box)}, (11,r_3), (12,r_4), (121,r_5)\}) \\
\vdash_{\substack{\mathrm{resume}(r_1,1,4,r_4)\\ \mathrm{read}(r_4,2,1)}} & (\langle r_4,2,1\rangle, \{(\varepsilon, @), (1,\Box), (11,r_3), \underline{(12,\langle r_1,1,4\rangle)}, (121,r_5)\}) \\
\vdash_{\mathrm{resume}(r_4,2,2,r_5)} & (\langle r_5,2,0\rangle, \{(\varepsilon, @), (1,\Box), (11,r_3), (12,\langle r_1,1,4\rangle), \underline{(121,\langle r_4,2,2\rangle)}\}) \\
\vdash_{\mathrm{suspend}(r_5,2,\langle r_4,2,2\rangle)} & (\langle r_4,2,2\rangle, \{(\varepsilon, @), (1,\Box), (11,r_3), \underline{(12,\langle r_1,1,4\rangle)}, (121,r_5)\}) \\
\vdash_{\mathrm{suspend}(r_4,2,\langle r_1,1,4\rangle)} & (\langle r_1,1,4\rangle, \{(\varepsilon, @), \underline{(1,\Box)}, (11,r_3), (12,r_4), (121,r_5)\}) \\
\vdash_{\mathrm{suspend}(r_1,1,\Box)} & (\Box, \{\underline{(\varepsilon, @)}, (1,r_1), (11,r_3), (12,r_4), (121,r_5)\})
\end{array}
$$

Fig. 1. Run of $\mathcal{M}(G)$ that recognises bd (cf. Example 8). The symbols b and d are read by $\mathrm{read}(r_4, 1, 1)$ and $\mathrm{read}(r_4, 2, 1)$, respectively, all other transitions in this run read ε.

$u_\kappa = \omega_1 \cdots \omega_\ell$ where $\omega_1, \ldots, \omega_\ell \in \Sigma \cup X$. We define $\theta_\kappa = \omega'_1 \cdots \omega'_\ell$ as the run in $\mathcal{M}(G)$ such that for every $\kappa' \in [\ell]$, we have that $\omega'_{\kappa'} = \mathrm{read}(r, \kappa, \kappa')$ if $\omega_{\kappa'} \in \Sigma$, $\omega'_{\kappa'} = \mathrm{call}(r, \kappa, \kappa', r')\, \theta_i^1\, \mathrm{suspend}(r', 1, \langle r, \kappa, \kappa' \rangle)$ if $\omega_{\kappa'} = x_i^1$ for some $i \geq 1$, and $\omega'_{\kappa'} = \mathrm{resume}(r, \kappa, \kappa', r')\, \theta_i^j\, \mathrm{suspend}(r', j, \langle r, \kappa, \kappa' \rangle)$ if $\omega_{\kappa'} = x_i^j$ for some $i \geq 1$ and $j \geq 2$, where $r' = d_i(\varepsilon)$. We can prove by structural induction on d that $[\![d]\!] = ([\![\theta^1]\!], \ldots, [\![\theta^s]\!])$. If $d \in \mathrm{D}_G^{\mathrm{c}}$, then s is 1 and hence the valid run $\mathrm{init}(r)\, \theta^1\, \mathrm{suspend}(r, 1, \Box)$ recognises exactly $[\![d]\!]$. $\Box$

Lemma 11. *Let $\tau_1, \ldots, \tau_n \in \delta$ with $\theta = \tau_1 \cdots \tau_n \in R_{\mathcal{M}(G)}$ and let $\rho \in \mathbb{N}_+^* \setminus \{\varepsilon\}$. There is a rule $\varphi_\theta(\rho)$ in G such that, during the run θ, the automaton $\mathcal{M}(G)$ is in some state $\langle \varphi_\theta(\rho), i, j \rangle \in \bar{R}$ whenever the stack pointer is at ρ.*

Proof. The rule $\varphi_\theta(\rho)$ is selected when ρ is first reached (with call). Then whenever we enter ρ with resume, a previous $\mathrm{suspend}_1$ has stored $\varphi_\theta(\rho)$ at position ρ and resume_2 enforces the claimed property. The claimed property is preserved by read. And whenever we enter ρ with suspend, a previous call or resume_2 has stored an appropriate state in the stack and suspend merely jumps back to that state, observing the claimed property. $\Box$

Examining the form of runs in $\mathcal{M}(G)$ (Construction 7) and using Lemma 11 we observe:

Observation 12. Let $\tau, \tau' \in \delta$, $q, q', q'' \in Q$, $\xi, \xi', \xi'' \in \mathrm{TS}(\Gamma)$, $\rho \in \mathbb{N}_+^*$, $i \in \mathbb{N}_+$, and $\varphi_\theta(\rho i)$ be of the form $A \to [u_1, \ldots, u_s](A_1, \ldots, A_\ell)$. Then:

1. If $(q', (\xi', \rho)) \vdash_\tau (q, (\xi, \rho i))$ with $q \in \bar{R}$, then $q = \langle \varphi_\theta(\rho i), j, 0 \rangle$ for some $j \in [s]$ and τ must be either an init- or call-transition.
2. If $(q'', (\xi'', \rho)) \vdash_\tau (q', (\xi', \rho i)) \vdash_{\tau'} (q, (\xi, \rho i))$ with $q' \in \{q_+ \mid q \in \bar{R}\}$, then $q = \langle \varphi_\theta(\rho i), j, 0 \rangle$ for some $j \in [s]$, τ is a resume$_1$-transition, and τ' is a resume$_2$-transition.
3. If $(q, (\xi, \rho i)) \vdash_\tau (q', (\xi', \rho i)) \vdash_{\tau'} (q'', (\xi'', \rho))$, then $q = \langle \varphi_\theta(\rho i), j, |u_j| \rangle$ for some $j \in [s]$, τ is a suspend$_1$-transition, and τ' is a suspend$_2$-transition. □

Lemma 13. *$L(G) \supseteq L(\mathcal{M}(G))$ if G only has productive non-terminals.*

Proof. For every run $\theta \in R_{\mathcal{M}(G)}$ we define $\varphi'_\theta \colon \mathbb{N}^* \to R$ by $\varphi'_\theta(\rho) = \varphi_\theta(1\rho)$ for every $\rho \in \mathbb{N}_+^*$ with $1\rho \in \mathrm{dom}(\varphi_\theta)$ (cf. Lemma 11). Then φ'_θ is a tree. One could show for every $d \in \mathrm{D}_G$ with $d \supseteq \varphi'_\theta$ by structural induction on φ'_θ that for every $\rho \in \mathrm{dom}(\varphi'_\theta)$ and every maximal interval $[a, b]$ where $\rho_a, \ldots, \rho_b$ have prefix ρ, we have $[\![\tau_a \cdots \tau_b]\!] = [\![d|_\rho]\!]_m$ with $q_a = \langle \varphi'_\theta(\rho), m, 0 \rangle$ for some $m \in \mathbb{N}_+$. Let us call this property (†). Let $\tau_1, \ldots, \tau_n \in \delta$ with $\theta = \tau_1 \cdots \tau_n \in R^{\mathrm{v}}_{\mathcal{M}(G)}$. Consider the run $(\square, (@, \varepsilon)) \vdash_{\tau_1} (q_1, (\xi_1, 1\rho_1)) \vdash_{\tau_2} \ldots \vdash_{\tau_{n-1}} (q_{n-1}, (\xi_{n-1}, 1\rho_{n-1})) \vdash_{\tau_n} (\square, (\xi_n, \varepsilon))$. By (†) we obtain that $[\![\tau_2 \cdots \tau_{n-1}]\!] = [\![d]\!]$. By Observation 12 and the fact that only an init-transition may start from $\square$ we obtain that τ_1 is an init-transition and τ_n is a suspend$_2$-transition. Thus $[\![\tau_1]\!] = \varepsilon = [\![\tau_n]\!]$ and therefore $[\![\theta]\!] = [\![d]\!]$. □

Proposition 14. *$L(G) = L(\mathcal{M}(G))$ if G only has productive non-terminals.*

Proof. The claim follows directly from Lemmas 10 and 13. □

$\mathcal{M}(G)$ is almost a parser for G. Let (ξ, ε) be a storage configuration of $\mathcal{M}(G)$ after recognising some word w and let $\xi|_1$ be the first subtree of ξ, defined by the equation $\xi|_1(\rho) = \xi(1\rho)$. Then every complete derivation d in G with $\xi|_1 \subseteq d$ generates w. If G only contains rules with non-deleting composition functions, we even have that $\xi|_1$ *is* a derivation in G generating w. In Fig. 1, for example, we see that $r_1(r_3, r_4(r_5))$ is a derivation of bd in G (cf. Example 8).

4.2 Every Restricted TSA Has an Equivalent MCFG

We construct an MCFG $G'(\mathcal{M})$ that recognises the valid runs of a given automaton $\mathcal{M}$, and then use the closure of MCFGs under homomorphisms. A tuple of runs $(\theta_1, \ldots, \theta_m)$ can be derived from non-terminal $\langle q_1, q'_1, \ldots, q_m, q'_m; \gamma_0, \ldots, \gamma_m \rangle$ iff the runs $\theta_1, \ldots, \theta_m$ all return to the stack position they started from and never go below it, and θ_i starts from state q_i and stack symbol γ_{i-1} and ends with q'_i and γ_i for every $i \in [m]$. We start with an example.

$$
\begin{aligned}
\delta\colon \tau_1 &= (1, \mathrm{a}, \mathrm{TS}(\Gamma), \mathrm{push}_1(*), 1)\\
\tau_2 &= (1, \varepsilon, \mathrm{TS}(\Gamma), \mathrm{push}_1(\#), 2)\\
\tau_3 &= (2, \varepsilon, \mathrm{equals}(\#), \mathrm{down}, 2)\\
\tau_4 &= (2, \mathrm{b}, \mathrm{equals}(*), \mathrm{down}, 2)\\
\tau_5 &= (2, \varepsilon, \mathrm{bottom}, \mathrm{up}_1, 3)\\
\tau_6 &= (3, \mathrm{c}, \mathrm{equals}(*), \mathrm{up}_1, 3)\\
\tau_7 &= (3, \varepsilon, \mathrm{equals}(\#), \mathrm{down}, 4)\\
\tau_8 &= (4, \mathrm{d}, \mathrm{equals}(*), \mathrm{down}, 4)\\
\tau_9 &= (4, \varepsilon, \mathrm{bottom}, \mathrm{id}, 5)
\end{aligned}
$$

$$
\begin{aligned}
& (1, \{\underline{(\varepsilon, @)}\}, \mathrm{abcd})\\
\vdash_{\tau_1}\ & (1, \{(\varepsilon, @), \underline{(1, *)}\}, \mathrm{bcd})\\
\vdash_{\tau_2}\ & (2, \{(\varepsilon, @), (1, *), \underline{(11, \#)}\}, \mathrm{bcd})\\
\vdash_{\tau_3}\ & (2, \{(\varepsilon, @), \underline{(1, *)}, (11, \#)\}, \mathrm{bcd})\\
\vdash_{\tau_4}\ & (2, \{\underline{(\varepsilon, @)}, (1, *), (11, \#)\}, \mathrm{cd})\\
\vdash_{\tau_5}\ & (3, \{(\varepsilon, @), \underline{(1, *)}, (11, \#)\}, \mathrm{cd})\\
\vdash_{\tau_6}\ & (3, \{(\varepsilon, @), (1, *), \underline{(11, \#)}\}, \mathrm{d})\\
\vdash_{\tau_7}\ & (4, \{(\varepsilon, @), \underline{(1, *)}, (11, \#)\}, \mathrm{d})\\
\vdash_{\tau_8}\ & (4, \{\underline{(\varepsilon, @)}, (1, *), (11, \#)\}, \varepsilon)\\
\vdash_{\tau_9}\ & (5, \{\underline{(\varepsilon, @)}, (1, *), (11, \#)\}, \varepsilon)
\end{aligned}
$$

Fig. 2. Set of transitions and a valid run in $\mathcal{M}$ (cf. Example 2).

Example 15. Recall the TSA $\mathcal{M}$ from Example 2 (also cf. Fig. 2). Note that $\mathcal{M}$ is cycle-free and in stack normal form. Let us consider position ε of the stack. The only transitions applicable there are τ_1, τ_2, τ_5, and τ_9. Clearly, every valid run in $\mathcal{M}$ starts with τ_1 or τ_2 and ends with τ_9, every τ_5 must be preceded by τ_4 or τ_3, and every τ_9 must be preceded by τ_8 or τ_7. Thus each valid run in $\mathcal{M}$ is either of the form $\theta = \tau_1\theta_1\tau_4\tau_5\theta_2\tau_8\tau_9$ or $\theta' = \tau_2\theta'_1\tau_3\tau_5\theta'_2\tau_7\tau_9$ for some runs θ_1, θ_2, θ'_1, and θ'_2. The target state of τ_1 is 1 and the source state of τ_4 is 2. Also τ_1 pushes a $*$ to position 1 and the predicate of τ_4 accepts only $*$. Thus θ_1 must go from state 1 to 2 and from stack symbol $*$ to $*$ at position 1. Similarly, we obtain that θ_2, θ'_1, and θ'_2 go from state 3 to 4, 2 to 2, and 3 to 3, respectively, and from stack symbol $*$ to $*$, $\#$ to $\#$, and $\#$ to $\#$, respectively, at position 1. The runs θ_1 and θ_2 are linked since they are both executed while the stack pointer is in the first subtree of the stack; the same holds for θ'_1 and θ'_2.

Clearly, linked runs need to be produced by the same non-terminal. For the pair (θ_1, θ_2) of linked runs, we have the non-terminal $\langle 1, 2, 3, 4; *, *, * \rangle$ and for (θ'_1, θ'_2) we have $\langle 2, 2, 3, 3; \#, \#, \# \rangle$. Since θ and θ' go from state 1 to 5 and from storage symbol @ to @, we have the rules

$$\langle 1, 5; @, @\rangle \to \big[\tau_1 x_1^1 \tau_4 \tau_5 x_1^2 \tau_8 \tau_9\big]\big(\langle 1, 2, 3, 4; *, *, *\rangle\big) \text{ and}$$
$$\langle 1, 5; @, @\rangle \to \big[\tau_2 x_1^1 \tau_3 \tau_5 x_1^2 \tau_7 \tau_9\big]\big(\langle 2, 2, 3, 3; \#, \#, \#\rangle\big) \text{ in } G'(\mathcal{M}).$$

Next, we explore the non-terminal $\langle 1, 2, 3, 4; *, *, * \rangle$, i.e. we need a run that goes from state 1 to 2 and from storage symbol $*$ to $*$ and another run that goes from state 3 to 4 and from storage symbol $*$ to $*$. There are only two kinds of suitable pairs of runs: $(\tau_1\theta_1\tau_4, \tau_6\theta_2\tau_8)$ and $(\tau_2\theta'_1\tau_3, \tau_6\theta'_2\tau_7)$ for some runs θ_1, θ_2, θ'_1, and θ'_2. The runs θ_1, θ_2, θ'_1, and θ'_2 of this paragraph then have the same state and storage behaviour as in the previous paragraph and we have rules

$$\langle 1,2,3,4;*,*,*\rangle \to \big[\tau_1 x_1^1 \tau_4, \tau_6 x_1^2 \tau_8\big]\big(\langle 1,2,3,4;*,*,*\rangle\big) \text{ and}$$
$$\langle 1,2,3,4;*,*,*\rangle \to \big[\tau_2 x_1^1 \tau_3, \tau_6 x_1^2 \tau_7\big]\big(\langle 2,2,3,3;\#,\#,\#\rangle\big) \text{ in } G'(\mathcal{M}).$$

For non-terminal $\langle 2,2,3,3;\#,\#,\#\rangle$, we may only take the pair of empty runs and thus have the rule $\langle 2,2,3,3;\#,\#,\#\rangle \to \big[\varepsilon,\varepsilon\big]()$ in $G'(\mathcal{M})$. □

For all $q, q' \in Q$, $\gamma, \gamma' \in \Gamma$, and $j \in \mathbb{N}_+$ we define the following sets:

$$\delta(q,q')|_{\mathrm{up}_j}^{\gamma\nearrow\bullet} = \{(q,\omega,p,\mathrm{up}_j,q') \in \delta \mid \gamma \in p\},$$
$$\delta(q,q')|_{\mathrm{push}_j}^{\gamma\nearrow\gamma'} = \{(q,\omega,p,\mathrm{push}_j(\gamma'),q') \in \delta \mid \gamma \in p\}, \text{ and}$$
$$\delta(q,q')|_{\mathrm{down}}^{\gamma\searrow\bullet} = \{(q,\omega,p,\mathrm{down},q') \in \delta \mid \gamma \in p\}.$$

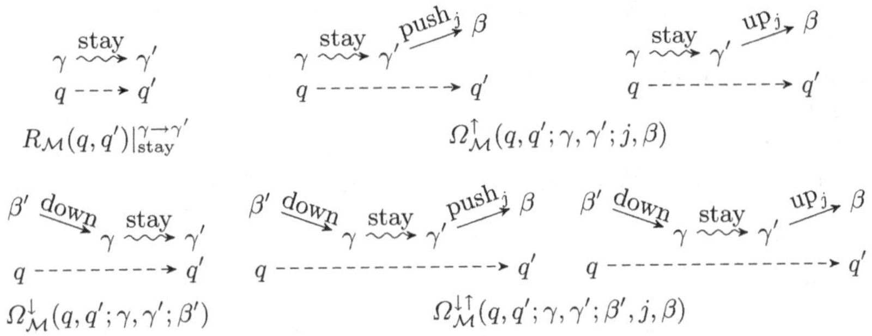

Fig. 3. Groups of runs in $\mathcal{M}$ where dashed arrows signify the change of states and continuous arrows signify the change in the storage.

For every $q, q' \in Q$, $\gamma, \gamma' \in \Gamma \cup \{@\}$, $\beta, \beta' \in \Gamma$, and $j \in \mathbb{N}_+$ we distinguish the following groups of runs (to help the intuition, they are visualised in Fig. 3):

1. A sequence of id- or set-instructions followed by an up- or push-instruction:

$$\Omega^{\uparrow}_{\mathcal{M}}(q,q';\gamma,\gamma';j,\beta) = \bigcup_{\bar{q}\in Q} R_{\mathcal{M}}(q,\bar{q})|_{\mathrm{stay}}^{\gamma\to\gamma'} \cdot \big(\delta(\bar{q},q')|_{\mathrm{push}_j}^{\gamma'\nearrow\beta} \cup \delta(\bar{q},q')|_{\mathrm{up}_j}^{\gamma'\nearrow\bullet}\big)$$

2. A down-instruction followed by id- or set-instructions:

$$\Omega^{\downarrow}_{\mathcal{M}}(q,q';\gamma,\gamma';\beta') = \bigcup_{\bar{q}\in Q} \delta(q,\bar{q})|_{\mathrm{down}}^{\beta'\searrow\bullet} \cdot R_{\mathcal{M}}(\bar{q},q')|_{\mathrm{stay}}^{\gamma\to\gamma'}$$

3. A down-instruction, then a sequence of id- or set-instructions and finally an up- or push-instruction:

$$\Omega^{\downarrow\uparrow}_{\mathcal{M}}(q,q';\gamma,\gamma';\beta',j,\beta) = \bigcup_{\bar{q}\in Q} \delta(q,\bar{q})|_{\mathrm{down}}^{\beta'\searrow\bullet} \cdot \Omega^{\uparrow}_{\mathcal{M}}(\bar{q},q';\gamma,\gamma';j,\beta)$$

The arguments of $\Omega^{\uparrow}_{\mathcal{M}}$, $\Omega^{\downarrow}_{\mathcal{M}}$, and $\Omega^{\downarrow\uparrow}_{\mathcal{M}}$ are grouped using semicolons. The first group describes the state behaviour of the run; the second group describes the storage behaviour at the parent position (i.e. the position of the set- and id-instructions), and the third group describes the storage behaviour at the child positions (i.e. the positions immediately above the parent position).

We build tuples of runs from the three groups above by matching the storage behaviour of neighbouring runs at the parent position. A tuple $t = (\theta_0, \ldots, \theta_\ell)$ of runs is *admissible* if $\ell = 0$ and θ_0 only uses id- and set-instructions; or if $\ell \geq 1$, θ_0 is in group 1, θ_ℓ is in group 2, and for every $i \in [\ell]$, we have

$$\theta_{i-1} \in \Omega^{\uparrow}_{\mathcal{M}}(q, q'; \gamma, \bar{\gamma}; j, \beta) \cup \Omega^{\downarrow\uparrow}_{\mathcal{M}}(q, q'; \gamma, \bar{\gamma}; \beta', j, \beta) \qquad \text{and}$$
$$\theta_i \in \Omega^{\downarrow}_{\mathcal{M}}(q'', q'''; \bar{\gamma}, \gamma'; \beta''') \cup \Omega^{\downarrow\uparrow}_{\mathcal{M}}(q'', q'''; \bar{\gamma}, \gamma'; \beta''', j', \beta'')$$

for some $\gamma, \bar{\gamma}, \gamma' \in \Gamma \cup \{@\}$, $\beta, \beta', \beta'', \beta''' \in \Gamma$, $q, q', q'', q''' \in Q$, and $j, j' \in \mathbb{N}_+$. Note that only the $\bar{\gamma}$ has to match. Then $\theta_{i-1}\theta_i$ may *not* be a run in $\mathcal{M}$ since it is not guaranteed that $q' = q''$ and $\beta = \beta'$. We therefore say that there is a *$(q', q''; j, \beta, \beta')$-gap between θ_{i-1} and θ_i*. Let $q_1, q_2 \in Q$ and $\gamma_1, \gamma_2 \in \Gamma \cup \{@\}$. We say that t *has type* $\langle q_1, q_2; \gamma_1, \gamma_2 \rangle$ if $\ell = 0$ and $\theta_0 \in R_{\mathcal{M}}(q_1, q_2)|^{\gamma_1 \to \gamma_2}_{\text{stay}}$; or if $\ell \geq 1$, the first transition in θ_0 has source state q_1 and its predicate contains γ_1, the last transition of θ_ℓ has target state q_2, the last set-instruction occurring in t, if there is one, is $\text{set}(\gamma_2)$, and $\gamma_1 = \gamma_2$ if no set-instruction occurs in t. The set of admissible tuples in $\Omega^{*}_{\mathcal{M}}$ is denoted by $\Omega^{\star}_{\mathcal{M}}$. We define $t[y_1, \ldots, y_\ell] = \theta_0 y_1 \theta_1 \cdots y_\ell \theta_\ell$ for every $y_1, \ldots, y_\ell \in X$ to later fill the gaps with variables.

Let $T = (t_1, \ldots, t_s) \in (\Omega^{\star}_{\mathcal{M}})^*$ and $\ell_1, \ldots, \ell_s$ be the counts of gaps in $t_1, \ldots, t_s$, respectively. For every $i \in [s]$ and $\kappa \in [\ell_i]$ we set $q_{(i,\kappa)}, q'_{(i,\kappa)} \in Q$, $\beta_{(i,\kappa)}, \beta'_{(i,\kappa)} \in \Gamma$, and $j_{(i,\kappa)} \in \mathbb{N}_+$ such that the κ-th gap in t_i is a $(q_{(i,\kappa)}, q'_{(i,\kappa)}; j_{(i,\kappa)}, \beta_{(i,\kappa)}, \beta'_{(i,\kappa)})$-gap. Let $\varphi_T, \psi_T \colon \mathbb{N}_+ \times \mathbb{N}_+ \to \mathbb{N}_+$ and $\pi_T \colon \mathbb{N}_+ \times \mathbb{N}_+ \to \mathbb{N}_+ \times \mathbb{N}_+$ be partial functions such that for every $i \in [s]$ and $\kappa \in [\ell_i]$, the number $j_{(i,\kappa)}$ is the $\varphi_T(i,\kappa)$-th distinct number occurring in $J = j_{(1,1)} \cdots j_{(1,|t_1|)} \cdots j_{(s,1)} \cdots j_{(s,|t_s|)}$ when read left-to-right, $j_{(i,\kappa)}$ occurs for the $\psi_T(i,\kappa)$-th time at the element with index (i,κ) in J, and $\pi_T(i,\kappa) = (\varphi_T(i,\kappa), \psi_T(i,\kappa))$. Moreover let m be the count of distinct numbers in J. We call T *admissible* if

- the κ-th run in t_i ends with a push-instruction whenever $\varphi_T(i,\kappa) = 1$,
- $\beta'_{\pi_T^{-1}(\kappa',\kappa)} = \beta_{\pi_T^{-1}(\kappa',\kappa+1)}$ for every $\kappa' \in [m]$ and $\kappa \in [\ell_{\kappa'} - 1]$, and
- there are $q_1, \bar{q}_1, \ldots, q_s, \bar{q}_s \in Q$ and $\gamma_0, \ldots, \gamma_s \in \Gamma \cup \{@\}$ such that for every $\kappa \in [s]$, we have that t_κ is of type $\langle q_\kappa, \bar{q}_\kappa; \gamma_{\kappa-1}, \gamma_\kappa \rangle$.

We then say that T *has type* $(A; B_1, \ldots, B_m)$, denoted by $\text{type}(T) = (A; B_1, \ldots, B_m)$, where $A = \langle q_1, \bar{q}_1, \ldots, q_s, \bar{q}_s; \gamma_0, \ldots, \gamma_s \rangle$, and for every $\kappa' \in [m]$:

$$B_{\kappa'} = \langle q_{\pi_T^{-1}(\kappa',1)}, q'_{\pi_T^{-1}(\kappa',1)}, \ldots, q_{\pi_T^{-1}(\kappa',\ell_{\kappa'})}, q'_{\pi_T^{-1}(\kappa',\ell_{\kappa'})};$$
$$\beta_{\pi_T^{-1}(\kappa',1)}, \beta'_{\pi_T^{-1}(\kappa',1)}, \ldots, \beta'_{\pi_T^{-1}(\kappa',\ell_{\kappa'})} \rangle .$$

The set of admissible elements of $(\Omega^{\star}_{\mathcal{M}})^*$ is denoted by $\Omega^{\star\star}_{\mathcal{M}}$.

Construction 16. Let $\mathcal{M} = (Q, \mathrm{TS}(\Gamma), \Sigma, q_{\mathrm{i}}, \{\underline{(\varepsilon, @)}\}, \delta, Q_{\mathrm{f}})$ be a cycle-free k-restricted TSA in stack normal form. Define the k-MCFG $G'(\mathcal{M}) = (N, \Sigma, I, R')$ where $N = \{A, B_1, \ldots, B_m \mid \langle A; B_1, \ldots, B_m\rangle \in \mathrm{type}(\Omega^{\star\star}_{\mathcal{M}})\}$, $I = \{\langle q_{\mathrm{i}}, q; @, @\rangle \mid q \in Q_{\mathrm{f}}\}$, and R' contains for every $T = (t_1, \ldots, t_s) \in \Omega^{\star\star}_{\mathcal{M}}$ the rule $A \to [u_1, \ldots, u_s](B_1, \ldots, B_m)$ where $(A; B_1, \ldots, B_m)$ is the type of T and $u_\kappa = t_\kappa[x^{\psi_T(\kappa,1)}_{\varphi_T(\kappa,1)}, \ldots, x^{\psi_T(\kappa,\ell_\kappa)}_{\varphi_T(\kappa,\ell_\kappa)}]$ for every $\kappa \in [s]$. Let $G(\mathcal{M})$ be a k-MCFG recognising $\{[\![\theta]\!] \mid \theta \in L(G'(\mathcal{M}))\}$.[2] □

Proposition 17. *$L(\mathcal{M}) = L(G(\mathcal{M}))$ for every cycle-free k-restricted TSA $\mathcal{M}$ in stack normal form.*

Proof. We can show by induction that $G'(\mathcal{M})$ generates exactly the valid runs of $\mathcal{M}$. Our claim then follows from the definition of $G(\mathcal{M})$. □

4.3 The Main Theorem

Theorem 18. *Let $L \subseteq \Sigma^*$ and $k \in \mathbb{N}_+$. The following are equivalent:*

1. *There is a k-MCFG G with $L = L(G)$.*
2. *There is a k-restricted tree stack automaton $\mathcal{M}$ with $L = L(\mathcal{M})$.*

Proof. We get the implication ($1 \Longrightarrow 2$) from Observation 9 and Proposition 14 and the implication ($2 \Longrightarrow 1$) from Lemmas 4 and 6 and Proposition 17. □

5 Conclusion

The automata characterisation of multiple context-free languages presented in this paper is achieved through tree stack automata that possess, in addition to the usual finite state control, the ability to manipulate a tree-shaped stack; tree stack automata are then restricted by bounding the number of times that the stack pointer enters any position of the stack from below (cf. Sect. 3). The proofs for the inclusions of multiple context-free languages in restricted tree stack languages and vice versa are both constructive; the former even works for parallel multiple context-free grammars, although the resulting automaton may then no longer be restricted (cf. Sect. 4). Theorem 18 closes a gap in formal language theory open since the introduction of MCFGs [13]. The proof allows for the easy implementation of a parser for parallel multiple context-free grammars.

References

1. Aho, A.V.: Nested stack automata. JACM **16**(3), 383–406 (1969)
2. Chomsky, N.: Formal properties of grammars. In: Luce, R.D., Bush, R.R., Galanter, E. (eds.) Handbook of Mathematical Psychology, vol. 2. Wiley, New York (1962)

[2] The k-MCFG $G(\mathcal{M})$ exists since $[\![\cdot]\!]$ is a homomorphism and k-MCFLs are closed under homomorphisms [13, Theorem 3.9].

3. Villemonte de la Clergerie, É.: Parsing MCS languages with thread automata. In: Proceedings of TAG+02, pp. 101–108 (2002)
4. Villemonte de la Clergerie, É.: Parsing mildly context-sensitive languages with thread automata. In: Proceedings of COLING 2002, vol. 1, pp. 1–7. ACL (2002)
5. Engelfriet, J.: Context-free grammars with storage. CoRR (2014)
6. Guessarian, I.: Pushdown tree automata. Math. Syst. Theor. **16**(1), 237–263 (1983)
7. Herrmann, L., Vogler, H.: A Chomsky-Schützenberger theorem for weighted automata with storage. In: Maletti, A. (ed.) CAI 2015. LNCS, vol. 9270, pp. 115–127. Springer, Heidelberg (2015). doi:10.1007/978-3-319-23021-4_11
8. Kennedy, K., Warren, S.K.: Automatic generation of efficient evaluators for attribute grammars. In: Proceedings of POPL 1976 (1976)
9. Kuhlmann, M., Satta, G.: Treebank grammar techniques for non-projective dependency parsing. In: Proceedings of EACL 2009, pp. 478–486. ACL (2009)
10. Maier, W.: Direct parsing of discontinuous constituents in German. In: Proceedings of SPMRL 2010, pp. 58–66. ACL (2010)
11. Schützenberger, M.P.: On context-free languages and push-down automata. Inf. Control **6**(3), 246–264 (1963)
12. Scott, D.: Some definitional suggestions for automata theory. J. Comput. Syst. Sci. **1**(2), 187–212 (1967)
13. Seki, H., Matsumura, T., Fujii, M., Kasami, T.: On multiple context-free grammars. TCS **88**(2), 191–229 (1991)
14. Vijay-Shanker, K.: A study of tree adjoining grammars. Ph.D. thesis (1988)

Weighted Automata and Logics on Infinite Graphs

Stefan Dück[✉]

Institute of Computer Science, Leipzig University, 04109 Leipzig, Germany
dueck@informatik.uni-leipzig.de

Abstract. We show a Büchi-like connection between graph automata and logics for infinite graphs. Using valuation monoids, a very general weight structure able to model computations like average or discounting, we extend this result to the quantitative setting. This gives us the first general results connecting automata and logics over infinite graphs in the qualitative and the quantitative setting.

Keywords: Quantitative automata · Infinite graphs · Graphs · Quantitative logic · Valuation monoids

1 Introduction

The coincidence between the languages recognizable by a finite state machine and the languages definable in monadic second order theory is one of the most fruitful results in theoretical computer science. Since Büchi, Elgot, and Trakhtenbrot [6,18,36] established this fundamental result, it has not only been the corner stone of multiple applications, like verification of finite-state programs, but also lead to multiple extensions covering finite and infinite trees [28,31], traces [32], pictures [22], (infinite) nested words [1], and texts [24]. A general result for finite graphs was given by Thomas [33].

It has remained an open question whether it is possible to get such a result for infinite graphs. In particular, this question is unanswered in the case of infinite pictures. The main contributions of this paper are the following:

- We show a Büchi-like equivalence between infinite graph acceptors and an EMSO-logic for infinite graphs.
- We establish a valuation-weighted automata model over graphs, which generalizes semiring-weighted automata and comprises previous automata models over special classes of graphs.
- Using methods of weighted logics, we extend our Büchi-like result from the qualitative to the quantitative setting, i.e., we show the equivalence of weighted infinite graph automata to a restricted weighted MSO-logic.

S. Dück—Supported by Deutsche Forschungsgemeinschaft (DFG) Graduiertenkolleg 1763 (QuantLA).

S. Brlek and C. Reutenauer (Eds.): DLT 2016, LNCS 9840, pp. 151–163, 2016.
DOI: 10.1007/978-3-662-53132-7_13

Formally introduced by Schützenberger [30], the study of quantitative questions (How often does an event arise?; What is the cost of this solution?; etc.) is another flourishing theory (see e.g. [2,17] and the recent handbook [13]). Quantitative automata modeling the long-time average or discounted behavior of systems were investigated, e.g., by Chatterjee et al. [7].

Recently, Bollig and Kuske [4] considered a logic FO^∞ featuring a first-order quantifier expressing that there are infinitely many elements satisfying a formula. In a different context than ours (for Muller message-passing automata), they were able to relate an extended Ehrenfeucht-Fraïssé game and k-equivalence of two formulas of FO^∞, thus developing a Hanf-like theorem [23] for this logic. We show how this result can be applied to infinite graphs to connect EMSO^∞ and infinite graph automata, yielding our first main result.

Using weighted MSO-logic [11], its extension to graphs [10], and valuation monoids [14], we generalize our graph automata model and our Büchi-like result to a quantitative setting. Here, one crucial part is the closure under the (restricted) weighted universal quantification (the *valuation-quantification*). An essential part of proving this closure is utilizing [4] to show that FO^∞ corresponds to one-state infinite graph acceptors.

To enhance readability, we first develop our weighted results in the finite case. Note that using valuation monoids, this model and the results are also new for finite graphs and enable us to consider examples using average or discounting in this general setting, as well as classical (possibly non-commutative) semirings. Furthermore, our approach is designed in an adaptable way, thereby facilitating the later extensions to infinite graphs.

2 Graphs and Graph Acceptors

In this section, we introduce the basic concepts around graphs and graph acceptors. Following [10,34], we define a *(directed) unpointed graph* as a relational structure $G = (V, (P_a)_{a\in A}, (E_b)_{b\in B})$ over two finite alphabets A and B, where V is the set of *vertices*, the sets P_a, $a \in A$, form a partition of V, and the sets E_b, $b \in B$, are pairwise disjoint irreflexive binary relations on V, called *edges*. We denote by $E = \bigcup_{b\in B} E_b$ the set of all edges. Then the elements of A are the vertex labels, and the elements of B are the edge labels. A graph is *bounded by* t if every vertex has an (in- plus out-) degree less than or equal to t.

We call a class of graphs *pointed* if every graph G of this class has a distinguished vertex. Formally, this assumption can be defined by adding a unary relation *root* to G with $\mathrm{root} = \{u\}$.

We consider subgraphs of a pointed graph (G, u) around a vertex v as follows. We call $\tau = (H, v, w)$ a *tile* if (H, v) is a pointed graph and either w is an additionally distinguished vertex of H or $w = empty$. Let $r \geq 0$. We denote by $\mathrm{dist}(x, y) \leq r$ that there exists a path $(x = x_0, x_1, ..., x_j = y)$ with $j \leq r$ and $(x_i, x_{i+1}) \in E$ or $(x_{i+1}, x_i) \in E$ for all $i < j$. We call (H, v, u) an *r-tile* if for every vertex x of H, it holds that $\mathrm{dist}(x, v) \leq r$. We denote by $\mathrm{sph}^r((G, u), v)$ the unique r-tile (H, v, w) consisting of all vertices x of G with $\mathrm{dist}(x, v) \leq r$

together with their edges and $w = u$ if $\mathrm{dist}(u, v) \leq r$ and $w = empty$, otherwise. We say v is the *center* of $\tau = (H, v, w)$, resp. of $\tau = (H, v) = (H, v, empty)$.

In this work, we assume all *graphs* to be pointed. We may omit the explicit root u of a graph and the radius r of a tile if the context is clear. Moreover, our results not explicitly utilizing the root also hold for unpointed graphs G.

We denote by $\mathrm{Lab}_G(v)$ the label of the vertex v of the graph G. We denote by $\mathrm{DG}_t(A, B)$ the class of all finite, directed, and pointed graphs over A and B, bounded by t. We denote by $\mathrm{DG}_t^{\infty}(A, B)$ the class of all infinite, directed, and pointed graphs over A and B, bounded by t. Note that r-tiles of finite or infinite graphs are finite structures, and there exist only finitely many non-isomorphic r-tiles since the degree of every considered graph is bounded.

Definition 1 ([33,34]). *A* graph acceptor (GA) $\mathcal{A}$ *over* $\mathrm{DG}_t(A, B)$ *is defined as a quadruple* $\mathcal{A} = (Q, \Delta, \mathrm{Occ}, r)$ *where*

- Q *is a finite set of states,*
- $r \in \mathbb{N}$ *is the* tile-radius,
- Δ *is a finite set of pairwise non-isomorphic r-tiles over $A \times Q$ and B,*
- Occ, *the* occurrence constraint, *is a boolean combination of formulas* "$\mathrm{occ}(\tau) \geq n$", *where $n \in \mathbb{N}$ and $\tau \in \Delta$.*

Note that Thomas (cf. [33,34]) uses non-pointed graphs. Here, the pointing can be seen as optional additional information to distinguish tiles from each other.

Given a finite graph $G = (G, u)$ of $\mathrm{DG}_t(A, B)$ and a mapping $\rho : V \to Q$, we consider the graph $G_\rho = (G_\rho, u) \in \mathrm{DG}_t(A \times Q, B)$, which consists of the same vertices and edges as G and is additionally labeled with $\rho(v)$ at every vertex v.

We call ρ a *run (or tiling) of $\mathcal{A}$ on G* if for every $v \in V$, $\mathrm{sph}^r(G_\rho, v)$ is isomorphic to a tile in Δ. We say G_ρ *satisfies* $\mathrm{occ}(\tau) \geq n$ if there exist at least n distinct vertices $v \in V$ such that $\mathrm{sph}^r(G_\rho, v)$ is isomorphic to τ. The semantics of "G_ρ *satisfies* Occ" are then defined in the usual way.

We call a run ρ *accepting* if G_ρ satisfies Occ. We say that $\mathcal{A}$ *accepts* the graph $G \in \mathrm{DG}_t(A, B)$ if there exists an accepting run ρ of $\mathcal{A}$ on G. We define $L(\mathcal{A}) = \{G \in \mathrm{DG}_t(A, B) \mid \mathcal{A} \text{ accepts } G\}$, the *language accepted by* $\mathcal{A}$. We call a language $L \subseteq \mathrm{DG}_t(A, B)$ *recognizable* if $L = L(\mathcal{A})$ for some GA $\mathcal{A}$.

Next, we introduce the logic $\mathrm{MSO}(\mathrm{DG}_t(A, B))$, short MSO, cf. [34]. We denote by $x, y, ...$ first-order variables ranging over vertices and by $X, Y, ...$ second order variables ranging over sets of vertices. The formulas of MSO are defined inductively by

$$\varphi ::= P_a(x) \mid E_b(x, y) \mid \mathrm{root}(x) \mid x = y \mid x \in X \mid \neg\varphi \mid \varphi \vee \varphi \mid \exists x.\varphi \mid \exists X.\varphi$$

where $a \in A$ and $b \in B$. An *FO-formula* is a formula of MSO without set quantifications, i.e., without using $\exists X$. An *EMSO-formula* is a formula of the form $\exists X_1 ... \exists X_k.\varphi$ where φ is an FO-formula.

The satisfaction relation $\models$ for graphs and MSO-sentences is defined in the natural way. Then for a sentence $\varphi \in \mathrm{MSO}$, we define the *language of* φ as

$L(\varphi) = \{G \in \mathrm{DG}_t(A,B) \mid G \models \varphi\}$. We call a language $L \subseteq \mathrm{DG}_t(A,B)$ *MSO-definable* (resp. *FO-definable*) if $L = L(\varphi)$ for some MSO-sentence (resp. FO-sentence) φ.

Theorem 2 ([34]). *Let $L \subseteq \mathrm{DG}_t(A,B)$ be a set of graphs. Then:*

1. *L is recognizable by a one-state GA iff L is definable by an FO-sentence.*
2. *L is recognizable iff L is definable by an EMSO-sentence.*

3 Infinite Graph Acceptors

In the following, we extend Theorem 2 to the infinite setting, thus showing a Büchi-like result for infinite graphs. We introduce infinite graph acceptors with an extended acceptance condition and an EMSO^∞ logic featuring a first-order quantifier $\exists^\infty x.\varphi$ to express that there exist infinitely many vertices fulfilling φ.

Using the occurrence constraint as acceptance condition, the introduced graph acceptor for finite graphs could also be interpreted as a model for infinite graphs. However, every occurrence constraint only checks for occurrences up to a certain threshold, i.e., it cannot express that a tile occurs infinitely many often. This motivates the following definition.

Definition 3. *An* infinite graph acceptor (GA^∞) $\mathcal{A}$ *over* $\mathrm{DG}_t^\infty(A,B)$ *is defined as a quadruple* $\mathcal{A} = (Q, \Delta, \mathrm{Occ}, r)$ *where*

- *Q, Δ, and r are defined as before, and*
- Occ, *the* extended occurrence constraint, *is a boolean combination of formulas "$\mathrm{occ}(\tau) \geq n$" and "$\mathrm{occ}(\tau) = \infty$", where $n \in \mathbb{N}$ and $\tau \in \Delta$.*

The notions of an *accepting run* ρ of $\mathcal{A}$ on $G \in \mathrm{DG}_t^\infty(A,B)$ and a *recognizable language* $L = L(\mathcal{A}) \subseteq \mathrm{DG}_t^\infty(A,B)$ are defined as before.

Next, following [4], we introduce the logic $\mathrm{MSO}^\infty(\mathrm{DG}_t^\infty(A,B))$, short MSO^∞, by the following grammar

$$\varphi ::= P_a(x) \mid E_b(x,y) \mid \mathrm{root}(x) \mid x = y \mid x \in X \mid \neg\varphi \mid \varphi \vee \varphi \mid \exists x.\varphi \mid \exists^\infty x.\varphi \mid \exists X.\varphi$$

We denote by FO^∞, resp. EMSO^∞, the usual first-order, resp. existential fragment. Defining an *assignment* σ and an *update* $\sigma[x \to v]$ as usual, the satisfaction relation $\models$ is defined as before, together with $(G,\sigma) \models \exists^\infty x.\varphi$ iff $(G, \sigma[x \to v]) \models \varphi$ for infinitely many $v \in V$.

Using an extended Ehrenfeucht-Fraïssé game, Bollig and Kuske [4] succeeded in proving a Hanf-like result for these structures. It says that for a given $k \in \mathbb{N}$ and a fixed maximal degree, there exists a sufficiently large tile-radius r and a threshold h such that two graphs which cannot be distinguished by an extended occurrence constraint over r and h are also indistinguishable by any FO^∞-formula up to quantifier depth k.

From this result, which was originally developed in a different context, namely Muller message-passing automata, we can deduce the following corollary.

Corollary 4. *Let φ be an FO^{∞}-sentence. Then there exists an extended occurrence constraint* Occ *such that $G \models \varphi$ iff $G \models$ Occ for all $G \in \mathrm{DG}_t^{\infty}(A, B)$.*

This result provides us with the means to prove our first main theorem.

Theorem 5. *Let $L \subseteq \mathrm{DG}_t^{\infty}(A, B)$ be a set of infinite graphs. Then:*

1. *L is recognizable by a one-state GA^{∞} iff L is definable by an* FO^{∞}*-sentence.*
2. *L is recognizable iff L is definable by an* EMSO^{∞}*-sentence.*

4 Weighted Graph Automata

In this section, we introduce and investigate a quantitative version of graph acceptors for finite graphs. We follow the approach of [10], but use more general structures than semirings, the *(graph-) valuation monoid* (cf. [14] for valuation monoids over words), which are able to model aspects like average, discounting, and other long-time behaviors of automata.

By abuse of notation, we also consider finite graphs $\mathrm{DG}_t(M, B)$ over an infinite set M. Note that we use this notation only in our weight assignments of the weighted automaton and never as part of the input or within a tile.

Definition 6. *A* (graph-) valuation monoid $\mathbb{D} = (D, +, \mathrm{Val}, 0)$ *consists of a commutative monoid $(D, +, 0)$ together with an absorptive valuation function* $\mathrm{Val} : \mathrm{DG}_t(D, B) \to D$*, i.e.,* $\mathrm{Val}(G) = 0$ *if at least one vertex of G is labeled* 0.

In the following, $\mathbb{D}$ will always refer to a valuation monoid[1].

Note that we do not enforce distributivity or another form of compatibility between $+$ and Val. The choice of valuation monoids is a natural one when you want to consider strictly more general structures than semirings and incorporate examples like average or discounting, as follows. In the context of trees, another closely related structure are multi-operator monoids (see e.g. [20]).

Example 7. Let $\mathrm{dia}(G)$ be the diameter of $G = (G, u) \in \mathrm{DG}_t(A, B)$. We define $\mathrm{avg}(G) = \frac{1}{|V|} \sum_{v \in V} \mathrm{Lab}_G(v)$ and $\mathrm{disc}_\lambda(G, u) = \sum_{r=0,\ldots,\mathrm{dia}(G)} \sum_{\mathrm{dist}(v,u)=r} \lambda^r \, \mathrm{Lab}_G(v)$. Then $\mathbb{D}_1 = (\mathbb{R} \cup \{-\infty\}, \sup, \mathrm{avg}, -\infty)$ and $\mathbb{D}_2 = (\mathbb{R} \cup \{-\infty\}, \sup, \mathrm{disc}_\lambda, -\infty)$ are two valuation monoids. Note that $\mathbb{D}_1$ does not use the root of the graph; therefore, we can omit it. In contrast, $\mathbb{D}_2$ is only utilizable for pointed graphs.

Definition 8. *A* weighted graph automaton (wGA) *over* $\mathrm{DG}_t(A, B)$ *and $\mathbb{D}$ is a tuple $\mathcal{A} = (Q, \Delta, \mathrm{wt}, \mathrm{Occ}, r)$ where*

- *$\mathcal{A}' = (Q, \Delta, \mathrm{Occ}, r)$ is a graph acceptor over* $\mathrm{DG}_t(A, B)$,
- *$\mathrm{wt} : \Delta \to D$ is the weight function assigning to every tile of Δ a value of D.*

[1] [14] enforced $\mathrm{Val}(d) = d$, which was later shown to be not required even in the word case, see e.g. [21].

An *accepting run* $\rho : V \to Q$ *of* $\mathcal{A}$ *on* $G \in \mathrm{DG}_t(A, B)$ is defined as an accepting run of $\mathcal{A}'$ on G. As in the unweighted case, the pointing of $G = (G, u)$ is optional.

For an accepting run ρ, we consider the graph G_ρ^D, where every vertex is labeled with the weight of the tile the run ρ defines around this vertex. More precisely, for a vertex v of G, let $\tau_\rho(v)$ be the r-tile of Δ which is isomorphic to $\mathrm{sph}^r(G_\rho, v)$. Then G_ρ^D is defined as the unique graph over $\mathrm{DG}_t(D, B)$ resulting from the graph G where for all vertices v, $\mathrm{Lab}_{G_\rho^D}(v) = \mathrm{wt}(\tau_\rho(v))$.

We denote by $\mathrm{acc}_\mathcal{A}(G)$ the set of all accepting runs of $\mathcal{A}$ on G. The *behavior* $[\![\mathcal{A}]\!] : \mathrm{DG}_t(A, B) \to D$ of a wGA $\mathcal{A}$ is defined, for each $G \in \mathrm{DG}_t(A, B)$, as

$$[\![\mathcal{A}]\!](G) = \sum_{\rho \in \mathrm{acc}_\mathcal{A}(G)} \mathrm{Val}(G_\rho^D).$$

We call any function $S : \mathrm{DG}_t(A, B) \to D$ a *series*. Then S is *recognizable* if $S = [\![\mathcal{A}]\!]$ for some wGA $\mathcal{A}$. By the usual identification of languages with functions assuming values in $\{0, 1\}$, we see that graph acceptors are expressively equivalent to wGA over the Boolean semiring $\mathbb{B}$.

Following [14], we call $\mathbb{D}$ *regular* if all constant series of D are recognizable, i.e., for every $d \in D$, there exists a wGA $\mathcal{A}_d$ with $[\![\mathcal{A}_d]\!](G) = d$ for every $G \in \mathrm{DG}_t(A, B)$.

Example 9. Let $A = \{a, b\}$ and $B = \{x\}$. For a given graph, we are interested in the value $\max_{a \in A} |V|_{a\&\mathrm{no\_outgoing}}/|V|$ which is the maximal proportion of nodes which are labeled with the same symbol and have no outgoing edges. For instance, in a tree the numerator would refer to the number of leafs labeled with a. We can compute this value with the following wGA over $\mathbb{D}_1 = (\mathbb{R} \cup \{-\infty\}, \sup, \mathrm{avg}, -\infty)$.

Set $\mathcal{A} = (\{q_1, q_2\}, \Delta, \mathrm{wt}, \mathrm{Occ}, r)$, with $r = 1$, $\Delta = \{\tau \mid \tau \text{ is a 1-tile}\}$, and

$$\mathrm{Occ} = \bigwedge_{\{\tau \mid \mathrm{center}(\tau) \in \{(a,q_1),(b,q_2)\}\}} \mathrm{occ}(\tau) = 0 \vee \bigwedge_{\{\tau \mid \mathrm{center}(\tau) \in \{(a,q_2),(b,q_1)\}\}} \mathrm{occ}(\tau) = 0.$$

Furthermore, we define $\mathrm{wt}(\tau) = 1$ if the center v of τ is labeled with q_1 and the center has no outgoing edges. Then $[\![\mathcal{A}]\!](G)$ is the desired proportion. □

Example 10. Let us assume our graph represents a social network. Now, we are interested into the affinity of a person to a certain characteristic (a hobby, a political tendency, an attribute, etc.) be it to use this information in a matching process or for personalized advertising. We assume that this affinity is closely related to the social environment of a person (e.g., I am more inclined to watch soccer if I play soccer myself, or I have friends who are interested into it).

We define a one-state wGA $\mathcal{A} = (\{q\}, \{\tau \mid \tau \text{ is a 1-tile}\}, \mathrm{wt}, \mathit{true}, 1)$ over $A = \{a, b\}$, $B = \{x\}$, and $\mathbb{D}_2 = (\mathbb{R} \cup \{-\infty\}, \sup, \mathrm{disc}_\lambda, -\infty)$, with $\mathrm{wt}(\tau) = \#_a(\tau)$, where $\#_a(\tau)$ is the number of vertices of τ labeled with a. Then depending on λ, $\mathcal{A}$ computes for a pointed graph (G, u) the affinity of u to the characteristic a.

Additionally introducing a nondeterministic choice for the center vertex u into the wGA, modifying the valuation function accordingly, and taking the supremum of all resulting runs, we can construct a nondeterministic automaton computing the maximal affinity of all vertices of a non-pointed graph. □

In the following, we give some results using ideas of [10]. These statements utilize the following formula. Let $\tau^* = \{\tau_1, ..., \tau_m\}$ be a finite set of tiles. For $N \in \mathbb{N}$, we shall write

$$\Big(\sum_{\tau \in \tau^*} \mathrm{occ}(\tau)\Big) \geq N \quad \text{short for} \quad \bigvee_{\substack{\sum_{i=1}^{m} n_i = N \\ n_i \in \{0,...,N\}}} \ \bigwedge_{i=1,...,m} \mathrm{occ}(\tau_i) \geq n_i. \tag{1}$$

We can interpret τ^* as a set of tiles matching a certain pattern. Then this formula is true iff the occurrence number of all tiles matching this pattern is at least N.

Let $S : \mathrm{DG}_t(A, B) \to D$ be a series recognizable by a wGA $\mathcal{A}$ with tile-radius s. Then we can show that for all $r \geq s$, S is recognizable by a wGA $\mathcal{B}$ with tile-radius r.

We extend the operation $+$ of our valuation monoid to series by means of point-wise definition, i.e., $(S + T)(G) = S(G) + T(G)$ for each $G \in \mathrm{DG}_t(A, B)$.

Proposition 11. *The class of recognizable series is closed under* $+$.

Let $S : \mathrm{DG}_t(A, B) \to D$ and $L \subseteq \mathrm{DG}_t(A, B)$. We define the *restriction* $S \cap L : \mathrm{DG}_t(A, B) \to D$ by letting $(S \cap L)(G) = S(G)$ if $G \in L$ and $(S \cap L)(G) = 0$, otherwise.

Proposition 12. *Let* $S : \mathrm{DG}_t(A, B) \to D$ *be a recognizable series and* $L \subseteq$ $DG_t(A, B)$ *be recognizable by a one-state GA. Then* $S \cap L$ *is recognizable.*

Proof (sketch). We build the wGA recognizing $S \cap L$ as a product-automaton from the wGA $\mathcal{A}$ recognizing S and the GA $\mathcal{B}$ recognizing L. The occurrence-constraint is combined by conjugating the projections to the constraints of $\mathcal{A}$ and $\mathcal{B}$ together with formula (1). Since $\mathcal{B}$ has exactly one state, we can control the number of runs of $\mathcal{C}$.

In the following, we show that recognizable series are closed under projection. Let $h : A' \to A$ be a mapping between two alphabets. Then h naturally defines a relabeling of graphs from $\mathrm{DG}_t(A', B)$ into graphs from $\mathrm{DG}_t(A, B)$, also denoted by h. Let $S : \mathrm{DG}_t(A', B) \to D$ be a series. We define $h(S) : \mathrm{DG}_t(A, B) \to D$ by

$$h(S)(G) = \sum_{\substack{G' \in \mathrm{DG}_t(A', B) \\ h(G') = G}} S(G'). \tag{2}$$

Proposition 13. *Let* $S : \mathrm{DG}_t(A', B) \to D$ *be a recognizable series and* $h : A' \to A$. *Then* $h(S) : \mathrm{DG}_t(A, B) \to D$ *is recognizable.*

5 Weighted Logics for Graphs

In the following, we introduce a weighted MSO-Logic for finite graphs, following the approach of Droste and Gastin [11] for words. We also incorporate an idea of Bollig and Gastin [3] to consider unweighted MSO-formulas as explicit fragment of our logic. We utilize an idea of Gastin and Monmege [21] to consider formulas with an 'if..then..else'-operator $\beta?\varphi_1 : \varphi_2$ instead of a weighted conjunction $\varphi_1 \otimes \varphi_2$. This operator is able to model the essential step-functions (resp. the almost FO-boolean fragment) without the need to add a second operation to the valuation monoid (the product $\diamond$).

Note that our underlying structure may still provide a product (e.g. as in the case of semirings). In this case, it remains possible to enrich our logic with a second operation (previously denoted by $\otimes$), therefore getting a direct connection to previous works [10,11,14].

In both cases, we are able to prove a Büchi-like connection between our introduced weighted graph automata and the (restricted) weighted MSO logic. Since the second operation enforces additional technical restrictions, we omit the details for this case here.

Definition 14. *We define the weighted logic* $\mathrm{MSO}(\mathbb{D}, \mathrm{DG}_t(A, B)), \mathrm{MSO}(\mathbb{D})$, *as*

$$\beta ::= P_a(x) \mid E_b(x,y) \mid \mathrm{root}(x) \mid x = y \mid x \in X \mid \neg\beta \mid \beta \vee \beta \mid \exists x.\beta \mid \exists X.\beta$$
$$\varphi ::= d \mid \varphi \oplus \varphi \mid \beta?\varphi : \varphi \mid \bigoplus_x \varphi \mid \bigoplus_X \varphi \mid \mathrm{Val}_x \varphi$$

where $d \in D$*;* x, y *are first-order variables; and* X *is a second order variable.*

Let $G \in \mathrm{DG}_t(A, B)$ and $\varphi \in \mathrm{MSO}(\mathbb{D})$. We follow classical approaches for logics and semantics. Let $\mathrm{free}(\varphi)$ be the set of all free variables in φ, and let $\mathcal{V}$ be a finite set of variables containing $\mathrm{free}(\varphi)$. A $(\mathcal{V}, G)$-*assignment* σ is a function assigning to every first-order variable of $\mathcal{V}$ an element of V and to every second order variable a subset of V. We define the *update* $\sigma[x \to v]$ as the $(\mathcal{V} \cup \{x\}, G)$-assignment mapping x to v and equaling σ everywhere else. The assignment $\sigma[X \to I]$ is defined analogously.

We represent the graph G together with the assignment σ as a graph (G, σ) over the vertex alphabet $A_{\mathcal{V}} = A \times \{0,1\}^{\mathcal{V}}$ where 1 denotes every position where x resp. X holds. A graph over $\mathcal{A}_{\mathcal{V}}$ is called *valid* if every first-order variable is assigned to exactly one position.

We define the *semantics* of $\varphi \in \mathrm{MSO}(\mathbb{D})$ as a function $[\![\varphi]\!]_{\mathcal{V}} : \mathrm{DG}_t(A_{\mathcal{V}}, B) \to D$ inductively for all valid $(G, \sigma) \in \mathrm{DG}_t(A_{\mathcal{V}}, B)$, as seen in Fig. 1. For not valid (G, σ), we set $[\![\varphi]\!]_{\mathcal{V}}(G, \sigma) = 0$. We write $[\![\varphi]\!]$ for $[\![\varphi]\!]_{\mathrm{free}(\varphi)}$.

Whether a graph is valid can be checked by an FO-formula, hence the language of all valid graphs over $A_{\mathcal{V}}$ is recognizable. For the Boolean semiring $\mathbb{B}$, the unweighted MSO is expressively equivalent to $\mathrm{MSO}(\mathbb{B})$.

The following lemma shows that for each finite set of variables containing $\mathrm{free}(\varphi)$, the semantics $[\![\varphi]\!]_{\mathcal{V}}$ are consistent with each other (cf. [11]).

$$\begin{aligned}
[\![d]\!]_{\mathcal{V}}(G,\sigma) &= d \quad \text{for all } d \in D\\
[\![\varphi \oplus \psi]\!]_{\mathcal{V}}(G,\sigma) &= [\![\varphi]\!]_{\mathcal{V}}(G,\sigma) + [\![\psi]\!]_{\mathcal{V}}(G,\sigma)\\
[\![\beta ? \varphi : \psi]\!]_{\mathcal{V}}(G,\sigma) &= \begin{cases} [\![\varphi]\!]_{\mathcal{V}}(G,\sigma)\ , & \text{if } (G,\sigma) \models \beta \\ [\![\psi]\!]_{\mathcal{V}}(G,\sigma)\ , & \text{otherwise}\end{cases}\\
[\![\textstyle\bigoplus_x \varphi]\!]_{\mathcal{V}}(G,\sigma) &= \sum_{v \in V} [\![\varphi]\!]_{\mathcal{V}\cup\{x\}}(G,\sigma[x \to v])\\
[\![\textstyle\bigoplus_X \varphi]\!]_{\mathcal{V}}(G,\sigma) &= \sum_{I \subseteq V} [\![\varphi]\!]_{\mathcal{V}\cup\{X\}}(G,\sigma[X \to I])\\
[\![\mathrm{Val}_x\, \varphi]\!]_{\mathcal{V}}(G,\sigma) &= \mathrm{Val}((G,\sigma)_\varphi) \text{ where } (G,\sigma)_\varphi \text{ is the graph } (G,\sigma) \text{ where every}\\
&\quad\ \text{vertex } v \text{ is labeled with } [\![\varphi]\!]_{\mathcal{V}\cup\{x\}}(G,\sigma[x \to v])
\end{aligned}$$

Fig. 1. Semantics

Lemma 15. *Let $\varphi \in \mathrm{MSO}(\mathbb{D})$ and $\mathcal{V}$ be a finite set of variables with $\mathcal{V} \supseteq \mathrm{free}(\varphi)$. Then $[\![\varphi]\!]_{\mathcal{V}}(G,\sigma) = [\![\varphi]\!](G,\sigma{\restriction}_{\mathrm{free}(\varphi)})$ for each valid $(G,\sigma) \in \mathrm{DG}_t(A_{\mathcal{V}},B)$. Furthermore, if $[\![\varphi]\!]$ is recognizable, then $[\![\varphi]\!]_{\mathcal{V}}$ is recognizable.*

Now, we show that recognizable series are closed under $\bigoplus_x$ and $\bigoplus_X$ quantification (in previous papers called the weighted existential quantification).

Lemma 16. *Let $[\![\varphi]\!]$ be recognizable. Then $[\![\bigoplus_x \varphi]\!]$ and $[\![\bigoplus_X \varphi]\!]$ are recognizable.*

The interesting case is the Val_x-quantification (previously called the weighted universal quantification [11]). Similarly to [11], our unrestricted logic is strictly more powerful than our automata model. Therefore, inspired by [14,21], we introduce the following fragment.

We call a formula $\varphi \in \mathrm{MSO}(\mathbb{D})$ *almost FO-boolean* if φ is built up inductively from the grammar, $\varphi ::= d \mid \beta ? d : \varphi$, where $d \in D$ and β is an unweighted FO-formula.

This fragment is equivalent to all formulas φ such that $[\![\varphi]\!]$ is an *FO-step function*, i.e., it takes only finitely many values and for each value its preimage is FO-definable. Denoting the constant series $d(G) = d$ for all $G \in DG_t(A,B)$ also with d, we get the following. If φ is almost FO-boolean, then $[\![\varphi]\!]$ has a representation $[\![\varphi]\!] = \sum_{i=1}^m d_i \mathbb{1}_{L_i} = \sum_{i=1}^m d_i \cap L_i$, where $m \in \mathbb{N}$, $d_i \in D$, L_i are languages definable by an unweighted FO-formula, and $(L_i)_{i=1\ldots m}$ form a partition of $\mathrm{DG}_t(A,B)$.

Proposition 17. *Let $\varphi \in \mathrm{MSO}(\mathbb{D})$ such that $[\![\varphi]\!]$ is an FO-step function. Then $[\![\mathrm{Val}_x\, \varphi]\!]$ is recognizable.*

Proof (sketch). Let $\mathcal{V} = \mathrm{free}(\mathrm{Val}_x\, \varphi)$ and $\mathcal{W} = \mathcal{V} \cup \{x\}$. Then $[\![\varphi]\!] = \sum_{i=1}^m d_i \mathbb{1}_{L_i}$, where L_i are FO-definable languages forming a partition of all of $\mathrm{DG}_t(A_{\mathcal{W}},B)$.

Now, we can encode the information in which language a given graph falls into an FO-formula $\tilde{L}$ over an extended alphabet. Using Theorem 2 yields a one-state GA $\tilde{\mathcal{A}}$ with $L(\tilde{\mathcal{A}}) = \tilde{L}$. Finally, we define a wGA $\mathcal{A}$ by adding weights to every tile depending on the state-label at its center and taking special care of the occurrence constraint. Then we can show that $[\![\mathcal{A}]\!] = [\![\mathrm{Val}_x\, \varphi]\!]$.

Let $\varphi \in \mathrm{MSO}(\mathbb{D})$. We call φ *FO-restricted* if all unweighted subformulas β are FO-formulas and for all subformulas $\mathrm{Val}_x\, \psi$ of φ, ψ is almost FO-boolean.

These restrictions are motivated in [11] (restriction of $\mathrm{Val}_x\,\psi$) and [19] (restriction to FO) where it is shown that the unrestricted versions of the logic are strictly more powerful than weighted automata on words, resp. pictures. For graphs this is also true, even for the Boolean semiring. We summarize our results.

Proposition 18. *If $\mathbb{D}$ is regular, then for every FO-restricted* MSO($\mathbb{D}$)*-sentence φ, there exists a wGA $\mathcal{A}$ with $[\![\mathcal{A}]\!] = [\![\varphi]\!]$.*

Proof (sketch). We use structural induction on φ. One new case is $[\![\beta?\varphi_1 : \varphi_2]\!] = [\![\varphi_1]\!] \cap L(\beta) + [\![\varphi_2]\!] \cap L(\neg\beta)$, which is recognizable by Propositions 11 and 12, because $L(\beta)$ and $L(\neg\beta)$ are recognizable by a one-state GA, since β is an FO-formula. The other cases are covered by regularity of $\mathbb{D}$ and the proven closure results (Lemma 16 and Proposition 17 together with Lemma 15).

Now, we show that every wGA can be simulated by an MSO($\mathbb{D}$)-sentence.

Proposition 19. *For every wGA $\mathcal{A}$, there exists an FO-restricted* MSO($\mathbb{D}$)*-sentence φ with $[\![\mathcal{A}]\!] = [\![\varphi]\!]$.*

Together with Proposition 18, this gives our second main result, a Büchi-like connection of the introduced weighted graph automata and the restricted weighted logic.

Theorem 20. *Let $\mathbb{D} = (D, +, \mathrm{Val}, 0)$ be a regular valuation monoid and let $S : \mathrm{DG}_t(A, B) \to D$ be a series. Then the following are equivalent:*

1. *S is recognizable.*
2. *S is definable by an FO-restricted* MSO($\mathbb{D}$)*-sentence.*

Examples of a regular valuation monoid are the introduced valuation monoids using average or discounting and all semirings.

6 Weighted Automata and Logics for Infinite Graphs

In the following, we extend our results in the weighted setting to infinite graphs. We utilize *∞-valuation monoids* to introduce weighted infinite graph automata.

We call a commutative monoid $(D, +, 0)$ *complete* if it has infinitary sum operations $\sum_I : D^I \to D$ for any index set I such that

- $\sum_{i\in\emptyset} d_i = 0$, $\sum_{i\in\{k\}} d_i = d_k$, $\sum_{i\in\{j,k\}} d_i = d_j + d_k$ for $j \neq k$,
- $\sum_{j\in J}(\sum_{i\in I_j} d_i) = \sum_{i\in I} d_i$ if $\bigcup_{j\in J} I_j = I$ and $I_j \cap I_k = \emptyset$ for $j \neq k$.

Definition 21. *An* ∞-(graph)-valuation monoid $(D, +, \mathrm{Val}^\infty, 0)$ *consists of a complete monoid $(D, +, 0)$ together with an absorptive* ∞-valuation function $\mathrm{Val}^\infty : \mathrm{DG}_t^\infty(D, B) \to D$.

Example 22. Let $\bar{\mathbb{R}}_+ = \{x \in \mathbb{R} \mid x \geq 0\} \cup \{\infty, -\infty\}$. Let $t > 1$ be the maximal degree of our graphs and $0 < \lambda < \frac{1}{t-1}$. Then $\mathbb{D} = (\bar{\mathbb{R}}_+, \sup, \mathrm{disc}_\lambda^\infty, -\infty)$, with $\mathrm{disc}_\lambda^\infty(G, u) = \lim_{n\to\infty} \sum_{r=0}^{n} \sum_{\mathrm{dist}(v,u)=r} \lambda^r \, \mathrm{Lab}_G(v)$, is an ∞-valuation monoid.

Definition 23. *A weighted infinite graph automaton* (wGA$^\infty$) *over* $\mathrm{DG}_t^\infty(A,B)$ *and* $\mathbb{D}$ *is a tuple* $\mathcal{A} = (Q, \Delta, \mathrm{wt}, \mathrm{Occ}, r)$ *where*

- $\mathcal{A}' = (Q, \Delta, \mathrm{Occ}, r)$ *is an infinite graph acceptor over* $\mathrm{DG}_t^\infty(A,B)$,
- $\mathrm{wt} : \Delta \to D$ *is the weight function assigning to every tile of* Δ *a value of* D.

We transfer the previous notions of *accepting run* and *recognizable series.*

The weighted MSO$^\infty$-logic for infinite graphs and its fragments is defined as extensions of MSO$^\infty$ as in the finite case (using Val$^\infty$ instead of Val) and is denoted by MSO$^\infty(\mathbb{D})$. Again, the significant difference is that we have the operator $\exists^\infty x$ in our underlying unweighted fragment. Adapting our previous notations and results to the infinite setting, we get our third main result.

Theorem 24. *Let* $\mathbb{D}$ *be a regular* ∞*-valuation monoid and let* $S : \mathrm{DG}_t^\infty(A,B) \to D$ *be a series. Then* S *is recognizable by a wGA*$^\infty$ *if and only if* S *is definable by an FO*$^\infty$*-restricted* MSO$^\infty(\mathbb{D})$*-sentence.*

The proof mainly follows the proof of Theorem 20. A notably difference is found in the closure under $\mathrm{Val}_x\,\varphi$ (in previous papers the weighted universal quantification). Since we have to deal with the additional quantifier $\exists^\infty x$, we cannot apply Theorem 2. However, Theorem 5 gives us one-state infinite graph acceptors $\mathcal{A}_i$ recognizing L_i. Then the automata constructions of Proposition 17 give us a wGA$^\infty$ $\mathcal{A}$ with $[\![\mathcal{A}]\!] = [\![\mathrm{Val}_x\,\varphi]\!]$.

7 Conclusion

Utilizing Bollig and Kuske [4] and a Hanf-like theorem for a first-order logic together with an infinity operator, we have proven a Büchi-like theorem for infinite graphs.

We introduced a weighted automata model over graphs which is robust enough to compute very general weight functions and is adaptable to infinite graphs. We gave new examples for this model, employing average and discounting. Introducing a suitable weighted MSO-logic, we successfully generalized Büchi-like results from the unweighted setting [35] to the weighted setting, from words [11] to graphs and from finite graphs [10] to infinite graphs.

Similar to [10], it can be shown that weighted word, tree, picture, and nested word automata are special instances of these weighted graph automata, which gives us, e.g., results of [11,16,19,26] and [12,14] as corollaries. Note that these lists are not exhaustive, as graphs are a very general structure comprising many other structures like traces [27], texts [25], distributed systems [5], and others.

Infinite graphs cover for example infinite words [15], infinite trees [29], infinite traces [8], and infinite nested words [9] and it would be interesting to study the expressive power of weighted infinite graph automata over these special classes.

Acknowledgments. I want to thank Manfred Droste and Tobias Weihrauch for helpful discussions and insightful remarks on earlier drafts of this paper.

References

1. Alur, R., Madhusudan, P.: Adding nesting structure to words. J. ACM **56**(3), 16:1–16:43 (2009)
2. Berstel, J., Reutenauer, C.: Rational Series and Their Languages. EATCS Monographs in Theoretical Computer Science, vol. 12. Springer, Heidelberg (1988)
3. Bollig, B., Gastin, P.: Weighted versus probabilistic logics. In: Diekert, V., Nowotka, D. (eds.) DLT 2009. LNCS, vol. 5583, pp. 18–38. Springer, Heidelberg (2009)
4. Bollig, B., Kuske, D.: Muller message-passing automata and logics. In: LATA 2007, Report 35/07, pp. 163–174. Universitat Rovira i Virgili, Tarragona (2007)
5. Bollig, B., Meinecke, I.: Weighted distributed systems and their logics. In: Artemov, S., Nerode, A. (eds.) LFCS 2007. LNCS, vol. 4514, pp. 54–68. Springer, Heidelberg (2007)
6. Büchi, J.R.: Weak second-order arithmetic and finite automata. Z. Math. Logik und Grundlagen Math. **6**, 66–92 (1960)
7. Chatterjee, K., Doyen, L., Henzinger, T.A.: Quantitative languages. In: Kaminski, M., Martini, S. (eds.) CSL 2008. LNCS, vol. 5213, pp. 385–400. Springer, Heidelberg (2008)
8. Diekert, V., Gastin, P.: LTL is expressively complete for Mazurkiewicz traces. In: Welzl, E., Montanari, U., Rolim, J.D.P. (eds.) ICALP 2000. LNCS, vol. 1853, pp. 211–222. Springer, Heidelberg (2000)
9. Droste, M., Dück, S.: Weighted automata and logics for infinite nested words. In: Dediu, A.-H., Martín-Vide, C., Sierra-Rodríguez, J.-L., Truthe, B. (eds.) LATA 2014. LNCS, vol. 8370, pp. 323–334. Springer, Heidelberg (2014)
10. Droste, M., Dück, S.: Weighted automata and logics on graphs. In: Italiano, G.F., Pighizzini, G., Sannella, D.T. (eds.) MFCS 2015. LNCS, vol. 9234, pp. 192–204. Springer, Heidelberg (2015)
11. Droste, M., Gastin, P.: Weighted automata and weighted logics. Theor. Comput. Sci. **380**(1–2), 69–86 (2007)
12. Droste, M., Götze, D., Märcker, S., Meinecke, I.: Weighted tree automata over valuation monoids and their characterization by weighted logics. In: Kuich, W., Rahonis, G. (eds.) Algebraic Foundations in Computer Science. LNCS, vol. 7020, pp. 30–55. Springer, Heidelberg (2011)
13. Droste, M., Kuich, W., Vogler, H. (eds.): Handbook of Weighted Automata. EATCS Monographs in Theoretical Computer Science. Springer, Heidelberg (2009)
14. Droste, M., Meinecke, I.: Weighted automata and weighted MSO logics for average and long-time behaviors. Inf. Comput. **220**, 44–59 (2012)
15. Droste, M., Rahonis, G.: Weighted automata and weighted logics on infinite words. In: Ibarra, O.H., Dang, Z. (eds.) DLT 2006. LNCS, vol. 4036, pp. 49–58. Springer, Heidelberg (2006)
16. Droste, M., Vogler, H.: Weighted tree automata and weighted logics. Theor. Comput. Sci. **366**(3), 228–247 (2006)
17. Eilenberg, S.: Automata, Languages, and Machines, Pure and Applied Mathematics, vol. 59-A. Academic Press, New York (1974)
18. Elgot, C.C.: Decision problems of finite automata design and related arithmetics. Trans. Am. Math. Soc. **98**(1), 21–52 (1961)
19. Fichtner, I.: Weighted picture automata and weighted logics. Theory Comput. Syst. **48**(1), 48–78 (2011)

20. Fülöp, Z., Stüber, T., Vogler, H.: A Büchi-like theorem for weighted tree automata over multioperator monoids. Theory Comput. Syst. **50**(2), 241–278 (2012)
21. Gastin, P., Monmege, B.: A unifying survey on weighted logics and weighted automata. Soft Comput. (2015). http://dx.doi.org/10.1007/s00500-015-1952-6
22. Giammarresi, D., Restivo, A., Seibert, S., Thomas, W.: Monadic second-order logic over rectangular pictures and recognizability by tiling systems. Inf. Comput. **125**(1), 32–45 (1996)
23. Hanf, W.: Model-theoretic methods in the study of elementary logic. In: Addison, J., Henkin, L., Tarski, A. (eds.) The Theory of Models, pp. 132–145. North-Holland, Amsterdam (1965)
24. Hoogeboom, H.J., ten Pas, P.: Monadic second-order definable text languages. Theory Comput. Syst. **30**(4), 335–354 (1997)
25. Mathissen, C.: Definable transductions and weighted logics for texts. Theory Comput. Sci. **411**(3), 631–659 (2010)
26. Mathissen, C.: Weighted logics for nested words and algebraic formal power series. In: Aceto, L., Damgård, I., Goldberg, L.A., Halldórsson, M.M., Ingólfsdóttir, A., Walukiewicz, I. (eds.) ICALP 2008, Part II. LNCS, vol. 5126, pp. 221–232. Springer, Heidelberg (2008)
27. Meinecke, I.: Weighted logics for traces. In: Grigoriev, D., Harrison, J., Hirsch, E.A. (eds.) CSR 2006. LNCS, vol. 3967, pp. 235–246. Springer, Heidelberg (2006)
28. Rabin, M.O.: Decidability of second order theories and automata on infinite trees. Trans. Am. Math. Soc. **141**, 1–35 (1969)
29. Rahonis, G.: Weighted muller tree automata and weighted logics. Int. J. Autom. Lang. Comb. **12**(4), 455–483 (2007)
30. Schützenberger, M.P.: On the definition of a family of automata. Inf. Control **4**(2–3), 245–270 (1961)
31. Thatcher, J.W., Wright, J.B.: Generalized finite automata theory with an application to a decision problem of second-order logic. Math. Syst. Theory **2**(1), 57–81 (1968)
32. Thomas, W.: On logical definability of trace languages. In: Diekert, V. (ed.) Proceedings of Workshop on ASMICS 1989, pp. 172–182. Technical University of Munich (1990)
33. Thomas, W.: On logics, tilings, and automata. In: Albert, J.L., Monien, B., Artalejo, M.R. (eds.) ICALP 1991. LNCS, vol. 510, pp. 441–454. Springer, Heidelberg (1991)
34. Thomas, W.: Elements of an automata theory over partial orders. In: Proceedings of DIMACS Workshop, POMIV 1996, pp. 25–40. AMS Press Inc., New York (1996)
35. Thomas, W.: Languages, automata, and logic. In: Rozenberg, G., Salomaa, A. (eds.) Handbook of Formal Languages, vol. 3, pp. 389–455. Springer, New York (1997)
36. Trakhtenbrot, B.A.: Finite automata and logic of monadic predicates. Doklady Akademii Nauk SSR **140**, 326–329 (1961). (in Russian)

Degrees of Infinite Words, Polynomials and Atoms

Jörg Endrullis[1(✉)], Juhani Karhumäki[2], Jan Willem Klop[1,3], and Aleksi Saarela[2]

[1] Department of Computer Science, VU University Amsterdam, Amsterdam, The Netherlands
{j.endrullis,j.w.klop}@vu.nl
[2] Department of Mathematics and Statistics & FUNDIM, University of Turku, Turku, Finland
{karhumak,amsaar}@utu.fi
[3] Centrum Wiskunde & Informatica (CWI), Amsterdam, The Netherlands

Abstract. Our objects of study are finite state transducers and their power for transforming infinite words. Infinite sequences of symbols are of paramount importance in a wide range of fields, from formal languages to pure mathematics and physics. While finite automata for recognising and transforming languages are well-understood, very little is known about the power of automata to transform infinite words.

We use methods from linear algebra and analysis to show that there is an infinite number of atoms in the transducer degrees, that is, minimal non-trivial degrees.

1 Introduction

The transformation realised by finite state transducers induces a partial order of degrees of infinite words: for words $v, w \in \Delta^{\mathbb{N}}$, we write $v \geq w$ if v can be transformed into w by some finite state transducer. If $v \geq w$, then v can be thought of as *at least as complex as* w. This complexity comparison induces equivalence classes of words, called *degrees*, and a partial order on these degrees, that we call *transducer degrees*.

The ensuing hierarchy of degrees is analogous to the recursion theoretic *degrees of unsolvability*, also known as *Turing degrees*, where the transformational devices are Turing machines. The Turing degrees have been widely studied in the 60's and 70's. However, as a complexity measure, Turing machines are too strong: they trivialise the classification problem by identifying all computable infinite words. Finite state transducers give rise to a much more fine-grained hierarchy.

We are interested in the structural properties of the hierarchy of transducer degrees. In this paper, we investigate the existence of atom degrees. An *atom degree* is a minimal non-trivial degree, that is, a degree that is directly above the bottom degree without interpolant degree.

This research has been supported by the Academy of Finland under the grant 257857.

S. Brlek and C. Reutenauer (Eds.): DLT 2016, LNCS 9840, pp. 164–176, 2016.
DOI: 10.1007/978-3-662-53132-7_14

Our Contribution. In [4,7] it has been proven that the degree of the words $\langle n \rangle$ and $\langle n^2 \rangle$ are atoms. Surprisingly, we find that this does not hold for $\langle n^3 \rangle$. In particular, we show that the degree of $\langle n^k \rangle$ is *never* an atom for $k \geq 3$ (see Theorem 22). On the other hand, we prove that for every $k > 0$, there exists a unique atom among the degrees of words $\langle p(n) \rangle$ for polynomials $p(n)$ of order k (see Theorem 31). (To avoid confusion between two meanings of *degrees*, namely *degrees of words* and *degrees of polynomials*, we speak of the *order* of a polynomial.) We moreover show that this atom is the infimum of all degrees of polynomials $p(n)$ of order k.

Further Related Work. The paper [11] discusses complexity hierarchies derived from notions of reduction. The paper [9] gives an overview over the subject of transducer degrees and compares them with the well-known Turing degrees [12, 15]. Restricting the transducers to output precisely one letter in each step, we arrive at Mealy machines. These gives rise to an analogous hierarchy of Mealy degrees that has been studied in [2,13]. The structural properties of this hierarchy are very different from the transducer degrees, see further [9].

2 Preliminaries

Let Σ be an alphabet. We write ε for the empty word, Σ^* for the set of finite words over Σ, and let $\Sigma^+ = \Sigma^* \backslash \{\varepsilon\}$. The set of infinite words over Σ is $\Sigma^{\mathbb{N}} = \{\sigma \mid \sigma : \mathbb{N} \to \Sigma\}$ and we let $\Sigma^\infty = \Sigma^* \cup \Sigma^{\mathbb{N}}$. Let $u, w \in \Sigma^\infty$. Then u is called a *prefix of* w, denoted $u \sqsubseteq w$, if $u = w$ or there exists $u' \in \Sigma^\infty$ such that $uu' = w$.

A *sequential finite state transducer* (FST) [1,14], a.k.a. *deterministic generalised sequential machine (DGSM)*, is a finite automaton with input letters and finite output words along the edges.

Definition 1. A *sequential finite state transducer* $A = \langle \Sigma, \Gamma, Q, q_0, \delta, \lambda \rangle$ consists of a finite *input alphabet* Σ, a finite *output alphabet* Γ, a finite set of *states* Q, an *initial state* $q_0 \in Q$, a *transition function* $\delta : Q \times \Sigma \to Q$, and an *output function* $\lambda : Q \times \Sigma \to \Gamma^*$. Whenever the alphabets Σ and Γ are clear from the context, we write $A = \langle Q, q_0, \delta, \lambda \rangle$.

We only consider sequential transducers and will simply speak of finite state transducers henceforth.

Definition 2. Let $A = \langle \Sigma, \Gamma, Q, q_0, \delta, \lambda \rangle$ be a finite state transducer. We homomorphically extend the transition function δ to $Q \times \Sigma^* \to Q$ by: for $q \in Q$, $a \in \Sigma$, $u \in \Sigma^*$ let $\delta(q, \varepsilon) = q$ and $\delta(q, au) = \delta(\delta(q, a), u)$. We extend the output function λ to $Q \times \Sigma^\infty \to \Gamma^\infty$ by: for $q \in Q$, $a \in \Sigma$, $u \in \Sigma^\infty$, let $\lambda(q, \varepsilon) = \varepsilon$ and $\lambda(q, au) = \lambda(q, a) \cdot \lambda(\delta(q, a), u)$.

We note that finite state transducers can be viewed as productive term rewrite systems [6] and the transduction of infinite words as infinitary rewriting [5].

3 Transducer Degrees

In this section, we explain how finite state transducers give rise to a hierarchy of degrees of infinite words, called transducer degrees. First, we formally introduce the transducibility relation $\geq$ on words as realised by finite state transducers.

Definition 3. Let $w \in \Sigma^{\mathbb{N}}$, $u \in \Gamma^{\mathbb{N}}$ for finite alphabets Σ, Γ. Let $A = \langle \Sigma, \Gamma, Q, q_0, \delta, \lambda \rangle$ be a finite state transducer. We write $w \geq_{\mathrm{A}} u$ if $u = \lambda(q_0, w)$. We write $w \geq u$, and say that u is a *transduct* of w, if there exists a finite state transducer A such that $w \geq_{\mathrm{A}} u$.

Note that the transducibility relation $\geq$ is a pre-order. It thus induces a partial order of 'degrees', the equivalence classes with respect to $\geq \cap \leq$. We denote equivalence using $\equiv$. It is not difficult to see that every word over a finite alphabet is equivalent to a word over the alphabet $\mathbf{2} = \{0, 1\}$. For the study of transducer degrees it suffices therefore to consider words over the latter alphabet.

Definition 4. Define the equivalence relation $\equiv\; = \;(\geq \cap \leq)$. The *(transducer) degree* $w^{\equiv}$ of an infinite word w is the equivalence class of w with respect to $\equiv$, that is, $w^{\equiv} = \{u \in \mathbf{2}^{\mathbb{N}} \mid w \equiv u\}$. We write $\mathbf{2}^{\mathbb{N}}/_{\equiv}$ to denote the set of degrees $\{w^{\equiv} \mid w \in \mathbf{2}^{\mathbb{N}}\}$.

The *transducer degrees* form the partial order $\langle \mathbf{2}^{\mathbb{N}}/_{\equiv}, \geq \rangle$[1] induced by the pre-order $\geq$ on $\mathbf{2}^{\mathbb{N}}$, that is, for words $w, u \in \mathbf{2}^{\mathbb{N}}$ we have $w^{\equiv} \geq u^{\equiv} \iff w \geq u$.

The *bottom degree* $\mathbf{0}$ of the transducer degrees is the least degree of the hierarchy, that is, the unique degree $\mathfrak{a} \in \mathbf{2}^{\mathbb{N}}/_{\equiv}$ such that $\mathfrak{a} \leq \mathfrak{b}$ for every $\mathfrak{b} \in \mathbf{2}^{\mathbb{N}}/_{\equiv}$. The bottom degree $\mathbf{0}$ consists of the ultimately periodic words, that is, words of the form $uvvv\cdots$ for finite words u, v where $v \neq \varepsilon$.

An atom is a degree that has only $\mathbf{0}$ below itself.

Definition 5. An *atom* is a minimal non-bottom degree, that is, a degree $\mathfrak{a} \in \mathbf{2}^{\mathbb{N}}/_{\equiv}$ such that $\mathbf{0} < \mathfrak{a}$ and there exists no $\mathfrak{b} \in \mathbf{2}^{\mathbb{N}}/_{\equiv}$ with $\mathbf{0} < \mathfrak{b} < \mathfrak{a}$.

4 Spiralling Words

We now consider *spiralling words* over the alphabet $\mathbf{2} = \{0, 1\}$ for which the distance of consecutive 1's in the word grows to infinity. We additionally require that the sequence of distances of consecutive 1's is ultimately periodic modulo every natural number. The class of spiralling words allows for a characterisation of their transducts in terms of weighted products.

For a function $f : \mathbb{N} \to \mathbb{N}$, we define $\langle f \rangle \in \mathbf{2}^{\mathbb{N}}$

$$\langle f \rangle = \textstyle\prod_{i=0}^{\infty} 10^{f(i)} = 10^{f(0)}\, 10^{f(1)}\, 10^{f(2)} \ldots \;.$$

We write $\langle f(n) \rangle$ as shorthand for $\langle n \mapsto f(n) \rangle$.

[1] We note that finite state transducers transform infinite words to finite or infinite words. The result of the transformation is finite if the transducer outputs the empty word ε for all except a finite number of letters of the input word. We are interested in infinite words only, since the set of finite words would merely entail two spurious extra sub-bottom degrees in the hierarchy of transducer degrees.

Definition 6. A function $f : \mathbb{N} \to \mathbb{N}$ is called *spiralling* if

(i) $\lim_{n\to\infty} f(n) = \infty$, and
(ii) for every $m \geq 1$, the function $n \mapsto f(n) \bmod m$ is ultimately periodic.

A word $\langle f \rangle$ is called *spiralling* whenever f is spiralling.

For example, $\langle p(n) \rangle$ is spiralling for every polynomial $p(n)$ with natural numbers as coefficients. Spiralling functions are called 'cyclically ultimately periodic' in the literature [3]. For a tuple $\boldsymbol{\alpha} = \langle \alpha_0, \ldots, \alpha_m \rangle$, we define

- the *length* $|\boldsymbol{\alpha}| = m + 1$, and
- its *rotation* by $\boldsymbol{\alpha}' = \langle \alpha_1, \ldots, \alpha_m, \alpha_0 \rangle$.

Let A be a set and $f : \mathbb{N} \to A$ a function. We write $\mathcal{S}^k(f)$ for the *k-th shift* of f defined by $\mathcal{S}^k(f)(n) = f(n+k)$.

We use 'weights' to represent linear functions.

Definition 7. A *weight* $\boldsymbol{\alpha}$ is a tuple $\langle a_0, \ldots, a_{k-1}, b \rangle \in \mathbb{Q}^{k+1}$ of rational numbers such that $k \in \mathbb{N}$ and $a_0, \ldots, a_{k-1} \geq 0$. The weight $\boldsymbol{\alpha}$ is called

- *non-constant* if $a_i \neq 0$ for some $i < k$, else *constant*,
- *strongly non-constant* if $a_i, a_j \neq 0$ for some $i < j < k$.

Now, let us also consider a tuple of tuples. For a tuple $\boldsymbol{\alpha} = \langle \boldsymbol{\alpha_0}, \ldots, \boldsymbol{\alpha_{m-1}} \rangle$ of weights we define $||\boldsymbol{\alpha}|| = \sum_{i=0}^{m-1} (|\boldsymbol{\alpha_i}| - 1)$.

Definition 8. Let $f : \mathbb{N} \to \mathbb{Q}$ be a function. For a weight $\boldsymbol{\alpha} = \langle a_0, \ldots, a_{k-1}, b \rangle$ we define $\boldsymbol{\alpha} \cdot f \in \mathbb{Q}$ by $\boldsymbol{\alpha} \cdot f = a_0 f(0) + a_1 f(1) + \cdots + a_{k-1} f(k-1) + b$. For a tuple of weights $\boldsymbol{\alpha} = \langle \boldsymbol{\alpha_0}, \boldsymbol{\alpha_1}, \ldots, \boldsymbol{\alpha_{m-1}} \rangle$, we define the *weighted product* $\boldsymbol{\alpha} \otimes f : \mathbb{N} \to \mathbb{Q}$ by induction on n:

$$(\boldsymbol{\alpha} \otimes f)(0) = \boldsymbol{\alpha_0} \cdot f$$
$$(\boldsymbol{\alpha} \otimes f)(n+1) = (\boldsymbol{\alpha}' \otimes \mathcal{S}^{|\boldsymbol{\alpha_0}|-1}(f))(n) \qquad (n \in \mathbb{N})$$

We say that $\boldsymbol{\alpha} \otimes f$ is a *natural weighted product* if $(\boldsymbol{\alpha} \otimes f)(n) \in \mathbb{N}$ for all $n \in \mathbb{N}$.

Weighted products are easiest understood by an example.

Example 9. Let $f(n) = n^2$ be a function and $\boldsymbol{\alpha} = \langle \boldsymbol{\alpha_1}, \boldsymbol{\alpha_2} \rangle$ a tuple of weights with $\boldsymbol{\alpha_1} = \langle 1, 2, 3, 4 \rangle$, $\boldsymbol{\alpha_2} = \langle 0, 1, 1 \rangle$. Then the weighted product $\boldsymbol{\alpha} \otimes f$ can be visualised as follows

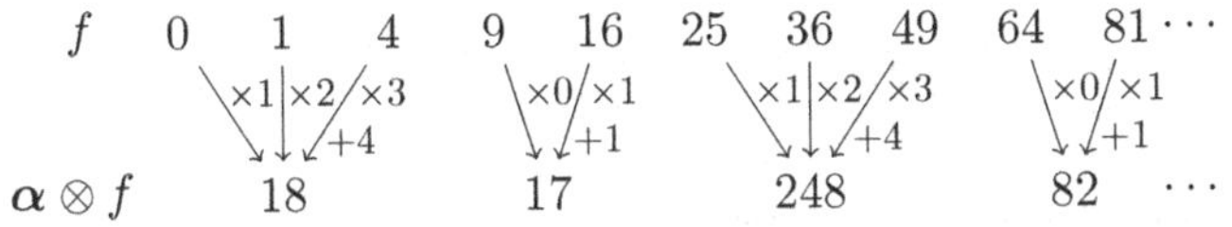

Intuitively, the weight $\boldsymbol{\alpha_1} = \langle 1, 2, 3, 4 \rangle$ means that 3 consecutive entries are added while being multiplied by 1, 2 and 3, respectively, and 4 is added to the result.

We introduce a few operations on weights. We define scalar multiplication of weights in the obvious way. We also introduce a multiplication $\odot$ that affects only the last entry of weights (the constant term).

Definition 10. Let $c \in \mathbb{Q}_{\geq 0}$, $\boldsymbol{\alpha} = \langle a_0, \dots, a_{\ell-1}, b \rangle$ a weight, $\boldsymbol{\beta} = \langle \boldsymbol{\beta_0}, \dots, \boldsymbol{\beta_{m-1}} \rangle$ a tuple of weights. We define

$$c\boldsymbol{\alpha} = \langle ca_0, \dots, ca_{\ell-1}, cb \rangle \qquad \boldsymbol{\alpha} \odot c = \langle a_0, \dots, a_{\ell-1}, bc \rangle$$
$$c\boldsymbol{\beta} = \langle c\boldsymbol{\beta_0}, \dots, c\boldsymbol{\beta_{m-1}} \rangle \qquad \boldsymbol{\beta} \odot c = \langle \boldsymbol{\beta_0} \odot c, \dots, \boldsymbol{\beta_{m-1}} \odot c \rangle$$

The following lemma follows directly from the definitions.

Lemma 11. *Let $c \in \mathbb{Q}_{\geq 0}$, $\boldsymbol{\alpha}$ a tuple of weights, and $f : \mathbb{N} \to \mathbb{Q}$ a function. Then $c(\boldsymbol{\alpha} \otimes f) = (c\boldsymbol{\alpha}) \otimes f = (\boldsymbol{\alpha} \odot c) \otimes (cf)$.* □

It is straightforward to define a *composition* of tuples of weights such that $\boldsymbol{\beta} \otimes (\boldsymbol{\alpha} \otimes f) = (\boldsymbol{\beta} \otimes \boldsymbol{\alpha}) \otimes f$ for every function $f : \mathbb{N} \to \mathbb{Q}$. Note that $\boldsymbol{\alpha} \otimes f$ is already defined. For the precise definition of $\boldsymbol{\beta} \otimes \boldsymbol{\alpha}$, we refer to [8]. It involves many details whose explicitation would not be illuminating. We will employ the following two properties of composition.

Lemma 12. *Let $\boldsymbol{\alpha}, \boldsymbol{\beta}$ be tuples of weights. Then we have that $\boldsymbol{\beta} \otimes (\boldsymbol{\alpha} \otimes f) = (\boldsymbol{\beta} \otimes \boldsymbol{\alpha}) \otimes f$ for every function $f : \mathbb{N} \to \mathbb{Q}$.* □

Lemma 13. *Let $\boldsymbol{\alpha}$ be tuple of weights, and $\boldsymbol{\beta}$ a tuple of strongly non-constant weights. Then $\boldsymbol{\alpha} \otimes \boldsymbol{\beta}$ is of the form $\langle \gamma_0, \dots, \gamma_{k-1} \rangle$ such that for every $i \in \mathbb{N}_{<k}$, the weight γ_i is either constant or strongly non-constant.* □

We need a few results on weighted products from [4].

Lemma 14 ([4]). *Let $f : \mathbb{N} \to \mathbb{N}$, and $\boldsymbol{\alpha}$ a tuple of weights. If $\boldsymbol{\alpha} \otimes f$ is a natural weighted product (i.e. $\forall n \in \mathbb{N}.\ (\boldsymbol{\alpha} \otimes f)(n) \in \mathbb{N}$), then $\langle f \rangle \geq \langle \boldsymbol{\alpha} \otimes f \rangle$.* □

For the proof of Theorem 21, below, we use the following auxiliary lemma. The lemma gives a detailed structural analysis, elaborated and explained in [4], of the transducts of a spiralling word $\langle f \rangle$.

Lemma 15 ([4]). *Let $f : \mathbb{N} \to \mathbb{N}$ be a spiralling function, and let $\sigma \in \mathbf{2}^{\mathbb{N}}$ be such that $\langle f \rangle \geq \sigma$ and $\sigma \notin \mathbf{0}$. Then there exist $n_0, m \in \mathbb{N}$, a word $w \in \mathbf{2}^*$, a tuple of weights $\boldsymbol{\alpha}$, and tuples of finite words $\mathbf{p}$ and $\mathbf{c}$ with $|\boldsymbol{\alpha}| = |\mathbf{p}| = |\mathbf{c}| = m > 0$ such that $\sigma = w \cdot \prod_{i=0}^{\infty} \prod_{j=0}^{m-1} p_j \, c_j^{\varphi(i,j)}$ where $\varphi(i,j) = (\boldsymbol{\alpha} \otimes \mathcal{S}^{n_0}(f))(mi + j)$, and*

(i) $c_j^\omega \neq p_{j+1} c_{j+1}^\omega$ for every j with $0 \leq j < m-1$, and $c_{m-1}^\omega \neq p_0 c_0^\omega$, and
(ii) $c_j \neq \varepsilon$, and α_j is non-constant, for all $j \in \mathbb{N}_{<m}$. □

Example 16. We continue Example 9. We have $\alpha = \langle \alpha_0, \alpha_1 \rangle$. Accordingly, we have prefixes $p_0, p_1 \in \mathbf{2}^*$ and cycles $c_0, c_1 \in \mathbf{2}^*$. Then the transduct σ in Lemma 15, defined by the double product, can be derived as follows:

f	0	1	4	9	16	25	36	49	64	81	$\cdots$
		α_0			α_1		α_0			α_1	
$\boldsymbol{\alpha} \otimes f$		18			17		248			82	$\cdots$
$\sigma = w$	$\cdot$	$p_0\, c_0^{18}$	$\cdot$		$p_1\, c_1^{17}$	$\cdot$	$p_0\, c_0^{248}$	$\cdot$		$p_1\, c_1^{82}$	$\cdots$

The infinite word σ is the infinite concatenation of w followed by alternating $p_0 c_0^{e_0}$ and $p_1 c_1^{e_1}$, where the exponents e_0 and e_1 are the result of applying weights α_0 and α_1, respectively.

The following theorem characterises the transducts of spiralling words up to equivalence ($\equiv$).

Theorem 17 ([4]). *Let $f : \mathbb{N} \to \mathbb{N}$ be spiralling, and $\sigma \in \mathbf{2}^{\mathbb{N}}$. Then $\langle f \rangle \geq \sigma$ if and only if $\sigma \equiv \langle \boldsymbol{\alpha} \otimes \mathcal{S}^{n_0}(f) \rangle$ for some $n_0 \in \mathbb{N}$, and a tuple of weights $\boldsymbol{\alpha}$.*

Roughly speaking, the next proposition states that polynomials of order k are closed under transduction.

Proposition 18 ([4]). *Let $p(n)$ be a polynomial of order k with non-negative integer coefficients, and let σ be a transduct of $\langle p(n) \rangle$ with $\sigma \notin \mathbf{0}$. Then $\sigma \geq \langle q(n) \rangle$ for some polynomial $q(n)$ of order k with non-negative integer coefficients.*

5 The Degree of $\langle n^k \rangle$ is Not an Atom for $k \geq 3$

We show that the degree of $\langle n^k \rangle$ is not an atom for $k \geq 3$. For this purpose, we prove a strengthening of Theorem 17, a lemma on weighted products of strongly non-constant weights, and we employ the power mean inequality.

First, we recall the power mean inequality [10].

Definition 19. For $p \in \mathbb{R}$, the *weighted power mean* $M_p(\boldsymbol{x})$ of $\boldsymbol{x} = \langle x_1, x_2, \ldots, x_n \rangle \in \mathbb{R}^n_{>0}$ with respect to $\boldsymbol{w} = \langle w_1, w_2, \ldots, w_n \rangle \in \mathbb{R}^n_{>0}$ with $\sum_{i=1}^n w_i = 1$ is

$$M_{\boldsymbol{w},0}(\boldsymbol{x}) = \prod_{i=1}^n x_i^{w_i} \qquad\qquad M_{\boldsymbol{w},p}(\boldsymbol{x}) = \left(\sum_{i=1}^n w_i x_i^p\right)^{1/p}.$$

Proposition 20 (Power mean inequality). *For all $p, q \in \mathbb{R}$, $\boldsymbol{x}, \boldsymbol{w} \in \mathbb{R}^n_{>0}$:*

$$p < q \implies M_{\boldsymbol{w},p}(\boldsymbol{x}) \leq M_{\boldsymbol{w},q}(\boldsymbol{x})$$
$$(p = q \vee x_1 = x_2 = \cdots = x_n) \iff M_{\boldsymbol{w},p}(\boldsymbol{x}) = M_{\boldsymbol{w},q}(\boldsymbol{x}).$$

Theorem 17 characterises transducts of spiralling sequences only up to equivalence. This makes it difficult to employ the theorem for proving non-transducibility. We improve the characterisation for the case of spiralling transducts as follows.

Theorem 21. *Let $f, g : \mathbb{N} \to \mathbb{N}$ be spiralling functions. Then $\langle g \rangle \geq \langle f \rangle$ if and only if some shift of f is a weighted product of a shift of g, that is:*

$$\mathcal{S}^{n_0}(f) = \boldsymbol{\alpha} \otimes \mathcal{S}^{m_0}(g)$$

for some $n_0, m_0 \in \mathbb{N}$ and a tuple of weights $\boldsymbol{\alpha}$.

Theorem 21 is a strengthening of Theorem 17 in the sense that the characterisation uses equality (= and shifts) instead of equivalence ($\equiv$). We will employ the gained precision to show that certain spiralling transducts of $\langle n^k \rangle$ cannot be transduced back to $\langle n^k \rangle$, and conclude that $\langle n^k \rangle$ is not an atom for $k \geq 3$. See further Theorem 22. Note, however, that Theorem 21 only characterises spiralling transducts whereas Theorem 17 characterises all transducts.

Proof (Theorem 21). For the direction '$\Leftarrow$', assume that $\mathcal{S}^{n_0}(f) = \boldsymbol{\alpha} \otimes \mathcal{S}^{m_0}(g)$. Then we have $\langle g \rangle \equiv \langle \mathcal{S}^{m_0}(g) \rangle \geq \langle \boldsymbol{\alpha} \otimes \mathcal{S}^{m_0}(g) \rangle = \langle \mathcal{S}^{n_0}(f) \rangle \equiv \langle f \rangle$ by invariance under shifts and by Lemma 14.

For the direction '$\Rightarrow$', assume that $\langle g \rangle \geq \langle f \rangle$. Then by Lemma 15 there exist $m_0, m \in \mathbb{N}$, $w \in \mathbf{2}^*$, $\boldsymbol{\alpha}$, $\boldsymbol{p}$ and $\boldsymbol{c}$ with $|\boldsymbol{\alpha}| = |\boldsymbol{p}| = |\boldsymbol{c}| = m > 0$ such that:

$$\langle f \rangle = w \cdot \prod_{i=0}^{\infty} \prod_{j=0}^{m-1} p_j \, c_j^{\varphi(i,j)} \tag{1}$$

where $\varphi(i,j) = (\boldsymbol{\alpha} \otimes \mathcal{S}^{m_0}(g))(mi + j)$ such that the conditions (i) and (ii) of Lemma 15 are fulfilled.

Note that, as $\lim_{n\to\infty} f(n) = \infty$, the distance of ones in the sequence $\langle g \rangle$ tends to infinity. For every $j \in \mathbb{N}_{<m}$, the word p_j occurs infinitely often in $\langle f \rangle$ by (1), and hence p_j can contain at most one occurrence of the symbol 1.

By condition (ii), we have for every $j \in \mathbb{N}_{<m}$ that $c_j \neq \varepsilon$, and the weight $\boldsymbol{\alpha_j}$ is not constant. As $\lim_{n\to\infty} g(n) = \infty$, it follows that c_j^2 appears infinitely often in $\langle f \rangle$ by (1). Hence c_j consists only of 0's, that is, $c_j \in \{0\}^+$ for every $j \in \mathbb{N}_{<m}$.

By condition (i) we never have $c_j^\omega = p_{j+1} c_{j+1}^\omega$ for $j \in \mathbb{N}_{<m}$ (where addition is modulo m). As $c_j^\omega = 0^\omega$ and $p_{j+1} 0^\omega = p_{j+1} c_{j+1}^\omega$, we obtain that p_{j+1} must contain a 1. Hence, for every $k \in \mathbb{N}_{<m}$, the word p_j contains precisely one 1.

Finally, we apply the following transformations to ensure $p_j = 1$ and $c_j = 0$ for every $j \in \mathbb{N}_{<m}$:

(i) For every $j \in \mathbb{N}_{<m}$ such that $c_j = 0^h$ for some $h > 1$, we set $c_j = 0$ and replace the weight $\boldsymbol{\alpha_j}$ in $\boldsymbol{\alpha}$ by $h\boldsymbol{\alpha_j}$.

(ii) For every $j \in \mathbb{N}_{<m}$ such that $p_j = 0^h 1 0^\ell$ for some $h \geq 1$ or $\ell \geq 1$, we set $p_j = 1$ and replace the weight $\boldsymbol{\alpha_j}$ in $\boldsymbol{\alpha}$ by $(\boldsymbol{\alpha_j} + \ell)$ and the weight $\boldsymbol{\alpha_{j-1}}$ by $(\boldsymbol{\alpha_{j-1}} + h)$. Here, for a weight $\boldsymbol{\gamma} = \langle x_0, \ldots, x_{\ell-1}, y \rangle$ and $z \in \mathbb{Q}$, we write $\boldsymbol{\gamma} + z$ for the weight $\langle x_0, \ldots, x_{\ell-1}, y + z \rangle$. If $j = 0$, we moreover append 0^h to the word w.

Note that both transformations leave Eq. (1) valid, they do not change the result of the double product.

Thus we now have $p_j = 1$ and $c_j = 0$ for every $j \in \mathbb{N}_{<m}$. It follows from (1) that $\langle f \rangle = w \langle \boldsymbol{\alpha} \otimes \mathcal{S}^{m_0}(g) \rangle$. Hence $\mathcal{S}^{n_0}(f) = \boldsymbol{\alpha} \otimes \mathcal{S}^{m_0}(g)$ for some $n_0 \in \mathbb{N}$. $\square$

Theorem 22. *For $k \geq 3$, the degree of $\langle n^k \rangle$ is not an atom.*

Proof. Define $f : \mathbb{N} \to \mathbb{N}$ by $f(n) = n^k$. We have $\langle f \rangle \geq \langle g \rangle$ where $g : \mathbb{N} \to \mathbb{N}$ is defined by $g(n) = (2n)^k + (2n+1)^k$. Assume that we had $\langle g \rangle \geq \langle f \rangle$. Then, by Theorem 21 we have $\mathcal{S}^{n_0}(f) = \boldsymbol{\alpha} \otimes \mathcal{S}^{m_0}(g)$ for some $n_0, m_0 \in \mathbb{N}$ and a tuple of weights $\boldsymbol{\alpha}$. Note that $g = \langle\langle 1,1,0\rangle\rangle \otimes f$ and

$$\begin{aligned}\mathcal{S}^{n_0}(f) &= \boldsymbol{\alpha} \otimes \mathcal{S}^{m_0}(\langle\langle 1,1,0\rangle\rangle \otimes f)\\ &= \boldsymbol{\alpha} \otimes (\langle\langle 1,1,0\rangle\rangle \otimes \mathcal{S}^{2m_0}(f)) = \boldsymbol{\beta} \otimes \mathcal{S}^{2m_0}(f)\end{aligned}$$

where $\boldsymbol{\beta} = \boldsymbol{\alpha} \otimes \langle\langle 1,1,0\rangle\rangle$. By Lemma 13 every weight in $\boldsymbol{\beta}$ is either constant or strongly non-constant. As $\mathcal{S}^{n_0}(f)$ is strictly increasing (and hence contains no constant subsequence), each weight in $\boldsymbol{\beta}$ must be strongly non-constant.

Let $\boldsymbol{\beta} = \langle \boldsymbol{\beta_0}, \dots, \boldsymbol{\beta_{\ell-1}}\rangle$. For every $n \in \mathbb{N}$ we have:

$$\mathcal{S}^{n_0}(f)(\ell n) = (\boldsymbol{\beta} \otimes \mathcal{S}^{2m_0}(f))(\ell n) = \boldsymbol{\beta_0} \cdot \mathcal{S}^{2m_0+||\boldsymbol{\beta}||\cdot n}(f)\ . \tag{2}$$

Then we have

$$\begin{aligned}\mathcal{S}^{n_0}(f)(\ell n) &= (n_0 + \ell n)^k = \textstyle\sum_{i=0}^{k} \binom{k}{i} n_0^i \ell^{k-i} n^{k-i}\\ &= \ell^k n^k + k n_0 \ell^{k-1} n^{k-1} + \cdots + k n_0^{k-1} \ell n + n_0^k\ .\end{aligned} \tag{3}$$

Let $\boldsymbol{\beta_0} = \langle a_0, a_1, \dots, a_{h-1}, b\rangle$. We define $c_i = a_i||\boldsymbol{\beta}||^k$ and $d_i = (2m_0 + i)/||\boldsymbol{\beta}||$. We obtain

$$\begin{aligned}\boldsymbol{\beta_0} \cdot \mathcal{S}^{2m_0+||\boldsymbol{\beta}||\cdot n}(f) &= b + \textstyle\sum_{i=0}^{h-1} a_i f(2m_0 + ||\boldsymbol{\beta}|| \cdot n + i)\\ &= b + \textstyle\sum_{i=0}^{h-1} a_i f(||\boldsymbol{\beta}||(n + \frac{2m_0+i}{||\boldsymbol{\beta}||}))\\ &= b + \textstyle\sum_{i=0}^{h-1} a_i ||\boldsymbol{\beta}||^k (n + d_i)^k = b + \sum_{i=0}^{h-1} c_i (n + d_i)^k\\ &= b + \textstyle\sum_{i=0}^{h-1} c_i (n^k + k d_i n^{k-1} + \cdots + k d_i^{k-1} n + d_i^k)\ .\end{aligned} \tag{4}$$

Recall Eq. (2). Comparing the coefficients of n^k, n^{k-1} and n in (3) and (4) we obtain

$$\ell^k = \sum_{i=0}^{h-1} c_i \qquad k n_0 \ell^{k-1} = \sum_{i=0}^{h-1} c_i k d_i \qquad k n_0^{k-1} \ell = \sum_{i=0}^{h-1} c_i k d_i^{k-1}\ , \text{ and hence}$$

$$1 = \sum_{i=0}^{h-1} \frac{c_i}{\ell^k} \qquad \frac{n_0}{\ell} = \sum_{i=0}^{h-1} \frac{c_i}{\ell^k} d_i \qquad \frac{n_0^{k-1}}{\ell^{k-1}} = \sum_{i=0}^{h-1} \frac{c_i}{\ell^k} d_i^{k-1}\ .$$

This is in contradiction with the weighted power means inequality (Proposition 20). Clearly all d_i are distinct, and, as a consequence of $\boldsymbol{\beta_0}$ being strongly non-constant, there are at least two $i \in \mathbb{N}_{<h}$ for which $c_i \neq 0$. Thus our assumption $\langle g\rangle \geq \langle f\rangle$ must have been wrong. Hence the degree of $\langle n^k\rangle$ is not an atom. □

6 Atoms of Every Polynomial Order

In the previous section, we have seen that $\langle n^k\rangle$ is not an atom for $k \geq 3$. In this section, we show that for every order $k \in \mathbb{N}$ there exists a polynomial $p(n)$ of order k such that the degree of the word $\langle p(n) \rangle$ is an atom. As a consequence, there are at least $\aleph_0$ atoms in the transducer degrees.

As we have seen in the proof of Theorem 22, whenever $k \geq 3$, we have that $\langle n^k \rangle \geq \langle g(n) \rangle$, but not $\langle g(n) \rangle \geq \langle n^k \rangle$ for $g(n) = (2n)^k + (2n+1)^k$. Thus there exist polynomials $p(n)$ of order k for which $\langle p(n) \rangle$ cannot be transduced to $\langle n^k \rangle$. However, the key observation underlying the construction in this section is the following: Although we may not be able to reach $\langle n^k \rangle$ from $\langle p(n) \rangle$, we can get arbitrarily close (Lemma 25, below). This enables us to employ the concept of *continuity*.

In order to have continuous functions over the space of polynomials to allow limit constructions, we now permit rational coefficients. For $k \in \mathbb{N}$, let $\mathfrak{Q}_k$ be the set of polynomials of order k with non-negative rational coefficients. We also use polynomials in $\mathfrak{Q}_k$ to denote spiralling sequences. However, we need to give meaning to $\langle q(n) \rangle$ for the case that the block sizes $q(n)$ are not natural numbers. For this purpose, we make use of the fact that the degree of a word $\langle f(n) \rangle$ is invariant under multiplication of the block sizes by a constant, as is easy to see. More precisely, for $f : \mathbb{N} \to \mathbb{N}$, we have $\langle f(n) \rangle \equiv \langle d \cdot f(n) \rangle$ for every $d \in \mathbb{N}$ with $d \geq 1$. So to give meaning to $\langle q(n) \rangle$, we multiply the polynomial by the least natural number $d > 0$ such that $d \cdot q(n)$ is a natural number for every $n \in \mathbb{N}$.

Definition 23. We call a function $f : \mathbb{N} \to \mathbb{Q}$ *naturalisable* if there exists a natural number $d \geq 1$ such that for all $n \in \mathbb{N}$ we have $(d \cdot f(n)) \in \mathbb{N}$.

For naturalisable $f : \mathbb{N} \to \mathbb{Q}$ we define $\langle f \rangle = \langle d \cdot f \rangle$ where $d \in \mathbb{N}$ is the least number such that $d \geq 1$ where for all $n \in \mathbb{N}$ we have $(d \cdot f(n)) \in \mathbb{N}$. (Note that, for $f : \mathbb{N} \to \mathbb{N}$, $\langle f(n) \rangle$ has been defined in Sect. 4.)

Observe that every $q(n) \in \mathfrak{Q}_k$ is naturalisable (multiply by the least common denominator of the coefficients). Also, naturalisable functions are preserved under weighted products.

Now, Lemma 14 can be generalised as follows. There is no longer need to require that the weighted product is natural. All weighted products of naturalisable functions can be realised by finite state transducers.

Lemma 24. *Let* $f : \mathbb{N} \to \mathbb{Q}$ *be naturalisable, and* $\boldsymbol{\alpha}$ *a tuple of weights. Then* $\boldsymbol{\alpha} \otimes f$ *is naturalisable and* $\langle f \rangle \geq \langle \boldsymbol{\alpha} \otimes f \rangle$.

Proof. Let $\boldsymbol{\alpha} = \langle \alpha_0, \ldots, \alpha_{m-1} \rangle$ for some $m \geq 1$. Let $c \in \mathbb{N}$ with $c \geq 1$ be minimal such that all entries of $c\boldsymbol{\alpha}$ are natural numbers. Let $d \in \mathbb{N}$ with $d \geq 1$ be the least natural number such that $\forall n \in \mathbb{N}\ (d \cdot f(n)) \in \mathbb{N}$.

Then we obtain $((dc\boldsymbol{\alpha}) \otimes f)(n) \in \mathbb{N}$ for ever $n \in \mathbb{N}$. By the definition of weighted products it follows immediately that $(dc\boldsymbol{\alpha}) \otimes f = dc(\boldsymbol{\alpha} \otimes f)$, and hence $\boldsymbol{\alpha} \otimes f$ is naturalisable. Let $e \in \mathbb{N}$ with $e \geq 1$ be the least natural number such that $\forall n \in \mathbb{N}\ (e \cdot (\boldsymbol{\alpha} \otimes f)(n)) \in \mathbb{N}$.

We have the following transduction

$$\begin{aligned}
\langle f \rangle &= \langle df \rangle && \text{by Definition 23} \\
&\geq \langle ((c\boldsymbol{\alpha}) \odot d) \otimes (df) \rangle && \text{by Lemma 14} \\
&= \langle (dc\boldsymbol{\alpha}) \otimes f \rangle = \langle dc(\boldsymbol{\alpha} \otimes f) \rangle && \text{by Lemma 11} \\
&\geq \langle \langle \langle \frac{e}{dc}, 0 \rangle \rangle \otimes (dc(\boldsymbol{\alpha} \otimes f)) \rangle && \text{by Lemma 14} \\
&= \langle e(\boldsymbol{\alpha} \otimes f) \rangle = \langle \boldsymbol{\alpha} \otimes f \rangle && \text{by Definition 23}
\end{aligned}$$

This conclude the proof. □

The following lemma states that every word $\langle q(n) \rangle$, for a polynomial $q(n) \in \mathfrak{Q}_k$ of order k, can be transduced arbitrarily close to $\langle n^k \rangle$.

Lemma 25. *Let $k \geq 1$ and let $q(n) \in \mathfrak{Q}_k$ be a polynomial of order k. For every $\varepsilon > 0$ we have $\langle q(n) \rangle \geq \langle n^k + b_{k-1}n^{k-1} + \cdots + b_1 n \rangle$ for some rational coefficients $0 \leq b_{k-1}, \ldots, b_1 < \varepsilon$.*

Proof. Let $q(n) = a_k n^k + a_{k-1}n^{k-1} + \cdots + a_1 n + a_0$, and let $\varepsilon > 0$ be arbitrary. Then for every $d \in \mathbb{N}$, we have

$$\begin{aligned}
\langle q(n) \rangle \geq \langle q(dn) \rangle \geq \langle \frac{q(dn)}{a_k d^k} \rangle &= \langle n^k + \frac{a_{k-1}}{a_k d} n^{k-1} + \ldots + \frac{a_1}{a_k d^{k-1}} n^1 + \frac{a_0}{a_k d^k} \rangle \\
&\geq \langle n^k + \frac{a_{k-1}}{a_k d} n^{k-1} + \ldots + \frac{a_1}{a_k d^{k-1}} n^1 \rangle
\end{aligned}$$

The first transduction is picking a subsequence of the blocks. The second transduction is a division of the size of each block (application of Lemma 24 with the weight $\langle \langle 1/a_k d^k, 0 \rangle \rangle$). The last transduction amounts to removing a constant number of zeros from each block (application of Lemma 24 with the weight $\langle \langle 1, -a_0/(a_k d^k) \rangle \rangle$). Finally, note that the last polynomial in the transduction is of the desired form if $d \in \mathbb{N}$ is chosen large enough. □

For polynomials $p(n) \in \mathfrak{Q}_k$, we want to express weighted products $\langle \boldsymbol{\alpha} \rangle \otimes p$ in terms of matrix products. For that purpose we need a couple of definitions.

Definition 26. For weights $\boldsymbol{\alpha} = \langle a_0, \ldots, a_{k-1}, b \rangle$ we define a column vector $U(\boldsymbol{\alpha}) = (a_0, \ldots, a_{k-1})^T$.

Definition 27. If $p(n) = \sum_{i=0}^{k} c_i n^i$ is a polynomial of order k, we define a column vector $V(p(n)) = (c_1, \ldots, c_k)^T$ and a square matrix

$$M(p(n)) = (V(p(kn + 0)), \ \ldots, \ V(p(kn + k - 1))) \, .$$

We also write $V(p)$ short for $V(p(n)$ and $M(p)$ for $M(p(n))$.

Note that we have omitted the constant term c_0 from the definition of $V(p)$. The reason is that for every $f : \mathbb{N} \to \mathbb{N}$ and $c \in \mathbb{N}$ we have $\langle f(n) \rangle \equiv \langle f(n) + c \rangle$. These words are of the same degree because a finite state transducer can add (or remove) a constant number of symbols 0 to (from) every block of 0's. For the same reason, b was omitted from the definition of $U(\boldsymbol{\alpha})$.

Example 28. Consider the polynomial n^3:

$$V(n^3) = \begin{pmatrix} 0 \\ 0 \\ 1 \end{pmatrix} \quad \text{and} \quad M(n^3) = \begin{pmatrix} 0 & 9 & 36 \\ 0 & 27 & 54 \\ 27 & 27 & 27 \end{pmatrix}$$

where the columns vectors of the matrix $M(n^3)$ are given by $V((3n)^3)$, $V((3n+1)^3)$ and $V((3n+2)^3)$.

Lemma 29. *Let $k \geq 1$. Let $\boldsymbol{\alpha} = \langle a_0, \ldots, a_{k-1}, b\rangle$ be a weight and $p(n) \in \mathfrak{Q}_k$. Then $M(p)\,U(\boldsymbol{\alpha}) = V(\langle\boldsymbol{\alpha}\rangle \otimes p)$.*

Proof. A direct calculation shows that

$$\begin{aligned} M(p)\,U(\boldsymbol{\alpha}) &= \sum_{i=0}^{k-1} a_i V(p(kn+i)) = V\big(\sum_{i=0}^{k-1} a_i p(kn+i)\big) \\ &= V\big(\sum_{i=0}^{k-1} a_i p(kn+i) + b\big) = V(\langle\boldsymbol{\alpha}\rangle \otimes p)\,, \end{aligned}$$

which proves the lemma. □

Let us take a closer look at the matrix $M(n^k)$. The element on the ith row and jth column is $M_{i,j} = \binom{k}{i} k^i (j-1)^{k-i}$. Dividing the ith row by $\binom{k}{i} k^i$ for each i gives a Vandermonde-type matrix, which is invertible. Thus also $M(n^k)$ is invertible.

Lemma 30. *For $k \geq 1$, $M(n^k)$ is invertible.* □

Theorem 31. *Let $k \geq 1$. Let $a_0, \ldots, a_{k-1}$ be positive rational numbers, $\boldsymbol{\alpha} = \langle a_0, \ldots, a_{k-1}, 0\rangle$, and*

$$p(n) = (\langle\boldsymbol{\alpha}\rangle \otimes n^k)(n) = \sum_{i=0}^{k-1} a_i (kn+i)^k.$$

Then $\langle q(n)\rangle \geq \langle p(n)\rangle$ for all $q(n) \in \mathfrak{Q}_k$. Moreover, the degree $\langle p(n)\rangle^{\equiv}$ is an atom. Note that the degree $\langle p(n)\rangle^{\equiv}$ is the infimum of all degrees of words $\langle q(n)\rangle$ with $q(n) \in \mathfrak{Q}_k$.

Proof. By Lemma 29, $M(n^k)\,U(\boldsymbol{\alpha}) = V(p)$. By Lemma 30, $M(n^k)$ is invertible and we can write $U(\boldsymbol{\alpha}) = M(n^k)^{-1}V(p)$. By Lemma 25, for every $\varepsilon > 0$ there exists $q_\varepsilon \in \mathfrak{Q}_k$ such that $\langle q(n)\rangle \geq \langle q_\varepsilon(n)\rangle$ and

$$q_\varepsilon(n) = n^k + b_{k-1}n^{k-1} + \cdots + b_1 n$$

with $0 \leq b_i \leq \varepsilon$ for every $i \in \{1, \ldots, k-1\}$. We will show that if ε is small enough, then $\langle q_\varepsilon(n)\rangle \geq \langle p(n)\rangle$.

We have $\lim_{\varepsilon\to 0} M(q_\varepsilon) = M(n^k)$. As $\det(M(n^3)) \neq 0$ and the determinant function is continuous, also $\det(M(q_\varepsilon)) \neq 0$ for all sufficiently small ε. Then

$M(q_\varepsilon)$ is invertible, and we define $U_\varepsilon = M(q_\varepsilon)^{-1}V(p)$. We would like to have $U_\varepsilon = U(\gamma)$ for some weight γ. This is not always possible, because some elements of U_ε might be negative. However, by the continuity of matrix inverse and product,

$$\lim_{\varepsilon \to 0} U_\varepsilon = \lim_{\varepsilon \to 0}(M(q_\varepsilon)^{-1}V(p)) = (\lim_{\varepsilon \to 0} M(q_\varepsilon))^{-1}V(p) = M(n^k)^{-1}V(p) = U(\boldsymbol{\alpha})$$

Since every element of $U(\boldsymbol{\alpha})$ is positive, we can fix a small enough ε so that every element of U_ε is positive. Then we have $U_\varepsilon = U(\gamma)$ for some weight γ.

We have $M(q_\varepsilon)\,U(\gamma) = V(\langle\gamma\rangle \otimes q_\varepsilon)$ by Lemma 29, and $M(q_\varepsilon)\,U(\gamma) = V(p)$ by the definition of U_ε. As a consequence $(\langle\gamma\rangle \otimes q_\varepsilon)(n) = p(n) + c$ for some constant c. By Lemma 24, we obtain $\langle q_\varepsilon(n)\rangle \geq \langle p(n)\rangle$.

It remains to show that the degree $\langle p(n)\rangle^{\equiv}$ is an atom. Assume that $\langle p(n)\rangle \geq w$ and $w \notin \mathbf{0}$. By Proposition 18 we have $w \geq \langle q(n)\rangle$ for some $q(n) \in \mathfrak{Q}_k$. As shown above, $\langle q(n)\rangle \geq \langle p(n)\rangle$, thus $w \geq \langle p(n)\rangle$. Hence $\langle p(n)\rangle^{\equiv}$ is an atom. □

7 Future Work

Our results hint at an interesting structure of the transducer degrees of words $\langle p(n)\rangle$ for polynomials $p(n)$ of order $k \in \mathbb{N}$. Here, we have only scratched the surface of this structure. Many questions remain open, for example:

(i) What is the structure of 'polynomial spiralling' degrees (depending on $k \in \mathbb{N}$)? Is the number of degrees finite for every $k \in \mathbb{N}$?
(ii) Are there interpolant degrees between the degrees of $\langle n^k\rangle$ and $\langle p_k(n)\rangle$?
(iii) Are there continuum many atoms?
(iv) Is the degree of the Thue–Morse sequence an atom?

References

1. Allouche, J.P., Shallit, J.: Automatic Sequences: Theory, Applications Generalizations. Cambridge University Press, New York (2003)
2. Belov, A.: Some algebraic properties of machine poset of infinite words. ITA **42**(3), 451–466 (2008)
3. Berstel, J., Boasson, L., Carton, O., Petazzoni, B., Pin, J.E.: Operations preserving regular languages. Theor. Comput. Sci. **354**(3), 405–420 (2006)
4. Endrullis, J., Grabmayer, C., Hendriks, D., Zantema, H.: The degree of squares is an atom. In: Manea, F., Nowotka, D. (eds.) WORDS 2015. LNCS, vol. 9304, pp. 109–121. Springer, Heidelberg (2015)
5. Endrullis, J., Hansen, H.H., Hendriks, D., Polonsky, A., Silva, A.: A coinductive framework for infinitary rewriting and equational reasoning. In: Proceedings of Conference on Rewriting Techniques and Applications (RTA 2015). Schloss Dagstuhl (2015)
6. Endrullis, J., Hendriks, D.: Lazy productivity via termination. Theor. Comput. Sci. **412**(28), 3203–3225 (2011)
7. Endrullis, J., Hendriks, D., Klop, J.W.: Degrees of streams. J. Integers **11B**(A6), 1–40 (2011). Proceedings of the Leiden Numeration Conference 2010

8. Endrullis, J., Karhumäki, J., Klop, J., Saarela, A.: Degrees of infinite words, polynomials and atoms (extended version). CoRR (2016)
9. Endrullis, J., Klop, J.W., Saarela, A., Whiteland, M.: Degrees of transducibility. In: Manea, F., Nowotka, D. (eds.) WORDS 2015. LNCS, vol. 9304, pp. 1–13. Springer, Heidelberg (2015)
10. Hardy, G.H., Littlewood, J.E., Pólya, G.: Inequalities. Cambridge University Press, Cambridge (1988). Reprint of the 1952 edition
11. Löwe, B.: Complexity hierarchies derived from reduction functions. In: Löwe, B., Piwinger, B., Räsch, T. (eds.) Classical and New Paradigms of Computation and their Complexity Hierarchies. Trends in Logic, vol. 23, pp. 1–14. Springer, Amsterdam (2004)
12. Odifreddi, P.: Classical Recursion Theory. Studies in Logic and the Foundations of Mathematics. North-Holland Publishing Co., Amsterdam (1999)
13. Rayna, G.: Degrees of finite-state transformability. Inf. Control **24**(2), 144–154 (1974)
14. Sakarovitch, J.: Elements of Automata Theory. Cambridge University Press, Cambridge (2003)
15. Shoenfield, J.R.: Degrees of Unsolvability. North-Holland, Elsevier, New York (1971)

Ternary Square-Free Partial Words with Many Wildcards

Daniil Gasnikov and Arseny M. Shur(✉)

Ural Federal University, Ekaterinburg, Russia
legionlon@gmail.com, arseny.shur@urfu.ru

Abstract. We contribute to the study of square-free words. The classical notion of a square-free word has a natural generalization to partial words, studied in several papers since 2008. We prove that the maximal density of wildcards in the ternary infinite square-free partial word is surprisingly big: 3/16. In addition, we introduce a related characteristic of infinite square-free words, called flexibility, and find its values for some interesting words and classes of words.

Keywords: Partial word · Square-free word · Letter density

1 Introduction

Partial words are a natural generalization of "ordinary" words. A partial word is a word with some positions undefined; more formally, a partial word over an alphabet Σ is a word over the alphabet $\Sigma \cup \{\diamond\}$, where the *wildcard* symbol $\diamond$ has a special meaning. Namely, when two words are being compared, a wildcard matches any symbol. Thus, the partial word $a\diamond bc$ matches $\diamond c\diamond c$. In the study of partial words, the matching relation replaces equality in such notions as periods, powers, etc. The main feature of the matching relation is its nontransitivity. This makes many problems on partial words hard. For example, the pattern matching problem, studied for partial words since 1974 [7], is at least as hard as the boolean multiplication [10].

Combinatorics of partial words is much younger than their algorithmics; it began with the paper by Berstel and Boasson [2] and subsequent works [4,15]. These and other early papers focused on periodicity properties. The study of avoidability began with the paper [9], which focused mostly on cube-free partial words. A suitable definition of a square-free partial word was proposed in [8] and independently in [5]; in both these papers the existence of infinite ternary square-free partial words with infinite number of wildcards was proved. In fact, it was demonstrated that wildcards in such a word can have nonzero density: the construction from [8] gives a word with the density 1/39. Thus, a natural question arises: *what is the maximum possible density of wildcards in an infinite ternary square-free partial word?* A related question about the minimum k such that any factor of length k of some infinite ternary square-free partial word contains

S. Brlek and C. Reutenauer (Eds.): DLT 2016, LNCS 9840, pp. 177–189, 2016.
DOI: 10.1007/978-3-662-53132-7_15

a wildcard was answered in [3]; the answer is $k = 7$. Moreover, a thorough analysis of the word from [3] shows that the density of wildcards in it is 7/39.

We study the density of wildcards in a more general context. Every square-free partial word can be obtained by taking a square-free word and replacing some of its letters with wildcards. Thus, for infinite square-free words we have a natural characteristic which we call *flexibility*: the maximum density of the set of positions, in which letters can be simultaneously replaced by wildcards preserving square-freeness (we refer to such sets as *wildcard sets*). Our main result is the exact value of the maximal flexibility of infinite ternary square-free words: 3/16. First, we prove that not only density, but even the upper density of a wildcard set for an infinite ternary square-free word cannot exceed 3/16. Second, we construct a square-free word G (which probably never appeared in the study of square-free words before) with flexibility 3/16. Moreover, the wildcard set for G is periodic with period 16. Additional results include the flexibility of the Arshon word (1/9) and the Dejean word (2/19), and also a series of "rigid" square-free words, which have no room for wildcards at all. Our technique is based on the encoding of ternary square-free words by the walks in the weighted K_{33} graph. This encoding was proposed by the second author [14] and proved useful in solving different problems on ternary square-free words [11–13].

The further text consists of preliminary Sect. 2, technical Sect. 3, and proofs of the main results in Sect. 4.

2 Preliminaries

Definitions and Notation. We study words over the ternary alphabet $\Sigma = \{a, b, c\}$; by default, the letters x, y, z denote variable symbols from Σ. Finite (infinite) words over Σ are treated as functions $w : \{1, \ldots, n\} \to \Sigma$ (resp., $w : \mathbb{N} \to \Sigma$); the numbers from the domain of such a function are *positions* (in w). In this setting *partial words* are partial functions; the *wildcard* symbol $\diamond$ is used to fill undefined positions.

Standard notions of factor, prefix, and suffix are used for both words and partial words. We write λ for the empty word, $|w|$ for the length of w, $w[i]$ for the ith letter of w and $w[i..j]$ for the factor of w occupying the positions $i, i{+}1, \ldots, j$. A factor $v = w[i..j]$ is referred to as the *occurrence* of v in w at the ith position. Two partial words u and v *match* if $|u| = |v|$ and for each $i = 1, \ldots, |u|$ either $u[i] = v[i]$ or at least one of $u[i], v[i]$ is a wildcard.

A finite word w has *period* $p < |w|$ if $w[1..|w|{-}p] = w[p{+}1..w]$. The *exponent* $\exp(w)$ of w is the ratio between its length and its minimal period. The *local exponent* $\operatorname{lexp}(w)$ of a finite or infinite word w is the supremum of the exponents of the finite factors of w. The *extension* of a factor $v = w[i..j]$ is the factor $u = w[i'..j']$ such that $i' \le i$, $j' \ge j$, u has period $|v|$ but the factors $w[i'..j'{+}1]$, $w[i'{-}1..j']$ has not.

A *square* is a nonempty word of the form uu. A word is *square-free* if it has no squares as factors. The set of ternary square-free words (both finite and infinite) is denoted by SF. A *partial square* is a word of the form uu' such that

u matches u'. A (partial or not) square is called *p-square* if $|u| = p$. Partial 1-squares of the form $\diamond x$ or $x\diamond$ occur in every partial word u such that $|u| > 1$ and u contains a wildcard, so we regard them as trivial. A partial word containing no 1-squares and no *partial* p-squares for any $p > 1$, is called *square-free*. Ternary square-free infinite partial words exist [5,8]; we write PSF for the partial counterpart of SF.

Words of the form uv and vu are *conjugates*; conjugacy is an equivalence relation. Linking up the ends of a finite word, we obtain a *circular word*. A circular word represents a conjugacy class in an obvious way. The factors of a circular word are just words, so one can speak about square-free circular words.

Let $P \subset \mathbb{N}$ and $d_n = |P \cap \{1, \ldots, n\}|/n$. Then the *density* of P is the limit $d = \lim_{n\to\infty} d_n$ if it exists; otherwise, we speak about *upper* and *lower* density, meaning $\limsup d_n$ and $\liminf d_n$, respectively.

Basic Properties. The following basic property of partial words is important.

Lemma 1. *If partial words u and v match and u' is obtained from u by replacing a letter with a wildcard, then u' and v match.*

Lemma 2. *Let a partial word u be square-free and $u[i] = \diamond$. The partial word u' obtained from u by replacing $u[i]$ with a letter distinct from the adjacent letters is square-free.*

Proof. Assume that u' contains a square. It is not a 1-square by construction and it contains $u'[i]$. Then u must contain a square at the same position by Lemma 1. This contradicts the square-freeness of u. □

Proposition 3. *Every finite or infinite square-free partial word matches some square-free word. In the case of ternary alphabet, such a matching word is unique up to the first and the last letter.*

Proof. The existence of a matching square-free word follows by repeated application of Lemma 2. Further, a square-free partial word has no factors of the form $x\diamond x$, because such a factor forms a 2-square with any subsequent/preceding symbol. Thus, the letters adjacent to a wildcard are distinct. If the alphabet is ternary, there is only one possibility to replace this wildcard. □

Due to Proposition 3, any element of PSF can be seen as a word from SF in which the letters in some positions are replaced by wildcards. Let $u \in \mathsf{SF}$, $P \subset \mathbb{N}$. We denote by u_P the partial word obtained by replacing the letters in u at the positions from P by wildcards. We call P *a wildcard set for* u if $u_P \in \mathsf{PSF}$. The maximum density of a wildcard set for u is a natural combinatorial characteristic of u; we call it *flexibility*[1]. The original question about the density of wildcards can be reformulated as

[1] As mentioned above, some sets do not have density; to avoid additional notions we postulate that upper (lower) bounds on flexibility should work for upper (resp., lower) densities of the corresponding wildcard sets.

– *What is the maximum flexibility of an infinite ternary square-free word?*

We give an upper bound in Sect. 4.1 and a matching lower bound in Sect. 4.2. Another natural question is about words of zero flexibility. We say that an infinite word $u \in \mathsf{SF}$ is *rigid* (resp., *almost rigid*) if it has no nonempty wildcard sets (resp., only finite wildcard sets). In Sect. 3.1, we characterize a class of almost rigid words and find a series of rigid words in it.

Codewalks. The representation of ternary square-free words described in this subsection was proposed in [14]. These words contain three-letter factors of the form xyx, called *jumps* (of one letter over another). Jumps occur quite often: if $u[i..i{+}2]$ is a jump in $u \in \mathsf{SF}$, then the next jump in u occurs at one of the positions $i{+}2$, $i{+}3$, $i{+}4$. (A jump at position $i{+}1$ produces a 2-square at position i, while no jump up to position $i{+}5$ leads to a 3-square at position $i{+}1$.) Moreover, a jump in a word can be uniquely reconstructed from the previous (or the next) jump and the distance between them. Indeed, let $u[i..i{+}2] = xyx$. If the next jump is in the position $i{+}2$ (resp., $i{+}3$, $i{+}4$), then $u[i..i{+}4] = xyxzx$ (resp., $u[i..i{+}5] = xyxzyz$, $u[i..i{+}6] = xyxzyxy$). Thus,

($\star$) a word $u \in \mathsf{SF}$ can be uniquely reconstructed from the following information: the leftmost jump, its position, the sequence of distances between successive jumps, the number of positions after the last jump (for finite words).

The property ($\star$) allows one to encode square-free words by walks in the weighted K_{33} graph shown in Fig. 1. A word $u \in \mathsf{SF}$ is encoded by the walk visiting the vertices in the order in which jumps occur when reading u left to right. If the leftmost jump occurs in u at position $i > 1$, then we add the edge of length $i{-}1$ to the beginning of the walk; note that in this case the walk begins with an edge, not a vertex. A symmetric procedure applies to the end of u if u is finite. By ($\star$), we can omit the vertices (except for the first one), keeping just the lengths of edges and marking the "hanging" edges in the beginning and/or the end. Due to symmetry, we can omit even the first vertex, retaining all information about u up to renaming the letters. For example, $u = abcbabcacbacabc$ has the jumps (left to right) bcb, bab, cac, aca and is encoded, according to Fig. 1, by $\underline{1}123\underline{2}$.

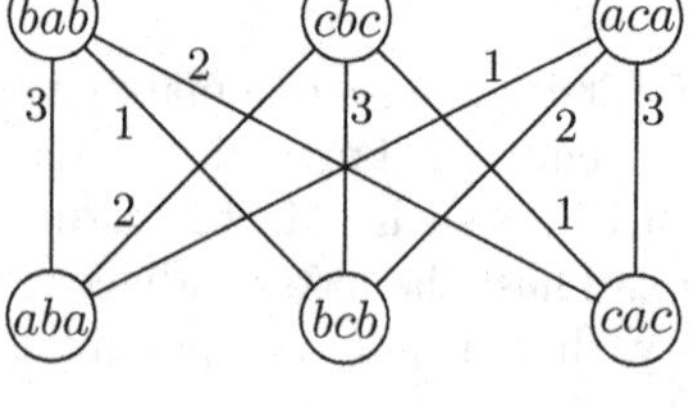

Fig. 1. The graph of jumps in ternary square-free words. Vertices are jumps; two jumps that can follow each other in a square-free word are connected by an edge of length i, where i is the number of positions between the starting positions of these jumps. Due to symmetry, the graph is undirected.

Such a code is called *codewalk* and denoted by $\mathsf{cwk}(u)$. The codewalk $\underline{1}123\underline{2}$ is decoded by any word $xyzyxyzxzyxzxyz$, where $\{x, y, z\} = \{a, b, c\}$. Note that the choice of decoding does not affect the properties concerning periods and squares. A codewalk is *closed* if it marks a closed walk without hanging edges in $K_{3,3}$; e.g., 121212 is closed and 1212 is not. For a codewalk w without hanging edges, its *literal length* $\ell(w)$ is the distance between the positions of the last and the first jumps in the decoded word; it can be computed by adding $|w|$ to the sum of digits of w. Note that if a codewalk $wv = \mathsf{cwk}(u)$ has period $|w|$ and w is closed, then u has period $\ell(w)$.

Clearly, not all walks in the weighted K_{33} graph encode square-free words.

Remark 4. The codewalk 11 is decoded by a word of the form $xyxzxyx$, which is square-free but cannot be extended to a square-free word by any letter. This property is shared by the codewalk 333 and, moreover, by any codewalk of the form $v3v$, where $v3$ is closed. The codewalks of the form vxv, where $x \in \{1, 2\}$ and vx is closed, encode words containing squares. The codewalks 223 and 322 decode to square-free words that cannot be extended by any letter to the left (resp., to the right).

A sufficient condition for square-freeness was proved in [14].

Lemma 5. *A codewalk having (a) no factors* 11, 222, 223, 322, 333, *and (b) no factors of the form $vxyv$, where x, y are symbols and the codewalk vxy is closed, encodes a square-free word.*

3 White and Black Positions

Assume that $u \in \mathsf{SF}$ is fixed. We say that a position i is *white* if it belongs to some wildcard set for u and *black* otherwise. Usually, the set of *all* white positions is not a wildcard set for u, because some white positions "interact" in the sense that a wildcard set can contain some of them but not all of them; the simplest interactions are considered in Lemma 10 below.

We start the study of white and black positions with the following criterion.

Proposition 6. *Given a fixed $u \in \mathsf{SF}$, a position i is black if for some factor vxv of u, where $v \neq \lambda$ and $x \in \Sigma$, i is either the position of x, or the position preceding the left v, or the position following the right v. Otherwise, i is white.*

Proof. The set of wildcard sets for u is closed downwards by Lemma 1. Then a position i is white if and only if $\{i\}$ is a wildcard set for u. Placing a wildcard in a position described in the conditions of the proposition gives us a $(|v|+1)$-square, so all such positions are black. Conversely, let some position i be black and $x = u[i]$. Since $\{i\}$ is not a wildcard set, u has a factor of the form $vxwvyw$, turning into a square when x is replaced by a wildcard. If either v or w is empty, the position of x satisfies the conditions of the proposition. Otherwise, note that $x \neq y$ and $w[1] = v[|v|] = z$, where $z \neq x$, $z \neq y$ since u is square-free. Hence $u[i-1..i+1] = zxz$, and the position of x satisfies the conditions again. □

Example 7. The word $u = abcbabcacb$ has two white positions: 2 and 9. All other positions are black by Proposition 6; e.g., the factor $u[2..4] = bcb$ makes black the positions 1, 3, and 5, while 4 and 8 are black due to $u[1..7] = abcbabc$. If we consider $u' = u[2..10]$ instead of u, the position of the second b (now position 3) will be white. More generally,

(∗) in some words with the prefix $xyxzx$ the position 3 is white.

The distribution of white positions in infinite square-free words (and in their finite factors) is densely related to jumps.

Proposition 8. *Let $u \in \mathsf{SF}$ be an infinite word.*
(1) Every white position in u is either the first or the last position of a jump.
(2) Modulo the single exclusion (∗), every jump in u contains at most one white position, and every white position belongs to exactly one jump.

Proof. The middle position of a jump and a position adjacent to a jump are black by Proposition 6. Since two consecutive jumps are separated by at most one position, statement 1 is proved.

Consider two consecutive jumps in u. Depending on the length of the edge between them, they form a factor $v_1 = xyxzx$, $v_2 = xyxzyz$, or $v_3 = xyxzyxy$ (see Fig. 1). The position of the middle x in v_1 is black if v_1 is not a prefix of u (cf. Example 7); if v_1 is a prefix, then we have the exclusion (∗). Applying Proposition 6 to v_2 and v_3 we see that all positions except for $1, 6$ in v_2 and for $3, 5$ in v_3 are black. Statement 2 now follows. □

So, the further study of the distribution of the white positions in a word should clarify which jumps contain white positions and which do not. From the proof of Proposition 8 we have the following basic picture (see Fig. 2). We call the positions of question marks *potentially white*.

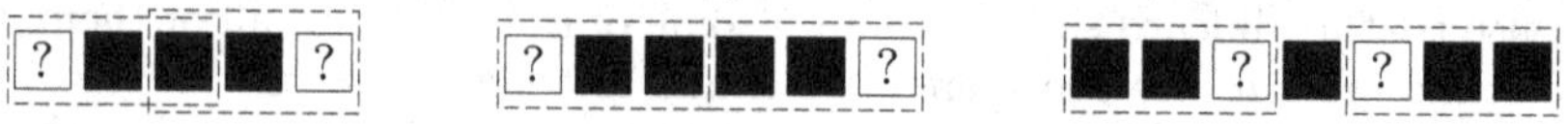

Fig. 2. Black and potentially white positions in the pairs of consecutive jumps. The jumps in the left (resp., middle, right) picture are connected in the K_{33} graph by the edge of length 1 (resp., 2, 3).

Our main interest is in the asymptotic distribution of white positions in infinite words. So we pay little attention to special cases concerning prefixes of these words (like (∗) and "almost squares" described in Remark 4). We call a factor of a codewalk *regular* if it is not its prefix. The following three lemmas form the basis for the proof of our main results.

Lemma 9. *A jump in $u \in \mathsf{SF}$ contains no white position if it is located in the place indicated by a dot in any of the following regular factors of* $\mathsf{cwk}(u)$*:*

$$1.2,\ 2.1,\ 2.2,\ 3.3,\ 13.,\ .31,\ .1221.,\ .2332., \tag{1}$$

$$.1212.321,\ 123.2121.,\ .13132312.,\ .21323131.,\ 323.1321.,\ .1231.323 \tag{2}$$

Proof. For the first four factors in (1), it is enough to look at Fig. 2. In the jump that follows 1 in the codewalk, the last position is potentially white; but if the next edge has length 2, this position is black. The same argument works for $21, 22$ and 33.

For the remaining factors we use Proposition 6. Decoding each of these factors (together with its unique extension if necessary), we get a factor of u of the form vxv; for convenience, the v's are overlined:

$$\begin{array}{ll}
\underline{1}13 \to \overline{zxy}x\overline{zxy}\boldsymbol{z}xz & 1212321\underline{2} \to \boldsymbol{x}\overline{yxzxyzyxyzx}\boldsymbol{z}\overline{yxzxyzyxyzx} \\
31\underline{1} \to xy\boldsymbol{x}\overline{zyx}y\overline{zyx} & \underline{2}1232121 \to \overline{xzyxyzyxzxy}\boldsymbol{z}\overline{xzyxyzyxzxy}\boldsymbol{x} \\
1221 \to \boldsymbol{x}\overline{yxzxy}z\overline{yxzxy}\boldsymbol{x} & 13132312 \to \boldsymbol{x}\overline{yxzxyzxzyzxy}z\overline{yxzxyzxzyzxy}\boldsymbol{x} \\
2332 \to \boldsymbol{x}\overline{yxzyzxy}z\overline{yxzyzxy}\boldsymbol{x} & 21323131 \to \boldsymbol{x}\overline{yxzyzxzyxzxy}z\overline{yxzyzxzyxzxy}\boldsymbol{x} \\
 & \underline{1}3231321 \to \overline{xzyzxyzyxzxy}\boldsymbol{z}\overline{xzyzxyzyxzxy}\boldsymbol{x} \\
 & 1231323\underline{1} \to \boldsymbol{x}\overline{yxzxyzyxzyzx}\boldsymbol{z}\overline{yxzxyzyxzyzx}
\end{array}$$

In the jumps indicated by dots, the potentially white positions are those with boldface letters (see Fig. 2); all these positions are black by Proposition 6. □

Lemma 10. *For every $u \in \mathsf{SF}$ and every $i > 1$, a wildcard set for u cannot contain simultaneously i and $i+5$, or i and $i+6$.*

Proof. Assume that both positions i and $i+5$ are white (otherwise, there is nothing to prove). Examining the location of white positions in jumps (Fig. 2), we see the only possibility: these positions are located in consecutive jumps connected by an edge of length 2 (the middle picture). This 2 in the codewalk gives a factor of the form $zxyxzyzx$ (note that $i > 1$, so the initial z exists). Placing wildcards in both white positions, we get a partial 4-square $z\diamond yx\, zy\diamond x$.

A slightly longer analysis shows that the white positions i and $i+6$ are always in the jumps connected by the codewalk 33: they cannot be connected by 11 by Remark 4 and by 13 or 31 by Lemma 9. The codewalk 33 gives us a factor of the form $xyxzyxyzxyx$. If we place wildcards in both white positions, we get a partial word starting with a partial 5-square $xy\diamond zy\, xyz\diamond y$. □

Lemma 11. *Let P be a wildcard set for an infinite word $u \in \mathsf{SF}$, j, $j+k \in P$ and $j+1, \ldots, j+k-1 \notin P$. Then (1) $k \notin \{1, 3, 5, 6, 8\}$, (2) if $k = 2$ (resp., $k = 4$; $k = 7$) then j, $j+k$ are located in jumps connected by 3 (resp., by 1; by one of the paths 12, 21, 23, or 32) in $\mathsf{cwk}(u)$.*

Proof. The statement readily follows from Proposition 8 and Fig. 2 for $k = 1, 2, 3, 4$ and from Lemma 10 for $k = 5, 6$. Further, it is easy to check that the distance between potentially white positions in jumps connected by a codewalk of length 3 is at least 9. Thus, for $k = 7, 8$ the jumps containing j and $j+k$ are connected by a two-edge codewalk. This codewalk is not 11 by Remark 4, not 13 or 31 by Lemma 9, and not 33 by Lemma 10. If it equals $12, 21, 23$, or 32, we have $k = 7$. If it equals 22, we have $k = 8$. But 22 is followed by 1 in $\mathsf{cwk}(u)$, which means that the position $j+k$ is black by Lemma 9, contradicting the condition $j+k \in P$. Hence, $k \neq 8$. The lemma is proved. □

3.1 Rigid Words

Lemma 9 implies the following result about almost rigid words.

Proposition 12. *Let $u \in \mathsf{SF}$ be an infinite word. If $\mathsf{cwk}(u)$ contains finitely many 3's, then u is almost rigid.*

It is not a priori clear whether a word required in Proposition 12 (or, equivalently, an infinite word $u \in \mathsf{SF}$ such that $\mathsf{cwk}(u)$ contains no 3's) exists. Fortunately, this is the case. Consider the alphabet $\{1,2\}$ and take the *Fibonacci word* F which is the fixed point of the morphism ϕ:

$$\phi(2) = 21,\ \phi(1) = 2; \qquad \mathsf{F} = 2122121221221\cdots$$

The *1-2-bonacci word* is the ternary word $\mathsf{F}_{12} = ab\cdots$ such that $\mathsf{cwk}(\mathsf{F}_{12}) = \mathsf{F}$. As was proved in [13], $\mathrm{lexp}(\mathsf{F}_{12}) = 11/6$. This means the existence of an infinite set of rigid square-free words, as the next proposition shows.

Proposition 13. *Any suffix u of the 1-2-bonacci word such that $\mathsf{cwk}(u) = \underline{1}1221\cdots$ is a rigid square-free word.*

Proof. By Proposition 8(2), every jump in u has at most one white position: since $u = xyz\cdots$, we avoid the case $(*)$. By Lemma 9, all positions in the jumps are black. The first position of u precedes a jump, so it is also black. □

4 Proofs of Main Results

4.1 Upper Bound on Flexibility

Theorem 14. *The flexibility of any infinite word $w \in \mathsf{SF}$ is at most 3/16.*

Proof. Let P be any infinite wildcard set for w. We aim at building a factorization $w = u_0u_1\cdots u_n\cdots$ such that the lengths of all factors u_i are bounded and u_i contains $p_i \le \frac{3|u_i|}{16}$ positions from P for any $i > 0$. The existence of such a factorization implies the upper bound 3/16 on the upper density of P, thus proving the theorem.

We factorize w greedily from left to right, checking that each factor u_i, $i > 0$, satisfies the conditions $|u_i| \le 22$, $p_i \le \frac{3|u_i|}{16}$, and begins with a position from P whenever $p_i > 0$. To define the position of u_1, consider the third from the left jump in w. By Proposition 8, it contains at most one position from P. If it contains a position from P, u_1 begins at this position; otherwise, it begins at any position of this jump. Now assume that all factors up to u_{i-1} are built and u_i begins at the jth position of w. We should define $k = |u_i|$. Let $l < l' < l'' < l'''$ be the first four positions from P on the right of j. If $j \notin P$, then put $k = \min\{22, l-j\}$. If $j \in P$, the choice of k depends on the distances between j, l, l', l'', l''' and is described in Table 1. In all possible cases u_i satisfies the prescribed conditions; the desired factorization is thus constructed. □

Table 1. Choosing the length of u_i in the proof of Theorem 14. The possible distances between consecutive positions from P as well as the corresponding fragments of $\mathsf{cwk}(w)$ are taken from Lemma 11. For impossible sets of distances, the contradictions are given.

$l-j$	$l'-l$	$l''-l'$	$l'''-l''$	Fragment of $\mathsf{cwk}(w)$	k	p_i	$\frac{3k}{16}$
≥ 7	any	any	any	irrelevant	$\min\{22, l-j\}$	1	$\geq \frac{21}{16}$
4	≥ 7	any	any	irrelevant	$\min\{22, l'-j\}$	2	$\geq \frac{33}{16}$
4	4	any	any	$j\ l\ l'$ $\|1\|1\|$	$w \notin \mathsf{SF}$ by Remark 4		
4	2	any	any	$j\ l\ l'$ $\|1\|3\|$	$l' \notin P$ by Lemma 9		
2	≥ 9	any	any	irrelevant	$\min\{22, l'-j\}$	2	$\geq \frac{33}{16}$
2	7	≥ 7	any	irrelevant	$\min\{22, l''-j\}$	3	≥ 3
2	7	4	≥ 9	irrelevant	22	4	$\frac{66}{16}$
2	7	4	7	$j\ l\ l'\ l''\ l'''$ $\|3\|R\|1\|Q\|$	$R \in \{12, 32\}$: $l' \notin P$ by Lemma 9		
					$Q \in \{21, 23\}$: $l'' \notin P$ by Lemma 9		
					$R = 21$ or $Q = 12$: $w \notin \mathsf{SF}$ by Remark 4		
					$R = 23$, $Q = 32$: $l' \notin P$ by Lemma 9(2)		
2	7	4	4	$j\ l\ l'\ l''\ l'''$ $\|3\|R\|1\|1\|$	$w \notin \mathsf{SF}$ by Remark 4		
2	7	4	2	$j\ l\ l'\ l''\ l'''$ $\|3\|R\|1\|3\|$	$l''' \notin P$ by Lemma 9		
2	7	2	any	$j\ l\ l'\ l''$ $\|3\|R\|3\|$	j [resp. $l; l'; l''$] $\notin P$ by Lemma 9		
					for $R = 12$ [resp. 32; 23; 21]		
2	4	any	any	$j\ l\ l'$ $\|3\|1\|$	$j \notin P$ by Lemma 9		
2	2	any	any	$j\ l\ l'$ $\|3\|3\|$	$l \notin P$ by Lemma 9		

4.2 Word of Maximal Flexibility

Return to the 1-2-bonacci word from Sect. 3.1 and consider the codewalk $\mathsf{H} = \eta(\mathsf{F}_{12})$, where the morphism $\eta : \Sigma^* \to \{1,2,3\}^*$ is defined by

$$\eta(a) = 1232\,132323$$
$$\eta(b) = 1232\,13232\,132323$$
$$\eta(c) = 1232\,132323\,12323$$

Theorem 15. *The word* $\mathsf{G} = ab\cdots$ *with the codewalk* H *is square-free and has flexibility 3/16.*

Proof. To prove square-freeness of G, it suffices to show that H satisfies the conditions of Lemma 5. The condition (a) is obviously satisfied; let us check (b). Note that any conjugate of a closed codewalk is closed; hence if g is a codewalk with period p and some factor of g of length p is a closed codewalk, then all such factors are closed. According to the condition (b), our aim is to prove that for any closed codewalk h its extension in H, denoted below by g, has length $< 2|h| - 2$. Since $K_{3,3}$ is bipartite, closed codewalks have even lengths.

It is easy to check (b) for closed walks of length 4 (1232 and 1323) and 6 (no such codewalks in H). So let $|h| \geq 8$. Assume to the contrary that $|g| \geq 2|h| - 2$. Then g has a factor of length $|h|$ beginning with 1. So we assume w.l.o.g. that h begins with 1. We call the codewalks 1232, 12323, 13232, 132323 *miniblocks*; they constitute the blocks $\eta(a), \eta(b), \eta(c)$. Let $h = u_1 \cdots u_n u_{n+1}$, where $u_1, \ldots, u_n$ are miniblocks, while u_{n+1} is a prefix of a miniblock but not a miniblock itself. Then h is followed in H by a symbol distinct from $1 = h[1]$. Hence g ends at the same position in H as h. On the other hand, g extends h to the left by less than $|u_{n+1}| < 6$ symbols. Since $|h| \geq 8$, this contradicts our assumption on $|g|$. Therefore, h is a product of miniblocks.

To know which codewalks are closed, we partition them into six *types*: λ, $1, 2, 3, 12$, and 13. A codewalk u has type $t = \mathsf{type}(u)$ if the paths in $K_{3,3}$ with a common starting point and labels u and t, respectively, have a common endpoint. In particular, the codewalks of type λ, and only they, are closed. The concatenation of codewalks has the same type as the concatenation of their types, and the latter can be easily computed by Fig. 1. For miniblocks, one has $\mathsf{type}(1232) = \lambda$, $\mathsf{type}(12323) = 3$, $\mathsf{type}(13232) = 2$, $\mathsf{type}(132323) = 12$.

A direct check of types of concatenations of miniblocks shows that h, which is closed, is long enough; in particular, h contains at least two occurrences of the miniblock 1232, each followed by 1 (not by 3). Then some factor of g of length $|h|$ starts with the leftmost factor 12321 in g (otherwise g would be too short); we assume w.l.o.g. that h begins with this leftmost 12321. Then $h = \eta(w)u$, where $w \neq \lambda$ and u is a proper prefix of a block, but not a block. If u is nonempty, h is not followed by 12321 in H, so g extends h to the right by at most four symbols. At the same time, g extends h to the left by less than $|u|$ symbols. This contradicts our assumption on $|g|$. Therefore, $h = \eta(w)$ is a product of blocks. Note that $\mathsf{type}(\eta(a)) = 12$, $\mathsf{type}(\eta(b)) = 3$, $\mathsf{type}(\eta(c)) = 2$. Since h is closed, $|w| \geq 3$; e.g., $\mathsf{type}(\eta(abc)) = \lambda$. Let $w = x_1 \cdots x_n$.

By the choice of h and square-freeness of F_{12}, the extension of w in F_{12} equals $x_1 \cdots x_n x_1 \cdots x_{n-i}$ for some $i \geq 1$. Hence this extension occurs in F_{12} inside the factor $\hat{w} = \bar{x} x_1 \cdots x_n x_1 \cdots x_{n-i} \hat{x}$, where $\bar{x} \neq x_1, x_n$; $\hat{x} \neq x_{n-i}, x_{n-i+1}$. Then $|g| = 2|h| - |\eta(x_{n-i+1} \cdots x_n)| + M + N$, where M is the length of the common suffix of $\eta(\bar{x})$ and $\eta(x_n)$, while N is the length of the common prefix of $\eta(\hat{x} \cdots)$ and $\eta(x_{n-i+1} \cdots)$. One has $M = 9$ for the pair (a, b) and $M = 4$ otherwise; $N = 14$ for the pair (a, c) and $N = 9$ otherwise. If $i > 2$, then clearly $|g| < 2|h| - 2$; hence $i \leq 2$.

Case $i = 1$. Since $\mathrm{lexp}(\mathsf{F}_{12}) = 11/6$, we have $|w| \leq 6$. Note that F_{12} has no factors of the form $xyzxy$, because F contains no 3's (this is why we have chosen F_{12} as the argument of η); hence $|w| \geq 4$. It is easy to check that the only candidates to w have the form $xyzy$ or $xyzyxz$, but then $\eta(w)$ is not closed for any values of x, y, and z.

Case $i = 2$. We have $M = 9$, $N = 14$, $|\eta(x_{n-1} x_n)| = 25$, and, respectively, $|g| = 2|h| - 2$. Then $\bar{x} \in \{a, b\}$, $\hat{x} \in \{a, c\}$, $a \in \{x_{n-1}, x_n\}$. By square-freeness of F_{12} we get

$$\hat{w} = b\,\underline{c\,u\,b\,c\,a}\,c\,u\,b\,a \quad \text{or} \quad \hat{w} = a\,\underline{c\,u\,b\,a\,b}\,c\,u\,b\,c,$$

where w is underlined, $u \in \Sigma^*$. We have $\exp(\hat{w}) \leq 11/6$ and then $|w| \leq 12$. It appears that the only possibility for u, giving a square-free word $\hat{w}$ such that $\mathsf{cwk}(\hat{w})$ contains no 3's, is $u = ba$ (resp., $u = ca$) for the left (resp., right) case. But then $\eta(w)$ is not closed. This finishes the proof of condition (b); so G is square-free by Lemma 5.

To find the flexibility of G, we need a technical lemma.

Lemma 16. *All factors of* G *of the forms* vxv *and* $vxyv$ *have periods* ≤ 10.

Proof. Consider a factor $w = vuv$ of G with the minimal period $p = |vu| \geq 11$ and minimal possible $|u|$. Then w is the extension of vu. Aiming at a contradiction, assume that $|u| \leq 2$. Consider the codewalk w' starting at the leftmost jump and ending at the rightmost jump in w. Since v is long enough to contain at least two jumps, the walk between the leftmost and the rightmost jump in v repeats twice, so $w' = v'u'v'$, where $v'u'$ is closed and $\ell(v'u') = p$. Now compute $|w|$. By the definition of extension, the left and right v in w are preceded (and also followed) in G by different letters. Hence the first (resp., last) jump in w is preceded (resp., followed) by at most two letters. The length of the last jump is 3, and the remaining part of w has length $\ell(v'u'v')$. Thus, $|w| \leq 2p - \ell(u') + 7$. Since w' is a factor of H, we know that $|u'| \geq 3$ by condition (b). Then $\ell(u') \geq 9$, where the equality takes place for $u' = 123, 132, 213, 231, 312, 321$. If $\ell(u') \geq 10$, we obtain $|w| \leq 2p - 3$ and $|u| \geq 3$, contradicting our assumption. Let $\ell(u') = 9$. Then $v'[|v'|] \in \{2, 3\}$. Since H has no factors 22 and 33, either $u[1]$ or the symbol following w' in H is 1. Hence the last jump in w is followed by just one letter, not two; we again have $|w| \leq 2p - 3$, contradicting the assumption $|u| \leq 2$. □

By Lemma 16 and Proposition 6, the potentially white positions in all jumps in G, except for those described in Lemma 9(1), are white. The set P of all white positions in G is periodic with period 16 and has density 3/16. Indeed, white positions in the jumps connected by 3 (resp., 12, 23, 21, 32) in the codewalk, are at distance 2 (resp., 7) by Lemma 11:

$$\mathsf{G} = 12^{|}3^{|}21^{|}32^{|}3^{|}23^{|}12^{|}3^{|}21^{|}32^{|}3^{|}21^{|}32^{|}3^{|}23^{|}12^{|}3^{|}21^{|}32^{|}3^{|}23^{|}12^{|}3^{|}23^{|}\cdots$$

It remains to prove that P is a wildcard set. Assume to the contrary that the partial word G_P contains a partial square vv', $m = |v|$. A direct check shows that $m > 10$. We transform vv', whenever possible, replacing back wildcards with the letters from the same positions of G. The replacement rules are as follows. Consider all pairs $(v[i], v'[i])$. If one symbol is x and the other is a wildcard obtained from x, replace the wildcard; if both are wildcards obtained from the same x, replace both; if both are wildcards obtained from different letters, replace one of them. Let uu' be the resulting partial square. At least one of the words $u[2..m{-}1]$, $u'[2..m{-}1]$ contains a wildcard; otherwise G has the factor $u[2..m{-}1]u[m]u'[1]u[2..m{-}1]$ with period m, contradicting Lemma 16. W.l.o.g., $u[i]$ is a wildcard obtained from a letter z and $1 < i < m$. Then $u'[i] = y \neq z$. One of the letters $u[i{-}1], u[i{+}1]$ is y; the corresponding letter of u' cannot be y, hence it is a wildcard. Thus, the wildcard $u[i]$ matches a letter adjacent to another wildcard. This condition is quite restrictive. A case analysis shows that only short factors of G_P match under this condition, and the square uu' cannot exist. This finishes the proof of square-freeness of G_P and thus of the theorem.

Some details of the case analysis follow. Up to symmetry, there are two possible fragments of G that can contain the position of $u[i]$ (this position is indicated by a wildcard replacing z):

$$2^{|}32:\quad \begin{array}{ccccccccccccc} x & y & x & z & y & \diamond & x & \boldsymbol{y} & z & y & x & z & x \\ \hline & & \boldsymbol{y} & z & \boldsymbol{x} & y & x & z & & & & & \\ & & & & \boldsymbol{z} & y & x & z & & & & & \\ & y & x & z & \boldsymbol{x} & y & \boldsymbol{z} & x & z & y & x & & \\ & & x & z & y & x & \boldsymbol{y} & z & \boldsymbol{x} & y & x & z & \\ & & & z & y & x & \boldsymbol{z} & y & z & & & & \\ & & & & \boldsymbol{z} & x & \boldsymbol{y} & & & & & & \end{array}$$

$$21^{|}3:\quad \begin{array}{ccccccccccccc} z & \boldsymbol{x} & y & x & z & y & z & x & \diamond & y & x & z & x \\ \hline & z & y & x & z & y & z & x & y & \boldsymbol{x} & & & \\ & & & & & & z & x & y & \boldsymbol{z} & x & z & \\ & & & & & & & \boldsymbol{z} & y & \boldsymbol{x} & & & \\ & & & & & & z & \boldsymbol{y} & x & y & & & \\ & & & & & & & \boldsymbol{z} & x & y & & & \\ & & & & & y & z & \boldsymbol{y} & x & \boldsymbol{z} & & & \end{array}$$

Below each line, the maximal factors of G that match the top word are given; here all boldface symbols occupy white positions and can be replaced by wildcards. These factors are computed to satisfy both Fig. 2 and the structure of the codewalk H. For example, the factor $w = yxz\boldsymbol{x}y\boldsymbol{z}xzyx$ in the left column contains two white positions at distance 2, so the corresponding jumps are surrounded by 2's in H: $2^{|}3^{|}2$. Then w is preceded by z and followed by y, the letters mismatching their counterparts in the top word. □

4.3 Morphic and Substitutional Flexible Words

The flexibility of well-known square-free words is of certain interest. The proofs of the next results are similar to Theorem 15 and omitted due to space constraints. The result of Theorem 17 is the best we have for purely morphic words.

Theorem 17. *The Dejean word [6] has flexibility 2/19.*

Theorem 18. *The Arshon word [1] has flexibility 1/9.*

5 Conclusion and Open Problems

The problem of finding the maximum density of wildcards in a ternary infinite square-free partial word can be conveniently reformulated in terms of placing wildcards at some positions in square-free words. We developed a technique to find the appropriate sets of positions (wildcard sets) and define flexibility of a square-free word as the maximum density of its wildcard set. We proved that the maximum flexibility of a ternary square-free word is 3/16. Besides that, we proved the existence of rigid words, having no positions for wildcards at all. Two open problems can direct further development of this topic:

1. What is the maximum flexibility of a morphic/purely morphic ternary square-free word?
2. What is the minimum local exponent of a rigid word?

References

1. Arshon, S.E.: Proof of the existence of asymmetric infinite sequences. Mat. Sbornik **2**, 769–779 (1937). in Russian, with French abstract
2. Berstel, J., Boasson, L.: Partial words and a theorem of Fine and Wilf. Theoret. Comput. Sci. **218**, 135–141 (1999)
3. Blanchet-Sadri, F., Black, K., Zemke, A.: Unary pattern avoidance in partial words dense with holes. In: Dediu, A.-H., Inenaga, S., Martín-Vide, C. (eds.) LATA 2011. LNCS, vol. 6638, pp. 155–166. Springer, Heidelberg (2011)
4. Blanchet-Sadri, F., Hegstrom, R.A.: Partial words and a theorem of Fine and Wilf revisited. Theor. Comput. Sci. **270**(1–2), 401–419 (2002)
5. Blanchet-Sadri, F., Mercaş, R., Scott, G.: A generalization of Thue freeness for partial words. Theoret. Comput. Sci. **410**, 793–800 (2009)
6. Dejean, F.: Sur un théorème de Thue. J. Combin. Theory. Ser. A **13**, 90–99 (1972)
7. Fischer, M., Paterson, M.: String matching and other products. SIAM-AMS Proc. **7**, 113–125 (1974)
8. Halava, V., Harju, T., Kärki, T.: Square-free partial words. Inform. Process. Lett. **108**(5), 290–292 (2008)
9. Manea, F., Mercaş, R.: Freeness of partial words. Theoret. Comput. Sci. **389**(1–2), 265–277 (2007)
10. Muthukrishnan, S., Ramesh, H.: String matching under a general matching relation. In: Shyamasundar, R.K. (ed.) FSTTCS 1992. LNCS, vol. 652, pp. 356–367. Springer, Heidelberg (1992)
11. Petrova, E.A., Shur, A.M.: Constructing premaximal ternary square-free words of any level. In: Rovan, B., Sassone, V., Widmayer, P. (eds.) MFCS 2012. LNCS, vol. 7464, pp. 752–763. Springer, Heidelberg (2012)
12. Petrova, E.A., Shur, A.M.: On the tree of ternary square-free words. In: Manea, F., Nowotka, D. (eds.) WORDS 2015. LNCS, vol. 9304, pp. 223–236. Springer, Heidelberg (2015)
13. Petrova, E.A.: Avoiding letter patterns in ternary square-free words. Electr. J. Comb. **23**(1), P1.18 (2016)
14. Shur, A.M.: On ternary square-free circular words. Electronic J. Combinatorics **17**, R140 (2010)
15. Shur, A.M., Konovalova, Y.V.: On the periods of partial words. In: Sgall, J., Pultr, A., Kolman, P. (eds.) MFCS 2001. LNCS, vol. 2136, pp. 657–665. Springer, Heidelberg (2001)

Alternating Demon Space Is Closed Under Complement and Other Simulations for Sublogarithmic Space

Viliam Geffert(✉)

Department of Computer Science,
P.J. Šafárik University, Jesenná 5, 04001 Košice, Slovakia
viliam.geffert@upjs.sk

Abstract. We present new simulations for $\text{ASPACE}^{\text{dm}}(s(n))$, the class of languages that can be accepted by alternating Turing machines starting with $s(n)$ worktape cells delimited initially. Under weak constructibility assumptions, not excluding monotone functions below $\log n$, we show: *(i)* $\text{ASPACE}^{\text{dm}}(s(n)) \subseteq \text{DTIME}(n \cdot 2^{O(s(n))})$. This extends, to sublogarithmic space, the classical simulation of alternating space by deterministic time. *(ii)* $\text{ASPACE}^{\text{dm}}(s(n)) \subseteq \text{NTIMESPACE}(n \cdot 2^{O(s(n))}, 2^{O(s(n))})$, a simulation with simultaneous bounds on time and space. This improves the known inclusion, stating that $\text{ASPACE}^{\text{dm}}(s(n)) \subseteq \text{NSPACE}(2^{O(s(n))})$. *(iii)* $\text{ASPACE}^{\text{dm}}(s(n)) = \text{co-ASPACE}^{\text{dm}}(s(n)))$, i.e., the alternating space is closed under complement. This simulation does not depend on whether $s(n)$ is above $\log n$ nor on whether the original machine gets into infinite loops, which solves a long-standing open problem.

Keywords: Computational complexity · Alternation · Sublogarithmic space

1 Introduction

Space complexity of a computation, introduced in [10] in 1965, is the second in importance among various computational complexity measures, right after the time complexity. It turns out that $\log n$ is the most significant boundary among all space complexity bounds, since the space complexity classes below $\log n$ are radically different from those above.

For example, if $s(n) \geq \Omega(\log n)$, it is trivial to show that $\text{DSPACE}(s(n))$ is closed under complement. However, the trivial argument does not work below $\log n$, because the machine may reject by getting into an infinite loop and we do not have enough space to detect such loops by counting executed steps, up to $n \cdot 2^{\Omega(s(n))}$. To show that $\text{DSPACE}(s(n)) = \text{co-DSPACE}(s(n))$ without any assumption on $s(n)$, a more sophisticated simulation was necessary [18]. In the nondeterministic case, we have that $\text{NSPACE}(s(n))$ is closed under complement

Supported by the Slovak grant contracts VEGA 1/0142/15 and APVV-15-0091.

S. Brlek and C. Reutenauer (Eds.): DLT 2016, LNCS 9840, pp. 190–202, 2016.
DOI: 10.1007/978-3-662-53132-7_16

for $s(n) \geq \Omega(\log n)$ [14,20] but, for $s(n)$ below $\log n$, the problem is still open. The problem stays open even if we consider another reasonable way to define space complexity, studied, e.g., in [1,5]: the classes[1] $\text{NSPACE}^{\text{dm}}(s(n))$.

The same problem arises for the alternating machines, introduced in [4] by generalization of nondeterminism and parallelism. It is trivial to invert the roles of existential and universal decisions and of accepting and rejecting states, which gives a machine for the complement of the original language, *if the original machine never gets into an infinite loop.* Thus, both $\text{ASPACE}(s(n))$ and $\text{ASPACE}^{\text{dm}}(s(n))$ are closed under complement for $s(n) \geq \Omega(\log n)$, since we can force the machine to halt [4, Theorem 2.6]. This does not imply anything for $s(n)$ below $\log n$. For example, by inductive counting [14,20] (see also [21]), the hierarchy of $s(n)$ space bounded machines making a constant number of alternations collapses and hence Σ_k- and Π_k-$\text{SPACE}(s(s))$ are closed under complement for $s(n) \geq \Omega(\log n)$, but they are provably not closed, if $s(n) \leq o(\log n)$ [3,7,16].

The importance of even the lowest levels of space bounded computations is established by several results. For example, we know that $\text{NSPACE}(\log n)$ separates from $\text{DSPACE}(\log n)$ if and only if there exists a unary language in $\text{NSPACE}(\log\log n) - \text{DSPACE}(\log\log n)$ [8]. The sublogarithmic alternating space classes may actually be quite strong, e.g., there exists a binary NP-complete language such that its unary coded version is in $\text{ASPACE}(\log\log n)$ [9].

In this paper, we first provide a new time efficient simulation of alternating machines with small space by deterministic machines. Namely, we show that

$$\text{ASPACE}^{\text{dm}}(s(n)) \subseteq \text{DTIME}(n \cdot 2^{O(s(n))}),$$

for each $s(n)$ such that $\lfloor s(n) \rfloor$ can be computed by a deterministic multi-tape Turing machine in $n \cdot 2^{O(s(n))}$ time. Such constructibility condition is very weak and does not exclude[2] even functions below $\log\log n$. This extends, to sublogarithmic space bounds, the classical result [4] stating that $\text{ASPACE}(s(n)) \subseteq \text{DTIME}(2^{O(s(n))})$ for $s(n) \geq \log n$.

Our deterministic simulation within $n \cdot 2^{O(s(n))}$ time uses superlinear space, namely, $n \cdot 2^{\Omega(s(n))}$. However, it turns out that the simulating machine has several additional special properties, and hence it can be simulated space efficiently by more powerful machine models. Based on this, we shall derive that

$$\text{ASPACE}^{\text{dm}}(s(n)) \subseteq \text{1-NTIMESPACE}^{\text{dm}}(n \cdot 2^{O(s(n))}, 2^{O(\lfloor s(n) \rfloor)}),$$

[1] By $X\text{SPACE}^{\text{dm}}(s(n))$, for $X \in \{\text{D}, \text{N}, \text{A}\}$, we denote the classes of languages accepted by deterministic, nondeterministic, and alternating Turing machines starting with a worktape consisting of $\lfloor s(n) \rfloor$ blank cells delimited by endmarkers (here $\lfloor x \rfloor$ denotes the largest integer satisfying $i \leq x$, for the given real value x), as opposed to the more common complexity classes $X\text{SPACE}(s(n))$ where the worktape is initially empty and the machine must use its own computational power to make sure that it respects, along each computation path on each input of length n, the space bound of $s(n)$. The notation "dm" derives from "Demon" Turing Machines [5].

[2] It is known that $\text{ASPACE}(o(\log\log n))$ contains only regular languages [15]. However, it is still possible to accept some nonregular languages, if $\lfloor s(n) \rfloor \leq o(\log\log n)$ worktape cells are delimited automatically at the very beginning. As an example [2], take $\mathcal{L} = \{1^n : n \mod \lceil \log\log\log n \rceil = 0\}$, contained in $\text{DSPACE}^{\text{dm}}(\log\log\log\log n)$.

which represents a simulation by one-way nondeterministic machines starting with a delimited worktape of size $2^{O(\lfloor s(n)\rfloor)}$ and executing at most $n{\cdot}2^{O(s(n))}$ steps along each computation path, without any assumptions on $s(n)$. This improves the known inclusion $\text{ASPACE}(s(n)) \subseteq \text{NSPACE}(2^{O(s(n))})$ that was proved for $s(n)$ above $\Omega(\log\log n)$ under some weak constructibility assumptions [19]. If $\lfloor s(n)\rfloor$ can be computed by a deterministic machine in $n{\cdot}2^{O(s(n))}$ time and, simultaneously, in $2^{O(s(n))}$ space, we obtain a nondeterministic simulation using worktapes that are initially empty. The new machine is no longer one-way:

$$\text{ASPACE}^{\text{dm}}(s(n)) \subseteq \text{NTIMESPACE}(n{\cdot}2^{O(s(n))}, 2^{O(s(n))}).$$

Finally, we convert a two-way alternating machine into a machine for the complement of the original language, keeping the same amount of space. The conversion does not depend on whether $s(n)$ is above $\log n$ nor on whether the original machine gets into infinite loops:

$$\text{ASPACE}^{\text{dm}}(s(n)) = \text{co-ASPACE}^{\text{dm}}(s(n)) \quad \text{for each } s(n).$$

This solves a long-standing open problem [3]. Quite surprisingly, this complementing does not eliminate infinite loops — the new machine itself rejects by going into infinite loops along some computation paths.

We assume the reader is familiar with standard deterministic, nondeterministic, and alternating Turing machines, equipped with a finite-state control, a two-way read-only input tape, and a fixed number of two-way read-write worktapes. (See, e.g., [4,13,21].) Throughout the paper, $\text{DTIME}(n{\cdot}2^{O(s(n))})$ is a shorthand notation representing $\bigcup_{k\geq 1} \text{DTIME}(n{\cdot}2^{ks(n)})$; the same kind of notation is used for other classes. Because of the page limit, all proofs are sketched.

2 Simulations

The first lemma presents a little bit artificial deterministic machine, with many additional special properties not stated here explicitly. The lemma serves as a basis for all subsequent simulations by more natural machine models.

Lemma 1. *If a language $\mathcal{L}$ is accepted by an alternating Turing machine* A *using an initially delimited worktape of size $\lfloor s(n)\rfloor$, then $\mathcal{L}$ is accepted by a deterministic Turing machine* A′ *working in time $n{\cdot}2^{O(s(n))}$, equipped, besides a two-way read-only input tape, with two worktapes: a so-called* primary *worktape that is initially empty, containing the left endmarker followed by infinitely many blank symbols, and a* secondary *worktape, containing initially $\lfloor s(n)\rfloor$ blank symbols delimited in between two endmarkers. This holds for each $s(n) \geq 1$.*

Proof. Let Δ be worktape alphabet of A and Q the set of its states. A *memory state* of A is a triple $p = \langle q, x, h\rangle$, where $q \in Q$ is a finite control state, $x \in \Delta^{\lfloor s(n)\rfloor}$ a content of the worktape, and $h \in \{0, \ldots, \lfloor s(n)\rfloor{+}1\}$ a position of the worktape head. A *configuration* is a pair $P = \langle p, i\rangle$, where p is a memory state and $i \in \{0,\ldots,n+1\}$ a position of the input head. We shall encode a memory state to a number $j \in \{0,\ldots,2^{\psi(n)}-1\}$ by the use of $\psi(n)$ bits, where

$$\psi(n) = \lceil \log \|Q\| \rceil + \lfloor s(n) \rfloor \times \lceil 1 + \log \|\Delta\| \rceil + 1 \leq O(s(n)). \tag{1}$$

That is, we code the finite control state, each symbol on the worktape, and, for each worktape position, the presence/absence of the head by one extra bit. By coding the initial state by $0^{\lceil \log \|Q\| \rceil}$ and the worktape blank symbol with absent head by $0^{\lceil 1+\log \|\Delta\| \rceil}$, we achieve that the initial memory state is coded by $j = 0$. We are now ready to simulate A by a deterministic Turing machine A′.

Phase I. Starting with the delimited $\lfloor s(n) \rfloor$ blank cells, A′ computes $2^{\psi(n)}-1$ and saves this value in a separate track of the secondary worktape. After that, consulting the given input tape $\vdash w \dashv = \vdash a_1 \ldots a_n \dashv$, the machine prepares, on its primary worktape, the string in the form

$$\widehat{w} = \$\mathbb{P}_0 \ldots \$\mathbb{P}_i \ldots \$\mathbb{P}_{n+1}, \qquad \text{where}$$
$$\mathbb{P}_i = a_i \text{¢} P_{i,0} \text{¢} \ldots \text{¢} P_{i,j} \text{¢} \ldots \text{¢} P_{i,2^{\psi(n)}-1} \text{¢} a_i.$$

That is, $\widehat{w}$ consists of blocks corresponding to the $n+2$ input tape positions. The i-th block, apart from the input symbol a_i at the very beginning and at the very end,[3] consists of records corresponding to the $2^{\psi(n)}$ memory states. The record $P_{i,j}$ represents the configuration with the input head position i and the memory state binary coded by j. The record itself is of constant length, namely,

$$P_{i,j} \in \{0,1\}^{\varrho}, \text{ where } \varrho = 7 + \lceil \log \|\Delta\| \rceil + \lceil \log \|Q\| \rceil.$$

The first 3 bits in $P_{i,j}$ encode a *mode* of the given configuration, which is a value $m_{i,j} \in \{\mathsf{root}, \mathsf{accept}, \mathsf{reject}, \mathsf{unknown}, \mathsf{lock_L}, \mathsf{lock_R}\}$. Initially, we set the mode to root in the initial configuration $P_{0,0}$, to accept and reject in configurations that halt in accepting and rejecting states, respectively, and to unknown in all remaining configurations. In all these cases, the remaining $\varrho-3$ bits in the corresponding record $P_{i,j}$ are initially cleared to zero. Later, in Phase II, they will be utilized to save a *backup link* $b_{i,j} \in \{-1,0,+1\} \times \{-1,0,+1\} \times \Delta \times Q$. After creating $\widehat{w}$, A′ clears all intermediate data and returns both worktape heads to the left.

Phase II. Using $\widehat{w}$ on the primary worktape, A′ traverses the directed graph the nodes of which are configurations and the edges are single computation steps of A on the given input w. The depth-first search starts in $P_{0,0}$, the initial configuration of A. (Without loss of generality, no node has more than 2 sons and the root has exactly 1 son.) For each explored configuration $P_{i,j}$, A′ evaluates whether the subtree of all computation paths rooted in $P_{i,j}$ is accepting or rejecting, and saves this information in the structure $\widehat{w}$. Every time A′ is visiting $P_{i,j}$ on the primary worktape, the memory state j is loaded in a secondary worktape track and the current input symbol a_i is loaded in the finite state control.

[3] For $i \in \{0, n+1\}$, we take $a_i \in \{\triangleright, \triangleleft\}$, two new symbols representing the respective endmarkers.

Moreover, if A' arrived to $P_{i,j}$ from some of its "parents", by following a single-step edge from some $P_{i',j'}$, the machine A' keeps, in the finite state control, a *backup link* $b = \langle d_{\mathsf{I}}, d_{\mathsf{W}}, \sigma, q\rangle \in \{-1,0,+1\}\times\{-1,0,+1\}\times\Delta\times Q$. This link can be used to restore the original configuration $P_{i',j'}$ by "undo" operations on $P_{i,j}$. Namely, d_{I} and d_{W} describe the reversed directions for the input and the worktape head movements, after which the original worktape symbol under the head and the finite control state can be restored by the use of σ and q.

Conversely, if A' arrived to $P_{i,j}$ from some of its "sons", i.e., by backing up against the direction of an edge from $P_{i,j}$ to some $P_{i',j'}$, the machine A' keeps, in the finite state control, a *result* $r \in \{\mathsf{accept}, \mathsf{reject}\}$. This result depends on whether the subtree of all computation paths rooted in $P_{i',j'}$ is accepting or rejecting. Now we are ready to present details for this depth-first search.

(a) If A' arrives to $P_{i,j}$ from some of its parents, it checks the mode $m_{i,j}$ in $P_{i,j}$ on the primary worktape, after which we have the following cases:

(a.1) $m_{i,j} = \mathsf{unknown}$, i.e., $P_{i,j}$ has not been explored yet. First, A' saves the current backup link b as $b_{i,j}$ in the record $P_{i,j}$ on the primary worktape. Then, by inspecting j on the secondary worktape and the current input symbol a_i in the finite state control, A' can determine the *first* executable instruction of A, together with the new backup link b and the input head movement $d \in \{-1,0,+1\}$ related to this instruction. By applying this instruction, the memory state changes from j to some j'. Then the mode in $P_{i,j}$ is updated; from $m_{i,j} = \mathsf{unknown}$ to $m_{i,j} = \mathsf{lock}_{\mathsf{L}}$, if $P_{i,j}$ has two sons, but to $m_{i,j} = \mathsf{lock}_{\mathsf{R}}$, if it has one son.

It remains to position the primary worktape head on $P_{i+d,j'}$. Consider first the case of $d = -1$. By moving to the left, A' finds the \$-symbol in between $\mathbb{P}_{i-1}$ and $\mathbb{P}_i$, updates the current input tape symbol in the finite state control from a_i to a_{i-1}, and, in an auxiliary track on the secondary worktape, it writes down $\tilde{\jmath} = 2^{\psi(n)}-1$. Then, counting down in $\tilde{\jmath}$, it moves along $\mathbb{P}_{i-1}$ to the left until it gets to the record with $\tilde{\jmath} = j'$. The cases of $d = 0$ and $d = +1$ are similar.

(a.2) $m_{i,j} \in \{\mathsf{accept}, \mathsf{reject}\}$, i.e., $P_{i,j}$ has been explored already (we are just visiting $P_{i,j}$ from another parent) or $P_{i,j}$ represents a halting configuration. In either case, A' loads $m_{i,j} \in \{\mathsf{accept}, \mathsf{reject}\}$ to the finite state control, as the result r. Then, using the backup link $b = \langle d_{\mathsf{I}}, d_{\mathsf{W}}, \sigma, q\rangle$ in the finite state control, A' backs up to the recent parent $P_{i',j'}$, against the direction of the edge. Namely, in j on the secondary worktape, A' updates the head position by the difference d_{W}, after which it restores the worktape symbol under the head to σ and the finite control state to q. This changes j to j'. Then A' places the primary worktape head on $P_{i+d_{\mathsf{I}},j'}$, in the same way as in (a.1).

(a.3) $m_{i,j} \in \{\mathsf{lock}_{\mathsf{L}}, \mathsf{lock}_{\mathsf{R}}, \mathsf{root}\}$, i.e., in the course of exploring the subtree of all computation paths rooted in $P_{i,j}$, the machine A' visits $P_{i,j}$ again, having followed a computation path of A that enters a loop. Therefore, A' backs up to the recent parent $P_{i',j'}$ with the result $r = \mathsf{reject}$, in the same way as if, in (a.2), the mode $m_{i,j}$ were equal to reject.

(b) If A' arrives to $P_{i,j}$ from some of its sons, it checks the mode $m_{i,j}$, after which we have the following cases:

(b.1) $m_{i,j} = \mathsf{lock_L}$, i.e., A' comes with a result $r \in \{\mathsf{accept}, \mathsf{reject}\}$ from the first son. First, A' inspects whether the memory state j on the secondary worktape is existential or universal.

If $r = \mathsf{reject}$ and j is existential, or $r = \mathsf{accept}$ and j is universal, A' changes the mode on the primary worktape from $m_{i,j} = \mathsf{lock_L}$ to $m_{i,j} = \mathsf{lock_R}$. Then, by inspecting j on the secondary worktape and the current input symbol a_i in the finite state control, A' determines the *second* executable instruction, together with the related backup link b and the input head movement $d \in \{-1, 0, +1\}$. By applying this instruction, A' changes j to some j' and traverses to the second son, in the same way as described earlier for the first son, in (a.1).

If $r = \mathsf{accept}$ and j is existential, or $r = \mathsf{reject}$ and j is universal, A' changes the mode from $m_{i,j} = \mathsf{lock_L}$ to $m_{i,j} = r$. Then A' loads $b_{i,j}$ from the primary worktape to the finite state control and, using this updated backup link b, it traverses back to the parent, as described earlier, in (a.2).

(b.2) $m_{i,j} = \mathsf{lock_R}$, i.e., A' comes with a result $r \in \{\mathsf{accept}, \mathsf{reject}\}$ from the second son. (This covers also the case of $P_{i,j}$ with only one son.) A' changes the mode on the primary worktape from $m_{i,j} = \mathsf{lock_R}$ to $m_{i,j} = r$, loads $b_{i,j}$ from the primary worktape to the finite state control, and, using the updated backup link b, it traverses to the parent configuration, as in (a.2).

(b.3) $m_{i,j} = \mathsf{root}$, i.e., A' comes with a result $r \in \{\mathsf{accept}, \mathsf{reject}\}$ to the initial configuration $P_{0,0}$, from its only son. After erasing all data on the secondary worktape and parking the primary worktape head at the leftmost symbol in $\widehat{w}$, A' halts and accepts or rejects, in accordance with the value r.

(c) Initially, after moving the primary worktape head from the left endmarker to $P_{0,0}$, the depth-first search is activated by traversing along the edge from the initial configuration to its only son, keeping the mode $m_{0,0} = \mathsf{root}$ unchanged.

Consider now the time requirements for Phase II. In each of the above cases, we charge at most $O(\psi(n)^2)$ steps of A' per each traversal along one edge in the configuration graph of A and per each position on the primary worktape visited by this traversal. Since the traversal along one edge visits $O(2^{\psi(n)})$ primary worktape positions, none of the edges is traversed more than twice (backup included), the number of configurations of A is bounded by $(n+2)\cdot 2^{\psi(n)}$, and no configuration has more than 2 sons, the total time for Phase II is bounded by $O(n\cdot 2^{\psi(n)} \times 2^{\psi(n)} \times \psi(n)^2) \leq n\cdot 2^{O(\psi(n))} \leq n\cdot 2^{O(s(n))}$, using (1). □

It is well known that $\text{ASPACE}(s(n)) \subseteq \text{DTIME}(2^{O(s(n))})$ for $s(n) \geq \log n$ [4]. The next theorem is an extension to sublogarithmic space bounds. (For example, the functions like $\log\log\log n$ or $\log^* n$ satisfy the weak constructibility assumptions of this theorem—see also Footnote 2.)

Theorem 2. $\text{ASPACE}^{\text{dm}}(s(n)) \subseteq \text{DTIME}(n\cdot 2^{O(s(n))})$, *provided that* $\lfloor s(n) \rfloor$ *can be computed by a deterministic multi-tape Turing machine in* $n\cdot 2^{O(s(n))}$ *time.*

Proof. The standard deterministic machine computes $\lfloor s(n) \rfloor$, marks a segment of size $\lfloor s(n) \rfloor$ on one of its worktapes, and then it simulates A' from Lemma 1. □

Next, we present a simulation with simultaneous bounds on time and space by nondeterministic one-way machines, with no assumptions on $s(n)$, but starting with delimited $2^{O(\lfloor s(n)\rfloor)}$ space. The simulation is based on crossing sequence techniques [6,12,17]—introduced in [11]—with several innovations: crossing sequences are considered for one of the worktapes of a machine equipped with an input tape and two worktapes and, moreover, the simulation of the first phase of the original computation (in which the input head is used) is skipped.

Theorem 3. $\mathrm{ASPACE}^{\mathrm{dm}}(s(n)) \subseteq \text{1-}\mathrm{NTIMESPACE}^{\mathrm{dm}}(n\cdot 2^{O(s(n))}, 2^{O(\lfloor s(n)\rfloor)})$, *for each* $s(n) \geq 1$.

Proof. Let A be the original alternating Turing machine, and let A' be the equivalent deterministic machine constructed in Lemma 1. We shall devise an equivalent nondeterministic one-way machine A'' using three worktapes, starting with a delimited worktape space of size $2^{k\cdot\lfloor s(n)\rfloor}$, for a fixed integer constant $k \geq 1$.

Let us recall some details about A'. In Phase I, for the given input $w = a_1 \ldots a_n$, A' constructs $\widehat{w}$ on the primary worktape, of length $O(n\cdot 2^{\psi(n)})$. In Phase II, A' traverses the configuration graph of A, using the secondary worktape of size $\lfloor s(n)\rfloor$ and storing partial results in $\widehat{w}$ on the primary worktape.

Let a *secondary memory state* of A' be a triple $\pi = \langle q, x_{\mathsf{S}}, h_{\mathsf{S}}\rangle$, where q is a finite control state, x_{S} is a content of the secondary worktape, and h_{S} a position of the secondary worktape head. Clearly, we can write π with $O(s(n))$ bits. A' starts Phase II at the left end of the primary worktape in the unique secondary memory state π'_{II} and, depending on the outcome, it halts in the unique accepting/rejecting secondary memory state π'_{A} or π'_{R}, respectively.

Next, let $h \in \{1, \ldots, |\widehat{w}|\}$ be a position along the primary worktape, and let t_h denote the number of times the head of A' is placed on h in the course of Phase II. As pointed out in Lemma 1, we charge at most $O(\psi(n)^2)$ steps of A' per each traversal along one edge in the configuration graph of A and per each primary worktape position h visited by this traversal. If the position h is located on a block $\mathbb{P}_i$, for some i, only traversals of those edges that start in the configurations of A with the input head positions in the range $\{i-1, i, i+1\}$ can make contributions to t_h. There are at most $O(2^{\psi(n)})$ such edges, and hence

$$t_h \leq O(2^{\psi(n)} \times \psi(n)^2) \leq 2^{O(\psi(n))}, \text{ for each } h \in \{1, \ldots, |\widehat{w}|\}. \qquad (2)$$

Finally, consider a computation of A' in Phase II, and a boundary between h and $h+1$. Let $\bar{\pi}_{h,g}$ denote the secondary memory state when the primary worktape head crosses this boundary for the g-th time. Then the list $\Pi_h = (\bar{\pi}_{h,1}, \bar{\pi}_{h,2}, \ldots, \bar{\pi}_{h,g}, \ldots)$ will be called a *secondary crossing sequence* for the given boundary. Using (2), $s(n) \geq 1$, and fixing a sufficiently large constant k, the list Π_h can be written with $(t_h + t_{h+1}) \times O(s(n)) \leq 2^{O(s(n))} \leq 2^{k\cdot\lfloor s(n)\rfloor}$ bits.

The machine A'' simulates the computation of A' in the course of Phase II. The machine A'' prepares the prerequisites for Phase II in a different way by itself, so Phase I is skipped. In a loop running for $h = 1, \ldots, |\widehat{w}|$, the machine A'' nondeterministically guesses secondary crossing sequences $\Pi_2, \ldots, \Pi_{|\widehat{w}|-1}$ and verifies if they correspond to a valid accepting computation of A' on $\widehat{w}$ in Phase II.

In the body of this loop, A'' checks whether Π_{h-1}, Π_h are compatible with respect to $\widehat{w}_h$, the h-th symbol of $\widehat{w}$. The machine A'' keeps the two adjacent lists Π_{h-1}, Π_h on two separate worktapes, using the third worktape for π, the current secondary memory state in the course of simulation. In the finite state control, A'' keeps also σ, the current symbol at the position h on the primary *read-write* worktape. There are two special cases. If $h = 1$, A'' works with $\Pi_0 = (\pi'_{\mathsf{II}})$ and $\Pi_1 = (\pi'_{\mathsf{II}}, \pi'_{\mathsf{A}})$. If $h = |\widehat{w}|$, A'' works with the empty list $\Pi_{|\widehat{w}|}$.

Since $h \leq |\widehat{w}| \leq O(n \cdot 2^{\psi(n)})$ and, by (2), we simulate at most $t_h \leq 2^{O(\psi(n))}$ steps of A' at each position h, taking time $O(\psi(n)^2)$ per each simulated step, the cost of the simulation can be bounded by $n \cdot 2^{O(\psi(n))} \leq n \cdot 2^{O(s(n))}$.

The only problem is that A'' does not have enough space to keep the primary *read-write* worktape of A' which, at the beginning of Phase II, contains $\widehat{w}$. For this reason, the initial primary worktape containing $\widehat{w}$ is manipulated "on demand". More precisely, the current position h in $\widehat{w}$ is represented by *(i)* a block $\mathbb{P}_i$ in $\widehat{w}$, given implicitly by the input head of A'' pointing to the symbol a_i on the input tape, *(ii)* a record $P_{i,j}$ in $\mathbb{P}_i$, given by $j \in \{0, \ldots, 2^{\psi(n)}-1\}$ kept in a separate track of the third worktape, and *(iii)* a relative position d of the current symbol in the string $¢P_{i,j}$, kept in the finite state control. For $j = 0$, the position d points to a symbol in $\$a_i¢P_{i,0}$ and, for $j = 2^{\psi(n)}-1$, the position d points to a symbol in $¢P_{i,2^{\psi(n)}-1}¢a_i$.

Each time A'' needs to *read the current symbol* in $\widehat{w}$, it uses the d-th symbol in $¢P_{i,j}$ (with obvious differences for $j \in \{0, 2^{\psi(n)}-1\}$). For the first 3 bits in $P_{i,j}$, this requires to determine whether j represents a memory state in which A halts and accepts/rejects and whether $P_{i,j} = P_{0,0}$.

Each time A'' needs to *move forward* along $\widehat{w}$, it increases d modulo $|¢|P_{i,j}$ (or modulo $|¢|P_{i,j}+2$, if $j \in \{0, 2^{\psi(n)}-1\}$). If, after that, $d = 0$, A'' increases j modulo $2^{\psi(n)}$ and then, if $j = 0$, it moves the input head to the right. □

With an additional very weak constructibility assumption, we can obtain a simulation by a nondeterministic machine using worktapes that are initially empty. The price we pay is that the new machine is no longer one-way.

Theorem 4. $\text{ASPACE}^{\text{dm}}(s(n)) \subseteq \text{NTIMESPACE}(n \cdot 2^{O(s(n))}, 2^{O(s(n))})$, *provided that* $\lfloor s(n) \rfloor$ *can be computed by a deterministic multi-tape Turing machine in* $n \cdot 2^{O(s(n))}$ *time and, simultaneously, in* $2^{O(s(n))}$ *space.*

Proof. We can devise a new machine that marks off a worktape space of size $2^{k \cdot \lfloor s(n) \rfloor}$ by itself and then it simulates A'' presented in Theorem 3. □

The deterministic two-way machine constructed in Lemma 1 uses large amount of space, namely, $n \cdot 2^{\Omega(s(n))}$. Now we shall convert it back into an alternating machine using $O(s(n))$ space, this time accepting the complement of the original language. The conversion works with no assumptions on $s(n)$, even if the original machine rejects by going into infinite loops.

Theorem 5. $\text{ASPACE}^{\text{dm}}(s(n)) = \text{CO-ASPACE}^{\text{dm}}(s(n))$, *for each* $s(n) \geq 1$.

Proof. Let A be the original alternating machine, and let A' be the equivalent deterministic machine from Lemma 1. This time we shall construct an alternating machine A''' equipped with a single worktape, starting with a delimited worktape space of size $\lfloor s(n) \rfloor$, such that A''' accepts if and only if A' rejects.

Similarly as A'' in Theorem 3, A''' simulates A' in the course of Phase II only, skipping Phase I and using the virtual string $\widehat{w}$ instead. The current position h in $\widehat{w}$ is represented as described in Theorem 3, with $O(s(n))$ space. However, besides *read-current-symbol* and *move-forward*, implemented in the same way as in Theorem 3, the machine A''' shall also use a *move-backward* operation, implemented as the "undo" operation of *move-forward*.

Recall that A' starts Phase II at the left end of the primary worktape in the secondary memory state π'_{II} and, depending on the outcome, it halts there in the accepting/rejecting secondary memory state π'_{A} or π'_{R}, respectively. Moreover, by (2) and (1), the machine A' does not visit any position h on the primary worktape more than $t_h \leq 2^{O(\psi(n))} \leq 2^{O(s(n))}$ times. By fixing a sufficiently large constant k', we obtain $t_h < 2^{k' \cdot \lfloor s(n) \rfloor}$.

Now, for $h \in \{1, \ldots, |\widehat{w}|\}$ and $g \in \{1, \ldots, 2^{k' \cdot \lfloor s(n) \rfloor} - 1\}$, let $\dot{\pi}_{h,g}$ denote the secondary memory state when, in the course of Phase II, the machine A' visits the position h for the g-th time. If $g > t_h$, we take $\dot{\pi}_{h,g} = \mathsf{undefined}$. Note that $\dot{\pi}_{1,1} = \pi'_{\mathsf{II}}$ and $\dot{\pi}_{1,2} \in \{\pi'_{\mathsf{A}}, \pi'_{\mathsf{R}}\}$, for each input w. Moreover, $w \notin L(\mathsf{A})$ if and only if $w \notin L(\mathsf{A}')$ which, in turn, holds if and only if $\dot{\pi}_{1,2} = \pi'_{\mathsf{R}}$.

In order to decide whether $w \notin L(\mathsf{A})$, A''' decides whether $\pi'_{\mathsf{R}} = \dot{\pi}_{1,2}$. To this aim, consider a more general task, testing whether $\pi = \dot{\pi}_{h,g}$, for any given π, h, g. Testing this predicate is implemented in the form of an alternating procedure $\mathsf{test}(\pi, h, g)$. Thus, A''' calls the procedure $\mathsf{test}(\pi'_{\mathsf{R}}, 1, 2)$.

The procedure $\mathsf{test}(\pi, h, g)$ starts in a special finite control state q_{test} with π and g written in two tracks of the worktape of A''', and the primary worktape head placed at the position h on $\widehat{w}$. (Actually, the string $\widehat{w}$ on the primary worktape of A' is *virtual*, we only imitate a two-way read-only access to it.) Depending on whether $\pi = \dot{\pi}_{h,g}$, the alternating subtree of all computation paths rooted in the point of activation of $\mathsf{test}(\pi, h, g)$ will be accepting or rejecting. The procedure test is allowed to reject by going into infinite loops.

In the body of this procedure, the machine A''' runs a "local" simulation of A' on the segment of the primary worktape cells at the positions $h-1, h, h+1$, in the course of Phase II. This way A''' obtains the value $\dot{\pi}_{h,g}$, which is then compared with π. (An example of a local simulation is displayed in Fig. 1.) During the local simulation, A''' maintains the following data about A':

(a) $\sigma_{-1}, \sigma_0, \sigma_{+1}$, the current contents in the primary worktape cells at the respective positions $h-1, h, h+1$, kept in the finite state control. Initially, A''' loads the respective symbols $\widehat{w}_{h-1}, \widehat{w}_h, \widehat{w}_{h+1}$ by the use of the operations that handle the virtual string $\widehat{w}$, namely, by the use of *read-current-symbol*, *move-forward*, and *move-backward*, so that the original position h on $\widehat{w}$ is preserved.

(b) g_{-1}, g_0, g_{+1}, the local time counters for the positions $h-1, h, h+1$, kept in separate worktape tracks, in which A''' counts the number of times the primary

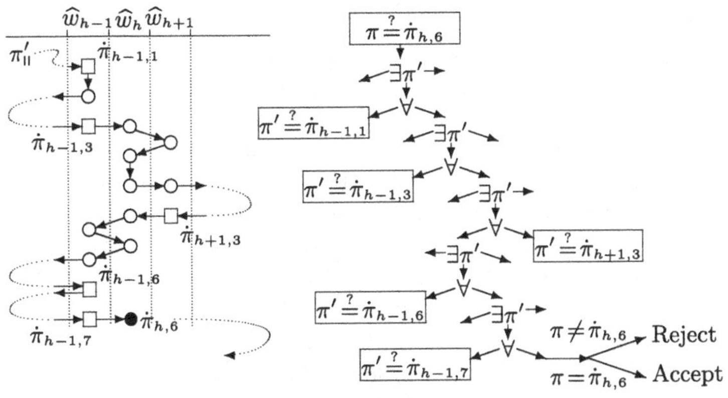

Fig. 1. An example of a local simulation (left), testing whether $\pi = \dot{\pi}_{h,6}$ by calling $\mathsf{test}(\pi, h, 6)$, and the structure of existential and universal decisions in the corresponding fragment of the computation tree (right).

worktape head of A' visits the corresponding position in the course of the local simulation. Initially, A''' assigns $g_{-1} := 1$, $g_0 := 0$, and $g_{+1} := 0$.

(c) $h' \in \{-1, 0, +1\}$, the current position of the head of A' on the primary worktape, relative to h, kept the finite state control. Initially, $h' := -1$.

(d) π', the current secondary memory state of A', kept in a separate worktape track. The initial value $\pi' := \dot{\pi}_{h-1,1}$ is guessed existentially. After guessing, A''' branches universally. The first branch starts the local simulation, assuming π' is correct. The second branch verifies whether $\pi' = \dot{\pi}_{h-1,1}$, that is, it "recursively" activates $\mathsf{test}(\pi', h-1, 1)$, running in parallel with the first branch. Thus, A''' replaces the "old" values π and g, written in the corresponding worktape tracks, by the "new" ones, namely, by π' and 1, respectively. After that, A''' moves backward along the virtual string $\widehat{w}$, by the use of the operation *move-backward*. This replaces the "old" value h by $h-1$. Finally, A''' switches to the state q_{test}. This restarts $\mathsf{test}(\pi', h-1, 1)$ for testing whether $\pi' = \dot{\pi}_{h-1,1}$.

Now, while A' does not leave the primary worktape cells at the positions $h-1, h, h+1$, the simulation is straightforward. In accordance with the progress of the computation of A', the machine A''' updates π', h', and $\sigma_{-1}, \sigma_0, \sigma_{+1}$. Before each simulated step, A''' increments the corresponding local time counter $g_{h'}$.

If A' leaves the local area on the primary worktape to the left, in some secondary memory state π' with $h' = -1$ and some $g_{-1} \geq 1$, the machine A''' first increments g_{-1}. Now A''' resumes the local simulation by guessing existentially $\pi' := \dot{\pi}_{h-1,g_{-1}}$. After guessing, A''' branches universally. The first branch resumes the local simulation, assuming π' is correct. The second branch verifies whether $\pi' = \dot{\pi}_{h-1,g_{-1}}$, that is, it activates $\mathsf{test}(\pi', h-1, g_{-1})$, running in parallel with the first branch. Namely, A''' replaces π and g in the corresponding worktape tracks by π' and g_{-1}, after which A''' moves backward along the virtual string $\widehat{w}$, by the use of the operation *move-backward*. Then A''' switches to the state q_{test}.

Thus, we have a parallel alternating subtree rooted in the point of activation of $\mathsf{test}(\pi', h-1, g_{-1})$, testing whether $\pi' = \dot{\pi}_{h-1,g_{-1}}$.

Similarly, if A' leaves the local area on the primary worktape to the right, in some secondary memory state π' with $h' = +1$ and some $g_{+1} \geq 1$, the machine A''' proceeds in the same way, using *move-forward* instead of *move-backward* and activating $\mathsf{test}(\pi', h+1, g_{+1})$ instead of $\mathsf{test}(\pi', h-1, g_{-1})$, in a parallel path.

When, in the course of the local simulation, the local time counter g_0 reaches the value g, the simulation is stopped and, depending on whether $\pi = \pi'$, the machine A''' accepts or rejects. (If all existential guesses were correct, we have $\pi' = \dot{\pi}_{h,g}$.) Conversely, when one of the counters g_{-1}, g_{+1} overflows, i.e., it reaches $2^{k' \cdot \lfloor s(n) \rfloor}$, the machine A''' rejects — wrong guess in the past.

In the special case of $h = 1$, the local simulation does not start with an existentially guessed $\pi' = \dot{\pi}_{h-1,1}$ and $h' = -1$ but, rather, with $\pi' := \pi'_{\shortparallel}$ (equal to $\dot{\pi}_{1,1}$), $h' := 0$, $g_0 := 1$, and $\sigma_{-1} := \vdash$. Note also that the computation of $\mathsf{test}(\pi, 1, 1)$ is deterministic: if, in addition, $g = 1$, the local simulation is stopped immediately after initialization, since then $g_0 = g$ at the very beginning. After that, depending on whether $\pi = \pi'_{\shortparallel}$, the machine A''' accepts or rejects.

It can be shown that, depending on whether $\pi = \dot{\pi}_{h,g}$, the alternating subtree of all computation paths rooted in the point of activation of $\mathsf{test}(\pi, h, g)$ is accepting or rejecting. □

3 Conclusion

By combining Theorems 2 and 4 with $\text{ASPACE}^{\text{dm}}(s(n)) \subseteq \text{DSPACE}(2^{O(s(n))})$, shown in [19], we have, for each $s(n) \geq \log\log n$ such that $\lfloor s(n) \rfloor$ can be computed deterministically in $n \cdot 2^{O(s(n))}$ time and, simultaneously, in $2^{O(s(n))}$ space, that

$$\text{ASPACE}^{\text{dm}}(s(n)) \subseteq \text{DTIME}(n \cdot 2^{O(s(n))}) \cap \text{DSPACE}(2^{O(s(n))}) \text{ and}$$
$$\text{ASPACE}^{\text{dm}}(s(n)) \subseteq \text{NTIMESPACE}(n \cdot 2^{O(s(n))}, 2^{O(s(n))}).$$

It should be pointed out that even though this gives $\text{ASPACE}(\log\log n) \subseteq \text{ASPACE}^{\text{dm}}(\log\log n) \subseteq \text{DTIME}(n \cdot (\log n)^{O(1)}) \cap \text{DSPACE}((\log n)^{O(1)})$, we do not know whether each language in $\text{ASPACE}(\log\log)$ can be accepted deterministically in polynomial time and, *simultaneously*, in polylogarithmic space, i.e., whether $\text{ASPACE}(\log\log) \subseteq \text{SC}$. So far, by Theorem 4, we have obtained such inclusion only for the nondeterministic counterpart of SC.

We have also shown that $\text{ASPACE}^{\text{dm}}(s(n))$ is closed under complement even for sublogarithmic space, which solved a long-standing open problem [3]:

$$\text{ASPACE}^{\text{dm}}(s(n)) = \text{co-ASPACE}^{\text{dm}}(s(n)), \text{ for each } s(n). \tag{3}$$

As a future work, we are going to show that (3) holds also for $\text{ASPACE}(s(n))$; the problem of complement stays open for both $\text{NSPACE}(s(n))$ and $\text{NSPACE}^{\text{dm}}(s(n))$; the answer to the analogous question for $\text{DSPACE}(s(n))$ and $\text{DSPACE}^{\text{dm}}(s(n))$ is affirmative, by the Sipser's simulation [18].

It is somewhat annoying that though we have a construction of a complementary machine A''' that does not depend on whether $s(n)$ is below $\log n$ and works even if the original machine A can reject by going into infinite loops, the machine A''' itself rejects by going into infinite loops along some computation paths. This, despite (3), leaves us with a fascinating question:

> Is it possible to replace each $O(s(n))$ space bounded alternating Turing machine by an equivalent alternating machine using the same amount of space, such that it halts along each computation path on every input?

This kind of problem is open for both $\mathrm{ASPACE}(s(n))$ and $\mathrm{ASPACE}^{\mathrm{dm}}(s(n))$, for space complexity classes *not* satisfying $s(n) \in \Omega(\log n)$.

References

1. Allender, E., Mix Barrington, D., Hesse, W.: Uniform circuits for division: consequences and problems. In: Proceedings of IEEE Conference on Computer Complexity, pp. 150–159 (2001)
2. Bertoni, A., Mereghetti, C., Pighizzini, G.: On languages accepted with simultaneous complexity bounds and their ranking problem. In: Prívara, I., Ružička, P., Rovan, B. (eds.) MFCS 1994. LNCS, vol. 841, pp. 245–255. Springer, Heidelberg (1994)
3. Braunmühl, B., Gengler, R., Rettinger, R.: The alternation hierarchy for sublogarithmic space is infinite. Comput. Complex. **3**, 207–230 (1993)
4. Chandra, A., Kozen, D., Stockmeyer, L.: Alternation. J. Assoc. Comput. Mach. **28**, 114–133 (1981)
5. Chang, R., Hartmanis, J., Ranjan, D.: Space bounded computations: review and new separation results. Theoret. Comput. Sci. **80**, 289–302 (1991)
6. Geffert, V.: A speed-up theorem without tape compression. Theoret. Comput. Sci. **118**, 49–65 (1993)
7. Geffert, V.: A hierarchy that does not collapse: alternations in low level space. RAIRO Inform. Théor. Appl. **28**, 465–512 (1994)
8. Geffert, V.: Bridging across the $\log(n)$ space frontier. Inform. Comput. **142**, 127–158 (1998)
9. Geffert, V., Pardubská, D.: Unary coded NP-complete languages in $\mathrm{ASPACE}(\log\log n)$. Int. J. Found. Comput. Sci. **24**, 1167–1182 (2013)
10. Hartmanis, J., Lewis II, P., Stearns, R.: Hierarchies of memory limited computations. In: IEEE Conference on Record on Switching Circuit Theory and Logical Design, pp. 179–190 (1965)
11. Hennie, F.: One-tape, off-line Turing machine computations. Inform. Control **8**, 553–578 (1965)
12. Hong, J.: A tradeoff theorem for space and reversal. Theoret. Comput. Sci. **32**, 221–224 (1984)
13. Hopcroft, J., Motwani, R., Ullman, J.: Introduction to Automata Theory, Languages, and Computation. Addison Wesley, Reading (2001)
14. Immerman, N.: Nondeterministic space is closed under complementation. SIAM J. Comput. **17**, 935–938 (1988)
15. Iwama, K.: $\mathrm{ASPACE}(o(\log\log n))$ is regular. SIAM J. Comput. **22**, 136–146 (1993)

16. Liśkiewicz, M., Reischuk, R.: The sublogarithmic alternating space world. SIAM J. Comput. **25**, 828–861 (1996)
17. Mereghetti, C.: Testing the descriptional power of small Turing machines on non-regular language acceptance. Int. J. Found. Comput. Sci. **19**, 827–843 (2008)
18. Sipser, M.: Halting space bounded computations. Theoret. Comput. Sci. **10**, 335–338 (1980)
19. Sudborough, I.: Efficient algorithms for path system problems and applications to alternating and time-space complexity classes. In: Proceedings of the IEEE Symposium on Foundations of Computer Science, pp. 62–73 (1980)
20. Szelepcsényi, R.: The method of forced enumeration for nondeterministic automata. Acta Inform. **26**, 279–284 (1988)
21. Szepietowski, A.: Turing Machines with Sublogarithmic Space. LNCS, vol. 843. Springer, Heidelberg (1994)

Weighted Symbolic Automata with Data Storage

Luisa Herrmann(✉) and Heiko Vogler

Faculty of Computer Science, Technische Universität Dresden,
Nöthnitzer Str. 46, 01062 Dresden, Germany
{Luisa.Herrmann,Heiko.Vogler}@tu-dresden.de

Abstract. We introduce weighted symbolic automata with data storage, which combine and generalize the concepts of automata with storage types, weighted automata, and symbolic automata. By defining two particular data storages, we show that this combination is rich enough to capture symbolic visibly pushdown automata and weighted timed automata. We introduce a weighted MSO-logic and prove a Büchi-Elgot-Trakhtenbrot theorem, i.e., the new logic and the new automaton model are expressively equivalent.

1 Introduction

Finite-state (string) automata have been generalized in at least three directions.

Due to the introduction of a wealth of new automata models, like pushdown automata, stack automata, nested stack automata, and counter automata, Scott proposed a homogeneous framework [16]. Using the notions of [9], such an *automaton with storage* consists of an automaton and a storage type; in each transition, the automaton can test the current storage configuration by a predicate (like: top $= \gamma$?) and transform it by an instruction (like: push(δ) or pop).

In a second generalization, each transition of a finite-state automaton was equipped with a weight taken from some semiring in order to analyse quantitative aspects of the recognition process. This led to the concept of *weighted automata* and its well investigated theory, cf. e.g. [3,8,13,15]. In recent work, unital valuation monoids were used as weight algebras [4] in order to calculate along a run of an automaton also with non-sequential operations, like average. In the literature, combinations of the first two generalizations were investigated: weighted pushdown automata over semirings [13] and unital valuation monoids [6], and weighted automata over arbitrary storage types and unital valuation monoids [11,20].

In a third generalization, finite-state automata were allowed to process input strings over an arbitrary, not necessarily finite set. This extension is relevant, e.g., when dealing with XML-documents involving data. An example of such automata are symbolic automata [17,19] in which each transition τ involves a unary predicate π; τ is applicable to the current input symbol if it satisfies π.

In this paper, we introduce a new automaton model, called *weighted symbolic automata with data storage.* It captures all three mentioned generalizations and it

L. Herrmann—Supported by DFG Graduiertenkolleg 1763 (QuantLA).

S. Brlek and C. Reutenauer (Eds.): DLT 2016, LNCS 9840, pp. 203–215, 2016.
DOI: 10.1007/978-3-662-53132-7_17

is defined in the same modular style as automata with storage introduced in [9,16]. As weight structure we choose unital valuation monoids. We extend the concept of storage type to that of a *data storage type*. There, predicates and instructions do not only depend on the current storage configuration, but also on storage inputs. In each transition, a predicate checks a property of the current data symbol of the input string (as in symbolic automata); via an encoding function specified in the automaton, this data symbol is transformed into a storage input. In this sense, predicates and instructions become 'sensitive' to the input string.

It turns out that our combination of symbolicalness of input strings and of sensitivity of predicates and instructions is rich enough to capture two recently introduced classes of automata which can process words over infinite sets. We define the data storage types VP(N) and TIME($\mathcal{C}$) and show that weighted automata over VP(N) and TIME($\mathcal{C}$) are exactly the (weighted version of) symbolic visibly pushdown automata [1] and weighted timed automata [7], respectively.

Moreover, we introduce a *weighted MSO-logic over data storage types* extending [20] by employing an infinite set of input symbols and a data storage type. Each formula of this logic has the form $\sum_B^\eta e$ where η is an encoding of input symbols into storage inputs, and $\sum_B$ represents the weighted version of a second-order existential quantification over the second-order behavior variable B; it ranges over behaviors of the underlying data storage type. Intuitively, a behavior is an executable string $(p_1, f_1) \ldots (p_n, f_n)$ of pairs of predicates p_i and instructions f_i. The subformula e is an expression as defined in [20] (also cf. [10, Definition 3.1]); we note that, for semirings, such expressions are equivalent to the fragment of restricted weighted MSO-logic introduced in [2] (cf. [10, Proposition 5.14]).

We prove a Büchi-Elgot-Trakhtenbrot (BET) theorem (cf. Theorems 13 and 16) stating that weighted symbolic automata over data storage types are expressively equivalent to weighted MSO-logic over data storage types. In particular, we obtain the BET theorem for weighted symbolic visibly pushdown automata (which is new) and for weighted timed automata (which is an alternative to [14, Theorem 41]). As a consequence of our BET theorem we obtain that, for each bounded lattice, the satisfiability problem of weighted MSO-logic over VP(N) is decidable.

2 Preliminaries

Notations and Notions. We denote the set of natural numbers including zero by $\mathbb{N}$. For $n \in \mathbb{N}$ we let $[n]$ denote the set $\{i \in \mathbb{N} \mid 1 \leq i \leq n\}$. Thus $[0] = \emptyset$. In the following let $A, A_1, \ldots, A_n$, and B be sets. The set of all words over A is denoted by A^*. For each $w \in A^*$, $|w|$ is the length of w, $\mathrm{pos}(w) = \{1, \ldots, |w|\}$ is the set of *positions of* w, and w_i is the label at the i-th position of w. The *empty word* (of length 0) is denoted by ε. We let $A^+ = A^* \setminus \{\varepsilon\}$. A *relabeling* is a mapping $\rho\colon A \to \mathcal{P}(B)$. We denote the unique extension of ρ to the morphism from the free monoid $(A^*, \cdot, \varepsilon)$ to the monoid $(\mathcal{P}(B^*), \circ, \{\varepsilon\})$, where $\circ$ denotes language concatenation, also by ρ. For each $L \subseteq A^*$ we define $\rho(L) = \bigcup_{w \in L} \rho(w)$. For each $i \in [n]$ we define the *i-th projection* as function $(.)_i\colon A_1 \times \ldots \times A_n \to A_i$ such that for each $(a_1, \ldots, a_n) \in A_1 \times \ldots \times A_n$ we have $(a_1, \ldots, a_n)_i = a_i$. We require that each function which we consider in this work is computable.

Unital Valuation Monoids. The concept of valuation monoid was introduced in [4] and extended in [6] to unital valuation monoid. A *unital valuation monoid* is a tuple $(K, +, \mathrm{val}, 0, 1)$ where $(K, +, 0)$ is a commutative monoid and $\mathrm{val}\colon K^* \to K$ is a mapping such that (i) $\mathrm{val}(k) = k$ for each $k \in K$, (ii) $\mathrm{val}(k) = 0$ for each $k \in K^*$ whenever $k_i = 0$ for some $i \in [|k|]$, (iii) $\mathrm{val}(k1k') = \mathrm{val}(kk')$ for every $k, k' \in K^*$, and (iv) $\mathrm{val}(\varepsilon) = 1$. Moreover, K is called *zero-sum-free*, if $k + k' = 0$ implies $k = k' = 0$. *In the rest of this paper, we let K denote an arbitrary unital valuation monoid $(K, +, \mathrm{val}, 0, 1)$ unless specified otherwise.*

Example 1. Recall that a strong bimonoid [5] is a structure $(K, +, \cdot, 0, 1)$, where $(K, +, 0)$ is a commutative monoid, $(K, \cdot, 1)$ is a monoid, and $a \cdot 0 = 0 \cdot a = 0$ for every $a \in K$. A *semiring* is a strong bimonoid in which $\cdot$ distributes over $+$ from both sides. A particular semiring is the Boolean semiring $(\mathbb{B}, \vee, \wedge, 0, 1)$ with $\mathbb{B} = \{0, 1\}$. A *bounded lattice* is a strong bimonoid in which both operations are idempotent (and other laws are satisfied). It is clear that each strong bimonoid is a unital valuation monoid $(K, +, \mathrm{val}, 0, 1)$, where for every $n \in \mathbb{N}$ and $k_1, \ldots, k_n \in K$ we let $\mathrm{val}(k_1 \ldots k_n) = k_1 \cdot \ldots \cdot k_n$ (see [6]). Unital valuation monoids can be used to compute averages. For this consider $K_{\mathrm{avg}} = (\mathbb{R} \cup \{\infty, -\infty\}, \sup, \mathrm{avg}, -\infty, \infty)$ with $\mathrm{avg}(a_1 \ldots a_n) = \frac{1}{n} \cdot \sum_{1 \leq i \leq n} a_i$ for every $a_1, \ldots, a_n \in \mathbb{R}$.

Weighted Languages. Let D be a non-empty set. A *K-weighted language (over D)* is a mapping of the form $r\colon D^* \to K$. We denote the set of all such mappings by $K\langle\langle D^* \rangle\rangle$. Let $r \in K\langle\langle D^* \rangle\rangle$. We denote the set $\{w \in D^* \mid r(w) \neq 0\}$ by $\mathrm{supp}(r)$ (*support of* r). Moreover, let $L \subseteq D^*$ and $w \in D^*$. We define the weighted language $(r \cap L) \in K\langle\langle D^* \rangle\rangle$ by $(r \cap L)(w) = r(w)$ if $w \in L$, and 0 otherwise. Now let $r' \in K\langle\langle D^* \rangle\rangle$. We define the *sum of r and r'* as the weighted language $r + r' \in K\langle\langle D^* \rangle\rangle$ by $(r + r')(w) = r(w) + r'(w)$.

Label Structure. Let D be a non-empty set. A *predicate over D* is a mapping $\pi\colon D \to \{0, 1\}$ and we identify π with $\{a \in D \mid \pi(a) = 1\}$. We say that π *is decidable* if it is decidable whether $\pi \neq \emptyset$. We denote by $\mathrm{Pred}(D)$ the set of all decidable predicates over D. For every $\Pi \subseteq \mathrm{Pred}(D)$ we define the Boolean closure $\mathrm{BC}(\Pi)$ as usual and we denote the always true predicate by T. Obviously, if Π is recursively enumerable, then so is $\mathrm{BC}(\Pi)$. A *label structure (over D)* is a tuple (D, Π), where $\Pi \subseteq \mathrm{Pred}(D)$ is a recursively enumerable set of predicates such that $\mathrm{BC}(\Pi) = \Pi$. If D is clear from the context, then we only write Π instead of (D, Π).

3 Weighted Symbolic Automata with Data Storage

Data Storage Types and Behavior. We extend the notion of storage type [9,16] in such a way that the predicates and instructions do not only depend on the current configuration, but also on storage inputs (which, in their turn, are encodings of the input of an automaton).

A *data storage type* is a tuple $S = (C, M, P, F, c_0)$ where C is a set (*configurations*), M is a set (*storage inputs*), P is a set of functions each of the type

$p: C \times M \to \{\text{true}, \text{false}\}$ (*predicates*), F is a set of partial functions each of the type $f: C \times M \to C$ (*instructions*), and $c_0 \in C$ (*initial configuration*).

If M is a singleton, then we reobtain the concept of storage type as introduced in [11,20]. *Throughout this paper we let S denote an arbitrary data storage type (C, M, P, F, c_0) unless specified otherwise.*

Example 2. (1) For some fixed elements c and m we define the *trivial storage type* as the data storage type $\text{TRIV} = (\{\text{c}\}, \{\text{m}\}, \{\text{p}_{\text{true}}\}, \{\text{f}_{\text{id}}\}, \text{c})$ where $\text{p}_{\text{true}}(\text{c}, \text{m}) =$ true and $\text{f}_{\text{id}}(\text{c}, \text{m}) = \text{c}$.

(2) Let COUNT be the data storage type $(\mathbb{N}, \mathbb{N}, \{\text{T?}, 0?\}, \{+, -\}, 0)$ where for every $c, d \in \mathbb{N}$ we let $\text{T?}(c, d) = \text{true}$, $0?(c, d) = \text{true}$ iff $c = 0$, $+(c, d) = c + d$, and $-(c, d) = c - d$ if $c \geq d$ and undefined otherwise.

A central notion of our MSO-logic is the concept of storage behavior. Let Ω be a finite subset of $P \times F$. Also, let $n \in \mathbb{N}$, $m_1, \ldots, m_n \in M$, and $b = (p_1, f_1) \ldots (p_n, f_n) \in \Omega^*$. We call b an $m_1 \ldots m_n$*-behavior (over Ω)* if for every $i \in [n]$ we have $p_i(c', m_i) = \text{true}$ and $f_i(c', m_i)$ is defined where $c' = f_{i-1}(\ldots f_1(c_0, m_1) \ldots, m_{i-1})$. Note that $c' = c_0$ for $i = 1$. We denote the set of all $m_1 \ldots m_n$-behaviors over Ω by $\text{B}(\Omega, m_1 \ldots m_n)$.

Example 3. Consider Count and $w = 284770 \in \mathbb{N}^*$ with $|w| = 6$. Then the word $(\text{T?}, +)^3(\text{T?}, -)^2(0?, +)$ is a w-behavior over $\{\text{T?}, 0?\} \times \{+, -\}$.

Weighted Symbolic Automata with Data Storage. Let D be a set. A *K-weighted symbolic automaton with data storage type S and input D* (short: *(S, D, K)-automaton*) is a tuple $\mathcal{A} = (Q, \Pi, Q_0, Q_f, T, \text{wt}, \eta)$ where Q is a finite set (*states*), Π is a label structure over D, $Q_0 \subseteq Q$ (*initial states*), $Q_f \subseteq Q$ (*final states*), $T \subseteq Q \times \Pi \times P \times Q \times F$ is a finite set (*transitions*), $\text{wt}: T \times D \to K$ is a function (*weight assignment*), and $\eta: D \to M$ is a relabeling (*storage encoding*). We call $\mathcal{A}$ *projective* if η is a projection. Moreover, we call $\mathcal{A}$ *homogeneous* if for each transition $\tau \in T$ and for every $d_1, d_2 \in (\tau)_2$ we have $\text{wt}(\tau, d_1) = \text{wt}(\tau, d_2)$. In this case we view wt as function of type $T \to K$.

The set of $\mathcal{A}$*-configurations* is the set $Q \times D^* \times C$. For each transition $\tau = (q, \pi, p, q', f)$ in T we define the binary relation $\vdash^\tau$ on the set of $\mathcal{A}$-configurations as follows: for every $d \in D$, $w \in D^*$, and $c \in C$, we let $(q, dw, c) \vdash^\tau (q', w, f(c, \eta(d)))$ if $\pi(d)$ is true, $p(c, \eta(d))$ is true, and $f(c, \eta(d))$ is defined. The *computation relation of $\mathcal{A}$* is the binary relation $\vdash = \bigcup_{\tau \in T} \vdash^\tau$.

A *computation* is a sequence $\xi_0 \vdash^{\tau_1} \xi_1 \cdots \vdash^{\tau_n} \xi_n$ such that $n \in \mathbb{N}$, $\tau_1, \ldots, \tau_n$ are transitions, $\xi_0, \ldots, \xi_n$ are $\mathcal{A}$-configurations, and $\xi_{i-1} \vdash^{\tau_i} \xi_i$ for each $i \in [n]$. Sometimes we abbreviate this computation by $\xi_0 \vdash^{\tau_1 \cdots \tau_n} \xi_n$. Let $w = d_1 \ldots d_n \in D^*$ with $d_i \in D$. A *successful computation on w* is a computation $\theta = ((q_0, w, c_0) \vdash^{\tau_1 \cdots \tau_n} (q_f, \varepsilon, c'))$ for some $q_0 \in Q_0$, $q_f \in Q_f$, $c' \in C$, and $\tau_1, \ldots, \tau_n \in T$. The *weight of θ* is the element in K defined by

$$\text{wt}(\theta) = \text{val}(\text{wt}(\tau_1, d_1) \ldots \text{wt}(\tau_n, d_n))$$

and we denote the set of all successful computations on w by $\Theta_{\mathcal{A}}(w)$.

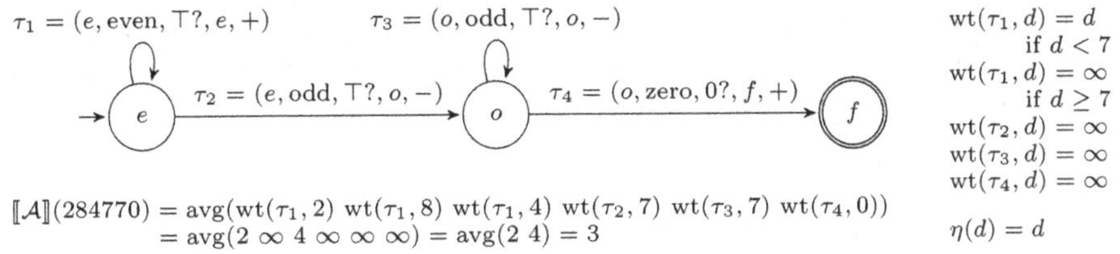

Fig. 1. The projective $(\text{Count}, \mathbb{N}, K_{\text{avg}})$-automaton $\mathcal{A}$ recognizing r.

The *weighted language recognized by* $\mathcal{A}$ is the K-weighted language $[\![\mathcal{A}]\!] : D^* \to K$ defined for every $w \in D^*$ by

$$[\![\mathcal{A}]\!](w) = \textstyle\sum_{\theta \in \Theta_{\mathcal{A}(w)}} \text{wt}(\theta).$$

A weighted language $r: D^* \to K$ is (S, D, K)-recognizable if there is an (S, D, K)-automaton $\mathcal{A}$ such that $r = [\![\mathcal{A}]\!]$. In the obvious way, we define projectively (S, D, K)-recognizable and homogeneously (S, D, K)-recognizable.

Example 4. Consider the language $L \subseteq \mathbb{N}^*$ consisting of words $u_1 \ldots u_n v_1 \ldots v_m 0$, $m, n \geq 1$, such that u_i is even and v_j is odd for each $i \in [n]$, $j \in [m]$, and $u_1 + \ldots + u_n = v_1 + \ldots + v_m$. We define the weighted language $r: \mathbb{N}^* \to K_{\text{avg}}$ with $\text{supp}(r) = L$; each word in L is mapped to the average value of all even symbols in $u_1 \ldots u_n$ which are smaller than 7. The projective, non-homogeneous $(\text{Count}, \mathbb{N}, K_{\text{avg}})$-automaton $\mathcal{A} = (\{e, o, f\}, \Pi, \{e\}, \{f\}, T, \text{wt}, \eta)$ shown in Fig. 1 recognizes r, where $\Pi = \text{BC}(\{\text{even}, \text{odd}, \text{zero}\})$ with the intuitive interpretations.

Lemma 5. *For each $(S, D, \mathbb{B})$-automaton $\mathcal{A}$ there is a homogeneous $(S, D, \mathbb{B})$-automaton $\mathcal{B}$ with $[\![\mathcal{B}]\!] = [\![\mathcal{A}]\!]$.*

Special Cases. (1) Let D be a finite set. Then it is easy to see that each weighted automaton with storage (in the manner of [20]) is a homogeneous (S, D, K)-automaton (for some data storage type S).

(2) Let $K = \mathbb{B}$. Since we can assume each $(S, D, \mathbb{B})$-automaton $\mathcal{A}$ to be homogeneous and, therefore, the weight assignment wt does not depend on its second argument we can presume that the set of transitions of $\mathcal{A}$ consists of those transitions which are mapped to 1. Thus, we can specify an $(S, D, \mathbb{B})$-automaton by a tuple $\mathcal{A} = (Q, \Pi, Q_0, Q_f, T, \eta)$ and define the *language recognized by* $\mathcal{A}$ as the set $L(\mathcal{A}) = \text{supp}([\![\mathcal{A}]\!])$.

(3) Let $S = \text{Triv}$. Then we drop all references to S from the concepts introduced for (S, D, K)-automata. Thus $T \subseteq Q \times \Pi \times Q$ and we speak about (D, K)-*automata* and (D, K)-*recognizable*. Note that homogeneous (D, K)-automata can be seen as a K-weighted version of symbolic automata.

(4) Let $S = \text{Triv}$ and $K = \mathbb{B}$. Then we use both conventions mentioned above and speak about D-*automata* and D-*recognizable*. Moreover, we say that

a D-automaton $\mathcal{A} = (Q, \Pi, Q_0, Q_f, T)$ is *deterministic* if $|Q_0| = 1$ and for every two transitions (q, π_1, q_1) and (q, π_2, q_2) in T with $\pi_1 \cap \pi_2 \neq \emptyset$ we have $q_1 = q_2$, and *total* if for each $q \in Q$, $d \in D$ there is a transition $(q, \pi, q') \in T$ with $d \in \pi$.

Closure Properties. In [18, Theorem 1] it was proved that symbolic tree automata can be made total and deterministic. As a special case we easily obtain:

Lemma 6 (cf. [18, Theorem 1]). *For every D-automaton $\mathcal{A}$ there is a total and deterministic D-automaton $\mathcal{B}$ such that $L(\mathcal{A}) = L(\mathcal{B})$.*

By slightly modifying usual constructions we obtain the following two results:

Lemma 7. Let r_1, r_2 be (S, D, K)-recognizable weighted languages and let L_1, L_2 be D-recognizable languages. Then the weighted languages $r_1 + r_2$ and $r_1 \cap L_1$ are (S, D, K)-recognizable. Moreover, $L_1 \backslash L_2$ is D-recognizable (cf. [19]).

Lemma 8. Let D, D' be sets, L a D-recognizable language, and $\rho\colon D \to \mathcal{P}(D')$ a relabeling. Then $\rho(L)$ is D'-recognizable.

4 Data Storage for Symbolic Visibly Pushdown Automata

A *nested set* is a set $N = N_i \cup N_c \cup N_r$, where N_i (*internal symbols*), N_c (*call symbols*), and N_r (*return symbols*) are pairwise disjoint sets. Let $\mathcal{M} = (Q, Q_0, \Gamma, \delta_i, \delta_c, \delta_r, \delta_b, Q_f)$ be a symbolic visibly pushdown automaton (svpda) as defined in [1] (with pushdown alphabet Γ). As explained there, $\mathcal{M}$ uses binary predicates over matching positions. A pair (i, j) of positions of an input word is *matching* if the pushdown cell pushed at i is popped at j. We introduce the data storage type $\mathrm{VP}(N)$ which simulates the pushdown part of an svpda and encodes these binary predicates as parameters of storage instructions.

Let N be a nested set. We define the data storage type $\mathrm{VP}(N) = (C, N, P, F, \varepsilon)$ where $C = (\Lambda \times N_c)^*$ and Λ is an infinite set of pushdown symbols, $P = \{\mathrm{true}\}$, and $F = \{\mathrm{push}_\gamma \mid \gamma \in \Lambda\} \cup \{\mathrm{pop}_{\gamma,\pi} \mid \gamma \in \Lambda, \pi \subseteq N_c \times N_r \text{ decidable}\} \cup \{\mathrm{stay}_i, \mathrm{stay}_r\}$ such that for each $\gamma \in \Lambda$, $\pi \subseteq N_c \times N_r$, $c \in C$, and $d \in N$ we have

- $\mathrm{push}_\gamma(c, d) = (\gamma, d)c$ if $d \in N_c$,
- $\mathrm{pop}_{\gamma,\pi}(c, d) = c'$ if $d \in N_r$, $c = (\gamma, a)c'$ for some $a \in N_c$, and $(a, d) \in \pi$,
- $\mathrm{stay}_i(c, d) = c$ if $d \in N_i$ and $\mathrm{stay}_r(c, d) = c$ if $d \in N_r$ and $c = \varepsilon$,

and undefined otherwise.

Theorem 9. *Let N be a nested set and $L \subseteq N^*$. Then L is recognizable by a symbolic visibly pushdown automaton with decidable label theory (introduced in [1]) if and only if L is projectively $(\mathrm{VP}(N), N, \mathbb{B})$-recognizable.*

Table 1. The svpda $\mathcal{M}$ (left) and the $(\mathrm{VP}(N), N)$-automaton $\mathcal{A}$ (right), both recognizing L.

$\mathcal{M} = (Q, Q_0, \Gamma, \delta_i, \delta_c, \delta_r, \delta_b, Q_f)$	$\mathcal{A} = (Q, \Pi, Q_0, Q_f, T, \eta)$
$Q = Q_0 = Q_f = \{q\}$	$Q = Q_0 = Q_f = \{q\}$
$\Gamma = \{e, o\}$	
$\delta_i = \{(q, \top, q)\}, \delta_b = \{(q, \top, q)\}$	$\{(q, \top, \mathrm{true}, q, \mathrm{stay}_i), (q, \top, \mathrm{true}, q, \mathrm{stay}_r)\} \subseteq T$
$\delta_c = \{(q, \mathrm{even}, q, e), (q, \mathrm{odd}, q, o)\}$	$\{(q, \mathrm{even}, \mathrm{true}, q, \mathrm{push}_e), (q, \mathrm{odd}, \mathrm{true}, q, \mathrm{push}_o)\} \subseteq T$
$\delta_r = \{(q, \sim, e, q), (q, \top, o, q)\}$	$\{(q, \top, \mathrm{true}, q, \mathrm{pop}_{e,\sim}), (q, \top, \mathrm{true}, q, \mathrm{pop}_{o,\top})\} \subseteq T$

Instead of a formal proof we demonstrate our construction by an example. For this let $N_i = \mathbb{N}$, $N_c = \{\langle x \mid x \in \mathbb{N}\}$, and $N_r = \{x\rangle \mid x \in \mathbb{N}\}$. We consider the language $L \subseteq N^*$ which consists of all words w such that for every two symbols $\langle x$ and $y\rangle$ at matching positions of w we have $x = y$ if x is even. For an example consider $w = 3\rangle\langle 2\ 4\ \langle 3\ 5\rangle\ 2\rangle \in L$ with $|w| = 6$ and matching positions $(2, 6)$ and $(4, 5)$. Clearly, L can be recognized by the svpda $\mathcal{M}$ shown in Table 1(left). Note that $\mathcal{M}$ uses a label theory (cf. [1]), which can be seen as the Boolean closure of the unary predicates even and odd with their intuitive interpretations, and of the binary predicate $\sim$ where $(\langle x, y\rangle) \in \sim$ iff $x = y$. Then we construct the projective $(\mathrm{VP}(N), N, \mathbb{B})$-automaton $\mathcal{A}$ as shown in Table 1(right), which uses the label structure $\Pi = \mathrm{BC}(\{\mathrm{even}, \mathrm{odd}\})$. Clearly, $L(\mathcal{A}) = L$.

Now Theorem 9 opens the possibility of considering weighted svpda. For example we can easily construct a $(\mathrm{VP}(N), N, K_{\mathrm{avg}})$-automaton $\mathcal{A}'$ which maps each word in L to the average value of all its even call symbols.

5 Data Storage for Weighted Timed Automata

Our definition of timed words and weighted timed automata closely resembles the one in [14]. The only difference is that in our definition of timed words, each symbol stores the time difference to its predecessor as in [7], while in [14] the corresponding point in time is recorded. Clearly, both views are isomorphic. In the course of this section let Σ be a finite set and K a semiring.

A *timed word (over Σ)* is a non-empty finite sequence $(a_1, t_1) \dots (a_n, t_n) \in (\Sigma \times \mathbb{R}_{\geq 0})^+$. The set of timed words over Σ is denoted by $T\Sigma^+$ and for some semiring K a mapping $r\colon T\Sigma^+ \to K$ is called a *timed series (over Σ and K)*.

A *clock variable* is a variable ranging over $\mathbb{R}_{\geq 0}$ and we denote the set of all clock variables by $\mathcal{C}$. The set of all *clock constraints over* $\mathcal{C}$ is denoted by $\Phi(\mathcal{C})$. A *clock valuation* is a function $\nu\colon \mathcal{C} \to \mathbb{R}_{\geq 0}$, we let $\nu_0(x) = 0$ for each $x \in \mathcal{C}$, and for $t \in \mathbb{R}_{\geq 0}$ and $\lambda \subseteq \mathcal{C}$ we let the clock valuations $\nu + t$ (where t is added to each clock) and $\nu[\lambda := 0]$ (where all clocks in λ are set to 0) be defined as in [14]. Moreover, the *satisfaction relation* $\models \subseteq {\mathbb{R}_{\geq 0}}^{\mathcal{C}} \times \Phi(\mathcal{C})$ is defined as usual.

A *K-weighted timed automaton* over Σ (and $\mathcal{C}$) is a tuple $\mathcal{A} = (Q, Q_i, Q_f, \mathcal{C}, E, \mathrm{ewt}, \mathrm{dwt})$, where Q is a finite set (*states*), $Q_i \subseteq Q$ (*initial states*) and $Q_f \subseteq Q$ (*final states*), $\mathcal{C}$ is a finite set of clock variables,

$E \subseteq Q \times \Sigma \times \Phi(\mathcal{C}) \times 2^{\mathcal{C}} \times Q$ is a finite set (*edges*), ewt: $E \to K$ is a function (*edge weights*), and dwt: $Q \times \mathbb{R}_{\geq 0} \to K$ is a function (*delay weights*). A *run* of $\mathcal{A}$ is a finite sequence

$$\rho = (q_0, \nu_0) \xrightarrow{t_1} \xrightarrow{e_1} \dots \xrightarrow{t_n} \xrightarrow{e_n} (q_n, \nu_n)$$

where $n \geq 1$, $q_0, \dots, q_n \in Q$, ν_i are clock valuations, $t_i \in \mathbb{R}_{\geq 0}$, and $e_i \in E$ satisfying the following conditions: $q_0 \in Q_i$, $q_n \in Q_f$, and $e_i = (q_{i-1}, a_i, \phi_i, \lambda_i, q_i)$ such that $\nu_{i-1} + t_i \models \phi_i$ and $\nu_i = (\nu_{i-1} + t_i)[\lambda_i := 0]$. The *label* of ρ is the timed word $\text{label}(\rho) = ((e_1)_2, t_1) \dots ((e_n)_2, t_n)$, and the *running weight* $\text{rwt}(\rho)$ of ρ is given by $\text{rwt}(\rho) = \prod_{i \in [n]} \text{dwt}(q_{i-1}, t_i) \cdot \text{ewt}(e_i)$. For any timed word $w \in T\Sigma^+$ let $\text{Run}_{\mathcal{A}}(w)$ denote the set of all runs ρ of $\mathcal{A}$ with $\text{label}(\rho) = w$. The *timed series recognized by* $\mathcal{A}$ is the mapping $[\![\mathcal{A}]\!]: T\Sigma^+ \to K$ such that $[\![\mathcal{A}]\!](w) = \sum_{\rho \in \text{Run}_{\mathcal{A}}(w)} \text{rwt}(\rho)$.

Now we define a data storage type $\text{TIME}(\mathcal{C})$ to simulate the clock behavior of weighted timed automata. Let $\mathcal{C}$ be a finite set of clock variables and let $\text{TIME}(\mathcal{C}) = (\mathbb{R}_{\geq 0}{}^{\mathcal{C}}, \mathbb{R}_{\geq 0}, P, F, \nu_0)$ where $P = \{p_\phi \mid \phi \in \Phi(\mathcal{C})\}$, $F = \{f_\lambda \mid \lambda \subseteq \mathcal{C}\}$, and for every $\phi \in \Phi(\mathcal{C})$, $\lambda \subseteq \mathcal{C}$, $\nu \in \mathbb{R}_{\geq 0}{}^{\mathcal{C}}$, and $t \in \mathbb{R}_{\geq 0}$ we let

- $p_\phi(\nu, t) = \text{true}$ iff $(\nu + t) \models \phi$, and
- $f_\lambda(\nu, t) = (\nu + t)[\lambda := 0]$.

Theorem 10. *Let K be a semiring and $r: T\Sigma^+ \to K$ a timed series. Then r is recognized by a K-weighted timed automaton over Σ and $\mathcal{C}$ if and only if r is projectively* $(\text{TIME}(\mathcal{C}), \Sigma \times \mathbb{R}_{\geq 0}, K)$*-recognizable.*

The formal proof uses the following ideas. In "⇒" each transition results from a given edge, and the weight amounts to the product of ewt and dwt; the resulting $(\text{TIME}(\mathcal{C}), \Sigma \times \mathbb{R}_{\geq 0}, K)$-automaton is (in general) not homogeneous. In "⇐" we code the transitions and the finite part of the input symbols into states, split up each transition, and then simulate the weight assignment with dwt.

6 Weighted Symbolic MSO-Logic with Storage Behavior

Our new logic is based on the concepts of M-expression [10, Definition 3.1] and B-expression [20, Definition 5]. Since these expressions depend on unweighted MSO formulas, we first extend unweighted MSO-logic to symbolic MSO-logic.

Symbolic MSO-Logic. As usual, we use first-order variables, like x, y, and second-order variables, like X, Y. Furthermore, we introduce one more variable B which we call *second-order behavior variable* and which ranges over behaviors of S.

Let D be a set, Π a label structure over D, and Ω a finite subset of $P \times F$. We define the set of *formulas of symbolic MSO-logic over Ω and Π*, denoted by $\text{MSO}(\Omega, \Pi)$, by the following EBNF:

$$\psi ::= P_\pi(x) \mid \text{next}(x, y) \mid x \in X \mid B(x) = (p, f)$$
$$\varphi ::= \psi \mid \neg\varphi \mid \varphi \wedge \varphi \mid \exists x.\varphi \mid \exists X.\varphi$$

where $\pi \in \Pi$ and $(p, f) \in \Omega$. Let $\varphi \in \mathrm{MSO}(\Omega, \Pi)$. The set of *free variables of* φ and *bound variables of* φ, denoted by $\mathrm{Free}(\varphi)$ and $\mathrm{Bound}(\varphi)$, resp., is defined as usual. In particular, we set $\mathrm{Free}(P_\pi(x)) = \{x\}$ and $\mathrm{Free}(B(x) = (p, f)) = \{x, B\}$.

Let $\mathcal{V}$ be a finite set of variables with $B \in \mathcal{V}$, let $\eta\colon D \to M$ be a relabeling, and let $w \in D^*$. A $(\mathcal{V}, \eta)$-*assignment for* w is a function with domain $\mathcal{V}$ which maps each first-order variable in $\mathcal{V}$ to an element of $\mathrm{pos}(w)$, each second-order variable in $\mathcal{V}$ to a subset of $\mathrm{pos}(w)$, and B to an $\eta(w)$-behavior over Ω. We let $\Phi_{(\mathcal{V},\eta),w}$ denote the set of all $(\mathcal{V}, \eta)$-assignments for w. In the usual way we define updates of $(\mathcal{V}, \eta)$-assignments. Let $\sigma \in \Phi_{(\mathcal{V},\eta),w}$ and $i \in \mathrm{pos}(w)$. By $\sigma[x \mapsto i]$ we denote the $(\mathcal{V} \cup \{x\}, \eta)$-assignment for w that agrees with σ on $\mathcal{V} \backslash \{x\}$ and that satisfies $\sigma[x \mapsto i](x) = i$. Similarly, we define the updates $\sigma[X \mapsto I]$ and $\sigma[B \mapsto b]$ for each set $I \subseteq \mathrm{pos}(w)$ and each behavior $b \in \mathrm{B}(\Omega, \eta(w))$, respectively.

Extending the usual technique we encode a pair (w, σ), where $w \in D^*$ and $\sigma \in \Phi_{(\mathcal{V},\eta),w}$, as a word over an extended set as follows. For each finite set $\mathcal{V}$ of variables with $B \in \mathcal{V}$ we let

$$D_{\mathcal{V}} = D \times \mathcal{P}(\mathrm{fo}(\mathcal{V}) \cup \mathrm{so}(\mathcal{V})) \times \Omega$$

where $\mathrm{fo}(\mathcal{V})$ and $\mathrm{so}(\mathcal{V})$ are the subsets of all first-order and second-order variables occurring in $\mathcal{V}$, respectively. Let $\zeta = \zeta_1 \dots \zeta_n \in D_{\mathcal{V}}^*$. We call ζ fo-*valid* if for each $x \in \mathrm{fo}(\mathcal{V})$ there is a unique $i \in \mathrm{pos}(\zeta)$ such that x occurs in the second component of ζ_i. We denote the set of all fo-valid words over $D_{\mathcal{V}}$ by $D_{\mathcal{V}}^{*\mathrm{fo}}$. Moreover, we call ζ η-*valid* if the word $(\zeta_1)_3 \dots (\zeta_n)_3$ is an $\eta((\zeta_1)_1 \dots (\zeta_n)_1)$-behavior over Ω. We denote the set of all η-valid words over $D_{\mathcal{V}}$ by $D_{\mathcal{V}}^{*\eta}$.

It is clear that, for each finite set $\mathcal{V}$ of variables with $B \in \mathcal{V}$ and relabeling $\eta\colon D \to M$, there is a one-to-one correspondence between the set $\{(w, \sigma) \mid w \in D^*, \sigma \in \Phi_{(\mathcal{V},\eta),w}\}$ and the set $D_{\mathcal{V}}^{*\mathrm{fo}} \cap D_{\mathcal{V}}^{*\eta}$. Thus, as usual, we will not distinguish between the pair (w, σ) and the corresponding word $\zeta \in D_{\mathcal{V}}^{*\mathrm{fo}} \cap D_{\mathcal{V}}^{*\eta}$.

Lemma 11. Let D be a set and let $\mathcal{V}$ be a finite set of variables with $B \in \mathcal{V}$. Then $D_{\mathcal{V}}^{*\mathrm{fo}}$ is $D_{\mathcal{V}}$-recognizable.

Let $\varphi \in \mathrm{MSO}(\Omega, \Pi)$ and $\mathcal{V}$ be a finite set of variables such that $\mathrm{Free}(\varphi) \subseteq \mathcal{V}$ and $B \in \mathcal{V}$. Moreover, let $\eta\colon D \to M$ be a relabeling. For every $(w, \sigma) \in D_{\mathcal{V}}^{*\mathrm{fo}} \cap D_{\mathcal{V}}^{*\eta}$ we define the relation $(w, \sigma) \models \varphi$ by extending the usual models operator of classical MSO-logic as shown for atoms in Fig. 2. Then we define the set of models of φ as the set

$$L_{\mathcal{V},\eta}(\varphi) = \{(w, \sigma) \mid w \in D^*,\ \sigma \in \Phi_{(\mathcal{V},\eta),w}, (w, \sigma) \models \varphi\}.$$

Thus, $L_{\mathcal{V},\eta}(\varphi) \subseteq D_{\mathcal{V}}^{*\mathrm{fo}} \cap D_{\mathcal{V}}^{*\eta}$.

Lemma 12. Let D be a set and Π a label structure over D, let Ω be a finite subset of $P \times F$, and let $\eta\colon D \to M$ be a relabeling. For each $\varphi \in \mathrm{MSO}(\Omega, \Pi)$ and each finite set $\mathcal{V} \supseteq \mathrm{Free}(\varphi)$ of variables with $\mathcal{V} \cap \mathrm{Bound}(\varphi) = \emptyset$ and $B \in \mathcal{V}$ there is a $D_{\mathcal{V}}$-recognizable language L such that $L_{\mathcal{V},\eta}(\varphi) = L \cap D_{\mathcal{V}}^{*\eta}$.

$$
\begin{array}{rcl}
(w,\sigma) \models P_\pi(x) \text{ is true} & \iff & w_{\sigma(x)} \in \pi \\
(w,\sigma) \models \mathrm{next}(x,y) \text{ is true} & \iff & \sigma(x)+1 = \sigma(y) \\
(w,\sigma) \models (x \in X) \text{ is true} & \iff & \sigma(x) \in \sigma(X) \\
(w,\sigma) \models (B(x) = (p,f)) \text{ is true} & \iff & \sigma(B)_{\sigma(x)} = (p,f).
\end{array}
$$

Fig. 2. Models operator for atoms.

Weighted Symbolic MSO-Logic. Here we introduce our new weighted MSO-logic over data storage types. This logic extends the one in [20, Definition 5] from a finite set Σ to an arbitrary set D.

Let D be a set, Π a label structure over D, and Ω a finite subset of $P \times F$. We define the set $\mathrm{BExp}(\Omega, \Pi, K)$ of *B-expressions over* (Ω, Π, K) to be the set generated by the EBNF:

$$
e ::= \mathrm{Val}_\kappa \mid (e+e) \mid (\varphi \rhd e) \mid \sum\nolimits_x e \mid \sum\nolimits_X e,
$$

where $\kappa\colon D_{\mathcal{U}} \to K$ is a relabeling for some finite set $\mathcal{U}$ of variables with $B \in \mathcal{U}$, and $\varphi \in \mathrm{MSO}(\Omega, \Pi)$. As in the unweighted case the sets $\mathrm{Free}(e)$ and $\mathrm{Bound}(e)$ for each B-expression e are defined as usual where we set $\mathrm{Free}(\mathrm{Val}_\kappa) = \mathcal{U}$.

We define the set $\mathrm{Exp}(\Omega, \Pi, K)$ of *MSO-expressions over* (Ω, Π, K) as the set of all expressions of the form $\sum_B^\eta e$ with $e \in \mathrm{BExp}(\Omega, \Pi, K)$, $\mathrm{Free}(e) = \{B\}$, and relabeling $\eta\colon D \to M$. An *MSO-expression over* (S, D, K) is an MSO-expression over (Ω, Π, K) for some finite $\Omega \subseteq P \times F$ and label structure Π over D.

$$
\begin{aligned}
[\![\mathrm{Val}_\kappa]\!]_{\mathcal{V},\eta}(\zeta) &= \mathrm{val}(\kappa(\zeta_{\mathcal{U}})) \text{ where } \zeta_{\mathcal{U}} \text{ is obtained from } \zeta \text{ by replacing each} \\
&\qquad \text{symbol } (a, V, \omega) \text{ by } (a, V \cap (\mathrm{fo}(\mathcal{U}) \cup \mathrm{so}(\mathcal{U})), \omega) \\
[\![e_1 + e_2]\!]_{\mathcal{V},\eta}(\zeta) &= [\![e_1]\!]_{\mathcal{V},\eta}(\zeta) + [\![e_2]\!]_{\mathcal{V},\eta}(\zeta) \\
[\![\varphi \rhd e]\!]_{\mathcal{V},\eta}(\zeta) &= [\![e]\!]_{\mathcal{V},\eta}(\zeta), \text{if } \zeta \in \mathcal{L}_{\mathcal{V},\eta}(\varphi), \text{ and } 0 \text{ otherwise} \\
[\![\textstyle\sum_x e]\!]_{\mathcal{V},\eta}(\zeta) &= \sum_{i \in \mathrm{pos}(\zeta)} [\![e]\!]_{\mathcal{V}\cup\{x\},\eta}(w, \sigma[x \mapsto i]) \\
[\![\textstyle\sum_X e]\!]_{\mathcal{V},\eta}(\zeta) &= \sum_{I \subseteq \mathrm{pos}(\zeta)} [\![e]\!]_{\mathcal{V}\cup\{X\},\eta}(w, \sigma[X \mapsto I])
\end{aligned}
$$

Fig. 3. Semantics of B-expressions (also cf. [20, Definition 6]).

Let $e \in \mathrm{BExp}(\Omega, \Pi, K)$, $\mathcal{V}$ be a finite set of variables containing $\mathrm{Free}(e)$, and $\eta\colon D \to M$ be a relabeling. The *semantics of e with respect to $\mathcal{V}$* and η is the weighted language $[\![e]\!]_{\mathcal{V},\eta}\colon D_{\mathcal{V}}^* \to K$ such that $\mathrm{supp}([\![e]\!]_{\mathcal{V},\eta}) \subseteq D_{\mathcal{V}}^{*\mathrm{fo}} \cap D_{\mathcal{V}}^{*\eta}$ and for each $\zeta = (w, \sigma) \in D_{\mathcal{V}}^{*\mathrm{fo}} \cap D_{\mathcal{V}}^{*\eta}$ we define $[\![e]\!]_{\mathcal{V},\eta}(\zeta)$ inductively as shown in Fig. 3. Let $e = \sum_B^\eta e'$ be an MSO-expression over (Ω, Π, K). We define the weighted language $[\![e]\!]\colon D^* \to K$ for each $w \in D^*$ by:

$$
[\![\textstyle\sum_B^\eta e']\!](w) = \sum\nolimits_{b \in \mathrm{B}(\Omega, \eta(w))} [\![e']\!]_{\{B\},\eta}(w, [B \mapsto b]).
$$

We say that a weighted language $r\colon D^* \to K$ is *definable by an MSO-expression over* (S, D, K) if there is an MSO-expression e over (S, D, K) such that $r = [\![e]\!]$.

From Automata to Logic. The proof of the claim that recognizability implies definability follows the standard construction idea and is exactly the same as the proof of Lemma 9 of [20] where D is a finite set (there denoted by Σ), except that (1) the atomic formula $P_a(x)$ (for $a \in \Sigma$) in ψ_1 has to be replaced by $P_\pi(x)$ and (2) $\kappa((d, V, \omega)) = \mathrm{wt}(\tau, d)$ if $V = \{X_\tau\}$ and 0 otherwise, where wt is the weight function of the given automaton.

Theorem 13. *Let $r\colon D^* \to K$. If r is (S, D, K)-recognizable, then r is definable by some MSO-expression over (S, D, K).*

From Logic to Automata. We can prove the following lemma by induction on the structure of the B-expression e. The proof of the cases are easy generalizations of Lemmas 11–14 of [20].

Lemma 14. Let $e \in \mathrm{BExp}(\Omega, \Pi, K)$ and $\mathcal{V} \supseteq \mathrm{Free}(e)$ a finite set of variables with $\mathcal{V} \cap \mathrm{Bound}(\varphi) = \emptyset$ and $B \in \mathcal{V}$. Moreover, let $\eta\colon D \to M$ be a relabeling. There is a $(D_\mathcal{V}, K)$-recognizable weighted language r such that $[\![e]\!]_{\mathcal{V},\eta} = r \cap D_\mathcal{V}^{*\eta}$.

Due to the symbolicness of the automata, the next lemma is slightly more complicated to prove than the corresponding Lemma 15 of [20].

Lemma 15. Let $e \in \mathrm{BExp}(\Omega, \Pi, K)$ with $\mathrm{Free}(e) = \{B\}$ and $\eta\colon D \to M$ be a relabeling. If $[\![e]\!]_{\{B\},\eta} = r \cap D_{\{B\}}^{*\eta}$ for some $(D_{\{B\}}, K)$-recognizable weighted language r, then $[\![\sum_B^\eta e]\!]$ is an (S, D, K)-recognizable weighted language.

Proof. Let $\mathcal{A} = (Q, \Pi, Q_0, Q_f, T, \mathrm{wt})$ be a $(D_{\{B\}}, K)$-automaton such that $[\![e]\!]_{\{B\},\eta} = [\![\mathcal{A}]\!] \cap D_{\{B\}}^{*\eta}$. We will construct the (S, D, K)-automaton $\mathcal{A}' = (Q', \mathrm{BC}(\Pi'), Q_0', Q_f, T', \mathrm{wt}', \eta)$ such that $[\![\mathcal{A}']\!] = [\![\sum_B^\eta e]\!]$, using the following idea. Since each predicate $\pi \subseteq D \times \{\emptyset\} \times \Omega$ occurring in T combines elements from D and Ω, we have to keep these combinations also in $\mathcal{A}'$. For this, we partition Π into a family $\Pi \times \Omega$ and keep track of elements of Ω in the state set of $\mathcal{A}'$.

Formally, we let $Q' = (Q \times \Omega) \cup Q_f$ and $Q_0' = Q_0 \times \Omega$. We let $\Pi' = \Pi \times \Omega$ such that for every $(\pi, (p, f)) \in \Pi'$ we have $(\pi, (p, f)) = \{d \mid (d, \emptyset, (p, f)) \in \pi\}$. If $\tau = (q, \pi, q')$ is in T, then for every $(p, f), (p', f') \in \Omega$ we let $\tau' = ((q, (p, f)), (\pi, (p, f)), p, (q', (p', f')), f)$ be in T'. Moreover, if $q' \in Q_f$, then also $\tau' = ((q, (p, f)), (\pi, (p, f)), p, q', f)$ is in T'. For every $\tau' \in T'$ and $d \in D$ we define $\mathrm{wt}'(\tau', d) = \mathrm{wt}(\tau, (d, \emptyset, (p, f)))$ if $(\tau')_2 = (\pi, (p, f))$ and $d \in (\pi, (p, f))$, and $\mathrm{wt}'(\tau', d) = 0$ otherwise (where τ is the transition from which τ' was constructed). It is not difficult to prove that $[\![\mathcal{A}']\!] = [\![\sum_B^\eta e]\!]$. □

Using Lemma 14 for $\mathcal{V} = \{B\}$ and Lemma 15 we obtain the following theorem.

Theorem 16. *Let $r\colon D^* \to K$. If r is definable by some MSO-expression over (S, D, K), then r is (S, D, K)-recognizable.*

Decidability Result. Based on the method introduced by Kirsten in [12] and the *zero generation problem* (ZGP), we can prove the following theorem (also cf. [20, Theorem 17]), where a strong bimonoid $(K, +, \cdot, 0, 1)$ is *commutative* if $\cdot$ is so.

Theorem 17 (cf. [12, Theorem 1]). *Let K be a zero-sum-free commutative strong bimonoid. Then, for each (S, D, K)-recognizable weighted language r, the support* $\mathrm{supp}(r)$ *is $(S, D, \mathbb{B})$-recognizable. Moreover, if $(K, \cdot, 1)$ has a decidable ZGP, then there is an effective construction of an $(S, D, \mathbb{B})$-automaton recognizing* $\mathrm{supp}(\llbracket\mathcal{A}\rrbracket)$ *from any given (S, D, K)-automaton $\mathcal{A}$.*

In particular, each bounded lattice satisfies the conditions of Theorem 17. An MSO-expression e over (S, D, K) is *satisfiable* if $\mathrm{supp}(r) \neq \emptyset$.

Corollary 18. *Let N be a nested set, K a zero-sum-free commutative strong bimonoid with a decidable ZGP, and r a projectively* $(\mathrm{VP}(N), N, K)$*-recognizable weighted language. (1) It is decidable whether* $\mathrm{supp}(r) = \emptyset$*. (2) The satisfiability problem of each MSO-expression over* $(\mathrm{VP}(N), N, K)$ *is decidable.*

Proof. (1) follows from Theorem 17, Theorem 9, and the decidability of the emptiness problem of symbolic visibly pushdown automata [1, Theorem 4]. (2) follows from Theorem 16 and (1).

References

1. D'Antoni, L., Alur, R.: Symbolic Visibly Pushdown Automata. In: Biere, A., Bloem, R. (eds.) CAV 2014. LNCS, vol. 8559, pp. 209–225. Springer, Heidelberg (2014)
2. Droste, M., Gastin, P.: Weighted Automata and Weighted Logics. In: Caires, L., Italiano, G.F., Monteiro, L., Palamidessi, C., Yung, M. (eds.) ICALP 2005. LNCS, vol. 3580, pp. 513–525. Springer, Heidelberg (2005)
3. Droste, M., Kuich, W., Vogler, H. (eds.): Handbook of Weighted Automata. EATCS Monographs in Theoretical Computer Science. Springer, Heidelberg (2009)
4. Droste, M., Meinecke, I.: Describing Average- and Longtime-Behavior by Weighted MSO Logics. In: Hliněný, P., Kučera, A. (eds.) MFCS 2010. LNCS, vol. 6281, pp. 537–548. Springer, Heidelberg (2010)
5. Droste, M., Stüber, T., Vogler, H.: Weighted finite automata over strong bimonoids. Inf. Sci. **180**(1), 156–166 (2010)
6. Droste, M., Vogler, H.: The Chomsky-Schützenberger Theorem for Quantitative Context-Free Languages. In: Béal, M.-P., Carton, O. (eds.) DLT 2013. LNCS, vol. 7907, pp. 203–214. Springer, Heidelberg (2013)
7. Droste, M., Perevoshchikov, V.: A Nivat Theorem for Weighted Timed Automata and Weighted Relative Distance Logic. In: Esparza, J., Fraigniaud, P., Husfeldt, T., Koutsoupias, E. (eds.) ICALP 2014, Part II. LNCS, vol. 8573, pp. 171–182. Springer, Heidelberg (2014)
8. Eilenberg, S.: Automata, Languages, and Machines. Pure and Applied Mathematics, vol. 59. Academic Press, New York (1974)
9. Engelfriet, J.: Context-free grammars with storage. Technical report 86–11, University of Leiden (1986), see also: arXiv:1408.0683 [cs.FL] (2014)

10. Fülöp, Z., Stüber, T., Vogler, H.: A Büchi-like theorem for weighted tree automata over multioperator monoids. Theor. Comput. Syst. **50**(2), 241–278 (2012)
11. Herrmann, L., Vogler, H.: A Chomsky-Schützenberger theorem for weighted automata with storage. In: Maletti, A. (ed.) CAI 2015. LNCS, vol. 9270, pp. 115–127. Springer, Switzerland (2015)
12. Kirsten, D.: The support of a recognizable series over a zero-sum free, commutative semiring is recognizable. Acta Cybern. **20**(2), 211–221 (2011)
13. Kuich, W., Salomaa, A.: Semirings, Automata, Languages. EATCS Monographs on Theoretical Computer Science, vol. 5. Springer, Heidelberg (1986)
14. Quaas, K.: MSO logics for weighted timed automata. Form. Methods Syst. Des. **38**(3), 193–222 (2011)
15. Sakarovitch, J.: Elements of Automata Theory. Cambridge University Press, Cambridge (2009)
16. Scott, D.: Some definitional suggestions for automata theory. J. Comput. Syst. Sci. **1**, 187–212 (1967)
17. Veanes, M.: Applications of Symbolic Finite Automata. In: Konstantinidis, S. (ed.) CIAA 2013. LNCS, vol. 7982, pp. 16–23. Springer, Heidelberg (2013)
18. Veanes, M., Bjørner, N.: Symbolic tree automata. Inf. Process. Lett. **115**(3), 418–424 (2015)
19. Veanes, M., Bjørner, N., de Moura, L.: Symbolic Automata Constraint Solving. In: Fermüller, C.G., Voronkov, A. (eds.) LPAR-17. LNCS, vol. 6397, pp. 640–654. Springer, Heidelberg (2010)
20. Vogler, H., Droste, M., Herrmann, L.: A weighted MSO logic with storage behaviour and its Büchi-Elgot-Trakhtenbrot theorem. In: Dediu, A.H., Janoušek, J., Martín-Vide, C., Truthe, B. (eds.) LATA 2016. LNCS, vol. 9618, pp. 127–139. Springer, Switzerland (2016)

On Families of Full Trios Containing Counter Machine Languages

Oscar H. Ibarra[1] and Ian McQuillan[2](✉)

[1] Department of Computer Science, University of California, Santa Barbara, CA 93106, USA
ibarra@cs.ucsb.edu

[2] Department of Computer Science, University of Saskatchewan, Saskatoon, SK S7N 5A9, Canada
mcquillan@cs.usask.ca

Abstract. We look at NFAs augmented with multiple reversal-bounded counters where, during an accepting computation, the behavior of the counters during increasing and decreasing phases is specified by some fixed "pattern". We consider families of languages defined by various pattern behaviors and show that some correspond to the smallest full trios containing restricted classes of bounded semilinear languages. For example, one such family is exactly the smallest full trio containing all the bounded semilinear languages. Another family is the smallest full trio containing all the bounded context-free languages. Still another is the smallest full trio containing all bounded languages whose Parikh map is a semilinear set where all periodic vectors have at most two non-zero coordinates. We also examine relationships between the families.

Keywords: Counter machines · Full trios · Semilinearity · Bounded languages

1 Introduction

A language L is bounded if $L \subseteq w_1^* \cdots w_k^*$, for non-empty words $w_1, \ldots, w_k$. Further, L is *bounded semilinear* if there exists a semilinear set $Q \subseteq \mathbb{N}_0^k$ such that $L = \{w \mid w = w_1^{i_1} \cdots w_k^{i_k}, (i_1, \ldots, i_k) \in Q\}$ [10]. It is known that every bounded semilinear language can be accepted by a one-way nondeterministic reversal-bounded multicounter machine (NCM, [9]). Also, every bounded language accepted by an NCM can be accepted by a deterministic reversal-bounded multicounter machine (DCM, [10]). Thus, every bounded semilinear language can be accepted by a DCM.

Recently, several families of languages that are both bounded and semilinear have been defined and studied [7]. The notion of bounded semilinear above is

The research of O. H. Ibarra was supported, in part, by NSF Grant CCF-1117708. The research of I. McQuillan was supported, in part, by Natural Sciences and Engineering Research Council of Canada Grant 327486-2010.

S. Brlek and C. Reutenauer (Eds.): DLT 2016, LNCS 9840, pp. 216–228, 2016.
DOI: 10.1007/978-3-662-53132-7_18

referred to as *bounded Ginsburg semilinear* to distinguish from other types. Two other interesting types are: a language $L \subseteq w_1^* \cdots w_k^*$ over alphabet Σ is *bounded Parikh semilinear* if $L = \{w \mid w = w_1^{i_1} \cdots w_k^{i_k}$, the Parikh map of w is in $Q\}$, where Q is a semilinear set with $|\Sigma|$ components; L is *bounded general semilinear* if L is both bounded and semilinear. It was shown that the family of bounded Parikh semilinear languages is a strict subset of the family of bounded Ginsburg semilinear languages, which is a strict subset of the family of bounded general semilinear languages. However, it was shown that in any language family $\mathcal{L}$ that is a semilinear trio (the family only contains semilinear languages, and is closed under λ-free homomorphism, inverse homomorphism, and intersection with regular languages), all bounded languages within $\mathcal{L}$ are bounded Ginsburg semilinear, and can therefore be accepted by machines in NCM and even DCM. This implies that the equality problem, containment problem, and disjointness problem are decidable for bounded languages in $\mathcal{L}$ since they are decidable for DCM. Furthermore, a criterion was developed for testing when the bounded languages within $\mathcal{L}$ and the family accepted by machines in DCM coincide; this occurs if and only if $\mathcal{L}$ contains all distinct-letter-bounded Ginsburg semilinear languages. This was shown to be the case for finite-index ET0L languages [12], and therefore the bounded languages within these families are the same.

In this paper, we attempt to restrict the operation of NCM in order to precisely characterize types of languages that are bounded and semilinear. Indeed, restricting the behavior of NCM can naturally capture several interesting families through the use of so-called *instruction languages*. Informally, a k-counter machine M is said to satisfy instruction language $I \subseteq \{C_1, D_1, \ldots, C_k, D_k\}^*$ if, for every accepting computation of M, replacing each increase of counter i with C_i, and decrease of counter i with D_i, gives a sequence in I. Then, for a family of instruction languages $\mathcal{I}$, $\mathsf{NCM}(\mathcal{I})$ is the family of NCM machines satisfying some $I \in \mathcal{I}$. Several interesting instruction language families are defined and studied. For example, if one considers $\mathsf{BD}_i\mathsf{LB}_d$, the family of instruction languages consisting of bounded increasing instructions followed by letter-bounded decreasing instructions, then we show that the family of languages accepted by machines in $\mathsf{NCM}(\mathsf{BD}_i\mathsf{LB}_d)$ is the smallest full trio containing all bounded Ginsburg semilinear languages (and therefore, the smallest full trio containing all bounded languages from any semilinear trio). It is also possible to characterize exactly the bounded context-free languages with a subfamily of counter languages. Several other families are also defined and compared. For each, characterizations are given such that the families are the smallest full trios containing the languages. Using these characterizations, we are able to give even simpler criteria than those in [7] for testing if the bounded languages within a semilinear full trio coincide with those accepted by machines in DCM. We then give applications to several interesting families, such as the multi-pushdown languages [1], and restricted types of Turing machines, and it is shown that the bounded languages within each are the same as those accepted by machines in DCM. In a future paper, we will examine closure and decision properties of the models.

All proofs in this paper are omitted due to space constraints and appear in a technical report [8].

2 Preliminaries

In this paper, we assume knowledge of automata and formal languages, and refer to [6] for an introduction. Let Σ be a finite alphabet. Then, Σ^* (resp. Σ^+) is the set of all words (non-empty words) over Σ. A word is any $w \in \Sigma^*$, and a language is any $L \subseteq \Sigma^*$. The empty word is denoted by λ. The complement of L with respect to Σ^* is $\overline{L} = \Sigma^* - L$. The shuffle of words $u, v \in \Sigma^*$ is $u \sqcup\!\sqcup v = \{u_1 v_1 \cdots u_n v_n \mid n \geq 1, u = u_1 \cdots u_n, v = v_1 \cdots v_n, u_i, v_i \in \Sigma^*, 1 \leq i \leq n\}$, extended to languages $L_1 \sqcup\!\sqcup L_2 = \{u \sqcup\!\sqcup v \mid u \in L_1, v \in L_2\}$.

A language $L \subseteq \Sigma^*$ is bounded if there exist $w_1, \ldots, w_k \in \Sigma^+$ such that $L \subseteq w_1^* \cdots w_k^*$, and is letter-bounded if $w_1, \ldots, w_k$ are letters. Furthermore, L is distinct-letter-bounded if each letter is distinct.

Let $\mathbb{N}$ be the set of positive integers and $\mathbb{N}_0 = \mathbb{N} \cup \{0\}$. A *linear set* is a set $Q \subseteq \mathbb{N}_0^m$ if there exist $\boldsymbol{v_0}, \boldsymbol{v_1}, \ldots, \boldsymbol{v_n}$ such that $Q = \{\boldsymbol{v_0} + i_1 \boldsymbol{v_1} + \cdots + i_n \boldsymbol{v_n} \mid i_1, \ldots, i_n \in \mathbb{N}_0\}$. The vector $\boldsymbol{v_0}$ is called the *constant*, and $\boldsymbol{v_1}, \ldots, \boldsymbol{v_n}$ are the *periods*. A *semilinear set* is a finite union of linear sets. Given an alphabet $\Sigma = \{a_1, \ldots, a_m\}$, the length of a word $w \in \Sigma^*$ is denoted by $|w|$. And, given $a \in \Sigma$, $|w|_a$ is the number of a's in w. Then, the *Parikh map* of w is $\psi(w) = (|w|_{a_1}, \ldots, |w|_{a_m})$, and the Parikh map of a language L is $\psi(L) = \{\psi(w) \mid w \in L\}$. Also, $\mathrm{alph}(w) = \{a \in \Sigma \mid |w|_a > 0\}$. We refer to Sect. 1 for the definitions of bounded Ginsburg semilinear and bounded Parikh semilinear languages.

For a class of machines $\mathcal{M}$, we let $\mathcal{L}(\mathcal{M})$ be the family of languages accepted by machines in $\mathcal{M}$. Let $\mathcal{L}(\mathsf{CFL})$ be the family of context-free languages. A *trio* (resp. *full trio*) is any family of languages closed under λ-free homomorphism (resp. homomorphism), inverse homomorphism, and intersection with regular languages. A *full semi-AFL* is a full trio closed under union [2]. Many well-known families of languages are trios, such as every family of the Chomsky hierarchy. Many important families are full trios and full semi-AFLs as well, such as the families of regular and context-free languages. Given a language family $\mathcal{L}$, $\mathcal{L}^{\mathrm{bd}}$ are the bounded languages in $\mathcal{L}$.

We only define one-way k-counter machines informally and refer to [8,9] for formal definitions. These machines are similar to pushdown automata, with k independent pushdowns that each have one symbol plus an end-marker. A configuration is a tuple $(q, w, i_1, \ldots, i_k)$ where q is the current state, w is the remaining input, and $i_1, \ldots, i_k \in \mathbb{N}_0$ are the contents of the k counters. The derivation relation $\vdash_M$ and its reflexive, transitive closure $\vdash_M^*$ are defined in the usual way [8]. The language accepted by M is denoted by $L(M)$.

Further, M is l-reversal-bounded if, in every accepting computation, the counter alternates between increasing and decreasing at most l times. We will often associate labels from an alphabet T to the transitions of M bijectively, and then write $\vdash_M^t$ to represent the changing of configurations via transition t. This is generalized to derivations over words in T^*.

Then $\mathsf{NCM}(k,l)$ is the class of one-way l-reversal-bounded k-counter machines, and NCM is all reversal-bounded multicounter languages, and replacing N with D gives the deterministic variant.

3 Instruction NCM Machines

It is known that all of the bounded languages in every semilinear trio are in $\mathcal{L}(\mathsf{NCM})$ [7]. We start this section by considering subclasses of $\mathcal{L}(\mathsf{NCM})$ in order to determine more restricted methods of computation where this property also holds. We are able to do this optimally. Furthermore, characterizations of the restricted families are also possible, and lead to even simpler methods to determine the bounded languages within semilinear full trios.

First, we define restrictions of NCM depending on the sequences of counter instructions that occur. These restrictions will only be defined on NCMs that we will call *well-formed*. A k-counter NCM M is well-formed if $M \in \mathsf{NCM}(k,1)$ whereby all transitions change at most one counter value per transition, and all counters decrease to zero before accepting. Indeed, an NCM (or DCM) can be assumed without loss of generality to be 1-reversal-bounded by increasing the number of counters [9]. It is also clear that all counters can be forced to change one counter value at a time, and decrease to zero before accepting without loss of generality. Thus, every language in $\mathcal{L}(\mathsf{NCM})$ can be accepted by a well-formed NCM. Let Δ be an infinite set of new symbols, $\Delta = \{C_1, D_1, C_2, D_2, \ldots\}$, and for $k \geq 1, \Delta_k = \{C_1, D_1, \ldots, C_k, D_k\}, \Delta_{(k,c)} = \{C_1, \ldots, C_k\}, \Delta_{(k,d)} = \{D_1, \ldots, D_k\}$.

Given a well-formed k-counter NCM machine M, let T be a set of labels in bijective correspondence with transitions of M. Then, define a homomorphism h_Δ from T^* to Δ_k that maps every transition label associated with a transition that increases counter i to C_i, maps every label associated with a transition that decreases counter i to D_i, and maps all labels associated with transitions that do not change any counter to λ. Also, define a homomorphism h_Σ that maps every transition that reads a letter $a \in \Sigma$ to a, and erases all others. Then, we say that M *satisfies instruction language* $I \subseteq \Delta_k^*$ if every sequence of transitions $\alpha \in T^*$ corresponding to an accepting computation — that is $(q_0, w, 0, \ldots, 0) \vdash_M^\alpha (q, \lambda, c_1, \ldots, c_k), q$ a final state — has $h_\Delta(\alpha) \in I$. This means that M satisfies instruction language I if I describes all possible counter increase and decrease instructions that can be performed in an accepting computation by M, with C_i occurring for every increase of counter i by one, and D_i occurring for every decrease of counter i by one.

Given a family of languages $\mathcal{I}$ with each $I \in \mathcal{I}$ over Δ_k, for some $k \geq 1$, let $\mathsf{NCM}(k,\mathcal{I})$ be the subset of well-formed k-counter NCM machines that satisfy I for some $I \in \mathcal{I}$ with $I \subseteq \Delta_k^*$; these are called the k-counter $\mathcal{I}$-instruction machines. The family of languages they accept, $\mathcal{L}(\mathsf{NCM}(k,\mathcal{I}))$, are called the k-counter $\mathcal{I}$-instruction languages. Furthermore, $\mathsf{NCM}(\mathcal{I}) = \bigcup_{k\geq 1} \mathsf{NCM}(k,\mathcal{I})$ (resp. $\mathcal{L}(\mathsf{NCM}(\mathcal{I})) = \bigcup_{k\geq 1} \mathcal{L}(\mathsf{NCM}(k,\mathcal{I}))$) are the $\mathcal{I}$-instruction machines (and languages). We will only consider instruction languages I where, for all $w \in I$, every occurrence of C_i occurs before any occurrence of D_i, for all i, $1 \leq i \leq k$, which is enough since every well-formed machine is 1-reversal-bounded.

First, we will study properties of these restrictions before examining some specific types.

Proposition 1. *Given any family of languages $\mathcal{I}$ over Δ_k, $\mathcal{L}(\mathsf{NCM}(k,\mathcal{I}))$ is a full trio. Furthermore, given any family of languages $\mathcal{I}$, where each $I \in \mathcal{I}$ is over some $\Delta_k, k \geq 1$, $\mathcal{L}(\mathsf{NCM}(\mathcal{I}))$ is a full trio.*

Next, we require another definition. Given a language I over Δ_k, let

$$I_{eq} = \{w \mid w \in I, |w|_{C_i} = |w|_{D_i}, \text{every } C_i \text{ occurs before any } D_i, \text{ for } 1 \leq i \leq k\}.$$

Further, given a language family $\mathcal{I}$ over Δ where each $I \in \mathcal{I}$ is over Δ_k, for some $k \geq 1$, then $\mathcal{I}_{eq}$ is the family of all languages I_{eq}, where $I \in \mathcal{I}$.

Proposition 2. *Let $\mathcal{I}$ be a family of languages where each $I \in \mathcal{I}$ is a subset of Δ_k^*, for some $k \geq 1$, and $\mathcal{I}$ is a subfamily of the regular languages. Then $\mathcal{L}(\mathsf{NCM}(\mathcal{I}))$ is the smallest full trio containing $\mathcal{I}_{eq}$.*

We will consider several instruction language families that define interesting subfamilies of $\mathcal{L}(\mathsf{NCM})$.

Definition 3. *We define instruction language families:*

- $\mathsf{LB}_i\mathsf{LB}_d = \{I = YZ \mid k \geq 1, Y = a_1^* \cdots a_m^*, a_i \in \Delta_{(k,c)}, 1 \leq i \leq m, Z = b_1^* \cdots b_n^*, b_j \in \Delta_{(k,d)}, 1 \leq j \leq n\}$,
  (letter-bounded-increasing/letter-bounded-decreasing instructions),
- $\mathsf{StLB}_{id} = \{I \mid k \geq 1, I = a_1^* \cdots a_m^*, a_i \in \Delta_k, 1 \leq i \leq m,$ *there is no* $1 \leq l < l' < j < j' \leq m$ *such that* $a_l = C_r, a_{l'} = C_s, a_j = D_r, a_{j'} = D_s, r \neq s\}$,
  (stratified-letter-bounded instructions),
- $\mathsf{LB}_{id} = \{I \mid k \geq 1, I = a_1^* \cdots a_m^*, a_i \in \Delta_k, 1 \leq i \leq m\}$,
  (letter-bounded instructions),
- $\mathsf{BD}_i\mathsf{LB}_d = \{I = YZ \mid k \geq 1, Y = w_1^* \cdots w_m^*, w_i \in \Delta_{(k,c)}^*, 1 \leq i \leq m, Z = a_1^* \cdots a_n^*, a_j \in \Delta_{(k,d)}, 1 \leq j \leq n\}$,
  (bounded-increasing/letter-bounded-decreasing instructions),
- $\mathsf{LB}_i\mathsf{BD}_d = \{I = YZ \mid k \geq 1, Y = a_1^* \cdots a_m^*, a_i \in \Delta_{(k,c)}, 1 \leq i \leq m, Z = w_1^* \cdots w_n^*, w_j \in \Delta_{(k,d)}^*, 1 \leq j \leq n\}$,
  (letter-bounded-increasing/bounded-decreasing instructions),
- $\mathsf{BD}_{id} = \{I \mid k \geq 1, I = w_1^* \cdots w_m^*, w_i \in \Delta_k^*, 1 \leq i \leq m\}$,
  (bounded instructions),
- $\mathsf{LB}_d = \{I \mid k \geq 1, I = Y \sqcup\!\sqcup Z, Y = \Delta_{(k,c)}^*, Z = a_1^* \cdots a_n^*, a_j \in \Delta_{(k,d)}, 1 \leq j \leq n\}$, *(letter-bounded-decreasing instructions)*,
- $\mathsf{LB}_i = \{I \mid k \geq 1, I = Y \sqcup\!\sqcup Z, Y = a_1^* \cdots a_m^*, a_i \in \Delta_{(k,c)}, 1 \leq i \leq m, Z = \Delta_{(k,d)}^*, \}$, *(letter-bounded increasing instructions)*,
- $\mathsf{LB}_\cup = \mathsf{LB}_d \cup \mathsf{LB}_i$,
  (either letter-bounded-decreasing or letter-bounded-increasing instructions),
- $\mathsf{ALL} = \{I \mid k \geq 1, I = \Delta_k^*\}$.

For example, every NCM machine M where the counters are increased and decreased according to some bounded language, then there is an instruction language I such that M satisfies I, and $I \in \mathsf{BD}_{id}$, and $L(M) \in \mathcal{L}(\mathsf{NCM}(\mathsf{BD}_{id}))$. Even though not all instructions in I are necessarily used, the instructions used will be a subset of I since the instructions used are a subset of a bounded language. It is also clear that $\mathcal{L}(\mathsf{NCM}) = \mathcal{L}(\mathsf{NCM}(\mathsf{ALL}))$.

Example 4. Let $L = \{ua^i vb^j wa^i xb^j y \mid i, j > 0, u, v, w, x, y \in \{0,1\}^*\}$. We can easily construct a well-formed 2-counter machine M to accept L where, on input $ua^i vb^j wa^{i'} xb^{j'} y$, M increases counter 1 i times, then increases counter 2 j times, then decreases counter 1 verifying that $i = i'$, then decreases counter 2 verifying that $j = j'$. This machine satisfies instruction language $C_1^* C_2^* D_1^* D_2^*$, which is a subset of some instruction language in every family in Definition 3 except for StLB_{id}, and therefore $L \in \mathcal{L}(\mathsf{NCM}(\mathcal{I}))$ for each of these families $\mathcal{I}$.

Example 5. Let $L = \{a^{2+i+2j} b^{3+2i+5j} \mid i, j \geq 0\}$. Note that the Parikh map of L is a linear set $Q = \{(2,3) + (1,2)i + (2,5)j \mid i, j \geq 0\}$. L can be accepted by a well-formed 4-counter NCM (see [8] for construction).

The instructions of M as constructed are a subset of $I = (C_1C_2)^*(C_3C_4)^* D_1^* D_3^* D_2^* D_4^*$. This is a subset of some language in each of $\mathsf{BD}_i\mathsf{LB}_d, \mathsf{BD}_{id}, \mathsf{LB}_d$ but not the other families, and therefore M is in each of $\mathsf{NCM}(\mathsf{BD}_i\mathsf{LB}_d), \mathsf{NCM}(\mathsf{BD}_{id}), \mathsf{NCM}(\mathsf{LB}_d)$. Even though M is not in the other classes of machines such as $\mathsf{NCM}(\mathsf{LB}_i)$, it is possible for $L(M)$ to be in $\mathcal{L}(\mathsf{NCM}(\mathsf{LB}_i))$ (using some other machine that accepts the same language). Indeed, we will see that $L(M)$ is also in $\mathcal{L}(\mathsf{NCM}(\mathsf{LB}_i\mathsf{BD}_d))$ and $\mathcal{L}(\mathsf{NCM}(\mathsf{LB}_i))$.

Example 6. Let $L_1 = \{w\#a^i b^j \mid |w|_a = i, |w|_b = j\}$. This can indeed be accepted by a machine $M_1 \in \mathsf{NCM}(\mathsf{LB}_d)$ (see [8] for construction).

Example 7. Let $L = \{w \mid w \in \{a,b\}^+, |w|_a = |w|_b > 0\}$. L can be accepted by an NCM which uses two counters that increments counter 1 (resp. counter 2) whenever it sees an a (resp., b). Then it decrements counter 1 and counter 2 simultaneously and accepts if they reach zero at the same time. This counter usage does not have a pattern in any of the restrictions above. It is quite unlikely that $L(M) \in \mathcal{L}(\mathsf{NCM}(\mathcal{I}))$ for any of the families in the definition above except the full $\mathcal{L}(\mathsf{NCM}(\mathsf{ALL})) = \mathcal{L}(\mathsf{NCM})$.

Every family $\mathcal{I}$ in Definition 3 is a subfamily of the regular languages. Therefore, by Proposition 2, the following can be shown by proving closure under union:

Proposition 8. *Let $\mathcal{I}$ be any family of instruction languages from Definition 3. Then $\mathcal{L}(\mathsf{NCM}(\mathcal{I}))$ is the smallest full trio (and full semi-AFL) containing $\mathcal{I}_{eq}$.*

As a corollary, if we consider the instructions languages of ALL (thus, the instructions are totally arbitrary), and for $i \geq 1$, let $L_i = \{C_i^n D_i^n \mid n \geq 0\}$, then $\mathsf{ALL}_{eq} = \{I \mid I = L_1 \sqcup\!\sqcup L_2 \sqcup\!\sqcup \cdots \sqcup\!\sqcup L_k, k \geq 1\}$. Hence, $\mathcal{L}(\mathsf{NCM})$ can be characterized as the smallest full trio containing ALL_{eq}, by Proposition 2. Or, it could be stated as follows (this is essentially already known, and follows from work in [4,5]).

Corollary 9. [4,5] $\mathcal{L}(\mathsf{NCM})$ is the smallest shuffle or intersection closed full trio containing $\{a^n b^n \mid n \geq 0\}$.

Indeed, it is known that $\mathcal{L}(\mathsf{NCM})$ is shuffle and intersection closed full trio [9]. For intersection, this follows since each instruction language I above can be represented by taking each L_i, and a homomorphism h_i that maps C_i and D_i to itself, and erases all other letters of Δ_k. Then let $L_i' = h_i^{-1}(L_i)$. Then, $L_1 ⧢ L_2 ⧢ \cdots ⧢ L_k = L_1' \cap L_2' \cap \cdots \cap L_k'$.

Since $\{a^n b^n \mid n \geq 0\}$ is in $\mathcal{L}(\mathsf{NCM}(\mathcal{I}))$ for all $\mathcal{I}$ in Definition 3, the following is also immediate from Corollary 9:

Corollary 10. *For all $\mathcal{I}$ in Definition 3, $\mathcal{L}(\mathsf{NCM})$ is the smallest shuffle or intersection closed full trio containing $\mathcal{L}(\mathsf{NCM}(\mathcal{I}))$.*

Thus, any instruction family $\mathcal{I}$ whereby $\mathcal{L}(\mathsf{NCM}(\mathcal{I})) \subsetneq \mathcal{L}(\mathsf{NCM})$ and $\{a^n b^n \mid n \geq 0\} \in \mathcal{L}(\mathsf{NCM}(\mathcal{I}))$ is immediately not closed under intersection and shuffle.

Next, we will prove the following lemma regarding many of the instruction language families showing that letter-bounded instructions can be assumed to be distinct-letter-bounded, and for bounded languages, for each letter in Δ_k in the words to only appear once.

First, we need a definition. For each of the instruction families of Definition 3, we place an underline below each LB if the letter-bounded language is forced to have each letter occur exactly once (and therefore be distinct-letter-bounded), and we place an underline below BD if each letter $a \in \Delta_k$ appears exactly once within the words $w_1, \ldots, w_m$. Thus, as an example, $\underline{\mathsf{LB}}_i\underline{\mathsf{BD}}_d$ is the subset of $\mathsf{LB}_i\mathsf{BD}_d$ equal to $\{I = YZ \mid k \geq 1, Y = a_1^* \cdots a_k^*, a_i \in \Delta_{(k,c)}, |a_1 \cdots a_k|_a = 1$, for all $a \in \Delta_{(k,c)}, Z = w_1^* \cdots w_n^*, w_i \in \Delta_{(k,d)}^*, 1 \leq j \leq n, |w_1 w_2 \cdots w_n|_a = 1$, for all $a \in \Delta_{(k,d)}\}$. Thus, each letter appears exactly once in the words or letters. The construction uses multiple new instruction letters and counters, in order to allow each letter to only appear once.

Lemma 11. *The following are true:*

$$\mathcal{L}(\mathsf{NCM}(\mathsf{LB}_i\mathsf{LB}_d)) = \mathcal{L}(\mathsf{NCM}(\underline{\mathsf{LB}}_i\underline{\mathsf{LB}}_d)),\ \mathcal{L}(\mathsf{NCM}(\mathsf{LB}_{id})) = \mathcal{L}(\mathsf{NCM}(\underline{\mathsf{LB}}_{id})),$$
$$\mathcal{L}(\mathsf{NCM}(\mathsf{LB}_i\mathsf{BD}_d)) = \mathcal{L}(\mathsf{NCM}(\underline{\mathsf{LB}}_i\underline{\mathsf{BD}}_d)),\ \mathcal{L}(\mathsf{NCM}(\mathsf{LB}_d)) = \mathcal{L}(\mathsf{NCM}(\underline{\mathsf{LB}}_d)),$$
$$\mathcal{L}(\mathsf{NCM}(\mathsf{BD}_i\mathsf{LB}_d)) = \mathcal{L}(\mathsf{NCM}(\underline{\mathsf{BD}}_i\underline{\mathsf{LB}}_d)),\ \mathcal{L}(\mathsf{NCM}(\mathsf{LB}_i)) = \mathcal{L}(\mathsf{NCM}(\underline{\mathsf{LB}}_i)).$$

The next goal is to separate some families of NCM languages with different instruction languages.

A (quite technical) lemma that is akin to a pumping lemma is proven, but is done entirely on derivations rather than words, so that it can be used twice starting from the same derivation within Proposition 13. Due to the length and technicality, the statement of the lemma and proof can be found in [8].

The next result follows from Lemma 11 and this new pumping lemma.

Proposition 12. $\{a^n b^n c^n \mid n > 0\} \notin \mathcal{L}(\mathsf{NCM}(\mathsf{LB}_{id}))$.

In addition, the following can be shown with Lemma 11 and two applications of the pumping lemma.

Proposition 13. $\{a^n b^n c^l d^l \mid n, l > 0\} \notin \mathcal{L}(\mathsf{NCM}(\mathsf{LB}_i\mathsf{LB}_d))$.

Therefore, the following is immediate:

Proposition 14. $\mathcal{L}(\mathsf{NCM}(\mathsf{LB}_i\mathsf{LB}_d)) \subsetneq \mathcal{L}(\mathsf{NCM}(\mathsf{LB}_{id})) \subsetneq \mathcal{L}(\mathsf{NCM}(\mathsf{BD}_{id}))$.

4 Generators for the Families

We will go through certain families individually while creating a more restricted set of generators than is provided by Proposition 8.

First, we will give two characterizations of $\mathcal{L}(\mathsf{NCM}(\mathsf{LB}_{id}))$.

Proposition 15. *$\mathcal{L}(\mathsf{NCM}(\mathsf{LB}_{id}))$ is the smallest full trio containing all distinct-letter-bounded languages of the form $\{a_1^{i_1} \cdots a_m^{i_m} \mid a_j = C_l, a_n = D_l \text{ imply } i_j = i_n\}$, where $a_1, \ldots, a_m$ is a permutation of Δ_k such that $a_j = C_l, a_n = D_l$ implies $j < n$.*

A similar characterization can be obtained with a single language for each k.

Proposition 16. *Let $k \geq 1$, and let $L_k^{\mathsf{LB}_{id}} = \{a_1^{i_1} a_2^{i_2} \cdots a_m^{i_m} \mid \{a_1, \ldots, a_m\}$ is a permutation of Δ_k, and ($C_j = a_l, D_j = a_n$ implies both $l < n$ and $i_l = i_n$), for each $j, 1 \leq j \leq k\}$.*

Then $\mathcal{L}(\mathsf{NCM}(\mathsf{LB}_{id}))$ is the smallest full trio containing $L_k^{\mathsf{LB}_{id}}$, for each $k \geq 1$.

Next, we will give characterizations for $\mathcal{L}(\mathsf{NCM}(\mathsf{LB}_i\mathsf{LB}_d))$, whose proof is similar to Proposition 15.

Proposition 17. *The family $\mathcal{L}(\mathsf{NCM}(\mathsf{LB}_i\mathsf{LB}_d))$ is the smallest full trio containing all distinct-letter-bounded languages of the form $\{a_1^{l_1} \cdots a_k^{l_k} b_1^{j_1} \cdots b_k^{j_k} \mid a_i = C_m, b_n = D_m \text{ imply } l_i = j_n\}$, where $a_1, \ldots, a_k$ is a permutation of $\Delta_{(k,c)}$ and $b_1, \ldots, b_k$ is a permutation of $\Delta_{(k,d)}$.*

This can similarly be turned into one language for each k, as follows with a proof similar to Proposition 16:

Proposition 18. *Let $k \geq 1$, and let $L_k^{\mathsf{LB}_i\mathsf{LB}_d} = \{a_1^{l_1} \cdots a_k^{l_k} b_1^{j_1} \cdots b_k^{j_k} \mid a_1, \ldots, a_k$ is a permutation of $\Delta_{(k,c)}$, $b_1, \ldots, b_k$ is a permutation of $\Delta_{(k,d)}$, and ($C_m = a_i, D_m = b_n$ implies $l_i = j_n$), for each $j, 1 \leq j \leq k\}$.*

Then $\mathcal{L}(\mathsf{NCM}(\mathsf{LB}_i\mathsf{LB}_d))$ is the smallest full trio containing $L_k^{\mathsf{LB}_i\mathsf{LB}_d}$, for each $k \geq 1$.

Next, we will provide an alternate interesting characterization for both families using properties of semilinear sets. Let $m \geq 1$. A linear set $Q \subseteq \mathbb{N}_0^n, n \geq 1$, is m-bounded if the periodic vectors of Q have at most m non-zero coordinates. (There is no restriction on the constant vector.) A semilinear set Q is m-bounded if it is a finite union of m-bounded linear sets.

Let $L \subseteq a_1^* \cdots a_n^*, a_1, \ldots, a_n \in \Sigma$ be a distinct-letter-bounded language. L is called a *distinct-letter-bounded 2-bounded semilinear language* if there exists a

2-bounded semilinear set Q such that $L = \{a_1^{i_1} \cdots a_n^{i_n} \mid (i_1, \ldots, i_n) \in Q\}$. L is called a *distinct-letter-bounded 2-bounded overlapped semilinear language* if there exists a 2-bounded semilinear set Q with the property that in any of the linear sets comprising Q, there are no periodic vectors v with non-zero coordinates at positions $i < j$, and v' with non-zero coordinates at positions $i' < j'$ such that $1 \leq i < j < i' < j' \leq n$, and $L = \{a_1^{i_1} \cdots a_n^{i_n} \mid (i_1, \ldots, i_n) \in Q\}$. (They overlap in the sense that, for any such Q, v, i, j, v', i', j', then the interval $[i, j]$ must overlap with $[i', j']$.)

Proposition 19.

1. *The family $\mathcal{L}(\mathsf{NCM}(\mathsf{LB}_{id}))$ is the smallest full trio containing all distinct-letter-bounded 2-bounded semilinear languages.*
2. *The family $\mathcal{L}(\mathsf{NCM}(\mathsf{LB}_i\mathsf{LB}_d))$ is the smallest full trio containing all distinct-letter-bounded 2-bounded overlapped semilinear languages.*

Next we will give a characterization of the smallest full trio containing all bounded $\mathcal{L}(\mathsf{CFL})$ languages. For that, we consider instruction family StLB_{id}. An example of an StLB_{id} language (counter behavior) is $C_1^* C_2^* C_3^* D_3^* C_2^* D_2^* C_1^* D_1^*$. But the counter behavior $C_1^* C_2^* C_3^* D_3^* C_2^* D_2^* C_1^* D_2^* C_1^* D_1^*$ is not an StLB_{id} language since C_2 appears, then C_1, then D_2, then D_1, violating the StLB_{id} definition.

The next results show that $\mathcal{L}(\mathsf{NCM}(\mathsf{StLB}_{id}))$ is the smallest full trio containing all bounded context-free languages. It has previously been found that there is no principal full trio (ie. generated by a single language [2]) accepting these languages [11] (this paper does not use the 'principal' notation). Our proof uses a known characterization of distinct-letter-bounded context-free languages (CFLs) from [3].

Proposition 20. *The family $\mathcal{L}(\mathsf{NCM}(\mathsf{StLB}_{id}))$ is the smallest full trio containing all bounded context-free languages.*

From this, the following can be determined:

Corollary 21.

1. $\mathcal{L}(\mathsf{NCM}(\mathsf{StLB}_{id})) \subsetneq \mathcal{L}(\mathsf{NCM}(\mathsf{LB}_{id}))$.
2. $\mathcal{L}(\mathsf{NCM}(\mathsf{StLB}_{id})$ *and* $\mathcal{L}(\mathsf{NCM}(\mathsf{LB}_i\mathsf{LB}_d))$ *are incomparable.*

Next, we will show that all bounded Ginsburg semilinear languages are in two of the language families (and therefore in all larger families).

Lemma 22. *All bounded Ginsburg semilinear languages are in $\mathcal{L}(\mathsf{NCM}(\mathsf{BD}_i\mathsf{LB}_d))$ and in $\mathcal{L}(\mathsf{NCM}(\mathsf{LB}_i\mathsf{BD}_d))$.*

From the definition, it is immediate that if $\mathcal{I} \subseteq \mathcal{I}'$, then $\mathcal{L}(\mathsf{NCM}(\mathcal{I})) \subseteq \mathcal{L}(\mathsf{NCM}(\mathcal{I}'))$. It is clear that all of $\mathsf{LB}_{id}, \mathsf{BD}_i\mathsf{LB}_d, \mathsf{LB}_i\mathsf{BD}_d$ are a subset of BD_{id}. We will show that three of these counter families coincide.

Proposition 23. $\mathcal{L}(\mathsf{NCM}(\mathsf{BD}_i\mathsf{LB}_i)) = \mathcal{L}(\mathsf{NCM}(\mathsf{LB}_i\mathsf{BD}_d)) = \mathcal{L}(\mathsf{NCM}(\mathsf{BD}_{id}))$ *is the smallest full trio containing all bounded Ginsburg semilinear languages, and the smallest full trio containing all bounded Parikh semilinear languages.*

Corollary 24. *For all* $\mathcal{I} \in \{\mathsf{BD}_i\mathsf{LB}_d, \mathsf{LB}_i\mathsf{BD}_d, \mathsf{BD}_{id}, \mathsf{LB}_d, \mathsf{LB}_i, \mathsf{LB}_\cup, \mathsf{ALL}\}$, *then* $\mathcal{L}(\mathsf{NCM}(\mathcal{I}))$ *contains all bounded Ginsburg semilinear languages, and all bounded languages in* $\mathcal{L}(\mathsf{NCM})$.

Next, we establish two simple sets of generators for $\mathcal{L}(\mathsf{NCM}(\mathsf{BD}_{id}))$. These languages will therefore be a simple mechanism to show whether or not a full trio $\mathcal{L}$ contains every bounded Ginsburg semilinear language, and therefore has exactly the same bounded languages as NCM, and has all bounded languages contained in any semilinear trio.

Proposition 25. *For* $k \geq 1$, *let*

$$\begin{aligned} L_k^{\mathsf{BD}_i\mathsf{LB}_d} = \{w_1^{x_1}\cdots w_m^{x_m} D_1^{y_1}\cdots D_k^{y_k} \mid\ & w_j \in \Delta^+_{(k,c)}, x_j > 0, 1 \leq j \leq m, \\ & \textit{for } 1 \leq i \leq k, |w_1 w_2 \cdots w_m|_{C_i} = 1, \\ & (C_i \in \mathrm{alph}(w_j) \textit{ implies } y_i = x_j)\}, \end{aligned}$$

$$\begin{aligned} L_k^{\mathsf{LB}_i\mathsf{BD}_d} = \{C_1^{y_1}\cdots C_k^{y_k} w_1^{x_1}\cdots w_m^{x_m} \mid\ & w_j \in \Delta^+_{(k,d)}, x_j > 0, 1 \leq j \leq m, \\ & \textit{for } 1 \leq i \leq k, |w_1 w_2 \cdots w_m|_{D_i} = 1, \\ & (D_i \in \mathrm{alph}(w_j) \textit{ implies } y_i = x_j)\}. \end{aligned}$$

Then $\mathcal{L}(\mathsf{NCM}(\mathsf{BD}_i\mathsf{LB}_d)) = \mathcal{L}(\mathsf{NCM}(\mathsf{LB}_i\mathsf{BD}_d)) = \mathcal{L}(\mathsf{NCM}(\mathsf{BD}_{id}))$ *is the smallest full trio containing* $L_k^{\mathsf{BD}_i\mathsf{LB}_d}$, *for each* $k \geq 1$, *and also the smallest full trio containing* $L_k^{\mathsf{LB}_i\mathsf{BD}_d}$, *for each* $k \geq 1$.

Then, by Proposition 23, Proposition 25, and [7], the following is true:

Proposition 26. *Let* $\mathcal{L}$ *be a full trio. Then, the following are equivalent:*

- $\mathcal{L}$ *contains all bounded Ginsburg semilinear languages,*
- $\mathcal{L}$ *contains all distinct-letter-bounded Ginsburg semilinear languages,*
- $\mathcal{L}$ *contains all bounded Parikh semilinear languages,*
- $\mathcal{L}(\mathsf{NCM})^{\mathrm{bd}} (= \mathcal{L}(\mathsf{DCM})^{\mathrm{bd}} = \mathcal{L}(\mathsf{NCM}(\mathsf{BD}_{id}))^{\mathrm{bd}})$ *is contained in* $\mathcal{L}$,
- $\mathcal{L}(\mathsf{NCM}(\mathsf{BD}_i\mathsf{LB}_d)) (= \mathcal{L}(\mathsf{NCM}(\mathsf{LB}_i\mathsf{BD}_d)) = \mathcal{L}(\mathsf{NCM}(\mathsf{BD}_{id})))$ *is contained in* $\mathcal{L}$,
- $\mathcal{L}$ *contains* $L_k^{\mathsf{BD}_i\mathsf{LB}_d}$, *for each* $k \geq 1$,
- $\mathcal{L}$ *contains* $L_k^{\mathsf{LB}_i\mathsf{BD}_d}$, *for each* $k \geq 1$.

Furthermore, if $\mathcal{L}$ *is also semilinear, then these conditions are equivalent to* $\mathcal{L}^{\mathrm{bd}} = \mathcal{L}(\mathsf{NCM})^{\mathrm{bd}} = \mathcal{L}(\mathsf{DCM})^{bd}$.

By Proposition 12 and Proposition 26, the following is immediate:

Corollary 27. $\mathsf{NCM}(\mathsf{LB}_{id})$ *and* $\mathcal{L}(\mathsf{NCM}(\mathsf{LB}_i\mathsf{LB}_d))$ *do not contain all bounded Ginsburg semilinear languages, or all bounded Parikh semilinear languages.*

There is another simple equivalent form of the family $\mathcal{L}(\mathsf{NCM}(\mathsf{BD}_{id}))$. Let SBD_{id} be the subset of BD_{id} that is the family

$$\{I \mid k \geq 1, I = w_1^* \cdots w_m^*, w_i \in \Delta_k^+, |w_i| \leq 2, 1 \leq i \leq m\}.$$

Proposition 28. *The family* $\mathcal{L}(\mathsf{NCM}(\mathsf{SBD}_{id}))$ *contains all bounded Ginsburg semilinear languages. Hence,* $\mathcal{L}(\mathsf{NCM}(\mathsf{SBD}_{id})) = \mathcal{L}(\mathsf{NCM}(\mathsf{BD}_{id}))$.

Thus, SBD_{id} is enough to generate all bounded Ginsburg semilinear languages, whereas LB_{id} is not.

Next, we will explore the language families $\mathsf{NCM}(\mathsf{LB}_d)$ and $\mathsf{NCM}(\mathsf{LB}_i)$.

Proposition 29. *$\mathcal{L}(\mathsf{NCM}(\mathsf{LB}_d))$ is the smallest full trio containing, for each $k \geq 1$, (here, $w_0 \in \Delta^*_{(k,c)}$, and $w_k \in D^*_k$),*

$$L_k^{\mathsf{LB}_d} = \{w_0 w_1 \cdots w_k \mid w_i \in \{C_{i+1}, C_{i+2}, \ldots, C_k, D_i\}^*, 0 \leq i \leq k, \\ |w_0 w_1 \cdots w_{j-1}|_{C_j} = |w_j|_{D_j} > 0, 1 \leq j \leq k\} \subseteq \Delta_k^*.$$

The next proposition follows with a similar proof.

Proposition 30. *$\mathcal{L}(\mathsf{NCM}(\mathsf{LB}_i))$ is the smallest full trio containing, for each $k \geq 1$, (here, $w_0 \in C_1^*$, and $w_k \in \Delta^*_{(k,d)}$),*

$$L_k^{\mathsf{LB}_i} = \{w_0 \cdots w_k \mid w_i \in \{D_1, \ldots, D_i, C_{i+1}\}^*, 0 \leq i \leq k, \\ |w_{j-1}|_{C_j} = |w_j w_{j+1} \cdots w_k|_{D_j} > 0, 1 \leq j \leq k\} \subseteq \Delta_k^*.$$

5 Applications to Existing Families

We will apply the results of this paper to quickly characterize the bounded languages inside known language families. It has been recently shown that finite-index $\mathsf{ET0L}$ languages contain all bounded Ginsburg semilinear languages [7]. This implies $\mathcal{L}(\mathsf{NCM}(\mathsf{BD}_{id})) \subseteq \mathcal{L}(\mathsf{ET0L}_{\mathsf{fin}})$ (the family of finite-index $\mathsf{ET0L}$ languages, which is a full trio [12]; we refer to this paper for the formal definitions of $\mathsf{ET0L}$ systems and languages), and the bounded languages within $\mathsf{DCM}, \mathsf{NCM}$, and $\mathsf{ET0L}_{\mathsf{fin}}$ coincide by Proposition 26. Here, we strengthen this result. First, it is shown that each $L_k^{\mathsf{LB}_d}$ and $L_k^{\mathsf{LB}_i}$ is in $\mathcal{L}(\mathsf{ET0L}_{\mathsf{fin}})$.

Lemma 31. *For each $k \geq 1$, $L_k^{\mathsf{LB}_d}, L_k^{\mathsf{LB}_i} \in \mathcal{L}(\mathsf{ET0L}_{\mathsf{fin}})$.*

It was also shown that that there are $\mathcal{L}(\mathsf{ET0L}_{\mathsf{fin}})$ languages that are not in $\mathcal{L}(\mathsf{NCM})$ [7]. Then, this sub-family of $\mathcal{L}(\mathsf{NCM})$ is strictly contained in $\mathcal{L}(\mathsf{ET0L}_{\mathsf{fin}})$.

Proposition 32. $\mathcal{L}(\mathsf{NCM}(\mathsf{LB}_{\cup})) \subsetneq \mathcal{L}(\mathsf{ET0L}_{\mathsf{fin}})$, $\mathcal{L}(\mathsf{ET0L}_{\mathsf{fin}})^{\mathrm{bd}} = \mathcal{L}(\mathsf{DCM})^{\mathrm{bd}}$.

We leave as an open problem whether there are languages accepted by NCM that cannot be generated by a finite-index $\mathsf{ET0L}$ system. We conjecture that over $\Sigma_k = \{a_1, \ldots, a_k\}$, $\{w \mid |w|_{a_1} = \cdots = |w|_{a_k}\}$ is not in $\mathcal{L}(\mathsf{ET0L}_{\mathsf{fin}})$, for some k. One might think that the (non-finite-index $\mathsf{ET0L}$) one-sided Dyck language on one letter is a candidate witness, but this language is not in $\mathcal{L}(\mathsf{NCM})$ [4].

Next, the class of TCA machines are Turing machines with a one-way read-only input tape, and a finite-crossing[1] read/write worktape. This language family is a semilinear full trio [5]. Therefore, $\mathcal{L}(\mathsf{TCA})^{\mathrm{bd}} \subseteq \mathcal{L}(\mathsf{NCM})^{\mathrm{bd}}$. To show

[1] There is a fixed c such that the number of times the boundary between any two adjacent input cells is crossed is at most c.

that there is equality, we will simulate $\mathsf{NCM}(\mathsf{BD}_i\mathsf{LB}_d)$. Let M be a well-formed k-counter machine satisfying instruction language $I \subseteq w_1^* w_2^* \cdots w_l^* D_{i_1}^* \cdots D_{i_n}^*$, where $w_i \in \Delta_{(k,c)}^*, 1 \le i \le l, D_j \in \Delta_{(k,d)}, 1 \le j \le n$. Then we build a TCA machine M' with worktape alphabet Δ_k that, on input w, simulates a derivation of M, whereby, if M increases from counters in the sequence $C_{j_1}, \ldots, C_{j_m}$, M' instead writes this sequence on the worktape. Then, M' simulates the decreasing transitions of M as follows: for every section of decreases in $D_{i_j}^*$, for $1 \le j \le n$, M' sweeps the worktape from right-to-left, and corresponding to every decrease, replaces the next C_{i_j} symbol with the symbol D_{i_j} (thereby marking the symbol). This requires n sweeps of the worktape, and M' accepts if all symbols end up marked and the simulated computation is in a final state.

Proposition 33. $\mathcal{L}(\mathsf{NCM}(\mathsf{BD}_i\mathsf{LB}_d)) \subseteq \mathcal{L}(\mathsf{TCA})$ *and* $\mathcal{L}(\mathsf{TCA})^{\mathrm{bd}} = \mathcal{L}(\mathsf{DCM})^{\mathrm{bd}}$.

Next, the family of multi-push-down automata and languages has been introduced [1]. We let MP be these machines. They have some number k of pushdowns, and allow to push to every pushdown, but only pop from the first non-empty pushdown. This can clearly simulate every machine in $\mathsf{NCM}(\underline{\mathsf{LB}}_d)$ (distinct-letter-bounded, which is enough to accept every language in $\mathcal{L}(\mathsf{NCM}(\mathsf{LB}_d))$ by Lemma 11). Furthermore, it follows from results within [1] that $\mathcal{L}(\mathsf{MP})$ is closed under reversal (since it is closed under homomorphic replication with reversal, and homomorphism). Therefore, $\mathcal{L}(\mathsf{MP})$ also contains $\mathcal{L}(\mathsf{NCM}(\mathsf{LB}_\cup))$. Also, this family only contains semilinear languages [1]. Therefore, the bounded languages within $\mathcal{L}(\mathsf{MP})$ coincide with those in $\mathcal{L}(\mathsf{NCM})$ and $\mathcal{L}(\mathsf{DCM})$.

Proposition 34. $\mathcal{L}(\mathsf{NCM}(\mathsf{LB}_\cup)) \subseteq \mathcal{L}(\mathsf{MP})$ *and* $\mathcal{L}(\mathsf{MP})^{\mathrm{bd}} = \mathcal{L}(\mathsf{DCM})^{\mathrm{bd}}$.

References

1. Breveglieri, L., Cherubini, A., Citrini, C., Reghizzi, S.: Multi-push-down languages and grammars. Int. J. Found. Comput. Sci. **7**(3), 253–291 (1996)
2. Ginsburg, S.: Algebraic and Automata-Theoretic Properties of Formal Languages. North-Holland Publishing Company, Amsterdam (1975)
3. Ginsburg, S.: The Mathematical Theory of Context-Free Languages. McGraw-Hill Inc., New York (1966)
4. Greibach, S.: Remarks on blind and partially blind one-way multicounter machines. Theoret. Comput. Sci. **7**, 311–324 (1978)
5. Harju, T., Ibarra, O., Karhumäki, J., Salomaa, A.: Some decision problems concerning semilinearity and commutation. J. Comput. Syst. Sci. **65**(2), 278–294 (2002)
6. Hopcroft, J.E., Ullman, J.D.: Introduction to Automata Theory, Languages, and Computation. Addison-Wesley, Reading, MA (1979)
7. Ibarra, O.H., McQuillan, I.: On bounded semilinear languages, counter machines, and finite-index ET0L. In: Han, Y.-S., Salomaa, K. (eds.) CIAA 2016. LNCS, vol. 9705, pp. 138–149. Springer, Heidelberg (2016). doi:10.1007/978-3-319-40946-7_12
8. Ibarra, O., McQuillan, I.: On families of full trios containing counter machine languages. Technical Report 2016–01, University of Saskatchewan (2016). http://www.cs.usask.ca/documents/technical-reports/2016/TR-2016-01.pdf

9. Ibarra, O.H.: Reversal-bounded multicounter machines and their decision problems. J. ACM **25**(1), 116–133 (1978)
10. Ibarra, O.H., Seki, S.: Characterizations of bounded semilinear languages by one-way and two-way deterministic machines. Int. J. Found. Comput. Sci. **23**(6), 1291–1306 (2012)
11. Kortelainen, J., Salmi, T.: There does not exist a minimal full trio with respect to bounded context-free languages. In: Mauri, G., Leporati, A. (eds.) DLT 2011. LNCS, vol. 6795, pp. 312–323. Springer, Heidelberg (2011)
12. Rozenberg, G., Vermeir, D.: On ET0L systems of finite index. Inf. Control **38**, 103–133 (1978)

Non-regular Maximal Prefix-Free Subsets of Regular Languages

Jozef Jirásek Jr.(✉)

Kuzmányho 27, 04001 Košice, Slovakia
jirasekjozef@gmail.com

Abstract. We investigate non-regular maximal prefix-free subsets (MPFS) of regular languages. We give a method to decide whether or not a regular language has any non-regular MPFS.

Next, we prove that if a regular language has any non-regular MPFS, then it also has a MPFS which is context-sensitive but not context-free, it has a MPFS which is recursive but not context-sensitive, and it has a MPFS which is not recursively enumerable.

We show that no regular language has a MPFS which is recursively enumerable but not recursive. Finally, for any regular language we can decide whether or not it has a context-free non-regular MPFS.

1 Introduction

A language is *prefix-free* if it does not contain two distinct strings such that one of them is a prefix of the other. Prefix-free languages are used in prefix-codes, for example variable-length Huffman codes or country calling codes. In a prefix-code, no codeword is a proper prefix of any other codeword. Hence, a receiver can identify each codeword without any special marker between words. Motivated by prefix-codes, the class of prefix-free regular languages has been recently investigated [2–4,6–10].

A subset M of a language L is called a *maximal prefix-free subset* (MPFS) of L if (informally) M is prefix-free, but adding any other string from L to M would make it no longer be prefix-free.

In [7], we have been interested in finding maximal prefix-free subsets of regular languages. We have shown that a regular MPFS of a regular language can be obtained from a minimal deterministic finite automaton (DFA) for this language by removing all the out-transitions from every final state. Other properties as well as descriptional complexity of regular MPFS of regular languages have been investigated in [7].

Here we further deepen the study of MPFS. A regular language can have a non-regular MPFS. We focus on these non-regular MPFS of regular languages. First we show how to decide whether or not a regular language has a non-regular MPFS. We describe a property P of a language such that a language has a non-regular MPFS if and only if it has the property P. The property P can be decided by examining the minimal DFA for the language.

S. Brlek and C. Reutenauer (Eds.): DLT 2016, LNCS 9840, pp. 229–242, 2016.
DOI: 10.1007/978-3-662-53132-7_19

Next, we describe a method for constructing some MPFS of a regular language L which has the property P. For a given set of integers S we can construct a MPFS M of L, such that two different starting sets result in two different MPFS. This means that the language has uncountably many MPFS. Therefore, for any countable class of languages (such as regular languages, context-free languages, context-sensitive languages, recursively enumerable languages, etc.) there exists a MPFS of L which does not belong to this class.

A natural question to ask is whether a MPFS belonging to a specific class of languages exists. Using a refinement of the above method, we answer this question for the layers of the Chomsky hierarchy: If a language has the property P, then it has a MPFS which is context-sensitive but not context-free, a MPFS which is recursive but not context-sensitive, and a MPFS which is not recursively enumerable. Next, no regular language has a MPFS which is recursively enumerable but not recursive. Finally, a regular language may or may not have a context-free MPFS which is not regular. We give a sufficient and necessary condition for minimal automata accepting languages which have one.

2 Preliminaries

We assume that the reader is familiar with basic concepts of regular languages and finite automata. For details or unexplained notions, we refer to [5,11].

For a finite alphabet Σ, let Σ^* be the set of all strings over Σ, including the empty string ε. Let Σ^+ be the set of all non-empty strings over Σ. Throughout the paper, we assume that $|\Sigma| \geq 2$, since every prefix-free subset of a unary language contains at most one string.

If u, v, w are strings in Σ^* and $w = uv$, then the string u is a *prefix* of the string w. If, moreover, $v \neq \varepsilon$, then u is a *proper prefix* of w. We denote by $\leq_p$ the partial order on Σ^* defined by $u \leq_p v$ iff u is a prefix of v. We also say that u and v are *comparable*. For a string w, let $[w] = \{u \in \Sigma^* \mid u \leq_p w\} \cup \{u \in \Sigma^* \mid w \leq_p u\}$ be the set of strings that are comparable to w. Note that $[w]$ is a regular language. Observe that if $x \notin [w]$, then x is incomparable with wy for any $y \in \Sigma^*$.

A language over Σ is *prefix-free* if it does not contain two distinct strings such that one is a prefix of the other. A subset $M \subseteq L$ is a *maximal prefix-free subset* (MPFS) of L if M is prefix-free, and for each u in L there exists some w in M such that $u \in [w]$. Equivalently, M is a MPFS of L if M is prefix-free, but for any $w \in L \setminus M$ the set $M \cup \{w\}$ is not prefix-free.

A MPFS is not necessarily unique, in fact it is unique if and only if L itself is prefix-free, then $M = L$ is the unique MPFS. A regular language may have a non-regular MPFS: the language a^*b^* has a MPFS $\{a^ib^i \mid i \geq 1\}$.

For a regular language L and $u \in L$, let L_u be the left quotient of L by u, that is, $L_u = \{w \in \Sigma^* \mid uw \in L\}$. Note that L_u is regular. For a DFA $A = (Q, \Sigma, \cdot, s, F)$ and for $q \in Q$, let L_q be the language accepted by A from q, that is, $L_q = \{w \in \Sigma^* \mid q \cdot w \in F\}$. Note that L_q is regular.

Let $q \in F$. If $L_q = \{\varepsilon\}$, we say that q is an *ε-state*. Otherwise we say that q is a final *non-ε-state*. It is known that the language accepted by a minimal DFA

$A = (Q, \Sigma, \cdot, s, F)$ is prefix-free if and only if A has only one final state, and this final state is an ε-state.

Proposition 1. *Let L be a language accepted by a DFA $A = (Q, \Sigma, \cdot, s, F)$. Let A' be an incomplete DFA obtained from A by removing all the out-transitions from every final state of A. Then the language accepted by A' is a regular maximal prefix-free subset of L.*

Proof. The set $L(A')$ is a prefix-free subset of L since every final state in A' is an ε-state. Let $u \in L$. Then u is accepted by A. Let p be the first final state on the computation of A on u. Thus $u = u'u''$ with $s \cdot u' = p$. Then u' is in $L(A')$ and $u' \leq_p u$. Hence $L(A')$ is maximal. □

Proposition 2. *Let L be a regular language and $w \in L$. Then we can find a DFA recognizing a regular maximal prefix-free subset $M \subseteq L$ with $w \in M$.*

Proof. Let $L' = L \setminus [w]$. Let M' be a regular MPFS of L' (obtained from Proposition 1). Let $M = M' \cup \{w\}$. Then M is prefix-free, since no string comparable with w is in M'. Finally M is maximal, since every string in L is either comparable with w, or in L' and therefore comparable with a string in M'. □

3 Characterization of Regular Languages with Non-regular Maximal Prefix-Free Subsets

The aim of this section is to give a sufficient and necessary condition for a regular language to have a non-regular maximal prefix-free subset.

Let L be a regular language accepted by a minimal DFA $A = (Q, \Sigma, \cdot, s, F)$. Since A is minimal, it may contain at most one ε-state, and we denote it by q_ε. We say that a state q in Q is *reachable from a cycle* in A if there is a state p in Q and two strings u and v in Σ^*, such that $p \cdot u = p$, $p \cdot v = q$, and u and v differ in the first symbol.

We are now ready to define the property P mentioned in the introduction:

Definition 3. *A DFA* has the property P *if it contains a final non-ε-state that is reachable from a cycle.*

For short, we also say that a regular language L has the property P *if the minimal DFA accepting L has the property P.*

See Fig. 1 for an illustration of this property.

Our next goal is to show that a language has a non-regular MPFS if and only if it has the property P. First we show the "only if":

Lemma 4. *Let L be a regular language. If L has a non-regular maximal prefix-free subset, then L has the property P.*

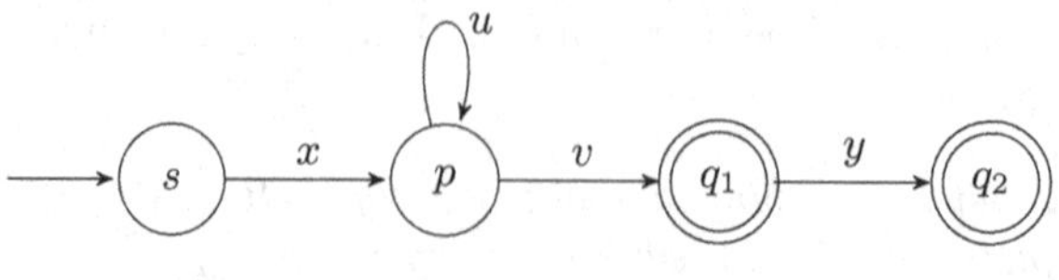

Fig. 1. A final non-ε-state q_1 that is reachable from a cycle

Proof. We prove the inverse, that is, if L does not have the property P, every MPFS of L is regular. Let L be accepted by a minimal DFA $A = (Q, \Sigma, \cdot, s, F)$. Assume that no final non-ε-state of A is reached from a cycle in A. Let M be a maximal prefix-free subset of L. We shall show that M is regular.

If A has a final ε-state, let this state be q_ε. Let $K = \{w \in \Sigma^* \mid s \cdot w = q_\varepsilon\}$ (the language of strings accepted by A in the state q_ε). Further let $M_1 = M \setminus K$ and $M_2 = M \cap K$. Note that M is a disjoint union of M_1 and M_2. If A does not have a final ε-state, let $M_1 = M$.

First we show that M_1 is regular. Let q be a final non-ε-state. Let F_q be the set of strings accepted by A in q, that is, $F_q = \{w \in \Sigma^* \mid s \cdot w = q\}$. Denote by U_q the set of strings u accepted by A in state q such that the state q does not occur inside the computation on u, that is,

$$U_q = \{u \in \Sigma^* \mid s \cdot u = q \text{ and } s \cdot v \neq q \text{ for each proper prefix } v \text{ of } u\}.$$

Notice that we must have $|u| \leq |Q|$ for each u in U_q because otherwise q would be reached from a cycle of A. Thus U_q is finite.

Next, let us show that there is at most one non-empty string w with $q \cdot w = q$ such that the computation on $q \xrightarrow{w} q$ does not pass through q. Suppose for a contradiction that w_1 and w_2 are two such distinct non-empty strings. Then w_1 and w_2 are not comparable, thus $w_1 = uax$ and $w_2 = uby$ for two distinct symbols a and b and some strings u, x, y. Let $p = q \cdot u$. Then we have $p \cdot axu = p$ and $p \cdot by = q$, so q is reached from a cycle of A, a contradiction.

Notice that $F_q = \{uw^i \mid u \in U_q, i \geq 0\}$ if such w exists, and $F_q = U_q$ otherwise. Further, for $u \in U_q$ at most one string in $\{uw^i \mid i \geq 0\}$ may be in the prefix-free set M_1. Since U_q is finite, M_1 may therefore contain only finitely many strings accepted in q, that is, $M_1 \cap F_q$ is finite. The set M_1 consists of strings accepted in non-ε states, so it is a finite union of these finite sets, and therefore M_1 is regular.

Observe that K is prefix-free and no string in K is a proper prefix of another string in L. Therefore if M is a MPFS, it must contain all the strings in K which do not have a proper prefix in M_1. Thus we have $M = M_1 \cup (K \setminus (M_1 \cdot \Sigma^*))$. Since both M_1 and K are regular, M must be regular as well. □

3.1 A Maximal Prefix-Free Subset C_S for a Set of Integers S

In this section we describe a method of obtaining MPFS of a regular language with particular properties. Given a regular language L with the property P and a set of non-negative integers $S \subseteq \mathbb{N}_0$, we want to create a MPFS of L. We want this method to create different MPFS when starting with two different sets S.

Let L be a regular language with the property P accepted by a minimal DFA $A = (Q, \Sigma, \cdot, s, F)$. Let q_1 be a final non-ε-state reachable from a cycle in A, and let $p \in Q$, $q_2 \in F$, $x \in \Sigma^*$, $u, v, y \in \Sigma^+$ be such that $p \cdot u = p$, $p \cdot v = q_1$, and u and v differ in the first symbol; finally $s \cdot x = p$ and $q_1 \cdot y = q_2$. The string x must exist since A is minimal and therefore p is reachable from s, and y must exist because q_1 is a non-ε-state. (See Fig. 1 for an illustration.) Let $S \subseteq \mathbb{N}_0$. Then:

- $L \setminus [x]$ is a regular language. Let C_0 be a regular MPFS of this language given by Proposition 1.
- $L_x \setminus [u]$ is a regular language. Both v and vy belong to this language, since u and v differ in the first symbol. Let C_1 be a regular MPFS of this language containing v, and let C_2 be a regular MPFS of this language containing vy. These can be obtained by Proposition 2.
- Let
$$C_S = C_0 \cup \bigcup_{i \in S} (xu^i \cdot C_1) \cup \bigcup_{i \in \mathbb{N}_0 \setminus S} (xu^i \cdot C_2)$$

Lemma 5. *Let L be a regular language with the property P, and let $S \subseteq \mathbb{N}_0$. Then the set C_S obtained as above is a maximal prefix-free subset of L. Furthermore, if $T \subseteq \mathbb{N}_0$ and $S \neq T$, then $C_S \neq C_T$.*

Proof. (1) *C_S is a subset of L*: $C_0 \subseteq L$. Next, for $i \geq 0$ we have $s \cdot xu^i = p = s \cdot x$, and both C_1 and C_2 are subsets of L_x. Therefore $xu^i \cdot C_1 \subseteq L$ and $xu^i \cdot C_2 \subseteq L$. So $C_S \subseteq L$.

(2) *C_S is prefix-free:* C_0 is prefix-free. C_0 does not contain any string from $[x]$, thus all strings from $xu^i \cdot C_1$ (resp. C_2) are incomparable with all strings in C_0. For a given $i \geq 0$, both $xu^i \cdot C_1$ and $xu^i \cdot C_2$ are prefix-free, since C_1 and C_2 are prefix-free and do not contain any string from $[u]$. Given $i < j$, $w_i = xu^i c_i$, and $w_j = xu^j c_j$, $w_i, w_j \in C_S$, and c_i, c_j in C_1 or C_2 as needed, we have $c_i \notin [u]$ and $w_j = xu^i uz$, thus w_i and w_j are incomparable. Therefore C_S is prefix-free.

(3) *C_S is maximal:* Consider a string $w \in L$. There are four possible cases:

- $w \notin [x]$. Then there is a w' comparable with w in C_0, since C_0 is a MPFS.
- $w \leq_p x$. Then $w \leq_p xv$ and $w \leq_p xvy$. One of these is in C_S, which one depends on whether $0 \in S$.
- $w = xu^i z$, $i \geq 0$, $z \leq_p u$. Then $w \leq_p xu^{i+1} v$, resp. $xu^{i+1} vy$, one of which is in C_S depending on whether $i + 1 \in S$.
- $w = xu^i z$, $i \geq 0$, $z \notin [u]$. We have $s \cdot xu^i = p = s \cdot x$, therefore $z \in L_x$. Since C_1 and C_2 are MPFS, there is a z' comparable with z in either C_1 or C_2, depending on whether $i \in S$. Then $w = xu^i z$ is comparable with $xu^i z' \in C_S$.

In every case we can find a string in C_S comparable with w, so C_S is a MPFS of L.

(4) $C_S \neq C_T$. Without loss of generality, let $i \in S$, $i \notin T$. Then we have $xu^i v \in C_S$ and $xu^i vy \in C_T$. Since C_S is prefix-free, $xu^i vy \notin C_S$, therefore $C_S \neq C_T$. This concludes the proof of the lemma. □

Corollary 6. *Since there are uncountably many sets of integers, Lemma 5 gives us uncountably many MPFS of L. Therefore, if L has the property P, for any countable class of languages (such as regular, context-free, recursively enumerable, etc.) there must exist a MPFS of L which does not belong in this class.*

We can now prove the main goal of this section:

Theorem 7. *A regular language L has a non-regular maximal prefix free subset if and only if L has the property P.*

Proof. Immediate from Lemma 4 and Corollary 6. □

4 Maximal Prefix-Free Subsets and Chomsky Hierarchy

In this section we further investigate non-regular maximal prefix-free subsets of regular languages. We have already shown that if a language has the property P, then it has uncountably many MPFS. However, it might be possible that all of these MPFS are incredibly complex, maybe not even recursively enumerable. In this section we show that we can always find MPFS of "reasonable" complexity, specifically those belonging to certain levels of the Chomsky hierarchy.

Let Reg, CF, CS, Rec, and RE denote the classes of regular, context-free, context-sensitive, recursive, and recursively enumerable languages, respectively. Recall that these classes are recognized by regular automata, context-free grammars, linear bounded automata, Turing machines that halt on every input, and Turing machines, respectively.

4.1 MPFS in CS \ CF, Rec \ CS, and Not in RE

We use the method given in Lemma 5 to find MPFS which are context-sensitive but not context-free, and recursive but not context-sensitive.

Theorem 8. *Let L be a regular language with the property P. Then L has a maximal prefix-free subset which is context-sensitive, but not context-free.*

Proof. We continue using the notation of Lemma 5 with $S = \{2^k \mid k \geq 0\}$. We shall show that C_S is context-sensitive, but not context-free.

Since the sets C_0, C_1, and C_2 are regular languages over Σ, there exist regular grammars $G_i = (N_i, \Sigma, S_i, P_i)$ such that $L(G_i) = C_i$ for $i = 0, 1, 2$. Next, the language $K = \{a^{2^k} \mid k \geq 0\}$ over a unary alphabet $\{a\}$ is context-sensitive, and its complement K^c is context-sensitive as well. Hence there are context-sensitive grammars $G_3 = (N_3, \{a\}, S_3, P_3)$ and $G_4 = (N_4, \{a\}, S_4, P_4)$ with $L(G_3) = K$ and $L(G_4) = K^c$.

We can assume that the sets of non-terminals N_i $(0 \leq i \leq 4)$ are disjoint. Let A and S be new non-terminals which are not in any N_i. In G_3 and G_4, we replace all the occurrences of the terminal a with the non-terminal A, and denote the corresponding set of new productions by P_3' and P_4'. Define a grammar $G = (N, \Sigma, S, P)$, where $N = N_0 \cup N_1 \cup N_2 \cup N_3 \cup N_4 \cup \{A, S\}$; $P = P_0 \cup P_1 \cup P_2 \cup P_3' \cup P_4' \cup P'$, where P' contains the following productions:

- $S \to S_0$;
- $S \to xS_3S_1$;
- $S \to xS_4S_2$; and
- $A \to u$.

Then $L(G) = C_S$, and since $u \neq \varepsilon$, the grammar G is context-sensitive. Hence the set C_S is context-sensitive.

Now we prove that C_S is not context-free. Assume for a contradiction that C_S is context-free. Recall that $v \notin C_2$ and u and v differ in the first symbol. Therefore we have $C_S \cap \{xu^\ell v \mid \ell \geq 0\} = \{xu^{2^k} v \mid k \geq 0\}$. Since CF is closed under intersection with regular languages, the latter set should be context-free. However, the pumping lemma for context-free languages gives a pumping constant p and a string w in this language with $|xu^{2^p} v| < |w| < |xu^{2^{p+1}} v|$, which is a contradiction. □

Theorem 9. *Let L be a regular language with the property P. Then L has a maximal prefix-free subset which is recursive, but not context-sensitive.*

Proof. Let Σ be an alphabet with $|\Sigma| \geq 2$ and $a \in \Sigma$. Let $A_1, A_2, A_3, \ldots$ be the list of halting Turing machines for context-sensitive languages over Σ; cf. [5, Theorem9.8]. Using the notation of Lemma 5, now let $S = \{k \geq 0 \mid a^k \notin L(A_k)\}$. Then the set $\{a^k \mid k \in S\}$ is accepted by the Turing decider T_S described by Algorithm 1:

Algorithm 1. Turing decider T_S for $\{a^k \mid k \in S\}$

if $w \notin a^*$ **then REJECT**
else
    $k \leftarrow |w|$
    simulate A_k on w
    if A_k on w accepts **then REJECT**
    else ACCEPT

Next, C_S is accepted by the Turing decider described by Algorithm 2:

Algorithm 2. Turing decider for C_S

if $w \in C_0$ **then ACCEPT**
if there is no $k \geq 0$ with $w = xu^k w'$ and $u \not\leq_p w'$ **then REJECT**
else execute T_S on a^k
    if T_S accepts a^k **then**
        if $w' \in C_1$ **then ACCEPT**
        else REJECT
    else
        if $w' \in C_2$ **then ACCEPT**
        else REJECT

Now we prove that C_S is not context-sensitive. Assume for a contradiction that C_S is context-sensitive. Recall that $v \notin C_2$ and u and v differ in the first symbol. Therefore we have $C_S \cap \{xu^\ell v \mid \ell \geq 0\} = \{xu^k v \mid k \in S\}$. Since CS is closed under intersection with regular languages, the latter set should also be context-sensitive. Let T be a linear bounded automaton for $\{xu^k v \mid k \in S\}$. We can describe a linear bounded automaton T' for $\{a^k \mid k \in S\}$:

Algorithm 3. Linear bounded automaton T' for $\{a^k \mid k \in S\}$

if $v \notin a^*$ **then REJECT**
else
  write the string x on the work tape
  for every a on input, write the string u on the work tape
  write the string v on the work tape
  execute the linear bounded automaton T on the string $xu^k v$
  if T accepts **then ACCEPT**
  else REJECT

Since x, u, and v are constant strings, the length of $xu^k v$ is linear in k. Next, the automaton T is linear bounded, so it uses space linear in $|xu^k v|$. It follows that T' is linear bounded. Therefore there is some i such that $L(T') = L(A_i)$. If A_i rejects a^i, then $i \in S$, so a^i should be accepted by T'. If A_i accepts a^i, then $i \notin S$, and T' should not accept a^i. In either case we arrive at a contradiction with $L(T') = L(A_i)$. It follows that the set C_S is not context-sensitive. □

Theorem 10. *Let L be a regular language with the property P. Then L has a maximal prefix-free subset which is not recursively enumerable.*

Proof. Immediate from Corollary 6. □

4.2 MPFS in RE \ Rec

In this subsection we prove that no regular language has a maximal prefix-free subset which is recursively enumerable but not recursive.

Theorem 11. *Let L be a regular language. Let M be a recursively enumerable maximal prefix-free subset of L. Then M is recursive.*[1]

Proof. We describe an algorithm for a Turing decider which accepts strings w such that $w \in M$.

First, if $w \notin L$, we immediately **REJECT**. Now let $w \in L$. Since M is recursively enumerable, we can enumerate all strings u of M until we find a string with $u \leq_p w$ or $w \leq_p u$. Such a string u must exist, since $w \in L$ and M is a MPFS of L. Therefore the computation halts after a finite time.

Then if $u = w$ we have $w \in M$, otherwise $w \notin M$ since M is prefix-free. □

[1] In the special case of $L = \Sigma^*$, this corresponds to Lemma 10 of [1].

4.3 MPFS in CF \ Reg

Finally, we investigate maximal prefix-free subsets belonging to CF \ Reg. We show that a regular language with the property P may or may not have a MPFS in this class. We give a sufficient and necessary condition for a language to have such a MPFS. This condition can be decided by examining the minimal DFA for the language.

Example 12. The regular language a^*b^* has a MPFS $\{a^ib^i \mid i \geq 1\}$. This set is context-free, but not regular. However, it can be shown that the language given by the regular expression $a^*(b+bb)$ has the property P, but does not have any MPFS in CF \ Reg. The proof can be obtained using Lemma 18 below.

Let us define a new property P_2.

Definition 13. *A DFA* $A = (Q, \Sigma, \cdot, s, F)$ has the property P_2, *if there exist* $p, q \in Q$, $x \in \Sigma^*$, $u, y, v \in \Sigma^+$, *such that all of the following holds:*

- $s \cdot x = p$, $p \cdot u = p$, $p \cdot y = q$, $q \cdot v = q$
- *u and y differ in the first symbol*
- *either q is a final state, or $L_q \setminus [v]$ is not prefix-free.*

We say that a regular language L has the property P_2, *if the minimal DFA accepting L has the property P_2.*

See Fig. 2 for an illustration.

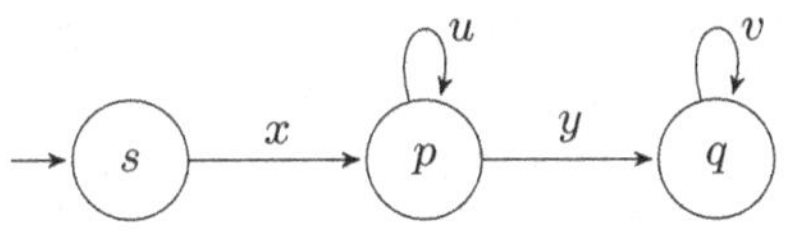

Fig. 2. Sketch of a DFA for a language with the property P_2

In the following we show that a regular language L has a MPFS which is context-free, but not regular, if and only if L has the property P_2.

Lemma 14. *Let L be a regular language with the property P_2. Then L has a maximal prefix-free subset which is context-free, but not regular.*

Proof. Let A, p, q, x, v, y, and u be as in Definition 13. Define the following:

- Let C_0 be a regular MPFS of $L \setminus [x]$.
- Let C_1 be a regular MPFS of $(L_p \setminus [u]) \setminus [y]$.
- (Case 1) $q \in F$
  - Let C_2 be a regular MPFS of $L_q \setminus [v]$.
  - Let

$$C = C_0 \cup \{xu^i \mid i \geq 0\} \cdot C_1 \cup \{xu^iyv^i \mid i \geq 0\} \cup \{xu^iyv^j \mid 0 \leq j < i\} \cdot C_2$$

- (Case 2) $q \notin F$, therefore $L_q \setminus [v]$ is not prefix-free.
  - Let z, z' be two strings in $L_q \setminus [v]$ such that z is a proper prefix of z'.
  - Let C_2 be a regular MPFS of $L_q \setminus [v]$ containing z, and let C_3 be a regular MPFS of $L_q \setminus [v]$ containing z'.
  - Let

$$C = C_0 \cup \{xu^i \mid i \geq 0\} \cdot C_1 \cup \{xu^i yv^i \mid i \geq 0\} \cdot C_2 \cup \{xu^i yv^j \mid i, j \geq 0; i \neq j\} \cdot C_3$$

We can obtain all the required MPFS using Propositions 1 and 2.

The proof that C is a MPFS of L is very similar to the proof of Lemma 5, thus we omit it here. See Appendix for the full proof. Here we show that C is in CF $\setminus$ Reg.

C is obtained as a combination of unions and concatenations of context-free languages, thus C is in CF.

Let us show that C is not regular.

(Case 1) Consider the set $C \cap \{xu^i yv^j \mid i, j \geq 0\}$. We have $\varepsilon \notin C_2$, so this set is equal to $\{xu^i yv^i \mid i \geq 0\}$, which is not regular.

(Case 2) Consider the set $C \cap \{xu^i yv^j z \mid i, j \geq 0\}$. We have $z \notin C_3$, so this set is equal to $\{xu^i yv^i z \mid i \geq 0\}$, which is not regular.

In both cases we have an intersection of C and a regular language. Since regular languages are closed under intersection, C is not regular. □

Lemma 15. *Let L be a regular language with a maximal prefix-free subset which is context-free, but not regular. Then L has the property P_2.*

Proof. For the entirety of this proof, let L be accepted by the minimal DFA $A = (Q, \Sigma, \cdot, s, F)$, and let M be a MPFS of L in CF $\setminus$ Reg. For a contradiction, assume that L does not have the property P_2.

If A has a final ε-state, let this state be q_ε. Let $K = \{w \in \Sigma^* \mid s \cdot w = q_\varepsilon\}$ (the language of strings accepted by A in the state q_ε). Further let $M_1 = M \setminus K$ and $M_2 = M \cap K$. Note that M is a disjoint union of M_1 and M_2, and both M_1 and M_2 are context-free. If A does not have a final ε-state, let $M_1 = M$.

First let us prove that M_1 is regular.

Let $w \in M_1$ and let the computation of A on $w = a_1 a_2 \cdots a_k; a_i \in \Sigma$, be $s = q_0 \xrightarrow{a_1} q_1 \xrightarrow{a_2} q_2 \xrightarrow{a_3} \cdots \xrightarrow{a_k} q_k$; $q_k \in F \setminus \{q_\varepsilon\}$.

Then we can prove the following:

Lemma 16. *For every state $q_i, 0 \leq i \leq k$, there is at most one non-empty string s_{q_i} such that $q_i \cdot s_{q_i} = q_i$, and the computation $q_i \xrightarrow{s_{q_i}} q_i$ does not pass through q_i except for the first and last state.*

Lemma 17. *Let $\mathrm{AC}(q_k) = \{w' \in \Sigma^* \mid s \cdot w' = q_k$, and the computation $s \xrightarrow{w'} q_k$ does not contain any state more than once$\}$. Let ℓ be the first index such that q_ℓ occurs in the computation on w more than once, if such a state exists. Then $w = r s_{q_\ell}^i t$, where $r, t \in \Sigma^*$, $i \geq 0$, and $rt \in \mathrm{AC}(q_k)$.*[2]

[2] AC for acyclic. The computation does not contain a cycle.

Lemma 18. *Let R, S, and T be finite languages. Let $L \subseteq \{rs^it \mid r \in R, s \in S, t \in T, i \geq 0\}$ be a context-free language. Then L is regular.*

See Appendix for a proof of these three lemmas.

Now we are ready to show that M_1 is regular. Let $w \in M_1$. If $w \in \mathrm{AC}(q_k)$, let $r = \varepsilon, t = w, i = 0$. Otherwise there is a state in the computation on w that occurs more than once, and let r, s_{q_ℓ}, t, i be as in Lemma 17. Every $w \in M_1$ can therefore be written as rs_p^it such that $rt \in \mathrm{AC}(q)$ for some $p, q \in Q$. Since $\mathrm{AC}(q)$ is finite for every state q, there can only be finitely many distinct strings r and t used among all $w \in M_1$. Since s_p is unique for every state p, there can be only finitely many distinct strings s_p used among all $w \in M_1$. We know that M_1 is context-free, and thus it fulfills all the conditions for Lemma 18. Therefore M_1 is regular. We can show that M is also regular in the same way as in Lemma 4.

We have arrived at a contradiction with the fact that $M \in \mathrm{CF} \setminus \mathrm{Reg}$, and therefore L must have the property P_2. □

Theorem 19. *Let L be a regular language. Then L has a maximal prefix-free subset which is context-free, but not regular, if and only if L has the property P_2.*

Proof. Immediate from Lemmas 14 and 15. □

5 Conclusions

We have investigated the existence of non-regular maximal prefix-free subsets of regular languages. We have defined a property P: *The minimal DFA for the regular language L has a final non-ε-state which is reachable from a cycle.* We have shown that if a language L does not have the property P, then L has only regular MPFS. On the other hand, if L has the property P, then it always has all of the following:

- A MPFS in the class CS \ CF.
- A MPFS in the class Rec \ CS.
- A MPFS which is not recursively enumerable.
- Uncountably many MPFS.

In the first two cases we have given a description of such MPFS as a context-sensitive grammar, resp. Turing decider.

Further we have shown that no regular language has a MPFS in the class RE \ Rec.

Finally, for languages which have the property P, we have defined a second property P_2, such that a language has the property P_2 if and only if it has a MPFS in the class CF \ Reg.

Both properties P and P_2 can be determined by examining the minimal DFA for the language.

Appendix

Lemma 14. Let L be a regular language with the property P_2. Then L has a maximal prefix-free subset which is context-free, but not regular.

Proof. Here we show that the set C obtained as described in Lemma 14 is a MPFS of L. The proof of the lemma in the article shows that $C \in \mathrm{CF} \setminus \mathrm{Reg}$.

Let $D_1 = \{xu^i \mid i \geq 0\} \cdot C_1$. In Case 1, let $D_2 = \{xu^iyv^i \mid i \geq 0\}$, $D_3 = \{xu^iyv^j \mid i \geq 1, 0 \leq j < i\} \cdot C_2$. In Case 2, let $D_2 = \{xu^iyv^i \mid i \geq 0\} \cdot C_2$, $D_3 = \{xu^iyv^j \mid i, j \geq 0; i \neq j\} \cdot C_3$.

(1) $C \subseteq L$: $C_0 \subseteq L$. We have $s \cdot xu^i = p$, and $C_1 \subseteq L_p$. Next, $s \cdot xu^iyv^j = q$ for $i, j \geq 0$. In Case 1, $q \in F$ and $C_2 \subseteq L_q$. In Case 2, $C_2, C_3 \subseteq L_q$. Therefore $C \subseteq L$.

(2) *C is prefix-free*: C_0 is prefix-free. C_0 does not contain any string in $[x]$, therefore strings in C_0 are incomparable with any string in $D_1 \cup D_2 \cup D_3$.

Since C_1 is prefix-free and does not contain any strings in $[u]$, D_1 is prefix-free as well. C_1 does not contain any string in $[u]$ or $[y]$, so strings in D_1 are incomparable with any string in $D_2 \cup D_3$.

Since C_2 and C_3 are prefix-free and do not contain any string in $[v]$, for a given $i, j \geq 0$ the languages $xu^iyv^j \cdot C_2$ and $xu^iv^j \cdot C_3$ are also prefix-free. Let $i_1 < i_2$. Then $xu^{i_1}yv$ and $xu^{i_2}yv$ are incomparable, since u and y differ in the first symbol. Let $j_1 < j_2$. Then any string in $ux^iyv^{j_1} \cdot C_2$, resp. C_3 is incomparable with any string in $ux^iyv^{j_2} \cdot C_2$, resp. C_3, since no string in C_2, resp. C_3 is in $[v]$.

(3) *C is maximal.* Let $w \in L$. Consider the following cases:

- $w \notin [x]$. Then w is comparable to a string in C_0.
- $w \leq_p x$. Then in Case 1 $w \leq_p xy \in D_2$, in Case 2 $w \leq_p xyz \in D_2$.
- $w = xu^iw_1, i \geq 0, w_1 \notin [u] \cup [y]$. Then w_1 is comparable to a string w_1' in C_1 and w is comparable to $xu^iw_1' \in D_1$.
- $w = xu^iw_1, i \geq 0, w_1 \leq_p u$. Then in Case 1 $w \leq_p xu^{i+1}yv^{i+1} \in D_2$, in Case 2 $w \leq_p xu^{i+1}yv^{i+1}z \in D_2$.
- $w = xu^iw_1, i \geq 0, w_1 \leq_p y$. Then in Case 1 $w \leq_p xu^iyv^i \in D_2$, in Case 2 $w \leq_p xu^iyv^iz \in D_2$.
- $w = xu^iyv^jw_2, i, j \geq 0, w_2 \notin [v]$.
  - Case 1: If $j \geq i$, then $w \geq_p xu^iyv^i \in D_2$. Otherwise w_2 is comparable to a $w_2' \in C_2$ and w is comparable to $xu^iyv^jw_2' \in D_3$.
  - Case 2: If $j = i$, then w_2 is comparable to a $w_2' \in C_2$ and w is comparable to $xu^iyv^iw_2' \in D_2$. Otherwise w_2 is comparable to a $w_2'' \in C_3$ and w is comparable to $xu^iyv^jw_2'' \in D_3$.
- $w = xu^iyv^jw_2, i, j \geq 0, w_2 \leq_p v$.
  - Case 1: If $j \geq i$, then $w \geq_p xu^iyv^i \in D_2$. Otherwise $w \leq_p xu^iyv^i \in D_2$.
  - Case 2: $w \leq_p xu^iyv^{i+j+1}z' \in D_3$.

Therefore for any $w \in L$ there is a $w' \in C$ such that w is comparable to w'. Then C is a MPFS. □

In the following, we use the notation introduced in Lemma 15.

Lemma 16. For every state $q_i, 0 \leq i \leq k$, there is at most one non-empty string s_{q_i} such that $q_i \cdot s_{q_i} = q_i$, and the computation $q_i \xrightarrow{s_{q_i}} q_i$ does not pass through q_i except for the first and last state.

Proof. For a contradiction, let $q_i \cdot s_1 = q_i$ and $q_i \cdot s_2 = q_i$ for non-empty $s_1 \neq s_2$, where the computations on s_1 and s_2 do not pass through q_i. Then s_1 and s_2 are not comparable, and we have $s_1 = s'as'_1$ and $s_2 = s'bs'_2$ for $s, s'_1, s'_2 \in \Sigma^*$ and $a, b \in \Sigma, a \neq b$. Let $q_0 \cdot x' = q_i$ and $q_i \cdot z' = q_k$.

Let $p = q = q_i \cdot s'$, $x = x's'$, , $u = v = as'_1 s'$, $y = bs'_2 s'$. Then L has the property P_2 since either $q = q_i$ is a final state if $i = k$ (Case 1), or $L_q \setminus [v]$ is not prefix-free, since $q \cdot bs'_2 s'z' = q_f$ and q_f is a final non-ε state (Case 2).

Lemma 17. Let $\mathrm{AC}(q_k) = \{w' \in \Sigma^* \mid s \cdot w' = q_k$, and the computation $s \xrightarrow{w'} q_k$ does not contain any state more than once$\}$. Let ℓ be the first index such that q_ℓ occurs in the computation on w more than once, if such a state exists. Then $w = rs^i_{q_\ell}t$, where $r, t \in \Sigma^*$, $i \geq 0$, and $rt \in \mathrm{AC}(q_k)$.

Proof. Let q_ℓ be the first and $q_{\ell'}$ be the last occurrence of the state q_ℓ in the computation on w. By Lemma 16, the only possible string that can be read between two consecutive passes through q_ℓ must be s_{q_ℓ}. The computation therefore looks like this:

$$q_0 \xrightarrow{a_1} q_1 \xrightarrow{a_2} \cdots \xrightarrow{a_\ell} q_\ell \xrightarrow{s_{q_\ell}} q_\ell \xrightarrow{s_{q_\ell}} \cdots \xrightarrow{s_{q_\ell}} q_{\ell'} \xrightarrow{a_{\ell'+1}} q_{\ell'+1} \xrightarrow{a_{\ell'+2}} \cdots \xrightarrow{a_k} q_k$$

Let $r = a_1 a_2 \cdots a_\ell$ and $t = a_{\ell'+1} a_{\ell'+2} \cdots a_k$. Let us show that $rt \in \mathrm{AC}(q_k)$; that is, the states $q_0, q_1, \ldots, q_{\ell-1}, q_{\ell'+1}, \ldots, q_k$ are all distinct.

We know that q_ℓ is the first state which occurs in the computation more than once, therefore if two states q_i and q_j for $i < j$ among the above are equivalent, it must be that $\ell' < i < j \leq k$.

Thus the computation goes through a cycle $q_\ell \xrightarrow{s_{q_\ell}} q_\ell$, potentially several times. After that, the computation goes through the cycle $q_i \xrightarrow{s_{q_i}} q_i$. This cycle does not contain q_ℓ, since q'_ℓ is the last occurrence of this state and $\ell' < i$. The computation must therefore "leave" the q_ℓ cycle at some point. Let q_p be the last state in the computation that follows this cycle. Without further technical details, let us observe that we have $p = q_p$, $q = q_i$, $x = a_1 a_2 \cdots a_p$, $u = s_{q_p}$, $y = a_{p+1} a_{p+2} \cdots a_i$, $v = s_{q_i}$ and L has the property P since either $q = q_i = q_k$ is final, or the final non-ε state q_k is reachable from q_i and thus L_{q_i} is not prefix-free.

This is a contradiction with the initial assumption that the language L does not have the property P. □

Lemma 18. Let R, S, and T be finite languages. Let $L \subseteq \{rs^i t \mid r \in R, s \in S, t \in T, i \geq 0\}$ be a context-free language. Then L is regular.

Proof. It holds that:

$$L = \bigcup_{\substack{r \in R \\ s \in S \\ t \in T}} \{rs^i t \mid rs^i t \in L\},$$

that is, L is a union of finitely many languages of the form $\{rs^i t \mid rs^i t \in L\}$ for some specific r, s, t. For each of these languages we have $\{rs^i t \mid rs^i t \in L\} = r \cdot \{s^i \mid rs^i t \in L\} \cdot t = r \cdot (r \backslash L/t) \cdot t$, where $\backslash$ and $/$ are the left and right quotient operation, respectively. This set is context-free, since L is context-free and CF is closed under concatenation and left and right quotients by regular languages.

It follows that $r \backslash \{rs^i t \mid rs^i t \in L\}/t = \{s^i \mid rs^i t \in L\}$ is also context-free, and since CF is closed under inverse homomorphism, the set $\{a^i \mid rs^i t \in L\}$ is context-free as well. However, the latter language is unary, and every unary context-free language is also regular. Therefore every language $\{rs^i t \mid rs^i t \in L\} = r \cdot \{s^i \mid rs^i t \in L\} \cdot t$ is regular as well, and L is a union of finitely many regular languages. Hence L is regular. □

References

1. Calude, C.S., Staiger, L.: On universal computably enumerable prefix codes. Math. Struct. Comput. Sci. **19**(1), 45–57 (2009)
2. Han, Y.-S., Salomaa, K., Wood, D.: Nondeterministic state complexity of basic operations for prefix-free regular languages. Fundam. Inform. **90**(1–2), 93–106 (2009)
3. Han, Y.-S., Salomaa, K., Wood, D.: Operational state complexity of prefix-free regular languages. In: Automata, Formal Languages, and Related Topics - Dedicated to Ferenc Gécseg on the Occasion of his 70th Birthday, pp. 99–115 (2009)
4. Han, Y.-S., Salomaa, K., Yu, S.: State complexity of combined operations for prefix-free regular languages. In: Dediu, A.H., Ionescu, A.M., Martín-Vide, C. (eds.) LATA 2009. LNCS, vol. 5457, pp. 398–409. Springer, Heidelberg (2009)
5. Hopcroft, J.E., Ullman, J.D.: Introduction to Automata Theory. Languages and Computation. Addison-Wesley, Reading (1979)
6. Jirásek, J., Jirásková, G.: Cyclic shift on prefix-free languages. In: Bulatov, A.A., Shur, A.M. (eds.) CSR 2013. LNCS, vol. 7913, pp. 246–257. Springer, Heidelberg (2013)
7. Jirásek, J.Š., Šebej, J.: Prefix-free subsets of regular languages and descriptional complexity. In: Shallit, J., Okhotin, A. (eds.) DCFS 2015. LNCS, vol. 9118, pp. 129–140. Springer, Heidelberg (2015)
8. Jirásková, G., Krausová, M.: Complexity in prefix-free regular languages. In: Proceedings Twelfth Annual Workshop on Descriptional Complexity of Formal Systems, DCFS 2010, Saskatoon, Canada, 8–10th, pp. 197–204, August 2010
9. Krausová, M.: Prefix-free regular languages: closure properties, difference, and left quotient. In: Kotásek, Z., Bouda, J., Černá, I., Sekanina, L., Vojnar, T., Antoš, D. (eds.) MEMICS 2011. LNCS, vol. 7119, pp. 114–122. Springer, Heidelberg (2012)
10. Palmovský, M., Šebej, J.: Star-complement-star on prefix-free languages. In: Shallit, J., Okhotin, A. (eds.) DCFS 2015. LNCS, vol. 9118, pp. 231–242. Springer, Heidelberg (2015)
11. Sipser, M.: Introduction to the Theory of Computation. PWS Publishing Company, Boston (1997)

Operations on Unambiguous Finite Automata

Jozef Jirásek Jr.[1,2], Galina Jirásková[1(✉)], and Juraj Šebej[2]

[1] Mathematical Institute, Slovak Academy of Sciences, Grešákova 6, 040 01 Košice, Slovakia
jirasekjozef@gmail.com, jiraskov@saske.sk
[2] Faculty of Science, Institute of Computer Science, P.J. Šafárik University, Jesenná 5, 040 01 Košice, Slovakia
juraj.sebej@gmail.com

Abstract. A nondeterministic finite automaton is unambiguous if it has at most one accepting computation on every input string. We investigate the complexity of basic regular operations on languages represented by unambiguous finite automata. We get tight upper bounds for intersection (mn), left and right quotients ($2^n - 1$), positive closure ($\frac{3}{4} \cdot 2^n - 1$), star ($\frac{3}{4} \cdot 2^n$), shuffle ($2^{mn} - 1$), and concatenation ($\frac{3}{4} \cdot 2^{m+n} - 1$). To prove tightness, we use a binary alphabet for intersection and left and right quotients, a ternary alphabet for star and positive closure, a five-letter alphabet for shuffle, and a seven-letter alphabet for concatenation. We also get some partial results for union and complementation.

1 Introduction

A nondeterministic machine is unambiguous if it has at most one accepting computation on every input string. Ambiguity was studied intensively mainly in connection with context-free languages and it is well known that the classes of ambiguous, unambiguous, and deterministic context-free languages are all different. Ambiguity in finite automata was first considered by Schmidt [21] in his unpublished thesis, where he obtained a lower bound $2^{\Omega(\sqrt{n})}$ on the conversion of unambiguous finite automata into deterministic finite automata, as well as for the conversion of nondeterministic finite automata into unambiguous finite automata. He also developed an interesting lower bound method for the size of unambiguous automata based on the rank of certain matrices.

Stearns and Hunt [23] provided polynomial-time algorithms for the equivalence and containment problems for unambiguous finite automata (UFAs), and they extended them to ambiguity bounded by a fixed integer k. Chan and Ibarra [5] provided a polynomial space algorithm to decide, given a nondeterministic finite automaton (NFA), whether it is finitely ambiguous. They also showed that it is PSPACE-complete to decide, given an NFA M and an integer k, whether M is k-ambiguous.

G. Jirásková—Research supported by VEGA grant 2/0084/15 and grant APVV-15-0091.
J. Šebej—Research supported by VEGA grant 1/0142/15 and grant APVV-15-0091.

S. Brlek and C. Reutenauer (Eds.): DLT 2016, LNCS 9840, pp. 243–255, 2016.
DOI: 10.1007/978-3-662-53132-7_20

Ibarra and Ravikumar [12] defined the ambiguity function $a_M(n)\colon N \to N$ of an NFA M such that $a_M(n)$ is the maximum number of distinct accepting computations of M on any string of length n, and they proved that the exponential ambiguity problem is decidable for NFAs. Weber and Seidl [24] showed that if an n-state NFA is finitely ambiguous, then it is at most $5^{n/2}n^n$-ambiguous. Allauzen et al. [1] considered ε-NFAs, and they showed that, given a trim ε-cycle-free NFA A, it is decidable in time that is cubic in the number of transitions of A, whether A is finitely, polynomially, or exponentially ambiguous.

Ravikumar and Ibarra [20] considered the relationship between different types of ambiguity of NFAs to the succinctness of their representations, and they provided a complete picture for unary and bounded languages. Exponentially and polynomially ambiguous NFAs were separated by Leung [15] by providing, for every n, an exponentially ambiguous n-state NFA such that every equivalent polynomially ambiguous NFA requires $2^n - 1$ states.

The UFA-to-DFA tradeoff was improved to the optimal bound 2^n by Leung [16]. He described, for every n, a binary n-state UFA with a unique initial state whose equivalent DFA requires 2^n states. A similar binary example with multiple initial states was given by Leiss [14], and a ternary one was presented already by Lupanov [17]; note that the reverse of Lupanov's ternary witness for NFA-to-DFA conversion is deterministic. Leung [16] elaborated Schmidt's lower bound method for the number of states in a UFA. He considered, for a language L, a matrix whose rows are indexed by strings x_i and columns by strings y_i, and the entry in row x_i and column y_j is 1 if $x_i y_j \in L$ and it is 0 otherwise. He showed that the rank of such a matrix provides a lower bound on the number of states in any UFA for L. Using this method, he was able to describe for every n an n-state finitely ambiguous NFA, whose equivalent UFA requires $2^n - 1$ states.

A lower bound method was further elaborated by Hromkovič et al. [11]. They used communication complexity to show that so-called exact cover of all 1's with monochromatic sub-matrices in a communication matrix of a language provides a lower bound on the size of any UFA for the language. This allowed them to simplify proofs presented in [21,23]. Using communication complexity methods, Hromkovič and Schnitger [10] showed a separation of finitely and polynomially ambiguous NFAs, and even proved a hierarchy for polynomial ambiguity.

A survey paper on unambiguity in automata theory was presented by Colcombet [6], where he considered word automata, tropical automata, infinite tree automata, and register automata. He showed that the notion of unambiguity is not well understood so far, and that some challenging problems, including complementation of UFAs, remain open.

Unary unambiguous automata were examined by Okhotin [19], who proved that the tight upper bound for UFA-to-DFA conversion in the unary case is given by a function in $e^{\Theta(\sqrt[3]{n(\ln n)^2})}$, while the trade-off for NFA-to-UFA conversion is $e^{\sqrt{n \ln n}(1+o(1))}$. He also considered the operations of star, concatenation, and complementation on unary UFA languages, and obtained the tight upper bound $(n-1)^2+1$ for star, an upper bound mn for concatenation which is tight if m, n are relatively prime, and a lower bound $n^{2-\epsilon}$ for complementation.

In this paper, we continue this research and study the complexity of basic regular operations on languages represented by unambiguous finite automata. First, we restate the lower bound method from [16,21]. Using the notions of reachable and so-called co-reachable states in an NFA N, we assign a matrix M_N to the NFA N in such a way that the rank of M_N provides a lower bound on the number of states in any UFA for the language $L(N)$. We use this to get all our lower bounds. To get upper bounds, we first construct an NFA for the language resulting from an operation, and then we apply the (incomplete) subset construction to this NFA to get an incomplete DFA, so also UFA, for the resulting language.

2 Preliminaries

We assume that the reader is familiar with basic notions in formal languages and automata theory. For details, the reader may refer to [22].

A *nondeterministic finite automaton* (NFA) is a 5-tuple $N = (Q, \Sigma, \Delta, I, F)$, where Q is a finite nonempty set of states, Σ is a finite nonempty input alphabet, $\Delta \subseteq Q \times \Sigma \times Q$ is the transition relation, $I \subseteq Q$ is the set of initial states, and $F \subseteq Q$ is the set of final states. Each element (p, a, q) of Δ is called a *transition* of N. A *computation* of N on an input string $a_1 \cdots a_n$ is a sequence of transitions $(q_0, a_1, q_1)(q_1, a_2, q_2) \cdots (q_{n-1}, a_n, q_n) \in \Delta^*$. The computation is *accepting* if $q_0 \in I$ and $q_n \in F$; in such a case we say that the string $a_1 \cdots a_n$ is accepted by N. The *language accepted by* the NFA N is the set of strings $L(N) = \{w \in \Sigma^* \mid w \text{ is accepted by } N\}$.

An NFA $N = (Q, \Sigma, \Delta, I, F)$ is *unambiguous* (UFA) if it has at most one accepting computation on every input string, and it is *deterministic* (DFA) if $|I| = 1$ and for each state p in Q and each symbol a in Σ, there is at most one state q in Q such that (p, a, q) is a transition of N. Let us emphasize that we allow NFAs to have multiple initial states, and DFAs to be incomplete.

The transition relation Δ may be viewed as a function from $Q \times \Sigma$ to 2^Q, which can be extended to the domain $2^Q \times \Sigma^*$ in the natural way. We denote this function by $\cdot$. Using this notation we get $L(N) = \{w \in \Sigma^* \mid I \cdot w \cap F \neq \emptyset\}$.

Every NFA $N = (Q, \Sigma, \cdot, I, F)$ can be converted to an equivalent incomplete DFA $N' = (2^Q \setminus \{\emptyset\}, \Sigma, \cdot', I, F')$, where $F' = \{R \in 2^Q \setminus \{\emptyset\} \mid R \cap F \neq \emptyset\}$, and for each R in $2^Q \setminus \{\emptyset\}$ and each a in Σ, the partial transition function $\cdot'$ is defined as follows: $R \cdot' a = R \cdot a$ if $R \cdot a \neq \emptyset$ and $R \cdot' a$ is undefined otherwise. We call the DFA N' the *incomplete subset automaton* of NFA N. Since every incomplete DFA is a UFA, we get the following observation.

Proposition 1. *If a language L is accepted by an n-state* NFA, *then L is accepted by a* UFA *of at most $2^n - 1$ states.* □

The *reverse* w^R of a string w is defined by $\varepsilon^R = \varepsilon$ and $(va)^R = av^R$ where $a \in \Sigma$ and $v \in \Sigma^*$. The *reverse of a language* L is the language L^R defined by $L^R = \{w^R \mid w \in L\}$. The *reverse of an automaton* $N = (Q, \Sigma, \cdot, I, F)$ is the NFA N^R obtained from N by swapping the role of initial and final states and by reversing all the transitions. Formally, we have $N^R = (Q, \Sigma, \cdot^R, F, I)$, where $q \cdot^R a = \{p \in Q \mid q \in p \cdot a\}$ for each state q in Q and each symbol a in Σ. The NFA N^R accepts the reverse of the language $L(N)$.

Let $N = (Q, \Sigma, \cdot, I, F)$ be an NFA. We say that a set S is *reachable* in N if there is a string w in Σ^* such that $S = I \cdot w$. Next, we say that a set T is *co-reachable* in N if T is reachable in N^R. In what follows we are interested in *non-empty* reachable and co-reachable sets, and we use the following notation:

$$\mathcal{R} = \{S \subseteq Q \mid S \text{ is reachable in } N \text{ and } S \neq \emptyset\}, \tag{1}$$
$$\mathcal{C} = \{T \subseteq Q \mid S \text{ is co-reachable in } N \text{ and } T \neq \emptyset\}. \tag{2}$$

The next observation uses the notions of reachable and co-reachable sets in an NFA to get a characterization of unambiguous automata.

Proposition 2. *Let $\mathcal{R}$ and $\mathcal{C}$ be the families of non-empty reachable and co-reachable sets in an* NFA *N. Then N is unambiguous if and only if $|S \cap T| \leq 1$ for each S in $\mathcal{R}$ and each T in $\mathcal{C}$.* □

If N^R is deterministic, then each co-reachable set in N is of size one, and we get the following result.

Corollary 3. *Let N be an* NFA. *If N^R is deterministic, then N is unambiguous.*

Recall that the *state complexity* of a regular language L, $\text{sc}(L)$, is the smallest number of states in any complete DFA accepting the language L. The state complexity of a regular operation is the maximal state complexity of languages resulting from the operation, considered as the function of state complexities of the arguments. The *nondeterministic state complexity* of languages and operations is defined analogously using NFA representation of languages. We define the *unambiguous state complexity* of a regular language L, $\text{usc}(L)$, as the smallest number of states in any UFA for L.

To prove that a DFA is minimal, we only need to show that all its states are reachable from the initial state, and that no two distinct states are equivalent. To prove minimality of NFAs, a fooling set lower bound method may be used [2,8]. To prove a lower bound for the size of a UFA, a method based on ranks of certain matrices was developed by Schmidt [21, Theorem 3.9], Leung [16, Theorem 2] and Hromkovič et al. [11]. We use it in the following statement.

Proposition 4 ([11,16,21]). *Let L be accepted by an NFA N. Let $\mathcal{R}$ and $\mathcal{C}$ be the families of non-empty reachable and co-reachable sets in N, respectively. Let M_N be the matrix in which the rows are indexed by sets in $\mathcal{R}$, the columns are indexed by sets in $\mathcal{C}$, and in the entry (S, T), we have 0/1 if S and T are/are not disjoint. Then* $\text{usc}(L) \geq \text{rank}(M_N)$. □

Proof. Let A be a minimal UFA accepting L. Consider a matrix M'_A, in which rows are indexed by the states of A and columns are indexed by strings generating the co-reachable sets in $\mathcal{C}$. The entry (q, w) is 1 if w^R is accepted by A from q, and it is 0 otherwise. Then every row of M_N is a sum of the rows of M'_A corresponding to the states in S: Notice that since A is a UFA, for every column there is at most one such row that contains a 1. Thus every row of M_N is a linear combination of rows in M'_A, and therefore $\text{rank}(M_N) \leq \text{rank}(M'_A) \leq \text{usc}(L)$. □

Throughout our paper, we use the following observation from [15] and its corollary stated in the proposition below.

Lemma 5 ([15, Lemma 3]). *Let $|Q| = n$ and M_n be a $2^n - 1 \times 2^n - 1$ matrix over the field with characteristic 2 with rows and columns indexed by a non-empty subsets of Q such that $M_n(S,T) = 1$ if $S \cap T \neq \emptyset$ and $M_n(S,T) = 0$ otherwise. Then the rank of M_n is $2^n - 1$.* □

Proposition 6. *Let L be accepted by an* NFA *N. Let $\mathcal{R}$ be the family of all non-empty reachable sets in N. If each non-empty set is co-reachable in* NFA *N, then* $\text{usc}(L) \geq |\mathcal{R}|$. □

Proof. Let $N = (Q, \Sigma, \cdot, I, F)$ be an NFA for L with $|Q| = n$. Consider the matrix M_N given by Proposition 4. Notice that M_N contains $|\mathcal{R}|$ rows of the matrix M_n given in Lemma 5. By Lemma 5, the rank of M_n is $2^n - 1$, so the rows of M_n are linearly independent. Therefore all the rows of M_N must be linearly independent, and we have $\text{rank}(M_N) = |\mathcal{R}|$. Hence $\text{usc}(L) \geq \text{rank}(M_N) = |\mathcal{R}|$ by Proposition 4. □

3 Operations on Unambiguous Finite Automata

We start with the reversal and intersection operations. Then we continue with left and right quotients. Notice that if an NFA N is unambiguous then N^R is also unambiguous. Hence we get the following result.

Theorem 7 (Reversal). *Let L be a regular language. Then* $\text{usc}(L^R) = \text{usc}(L)$.

Theorem 8 (Intersection). *Let K and L be languages over Σ with* $\text{usc}(K) = m$ *and* $\text{usc}(L) = n$. *Then* $\text{usc}(K \cap L) \leq mn$, *and the bound is tight if $|\Sigma| \geq 2$.*

The *left quotient* of a language L by a string w is $w \backslash L = \{x \mid wx \in L\}$, and the left quotient of a language L by a language K is the language $K \backslash L = \bigcup_{w \in K} w \backslash L$. The state complexity of the left quotient operation is $2^n - 1$ [25], and its nondeterministic state complexity is $n+1$ [13]. In both cases, the witness languages are defined over a binary alphabet. Our next result shows that the tight upper bound for UFAs is $2^n - 1$. To prove tightness we use a binary alphabet.

The *right quotient* of a language L by a string w is $L/w = \{x \mid xw \in L\}$, and the right quotient of a language L by a language K is $L/K = \bigcup_{w \in K} L/w$. If a language L is accepted by an n-state DFA or NFA A, then the language L/K is accepted by an automaton that is exactly the same as A, except for the set of final states that consists of all states of A, from which some string in K is accepted by A [25]. Thus $\text{sc}(L/K) \leq n$ and $\text{nsc}(L/K) \leq n$. The tightness of the first upper bound has been shown using binary languages in [25]. The second upper bound is met by unary languages $a^{\geq m-1}$ and $a^{\leq n-1}$. Our next aim is to show that the tight upper bound for unambiguous finite automata is $2^n - 1$, with witnesses defined over a binary alphabet.

Theorem 9 (Left and Right Quotient). *Let $K, L \subseteq \Sigma^*$, $\text{usc}(K) = m$, and $\text{usc}(L) = n$. Then*
(a) $\text{usc}(K \setminus L) \leq 2^n - 1$, *and the bound is tight if* $|\Sigma| \geq 2$;
(b) $\text{usc}(L / K) \leq 2^n - 1$, *and the bound is tight if* $|\Sigma| \geq 2$.

Proof. (a) To get an upper bound, let A be an n-state UFA for L. Construct an n-state NFA N for $K \setminus L$ from A by making initial all states of A that are reachable from the initial set by some string in K. By Proposition 1, $\text{usc}(K \setminus L) \leq 2^n - 1$.

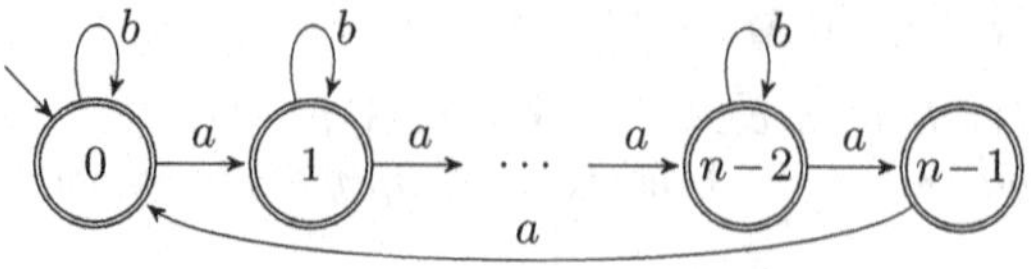

Fig. 1. The UFA of a language L with $\text{usc}(K \setminus L) = 2^n - 1$, where $K = a^{\geq m-1}$.

For tightness, let $K = \{a^k \mid k \geq m - 1\}$ and L be the language accepted by the n-state DFA $A = (\{0, 1, \dots, n-1\}, \{a, b\}, \{0\}, \{0, 1, \dots, n-1\})$ shown in Fig. 1. Notice that each state of A is reachable by some string in K. Construct an n-state NFA N for $K \setminus L$ from A by making all the states initial. Hence the initial set of N is $\{0, 1, \dots, n-1\}$. Next, we can shift every reachable subset right by one (modulo n) by reading a, and we can remove the state n from any subset containing state n by reading b. Therefore each non-empty set is reachable in N.

To construct N^R, we only need to reverse the transitions on a in N. The initial subset of N^R is $\{0, 1, \dots, n-1\}$, and we can again shift any subset and remove one state as before. It follows that every non-empty set is reachable in N^R, that is, co-reachable in NFA N. By Proposition 6, we have $\text{usc}(K \setminus L) \geq 2^n - 1$.

(b) To get an upper bound, let A be an n-state UFA for L. Construct an n-state NFA for L / K as described above. By Proposition 1, $\text{usc}(L / K) \leq 2^n - 1$.

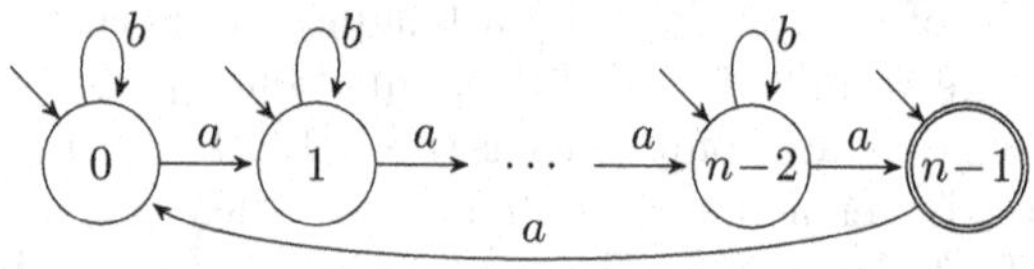

Fig. 2. The UFA of a language L with $\text{usc}(L / K) = 2^n - 1$, where $K = a^{\geq m-1}$.

To prove tightness, let $K = \{a^k \mid k \geq m-1\}$ and L be the language accepted by the n-state NFA $A = (\{0, 1, \dots, n-1\}, \{a, b\}, \{0, 1, \dots, n-1\}, \{n-1\})$ shown in Fig. 2. Since the automaton A^R is deterministic, the NFA A is unambiguous by Corollary 3. Since a string in K is accepted by A from each state of A, we construct an NFA N for L / K from A by making all the states of A final. Notice that we obtain the same NFA as in the proof of the previous lemma, thus by the same arguments $\text{usc}(L / K) \geq 2^n - 1$. □

Now let us continue with the shuffle and concatenation operations. The *shuffle* of two strings u and v over an alphabet Σ is defined as the set of strings $u \sqcup\!\sqcup\, v = \{u_1v_1 \cdots u_kv_k \mid u = u_1 \cdots u_k, v = v_1 \cdots v_k, u_1, \ldots, u_k, v_1, \ldots, v_k \in \Sigma^*\}$. The shuffle of languages K and L over Σ is defined by $K \sqcup\!\sqcup\, L = \bigcup_{u \in K, v \in L} u \sqcup\!\sqcup\, v$. The state complexity of the shuffle operation on languages represented by incomplete deterministic automata was studied by Câmpeanu et al. [3]. They proved that $2^{mn} - 1$ is a tight upper bound for that case. Here we show that the same upper bound is tight also for UFAs, and to prove tightness, we use almost the same languages as in [3, Theorem 1]. To the best of our knowledge, the problem is still open for complete deterministic automata.

Theorem 10 (Shuffle). *Let* $K, L \subseteq \Sigma^*$, $\mathrm{usc}(K) = m$, *and* $\mathrm{usc}(L) = n$. *Then* $\mathrm{usc}(L \sqcup\!\sqcup\, K) \leq 2^{mn} - 1$, *and the bound is tight if* $|\Sigma| \geq 5$.

Proof. Let $A = (Q_A, \Sigma, \cdot_A, I_A, F_A)$ and $B = (Q_B, \Sigma, \cdot_B, I_B, F_B)$ be m- and n-state UFAs for K and L respectively. Then $K \sqcup\!\sqcup\, L$ is accepted by an mn-state NFA $N = (Q_A \times Q_B, \Sigma, \cdot, I_A \times I_B, F_A \times F_B)$, where for each state (p, q) in $Q_A \times Q_B$ and each symbol a in Σ, we have $(p, q) \cdot a = (p \cdot_A a \times \{q\}) \cup (\{p\} \times q \cdot_B a)$. Hence $\mathrm{usc}(K \sqcup\!\sqcup\, L) \leq 2^{mn} - 1$ by Proposition 1.

To prove tightness, let $\Sigma = \{a, b, c, d, f\}$. Let K and L be the regular languages accepted by DFAs $A = (\{0, 1, \ldots, m-1\}, \Sigma, \cdot_A, \{0\}, \{m-1\})$ and $B = (\{0, 1, \ldots, n-1\}, \Sigma, \cdot_B, \{0\}, \{n-1\})$ shown in Fig. 3(left); notice that these DFAs are the same as in [3, Theorem 1] up to the position of final states. Construct an NFA N for $K \sqcup\!\sqcup\, L$ as described above. Figure 3(right) shows a sketch of the resulting NFA. It is shown in [3] that each non-empty set is reachable in N: The initial set $\{(0, 0)\}$ goes to the full set $\{0, 1, \ldots, m-1\} \times \{0, 1, \ldots, n-1\}$ by $c^m d^n$, and for each subset S with $(i, j) \in S$, we have $S \cdot a^{m-i} b^{n-j} f a^i b^j = S \setminus \{(i, j)\}$. Next, in N^R we have $\{(m-1, n-1)\} \cdot^R c^m d^n = \{0, 1, \ldots, m-1\} \times \{0, 1, \ldots, n-1\}$, and $S \cdot^R a^i b^j f a^{m-i} b^{n-j} = S \setminus \{(i, j)\}$ for each subset S with $(i, j) \in S$. It follows that each non-empty set is co-reachable in N, so $\mathrm{usc}(L) \geq 2^{mn} - 1$. □

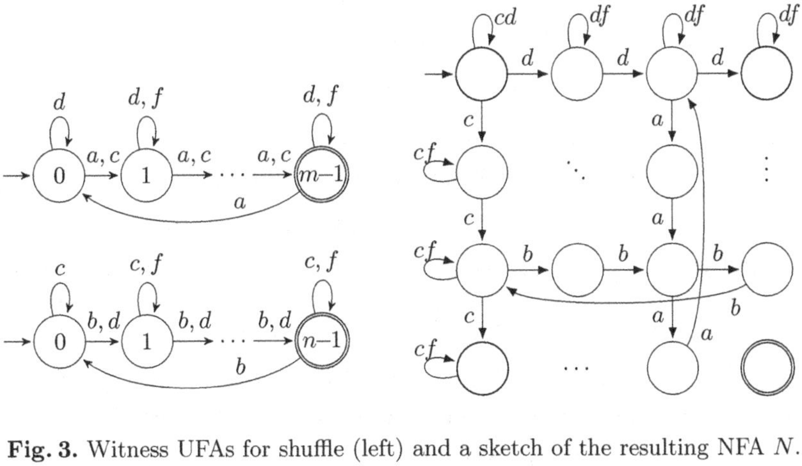

Fig. 3. Witness UFAs for shuffle (left) and a sketch of the resulting NFA N.

The *concatenation* of languages K and L is $KL = \{uv \mid u \in K \text{ and } v \in L\}$. The state complexity of concatenation is $m2^n - 2^{n-1}$, and its nondeterministic state complexity is $m + n$. In both cases, the witnesses are defined over a binary alphabet [9,13,18,25]. In the next theorem we get a tight upper bound for concatenation on UFAs. To prove tightness, we use a seven-letter alphabet.

Theorem 11 (Concatenation). *Let* $K, L \subseteq \Sigma^*$, $\text{usc}(K) = m$, *and* $\text{usc}(L) = n$, *where* $m, n \geq 2$. *Then* $\text{usc}(KL) \leq \frac{3}{4} \cdot 2^{m+n} - 1$, *and the bound is tight if* $|\Sigma| \geq 7$.

Proof. Let $A = (Q_A, \Sigma, \cdot_A, I_A, F_A)$ and $B = (Q_B, \Sigma, \cdot_B, I_B, F_B)$ be UFAs for languages K and L, respectively. Let $|Q_A| = m$, $|F_A| = k$, $|Q_B| = n$, $|I_B| = \ell$. Construct an NFA $N = (Q_A \cup Q_B, \Sigma, \cdot, I, F_B)$ for KL, where for each q in $Q_A \cup Q_B$ and each a in Σ,

$$q \cdot a = \begin{cases} q \cdot_A a, & \text{if } q \in Q_A \text{ and } q \cdot_A a \cap F_A = \emptyset; \\ q \cdot_A a \cup I_B, & \text{if } q \in Q_A \text{ and } q \cdot_A a \cap F_A \neq \emptyset; \\ q \cdot_B a, & \text{if } q \in Q_B, \end{cases}$$

and $I = I_A$ if $I_A \cap F_A = \emptyset$ and $I = I_A \cup I_B$ otherwise. Notice that if a set S is reachable in the NFA N and $S \cap F_A \neq \emptyset$, then $I_B \subseteq S$. It follows that the number of reachable sets is $2^{m-k}2^n + (2^m - 2^{m-k})2^{n-\ell}$, which is maximal if $\ell = 1$. In such a case, this number equals $(2^m + 2^{m-k})2^{n-1}$, which is maximal if $k = 1$. After excluding the empty set, we get the upper bound.

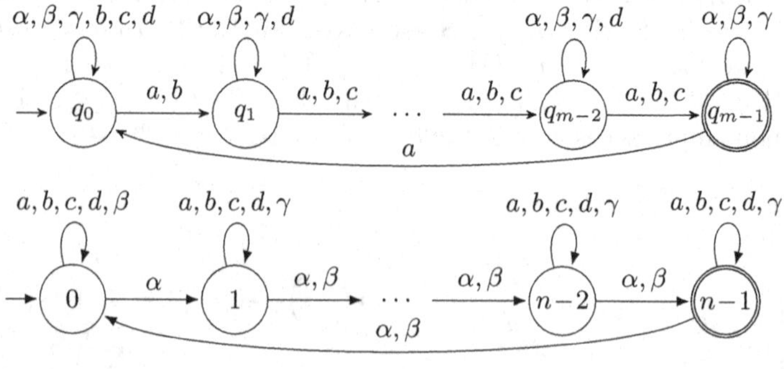

Fig. 4. Witness UFAs for concatenation meeting the upper bound $\frac{3}{4} \cdot 2^{m+n} - 1$.

For tightness, we can use K and L over $\{a, b, c, d, \alpha, \beta, \gamma\}$ accepted by automata A and B shown in Fig. 4, where $Q_A = \{q_0, q_1, \ldots, q_{m-1}\}$ and $Q_B = \{0, 1, \ldots, n-1\}$. Notice that A^R and B are deterministic. □

Now we consider the Kleene closure (star) and positive closure operations. For a language L, the *star* of L is the language $L^* = \bigcup_{i \geq 0} L^i$, where $L^0 = \{\varepsilon\}$ and $L^{i+1} = L^i L$. The *positive closure* of L is $L^+ = \bigcup_{i \geq 1} L^i$. The state complexity of the star operation is $\frac{3}{4} \cdot 2^n$ with binary witness languages [18,25]. In the unary case, the tight upper bound is $(n-1)^2 + 1$ [4,25]. The nondeterministic state complexity of star is $n+1$, with witnesses defined over a unary alphabet [9].

Theorem 12 (Positive Closure and Star). *Let L be a language over Σ with* $\text{usc}(L) = n$, *where $n \geq 2$. Then*
(a) $\text{usc}(L^+) \leq \frac{3}{4} \cdot 2^n - 1$, *and the bound is tight if $|\Sigma| \geq 3$;*
(b) $\text{usc}(L^*) \leq \frac{3}{4} \cdot 2^n$, *and the bound is tight if $|\Sigma| \geq 3$.*

Proof. (a) To get an upper bound, let $A = (Q, \Sigma, \cdot, I, F)$ be an n-state UFA for L. Construct an NFA $N = (Q, \Sigma, \cdot^+, I, F)$ for L^+ where the transition function $\cdot^+$ is defined as

$$q \cdot^+ a = \begin{cases} q \cdot a \cup I, & \text{if } q \cdot a \cap F \neq \emptyset; \\ q \cdot a, & \text{otherwise} \end{cases}$$

for each state q in Q and each symbol a in Σ. Notice that if a set S is reachable in N and $S \cap F \neq \emptyset$, then $I \subseteq S$. We can show that there are at most $\frac{3}{4} \cdot 2^n - 1$ reachable non-empty subsets in N. This proves the upper bound.

To prove tightness, let L be the language accepted by the ternary DFA A shown in Fig. 5(top). Construct the NFA N for L^+ as described above. Notice that the DFA A restricted to the alphabet $\{a, b\}$ is the same as the witness DFA for the star operation from [25, Theorem 3.3, Fig. 4] In particular, this means that N has $\frac{3}{4} \cdot 2^n - 1$ non-empty reachable subsets. We can show that each non-empty set is co-reachable in N.

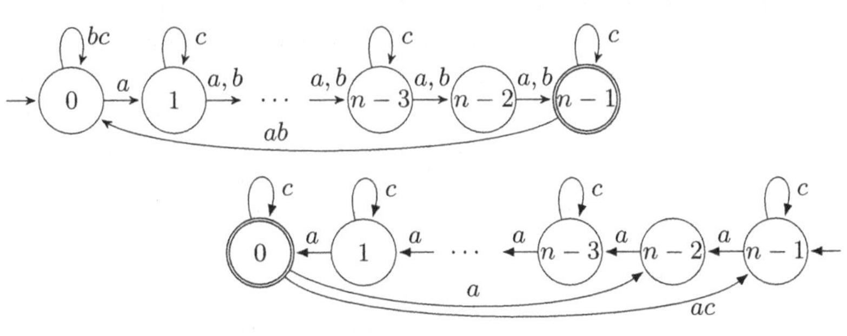

Fig. 5. The witness UFA for positive closure meeting the upper bound $\frac{3}{4} \cdot 2^n - 1$ (top), and the transitions on a, c in the NFA N^R (bottom).

(b) The upper bound follows from the case (a) since if $\varepsilon \in L$, then $L^+ = L^*$, and otherwise we only need to add one more initial and final state to the UFA for L^+ to accept the empty string. The resulting automaton is unambiguous since the new state accepts only the empty string which is not accepted by UFA for L^+. For tightness, consider the language L accepted by the UFA A shown in Fig. 5(top). Construct an NFA N for L^* from UFA A by adding a new initial and final state q_0, and by adding the transitions on a, b from $n-2$ to 0, and the transition by c from $n-1$ to 0. As shown in [25, Theorem 3.3] the NFA N has $\frac{3}{4} \cdot 2^n$ reachable sets: the initial set $\{q_0, 0\}$, all the subsets of $\{0, 1, \ldots, n-1\}$ containing state 0, and all the non-empty subsets of $\{1, 2, \ldots, n-2\}$. Consider the NFA N^R. The initial set of N^R is $\{q_0, n-1\}$. Next, as we have shown that each non-empty set is reachable in N^R. Now consider the $\frac{3}{4} \cdot 2^n \times 2^n$ matrix M_N, and show that its rank is $\frac{3}{4} \cdot 2^n \times 2^n$. □

4 Partial Results for Complementation and Union

In this section we present partial results for the complementation and union operations on UFA languages. The complement of a language L over Σ is the language $L^c = \Sigma^* \backslash L$. A language and its complement have the same state complexity since to get a DFA for the complement of L, we only need to interchange the sets of final and non-final states in a DFA for L. For NFAs, the tight upper bound for complementation is 2^n with witnesses defined over a binary alphabet [9,13]. For unary UFAs, the problem was studied by Okhotin, who provided a lower bound $n^{2-o(1)}$ for complementation of unary UFAs [19, Theorem 6]. In the next theorem we deal with an upper bound. Then we consider union.

Theorem 13 (Complementation: Upper Bound). *Let L be a regular language with* $\operatorname{usc}(L) = n$, *where* $n \geq 7$. *Then* $\operatorname{usc}(L^c) \leq 2^{0.79n + \log n}$.

Proof. Let A be an n-state UFA for L and $\mathcal{R}$ and $\mathcal{C}$ be the sets of non-empty reachable and co-reachable sets of A. First, we show that $\operatorname{usc}(L^c) \leq \min\{|\mathcal{R}|, |\mathcal{C}|\}$. We have $\operatorname{usc}(L^c) \leq |\mathcal{R}|$ since we can get a DFA for L^c by applying the subset construction to A and by interchanging the sets of final and non-final states in the resulting DFA that has $|\mathcal{R}|$ reachable states. Next, we have $\operatorname{usc}(L^c) \leq |\mathcal{C}|$ since the NFA A^R is unambiguous, so $\operatorname{usc}((L^R)^c) \leq |\mathcal{C}|$ which means that $\operatorname{usc}(L^c) \leq |\mathcal{C}|$ since complement and reversal commutes and the reverse of a UFA is a UFA.

Next, let $k = \max\{|X| \mid X \in \mathcal{R}\}$, and pick a set S in $\mathcal{R}$ of size k. Then each set in $\mathcal{R}$ has size at most k, and each set in $\mathcal{C}$ may have at most one element in S by Proposition 2. Thus

$$|\mathcal{R}| \leq \binom{n}{1} + \binom{n}{2} + \cdots + \binom{n}{k} \text{ and } |\mathcal{C}| \leq (k+1)2^{n-k}.$$

If $k \geq n/2$, then $|\mathcal{C}| \leq (n/2+1) \cdot 2^{n/2} \leq 2^{0.5n + \log n}$, and the theorem follows. Now assume that $k < n/2$. Then $|\mathcal{R}| \leq k\binom{n}{k} \leq n(\frac{en}{k})^k$ and $|\mathcal{C}| \leq n2^{n-k}$. Let $r(k) = n(\frac{en}{k})^k$ and $c(k) = n2^{n-k}$. Then $r(k)$ increases, while $c(k)$ decreases with k. It follows that if we pick a k_0 such that $k_0 < n/2$, then $\operatorname{usc}(L^c) \leq r(k_0)$ if $k \leq k_0$, and $\operatorname{usc}(L^c) \leq c(k_0)$ otherwise. By setting $k = nx$ and by solving $(\frac{en}{nx})^{nx} = 2^{n-nx}$, we get $x_0 = 0.2144$, $k_0 = 0.2144n$, $r(k_0) \leq 2^{0.7856n + \log n}$, and $c(k_0) \leq 2^{0.785629n + \log n}$. This completes our proof. □

Proposition 14 (Union). *Let K and L be languages over Σ with* $\text{usc}(K) = m$ *and* $\text{usc}(L) = n$, *where* $1 \le m \le n$. *Then*
(a) $\text{usc}(K \cup L) \le m + n \cdot \text{usc}(K^c) \le m + n2^{0.79n+\log n}$;
(b) the bound $mn + m + n$ *is met if* $|\Sigma| \ge 4$.

Proof. (a) The claim follows from the equality $K \cup L = K \dot{\cup} (L \cap K^c)$, where $\dot{\cup}$ denotes a disjoint union, since we have $\text{usc}(L \cap K^c) \le n \cdot \text{usc}(L^c)$ by Theorem 8, and, moreover, the NFA for a disjoint union of UFAs is unambiguous. The second inequality is given by Theorem 13. We can prove the lower bound in (b) using a four-letter alphabet. □

5 Conclusions

We investigated the complexity of operations on unambiguous finite automata. Since the reverse of an unambiguous automaton is unambiguous, a language and its reversal have the same complexity for UFAs. Next, we got tight upper bounds for intersection (mn), left and right quotients ($2^n - 1$), positive closure ($\frac{3}{4} \cdot 2^n - 1$), star ($\frac{3}{4} \cdot 2^n$), shuffle ($2^{mn} - 1$), and concatenation ($\frac{3}{4} \cdot 2^{m+n} - 1$).

To get upper bounds, we constructed an NFA for the language resulting from an operation, and applied the (incomplete) subset construction to it. For lower bounds, we defined witness languages in such a way that we were able to assign a matrix to a resulting language. The rank of this matrix provided a lower bound on the unambiguous state complexity of the resulting language. To prove tightness, we used a binary alphabet for intersection and left and right quotients, a ternary alphabet for star and positive closure, a five-letter alphabet for shuffle, and a seven-letter alphabet for concatenation. For complementation and union, we provided upper bounds $2^{0.79\,n+\log n}$ and $m+n2^{0.79\,n+\log n}$, respectively. Finally, we got a lower bound $mn + m + n$ for union.

In the case of complementation, we tried to use a fooling set lower bound method, but we were able to describe a fooling set for the complement of an n-state UFA language only of size $n + \log n$. Moreover, it seems that every such fooling set is of size which is quadratic in n [7]. Thus the fooling set technique cannot be used to get a larger lower bound. Neither the method based on the rank of matrices can be used here since the matrices of a language and its complement have the same rank, up to one. Therefore to get a larger lower bound for complementation, some other techniques should be developed.[1]

References

1. Allauzen, C., Mohri, M., Rastogi, A.: General algorithms for testing the ambiguity of finite automata and the double-tape ambiguity of finite-state transducers. Int. J. Found. Comput. Sci. **22**(4), 883–904 (2011). http://dx.doi.org/10.1142/S0129054111008477

[1] The full version can be found at http://im.saske.sk/~jiraskov/UFA/ufa.pdf.

2. Birget, J.: Partial orders on words, minimal elements of regular languages and state complexity. Theor. Comput. Sci. **119**(2), 267–291 (1993). http://dx.doi.org/10.1016/0304-3975(93)90160-U
3. Câmpeanu, C., Salomaa, K., Yu, S.: Tight lower bound for the state complexity of shuffle of regular languages. J. Autom. Lang. Comb. **7**(3), 303–310 (2002)
4. Čevorová, K.: Kleene star on unary regular languages. In: Jurgensen, H., Reis, R. (eds.) DCFS 2013. LNCS, vol. 8031, pp. 277–288. Springer, Heidelberg (2013). http://dx.doi.org/10.1007/978-3-642-39310-5_26
5. Chan, T., Ibarra, O.H.: On the finite-valuedness problem for sequential machines. Theor. Comput. Sci. **23**, 95–101 (1983). http://dx.doi.org/10.1016/0304-3975(88)90012-6
6. Colcombet, T.: Unambiguity in automata theory. In: Shallit, J., Okhotin, A. (eds.) DCFS 2015. LNCS, vol. 9118, pp. 3–18. Springer, Heidelberg (2015). http://dx.doi.org/10.1007/978-3-319-19225-3_1
7. Eliáš, P.: Fooling sets for complements of UFAs. Unpublished manuscript (2016)
8. Glaister, I., Shallit, J.: A lower bound technique for the size of nondeterministic finite automata. Inf. Process. Lett. **59**(2), 75–77 (1996). http://dx.doi.org/10.1016/0020-0190(96)00095-6
9. Holzer, M., Kutrib, M.: Nondeterministic descriptional complexity of regular languages. Int. J. Found. Comput. Sci. **14**(6), 1087–1102 (2003). http://dx.doi.org/10.1142/S0129054103002199
10. Hromkovič, J., Schnitger, G.: Ambiguity and communication. Theory Comput. Syst. **48**(3), 517–534 (2011). http://dx.doi.org/10.1007/s00224-010-9277-4
11. Hromkovič, J., Seibert, S., Karhumäki, J., Klauck, H., Schnitger, G.: Communication complexity method for measuring nondeterminism in finite automata. Inf. Comput. **172**(2), 202–217 (2002). http://dx.doi.org/10.1006/inco.2001.3069
12. Ibarra, O.H., Ravikumar, B.: On sparseness, ambiguity and other decision problems for acceptors and transducers. In: Monien, B., Vidal-Naquet, G. (eds.) STACS 1986. LNCS, vol. 210, pp. 171–179. Springer, Heidelberg (1986). http://dx.doi.org/10.1007/3-540-16078-7_74
13. Jirásková, G.: State complexity of some operations on binary regular languages. Theor. Comput. Sci. **330**(2), 287–298 (2005). http://dx.doi.org/10.1016/j.tcs.2004.04.011
14. Leiss, E.L.: Succint representation of regular languages by boolean automata. Theor. Comput. Sci. **13**, 323–330 (1981). http://dx.doi.org/10.1016/S0304-3975(81)80005-9
15. Leung, H.: Separating exponentially ambiguous finite automata from polynomially ambiguous finite automata. SIAM J. Comput. **27**(4), 1073–1082 (1998). http://dx.doi.org/10.1137/S0097539793252092
16. Leung, H.: Descriptional complexity of NFA of different ambiguity. Int. J. Found. Comput. Sci. **16**(5), 975–984 (2005). http://dx.doi.org/10.1142/S0129054105003418
17. Lupanov, O.B.: A comparison of two types of finite automata. Problemy Kibernetiki 9, Kibernetiki (1963), (in Russian) German translation: Über den Vergleich zweier Typen endlicher Quellen. Probleme der. Kybernetik **6**, 328–335 (1966)
18. Maslov, A.N.: Estimates of the number of states of finite automata. Soviet Math. Doklady **11**, 1373–375 (1970)
19. Okhotin, A.: Unambiguous finite automata over a unary alphabet. Inf. Comput. **212**, 15–36 (2012). http://dx.doi.org/10.1016/j.ic.2012.01.003

20. Ravikumar, B., Ibarra, O.H.: Relating the type of ambiguity of finite automata to the succinctness of their representation. SIAM J. Comput. **18**(6), 1263–1282 (1989). http://dx.doi.org/10.1137/0218083
21. Schmidt, E.M.: Succinctness of description of context-free, regular, and finite languages. Ph. D. thesis. Cornell University (1978)
22. Sipser, M.: Introduction to the Theory of Computation. PWS Publishing Company, Boston (1997)
23. Stearns, R.E., Hunt, H.B.: On the equivalence and containment problems for unambiguous regular expressions, regular grammars and finite automata. SIAM J. Comput. **14**(3), 598–611 (1985). http://dx.doi.org/10.1137/0214044
24. Weber, A., Seidl, H.: On the degree of ambiguity of finite automata. Theor. Comput. Sci. **88**(2), 325–349 (1991). http://dx.doi.org/10.1016/0304-3975(91)90381-B
25. Yu, S., Zhuang, Q., Salomaa, K.: The state complexities of some basic operations on regular languages. Theor. Comput. Sci. **125**(2), 315–328 (1994). http://dx.doi.org/10.1016/0304-3975(92)00011-F

The Trace Monoids in the Queue Monoid and in the Direct Product of Two Free Monoids

Dietrich Kuske and Olena Prianychnykova(✉)

Fachgebiet Automaten und Logik, Technische Universität Ilmenau,
Helmholtzplatz 5, 98684 Ilmenau, Germany
olena.prian@tu-ilmenau.de

Abstract. We prove that a trace monoid embeds into the queue monoid if and only if it embeds into the direct product of two free monoids. We also give a decidable characterization of these trace monoids.

Keywords: Trace monoid · Queue monoid · Coding problem

1 Introduction

Trace monoids model the behavior of concurrent systems whose concurrency is governed by the use of joint resources. They were introduced into computer science by Mazurkiewicz in his study of Petri nets [9]. Since then, much work has been invested on their structure, see [4] for comprehensive surveys. A basic fact about trace monoids is that they can be embedded into the direct product of free monoids [1]. Since the proof of this fact is constructive, an upper bound for the number of factors needed in such a free product is immediate (it is the number α of cliques needed to cover the dependence alphabet). If the dependence alphabet is a path on n vertices, than this upper bound equals the exact number, namely $n-1$. But there are cases where the exact number is considerably smaller (the examples are from [3]): if the independence alphabet is the disjoint union of two copies of C_4 (the cycle on four vertices), then $\alpha = 4$, but 3 factors suffice; if the independence alphabet is the disjoint union of n copies of K_k (the complete graph on k vertices), then $\alpha = k^n$, but k factors suffice.

The strongest result in this respect is due to Kunc [8]: Given a C_3- and C_4-free dependence alphabet and a natural number k, it is decidable whether the trace monoid embeds into the direct product of k free monoids. In this paper, we extend this positive result to all dependence alphabets (also those containing C_3 or C_4), but only for the case $k = 2$. More precisely, we give a complete and decidable characterization of all independence alphabets whose generated trace monoid embeds into the direct product of two free monoids.

Queue monoids, another class of monoids, have been introduced recently [6,7]. They model the behavior of a single fifo-queue. Intuitively, the basic actions (i.e., generators of the monoid) are the action of writing the letter a into the

Supported by the DFG-Project "Speichermechanismen als Monoide", KU 1107/9-1.

S. Brlek and C. Reutenauer (Eds.): DLT 2016, LNCS 9840, pp. 256–267, 2016.
DOI: 10.1007/978-3-662-53132-7_21

queue (denoted a) and reading the letter a from the queue (denoted $\bar{a}$). Sequences of actions are equivalent if they induce the same state change on any queue. For instance, writing a symbol into the queue and reading *another* symbol from the other end of the queue are two actions that can be permuted without changing the overall behavior, symbolically: $a\bar{b} \equiv \bar{b}a$. But there are also more complex equivalences that can be understood as "conditional commutativity", e.g., $ab\bar{b} \equiv a\bar{b}b$. The unconditional commutations allow to embed the direct product of two free monoids into the queue monoid [7]. In [7], it is conjectured that the monoid $\mathbb{N}^3$ cannot be embedded into the queue monoid. Note that these two monoids are special trace monoids and that any trace monoid embedding into the direct product of two free monoids consequently embeds into the queue monoid. In this paper, we prove the conjecture from [7] and characterize, more generally, the class of trace monoids that embed into the queue monoid.

In summary, this paper characterizes two classes of trace monoids defined by their embedability into $\{a,b\}^* \times \{c,d\}^*$ and into the queue monoid, respectively. As it turns out, these two classes are the same, i.e., a trace monoid embeds into the direct product of two free monoids if and only if it embeds into the queue monoid, and this property is decidable. Since this class is not closed under free or direct products, it follows that the classes of submonoids of $\{a,b\}^* \times \{c,d\}^*$ and of the queue monoid Q are not closed under these operations.

All missing proofs can be found in the complete version of this paper [5].

2 Preliminaries and Main Result

2.1 The Trace Monoid

Trace monoids are meant to model the behavior of concurrent systems whose concurrency is governed by the use of joint resources. Here, we take a slightly more abstract view and say that two actions are independent if they use disjoint resources. More formally, an *independence alphabet* is a pair (Γ, I) consisting of a countable (i.e., finite or of size $\aleph_0$) set Γ and an irreflexive and symmetric relation $I \subseteq \Gamma^2$ called the *independence relation.* By $D = \Gamma^2 \setminus I$, we denote the complementary *dependence relation.*

An independence alphabet (Γ, I) induces a trace monoid as follows: Let $\equiv_I$ denote the least congruence on the free monoid Γ^* with $ab \equiv_I ba$ for all pairs $(a,b) \in I$. Then the *trace monoid associated with* (Γ, I) is the quotient $\mathbb{M}(\Gamma, I) = \Gamma^*/{\equiv_I}$, the equivalence class containing $u \in \Gamma^*$ is denoted $[u]_I$. Thus the defining equations of the trace monoid are the equations $ab \equiv_I ba$ for some pairs of letters (a,b).

We only need the following very basic properties of the trace monoid from [1]:

Proposition 1. *Let (Γ, I) be an independence alphabet.*

(1) Let $\Gamma = \bigcup_{0 \leq i < n} C_i$ with $I = \Gamma^2 \setminus \bigcup_{0 \leq i < n} C_i \times C_i$ and $n \in \mathbb{N} \cup \{\omega\}$. Then the trace monoid $\mathbb{M}(\Gamma, I)$ embeds into the direct product of n free monoids with 2 generators each.

(2) The trace monoid $\mathbb{M}(\Gamma, I)$ *is cancellative, i.e.,* $uvw \equiv_I uv'w$ *implies* $v \equiv_I v'$ *for all words* $u, v, v', w \in \Gamma^*$.

In this paper, we will often use graph-theoretic terms to speak about an independence alphabet (Γ, I) – where we identify I with the set of edges $\{a, b\}$ for $(a, b) \in I$. In other words, we think of (Γ, I) as a symmetric and loop-free graph. We will also take the liberty to write (C, I) for the subgraph of (Γ, I) induced by $C \subseteq \Gamma$. We call a connected component C of (Γ, I) *nontrivial* if it is not an isolated vertex. The connected component C is *bipartite* if $I \cap C^2 \subseteq (C_1 \times C_2) \cup (C_2 \times C_1)$ for some partition $C_1 \uplus C_2$ of C. It is *complete bipartite* if $I \cap C^2 = (C_1 \times C_2) \cup (C_2 \times C_1)$. Finally, an independence alphabet (Γ, I) is *P_4-free* if no induced subgraph is isomorphic to P_4, i.e., if there are no four distinct vertices a, b, c, d with $(a, b), (b, c), (c, d) \in I$ and $(b, d), (d, a), (a, c) \in D$.

2.2 The Queue Monoid

The queue monoid models the behavior of a fifo-queue whose entries come from a finite set A. Consequently, the state of a valid queue is an element from A^*. In order to have a defined result even if a read action fails, we add the error state $\bot$. The basic actions are writing of the symbol $a \in A$ into the queue (denoted a) and reading the symbol $a \in A$ from the queue (denoted $\overline{a}$). Formally, $\overline{A}$ is a disjoint copy of A whose elements are denoted $\overline{a}$. Furthermore, we set $\Sigma = A \cup \overline{A}$. Hence, the free monoid Σ^* is the set of sequences of basic actions and it acts on the set $A^* \cup \{\bot\}$ by way of the function: $(A^* \cup \{\bot\}) \times \Sigma^* \to A^* \cup \{\bot\}$, which is defined, for $q \in A^*$, $a \in A$, and $u \in \Sigma^*$, as follows:

$$q.\varepsilon = q \qquad q.au = qa.u \qquad q.\overline{a}u = \begin{cases} q'.u & \text{if } q = aq' \\ \bot & \text{otherwise} \end{cases} \qquad \bot.u = \bot$$

Definition 2. *Two words* $u, v \in \Sigma^*$ *are* equivalent *if* $q.u = q.v$ *for all queues* $q \in A^*$. *In that case, we write* $u \equiv v$. *The equivalence class wrt.* $\equiv$ *containing the word* u *is denoted* $[u]$.

Since $\equiv$ *is a congruence on the free monoid* Σ^*, *we can define the quotient monoid* $Q_A = \Sigma^*/\equiv$ *that is called the* queue monoid.

Note that two queue monoids are not isomorphic if the generating sets have different size. But, for any finite generating set A, the queue monoid Q_A embeds into $Q_{\{a,b\}}$ [7, Corollary 5.5]. Since this paper is concerned with submonoids of Q_A, the concrete size of A does not matter (as long as it is finite and at least 2). Hence we will simply write Q for Q_A, no matter what the finite non-singleton set A is.

Theorem 3 ([7, Theorem 4.3]). *The equivalence relation* $\equiv$ *is the least congruence on the free monoid* Σ^* *satisfying the following for all* $a, b, c \in A$:

$$a\overline{b} \equiv \overline{b}a \text{ if } a \neq b; \quad a\overline{b}\overline{c} \equiv \overline{b}a\overline{c}; \quad ab\overline{c} \equiv a\overline{c}b$$

The second and third of these equations generalize nicely to words (we write $\overline{a_1 a_2 \dots a_n}$ for the word $\overline{a_1}\,\overline{a_2}\,\dots\,\overline{a_n}$ for any $a_1, \dots, a_n \in A$):

Lemma 4 ([7, Corollary 3.6]). *Let $u, v, w \in A^*$.*

- *If $|u| \leq |w|$, then $u\overline{v}\overline{w} \equiv \overline{v}u\overline{w}$.*
- *If $|u| \geq |w|$, then $uv\overline{w} \equiv u\overline{w}v$.*

Let $\pi\colon \Sigma^* \to A^*$ be the homomorphism defined by $\pi(a) = a$ and $\pi(\overline{a}) = \varepsilon$ for all $a \in A$. Similarly, define the homomorphism $\overline{\pi}\colon \Sigma^* \to A^*$ by $\pi(a) = \varepsilon$ and $\pi(\overline{a}) = a$ for all $a \in A$. Then, from Theorem 3, we immediately get

$$u \equiv v \implies \pi(u) = \pi(v) \text{ and } \overline{\pi}(u) = \overline{\pi}(v)$$

for all words $u, v \in \Sigma^*$. Hence the homomorphisms π and $\overline{\pi}$ define homomorphisms from Q to A^* by $[u] \mapsto \pi(u)$ and $[u] \mapsto \overline{\pi}(u)$. The words $\pi(u)$ and $\overline{\pi}(u)$ are called the *positive* and *negative projection* of u (or $[u]$).

Applying Lemma 4 iteratively to prefixes of uv, one gets

Corollary 5. *Let $u \in A^*$ and $v \in \Sigma^*$ with $|u| \geq |\overline{\pi}(v)|$. Then $uv \equiv u\overline{\pi}(v)\pi(v)$.*

Ordering the equations from Theorem 3 from left to right, we obtain a semi-Thue system. This semi-Thue system is confluent and terminating. Hence any equivalence class of $\equiv$ has a unique normal form. To describe these normal forms, we write $\langle a_1 a_2 \dots a_n, \overline{b_1 b_2 \dots b_n} \rangle$ for $a_1 \overline{b_1} a_2 \overline{b_2} \dots a_n \overline{b_n}$ (where $n \in \mathbb{N}$ and $a_i, b_i \in A$ for all $1 \leq i \leq n$). Then a word $u \in \Sigma^*$ is in normal form iff there are three words $u_1, u_2, u_3 \in A^*$ with $u = \overline{u_1}\langle u_2, \overline{u_2}\rangle u_3$. We write $\mathrm{nf}(u)$ for the unique word from the equivalence class $[u]$ in normal form. Furthermore, the mixed or central part of the word $\mathrm{nf}(u)$, i.e., the word u_2 with $\mathrm{nf}(u) = \overline{u_1}\langle u_2, \overline{u_2}\rangle u_3$ is denoted $\mu(u)$. The importance of this word $\mu(u)$ is described by the following observation: Let $u, v \in \Sigma^*$. Then the following are equivalent:

(1) $u \equiv v$ (2) $\mathrm{nf}(u) = \mathrm{nf}(v)$ (3) $\pi(u) = \pi(v)$, $\overline{\pi}(u) = \overline{\pi}(v)$, and $\mu(u) = \mu(v)$

Next, we describe the normal form of the product of two words. For this, we need the concept of the overlap of two words: Let $u, v \in A^*$. Then the *overlap* of u and v is the longest word x that is both, a suffix of u and a prefix of v. We write $\mathrm{ol}(u, v)$ for this overlap.

Theorem 6. ([7, Theorem 5.5]). *Let $u, v \in \Sigma^*$. Then $\mu(uv) = \mathrm{ol}(\mu(u)\overline{\pi}(v), \pi(u)\mu(v))$.*

With $s, t \in A^*$ such that $s\mu(uv) = \overline{\pi}(uv)$ and $\mu(uv)t = \pi(uv)$, we therefore have $\mathrm{nf}(uv) = \overline{s}\langle \mu(uv), \overline{\mu(uv)}\rangle t$.

In the following lemma we describe the normal form of the n-th power of an element of the queue monoid Q. Its inductive proof uses Theorem 6 as well as some tedious arguments about suffixes and prefixes of words.

Lemma 7. *Let $u \in A^*$. Then for every $n \geq 1$ we have*

$$\mu(u^n) = \mathrm{ol}(\mu(u)\overline{\pi}(u)^{n-1}, \pi(u)^{n-1}\mu(u)).$$

2.3 The Main Result

The results of this paper are summarised in the following theorem. It characterizes those trace monoids that can be embedded into the queue monoid as well as those that embed into the direct product of two free monoids. In particular, these two classes of trace monoids are the same. And, in addition, given a finite independence alphabet, it is decidable whether the generated trace monoid falls into this class.

Theorem 8. *The following are equivalent for any independence alphabet* (Γ, I)*:*

(1) The trace monoid $\mathbb{M}(\Gamma, I)$ *embeds into the queue monoid* $\mathcal{Q}$*.*
(2) The trace monoid $\mathbb{M}(\Gamma, I)$ *embeds into the direct product* $\{a,b\}^* \times \{c,d\}^*$ *of two free monoids.*
(3) One of the following conditions hold:
  (3.a) All nodes in (Γ, I) *have degree* ≤ 1*.*
  (3.b) The independence alphabet (Γ, I) *has only one non-trivial connected component and this component is complete bipartite.*

The implication "(2) implies (1)" follows immediately from [6, Proposition 8.2] since there, we showed that $\{a,b\}^* \times \{c,d\}^*$ embeds into the queue monoid $\mathcal{Q}$. In the following section, we present embeddings of $\mathbb{M}(\Gamma, I)$ whenever (Γ, I) satisfies condition (3). The main work here is concerned with independence alphabets satisfying (3.a). The subsequent section shows that any trace monoid that embeds into the queue monoid satisfies condition (3).

Remark 9. From this technically rather hard proof of the implication (1)⇒(3), one can extract a direct proof of the implication (2)⇒(3).

If a trace monoid $\mathbb{M}(\Gamma, I)$ embeds into the direct product of two free monoids, then there is a "weak coding", i.e., an embedding with certain additional properties, of $\mathbb{M}(\Gamma, I)$ into a direct product of two free monoids [8, Proposition 5.5]. As proposed by one of the reviewers, starting from this weak coding might simplify the extracted proof even further, but we do not see whether this is possible.

3 (3) Implies (2) in Theorem 8

Let (Γ, I) be an independence alphabet satisfying (3.a) or (3.b) of Theorem 8. We will prove that $\mathbb{M}(\Gamma, I)$ embeds into the direct product of two free monoids.

Lemma 10. *Let* (Γ, I) *be an independence alphabet such that all nodes in* (Γ, I) *have degree* ≤ 1*. Then* $\mathbb{M}(\Gamma, I)$ *embeds into the direct product of two countably infinite free monoids.*

Proof (sketch). Consider the independence alphabet (Σ, I) with $\Sigma = \{a_i, b_i \mid i \in \mathbb{N}\}$ and $I = \{(a_i, b_i), (b_i, a_i) \mid i \in \mathbb{N}\}$. Then (Γ, I) can be seen as a sub-alphabet of (Σ, I) so that $\mathbb{M}(\Gamma, I)$ embeds into $\mathbb{M}(\Sigma, I)$.

We embed $\mathbb{M}(\Sigma, I)$ into the direct product $M = \{c_i \mid i \in \mathbb{N}\} \times \{d_i \mid i \in \mathbb{N}\}$. In this monoid (c_i, d_i) and $(c_i, d_i d_i)$ commute. Hence there is a homomorphism

$\eta\colon \mathbb{M}(\Sigma, I) \to M$ with $\eta(a_i) = (c_i, d_i)$ and $\eta(b_i) = (c_i, d_i d_i)$ for all $i \in \mathbb{N}$. Using lexicographic normal forms we can show that this homomorphism is injective. Hence η embeds $\mathbb{M}(\Sigma, I)$ into M and we get $\mathbb{M}(\Gamma, I) \hookrightarrow \mathbb{M}(\Sigma, I) \hookrightarrow M$. □

Theorem 11. *Let (Γ, I) be an independence alphabet such that one of the following conditions holds:*

1. *all nodes in (Γ, I) have degree ≤ 1 or*
2. *(Γ, I) has only one non-trivial connected component and this component is complete bipartite*

Then $M(\Gamma, I)$ embeds into $\{a, b\}^* \times \{c, d\}^*$.

Proof. Let (Γ, I) be such that the first condition holds. Then by Lemma 10 there is an embedding of $M(\Gamma, I)$ into a direct product of two countably infinite free monoids.

Now let (Γ, I) be such that the second condition holds. Then the corresponding dependence alphabet (Γ, D) can be covered by two cliques. Consequently, [2, Corollary 1.4.5 (General Embedding Theorem), p. 26] implies that $M(\Gamma, I)$ is a submonoid of a direct product of two countably infinite free monoids.

Note that the countably infinite free monoid $\{a_i \mid i \in \mathbb{N}\}^*$ embeds into $\{a, b\}^*$ via $a_i \mapsto a^i b$. Hence, in any case, $\mathbb{M}(\Gamma, I)$ embeds into $\{a, b\}^* \times \{c, d\}^*$. □

4 (1) Implies (3) in Theorem 8

Definition 12. *Let (Γ, I) be an independence alphabet and $\eta\colon \mathbb{M}(\Gamma, I) \hookrightarrow Q$ be an embedding. We partition Γ into sets Γ_+, Γ_-, and $\Gamma_\pm$ according to the emptiness of the projections of $\eta(a)$:*
- $a \in \Gamma_+$ iff $\pi(\eta(a)) \neq \varepsilon$ and $\overline{\pi}(\eta(a)) = \varepsilon$ (i.e., $\eta(a)$ has no negative projection).
- $a \in \Gamma_-$ iff $\pi(\eta(a)) = \varepsilon$ and $\overline{\pi}(\eta(a)) \neq \varepsilon$ (i.e., $\eta(a)$ has no positive projection).
- $a \in \Gamma_\pm$ iff $\pi(\eta(a)) \neq \varepsilon$ and $\overline{\pi}(\eta(a)) \neq \varepsilon$ (i.e., $\eta(a)$ has both projections).

We will prove the following:

- $(\Gamma_+ \cup \Gamma_-, I)$ is complete bipartite (Proposition 13).
- Every node $a \in \Gamma_\pm$ has degree ≤ 1 (Corollary 22 which is the most difficult part of the proof).
- Any letter from $\Gamma_+ \cup \Gamma_-$ is connected to any edge (Proposition 15).
- The graph (Γ, I) is P_4-free (Proposition 24).

At the end of this section, we infer that the independence alphabet (Γ, I) has the required property from Theorem 8(3).

4.1 $(\Gamma_+ \cup \Gamma_-, I)$ Is Complete Bipartite

Proposition 13. *Let (Γ, I) be an independence alphabet, let $\eta\colon \mathbb{M}(\Gamma, I) \hookrightarrow Q$ be an embedding. Then (Γ_+, I) and (Γ_-, I) are discrete and $(\Gamma_+ \cup \Gamma_-, I)$ is complete bipartite.*

Proof. We first show that (Γ_+, I) is discrete.

Towards a contradiction, suppose there are $a, b \in \Gamma_+$ with $(a, b) \in I$. Consider the non-empty words $u = \mathrm{nf}(\eta(a))$ and $v = \mathrm{nf}(\eta(b))$. Since $\pi \circ \eta \colon \mathbb{M}(\Gamma, I) \to A^*$ is a homomorphism and since $[ab]_I = [ba]_I$, we get $uv = vu$. Hence u and v have a common root, i.e., there is a word p and there are $i, j > 0$ with $u = p^i$ and $v = p^j$. Hence

$$\eta(a)^j = u^j = v^i = \eta(b)^i.$$

Since η is injective, this implies $a^j \equiv_I b^i$ and therefore $a = b$, contradicting $(a, b) \in I$. Hence, there are no $a, b \in \Gamma_+$ with $(a, b) \in I$, i.e., (Γ_+, I) is discrete.

Symmetrically, also (Γ_-, I) is discrete.

It remains to be shown that $(a, b) \in I$ for any $a \in \Gamma_+$ and $b \in \Gamma_-$. There are words $u, v \in A^*$ with $\eta(a) = [u]$ and $\eta(b) = [\overline{v}]$ (note that u and v are nonempty since η is an injection). We have the following:

$$\begin{aligned}
\eta(abb^{|u|}) &= [u\overline{v}\overline{v}^{|u|}] \\
&= [\overline{v}u\overline{v}^{|u|}] \qquad \text{by the dual form of Corollary 5 since } |u| \le |v^{|u|}| \\
&= \eta(bab^{|u|})
\end{aligned}$$

Since η is injective, this implies $abb^{|u|} \equiv_I bab^{|u|}$ and therefore $ab \equiv_I ba$. Now $(a, b) \in I$ follows from $a \neq b$. □

4.2 Nodes from $\Gamma_+ \cup \Gamma_-$ Are Connected to Any Edge

Lemma 14. *Let $u, v, w \in \Sigma^+$ such that $\overline{\pi}(u) = \varepsilon$, $vw \equiv wv$ and $v \neq w$. Then there exist vectors $\overrightarrow{x} = (x_u, x_v, x_w)$ and $\overrightarrow{y} = (y_u, y_v, y_w)$ in $\mathbb{N}^3$ such that $x_v + x_w \neq 0$ and*

$$u^{x_u} v^{x_v} u w^{x_w} \equiv u^{y_u} w^{y_w} u v^{y_v}. \tag{1}$$

Proof (sketch). First note that since $vw \equiv wv$, there exist primitive words p and q and natural numbers a_v, a_w, b_v, b_w satisfying $\pi(v) = p^{a_v}$, $\overline{\pi}(v) = q^{b_v}$, $\pi(w) = p^{a_w}$, and $\overline{\pi}(w) = q^{b_w}$. Since $v, w \neq \varepsilon$, we get $a_v + b_v \neq 0 \neq a_w + b_w$.

The crucial step (to be found in the complete version [5] of this paper) is to show that there are natural numbers x_v, x_w, y_v, y_w (not all zero) that satisfy the following system of linear equations.

$$\left.\begin{aligned}
a_v x_v &= a_w y_w \\
a_w x_w &= a_v y_v \\
b_v x_v + b_w x_w &= b_w y_w + b_v y_v
\end{aligned}\right\} \tag{2}$$

Furthermore, let $x_u = y_u \in \mathbb{N}$ such that $|\overline{\pi}(v^{x_v} u w^{x_w})| \le |u| \cdot x_u = |u^{x_u}|$. Then we have the following:

$$\begin{aligned}
u^{x_u} v^{x_v} u w^{x_w} &\equiv u^{x_u}\, \overline{\overline{\pi}(v^{x_v} u w^{x_w})}\, \pi(v^{x_v} u w^{x_w}) && \text{by Corollary 5} \\
&= u^{x_u}\, \overline{q}^{b_v x_v + b_w x_w}\, p^{a_v x_v} u p^{a_w x_w} \\
&= u^{y_u}\, \overline{q}^{b_w y_w + b_v y_v}\, p^{a_w y_w} u p^{a_v y_v} \\
&= u^{y_u}\, \overline{\overline{\pi}(w^{y_w} u v^{y_v})}\, \pi(w^{y_w} u v^{y_v}) \\
&\equiv u^{y_u} w^{y_w} u v^{y_v} && \text{by Corollary 5}
\end{aligned}$$

Thus, we found $\overrightarrow{x}, \overrightarrow{y} \in \mathbb{N}^3$ satisfying Eq. (1) with $x_v + x_w \neq 0$. □

Proposition 15. *Let (Γ, I) be an independence alphabet and let $\eta \colon \mathbb{M}(\Gamma, I) \hookrightarrow Q$ be an embedding. Let $a \in \Gamma_+ \cup \Gamma_-$ and $b, c \in \Gamma$ with $(b, c) \in I$. Then $(a, b) \in I$ or $(a, c) \in I$.*

Proof. If $a \in \{b, c\}$, we get $(a, b) \in I$ or $(a, c) \in I$ from $(b, c) \in I$. So assume $a \notin \{b, c\}$. Let $u = \mathrm{nf}(\eta(a))$, $v = \mathrm{nf}(\eta(b))$, and $w = \mathrm{nf}(\eta(c))$. Since $(b, c) \in I$, we get $[vw] = \eta(bc) = \eta(cb) = [wv]$ and therefore $vw \equiv wv$. Furthermore, $[v] = \eta(b) \neq \eta(c) = [w]$ since η is injective and since $b \neq c$ follows from $(b, c) \in I$. Hence in particular $v \neq w$.

We first consider the case $a \in \Gamma_+$, i.e., $\overline{\pi}(u) = \varepsilon$. From Lemma 14, we find natural numbers $x_u, x_v, x_w, y_u, y_v, y_w$ with $u^{x_u} v^{x_v} u w^{x_w} \equiv u^{y_u} w^{y_w} u v^{y_v}$ and $x_v + x_w + y_v + y_w \neq 0$. Consequently,

$$\begin{aligned}\eta(a^{x_u} b^{x_v} a c^{x_w}) &= [u^{x_u} v^{x_v} u w^{x_w}] \\ &= [u^{y_u} w^{y_w} u v^{y_v}] \\ &= \eta(a^{y_u} c^{y_w} a b^{y_v}).\end{aligned}$$

Since η is injective, this implies $a^{x_u} b^{x_v} a c^{x_w} \equiv_I a^{y_u} c^{y_w} a b^{y_v}$.

If $x_v \neq 0$, then $(a, b) \in I$. Similarly, if $x_w \neq 0$, then $(a, c) \in I$. This settles the case $\overline{\pi}(u) = \varepsilon$.

Now let $\pi(u) = \varepsilon$. By duality, Lemma 14 yields natural numbers x_u, x_v, x_w, y_u, y_v, y_w with $x_v + x_w + y_v + y_w \neq 0$ and $v^{x_v} u w^{x_w} u^{x_u} \equiv w^{y_w} u v^{y_v} u^{y_u}$. Then we can derive $(a, b) \in I$ or $(a, c) \in I$ as above. □

4.3 Nodes from $\Gamma_\pm$ Have Degree ≤ 1

Let $a \in \Gamma_\pm$. Then there are nonempty primitive words p and q with $\pi(\eta(a)) \in p^+$ and $\overline{\pi}(\eta(a)) \in q^+$, i.e., p and q are the primitive roots of the two projections of $\eta(a)$. The proof of the fact that a has at most one neighbor in (Γ, I) distinguishes two cases: first, we handle the case that p and q are not conjugated (recall that p and q are *conjugated* if there are words $g \in A^*$ and $h \in A^+$ with $p = gh$ and $q = hg$). The second case, namely that p and q are conjugated, turns out to be far more difficult.

Non-conjugated Roots.

Proposition 16. *Let (Γ, I) be an independence alphabet and let $\eta \colon \mathbb{M}(\Gamma, I) \hookrightarrow Q$ be an embedding. Let furthermore $b \in \Gamma$ and $p, q \in A^+$ be primitive and not conjugated such that*

$$\pi(\eta(b)) \in p^+ \text{ and } \overline{\pi}(\eta(b)) \in q^+.$$

Then there is at most one letter $a \in \Gamma$ with $(a, b) \in I$.

Proof (sketch). By contradiction, let $a, c \in \Gamma$ be distinct with $(a, b), (b, c) \in I$. Choose $u = \mathrm{nf}(\eta([ab]_I))$, $v = \mathrm{nf}(\eta(b))$, and $w = \mathrm{nf}(\eta([bc]_I))$. First, one constructs a nontrivial solution to the equation

$$u^{x_u} v^{x_v} w^{x_w} \equiv u^{y_u} v^{y_v} w^{y_w} \tag{3}$$

in natural numbers as follows. Length conditions on the positive and negative projections yield the following system of linear equations

$$\left.\begin{aligned} a_u x_u + a_v x_v + a_w x_w &= a_u y_u + a_v y_v + a_w y_w \\ b_u x_u + b_v x_v + b_w x_w &= b_u y_u + b_v y_v + b_w y_w \end{aligned}\right\} \tag{4}$$

Since this system consists of two equations in the unknowns $x_u - y_u$, $x_v - y_v$ and $x_w - y_w$, it has an integer solution that can be increased by arbitrary natural numbers, i.e., there is a "sufficiently large" solution that makes the positive (and negative) projections of $u^{x_u} v^{x_v} w^{x_w}$ and $u^{y_u} v^{y_v} w^{y_w}$ equal. Using that this solution is "sufficiently large" and that p and q are not conjugated, we employ some combinatorics on words to prove that also the mixed parts of the normal forms of these two words are equal. Consequently, the normal forms of these two words coincide. Hence they are equivalent, i.e., as required, we found a non-trivial solution $\overrightarrow{x}$, $\overrightarrow{y}$ of Eq. (3). Injectivity of η implies $a^{x_u} b^{x_u + x_v + x_w} c^{x_w} \equiv_I a^{y_u} b^{y_u + y_v + y_w} c^{y_w}$. Since the letters a, b, and c are mutually distinct, this implies $\overrightarrow{x} = \overrightarrow{y}$, a contradiction. □

Conjugated Roots. We now want to prove a similar result in case p and q are conjugated. The proof, although technically more involved, will proceed similarly, i.e., we will determine and use a non-trivial solution of Eq. (3). But the use of the solution of (4) is more involved since words that are suffixes of q^m and prefixes of p^n can be arbitrary long. First, Lemma 19 describes the mixed part of the normal form of $u^{x_u} v^{x_v} w^{x_w}$. Then, Lemma 20 determines a nontrival solution to (some rotation of) Eq. (3), before, finally, Proposition 21 proves the analogue to Proposition 16 for conjugated roots.

The combinatorial lemma below describes words that are prefixes of some power of p and, at the same time, suffixes of some power of q (where p and q are conjugated).

Lemma 17. *Let $g \in A^*$, $h \in A^+$ such that $p = gh$ and $q = hg$ are both primitive words. Let furthermore y be some suffix of q^i and some prefix of p^j for some $i, j \geq 1$ such that $|y| \geq |q|$. Then $y = gq^k = p^k g$ where $k = \left\lfloor \frac{|y|}{|q|} \right\rfloor$.*

Using this combinatorial lemma, we can often determine the overlap of two words via the following corollary:

Corollary 18. *Let $g \in A^*$, $h \in A^+$ such that $p = gh$ and $q = hg$ are both primitive words. Furthermore, let p' be a suffix of p with $|p'| < |p|$ and let q' be a prefix of q with $|q'| < |q|$.*

Then for every $i, j \in \mathbb{N}$ we have $\mathrm{ol}(p' g q^i, p^j g q') = g q^{min(i,j)}$.

The following two lemmas are, technically, the centre of our proof of Proposition 21. The first one can be shown by straightforward but tedious calculations using Lemma 17 and Corollary 18.

Lemma 19. *Let $g \in A^*$, $h \in A^+$ such that $p = gh$ and $q = hg$ are primitive. Let $u, v, w \in Q$ such that the following holds for some $a_u, a_v, a_w, b_u, b_v, b_w \in \mathbb{N}\backslash\{0\}$ and $c_u, c_v, c_w \in \mathbb{Z}$:*

$$\pi(u) = p^{a_u} \qquad \overline{\pi}(u) = q^{b_u} \qquad c_u = \begin{cases} -1 & \text{if } |\mu(u)| < |g| \\ \left\lfloor \frac{|\mu(u)|}{|q|} \right\rfloor & \text{otherwise} \end{cases}$$

$$\pi(v) = p^{a_v} \qquad \overline{\pi}(v) = q^{b_v} \qquad c_v = \begin{cases} -1 & \text{if } |\mu(v)| < |g| \\ \left\lfloor \frac{|\mu(v)|}{|q|} \right\rfloor & \text{otherwise} \end{cases}$$

$$\pi(w) = p^{a_w} \qquad \overline{\pi}(w) = q^{b_w} \qquad c_w = \begin{cases} -1 & \text{if } |\mu(w)| < |g| \\ \left\lfloor \frac{|\mu(w)|}{|q|} \right\rfloor & \text{otherwise} \end{cases}$$

Let $\overrightarrow{x} = (x_u, x_v, x_w) \in \mathbb{N}^3$ with $x_u, x_v, x_w \geq 2$. Then $\mu(u^{x_u} v^{x_v} w^{x_w}) = gq^{X_{\overrightarrow{x}}} = p^{X_{\overrightarrow{x}}} g$ where $X_{\overrightarrow{x}}$ is the minimum of the three numbers

$$\begin{aligned}
&\min(a_u, b_u)x_u + b_v x_v + b_w x_w + c_u - \min(a_u, b_u), \\
&a_u x_u + \min(a_v, b_v)x_v + b_w x_w + c_v - \min(a_v, b_v), \quad \textit{and} \\
&a_u x_u + a_v x_v + \min(a_w, b_w)x_w + c_w - \min(a_w, b_w).
\end{aligned}$$

Lemma 20. *Let $g \in A^*$, $h \in A^+$ such that $p = gh$ and $q = hg$ are primitive. Let $u', v', w' \in \Sigma^+$ with $\pi(u'), \pi(v'), \pi(w') \in p^+$ and $\overline{\pi}(u'), \overline{\pi}(v'), \overline{\pi}(w') \in q^+$.*

Then there exist a rotation (u, v, w) of (u', v', w')[1] *and distinct vectors of non-negative integers $\overrightarrow{x} = (x_u, x_v, x_w)$ and $\overrightarrow{y} = (y_u, y_v, y_w)$ such that*

$$u^{x_u} v^{x_v} w^{x_w} \equiv u^{y_u} v^{y_v} w^{y_w}. \tag{5}$$

Proof (sketch). We choose the rotation (u, v, w) such that one of the following three conditions holds:

1. $|\pi(u)| = |\overline{\pi}(u)|$, $|\pi(v)| = |\overline{\pi}(v)|$, and $|\pi(w)| = |\overline{\pi}(w)|$ or
2. $|\pi(u)| > |\overline{\pi}(u)|$ or
3. $|\pi(w)| < |\overline{\pi}(w)|$.

Given this rotation, we define the natural numbers $a_u, a_v, a_w, b_u, b_v, b_w, c_u, c_v, c_w$ as in Lemma 19 and find a nontrival integer solution to the system of linear equations (4). Increasing all entries in this solution by the minimal entry plus 2 yields a nontrivial solution $\overrightarrow{x'} = (x'_u, x'_v, x'_w)$ and $\overrightarrow{y'} = (y'_u, y'_v, y'_w)$ with $\overrightarrow{x'}, \overrightarrow{y'} \in \mathbb{N}^3$ and $x'_u, x'_v, x'_w, y'_u, y'_v, y'_w \geq 2$. From this solution by Lemma 19 we then construct a nontrivial solution $\overrightarrow{x}$, $\overrightarrow{y}$ that, in addition, satisfies $X_{\overrightarrow{x}} = X_{\overrightarrow{y}}$. This is done by considering the three possible cases for the rotation (u, v, w) separately. We finally show that the two words $u^{x_u} v^{x_v} w^{x_w}$ and $u^{y_u} v^{y_v} w^{y_w}$ agree in their projections and their normal forms agree in their mixed part. Hence they are equivalent, i.e., as required, we found a non-trivial solution $\overrightarrow{x}$, $\overrightarrow{y}$ of Eq. (5). □

[1] i.e., (u, v, w) is one of the triples (u', v', w'), (v', w', u') and (w', u', v').

Using Lemma 20, we can now infer for conjugated roots a result similar to Proposition 16 for non-conjugated roots.

Proposition 21. *Let (Γ, I) be an independence alphabet and let $\eta\colon \mathbb{M}(\Gamma, I) \hookrightarrow Q$ be an embedding. Let furthermore $b \in \Gamma$ and $p, q \in A^+$ be primitive and conjugated such that*

$$\pi(\eta(b)) \in p^+ \text{ and } \overline{\pi}(\eta(b)) \in q^+.$$

Then there is at most one letter $a \in \Gamma$ with $(a, b) \in I$.

Proof (sketch). The proof is basically the same as the proof of Proposition 16. Towards a contradiction, suppose there are distinct letters a and c in Γ with $(a, b), (b, c) \in I$. Let $u' = \mathrm{nf}(\eta([ab]_I))$, $v' = \mathrm{nf}(\eta(b))$, and $w' = \mathrm{nf}(\eta([bc]_I))$.

The crucial point in the proof is that by Lemma 20 there exists a rotation (u, v, w) of (u', v', w') and distinct vectors $\overrightarrow{x}$, $\overrightarrow{y} \in \mathbb{N}^3$ satisfying Eq. (5). We consider the three possible rotations separately. We obtain that in all cases injectivity of η and commutation of b with a and with c yields

$$c^{x_v} b^{x_u+x_v+x_w} a^{x_w} \equiv_I c^{y_v} b^{y_u+y_v+y_w} a^{y_w}.$$

From the distinctness of a, b and c, we get $\overrightarrow{x} = \overrightarrow{y}$ which contradicts our choice of these two vectors as distinct. Thus there are no two distinct letters a and c with $(a, b), (b, c) \in I$. □

The following corollary is the main result of this section. Its proof is an immediate consequence of Propositions 16 and 21 (depending on whether the roots of the two projections of $\eta(a)$ are conjugated or not).

Corollary 22. *Let (Γ, I) be an independence alphabet, let $\eta\colon \mathbb{M}(\Gamma, I) \hookrightarrow Q$ be an embedding, and let $a \in \Gamma$. If $\pi(\eta(a)) \neq \varepsilon$ and $\overline{\pi}(\eta(b)) \neq \varepsilon$, then the degree of a is ≤ 1.*

4.4 (Γ, I) Is P_4-free

The proof of the next lemma is structurally similar to the proof of Lemma 14, but uses also Corollary 22.

Lemma 23. *Let $t, u, v, w \in \Sigma^+$ such that $\overline{\pi}(u) = \varepsilon$, $\pi(v) = \varepsilon$, $vw \equiv wv$, and $tu \equiv ut$. Then there exists a tuple $\overrightarrow{x} = (x_t, x_{u_1}, x_{u_2}, x_v, x_w)$ of natural numbers with $x_t, x_w \neq 0$ and*

$$u^{x_{u_1}} v^{x_v} w t^{x_t} w^{x_w} u^{x_{u_2}} \equiv u^{x_{u_1}} w u^{x_{u_2}} w^{x_w} t^{x_t} v^{x_v}. \tag{6}$$

The following proposition is a consequence of Lemma 23. Its proof is similar to the proof of Proposition 15.

Proposition 24. *Let (Γ, I) be an independence alphabet and let $\eta\colon \mathbb{M}(\Gamma, I) \hookrightarrow Q$ be an embedding. Then (Γ, I) is P_4-free.*

4.5 Proof of the Implication (1)⇒(3) in Theorem 8

Theorem 25. *Let* (Γ, I) *be an independence alphabet and* $\eta\colon \mathbb{M}(\Gamma, I) \to Q$ *be an embedding. Then one of the following conditions holds:*

1. *all nodes in* (Γ, I) *have degree* ≤ 1 *or*
2. (Γ, I) *has only one non-trivial connected component and this component is complete bipartite.*

Proof. Suppose (Γ, I) contains a node a of degree ≥ 2. Then, by Corollary 22, $a \in \Gamma_+ \cup \Gamma_-$. From Proposition 15, we obtain that a is connected to any edge, i.e., it belongs to the only nontrivial connected component C of (Γ, I). Now Proposition 15 implies $\Gamma_+ \cup \Gamma_- \subseteq C$. Note that all nodes in $C \setminus (\Gamma_+ \cup \Gamma_-)$ have degree 1 by Corollary 22. Hence, by Proposition 13, the connected graph (C, I) is a complete bipartite graph together with some additional nodes of degree 1. It follows that (C, I) is bipartite. By Proposition 24, it is a connected and P_4-free graph. Hence its complementary graph (C, D) is not connected [10]. But this implies that (C, I) is complete bipartite. □

References

1. Cori, P., Perrin, D.: Automates et commutations partielles. R.A.I.R.O. Informatique Théorique et Applications **19**, 21–32 (1985)
2. Diekert, V.: Combinatorics on Traces. LNCS, vol. 454. Springer, Heidelberg (1990)
3. Diekert, V., Muscholl, A., Reinhardt, K.: On codings of traces. In: Mayr, E.W., Puech, C. (eds.) STACS 1995. LNCS, vol. 900. Springer, Heidelberg (1995)
4. Diekert, V., Rozenberg, G.: The Book of Traces. World Scientific Publ. Co., River Edge (1995)
5. Kuske, D., Prianychnykova, O.: The trace monoids in the queue monoid and in the direct product of two free monoids. arXiv:1603.07217 (2016)
6. Huschenbett, M., Kuske, D., Zetzsche, G.: The monoid of queue actions. In: Csuhaj-Varjú, E., Dietzfelbinger, M., Ésik, Z. (eds.) MFCS 2014, Part I. LNCS, vol. 8634, pp. 340–351. Springer, Heidelberg (2014)
7. Huschenbett, M., Kuske, D., Zetzsche, G.: The monoid of queue actions (2016, submitted)
8. Kunc, M.: Undecidability of the trace coding problem and some decidable cases. Theor. Comput. Sci. **310**(1–3), 393–456 (2004)
9. Mazurkiewicz, A.: Concurrent program schemes and their interpretation. Technical report, DAIMI Report PB-78, Aarhus University (1977)
10. Seinsche, D.: On a property of the class of n-colorable graphs. J. Comb. Theory (B) **16**, 191–193 (1974)

On Ordered RRWW-Automata

Kent Kwee and Friedrich Otto(✉)

Fachbereich Elektrotechnik/Informatik, Universität Kassel, 34109 Kassel, Germany
{kwee,otto}@theory.informatik.uni-kassel.de

Abstract. It is known that the deterministic ordered restarting automaton accepts exactly the regular languages, while its nondeterministic variant accepts some languages that are not even growing context-sensitive. Here we study an extension of the ordered restarting automaton, the so-called ORRWW-automaton, which is obtained from the previous model by separating the restart operation from the rewrite operation. First we show that the deterministic ORRWW-automaton still characterizes just the regular languages. Then we prove that this also holds for the stateless variant of the nondeterministic ORRWW-automaton, which is obtained by splitting the transition relation into two parts, where the first part is used until a rewrite operation is performed, and the second part is used thereafter. Finally, we show that the nondeterministic ORRWW-automaton is even more expressive than the nondeterministic ordered restarting automaton.

Keywords: Restarting automaton · Ordered rewriting · Language class · Closure property

1 Introduction

The *restarting automaton* was introduced in [3] as a formal model for the linguistic technique of *analysis by reduction*, and since then many variants of restarting automata have been considered (see, e.g., [12]). Essentially, there are two main variants: those restarting automata that must restart immediately after performing a rewrite operation (denoted as *RWW-automata*), and those that have separate rewrite and restart operations and which therefore may continue scanning the tape after executing a rewrite operation (denoted as *RRWW-automata*). Under various restrictions, e.g., determinism and/or monotonicity, these two types of automata are equivalent [4,11], but it is still open whether (unrestricted) RRWW-automata are more expressive than RWW-automata. Here we address this question in the setting of ordered restarting automata.

The deterministic ordered restarting automaton (or det-ORWW-automaton) was introduced in [10] in the setting of picture languages. It is a very restricted form of the *shrinking restarting automaton* that was studied in [6]. In a shrinking restarting automaton, each rewrite operation is weight-reducing with respect to some predefined weight function, and it has been shown that for shrinking restarting automata, the RWW-variant is again equivalent to the RRWW-variant, both in the deterministic as well as in the nondeterministic case.

S. Brlek and C. Reutenauer (Eds.): DLT 2016, LNCS 9840, pp. 268–279, 2016.
DOI: 10.1007/978-3-662-53132-7_22

An *ordered restarting automaton* (or ORWW-automaton) has a finite-state control, a tape with end markers that initially contains the input, and a window of size three. Based on its state and the content of its window, the automaton can execute three types of operations: a *move-right step*, which shifts the window one position to the right and changes the state, a combined *rewrite/restart step*, which replaces the letter in the middle of the window by a letter that is strictly smaller with respect to a predefined ordering on the tape alphabet, moves the window back to the left end of the tape, and resets the automaton to its initial state, or an *accept step*, which causes the automaton to halt and accept. While the nondeterministic variant of this type of automaton accepts some languages that are not even growing context-sensitive [9], the deterministic variant accepts exactly the regular languages [10]. In addition, each det-ORWW-automaton can be simulated by an automaton of the same type that has only a single state, which means that for these automata, states are actually not needed. Accordingly, such an automaton is called a *stateless* det-ORWW-automaton (stl-det-ORWW-automaton). For such an automaton, the size of its tape alphabet can be taken as a complexity measure, and it has been shown [13] that these automata are polynomially related in size to the weight-reducing Hennie machines studied by Průša in [14] and that there is an exponential trade-off for converting a stl-det-ORWW-automaton into an NFA [8].

Here we introduce and study the ordered RRWW-automaton (ORRWW-automaton). It is obtained from the ORWW-automaton by looking at rewrite and restart operations as two separate operations, where, however, we still require that exactly one rewrite step is executed before the first and between any two successive restart operations. We will see that the det-ORRWW-automaton accepts exactly the regular languages, and so it is equivalent to the det-ORWW-automaton. We also consider the stateless variant of the ORRWW-automaton. However, here a problem arises, as an ORRWW-automaton must execute exactly one rewrite operation in each cycle of each computation. As in [7] we could either declare all computations that contain a cycle with none or with several rewrite steps as illegal, or we could clearly distinguish between the two phases of a cycle: the phase *up to* the rewrite step, and the phase *after* the rewrite step. Here we take the latter approach in studying stateless ORRWW-automata. Surprisingly, these automata still characterize the regular languages, both in the deterministic as well as in the nondeterministic case. Finally, we study the expressive power of the nondeterministic ORRWW-automaton with states. The class of languages that it accepts forms an abstract family of languages that properly contains the context-free languages. As the ORWW-automaton does not even accept all linear languages, this implies that the ORRWW-automaton is strictly more expressive than the ORWW-automaton. In fact, while the emptiness problem is decidable for ORWW-automata [9], it turns out that for ORRWW-automata, it is undecidable.

This paper is structured as follows. In the next section we present the definition of the ORRWW-automaton and study its deterministic variant. In Sect. 3 we consider the stateless variants of the ORRWW-automaton, and in Sect. 4 we study the class of languages that are accepted by nondeterministic ORRWW-automata. The paper closes with a short summary and some open problems.

2 Ordered RRWW-Automata

An *ordered RRWW-automaton* (ORRWW-automaton) is a one-tape machine that is described by an 8-tuple $M = (Q, \Sigma, \Gamma, \rhd, \lhd, q_0, \delta, >)$, where Q is a finite set of states, Σ is a finite input alphabet, Γ is a finite tape alphabet such that $\Sigma \subseteq \Gamma$, the letters $\rhd, \lhd \notin \Gamma$ serve as markers for the left and right border of the work space, respectively, $q_0 \in Q$ is the initial state, $>$ is a strict *partial ordering* on Γ, and

$$\delta : (Q \times ((\Gamma \cup \{\rhd\})^{\leq 1} \cdot \Gamma \cdot (\Gamma \cup \{\lhd\}))) \cup \{(q_0, \rhd\lhd)\} \rightarrow 2^{(Q \times (\{\mathsf{MVR}\} \cup \Gamma)) \cup \{\mathsf{Restart}, \mathsf{Accept}\}}$$

is the *transition relation*, which describes four different types of transition steps:

(1) A *move-right step* has the form $(q', \mathsf{MVR}) \in \delta(q, a_1a_2a_3)$, where $q, q' \in Q, a_1 \in \Gamma \cup \{\rhd\}$ and $a_2, a_3 \in \Gamma$. It causes M to shift the window one position to the right and to change to state q'. Observe that no move-right step is possible, if the window contains the right delimiter $\lhd$.
(2) A *rewrite step* has the form $(q', b) \in \delta(q, a_1a_2a_3)$, where $q, q' \in Q, a_1 \in \Gamma \cup \{\rhd\}$, $a_2, b \in \Gamma$, and $a_3 \in \Gamma \cup \{\lhd\}$ such that $a_2 > b$ holds. It causes M to replace the letter a_2 in the middle of its window by the letter b, to move the window one position to the right, and to change to state q'.
(3) A *restart step* has the form $\mathsf{Restart} \in \delta(q, a_1a_2a_3)$, where $q \in Q, a_1, a_2 \in \Gamma$, and $a_3 \in \Gamma \cup \{\lhd\}$, or $a_1 \in \Gamma, a_2 = \lhd$, and $a_3 = \lambda$ (the empty word). It causes M to restart, that is, the window is moved back to the left end of the tape, and M is reset to the initial state q_0.
(4) An *accept step* has the form $\mathsf{Accept} \in \delta(q, a_1a_2a_3)$, where $a_1 \in \Gamma \cup \{\rhd\}, a_2 \in \Gamma$, and $a_3 \in \Gamma \cup \{\lhd\}$, or $a_1 \in \Gamma$, $a_2 = \lhd$, and $a_3 = \lambda$. It causes M to halt and accept. In addition, we allow an accept step of the form $\delta(q_0, \rhd\lhd) = \{\mathsf{Accept}\}$.

If $\delta(q, u) = \emptyset$ for some pair (q, u), then M halts, when it is in state q with u in its window, and we say that M *rejects* in this situation. If $|\delta(q, u)| \leq 1$ for all pairs (q, u), then M is a *deterministic* ORRWW-automaton (det-ORRWW-automaton). Further, the letters in $\Gamma \smallsetminus \Sigma$ are called *auxiliary letters.*

Observe that for general RRWW-automata, a rewrite operation $(q', v) \in \delta(q, u)$ replaces the factor u by the word v, changes the state to q', and moves the window immediately to the right of v. In our case this would mean that a rewrite operation $(q', b) \in \delta(q, abc)$ should move the window *three* steps to the right, as it rewrites the factor abc into the word $ab'c$. However, for the stateless variant (that is, q_0 is the only state) this would mean that after a rewrite no information on the new letter would be available to the automaton, and therefore we have chosen the above interpretation for the rewrite step.

A *configuration* of an ORRWW-automaton M is a word of the form $\alpha q\beta$, where $q \in Q$ is a state, and $\alpha\beta \in \{\rhd\}\cdot\Gamma^*\cdot\{\lhd\}$ such that $|\beta| \geq 2$, and either $\alpha = \lambda$ and $\beta \in \{\rhd\}\cdot\Gamma^+\cdot\{\lhd\}$ or $\alpha \in \{\rhd\}\cdot\Gamma^*$ and $\beta \in \Gamma^+\cdot\{\lhd\}$; here $\alpha\beta$ is the current content of the tape, and it is understood that the window contains the first three letters of β or all of β, if $|\beta| \leq 3$. In addition, we admit the configuration $q_0 \rhd \lhd$. A *restarting configuration* has the form $q_0\rhd w\lhd$; if $w \in \Sigma^*$, then $q_0\rhd w\lhd$ is also called an *initial configuration.* Further, we use Accept to denote the *accepting configurations,* which are those configurations that M reaches by an accept step. A configuration of the form $\alpha q\beta$ such that $\delta(q, \beta_1) = \emptyset$, where β_1 is the current content of the window, is a *rejecting configuration.* A *halting configuration* is either an accepting or a rejecting configuration. By $\vdash_M$ we denote the *single-step computation relation* that M induces on the set of configurations, and the *computation relation* $\vdash_M^*$ of M is the reflexive and transitive closure of $\vdash_M$.

Any computation of an ORRWW-automaton M consists of certain phases. A phase, called a *cycle*, starts in a restarting configuration, the head is moved along the tape by MVR steps until a rewrite step is performed, which replaces a letter by a smaller one. After that further MVR steps may follow until, finally, a restart step is executed and thus, a new restarting configuration is reached. If no further restart operation is performed, any computation necessarily finishes in a halting configuration – such a phase is called a *tail.* It is required that each cycle contains *exactly* one rewrite step, and a tail may contain at most a single rewrite step. By $\vdash_M^c$ we denote the execution of a complete cycle, and $\vdash_M^{c^*}$ is the reflexive transitive closure of this relation. It can be seen as the *rewrite relation* that is realized by M on the set of restarting configurations.

An input $w \in \Sigma^*$ is accepted by M, if there is a computation of M which starts with the initial configuration $q_0\rhd w\lhd$ and ends with an accept step. By $L(M)$ we denote the language $L(M) = \{w \in \Sigma^* \mid q_0\rhd w\lhd \vdash_M^* \mathsf{Accept}\}$.

As each cycle contains a rewrite operation, which replaces a letter a by a letter b that is strictly smaller than a with respect to the given ordering $>$, we see that each computation of M on an input of length n consists of at most $(|\Gamma|-1)\cdot n$ many cycles. Thus, M can be simulated by a nondeterministic single-tape Turing machine in time $O(n^2)$. The following example illustrates the way in which the ORRWW-automaton works.

Example 1. For $m \geq 1$, let $L_{\mathrm{ch},m}$ be the following language:

$$L_{\mathrm{ch},m} = \{w_1 w_2 \dots w_n \in \{a,b\}^n \mid n \geq 2m \text{ and } w_m = w_{n+1-m} = w_n\},$$

that is, a word w of length $n \geq 2m$ belongs to this language iff the m-th letter and the m-th last letter both coincide with the last letter of w. We define a det-ORRWW-automaton $M = (Q, \Sigma, \Gamma, \rhd, \lhd, q_0, \delta, >)$ by taking $Q = \{q_0, q_1, \dots, q_{m-1}, q_a, q_b, q_r\}$, $\Sigma = \{a, b\}$ and $\Gamma = \Sigma \cup \{a_1, b_1, x_2, x_3, \dots, x_{m-1}\}$, by defining the partial ordering $>$ through $a > a_1 > x_i$ and $b > b_1 > x_i$ for all $i = 2, 3, \dots, m-1$, and by specifying the transition function through the following table, where $c, d, e, f \in \Sigma$:

$$\begin{array}{ll}\delta(q_0, \rhd dc) = (q_1, \mathsf{MVR}), & \delta(q_i, dce) = (q_{i+1}, \mathsf{MVR}), 1 \leq i \leq m-2,\\ \delta(q_{m-1}, dce) = (q_c, \mathsf{MVR}), & \delta(q_c, def) = (q_c, \mathsf{MVR}),\end{array}$$

$$\begin{array}{llll}
\delta(q_c, de\lhd) & = (q_r, e_1), & \delta(q_c, def_1) & = (q_r, x_2) \\
\delta(q_r, x_2c_1\lhd) & = \mathsf{Restart}, & \delta(q_c, dex_i) & = (q_r, x_{i+1}), 2 \le i \le m-2, \\
\delta(q_c, dcx_{m-1}) & = (q_c, \mathsf{MVR}), & \delta(q_c, x_{i+2}x_{i+1}x_i) & = (q_c, \mathsf{MVR}), 2 \le i \le m-3, \\
\delta(q_c, cx_{m-1}x_{m-2}) & = (q_c, \mathsf{MVR}) & \delta(q_c, x_3x_2c_1) & = (q_c, \mathsf{MVR}), \\
\delta(q_c, x_2c_1\lhd) & = \mathsf{Accept}, & \delta(q_r, x_{i+2}x_{i+1}x_i) & = (q_r, \mathsf{MVR}), 2 \le i \le m-3, \\
\delta(q_r, x_3x_2c_1) & = (q_r, \mathsf{MVR}), & \delta(q_r, c_1\lhd) & = \mathsf{Restart}.
\end{array}$$

Using its states M counts from left to right until it sees the m-th letter, say c, which it then remembers in its state. Then it rewrites the last $m-1$ letters from right to left, rewriting the last letter, say $w_n = d$, into d_1, and the letters $w_{n-1}, w_{n-2}, \ldots, w_{n+2-m}$ into $x_2, x_3, \ldots, x_{m-1}$. Finally, it checks whether w_{n+1-m}, which is the letter immediately before x_{m-1}, coincides with $w_m = c$. In the affirmative, M moves to the right, where it compare $w_m = w_{n+1-m} = c$ to the last letter d (or rather its encoding d_1). If a positive result is returned, then M accepts. It is easily seen that $L(M) = L_{\mathrm{ch},m}$ holds.

The ORWW-automaton studied in [8,9] differs from the ORRWW-automaton in that the rewrite and restart operations are combined into a joint operation. Obviously, (deterministic) ORWW-automata can be simulated by (deterministic) ORRWW-automata. Thus, it follows that all regular languages are accepted by det-ORRWW-automata. However, also the converse holds.

Theorem 2. $\mathcal{L}(\mathsf{det\text{-}ORRWW}) = \mathsf{REG}$.

Proof. Let $M = (Q, \Sigma, \Gamma, \rhd, \lhd, q_0, \delta, >)$ be a det-ORRWW-automaton, and let $L = L(M)$. Without loss of generality we can assume that M performs restart and accept operations only at the right delimiter $\lhd$. We present a det-ORWW-automaton $M' = (Q', \Sigma, \Gamma', \rhd, \lhd, q_0, \delta', >')$ that simulates M. Then $L(M') = L(M) = L$, which implies that L is a regular language. Each cycle of a computation of M is of the following form, where $u, v \in \Gamma^*, a, b, b', c, d, e \in \Gamma$, and $q_1, q_2, q_3 \in Q$:

$$\begin{aligned} q_0\rhd uabcvde\lhd &\vdash^{1+|u|}_{\mathrm{MVR}} \rhd uq_1abcvde\lhd \vdash_{\mathrm{Rewrite}} \rhd uaq_2b'cvde\lhd \\ &\vdash^{|v|+2}_{\mathrm{MVR}} \rhd uab'cvq_3de\lhd \vdash_{\mathrm{Restart}} q_0 \rhd uab'cvde\lhd, \end{aligned}$$

and in the next cycle M moves its window at least until it contains the newly written letter b' before the next rewrite step can be executed. In order for the det-ORWW-automaton M' to be able to correctly simulate the above cycle, M' must ensure (or verify) in some way that after the above rewrite operation M will eventually perform a restart. For this we let M' perform some kind of preprocessing during which it encodes certain additional information on its tape.

For each word $w \in \Gamma^+$, $|w| \ge 2$, and each letter $a \in \Gamma$, we define two sets $Q^{(a)}_{\mathrm{rs}}(w)$ and $Q^{(a)}_{+}(w)$ as follows, where $w = w_1bc$ for $b, c \in \Gamma$:

$$Q^{(a)}_{\mathrm{rs}}(w) = \{p \in Q \mid \rhd paw\lhd \vdash^{|w|-1}_{\mathrm{MVR}} \rhd aw_1p'bc\lhd \vdash_{\mathrm{Restart}} q_0\rhd aw\lhd\} \text{ and}$$
$$Q^{(a)}_{+}(w) = \{p \in Q \mid \rhd paw\lhd \vdash^{|w|-1}_{\mathrm{MVR}} \rhd aw_1p'bc\lhd \vdash_{\mathrm{Accept}} \mathsf{Accept}\}.$$

Now if the det-ORWW-automaton M' is to simulate the above cycle of M, then from the fact that $q_2 \in Q_{\text{rs}}^{(b')}(cvde)$ it sees that M will actually restart at the right end of the tape, and hence, it can safely perform the same rewrite operation and restart. Accordingly, we define a precomputation for M' that assigns, from right to left, the collection of sets $(Q_{\text{rs}}^{(a)}(z), Q_{+}^{(a)}(z))_{a\in\Gamma}$ with the first letter z_1 of each suffix z of the given input w. Thus, we define

$$\Gamma' = \Sigma \cup \{(A, (Q_1^{(a)}, Q_2^{(a)})_{a\in\Gamma}) \mid A \in \Gamma, Q_1^{(a)}, Q_2^{(a)} \subseteq Q\},$$

take $Q' = Q \cup \{q_C\}$, define $>'$ by taking $A >' (A, (Q_1^{(a)}, Q_2^{(a)})_{a\in\Gamma})$ for all $A \in \Gamma$ and all $Q_1^{(a)}, Q_2^{(a)} \subseteq Q$ and $(A, (Q_1^{(a)}, Q_2^{(a)})_{a\in\Gamma}) >' (B, (P_1^{(a)}, P_2^{(a)})_{a\in\Gamma})$ if $A > B$.

The transition function can now be defined in such a way that M' first encodes the information on the sets $(Q_{\text{rs}}, Q_+)_{a\in\Gamma}$ proceeding from right to left until it detects the position, say i, at which the next rewrite operation of M is to be simulated. Based on the information from the encoded sets of states, it then simulates this rewrite step, updating also the information on the stored sets of states of M at the current position. For this, it can extract the information on the corresponding sets from the symbol stored at position $i+1$. Observe that in the next cycle, M cannot execute a rewrite step until it has the newly written symbol in its window, that is, not to the left of position $i-1$. It can be shown that in this way M' can simulate M correctly, implying that $L(M') = L(M)$. Thus, $\mathcal{L}(\mathsf{det\text{-}ORRWW}) = \mathcal{L}(\mathsf{det\text{-}ORWW})$, which implies that $\mathcal{L}(\mathsf{det\text{-}ORRWW})$ coincides with the class REG of regular languages. □

3 On Stateless ORRWW-Automata

For restarting automata in general, each RR-variant is at least as powerful as the corresponding R-variant, but for stateless automata the situation is not that obvious. The feature of continuing to read the tape after a rewrite step has been executed is problematic for these automata, as they cannot distinguish between the phase of a cycle *before* the rewrite step and the phase *after* the rewrite step. Clearly, this distinction is important, since no rewrite steps may appear in the latter phase. For general restarting automata, this is avoided by using states, but how to deal with this situation for stateless RR-automata?

In [7] this problem has been addressed, and two options for dealing with it have been proposed. First, one can interpret any additional rewrite step within a cycle as a reject. However, this approach amounts to an *external* supervisor that aborts the computation in an unwanted situation. Here we rather follow the second option presented in [7] in which two *phases* of each cycle are distinguished: the first phase, which ends with the execution of a rewrite operation, and the second phase, which starts after the execution of a rewrite operation and ends with either a restart or an accept step. These two phases are realized by providing two separate transition functions. In [7] the corresponding stateless restarting automata are called *two-phase* restarting automata, but as we will only deal with this type of stateless ORRWW-automata, we just call them *stateless ORRWW-automata* (stl-ORRWW-automata). Formally these automata are defined as follows.

Definition 3. *A* stl-ORRWW-automaton *is described by a 7-tuple* $M = (\Sigma, \Gamma, \rhd, \lhd, \delta_1, \delta_2, >)$*, where* $\Sigma, \Gamma, \rhd, \lhd$*, and* $>$ *are defined as for ORRWW-automata, and*

$$\delta_1 : ((\Gamma \cup \{\rhd\}) \cdot \Gamma \cdot (\Gamma \cup \{\lhd\})) \cup \{\rhd\lhd\} \to 2^{\Gamma \cup \{\mathsf{MVR}, \mathsf{Accept}\}}$$

and

$$\delta_2 : (\Gamma^{\leq 2} \cdot (\Gamma \cup \{\lhd\})) \to 2^{\{\mathsf{MVR}, \mathsf{Restart}, \mathsf{Accept}\}}$$

are the transition relations. *Here it is required that* $b > b'$ *holds for each rewrite instruction* $b' \in \delta_1(abc)$*.*

A configuration of M is written as a pair (α, β), where $\alpha\beta$ is the current content of the tape and the window contains the prefix of β. Given a word $w \in \Sigma^+$ as input, the computation starts with the initial configuration $(\lambda, \rhd w \lhd)$. First, the transition relation δ_1 is used until either an accept instruction is reached, a rewrite instruction $b' \in \Gamma$ is reached, or the window contains a word for which δ_1 is undefined. In the first case, M accepts, in the second case the letter in the middle of the window is replaced by the letter b', the window is moved one step to the right, and the computation is continued by using the transition relation δ_2. Finally, in the third case M simply halts without accepting. The transition relation δ_2, which is used in the second phase of a cycle *after* the execution of a rewrite step, shifts the window to the right until either an accept instruction is executed, and then M accepts, until a restart instruction is executed, which resets the window to the left end of the tape and starts the next cycle, or until a window content is reached for which δ_2 is undefined. In the latter case M halts without accepting. For $w = \lambda$, there either is no applicable operation for the configuration $(\lambda, \rhd\lhd)$, or $\delta_1(\rhd\lhd) = \{\mathsf{Accept}\}$.

Theorem 4. $\mathcal{L}(\mathsf{stl\text{-}det\text{-}ORRWW}) = \mathcal{L}(\mathsf{stl\text{-}ORRWW}) = \mathsf{REG}$.

Proof. If L is a regular language, then there exists a stl-det-ORWW-automaton $M = (\Sigma, \Gamma, \rhd, \lhd, \delta, >)$ for L. From M we obtain an equivalent stl-det-ORRWW-automaton $M' = (\Sigma, \Gamma, \rhd, \lhd, \delta_1, \delta_2, >)$ by defining $\delta_1 = \delta$ and $\delta_2(abc) = \mathsf{MVR}$ and $\delta_2(de\lhd) = \mathsf{Restart}$ for all $a, b, c \in \Gamma$ and $de \in \Gamma^{\leq 2}$. Hence, $\mathsf{REG} \subseteq \mathcal{L}(\mathsf{stl\text{-}det\text{-}ORRWW}) \subseteq \mathcal{L}(\mathsf{stl\text{-}ORRWW})$ follows.

Conversely, let $M = (\Sigma, \Gamma, \rhd, \lhd, \delta_1, \delta_2, >)$ be a stl-ORRWW-automaton. We will prove that $L(M)$ is a regular language by showing that the Nerode relation $\sim$ of $L(M)$ has finite index (see, e.g., [2]). For doing so, we proceed as follows.

Let $w \in L(M)$, let m be an integer such that $1 \leq m \leq |w|$, and let C be an accepting computation of M on input w. During this computation M executes certain operations at position m, that is, when position m is in the middle of the window of M. These operations may include rewrite steps $b' \in \delta_1(abc)$, move-right steps $\mathsf{MVR} \in \delta_1(abc)$ or $\mathsf{MVR} \in \delta_2(abc)$, restart steps $\mathsf{Restart} \in \delta_2(abc)$, and accept steps Accept $\in \delta_1(abc)$ or Accept $\in \delta_2(abc)$. As by a rewrite step, the letter in the middle of the window is replaced by a smaller letter with respect to $>$, the rewrite steps that occur at position m are obviously ordered. Before the first rewrite step, and between two successive rewrite steps, a sequence of move-right steps may occur at position m. Here again an ordering is induced by $>$, if one of the letters at position $m-1$ or $m+1$ is rewritten during this part of the computation. It remains the case that a sequence of move-right steps occurs at position m that are all applied to the same window content abc. Some of them may be δ_1-steps, while others may be δ_2-steps. By associating the number 1 with a positive number of δ_1-steps and the number 2 with a positive number of δ_2-steps, we can assign a *type* $t \in (1 \cdot (2 \cdot 1)^* \cdot \{2, \lambda\}) \cup (2 \cdot (1 \cdot 2)^* \cdot \{1, \lambda\})$ to this sequence. By rearranging the corresponding cycles of the computation C, this computation can be transformed into a computation C' for which the type of any sequence of move-right steps with the same window content at position m is from the set $T = \{1, 2, 1 \cdot 2, 2 \cdot 1, 1 \cdot 2 \cdot 1, 2 \cdot 1 \cdot 2\}$. Additionally, we require that all cycles associated to the same number 1 appear uninterrupted within the computation. We call such a computation *normalized.*

Now let Ω be the following extended set of operations of M, where $a \in \Gamma \cup \{\rhd\}$, $b, b' \in \Gamma$ and $c \in \Gamma \cup \{\lhd\}$:

$$\begin{aligned}\Omega = \{(a, b, c, b') \mid b' \in \delta_1(abc)\} \cup \{(a, b, c, t) \mid t \in T\} \cup \\ \{(a, b, c, -) \mid \mathsf{Restart} \in \delta_2(abc)\} \cup \{(a, b, c, +_i) \mid i = 1, 2,\ \mathsf{Accept} \in \delta_i(abc)\}.\end{aligned}$$

Then the sequence of operations that are executed by M during the computation C' at position m can be described by a word $A_m^{C'}(w) \in \Omega^*$. In fact, as at most $|\Gamma| - 1$ rewrite steps can occur in this sequence, $A_m^{C'}(w)$ is of length $O(|\Gamma|)$.

With each word $w \in \Sigma^+$, we now associate two sets $S_1(w)$ and $S_2(w)$:

$$\begin{aligned}S_1(w) = \{A_{|w|}^{C}(wz) \mid\ & z \in \Sigma^* \text{ and } C \text{ is a normalized computation of } M \text{ for } wz \\ & \text{that accepts at a position} \leq |w|\}, \\ S_2(w) = \{\, A_{|w|}^{C}(wz) \mid\ & z \in \Sigma^* \text{ and } C \text{ is a normalized computation of } M \text{ for } wz \\ & \text{that accepts at a position} > |w|\}.\end{aligned}$$

We will show that there are no distinguishing extensions for $x, y \in \Sigma^+$ if $S_i(x) = S_i(y)$ for $i = 1, 2$. Accordingly, let $z \in \Sigma^*$ such that $xz \in L(M)$. Then there exists a number $i \in \{1, 2\}$ and a normalized accepting computation C_{xz} of M such that $A_{|x|}^{C_{xz}}(xz) = A \in S_i(x)$. As $S_i(y) = S_i(x)$, there exists a word $u \in \Sigma^*$ and a normalized accepting computation C_{yu} of M such that $A_{|y|}^{C_{yu}}(yu) = A$.

We claim that there is also an accepting computation C' of M for the word yz. We consider the sequences of cycles of C_{yu} and C_{xz} as working lists for constructing the cycles of C' that have their rewrite operations in the y-part and the z-part, respectively. We construct the computation C' for the word yz as follows. We divide the cycles into groups according to the different types of MVR-patterns. All consecutive cycles that contribute to the same number 1 in one pattern form a group of type 1. All consecutive cycles that contribute to the same number 2 belong to a group of type 2. Note that such a group may include short cycles that do not include a move-right step of δ_2 at the border. Additionally, each cycle that executes a rewrite at the border forms a rewrite group. We see that we have the same groups in C_{yu} and C_{xz}.

We now consider the group of cycles of C_{yu} one after the other. If we have a short cycle, we just append it to C' as the u-part is not involved. If we have a cycle of a rewrite group, we take the cycle up to position $|y|$ and complement it with the second part of the corresponding cycle of C_{xz}. If we have a group of type 1, we take the part up to position $|y| - 1$ of the first cycle and call it c_0. Then, we take all last parts starting at position $|x|$ of the cycles of the corresponding group of C_{xz} and call them $c_1, \ldots, c_k$. Finally, we append the cycles $c_0c_1, \ldots, c_0c_k$ to C'. The computation C' stays valid, as these new cycles do not make any changes in the y-part. Therefore, it is possible to execute $c_0c_1, \ldots, c_0c_k$ in this order. If we have a group of type 2, we take the last part of the last cycle of the corresponding group of C_{xz} starting at position $|x|$ and call it c_t. Then, we replace the last part of each cycle of the current group of C_{yu} starting at position $|y|$ by c_t if the cycle has length $> |y|$. Finally, we append all these cycles to C'.

This construction ends as soon as an accepting tail is encountered, which happens eventually as the computations C_{xz} and C_{yu} either both accept in the z- and u-parts, respectively, or they both accept to the left of these parts.

As there are only finitely many words $A_m^{C'}(w) \in \Omega^*$ that can occur as descriptions of sequences of operations of M, there are only finitely many different sets $S_1(x)$ and $S_2(x)$. Hence, the Nerode relation $\sim$ of $L(M)$ has finite index, which means that $L(M)$ is indeed a regular language. □

4 On Nondeterministic ORRWW-Automata

The class $\mathcal{L}(\mathsf{ORWW})$ of languages accepted by ORWW-automata is an abstract family of languages (see, e.g., [2]) that is closed under intersection, but that is not closed under reversal and complementation [9]. In addition, it contains a language that is not growing context-sensitive, while it does not even contain the deterministic linear language $\{a^nb^n \mid n \geq 1\}$. Thus, this class is incomparable to the (deterministic) linear, the (deterministic) context-free, and the growing context-sensitive languages. However, the inclusion $\mathcal{L}(\mathsf{ORWW}) \subsetneq \mathcal{L}(\mathsf{ORRWW})$ obviously holds. Actually, this inclusion is proper, as all context-free languages are accepted by ORRWW-automata.

Theorem 5. $\mathsf{CFL} \subsetneq \mathcal{L}(\mathsf{ORRWW})$.

Proof. Let $L \subseteq \Sigma^+$ be a context-free language that does not contain the empty word. Then there exists a grammar $G = (N, \Sigma, P, S)$ in quadratic Greibach normal form that generates L (see, e.g., [1]), that is, each production of P is of the form $A \to a$, $A \to aB$, or $A \to aBC$, where $a \in \Sigma$ and $A, B, C \in N$.

The language L is accepted by the ORRWW-automaton M that works as follows. Given an input $w \in \Sigma^+$, M guesses a leftmost derivation of w in G. In each step of this derivation the next symbol a of w must be produced by applying a corresponding production. Thus, the symbol a must be marked as having been read, and the nonterminals produced by that step are written in reverse order on the tape, where the rightmost of these nonterminals is marked as being 'active.' Now in each phase a production $A \rightarrow aT$ is chosen that has the active nonterminal A as its left-hand side and that produces the next symbol a of w on its right-hand side. Then the symbol a is replaced by an encoding of T^R, which is tagged. After that the marked nonterminal is deleted, the tag is removed, and the rightmost of the remaining nonterminals is marked as 'active.'

Formally the automaton M is defined as follows. We take the tape alphabet $\Gamma = \Sigma \cup \{[\lambda], [B], [BC], \overline{[B]}, \overline{[BC]}, \widetilde{[\lambda]}, \widetilde{[B]}, \widetilde{[BC]} \mid B, C \in N\}$ with the partial ordering $a > \widetilde{[CB]} > \widetilde{[C]} > \widetilde{[\lambda]} > [CB] > \overline{[CB]} > [C] > \overline{[C]} > [\lambda]$. The set of states Q and the transition relation δ are defined in such a way that, in each cycle, M scans its tape from left to right and executes one of the following steps depending on the form of the word on the tape. In the following we have $a \in \Sigma$, $B, C, D, E \in N$, and we use R to denote the set $R = \{[\lambda], [A], [AB] \mid A, B \in N\}$:

1. If the word α on the tape is of the form $\{a\} \cdot \Sigma^*$, then M can replace a by $[\lambda]$, if there is a production $S \rightarrow a$, by $[B]$, if there is a production $S \rightarrow aB$, or by $[CB]$, if there is a production $S \rightarrow aBC$.
2. If the word α is of the form $R^* \cdot \{[B], [CB]\} \cdot \{[\lambda]\}^* \cdot \Sigma^*$, then M replaces $[B]$ (or $[CB]$) by $\overline{[B]}$ (or $\overline{[CB]}$).
3. If the word α is of the form $R^* \cdot \{\overline{[B]}, \overline{[CB]}\} \cdot \{[\lambda]\}^* \cdot \{a\} \cdot \Sigma^*$, then M can replace a by $\widetilde{[\lambda]}$, if there is a production $B \rightarrow a$, by $\widetilde{D}$, if there is a production $B \rightarrow aD$, or by $\widetilde{ED}$, if there is a production $B \rightarrow aDE$.
4. If the word α is of the form $R^* \cdot \{\overline{[B]}, \overline{[CB]}\} \cdot \{[\lambda]\}^* \cdot \{\widetilde{[\lambda]}, \widetilde{[D]}, \widetilde{[ED]}\} \cdot \Sigma^*$, then M replaces $\overline{[B]}$ (or $\overline{[CB]}$) by $[\lambda]$ (or $[C]$).
5. If the word α is of the form $R^* \cdot \{[\lambda]\}^* \cdot \{\widetilde{[\lambda]}, \widetilde{[D]}, \widetilde{[ED]}\} \cdot \Sigma^*$, then M removes the tilde from $\widetilde{[\lambda]}$, $\widetilde{[D]}$, or $\widetilde{[ED]}$.
6. Finally, M halts and accepts, if the tape contains a word from $\{[\lambda]\}^*$.

It can easily be seen that $L(M) = L$, as the transitions of M are in close correspondence to the productions of G.

If the language L contains the empty word, we apply the above construction to the language $L \smallsetminus \{\lambda\}$ and then add the transition $\delta(q_0, \rhd\lhd) = \{\mathsf{Accept}\}$. □

Corollary 6. $\mathcal{L}(\mathsf{ORWW}) \cup \mathsf{CFL} \subsetneq \mathcal{L}(\mathsf{ORRWW})$.

In [9] it is shown that $\mathcal{L}(\mathsf{ORWW})$ is an abstract family of languages that is closed under intersection. The same proof can be used to show the following.

Theorem 7. $\mathcal{L}(\mathsf{ORRWW})$ *is closed under union, intersection, product, Kleene star, inverse morphisms, and non-erasing morphisms.*

However, in contrast to the situation for ORWW-automata, the class $\mathcal{L}(\mathsf{ORRWW})$ is closed under the operation of reversal, as the proof for general RRWW-automata also applies here [5].

Proposition 8. *For each ORRWW-automaton M, there exists an ORRWW-automaton M' such that $L(M') = L(M)^R$.*

Finally, the following result shows that ORRWW-automata can even accept some unary languages that are not context-free.

Proposition 9. *The unary language $L = \{a^n \mid \exists p, q > 1 : n = p \cdot q\}$ is accepted by an ORRWW-automaton.*

Proof. The languages $L_1 = \{a^n b^n \mid n > 1\}, L_2 = \{b^n a^n \mid n > 1\}, L_3 = \{a^n \mid n \geq 1\}$, and $L_4 = \{b^n \mid n \geq 1\}$ are all context-free and, therefore, they are accepted by ORRWW-automata. Now let $\varphi : \{a,b\}^* \to \{a\}^*$ denote the morphism that is defined by $\varphi(a) = \varphi(b) = a$. It is easily seen that

$$L = \varphi\left((L_1^* \cup L_1^* \cdot L_3) \cap (L_3 \cdot L_2^* \cup L_3 \cdot L_2^* \cdot L_4)\right).$$

Hence, by Theorem 7, L is accepted by an ORRWW-automaton. □

We complete this section by briefly looking at some decision problems for ORRWW-automata. It has been shown in [9] that the emptiness problem is decidable for ORWW-automata. However, as each context-free language is accepted by an ORRWW-automaton, and as the language class $\mathcal{L}(\mathsf{ORRWW})$ is closed under intersection, we obtain the following undecidabiliy result from the undecidability of the intersection-emptiness problem for context-free languages.

Corollary 10. *The emptiness problem for ORRWW-automata is undecidable.*

From an ORRWW-automaton M, we can construct an ORRWW-automaton M' for the language $L(M) \cdot \Sigma^+$, as the proof of Theorem 7 is actually constructive. Now $L(M')$ is finite, iff $L(M)$ is empty. Thus, from the undecidability of the emptiness problem, we immediately get the following.

Corollary 11. *The finiteness problem for ORRWW-automata is undecidable.*

Finally, from the corresponding results for context-free languages, it follows that for ORRWW-automata, also universality, regularity, inclusion, and equivalence are all undecidable.

5 Conclusion

We have introduced and studied the ORRWW-automaton, which is obtained from the ORWW-automaton by splitting the rewrite/restart operation of the latter into two separate operations, where, however, we still require that in any cycle of any computation, exactly one rewrite step is to be executed. We have seen that in the deterministic case, this change does not influence the expressive power of the model, and the same is true for the stateless variants. However, in the nondeterministic setting, the separation of the restart operation from the rewrite operation has quite a large impact. We still get an abstract family of languages that is closed under intersection, but in addition, we have closure under reversal. Furthermore, the class of languages that are accepted by ORRWW-automata extends the class of languages that are accepted by ORWW-automata substantially, as ORRWW-automata accept all context-free languages. Unfortunately, this entails that in this setting already the emptiness problem becomes undecidable. However, it remains open whether the language class $\mathcal{L}(\mathsf{ORRWW})$ is closed under the operation of complementation.

References

1. Autebert, J., Berstel, J., Boasson, L.: Context-free languages and pushdown automata. In: Rozenberg, G., Salomaa, A. (eds.) Handbook of Formal Languages, vol. 1, pp. 111–174. Springer, Heidelberg (1997)
2. Hopcroft, J., Ullman, J.: Introduction to Automata Theory, Languages, and Computation. Addison-Wesley, Reading (1979)
3. Jančar, P., Mráz, F., Plátek, M., Vogel, P.: Restarting automata. In: Reichel, H. (ed.) FCT 1995. LNCS, vol. 965, pp. 283–292. Springer, Heidelberg (1995)
4. Jančar, P., Mráz, F., Plátek, M., Vogel, J.: On monotonic automata with a restart operation. J. Auto. Lang. Comb. **4**, 287–311 (1999)
5. Jurdziński, T., Loryś, K., Niemann, G., Otto, F.: Some results on RWW- and RRWW-automata and their relation to the class of growing context-sensitive languages. J. Auto. Lang. Comb. **9**, 407–437 (2004)
6. Jurdziński, T., Otto, F.: Shrinking restarting automata. Intern. J. Found. Comp. Sci. **18**, 361–385 (2007)
7. Kutrib, M., Messerschmidt, H., Otto, F.: On stateless deterministic restarting automata. Acta Inform. **47**, 391–412 (2010)
8. Kwee, K., Otto, F.: On some decision problems for stateless deterministic ordered restarting automata. In: Shallit, J., Okhotin, A. (eds.) DCFS 2015. LNCS, vol. 9118, pp. 165–176. Springer, Heidelberg (2015)
9. Kwee, K., Otto, F.: On the effects of nondeterminism on ordered restarting automata. In: Freivalds, R.M., Engels, G., Catania, B. (eds.) SOFSEM 2016. LNCS, vol. 9587, pp. 369–380. Springer, Heidelberg (2016). doi:10.1007/978-3-662-49192-8_30
10. Mráz, F., Otto, F.: Ordered restarting automata for picture languages. In: Geffert, V., Preneel, B., Rovan, B., Štuller, J., Tjoa, A.M. (eds.) SOFSEM 2014. LNCS, vol. 8327, pp. 431–442. Springer, Heidelberg (2014)
11. Niemann, G., Otto, F.: Further results on restarting automata. In: Ito, M., Imaoka, T. (eds.) Proceedings of Words, Languages and Combinatorics III, pp. 352–369. World Scientific, Singapore (2003)
12. Otto, F.: Restarting automata. In: Ésik, Z., Martín-Vide, C., Mitrana, V. (eds.) Recent Advances in Formal Languages and Applications. SCI, vol. 25, pp. 269–303. Springer, Heidelberg (2006)
13. Otto, F.: On the descriptional complexity of deterministic ordered restarting automata. In: Jürgensen, H., Karhumäki, J., Okhotin, A. (eds.) DCFS 2014. LNCS, vol. 8614, pp. 318–329. Springer, Heidelberg (2014)
14. Průša, D.: Weight-reducing Hennie machines and their descriptional complexity. In: Dediu, A.-H., Martín-Vide, C., Sierra-Rodríguez, J.-L., Truthe, B. (eds.) LATA 2014. LNCS, vol. 8370, pp. 553–564. Springer, Heidelberg (2014)

Bispecial Factors in the Brun S-Adic System

Sébastien Labbé and Julien Leroy([✉])

Institut de Mathématique, Université de Liège,
Allée de la découverte 12 (B37), 4000 Liège, Belgium
{slabbe,j.leroy}@ulg.ac.be

Abstract. We study the bispecial factors in the S-adic system associated with the Brun Multidimensional Continued Fraction algorithm. More precisely, by describing how strong and weak bispecial words can appear, we get a sub-language of the Brun language for which all bispecial words are neutral.

Keywords: Substitutions · Brun · Factor complexity · Bispecial

1 Introduction

Sturmian sequences [14] are infinite sequences on a binary alphabet in which appear exactly $n+1$ distinct finite subsequences of consecutive n letters for each $n \in \mathbb{N}$. It is known that the symbolic dynamical system associated with a sturmian sequence (with the shift transformation) is minimal and is measure-theoretically isomorphic to an irrational rotation on the unit circle $\mathbb{T}_1$. The result was extended to higher dimensions when Rauzy proved in [15] that the symbolic dynamical system associated with the fixed point of the Tribonacci substitution $\sigma : 1 \mapsto 12, 2 \mapsto 13, 3 \mapsto 1$, which has $p(n) = 2n+1$ factors of length n, is measure-theoretically isomorphic to an irrational translation on the torus $\mathbb{T}_2$. Proving that Rauzy's result holds in a more general setting is still an open question known as the Pisot conjecture [1] in the case of all Pisot unimodular substitutions. However substitutive dynamical systems obtained from the iteration of *one* substitution is quite limited (frequencies of letters must be algebraic) and do not form a satisfactory generalization to larger alphabets of sturmian systems (achieving all irrational frequencies of letters).

A generalization of the Pisot conjecture was proposed in [7] in the case of S-adic symbolic dynamical systems. These shift spaces are obtained by iterating substitutions from a set S, generalizing the substitutive case where Card $S = 1$. Like it is the case for sturmian words, the sequence of substitutions is obtained from the continued fractions algorithm or some multidimensional version of it [8,17]. They proved using results from [3,4,10] that almost all S-adic shifts based on Brun's Multidimensional Continued Fraction Algorithm [9] are measurably

S. Labbé—Postdoctoral Marie Curie fellowship (BeIPD-COFUND).
J. Leroy—Postdoctoral FNRS fellowship.

S. Brlek and C. Reutenauer (Eds.): DLT 2016, LNCS 9840, pp. 280–292, 2016.
DOI: 10.1007/978-3-662-53132-7_23

conjugate to a translation on the torus $\mathbb{T}_2$. They also proved that these shifts provide a natural coding of almost all rotations on $\mathbb{T}_2$ providing a reverent generalization of sturmian systems to a three-letter alphabet.

One statement about Brun S-adic systems has remained unproved: the factor complexity. As mentioned in [7], it is believed that any Brun S-adic shift has a linear factor complexity and this is the subject of this contribution. In this work, we initiate a study of bispecial factors in the Brun S-adic systems pushing further methods already used in [6,12] and also in [5] where it was proved that $p(n) \leq \frac{5}{2}n + 1$ for Arnoux-Rauzy-Poincaré S-adic system. In the Brun system, it appears that left extensions of length 1 are not enough to study the evolution of bispecial factors. Also, some neutral bispecial factors can split into a pair of strong and weak bispecial factors which can later on merge again into a neutral bispecial factor. These phenomena are not possible in the case of the Arnoux-Rauzy-Poincaré algorithm and these are reasons why the linearity of the factor complexity for Brun S-adic systems has shown to be harder to prove.

2 Brun's Algorithm

Brun's algorithm [9] is a Multidimensional Continued Fraction Algorithm [8,17] which subtracts the second largest entry to the largest entry of a nonnegative vector in $\mathbb{R}^d_+$. In the most often used version of Brun's algorithm, the entries are sorted after each iteration. Keeping the entries sorted has the advantage of reducing the number of branches of the algorithm at each step but the disadvantage of losing the symmetry between them. In this work, we prefer to keep the symmetry and present below the unsorted version of Brun's algorithm which has 6 branches when $d = 3$. On $\Lambda = \mathbb{R}^3_+$, the *unsorted Brun's algorithm* is the map $F(x_1, x_2, x_3) = (x'_1, x'_2, x'_3)$ defined by

$$x'_{\pi 1} = x_{\pi 1}, \qquad x'_{\pi 2} = x_{\pi 2}, \qquad x'_{\pi 3} = x_{\pi 3} - x_{\pi 2}$$

where $\pi \in \mathcal{S}_3$ is the permutation of $\{1, 2, 3\}$ such that $x_{\pi 1} < x_{\pi 2} < x_{\pi 3}$. Equivalently, the map F on Λ can be defined as a linear application $F\mathbf{x} = M(\mathbf{x})^{-1}\mathbf{x}$ with $M(\mathbf{x}) = M_\pi$ if and only if $\mathbf{x} \in \Lambda_\pi$ where $\Lambda_\pi = \{(x_1, x_2, x_3) \in \Lambda \mid x_{\pi 1} < x_{\pi 2} < x_{\pi 3}\}$ defines a partition of the positive cone $\Lambda = \cup_{\pi \in \mathcal{S}_3} \Lambda_\pi$ up to a set of Lebesgue measure zero and M_π are the following elementary matrices:

$$M_{123} = \begin{pmatrix} 1 & 0 & 0 \\ 0 & 1 & 0 \\ 0 & 1 & 1 \end{pmatrix}, M_{132} = \begin{pmatrix} 1 & 0 & 0 \\ 0 & 1 & 1 \\ 0 & 0 & 1 \end{pmatrix}, M_{213} = \begin{pmatrix} 1 & 0 & 0 \\ 0 & 1 & 0 \\ 1 & 0 & 1 \end{pmatrix},$$
$$M_{231} = \begin{pmatrix} 1 & 0 & 1 \\ 0 & 1 & 0 \\ 0 & 0 & 1 \end{pmatrix}, M_{312} = \begin{pmatrix} 1 & 0 & 0 \\ 1 & 1 & 0 \\ 0 & 0 & 1 \end{pmatrix}, M_{321} = \begin{pmatrix} 1 & 1 & 0 \\ 0 & 1 & 0 \\ 0 & 0 & 1 \end{pmatrix}.$$

The algorithm F defines a *cocycle* $M_n : \Lambda \to SL(d, \mathbb{Z})$

$$M_0(\mathbf{x}) = I \quad \text{and} \quad M_n(\mathbf{x}) = M(\mathbf{x})M(F\mathbf{x}) \cdots M(F^{n-1}\mathbf{x})$$

with the cocycle property $M_{m+n}(\mathbf{x}) = M_m(\mathbf{x})M_n(F^m\mathbf{x})$. Since Brun's algorithm is strongly convergent almost everywhere when $d = 3$ [13] (also when $d = 4$ [16]), the columns of $M_n(\mathbf{x})$ are good rational approximations of $\mathbf{x}$. Indeed, an MCF algorithm is *strongly convergent* at $\mathbf{x} \in \Lambda$ with $\|\mathbf{x}\| = 1$ if for all i with $1 \leq i \leq d$, we have

$$\lim_{n\to\infty} M_n(\mathbf{x})\mathbf{e}_i - \|M_n(\mathbf{x})\mathbf{e}_i\|\mathbf{x} = 0. \tag{1}$$

3 Brun S-Adic System

3.1 S-Adic Words

Let S be a set of substitutions. A word $\mathbf{w} \in A^{\mathbb{N}}$ is said to be *S-adic* if there is a sequence $(\sigma_n : A_{n+1}^* \to A_n^*)_{n\in\mathbb{N}} \in S^{\mathbb{N}}$ and a sequence of letters $(a_n \in A_n)_{n\in\mathbb{N}}$ such that $A_0 = A$ and

$$\mathbf{w} = \lim_{n\to+\infty} \sigma_0\sigma_1\cdots\sigma_n(a_{n+1}).$$

For all $r \in \mathbb{N}$ we define the S-adic word

$$\mathbf{w}^{(r)} = \lim_{n\to+\infty} \sigma_r\sigma_{r+1}\cdots\sigma_{r+n}(a_{r+n+1}).$$

In our setting we will consider the alphabet $A = \{1, 2, 3\}$ and we usually use the set $\{i, j, k\}$ to represent A.

A directive sequence of substitutions $(\sigma_n : A_{n+1}^* \to A_n^*)_{n\in\mathbb{N}}$ is *primitive* if for all $r \in \mathbb{N}$, there exists $s \geq r$ such that for all $a \in A_r$ and all $b \in A_{s+1}$, the letter a occurs in $\sigma_r \cdots \sigma_s(b)$. Primitiveness of a directive sequence of substitutions implies the uniform recurrence of the associated S-adic word [11].

3.2 Brun Substitutions and Brun Words

For every totally irrational $\mathbf{x} \in \Lambda$, Brun's algorithm F defines a sequence of substitutions $(\sigma(F^n\mathbf{x}))_{n\in\mathbb{N}}$, where $\sigma(\mathbf{x}) = \beta_{jk}$ if and only if $\mathbf{x} \in \Lambda_{ijk}$ and β_{jk} : $i \mapsto i, j \mapsto jk, k \mapsto k$ is a substitution called *Brun substitution*. Note that the incidence matrix of β_{jk} is M_{ijk} for all $ijk \in \mathcal{S}_3$.

One can see that the allowed product of two consecutive Brun substitutions is restricted among the possibilities. One can show that after each β_{ij} only three of the six substitutions are allowed:

$$\{\sigma(\mathbf{x})\sigma(F\mathbf{x}) \mid \mathbf{x} \in \Lambda\} = \{\beta_{ij}\beta_{ij}, \beta_{ij}\beta_{ji}, \beta_{ij}\beta_{ki} \mid ijk \in \mathcal{S}_3\}.$$

Writing $\mathcal{S}_{\mathcal{B}} = \{\beta_{ij} \mid ijk \in \mathcal{S}_3\}$, the *Brun language* is:

$$\begin{aligned}\mathcal{L}_{\mathcal{B}} &= \{\sigma(\mathbf{x})\sigma(F\mathbf{x})\cdots\sigma(F^{n-1}\mathbf{x}) \mid \mathbf{x} \in \Lambda, n \in \mathbb{N}\}\\ &= \mathcal{S}_{\mathcal{B}}^* \setminus \mathcal{S}_{\mathcal{B}}^* \{\beta_{ij}\beta_{ik}, \beta_{ij}\beta_{jk}, \beta_{ij}\beta_{kj} \mid ijk \in \mathcal{S}_3\} \mathcal{S}_{\mathcal{B}}^*.\end{aligned}$$

It is a regular language accepted by the automaton represented in Fig. 1 where the label of an edge is β_{ij} whenever the edge goes to the state ij.

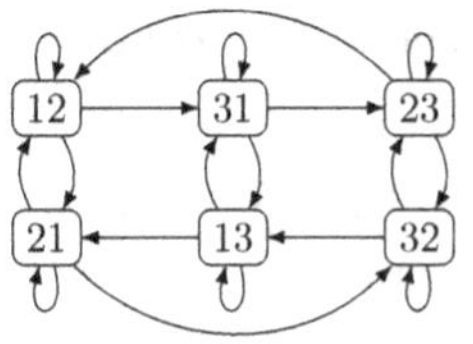

Fig. 1. The Brun language $\mathcal{L}_\mathcal{B}$ is regular.

If $\mathbf{x} \in \Lambda$ is totally irrational, then there are infinitely many $n \in \mathbb{N}$ such that $\sigma(F^n\mathbf{x})\sigma(F^{n+1}\mathbf{x}) \in \{\beta_{ij}\beta_{kj} \mid ijk \in \mathcal{S}_3\}$. This is equivalent to say that for all $ijk \in \mathcal{S}_3$, $\mathrm{Card}(\{n \in \mathbb{N} \mid \sigma(F^n\mathbf{x}) \in \{\beta_{ij}, \beta_{ik}\}) = +\infty$. This implies that $\lim_{n\to+\infty} \min_{i\in A} |\sigma(\mathbf{x})\sigma(F\mathbf{x})\cdots\sigma(F^n\mathbf{x})(i)| = +\infty$. As all Brun substitutions are prolongable on every letter, this allows us to define the S-adic infinite word

$$\lim_{n\to+\infty} \sigma(\mathbf{x})\sigma(F\mathbf{x})\cdots\sigma(F^{n-1}\mathbf{x})(1)$$

whose letter frequencies exist and are proportional to $\mathbf{x}$ by (1).

Definition 1 (**Brun word**). *A word* $\mathbf{w} \in A^\mathbb{N}$ *is a* Brun word *if it is an* $\mathcal{S}_\mathcal{B}$-*adic word whose directive sequence* $(\sigma_n)_{n\in\mathbb{N}} \in \mathcal{S}_\mathcal{B}^\mathbb{N}$ *is such that for all* $n \in \mathbb{N}$, $\sigma_0\sigma_1\cdots\sigma_n \in \mathcal{L}_\mathcal{B}$ *and for all* $ijk \in \mathcal{S}_3$, $\mathrm{Card}(\{n \in \mathbb{N} \mid \sigma_n \in \{\beta_{ij}, \beta_{ik}\}) = +\infty$.

Proposition 2. *If* $\mathbf{s} = (\sigma_n)_{n\in\mathbb{N}} \in \mathcal{S}_\mathcal{B}^\mathbb{N}$ *is the directive sequence of a Brun word* $\mathbf{w}$, *then* $(\sigma_n)_{n\in\mathbb{N}}$ *is primitive. In particular,* $\mathbf{w}$ *is uniformly recurrent.*

3.3 Relations with Arnoux-Rauzy and Poincaré Substitutions

The Brun substitutions share some relations with other well-known substitutions. For all $\{i,j,k\} = A$, we let α_i denote the *Arnoux-Rauzy substitution* [2] and π_{ij} denote the *Poincaré substitution* [5]:

$$\alpha_i : i \mapsto i, j \mapsto ji, k \mapsto ki, \qquad \pi_{ij} : i \mapsto ij, j \mapsto j, k \mapsto kij. \tag{2}$$

These are products of Brun substitutions. More precisely, for all $ijk \in \mathcal{S}_3$, we have $\pi_{ij} = \beta_{ij}\beta_{ki}$ and $\alpha_i = \beta_{ji}\beta_{ki} = \beta_{ki}\beta_{ji}$.

Note that a Poincaré substitution can appear as a product of two consecutive Brun substitutions in the Brun S-adic system, but not an Arnoux-Rauzy one. We let $\mathcal{S}_\mathcal{A}$ and $\mathcal{S}_\mathcal{P}$ respectively denote the set of Arnoux-Rauzy substitutions and the set of Poincaré substitutions: $\mathcal{S}_\mathcal{A} = \{\alpha_1, \alpha_2, \alpha_3\}$ and $\mathcal{S}_\mathcal{P} = \{\pi_{ij} \mid ijk \in \mathcal{S}_3\}$.

Now we show that Poincaré substitutions appear infinitely often as products of two consecutive Brun substitutions. This will be useful to study the extension type (defined in Sect. 4.1) of the empty word in Brun words.

Lemma 3. *Let* $\mathbf{w}$ *be a Brun word with directive sequence* $(\sigma_n)_{n\in\mathbb{N}} \in \mathcal{S}_\mathcal{B}^\mathbb{N}$. *There exist infinitely many integers* $n \in \mathbb{N}$ *such that* $\sigma_n\sigma_{n+1} \in \mathcal{S}_\mathcal{P}$. *Moreover, if* $\sigma_r = \beta_{ij}$, *then the smallest integer* $\ell \geq r$ *such that* $\sigma_\ell\sigma_{\ell+1} \in \mathcal{S}_\mathcal{P}$ *satisfies* $\sigma_\ell\sigma_{\ell+1} \in \{\pi_{ij}, \pi_{ji}\}$. *Finally, if* $\sigma_\ell\sigma_{\ell+1} = \pi_{xy}$ *with* $xy \in \{ij, ji\}$, *then* $(\sigma_n)_{n\geq\ell} \in L_{xy}$ *where*

$$L_{xy} = \pi_{xy}\beta_{yk}\mathcal{S}_\mathcal{B}^\mathbb{N} \cup \pi_{xy}\{\beta_{xk}, \beta_{kx}\}^*\{\pi_{xk}, \pi_{kx}\}\mathcal{S}_\mathcal{B}^\mathbb{N}. \tag{3}$$

3.4 Other Substitutions Used for Brun's Algorithm in the Litterature

In Jolivet's thesis or also in [4], they proposed the following substitutions for Brun's algorithm in its sorted version. Note that it was in the purpose of generating discrete planes. Therefore, their incidence matrix is dual to the matrix associated with the execution of the sorted Brun algorithm:

$$1 \mapsto 1, 2 \mapsto 2, 3 \mapsto 32, \qquad 1 \mapsto 1, 2 \mapsto 3, 3 \mapsto 23, \qquad 1 \mapsto 2, 2 \mapsto 3, 3 \mapsto 13.$$

In [10], they use the reversal of the above three substitutions for the sorted algorithm. For the unsorted one, they propose the six Brun substitutions:

$$\gamma_{ij} : j \mapsto ij, i \mapsto i, k \mapsto k, \qquad \text{for each } ijk \in \mathcal{S}_3,$$

with a language of allowed words of length two: $\{\gamma_{ij}\gamma_{ij}, \gamma_{ij}\gamma_{ji}, \gamma_{ij}\gamma_{jk} \mid ijk \in \mathcal{S}_3\}$. More recently, in [7], they present the Brun's algorithm in its sorted version using the following substitutions:

$$1 \mapsto 1, 2 \mapsto 23, 3 \mapsto 3, \qquad 1 \mapsto 1, 2 \mapsto 3, 3 \mapsto 23, \qquad 1 \mapsto 3, 2 \mapsto 1, 3 \mapsto 23.$$

One observes that any S-adic word obtained by the above substitutions of sorted Brun algorithm can be obtained as a Brun word with the unsorted algorithm.

4 Bispecial Factors Under Brun Substitutions

In this section we define the extension type of a word. We also describe the extensions of a word under the application of a Brun substitution.

4.1 Special Factors and Extension Type

Let $\mathbf{w}$ be a (infinite) word over A. We let $\mathrm{Fac}(\mathbf{w})$ denote the set of factors of $\mathbf{w}$:

$$\mathrm{Fac}(\mathbf{w}) = \{u \in A^* \mid \exists i \in \mathbb{N} : \mathbf{w}_i \cdots \mathbf{w}_{i+|u|-1} = u\}.$$

Let $u \in \mathrm{Fac}(\mathbf{w})$ and $\ell \in \mathbb{N}$. The *ℓ-extension set* of u is the set $E_\ell(u, \mathbf{w}) = \{(a, b) \in A^\ell \times A^1 \mid aub \in \mathrm{Fac}(\mathbf{w})\}$. We represent it by a tabular of the form

$$E_\ell(u, \mathbf{w}) = \begin{array}{c|ccc} u & 1 & \cdots & k \\ \hline v_1 & & \times & \times \\ \cdots & & & \\ v_n & \times & & \end{array},$$

where a symbol $\times$ in position (v_i, j) means that (v_i, j) belongs to $E_\ell(u, \mathbf{w})$. When the context is clear we omit the information on $\mathbf{w}$ and simply write $E_\ell(u)$. In this paper we will only work with $\ell \in \{1, 2\}$.

Two extension sets $E_\ell(u)$ and $E_\ell(v)$ are said to be *equivalent* if one can be obtained from the other by a permutation of the alphabet. The equivalent class of an extension set is called an *extension type*.

Given an extension set $E_\ell(u, \mathbf{w})$, we consider the corresponding set of *left extensions* $E_\ell^-(u, \mathbf{w}) = \pi_1(E_\ell(u, \mathbf{w}))$ (resp. *of right extensions* $E_\ell^+(u, \mathbf{w}) = \pi_2(E_\ell(u, \mathbf{w}))$), where π_1 (resp. π_2) represents the projection on the first (resp. second) component. We assume that the reader is familiar with the notion of left, right and bispecial words. For definitions, see [6, Chap. 4].

4.2 Antecedents, Extended Images and Their Extension Types

The next lemma allows to define the antecedents of a word under β_{ij}. It directly follows from the fact that the set $\{ij, j, k\}$ forms a prefix code.

Lemma 4 (Synchronization lemma). *Let i, j, k such that $\{i, j, k\} = A$. Consider a word $u \in A^*$ and let w be a factor of $\beta_{ij}(u)$.*

(i) If w is empty or belongs to $\{i, k\}A^*$, there exists a unique word $v \in A^*$ and a unique $s \in \{\varepsilon, i\}$ such that $w = \beta_{ij}(v) \cdot s$. We say that v is the antecedent *of w under β_{ij}.*

(ii) If $w \in jA^*$, there is a unique word $v \in A^*$ and a unique $s \in \{\varepsilon, i\}$ such that $w = j \cdot \beta_{ij}(v) \cdot s = \beta_{ij}(jv) \cdot s$. We say that v and jv are the two antecedents *of w under β_{ij}.*

Definition 5. *Suppose that v is an antecedent of w under σ as in Lemma 4. In this case, we say that w is an* extended image *of v. In particular, if w is a left special (resp. right special, bispecial) factor in $\sigma(u)$, then we say that it is a* left special *(resp.* right special, bispecial*)* extended image *of v under σ.*

The next lemma provides the link between the extensions of a word and those of its extended images.

Lemma 6 (Extensions). *Let i, j, k such that $\{i, j, k\} = A$. Let $u \in A^*$ and v be a factor of u. We assume that for all $(a, b) \in E_1(v)$, there exists a letter e such that $eavb$ is also a factor of u. The extensions of v in u are related to the extensions of $\beta_{ij}(v)$ and $j\beta_{ij}(v)$ considered as factors of $\beta_{ij}(u)$ as follows:*

$$\begin{array}{lll}
(i,b) \in E_1(v) & \iff (j,b) \in E_1(\beta_{ij}(v)) \ \ \textit{and} & (i,b) \in E_1(j\beta_{ij}(v)), \\
(ij,b) \ \textit{or}\ (jj,b) \in E_2(v) & \iff (j,b) \in E_1(\beta_{ij}(v)) \ \ \textit{and} & (j,b) \in E_1(j\beta_{ij}(v)), \\
(kj,b) \in E_2(v) & \iff (j,b) \in E_1(\beta_{ij}(v)) \ \ \textit{and} & (k,b) \in E_1(j\beta_{ij}(v)), \\
(k,b) \in E_1(v) & \iff (k,b) \in E_1(\beta_{ij}(v)), & \\
v = jv' \textit{with}\ (i,b) \in E_1(v') & \iff (i,b) \in E_1(\beta_{ij}(v)). &
\end{array}$$

Lemma 7 (Extended images). *Consider the same hypothesis as in Lemma 6.*

1. *If v is right special in u, then $\beta_{ij}(v)$ is a right special factor of $\beta_{ij}(u)$.*
2. *If v is left special in u, then v has at least one left special extended image in $\beta_{ij}(u)$.*
3. *If v is bispecial factor of u such that $\beta_{ij}(v)$ is not a left special factor of $\beta_{ij}(u)$, then $j\beta_{ij}(v)$ is a right special factor of $\beta_{ij}(u)$.*

Therefore, if v is a bispecial factor of u, then it has one or two bispecial extended images under β_{ij} in $\beta_{ij}(u)$; they are $\beta_{ij}(v)$ or $j \cdot \beta_{ij}(v)$.

Lemma 8 (Antecedents). *Let i, j, k such that $\{i, j, k\} = A$. Consider a word $u \in A^*$ and w a bispecial factor of $\beta_{ij}(u)$. Then at least one antecedent of w under β_{ij} is a bispecial factor of u. We call it* a bispecial antecedent *of w.*

5 Bispecial Words in the Brun System

In this section we study the set of bispecial words in a Brun word. We first show that any bispecial factor can be canonically desubstituted until the empty word is reached. We then define the descendants of a bispecial word and describe those of the empty word. At the end of the section, we give an example that illustrates some of the results that we obtained (see Fig. 2).

General Assumption. *In all what follows, we assume that* $\mathbf{w}$ *is a Brun word with directive sequence $(\sigma_n)_{n\in\mathbb{N}}$.*

5.1 Desubstitution of Bispecial Words

Definition 9 (nth-antecedent). *Let u be a bispecial factor of $\mathbf{w}$. Let $u^{(0)} = u$ and $u^{(i+1)}$ be the shortest bispecial antecedent of $u^{(i)}$ under σ_i for $i \geq 0$. We say that $u^{(n)}$ is the* n-th antecedent *of u. Observe that $u^{(n)}$ is a factor of $\mathbf{w}^{(n)}$.*

With Brun substitutions, as opposed to the Arnoux-Rauzy-Poincaré substitutions (2), we are unable to prove $|v| < |w|$ for any antecedent v of a bispecial word $w \neq \varepsilon$. This is not a problem since it holds for the n-th antecedent of w, for some $n \geq 1$, under the hypothesis that w is a factor of an S-adic Brun word.

Lemma 10. *If $u \neq \varepsilon$ is a bispecial word of $\mathbf{w}$, there is $s \geq 1$ such that $u^{(s)} = \varepsilon$.*

Definition 11 (Descendants). *Let u be a bispecial factor of $\mathbf{w}^{(s)}$ for some $s \in \mathbb{N}$. A bispecial factor v of $\mathbf{w}^{(r)}$, $r < s$, is called a* descendant *of u if there exists a sequence $(u_r, u_{r+1}, \ldots, u_s)$ such that $u_r = v$, $u_s = u$ and each u_ℓ, $r \leq \ell < s$ is a bispecial extended image of $u_{\ell+1}$. We let* $\mathrm{desc}(u)$ *denote the set of descendants of u and, for $r < s$, we let* $\mathrm{desc}_r(u)$ *denote the set of bispecial factors of $\mathbf{w}^{(r)}$ that are descendants of u.*

5.2 Extension Type of the Empty Word

The aim of this paper is to study bispecial words in $\mathbf{w}$. Since any such bispecial is a descendant of the empty word in some $\mathbf{w}^{(s)}$, the first step is to study the possible extension types of the empty word. The next result in particular ensures that the empty word is always a neutral bispecial factor. In the next section we will show that the extension type of a bispecial word essentially governs the extension type of any of its bispecial extended image. However, as seen in Lemma 6, we sometimes need to consider left extensions of length 2 to be able to describe those of a bispecial extended image. In the next result, we thus describe the 2-extension types of the empty word.

Theorem 12 *For all $s \in \mathbb{N}$, the empty word is a neutral bispecial factor of $\mathbf{w}^{(s)}$. More precisely, if $\sigma_s = \beta_{ij}$ and $\mathbf{s} = (\sigma_r)_{r \geq s}$ for some $\{i,j,k\} = A$, then the 2-extension type of the empty word is*

E_1 *if* $\mathbf{s} \in M_{ij}$,
E_2 *if* $\mathbf{s} \in (\beta_{ij}\beta_{ji}\{\beta_{ij},\beta_{ji}\}^* L_{ij}) \cup N_{ij}$,
E_3 *if* $\mathbf{s} \in (\beta_{ij}\beta_{ij}\{\beta_{ij},\beta_{ji}\}^* L_{ij}) \cup \beta_{ij} L_{ij}$,
E_4 *if* $\mathbf{s} \in (\beta_{ij}\beta_{ji}\{\beta_{ij}\beta_{ji}\}^* L_{ji}) \cup \beta_{ij} L_{ji}$,
E_5 *if* $\mathbf{s} \in \beta_{ij}\beta_{ij}\{\beta_{ij},\beta_{ji}\}^* L_{ji}$,

where L_{ij} as defined in Eq. (3) *and $L_{ij} = M_{ij} \cup N_{ij}$ with*

$$M_{ij} = \pi_{ij}\{\beta_{ik},\beta_{ki}\}^* \pi_{ki} \mathcal{S}_{\mathcal{B}}^{\mathbb{N}} \cup \pi_{ij}\beta_{jk}\mathcal{S}_{\mathcal{B}}^{\mathbb{N}},$$
$$N_{ij} = \pi_{ij}\{\beta_{ik},\beta_{ki}\}^* \pi_{ik} \mathcal{S}_{\mathcal{B}}^{\mathbb{N}},$$

and E_1, E_2, E_3, E_4 and E_5 are as follows:

E_1	i	j	k
ji		×	
ki		×	
ij	×	×	×
jj			×
jk	×		

,

E_2	i	j	k
ji		×	
ki		×	
ij	×	×	×
jj	×		
jk	×		

,

E_3	i	j	k
ji		×	
ki		×	
ij		×	
jj	×	×	×
jk	×		

,

E_4	i	j	k
ji		×	
ij	×	×	×
jj	×		
kj	×		
jk		×	

,

E_5	i	j	k
ji		×	
ij		×	
jj	×	×	×
kj	×		
jk		×	

.

5.3 Left Extensions of Length 2 Are Sufficient

As already stated and as seen in Lemma 6, we sometimes need to consider left extensions of length 2 to be able to determine the left extensions of the longer extended images. In this section, we show that considering 2-extensions is sufficient to recover 2-extensions of any descendant. For a word u and an integer $x \geq 1$, we let $u_{[-x:]}$ denote the suffix of length x of u.

Definition 13. *Assume that i,j,k are such that $A = \{i,j,k\}$. We define the function $\varphi_{ij} : A^2 \to A^2$ and the partial function $\psi_{ij} : A^2 \to A^2$ by*

$$\varphi_{ij}(x) = (\beta_{ij}(x))_{[-2:]} \quad \text{and} \quad \psi_{ij}(x) = \begin{cases} (j\beta_{ij}(x)j^{-1})_{[-2:]} & \text{if } \beta_{ij}(x) \in A^*j, \\ \text{undefined} & \text{otherwise.} \end{cases}$$

Proposition 14. *Let $s \in \mathbb{N}$ and assume that $\sigma_s = \beta_{ij}$ and that u is a factor of $\mathbf{w}^{(s+1)}$.*

1. $E_2(j\beta_{ij}(u), \mathbf{w}^{(s)}) = \{(\psi_{ij}(a), b) \mid (a,b) \in E_2(u, \mathbf{w}^{(s+1)}), \beta_{ij}(a) \in A^*j\}$.
2. $E_2(\beta_{ij}(u), \mathbf{w}^{(s)}) = \{(\varphi_{ij}(a), b) \mid (a,b) \in E_2(u, \mathbf{w}^{(s+1)})\}$ *if $u \in \{i,k\}A^*$ or $u = ju'$ for some $u' \in A^*$ which is not left special.*

Note that if $u = ju'$ for some left special $u' \in A^*$, then the equation in item 2. above does not hold but the equation $\beta_{ij}(u) = j\beta_{ij}(u')$ allows to use item 1.

5.4 First Descendants of the Empty Word

In this section we show that the first descendants of the empty word are always neutral bispecial words. We also show that if the empty word of $\mathbf{w}^{(s)}$ has no descendant which has 3 left extensions of length 1 and 3 left extensions of length 2, then all its descendants are neutral bispecial. Below, we denote the left valence of a factor v by $d_\ell^-(v) = \mathrm{Card}(E_\ell^-(v))$ for $\ell \in \{1, 2\}$. Given a bispecial word u of $\mathbf{w}^{(s)}$, for all $\ell < s$ we consider the multiset $D_\ell(u) = \{(d_1^-(v), d_2^-(v)) \mid v \in \mathrm{desc}_\ell(u)\}$.

Theorem 15. *Let $s \geq 1$ and consider ε as a bispecial factor of $\mathbf{w}^{(s)}$. One of the following occurs.*

1. *For all $r < s$, $(3, 3) \notin D_r(\varepsilon)$ and all bispecial words in $\mathrm{desc}(\varepsilon)$ are neutral.*
2. *There exists $r < s$ such that $(3, 3) \in D_r(\varepsilon)$ and one of the following occurs:*
   (a) $D_r(\varepsilon) = \{(3, 3)\}$. Furthermore, the bispecial word v such that $\mathrm{desc}_r(\varepsilon) = \{v\}$ is neutral.
   (b) $D_r(\varepsilon) = \{(2, 2), (3, 3)\}$. Furthermore, if v_1 and v_2 are the bispecial words such that $\mathrm{desc}_r(\varepsilon) = \{v_1, v_2\}$, with $d_1^-(v_1) = 3$ and $d_1^-(v_2) = 3$, then v_1 is neutral, v_2 is ordinary and the longest proper suffix of v_2 is not left special.

   Finally, if r is the greatest such integer, then all bispecial words in the set $\bigcup_{r<\ell<s} \mathrm{desc}_\ell(\varepsilon)$ are neutral.

Proposition 16. *Let $s \geq 1$ and assume that u is a non-empty bispecial factor of $\mathbf{w}^{(s)}$ such that $d_2^-(u) = 2$ and whose longest proper suffix is not left special. Then for all $r < s$, $\mathrm{desc}_r(u)$ contains a unique bispecial word and this word has the same bispecial multiplicity as u.*

5.5 Descendance of Bispecial Factors u with $d_2^-(u) = d_1^-(u) = 3$

By Theorem 15 and Proposition 16, strong and weak bispecial words can only occur as descendant of a neutral bispecial words u with $d_2^-(u) = d_1^-(u) = 3$. In this section we show that such a word u can have a descendant with the same property and we describe the sub-language of $\mathcal{L}_\mathcal{B}$ that makes this happen.

Definition 17. *For $ijk \in \mathcal{S}_3$, we define the regular language $\Gamma_{ijk} = \beta_{kj}\beta_{jk}^+\beta_{ij}$.*

By definition of $\mathcal{L}_\mathcal{B}$, we have $\Gamma_{xyz}\beta_{ij} \subset \mathcal{L}_\mathcal{B}$ if and only if $xyz \in \{ijk, jik, jki\}$. Furthermore, if $\sigma_0 \cdots \sigma_{s-1}\beta_{ij} \in \mathcal{L}_\mathcal{B}$, then there exists $r < s$ such that $\sigma_{[r,s)} = \sigma_r \cdots \sigma_{s-1}$ is a suffix of a word in $\Gamma_{ijk} \cup \Gamma_{jik} \cup \Gamma_{jki}$. The next result concerns the descendance of a bispecial word with left valence 3 under the application of a product of substitution in some Γ_{xyz}.

Theorem 18. *We assume that $\sigma_s = \beta_{ij}$ for some $s \geq 1$ and that u is a neutral bispecial factor of $\mathbf{w}^{(s)}$ such that $d_1^-(u) = d_2^-(u) = 3$. We also suppose that there exists $r < s$ such that $\sigma_{[r,s)} \in \Gamma_{ijk} \cup \Gamma_{jik} \cup \Gamma_{jki}$. We have the following.*

1. $\mathrm{desc}_r(u) = \{v\}$, *where* v *is a neutral bispecial factor of* $\mathbf{w}^{(r)}$ *such that* $d_1^-(v) = d_2^-(v) = 3$. *In particular,* v *is ordinary in* $\mathbf{w}^{(r)}$ *if and only if* u *is ordinary in* $\mathbf{w}^{(s)}$.
2. *for all* ℓ *such that* $r < \ell < s$, $\mathrm{Card}(\mathrm{desc}_\ell(u)) \in \{1, 2\}$ *and all bispecial words in* $\mathrm{desc}_\ell(u)$ *have left valence 2. Furthermore, if* $\mathrm{Card}(\mathrm{desc}_\ell(u)) = 2$, *then one is the longest proper suffix of the other.*

The previous result describes what happens for the descendants of a bispecial factor u of $\mathbf{w}^{(s)}$ with $d_1^-(u) = d_2^-(u) = 3$ and $\sigma_s = \beta_{ij}$ when some product of substitution $\sigma_{[r,s)}$ belongs to $\Gamma_{ijk} \cup \Gamma_{jik} \cup \Gamma_{jki}$. To describe what happens when this is not the case, we need the following notation: given a language L, $\mathrm{Suff}(L)$ is the set of suffixes of words in L. For a bispecial word v, we also let $m(v)$ denote its bispecial multiplicity.

Proposition 19. *We assume that* $\sigma_s = \beta_{ij}$ *for some* $s \in \mathbb{N}$ *and that* u *is a neutral bispecial factor of* $\mathbf{w}^{(s)}$ *such that* $d_1^-(u) = d_2^-(u) = 3$. *We also assume that there exists* $r < s - 1$ *such that* $\sigma_{[r+1,s)} \in \mathrm{Suff}(\Gamma_{ijk} \cup \Gamma_{jik} \cup \Gamma_{jki})$ *and* $\sigma_{[r,s)} \notin \mathrm{Suff}(\Gamma_{ijk} \cup \Gamma_{jik} \cup \Gamma_{jki})$. *Then for all* $\ell < s$, *we have the equality of multisets*

$$\{m(v) \mid v \in \mathrm{desc}_\ell(u)\} = \{m(v) \mid v \in \mathrm{desc}_{s-1}(u)\}.$$

5.6 Occurrences of Strong and Weak Bispecial Factors

Let u be a neutral bispecial factor of $\mathbf{w}^{(s)}$ such that $d_1^-(u) = d_2^-(u) = 3$. By Theorem 18 and Proposition 19, the bispecial multiplicity of the descendants of u is completely determined by what happens between two occurrences of a bispecial word v with $d_1^-(v) = d_2^-(v) = 3$ in the sequence $(\mathrm{desc}_r(u))_{0 \leq r \leq s}$. The first result of this section shows that strong and weak bispecial words can only appear when u is not ordinary.

Proposition 20. *Assume that* u *is an ordinary bispecial factor of* $\mathbf{w}^{(s)}$ *such that* $d_1^-(u) = d_2^-(u) = 3$. *All bispecial words in* $\mathrm{desc}(u)$ *are ordinary.*

If u is a neutral bispecial, then strong and weak bispecial words can appear in $\mathrm{desc}(u)$ depending on which letter $a \in A$ is such that $d^+(au) = 3$ and on which $\mathrm{Suff}(\Gamma_{xyz})$ the product $\sigma_{[r+1,s)}$ of Proposition 19 belongs to. This can be explained using Proposition 14 as follows. When we apply a Brun morphism β_{xy} on $\mathbf{w}^{(s)}$, the lines $L_x = E^+(xu)$ and $L_y = E^+(yu)$ are merged to one line in the extension set of $\beta_{xy}(u)$. For the other bispecial extended image $y\beta_{xy}(u)$, its extension set has two lines that are copies of L_x and L_y. Depending one whether $a = z$ or $a \in \{x, y\}$, we get a pair of strong and weak bispecial words or we get ordinary bispecial words.

Proposition 21. *Assume that* $\sigma_s = \beta_{ij}$ *for some* $s \geq 1$ *and that* u *is a neutral non-ordinary bispecial factor of* $\mathbf{w}^{(s)}$ *such that* $d_1^-(u) = d_2^-(u) = 3$. *Let also* $a \in A$ *such that* $d^+(au) = 3$. *Let finally* $r < s$ *such that* $\sigma_{[r,s)} \in \mathrm{Suff}(\Gamma_{xyz})$ *with* $xyz \in \{ijk, jik, jki\}$. *One of the following occurs:*

$$
\begin{pmatrix}
E(w) & 1 & 2 & 3 \\
21 & \times & \times & \times \\
32 & & & \times \\
23 & \times & & \\
\multicolumn{4}{l}{m(w)=0,\ \text{neutral}}
\end{pmatrix}
\xleftarrow{\ \beta_{32}\ }
\begin{pmatrix}
E(w) & 1 & 2 & 3 \\
12 & \times & \times & \times \\
32 & & & \times \\
23 & \times & & \\
\multicolumn{4}{l}{m(w)=0,\ \text{ord.}}
\end{pmatrix}
\xleftarrow{\ \beta_{23}\ }
\begin{pmatrix}
E(w) & 1 & 2 & 3 \\
21 & \times & & \\
31 & \times & & \\
12 & \times & \times & \times \\
22 & & & \times \\
23 & \times & & \\
\multicolumn{4}{l}{m(w)=0,\ \text{neutral}}
\end{pmatrix}
$$

β_{23} (downward), β_{21}, β_{31}

$$
\begin{pmatrix}
E(w) & 1 & 2 & 3 \\
31 & \times & \times & \times \\
23 & & & \times \\
33 & \times & & \\
\multicolumn{4}{l}{m(w)=1,\ \text{strong}} \\[4pt]
E(w) & 1 & & 3 \\
32 & & & \times \\
23 & \times & & \\
\multicolumn{4}{l}{m(w)=-1,\ \text{weak}}
\end{pmatrix}
\xrightarrow{\ \beta_{31}\ }
\begin{pmatrix}
E(w) & 1 & 2 & 3 \\
31 & \times & \times & \times \\
13 & \times & & \\
23 & & & \times \\
\multicolumn{4}{l}{m(w)=1,\ \text{strong}} \\[4pt]
E(w) & 1 & & 3 \\
31 & \times & & \\
12 & & & \times \\
\multicolumn{4}{l}{m(w)=-1,\ \text{weak}}
\end{pmatrix}
\xrightarrow{\ \beta_{31}\ }
\begin{pmatrix}
E(w) & 1 & 2 & 3 \\
31 & \times & \times & \times \\
13 & \times & & \\
23 & & & \times \\
\multicolumn{4}{l}{m(w)=1,\ \text{strong}} \\[4pt]
E(w) & 1 & & 3 \\
11 & \times & & \\
12 & & & \times \\
\multicolumn{4}{l}{m(w)=-1,\ \text{weak}}
\end{pmatrix}
\xrightarrow{\ \beta_{13}\ }
\begin{pmatrix}
E(w) & 1 & 2 & 3 \\
31 & \times & \times & \times \\
32 & & & \times \\
13 & \times & & \\
\multicolumn{4}{l}{m(w)=0,\ \text{neutral}}
\end{pmatrix}
$$

β_{13}

$$
\begin{pmatrix}
E(w) & 1 & 2 & 3 \\
11 & \times & \times & \times \\
21 & & & \times \\
13 & \times & & \\
\multicolumn{4}{l}{m(w)=0,\ \text{ord.}} \\[4pt]
E(w) & 1 & 2 & 3 \\
21 & \times & \times & \times \\
32 & & & \times \\
\multicolumn{4}{l}{m(w)=0,\ \text{ord.}}
\end{pmatrix}
\xrightarrow{\ \beta_{13}\ }
\begin{pmatrix}
E(w) & 1 & 2 & 3 \\
21 & & & \times \\
31 & \times & \times & \times \\
13 & \times & & \\
\multicolumn{4}{l}{m(w)=0,\ \text{ord.}} \\[4pt]
E(w) & 1 & 2 & 3 \\
32 & & & \times \\
13 & \times & \times & \times \\
\multicolumn{4}{l}{m(w)=0,\ \text{ord.}}
\end{pmatrix}
\xrightarrow{\ \beta_{13}\ }
\begin{pmatrix}
E(w) & 1 & 2 & 3 \\
21 & & & \times \\
31 & \times & \times & \times \\
13 & \times & & \\
\multicolumn{4}{l}{m(w)=0,\ \text{ord.}} \\[4pt]
E(w) & 1 & 2 & 3 \\
32 & & & \\
33 & \times & \times & \times \\
\multicolumn{4}{l}{m(w)=0,\ \text{ord.}}
\end{pmatrix}
\xrightarrow{\ \beta_{31}\ }
\begin{pmatrix}
E(w) & 1 & 2 & 3 \\
31 & \times & \times & \times \\
12 & & & \times \\
13 & \times & & \\
\multicolumn{4}{l}{m(w)=0,\ \text{neutral}}
\end{pmatrix}
$$

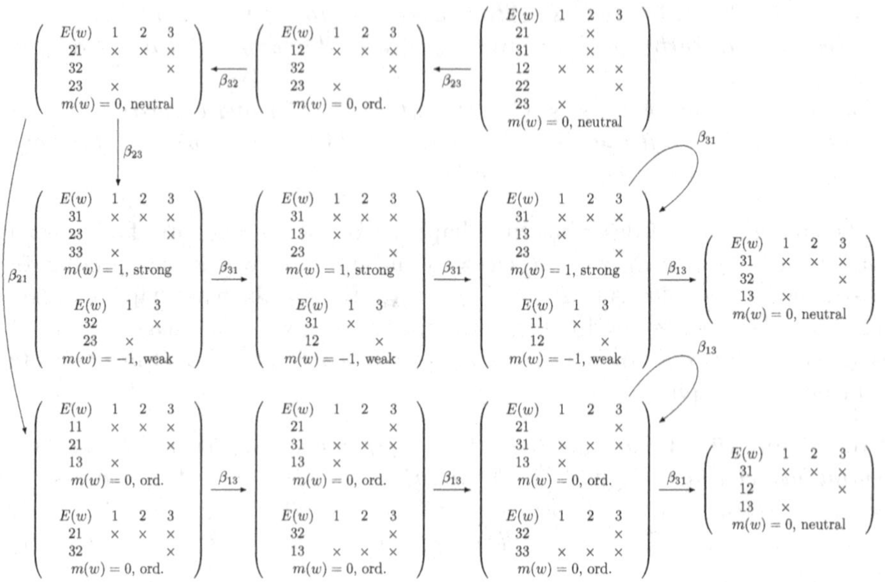

Fig. 2. Neutral bispecial words with left valence 3 can split either into two neutral bispecial words or, into a pair of a strong one and a weak one. Above $\beta_{32}\beta_{23}$ is applied on ε. Then we can apply morphisms in $\Gamma_{321}\cup\Gamma_{231}\cup\Gamma_{213}$. The figure illustrates $\Gamma_{213}=\beta_{31}\beta_{13}^{+}\beta_{21}$ and $\Gamma_{231}=\beta_{13}\beta_{31}^{+}\beta_{23}$ where strong and weak bispecial factors are created.

1. $a\in\{x,y\}$ and all bispecial words in $\mathrm{desc}_{s-1}(u)$ are ordinary;
2. $a=z$ and $\mathrm{desc}_{s-1}(u)=\{v_1,v_2\}$, where $m(v_1)=+1$ and $m(v_2)=-1$. In particular, we have $v_1=\beta_{xy}(u)$ and $v_2=y\beta_{xy}(u)$.

We now give an example that illustrates all results that we obtained (see Fig. 2). In that example, we describe the first elements of the sequence

$$(\mathrm{desc}_s(\varepsilon),\mathrm{desc}_{s-1}(\varepsilon),\mathrm{desc}_{s-2}(\varepsilon),\dots),$$

where ε is considered as a bispecial of $\mathbf{w}^{(s)}$ whose 2-extension set corresponds to E_1 in Theorem 12. In this example, the extension set on the top left is the one of a bispecial word u with $d_2^{-}(u)=d_1^{-}(u)=3$. We illustrate the fact that the multiplicity of its descendants depends on which Γ_{xyz} the product $\sigma_{[r,s)}$ belongs to.

6   Further Work

The results of the paper allow to understand how can appear strong and weak bispecial factors in a Brun word. These are preliminary results to perfectly understand the factor complexity of a Brun word. The missing information to complete this knowledge concerns the length of bispecial words. To ensure a linear complexity, we need to prove that strong and weak bispecial factors are "well distributed" in the sequence of bispecial factors ordered by length. Experimentally, strong and weak bispecial factors come by pairs and alternate. This is supported by Proposition 21 where we show that when they appear, the strong one is a suffix of the weak one. However this property is not preserved under the application of Brun morphisms so more work needs to be done. If strong and weak bispecial words indeed alternate, then the factor complexity of any Brun word is always between $2n + 1$ and $3n + 1$. Like in the Arnoux-Rauzy-Poincaré system, the upper bound should be improvable to $\frac{5}{2}n + 1$.

References

1. Akiyama, S., Barge, M., Berthé, V., Lee, J.Y., Siegel, A.: On the Pisot substitution conjecture. Preprint (2014)
2. Arnoux, P., Rauzy, G.: Représentation géométrique de suites de complexité 2n + 1. Bull. Soc. Math. Fr. **119**(2), 199–215 (1991)
3. Avila, A., Delecroix, V.: Some monoids of Pisot matrices. Preprint, June 2015. http://arXiv.org/abs/1506.03692
4. Berthé, V., Bourdon, J., Jolivet, T., Siegel, A.: A combinatorial approach to products of Pisot substitutions. Ergod. Theory Dyn. Syst. First View, 1–38. http://journals.cambridge.org/article_S0143385714001412
5. Berthé, V., Labbé, S.: Factor complexity of S-adic words generated by the Arnoux-Rauzy-Poincaré algorithm. Adv. Appl. Math. **63**, 90–130 (2015)
6. Berthé, V., Rigo, M. (eds.): Combinatorics, Automata and Number Theory, Encyclopedia of Mathematics and its Applications, vol. 135. Cambridge University Press, Cambridge (2010)
7. Berthé, V., Steiner, W., Thuswaldner, J.M.: Geometry, dynamics, and arithmetic of S-adic shifts. Preprint (2014). http://arXiv.org/abs/1410.0331
8. Brentjes, A.J.: Multidimensional Continued Fraction Algorithms. Mathematisch Centrum, Amsterdam (1981)
9. Brun, V.: Algorithmes euclidiens pour trois et quatre nombres. In: Treizième congrès des mathèmaticiens scandinaves, tenu à Helsinki 18–23 août 1957, pp. 45–64. Mercators Tryckeri, Helsinki (1958)
10. Delecroix, V., Hejda, T., Steiner, W.: Balancedness of Arnoux-Rauzy and Brun words. In: Karhumäki, J., Lepistö, A., Zamboni, L. (eds.) WORDS 2013. LNCS, vol. 8079, pp. 119–131. Springer, Heidelberg (2013)
11. Durand, F.: Corrigendum and addendum to: "Linearly recurrent subshifts have a finite number of non-periodic subshift factors". Ergod. Theory Dyn. Syst. **23**, 663–669 (2003)
12. Klouda, K.: Bispecial factors in circular non-pushy D0L languages. Theoret. Comput. Sci. **445**, 63–74 (2012). http://dx.doi.org/10.1016/j.tcs.2012.05.007
13. Lagarias, J.C.: The quality of the Diophantine approximations found by the Jacobi-Perron algorithm and related algorithms. Monatsh. Math. **115**(4), 299–328 (1993). http://dx.doi.org/10.1007/BF01667310
14. Morse, M., Hedlund, G.A.: Symbolic Dynamics II. Sturmian Trajectories. Am. J. Math. **62**(1), 1–42 (1940). http://www.jstor.org/stable/2371431
15. Rauzy, G.: Nombres algébriques et substitutions. Bulletin de la Société Mathématique de France **110**, 147–178 (1982)

16. Schratzberger, B.R.: The quality of approximation of Brun's algorithm in three dimensions. Monatsh. Math. **134**(2), 143–157 (2001). http://dx.doi.org/10.1007/s006050170004
17. Schweiger, F.: Multidimensional Continued Fraction. Oxford University Press, New York (2000)

Compositions of Tree-to-Tree Statistical Machine Translation Models

Andreas Maletti(✉)

Institute for Natural Language Processing, Universität Stuttgart,
Pfaffenwaldring 5b, 70569 Stuttgart, Germany
maletti@ims.uni-stuttgart.de

Abstract. Compositions of well-known tree-to-tree translation models used in statistical machine translation are investigated. Synchronous context-free grammars are closed under composition in both the unweighted as well as the weighted case. In addition, it is demonstrated that there is a close connection between compositions of synchronous tree-substitution grammars and compositions of certain tree transducers because the intermediate trees can encode finite-state information. Utilizing these close ties, the composition closure of synchronous tree-substitution grammars is identified in the unweighted and weighted case. In particular, in the weighted case, these results build on a novel lifting strategy that will prove useful also in other setups.

1 Introduction

Several different translation models are nowadays used in syntax-based statistical machine translation [17]. The translation model is the main component responsible for the transformation of the input into the translated output, and thus the expressive power of the translation model limits the possible translations. For example, the framework 'Moses' [18] provides implementations of synchronous context-free grammars (SCFGs) [1] and several variants of synchronous tree-substitution grammars (STSGs) [6]. The expressive power of SCFGs and STSGs is reasonably well-understood, and in particular, knowledge of the limitations of the models has helped many authors to preprocess [5,20,26] or post-process [4,25] their data and to achieve better translation results. Together with pre- or post-processing steps, the translation model is no longer solely responsible for the transformation process, but we rather obtain a composition of several models or simply a composition chain [23]. Occasionally, composition chains also appear because they ideally support a modular development of components for specific translation tasks [3] (e.g., translating numerals or geographic locations). However, it is often difficult to evaluate such composition chains efficiently especially when the pre- or post-processing steps are nondeterministic.

Supported by the German Research Foundation (DFG) grant MA/4959/1-1.

S. Brlek and C. Reutenauer (Eds.): DLT 2016, LNCS 9840, pp. 293–305, 2016.
DOI: 10.1007/978-3-662-53132-7_24

In the string-to-string setting the phrase-based models are essentially finite-state transducers and chains of them can be collapsed into a single transducer [24] because they are closed under composition. However, this is not true for several tree-to-tree models. Efficient on-the-fly evaluations for composition chains are presented in [23] along with the observation that the straightforward sequential evaluation of composition chains is terribly inefficient. Even in the on-the-fly evaluation the chains should be as short as possible. In this contribution, we will investigate the expressive power of composition chains of the established tree-to-tree translation models. The symmetric tree-to-tree setting, although typically worse in terms of translation quality than the string-to-tree or tree-to-string setting, is particularly convenient since it allows a clean notion of composition.

We first demonstrate that (unweighted and weighted) composition chains of SCFGs can always be reduced to just a single SCFG. In addition, we demonstrate how to utilize results for unweighted extended tree transducers [22] to obtain results for STSGs. The main insight in this part is that even local models like STSGs obtain a finite-state behavior in composition chains. Thus, a composition of two STSGs is as powerful as a composition of two corresponding tree transducers. This close connection allows us to show that two STSGs are necessary and sufficient for arbitrary composition chains of certain simple, yet commonly used STSGs. These results hold in the absence of weights. However, all translation models used in statistical machine translation are weighted, so as a second contribution we demonstrate how to lift the unweighted results into the weighted setting. Our novel lifting procedure, which we believe will be useful also in other setups, relies on a separation of the weights and several normalization procedures. Overall, we achieve the same results also in the weighted setting, which essentially shows that short chains of certain STSGs suffice.

2 Preliminaries

Let us start with some basic notions for trees, which we depict graphically whenever possible. Formally, our trees use a finite set N of internal labels and a finite set L of leaf labels. The internal labels can label any non-leaf node of the tree and such labeled nodes can have any positive number of children, whereas leaf labels only label leaves; i.e., nodes without children. Thus, our trees are inductively defined to be the smallest set $T_N(L)$ such that (i) every leaf node labeled $\ell \in L$ is a tree $\ell \in T_N(L)$ and (ii) $n(t_1, \dots, t_k) \in T_N(L)$ is a tree consisting of a root node labeled n and k direct subtrees for any given positive integer k, internal label $n \in N$, and trees $t_1, \dots, t_k \in T_N(L)$. A tree t that consists only of a (non-leaf) root node and leaf nodes is shallow, so a shallow tree is of the form $n(\ell_1, \dots, \ell_k)$ for some $n \in N$, $k \geq 1$, and $\ell_1, \dots, \ell_k \in L$. To easily access information in a tree, we use the following notation. For each node ν in a tree $t \in T_N(L)$ we write $t(\nu)$ for the label of the node ν. Occasionally, we are interested in the leaf nodes that are labeled by certain leaf labels $Q \subseteq L$. Consequently, the set of all nodes that are leaves and labeled by an element of Q is denoted by $\text{leaves}_Q(t)$, and the elements of $\text{leaves}_Q(t)$ are called anchors for

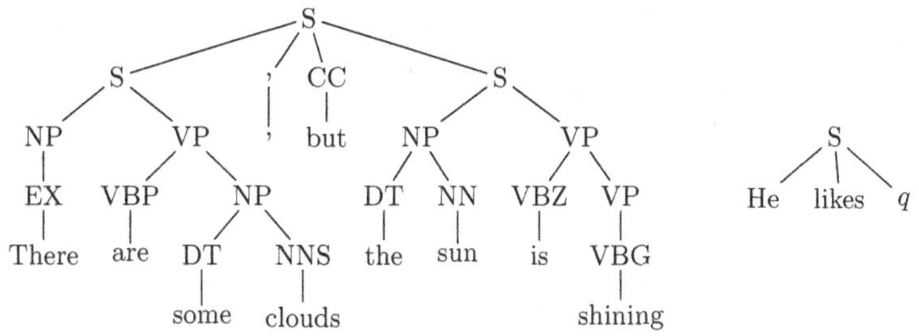

Fig. 1. The left tree t is not shallow, whereas the right tree u is. If $Q = \{q\}$, then $\text{leaves}_Q(t) = \emptyset$ and $\text{leaves}_Q(u) = \{\nu\}$, where ν is the q-labeled node in u. Obviously, its label is $u(\nu) = q$. Moreover, $u[\nu \leftarrow \text{her}] = \text{S}(\text{He}, \text{likes}, \text{her})$. Note that the q-labeled node vanishes in the substitution.

substitution. Moreover, given another tree $u \in T_N(L)$ and a leaf $\nu \in \text{leaves}_Q(t)$, we write $t[\nu \leftarrow u]$ for the tree obtained from t by replacing the leaf ν by the tree u. These notations are illustrated in Fig. 1, and we refer to [12,13] for an in-depth exposition.

Our weights will be taken from commutative semirings [14,16], which are algebraic structures $(A, +, \cdot, 0, 1)$ such that $(A, +, 0)$ and $(A, \cdot, 1)$ are commutative monoids and $(\sum_{i=1}^{k} a_i) \cdot a = \sum_{i=1}^{k} (a_i \cdot a)$ for all non-negative integers k and $a, a_1, \ldots, a_k \in A$. Typical examples of such semirings include the Boolean semiring $(\{0,1\}, \max, \min, 0, 1)$, the Viterbi semiring $([0,1], \max, \cdot, 0, 1)$ on the unit interval $[0,1]$, and the semiring $(\mathbb{Q}, +, \cdot, 0, 1)$ of rational numbers. In the following, let $(A, +, \cdot, 0, 1)$ be an arbitrary commutative semiring. Similarly, we fix the finite sets N and L of default internal labels and leaf labels, respectively.

A weighted (linear, nondeleting extended top-down) tree transducer [9,15] is a tuple $T = (Q, \Sigma, (q_1, q_2), R, \text{wt})$ consisting of (i) a finite set Q of states, (ii) a finite set Σ of internal labels for the trees generated, (iii) designated initial states $q_1, q_2 \in Q$, (iv) a finite set R of rules of the form $(q, t) \stackrel{\varphi}{—} (q', t')$ consisting of states $q, q' \in Q$, input and output tree fragments $t, t' \in T_\Sigma(L \cup Q)$,[1] and a bijective alignment $\varphi\colon \text{leaves}_Q(t) \to \text{leaves}_Q(t')$, and (v) a rule weight assignment $\text{wt}\colon R \to A$. The transducer T is a synchronous tree-substitution grammar (STSG) [6] if $Q = \Sigma$ and in each rule $(q, t) \stackrel{\varphi}{—} (q', t') \in R$ the root labels of t and t' are q and q', respectively.[2] Roughly speaking, an STSG replaces the "hidden" finite-state behavior by locality tests because the root labels (i.e., the states of a rule) are visible in the input and output tree fragments. Finally, T is a synchronous context-free grammar (SCFG) [1] if it is an STSG and in each rule $(q, t) \stackrel{\varphi}{—} (q', t') \in R$ the trees t and t' are shallow. In an SCFG, the input and the output tree are assembled like derivation trees of a context-free grammar (i.e., one level at a time). We recall two restrictions on tree transducer rules. A rule $(q, t) \stackrel{\varphi}{—} (q', t')$ is an ε-rule if $t \in Q$. Similarly, it is a non-strict rule

[1] For technical reasons we disallow that $\{t, t'\} \subseteq Q$.

[2] Note that in an STSG the elements of Σ can label internal nodes and leaves.

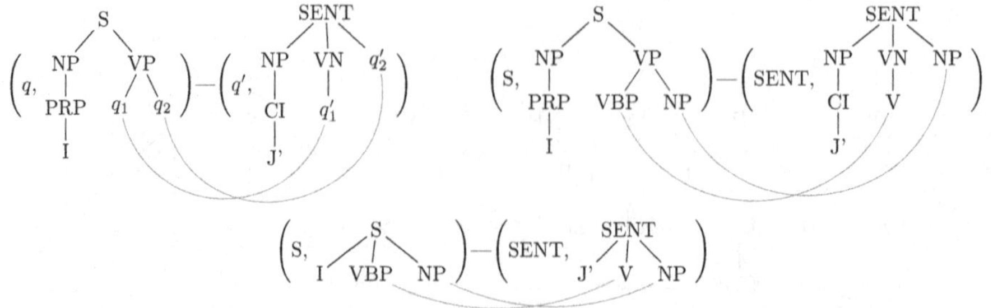

Fig. 2. Example rules of a tree transducer [top left], an STSG [top right], and an SCFG [bottom]

if $t' \in Q$. The tree transducer T is ε-free if it does not contain any ε-rules in R, and it is strict provided that it has no non-strict rules in R. Finally, simple tree transducers are both ε-free and strict. Note that an SCFG is always simple. We show a few example rules of each type in Fig. 2.

Let us recall the derivation semantics [10] of a weighted tree transducer $T = (Q, \Sigma, (q_1, q_2), R, \mathrm{wt})$. The derivations are defined over rule-like triples of the form (u, ψ, u') consisting of a (partial) input tree $u \in T_\Sigma(L \cup Q)$, a bijective alignment $\psi\colon \mathrm{leaves}_Q(u) \to \mathrm{leaves}_Q(u')$ linking synchronous states, and a (partial) output tree $u' \in T_\Sigma(L \cup Q)$. Note that the derivation forms are thus essentially rules without the states. Given two such forms (u, ψ, u') and (s, ψ', s') and a rule $\rho = (q, t) \overset{\varphi}{—} (q', t') \in R$, we say that (u, ψ, u') derives (s, ψ', s') via ρ, written $(u, \psi, u') \Rightarrow_T^\rho (s, \psi', s')$ if the least element $\nu \in \mathrm{leaves}_Q(u)$ with respect to some arbitrary linear order on nodes is such that (i) the node ν is labeled $u(\nu) = q$, (ii) its synchronized node $\psi(\nu)$ in u' has label $u'(\psi(\nu)) = q'$, (iii) $s = u[\nu \leftarrow t]$ is obtained from u by replacing ν by t, (iv) $s' = u'[\psi(\nu) \leftarrow t']$ is obtained from u' by replacing $\psi(\nu)$ by t', and (v) the synchronization ψ' is given for every $\nu' \in \mathrm{leaves}_Q(s)$ by $\psi'(\nu') = \varphi(\nu')$ if $\nu' \in \mathrm{leaves}_Q(t)$ and $\psi'(\nu') = \psi(\nu')$ otherwise. In other words, we keep the old synchronized states (except the replaced ones) and add the synchronized states of the rule ρ.[3] We illustrate a derivation step in Fig. 3. The derivation process starts with the initial form $\xi_0 = (q_1, \psi_0, q_2)$, in which the root nodes ν_1 and ν_2 of the trees q_1 and q_2, respectively, are synchronized (i.e., $\psi_0(\nu_1) = \nu_2$). Given trees $t, t' \in T_\Sigma(L)$, the transducer T assigns the weight

$$T(t, t') = \sum_{\xi_0 \Rightarrow_T^{\rho_1} \xi_1 \Rightarrow_T^{\rho_2} \cdots \Rightarrow_T^{\rho_n} (t, \emptyset, t')} \left(\prod_{i=1}^{n} \mathrm{wt}(\rho_i) \right)$$

to the pair (t, t'). We note that this sum always remains finite. Intuitively, we sum up the weights of all derivations of the tree pair (t, t'), where the weight of the derivation is obtained by multiplying the rule weights used in the derivation. In this manner, the transducer T computes a mapping $T\colon T_\Sigma(L) \times T_\Sigma(L) \to A$.

[3] For simplicity, we assume that nodes in different trees are disjoint.

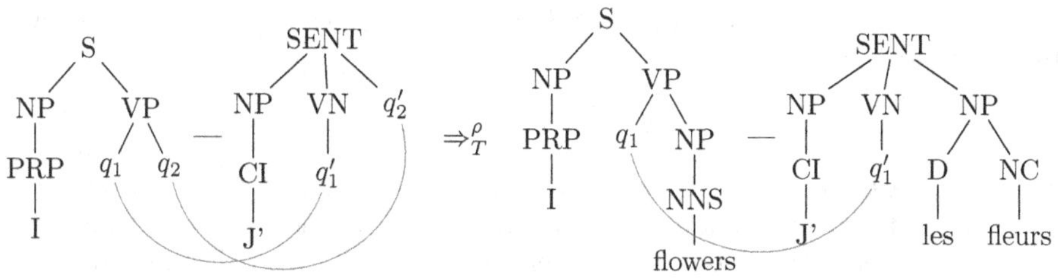

Fig. 3. The rule ρ: $(q_2, \mathrm{NP}(\mathrm{NNS}(\mathrm{flowers}))) \overset{\emptyset}{—} (q'_2, \mathrm{NP}(\mathrm{D}(\mathrm{les}), \mathrm{NC}(\mathrm{fleurs})))$ used in a derivation step, in which the states q_2 and q'_2 completely disappear

Finally, let us formally introduce compositions of weighted tree-to-tree translations. For all alphabets Σ and Δ, a mapping $\tau\colon T_\Sigma(L) \times T_\Delta(L) \to A$ is finitary, if for every $t \in T_\Sigma(L)$ there exist only finitely many $u \in T_\Delta(L)$ such that $\tau(t, u) \neq 0$. Similarly, it is co-finitary, if for every $u \in T_\Delta(L)$ there exist only finitely many $t \in T_\Sigma(L)$ such that $\tau(t, u) \neq 0$. Now let $\tau\colon T_\Sigma(L) \times T_\Delta(L) \to A$ and $\tau'\colon T_\Delta(L) \times T_\Gamma(L) \to A$ be such that τ is finitary or τ' is co-finitary. Then the composition τ followed by τ', written $\tau \,;\, \tau'$, is defined for every $t \in T_\Sigma(L)$ and $s \in T_\Gamma(L)$ by

$$(\tau \,;\, \tau')(t, s) = \sum_{u \in T_\Delta(L)} \tau(t, u) \cdot \tau'(u, s).$$

Note that this sum is well-defined because of the finitary or co-finitary restriction, which yields that only finitely many choices of u yield non-zero products. Roughly speaking, we sum over all potential intermediate trees u and take the product of the weights for the translation from t to u and the translation from u to s, which shows that composition corresponds to executing the second transducer on the output of the first transducer. Composition extends to classes $\mathcal{C}$ of weighted translations in the usual manner, and we use $\mathcal{C}^n$ for the composition $\mathcal{C} \,;\, \cdots \,;\, \mathcal{C}$ containing the class $\mathcal{C}$ exactly n times.

3 Unweighted Compositions

Let us first collect what is known about the unweighted case, which is obtained using the Boolean semiring $(\{0, 1\}, \max, \min, 0, 1)$ as weight structure. In this setting, tree-to-tree translations are essentially relations on trees. It is evident from the formal definitions that each SCFG is a special STSG, which in turn is a special tree transducer, so the expressive power increases from SCFGs to STSGs to tree transducers (TTs). Using the abbreviations as denotations for the classes of tree relations that can be generated by the corresponding translation model, we thus have SCFG $\subseteq$ STSG $\subseteq$ TT. Moreover, the key property that separates SCFGs and STSGs was identified in [6]. The relations computed by SCFGs only contain pairs of isomorphic trees (disregarding the labels and the order of the children). It is also easy to show that STSGs and TTs can be separated, so we obtain the strict hierarchy

$$\text{SCFG} \subset \text{STSG} \subset \text{TT}. \tag{1}$$

Composition essentially corresponds to running two translations consecutively, where the first translation translates the input into intermediate results and the second translation translates those intermediate results into the final results. Compositions of tree translations have been extensively investigated (see [11] for a survey). To avoid a careful distinction between transducers and their translations, we will conflate the class of models with the class of translations computable by it. Since all classes discussed here contain the identity relation, compositions of our classes $\mathcal{C}$ form a natural hierarchy; i.e., $\mathcal{C} \subseteq \mathcal{C}^2 \subseteq \mathcal{C}^3 \subseteq \cdots$. This hierarchy collapses at level n if $\mathcal{C}^n = \mathcal{C}^{n+1}$. We also say that the composition closure is obtained at level n provided that n is the least integer, for which the hierarchy collapses. Intuitively, if the closure is obtained at level n, then compositions of n translations of $\mathcal{C}$ are necessary and sufficient to generate any translation computable by any composition of $\mathcal{C}$. Provided that the composition closure for $\mathcal{C}$ is n, we thus have $\mathcal{C} \subset \cdots \subset \mathcal{C}^n = \mathcal{C}^{n+1} = \cdots$. We use ∞ to indicate that the hierarchy never collapses. We summarize the known results [2,7,8] on the composition closure in Table 1.

Table 1. Known results on composition closures.

Model	Composition closure	Reference
Top-down tree transducer	1	[7]
Simple tree transducer	2	[2]
Other tree transducer	∞	[8]

We start our investigation with SCFGs. Given two SCFGs T_1 and T_2 we can simply "join" rules of T_1 and T_2 that coincide on the intermediate tree. We illustrate this approach in Fig. 4. Such rules can certainly be executed consecutively in the on-the-fly approach [23]. A refined version of this approach taking the finite-state information and the non-shallow output into account is used to prove that (our linear and nondeleting) top-down tree transducers are closed under composition [7].

Theorem 1. *The composition closure of unweighted and weighted SCFGs is achieved at the first level.*

Proof. We prove the statement for arbitrary weighted SCFGs. Let

$$T = (\Sigma, \Sigma, (S_1, S_2), R, \text{wt}) \qquad \text{and} \qquad T' = (\Sigma', \Sigma', (S'_1, S'_2), R', \text{wt}')$$

be weighted SCFGs. If $S_2 \neq S'_1$, then the composition $T \,;\, T'$ is the constant 0 mapping, which can easily be computed by a single SCFG. Now suppose that

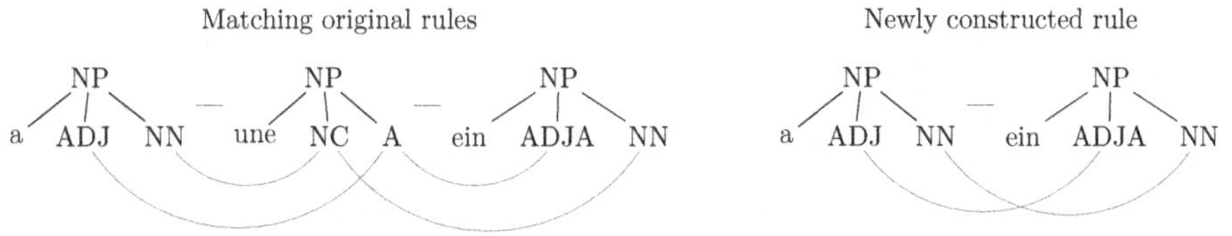

Fig. 4. Rule matching and joining in the composition of SCFGs. The left and middle part form a rule of T_1 and the middle and right part form a rule of T_2. The newly constructed rule will simply avoid the intermediate tree fragment.

$S_2 = S'_1$. We construct the weighted SCFG $T'' = (\Sigma'', \Sigma'', (S_1, S'_2), R'', \mathrm{wt}'')$, where $\Sigma'' = \Sigma \cup \Sigma'$ and the rules R'' and their weights wt'' are obtained as follows: For every rule $\rho = (\sigma, s) \stackrel{\varphi}{—} (\delta, t)$ of R and rule $\rho' = (\delta, t) \stackrel{\psi}{—} (\gamma, u)$ of R' we construct the rule $\rho'' = (\sigma, s) \stackrel{\varphi;\psi}{—} (\gamma, u)$ of R'' and set $\mathrm{wt}''(\rho'') = \mathrm{wt}(\rho) \cdot \mathrm{wt}'(\rho')$. No other rules are in R''. The correctness of this construction is straightforward. □

Next, we will show that the composition closure for simple STSGs can be obtained from the known results via a small insight. Recall that SCFGs and STSGs are both local, so they are missing the hidden finite-state behavior of general tree transducers. However, we can simulate the hidden finite-state behavior for both models in compositions with the help of the unknown (hidden) intermediate trees. Namely, we can annotate the desired finite-state information on the intermediate trees in the spirit of the representation of a regular tree language as the image of a local tree language under a relabeling [13]. We illustrate the approach in Fig. 5. Note that the first STSG encodes the states in its output (i.e., the intermediate tree), whereas the second STSG encodes them in its input (i.e., also the intermediate tree).

Lemma 2. *For all $n \geq 2$, compositions of n simple STSGs are as expressive as compositions of n simple tree transducers.*

Proof. We only provide the argument for compositions $T\,;T'$ of 2 simple STSGs T and T'. Assume that $(q, t) \stackrel{\varphi}{—} (q', t')$ is a rule of the first simple tree transducer $T = (Q, \Sigma, (q_1, q_2), R, \mathrm{wt})$. Since T is strict, we have $t' = \delta(t'_1, \ldots, t'_k)$ for some internal symbol $\delta \in \Sigma$ and subtrees $t'_1, \ldots, t'_k$. We will adjust the internal symbols to $\Sigma' = \Sigma \cup (\Sigma \times Q \times Q)$, which allows us to use combinations of internal symbols together with two states. For every state-labeled node $\nu \in \mathrm{leaves}_Q(t')$ in t' we additionally guess two internal symbols $\sigma_\nu, \delta_\nu \in \Sigma$. Then we construct the rule $(\sigma, u) \stackrel{\varphi}{—} ((\delta, q, q'), (\delta, q, q')(u'_1, \ldots, u'_k))$, where σ is the root label of t, $u = t[\varphi^{-1}(\nu) \leftarrow \sigma_\nu \mid \nu \in \mathrm{leaves}_Q(t')]$, and the subtrees $u'_1, \ldots, u'_k$ are obtained from the subtrees $t'_1, \ldots, t'_k$ by replacing each leaf node $\nu \in \mathrm{leaves}_Q(t')$ by the state-annotated variant $(\delta_\nu, t(\varphi^{-1}(\nu)), t'(\nu))$. In other words, we guess the internal symbols that will replace a state leaf in the input and output fragment t and t' and replace the state leaf by the guessed internal symbol in the input fragment

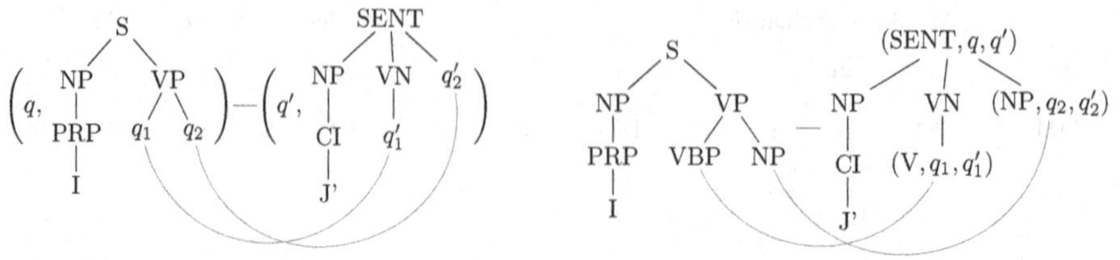

Fig. 5. Illustration of the state annotation on the intermediate tree. The left part shows the original tree transducer rule and the right STSG rule shows how the state annotation is performed on the output tree using the guessed nonterminal pairs (VBP, V) and (NP, NP) for (q_1, q_1') and (q_2, q_2'), respectively.

and the triple containing the guessed internal symbol and the two synchronized states. We construct a new rule for each original rule and all possible guesses. Similarly, we need to annotate the finite-state information of the second tree transducer T' in its input fragments, which works in essentially the same manner using the input fragments instead of the output fragments. This also shows that we actually need to annotate up to 4 states to each symbol in the intermediate tree, and we additionally need to guess the finite-state information (that can also occur in internal symbols) of the other tree transducer. We omit the technical details. □

Theorem 3. *The composition closure of simple STSGs is obtained at the second level.*

Proof. Simple tree transducers achieve the composition closure at level 2 [2]. Since the second levels of the composition hierarchy for simple tree transducers and simple STSGs coincide by Lemma 2 and simple STSGs are less expressive by (1), the composition closure of simple STSGs is achieved at level 2 as well. □

Finally, we examine the composition hierarchy of the remaining cases (non-strict STSGs and STSGs with ε-rules). In both cases, the corresponding hierarchy for tree transducers is infinite. Moreover, re-examining the counterexample translation τ provided in [8, Example 43], we can easily see that it does not utilize its finitely many states and can be generated by a non-strict STSG as well. Hence for every $n \geq 1$ we also obtain a translation τ^{n+1} that can be computed by $(n+1)$ STSGs, but not by n tree transducers according to [8, Lemma 44]. Since by (1) we have STSG $\subseteq$ TT, it follows that $\text{STSG}^n \subseteq \text{TT}^n$ and thus n STSGs also cannot implement τ^{n+1}. The analogous arguments using the inverse translation τ^{-1} can be used to prove the infiniteness of the composition hierarchy for STSGs with ε-rules. We summarize the results in Table 2.

Theorem 4. *The composition hierarchy of strict STSGs, ε-free STSGs, and general STSGs is infinite.*

Table 2. Composition closure results for unweighted and weighted SCFGs and STSGs. They mirror the corresponding results for tree transducers.

Model	Unweighted/weighted composition closure	Results
SCFGs	1	Theorem 1
Simple STSGs	2	Theorems 3 and 7
Other STSGs	∞	Theorems 4 and 8

4 Weighted Compositions

In the weighted setting, which is more relevant in statistical machine translation, the models assign a weight to each rule. During derivations the weights of the participating rules are multiplied, and if there are several ways to achieve the same input- and output-tree pair, then the derivation weights are summed up. To avoid infinite summations, we restrict ourselves to ε-free or strict models.

The goal of this section is to lift the unweighted results of the previous section into the weighted setting. In Theorem 1 we already proved that SCFGs are closed under composition also in the weighted case. Moreover, the result of Lemma 2 also holds in the weighted case, so the composition closure of simple weighted STSGs and that of simple weighted tree transducers again coincide. It only remains to establish the composition closure for simple weighted tree transducers. Roughly speaking, we will reduce the weighted problem to the unweighted setting by removing the weights from the tree transducer and moving them into a particularly simple type of translation, called weighted relabeling. For the ease of presentation we assume that no rule consists only of a leaf in the input or output tree fragment (i.e., $t \notin L$ and $t' \notin L$ for all considered rules $(q,t) \stackrel{\varphi}{—} (q',t')$). This is realistic in statistical machine translation since the parsers usually attach at least a part-of-speech tag to each lexical item. Moreover, we can easily adjust our approach and relabel leaf symbols as well.

For a given alphabet Σ, a *weighted relabeling* is a mapping $\kappa\colon \Sigma \times \Sigma \to A$. In other words, it is a weighted association between symbols. It extends to pairs of trees such that it assigns weight 0 to all pairs of trees of different shape. For trees of the same shape, it simply takes the product of the symbol-to-symbol weights given by κ for all corresponding nodes in the two trees. Formally, each such relabeling κ extends to a weighted tree translation $\overline{\kappa}\colon T_\Sigma(L) \times T_\Sigma(L) \to A$ inductively by (i) $\overline{\kappa}(\ell,\ell) = 1$ for every $\ell \in L$; i.e., we do not relabel leaf symbols, and (ii) for every $k \geq 1$, symbols $\sigma, \delta \in \Sigma$, subtrees $t_1, \ldots, t_k, u_1, \ldots, u_k \in T_\Sigma(L)$

$$\overline{\kappa}\big(\sigma(t_1,\ldots,t_k), \delta(u_1,\ldots,u_k)\big) = \kappa(\sigma,\delta) \cdot \prod_{i=1}^{k} \overline{\kappa}(t_i,u_i).$$

We relabel trees with an internal symbol as root by charging the weight for relabeling the root symbol to another symbol and then multiply the product of the weights of recursively relabeling the subtrees. In all remaining cases,

$\overline{\kappa}(\ldots,\ldots) = 0$. We use wREL for the class of all weighted translations computable by weighted relabelings and s-wTT for the corresponding class computed by simple weighted tree transducers. Since most devices in this section are weighted, we will drop the explicit mention that they are weighted and simply say 'relabeling' or 'simple tree transducer'.

Lemma 5. *For every composition of a simple tree transducer and a relabeling in either order, we can present an equivalent simple tree transducer.*

$$\text{s-wTT}; \text{wREL} \subseteq \text{s-wTT} \qquad \textit{and} \qquad \text{wREL}; \text{s-wTT} \subseteq \text{s-wTT}$$

Proof. The first statement is obtained by combining the decomposition of [9, Lemma 4.1] and the composition results of [19, Theorem 2.4]. Moreover, since all the involved models are symmetric, we also immediately obtain the second statement. □

The next lemma shows that we can separate the weights from a simple tree transducer leaving a composition of an essentially unweighted (i.e., unambiguous and Boolean[4]) simple tree transducer T' and a relabeling. Moreover, the tree relation computed by T' will be injective.[5] *Unambiguous* means that for each (successful) translation (t, u) containing an input and an output tree there exists exactly one derivation yielding (t, u). We use su-TT_{inj} for the injective translations computed by simple unambiguous tree transducers. Note that these weighted translations are essentially the characteristic functions of the translations of the corresponding unweighted tree transducers, which motivates the chosen abbreviation.

Lemma 6. *Every simple tree transducer T can be equivalently represented by a composition of a simple unambiguous Boolean tree transducer T' computing an injective translation followed by a relabeling κ.*

$$\text{s-wTT} \subseteq \text{su-TT}_{\text{inj}}; \text{wREL}$$

Proof. Let $T = (Q, \Sigma, (q_1, q_2), R, \text{wt})$. For every rule $\rho = (q, t) \xrightarrow{\varphi} (q', t')$ of R, we have $t' = \sigma(t'_1, \ldots, t'_k)$ for some integer k, symbol $\sigma \in \Sigma$, and subtrees $t'_1, \ldots, t'_k \in T_\Sigma(L \cup Q)$ because T is strict. For this rule ρ, we construct the rule $(q, t) \xrightarrow{\varphi} (q', \langle \sigma, \rho \rangle (t'_1, \ldots, t'_k))$ of T', which essentially records the rule application in the root of the output tree fragment. The weight of this new rule is 1 in T'. Finally, the relabeling κ is such that $\kappa(\sigma, \sigma) = 1$ and $\kappa(\langle \sigma, \rho \rangle, \sigma) = \text{wt}(\rho)$ for all $\sigma \in \Sigma$ and $\rho \in R$, and 0 otherwise. In other words, the relabeling removes the annotation and charges the weight of the annotated rule. Obviously, the constructed tree transducer is Boolean. In addition, since the derivation is completely visible in the output, the tree transducer T' is unambiguous. □

[4] Using only the weights 0 and 1.

[5] A tree translation $\tau\colon T_\Sigma(L) \times T_\Sigma(L) \to A$ is *injective* if for every output tree $u \in T_\Sigma(L)$ there exists at most one input tree $t \in T_\Sigma(L)$ such that $\tau(t, u) \neq 0$.

Using Lemmas 5 and 6 we can now separate the weights from a composition chain because

$$
\begin{aligned}
&\text{s-wTT} \,;\, \text{s-wTT}^2 \subseteq \text{su-TT}_{\text{inj}} \,;\, \text{wREL} \,;\, \text{s-wTT}^2 && \text{(Lemma 6)}\\
\subseteq\ &\text{su-TT}_{\text{inj}} \,;\, \text{s-wTT}^2 \subseteq \text{su-TT}_{\text{inj}}^2 \,;\, \text{wREL} \,;\, \text{s-wTT} && \text{(Lemmas 5 and 6)}\\
\subseteq\ &\text{su-TT}_{\text{inj}}^2 \,;\, \text{s-wTT} \subseteq \text{su-TT}_{\text{inj}}^3 \,;\, \text{wREL}. && \text{(Lemmas 5 and 6)}
\end{aligned}
$$

Now we can apply the result of [2] on the unweighted composition closure of s-TT. Note that the composition of injective translations is naturally again injective.

$$\text{su-TT}_{\text{inj}}^3 \,;\, \text{wREL} \subseteq \underbrace{\text{s-TT}^2}_{\text{injective}} \,;\, \text{wREL},$$

where s-TT is the class of translations computed by simple unweighted tree transducers. We cannot simply simulate those unweighted tree transducers directly by weighted tree transducers. We first use standard techniques (regular restrictions; see [22]) to make both unweighted translations injective. Moreover each injective translation can be made unambiguous using essentially the same techniques, so we obtain

$$\underbrace{\text{s-TT}^2}_{\text{injective}} \,;\, \text{wREL} \subseteq \text{su-TT}_{\text{inj}}^2 \,;\, \text{wREL} \subseteq \text{s-wTT}^2,$$

where the last step uses Lemma 5. Note that unambiguous unweighted tree transducers can easily be simulated by the corresponding weighted tree transducers. Thus, we derived the difficult part of the composition closure.

Theorem 7. *The composition closure of weighted simple STSGs is achieved at the second level.*

Proof. We showed that $\text{s-wTT}^3 \subseteq \text{s-wTT}^2$, so the composition hierarchy of the class s-wTT collapses at level 2. Moreover using the linking arguments of [21] we can also conclude that $\text{s-wTT} \subset \text{s-WTT}^2$. □

Finally, for the remaining classes (i.e., strict STSGs and ε-free STSGs), we can essentially import the infinite composition hierarchy from the unweighted case using the linking technique of [21]. We omit the details.

Theorem 8. *The composition hierarchy of strict weighted STSGs and ε-free weighted STSGs is infinite.*

5 Conclusion

We have investigated the expressive power of compositions of the well-established tree-to-tree translation models: SCFGs, STSGs, and tree transducers. In the unweighted case, the results for the local devices [i.e., SCFGs and STSGs] closely mirror the known composition results for tree transducers due to the fact that we can encode the finite-state information in the intermediate trees of a composition. The same picture presents itself in the weighted setting, for which we showed how to lift the corresponding results from the unweighted setting to the weighted setting. This uses a novel decomposition separating the weights from simple tree transducers and then constructions for the obtained unambiguous and injective tree transducers. Overall, we demonstrated that in the relevant cases, short (length 1 or 2) composition chains are necessary and sufficient to simulate arbitrarily long composition chains.

References

1. Aho, A.V., Ullman, J.D.: Syntax directed translations and the pushdown assembler. J. Comput. Syst. Sci. **3**(1), 37–56 (1969)
2. Arnold, A., Dauchet, M.: Morphismes et bimorphismes d'arbres. Theor. Comput. Sci. **20**(1), 33–93 (1982)
3. Chen, S., Matsumoto, T.: Translation of quantifiers in Japanese-Chinese machine translation. In: Isahara, H., Kanzaki, K. (eds.) JapTAL 2012. LNCS, vol. 7614, pp. 11–22. Springer, Heidelberg (2012)
4. Clifton, A., Sarkar, A.: Combining morpheme-based machine translation with post-processing morpheme prediction. In: Proceedings of ACL, pp. 32–42. ACL (2011)
5. Collins, M., Koehn, P., Kucerovǎ, I.: Clause re-structuring for statistical machine translation. In: Proceedings of ACL, pp. 531–540. ACL (2005)
6. Eisner, J.: Learning non-isomorphic tree mappings for machine translation. In: Proceedings of ACL, pp. 205–208. ACL (2003)
7. Engelfriet, J.: Bottom-up and top-down tree transformations: a comparison. Math. Syst. Theor. **9**(3), 198–231 (1975)
8. Engelfriet, J., Fülöp, Z., Maletti, A.: Composition closure of linear extended top-down tree transducers. Theor. Comput. Syst. (2016, to appear). doi:10.1007/s00224-015-9660-2
9. Fülöp, Z., Maletti, A., Vogler, H.: Weighted extended tree transducers. Fundam. Informaticae **111**(2), 163–202 (2011)
10. Fülöp, Z., Vogler, H.: Weighted tree transducers. J. Autom. Lang. Comb. **9**(1), 31–54 (2004)
11. Fülöp, Z., Vogler, H.: Weighted tree automata and tree transducers. In: Droste, M., Kuich, W., Vogler, H. (eds.) Handbook of Weighted Automata, Chap. 9, pp. 313–403. Springer, Heidelberg (2009)
12. Gécseg, F., Steinby, M.: Tree Automata. Akadémiai Kiadó, Budapest (1984)
13. Gécseg, F., Steinby, M.: Tree Automata. arXiv:1509.06233 (2015)
14. Golan, J.S.: Semirings and Their Applications. Springer, Dordrecht (1999)
15. Graehl, J., Knight, K.: Training tree transducers. In: Proceedings of HLT-NAACL, pp. 105–112. ACL (2004)
16. Hebisch, U., Weinert, H.J.: Semirings-Algebraic Theory and Applications in Computer Science. World Scientific, Singapore (1998)
17. Koehn, P.: Statistical Machine Translation. Cambridge University Press, Cambridge (2010)
18. Koehn, P., Hoang, H., Birch, A., Callison-Burch, C., Federico, M., Bertoldi, N., Cowan, B., Shen, W., Moran, C., Zens, R., Dyer, C., Bojar, O., Constantin, A., Herbst, E.: Moses: open source toolkit for statistical machine translation. In: Proceedings of ACL, pp. 177–180. ACL (2007)

19. Kuich, W.: Full abstract families of tree series I. In: Karhumäki, J., Maurer, H., Păun, G., Rozenberg, G. (eds.) Jewels are Forever, pp. 145–156. Springer, Heidelberg (1999)
20. Lerner, U., Petrov, S.: Source-side classifier preordering for machine translation. In: Proceedings of EMNLP, pp. 513–523. ACL (2013)
21. Maletti, A.: The power of weighted regularity-preserving multi bottom-up tree transducers. Int. J. Found. Comput. Sci. **26**(7), 987–1005 (2015)
22. Maletti, A., Graehl, J., Hopkins, M., Knight, K.: The power of extended top-down tree transducers. SIAM J. Comput. **39**(2), 410–430 (2009)
23. May, J., Knight, K., Vogler, H.: Efficient inference through cascades of weighted tree transducers. In: Proceedings of ACL, pp. 1058–1066. ACL (2010)
24. Mohri, M.: Finite-state transducers in language and speech processing. Comput. Linguist. **23**(2), 269–311 (1997)
25. Stymne, S.: Text harmonization strategies for phrase-based statistical machine translation. Ph.D. thesis, Linköping University (2012)
26. Xia, F., McCord, M.C.: Improving a statistical MT system with automatically learned rewrite patterns. In: Proceedings of CoLing, pp. 508–514 (2004)

On the Solvability Problem for Restricted Classes of Word Equations

Florin Manea[1], Dirk Nowotka[1], and Markus L. Schmid[2(✉)]

[1] Department of Computer Science, Kiel University, 24098 Kiel, Germany
flm@informatik.uni-kiel.de, nowotka@zs.uni-kiel.de
[2] Fachbereich IV – Abteilung Informatikwissenschaften,
Trier University, 54286 Trier, Germany
MSchmid@uni-trier.de

Abstract. We investigate the complexity of the solvability problem for restricted classes of word equations with and without regular constraints. For general word equations, the solvability problem remains NP-hard, even if the variables on both sides are ordered, and for word equations with regular constraints, the solvability problems remains NP-hard for variable disjoint (i. e., the two sides share no variables) equations with two variables, only one of which is repeated. On the other hand, word equations with only one repeated variable (but an arbitrary number of variables) and at least one non-repeated variable on each side, can be solved in polynomial-time.

Keywords: Word equations · Regular constraints · NP-hardness

1 Introduction

A *word equation* is an equation $\alpha = \beta$, such that α and β are words over an alphabet $\Sigma \cup X$, where Σ is a finite alphabet of *constants* and $X = \{x_1, x_2, x_3, \ldots\}$ is an enumerable set of *variables*. A *solution* to a word equation $\alpha = \beta$ is a morphism $h : (\Sigma \cup X)^* \to \Sigma^*$ that satisfies $h(\alpha) = h(\beta)$ and $h(b) = b$ for every $b \in \Sigma$. For example, $x\mathtt{ab}y = \mathtt{b}y x\mathtt{a}$ is a word equation with variables x, y, constants $\mathtt{a}, \mathtt{b}$ and h with $h(x) = \mathtt{bab}$, $h(y) = \mathtt{aba}$ is a solution, since $h(x\mathtt{ab}y) = \mathtt{babababa} = h(\mathtt{b}yx\mathtt{a})$.

The *solvability problem* for word equations, i. e., to decide whether or not a given word equation has a solution, has a long history with the most prominent landmark being Makanin's algorithm [11] from 1977, which showed the solvability problem to be decidable (see Chapter 12 of [10] for a survey). While the complexity of Makanin's original algorithm was very high, it is nowadays known that the solvability problem is in PSPACE (see [8,12]) and NP-hard (in fact, it is even believed to be in NP). Word equations with only a single variable can be solved in linear time [7] and equations with two variables can be solved in time $\mathcal{O}(n^5)$ [2]; it is not known whether there exist polynomial-time algorithms for solving word equations with at most k variables, for some $k \geq 3$.

S. Brlek and C. Reutenauer (Eds.): DLT 2016, LNCS 9840, pp. 306–318, 2016.
DOI: 10.1007/978-3-662-53132-7_25

If we require $\beta \in \Sigma^*$, i. e., only one side of the equation is allowed to contain variables, then we obtain the *pattern matching problem with variables* (or simply *matching problem*, for short), where the term *pattern* refers to the part α that can contain variables. The matching problem is NP-complete and, compared to the solvability problem for word equations, many more tractability and intractability results are known (see [4,5,13]). More precisely, while restrictions of numerical parameters (e. g., number of variables, number of occurrences per variable, length of substitution words, alphabet size, etc.) make the problem either polynomial-time solvable in a trivial way (e. g., if the number of variables is bounded by a constant) or result in strongly restricted, but still NP-complete variants (see [4]), structural restrictions of the pattern (e. g., of the order of the variables) are more promising and can yield rich classes of patterns for which the matching problem can be solved in polynomial-time (see [13]). For example, the matching problem remains NP-complete if $|\Sigma| = 2$, every variable has at most two occurrences in α and every variable can only be replaced by the empty word or a single symbol (or instead by non-empty words of size at most 3). On the other hand, non-trivial and efficient polynomial-time algorithms exist (see [3]), if the patterns are *regular* (i. e., every variable has at most one occurrence), the patterns are *non-cross* (i. e., between any two occurrences of the same variable x no other variable different from x occurs) or the patterns have a bounded *scope coincidence degree* (i. e., the maximum number of scopes of variables that overlap is bounded, where the scope of a variable is the interval in the pattern where it occurs).

Technically, all these results can be seen as tractability and intractability results for restricted variants of the solvability problem (in fact, as it seems, all NP-hardness lower bounds for restricted variants of the solvability problem in the literature are actually NP-hardness lower bounds for the matching problem). However, these results are disappointing in terms of how much they provide us with a better understanding of the complexity of word equations, since in the matching problem the most crucial feature of word equations is missing, which is the possibility of having variables on both sides.

The aim of this paper is to transfer the knowledge and respective techniques of the matching problem to variants of the solvability problem for word equations that are *not* just variants of the matching problem. In particular, we investigate whether the structural restrictions mentioned above, which are beneficial for the matching problem, can be extended, with a comparable positive impact, to classes of word equations that have variables on both sides. We pay special attention to *regular constraints*, i. e., each variable x is accompanied by a regular language L_x from which $h(x)$ must be selected in a solution h. While Makanin's algorithm still works in the presence of regular constraints, it turns out that for more restricted classes of equations, the addition of regular constraints can drastically increase the complexity of the solvability problem.

2 Definitions

Let Σ be a finite alphabet of *constants* and let $X = \{x_1, x_2, x_3, \ldots\}$ be an enumerable set of *variables*. For any word $w \in (\Sigma \cup X)^*$ and $z \in \Sigma \cup X$, we denote by $|w|_z$ the number of occurrences of z in w, by $\mathsf{var}(w)$ the set of variables occurring in w and, for every i, $1 \leq i \leq |w|$, $w[i]$ denotes the symbol at position i in w. A morphism $h : (\Sigma \cup X)^* \to \Sigma^*$ with $h(a) = a$ for every $a \in \Sigma$ is called a *substitution*. A *word equation* is a tuple $(\alpha, \beta) \in (\Sigma \cup X)^+ \times (\Sigma \cup X)^+$ (for the sake of convenience, we also write $\alpha = \beta$) and a *solution* to a word equation (α, β) is a substitution h with $h(\alpha) = h(\beta)$, where $h(\alpha)$ is the *solution word* (*of* h). A word equation is *solvable* if there exists a solution for it and the *solvability problem* is to decide for a given word equation whether or not it is solvable.

Let $\alpha \in (\Sigma \cup X)^*$. We say that α is *regular*[1], if, for every $x \in \mathsf{var}(\alpha)$, $|\alpha|_x = 1$; e. g., $\mathtt{a}x_1\mathtt{ba}x_2\mathtt{c}x_3\mathtt{bca}x_4\mathtt{a}x_5\mathtt{bb}$ is regular. The word α is *non-cross* if between any two occurrences of the same variable x no other variable different from x occurs, e. g., $\mathtt{a}x_1\mathtt{ba}x_1x_2\mathtt{a}x_2x_2x_3x_3\mathtt{b}x_4$ is non-cross, whereas $x_1\mathtt{b}x_1x_2\mathtt{ba}x_3x_3x_4x_4\mathtt{bc}x_2$ is not. A word equation (α, β) is regular or non-cross, if both α and β are regular or both α and β are non-cross, respectively. An equation (α, β) is *variable disjoint* if $\mathsf{var}(\alpha) \cap \mathsf{var}(\beta) = \emptyset$.

For a word equation $\alpha = \beta$ and an $x \in \mathsf{var}(\alpha\beta)$, a *regular constraint* (*for* x) is a regular language L_x and a solution h for $\alpha = \beta$ *satisfies* the regular constraint L_x if $h(x) \in L_x$. The solvability problem for word equations with regular constraints is to decide on whether an equation $\alpha = \beta$ with regular constraints L_x, $x \in \mathsf{var}(\alpha\beta)$, given as NFA, has a solution that satisfies all regular constraints. The *size* of the regular constraints is the sum of the number of states of the NFA. If the regular constraints are all of the form Γ^*, for some $\Gamma \subseteq \Sigma$, then we call them *word equations with individual alphabets*.

A word equation $\alpha = \beta$ along with an $m \in \mathbb{N}$ is a *bounded word equation*. The problem of *solving a bounded word equation* is then to decide on whether there exists a solution h for $\alpha = \beta$ with $|h(x)| \leq m$ for every $x \in \mathsf{var}(\alpha\beta)$.

For an $\alpha \in (\Sigma \cup X)^*$, $L(\alpha) = \{h(\alpha) \mid h \text{ is a substitution}\}$ is the *pattern language of* α.

3 Regular and Non-cross Word Equations

For the matching problem, the restriction of regularity implies that every variable has only one occurrence in the equation, which makes the solvability problem trivial (in fact, it boils down to the membership problem for a very simple regular language). However, word equations in which both sides are regular can still have repeated variables, although the maximum number of occurrences per variable is 2 (i. e., regular equations are restricted variants of quadratic equations (see, e. g., [14])) and these two occurrences must occur on different sides. Unfortunately, we are neither able to show NP-hardness nor to find a polynomial-time algorithm for the solvability problem of regular word equations.

[1] The use of the term regular in this context has historical reasons: the matching problem has been first investigated in terms of so-called *pattern languages*, i. e., the set of all words that match a given pattern $\alpha \in (\Sigma \cup X)^*$, which are regular languages if α is regular.

Open Problem 1. *Can regular word equations be solved in polynomial-time?*

As we shall see later, solving a system of two regular equations is NP-hard (Corollary 6), solving regular equations with regular constraints is even PSPACE-complete (Theorem 7), and solving bounded regular equations or regular equations with individual alphabets is NP-hard (Corollaries 17 and 19, respectively), as well.

On the positive side, it can be easily shown that regular word equations can be solved in polynomial-time, if we additional require them to be variable disjoint (which simply means that no variable is repeated in the whole equation). More precisely, in this case, we only have to check emptiness for the intersection of the pattern languages described by the two sides of the equations (which are regular languages).

Next, we show the stronger result that polynomial-time solvability is still possible if at most one variable is repeated, and each side contains at least one of the non-repeating variables.

Theorem 2. *Word equations with only one repeated variable, and each side containing at least one non-repeating variable, can be solved in polynomial time.*

If we allow an arbitrary number of occurrences of each variable, but require them to be sorted on both sides on the equation, where the sorting order might be different on the two sides, then we arrive at the class of non-cross word equations. As for the class of regular patterns, also for non-cross patterns the matching problem can be solved efficiently. However, as we shall see next, for non-cross equations, the solvability problem becomes NP-hard.

Theorem 3. *Solving non-cross word equations is* NP*-hard.*

We prove this theorem by a reduction[2] from a graph problem, for which we first need the following definition.

Let $\mathcal{G} = (V, E)$ be a graph with $V = \{t_1, t_2, \ldots, t_n\}$. A vertex s is the *neighbour* of a vertex t if $\{t, s\} \in E$ and the set $N_{\mathcal{G}}[t] = \{s \mid \{t, s\} \in E\} \cup \{t\}$ is called the (*closed*) *neighbourhood* of t. If, for some $k \in \mathbb{N}$, every vertex of $\mathcal{G}$ has exactly k neighbours, then $\mathcal{G}$ is k*-regular*. A *perfect code* for $\mathcal{G}$ is a subset $C \subseteq V$ with the property that, for every $t \in V$, $|N_{\mathcal{G}}[t] \cap C| = 1$. Next, we define the problem to decide whether or not a given 3-regular graph has a perfect code, which is NP-complete (see [9]):

3-Regular Perfect Code (3RPerCode)
Instance: A 3-regular graph $\mathcal{G}$.
Question: Does $\mathcal{G}$ contain a perfect code?

We now define a reduction from 3RPerCode. To this end, let $\mathcal{G} = (V, E)$ be a 3-regular graph with $V = \{t_1, t_2, \ldots, t_n\}$ and, for every i, $1 \leq i \leq n$, N_i is the neighbourhood of t_i. Since the neighbourhoods play a central role, we shall define them in a more convenient way. For every r, $1 \leq r \leq 4$, we use

[2] We will also use minor modifications later on of this reduction in order to conclude corollaries of Theorem 3.

a mapping $\wp_r : \{1,2\ldots,n\} \to \{1,2\ldots,n\}$ that maps an $i \in \{1,2\ldots,n\}$ to the index of the rth vertex of neighbourhood N_i, i.e., for every i, $1 \le i \le n$, $N_i = \{t_{\wp_1(i)}, t_{\wp_2(i)}, t_{\wp_3(i)}, t_{\wp_4(i)}\}$. Obviously, the mappings $\wp_r$, $1 \le r \le 4$, imply a certain order on the vertices in the neighbourhoods, but, since our constructions are independent of this actual order, any order is fine.

We transform $\mathcal{G}$ into a word equation with variables $\{x_{i,j} \mid 1 \le i,j \le n\} \cup \{y_i, y'_i \mid 1 \le i \le n\}$ and constants from $\Sigma = \{\star, \diamond, \bar{\diamond}, \odot, \#, \mathsf{a}\}$. For every i,j, $1 \le i,j \le n$, the variable $x_{i,j}$ represents $t_i \in N_j$. For every i, $1 \le i \le n$, we define

$$\alpha_i = x_{\wp_1(i),i} \ldots x_{\wp_4(i),i}, \qquad \alpha'_i = \#\,\mathsf{a}^8\,\#\,\#,$$
$$\beta_i = \mathsf{a}, \qquad \beta'_i = y_i\,\#(x_{i,\wp_1(i)})^2 \ldots (x_{i,\wp_4(i)})^2\,\#\,y'_i$$

and

$$u = \alpha_1 \star \ldots \star \alpha_n \star \odot \bar{\diamond}\, \alpha'_1 \diamond \ldots \diamond \alpha'_n,$$
$$v = \beta_1 \star \ldots \star \beta_n \star \odot \bar{\diamond}\, \beta'_1 \diamond \ldots \diamond \beta'_n.$$

Proposition 4. *The words u and v are non-cross and can be constructed from $\mathcal{G}$ in polynomial time.*

Lemma 5. *The graph $\mathcal{G}$ has a perfect code if and only if (u, v) has a solution.*

Proof. For the sake of convenience, let $u = u_1 \odot u_2$ and $v = v_1 \odot v_2$. We start with the *only if* direction. For a perfect code C of $\mathcal{G}$, we construct a substitution h with $h(u) = h(v)$ in the following way. For every i, $1 \le i \le n$, we define $h(x_{i,\wp_r(i)}) = \mathsf{a}$, $1 \le r \le 4$, if $t_i \in C$, and $h(x_{i,\wp_r(i)}) = \varepsilon$, otherwise. Thus, for every i, $1 \le i \le n$, $h((x_{i,\wp_1(i)})^2 \ldots (x_{i,\wp_4(i)})^2) \in \{\mathsf{a}^8, \varepsilon\}$, which implies that $h(y_i)$ and $h(y'_i)$ can be defined such that $h(\beta'_i) = h(\alpha'_i)$. Consequently, $h(v_2) = h(u_2)$. Since C is a perfect code, for every i, $1 \le i \le n$, there is an r, $1 \le r \le 4$, such that $t_{\wp_r(i)} \in C$ and $t_{\wp_{r'}(i)} \notin C$, $1 \le r' \le 4$, $r \ne r'$. Therefore, $h(x_{\wp_1(i),i}x_{\wp_2(i),i}x_{\wp_3(i),i}x_{\wp_4(i),i}) = h(x_{\wp_r(i),i}) = \mathsf{a}$, which means that $h(\alpha_i) = h(\beta_i)$. Since this particularly implies $h(u_1) = h(v_1)$, we can conclude $h(u) = h(v)$.

In order to prove the *if* direction, we assume that there exists a solution h.

Claim: If $h(u_1) = h(v_1)$ and $h(u_2) = h(v_2)$, then $\mathcal{G}$ has a perfect code.

Proof of Claim: From $h(u_1) = h(v_1)$, we can directly conclude that, for every i, $1 \le i \le n$, $h(\alpha_i) = \beta_i$, which means that exactly one of the variables $x_{\wp_1(i),i}, x_{\wp_2(i),i}, x_{\wp_3(i),i}, x_{\wp_4(i),i}$ is mapped to a, while the others are mapped to ε. From $h(v_2) = h(u_2)$ it follows that, for every i, $1 \le i \le n$, $h(\beta'_i) = \alpha'_i$. Next, we observe that, for every i, $1 \le i \le n$, due to the symbols $\#$ in β'_i and α'_i, $h((x_{i,\wp_1(i)})^2 \ldots (x_{i,\wp_4(i)})^2) \in \{\mathsf{a}^8, \varepsilon\}$. Since each of the variables $x_{i,\wp_1(i)}, x_{i,\wp_2(i)}, x_{i,\wp_3(i)}, x_{i,\wp_4(i)}$ are mapped to either a or ε, this implies that either all of these variables are erased or all of them are mapped to a. Let C be the set of exactly the vertices $t_i \in V$ for which $h(x_{i,\wp_1(i)}) = h(x_{i,\wp_2(i)}) = h(x_{i,\wp_3(i)}) = h(x_{i,\wp_4(i)}) = \mathsf{a}$. For every neighbourhood $V_j = \{t_{\wp_1(j)}, t_{\wp_2(j)}, t_{\wp_3(j)}, t_{\wp_4(j)}\}$,

$1 \le j \le n$, $h(x_{\wp_1(j),j}\, x_{\wp_2(j),j}\, x_{\wp_3(j),j}\, x_{\wp_4(j),j})$ is mapped to $\mathtt{a}$, which implies that for some r, $1 \le r \le 4$, $h(x_{\wp_r(j),j}) = \mathtt{a}$; thus, $t_{\wp_r(j)} \in C$. Furthermore, $h(x_{\wp_{r'}(j),j}) = \varepsilon$, $1 \le r' \le 4$, $r \ne r'$, which means that $t_{\wp_{r'}(j)} \notin C$, $1 \le r' \le 4$, $r \ne r'$. Consequently, C is a perfect code. *(Claim)* □

It remains to show that a solution h necessarily satisfies $h(u_1) = h(v_1)$ and $h(u_2) = h(v_2)$. Let w be the solution word of h. We first recall that, since $v_1, u_2 \in \Sigma^*$, $h(v_1) = v_1$ and $h(u_2) = u_2$, which particularly means that $v_1 \odot$ is a prefix and $\odot\, u_2$ is a suffix of w. If $|w|_\odot = 1$, then $w = v_1 \odot u_2$ and therefore $h(u_1) = h(v_1)$ and $h(u_2) = h(v_2)$. If, on the other hand, $|w|_\odot \ge 2$, then $w = v_1 \odot \gamma \odot u_2$. If $\gamma = \varepsilon$, then $w = v_1 \odot \odot u_2$, which is a contradiction, since w must contain the factor $\star \odot \overline{\diamond}$. From $h(u_2) = u_2$ and $h(v_1) = v_1$ it follows that $h(u_1) = v_1 \odot \gamma =$ and $h(v_2) = \gamma \odot u_2$. The factor v_2 starts with an occurrence of $\overline{\diamond}$ and since γ is a non-empty prefix of $h(v_2)$, this means that $|\gamma|_{\overline{\diamond}} = k \ge 1$. Moreover, γ is also a suffix of $h(u_1)$ and since $|u_1|_{\overline{\diamond}} = 0$, this implies that there are variables $z_1, z_2, \ldots, z_\ell \in \mathsf{var}(u_1)$, $1 \le \ell \le k$, with $\sum_{i=1}^{\ell} |h(z_i)|_{\overline{\diamond}} \ge k$. Since each of these variables z_i, $1 \le i \le \ell$, is repeated twice in v_2 and since $|v_2|_{\overline{\diamond}} = 1$, we can conclude that $|h(v_2)|_{\overline{\diamond}} \ge 2k + 1$. In the suffix $\odot\, u_2$ of $h(v_2)$, there is only one occurrence of $\overline{\diamond}$, which implies that $|\gamma|_{\overline{\diamond}} \ge 2k$. Since $k \ge 1$, this is clearly a contradiction to $|\gamma|_{\overline{\diamond}} = k$. □

The equation obtained by the reduction from above has the form $u_1 \odot u_2 = v_1 \odot v_2$, where in a solution h, $h(u_1) = h(v_1)$ and $h(u_2) = h(v_2)$. In order to achieve this synchronisation between the two left parts and between the two right parts, we need to repeat variables in v_2. However, we can as well represent $u_1 \odot u_2 = v_1 \odot v_2$ as a system of two equations $u_1 = v_1$ and $u_2 = v_2$ and, since the synchronisation of the left parts and the right parts is now enforced by the fact that we regard them as two separate equations, we can get rid of the repeated variables in v_2, which makes the two equations regular.

Corollary 6. *The problem of checking solvability of a system of* 2 *regular word equations* $\alpha_1 = \beta_1$, $\alpha_2 = \beta_2$ *with* $\beta_1, \beta_2 \in \Sigma^*$ *is* NP*-hard.*

We conclude this section by stressing the fact that the non-cross equation from the reduction above is "almost regular", i. e., one side is regular, while for the other the maximum number of occurrences per variable is 2. However, we were not able to get rid of these repeated variables, which suggests that a hardness reduction for the regular case needs to be substantially different or regular word equations can be solved in polynomial-time.

4 Word Equations with Regular Constraints

In practical scenarios, it seems rather artificial that we only want to find just any solution for a word equation and we are fine with whatever sequence of symbols the variables will be substituted with. It is often more realistic that the variables have a well-defined domain from which we want the solution to select the words.

This motivates the addition of regular constraints to word equations, as defined in Sect. 2, for which we investigate the solvability problem in this section.

As mentioned in Sect. 1, regular constraints can be easily incorporated into algorithms for the general solvability problem. However, while it is open whether solving general word equations is hard for PSPACE, for word equations with regular constraints, this can be easily shown, even for regular equations.

Theorem 7. *Solving word equations with regular constraints is* PSPACE-*complete, even for regular equations.*

Proof. We can reduce the PSPACE-hard intersection emptiness problem for NFA, i. e., deciding for given NFA M_i, $1 \leq i \leq n$, whether or not $\bigcap_{i=1}^{n} L(M_i) = \emptyset$. To this end, let $M_1, \ldots, M_n$ be NFA over some alphabet Σ with $\# \notin \Sigma$. We define $\alpha = x_1 \# x_2 \# \ldots \# x_{n-1}$ and $\beta = x_2 \# x_3 \# \ldots \# x_n$, and we define the regular constraints $L_{x_i} = L(M_i)$. We note that the equation $\alpha = \beta$ is regular.

If there exists a word $w \in \bigcap_{i=1}^{n} L(M_i)$, then h with $h(x_i) = w$, $1 \leq i \leq n$, is a solution for $\alpha = \beta$, since $h(\alpha) = (w\#)^{n-2}w = h(\beta)$, and, furthermore, h satisfies the regular constraints. Let h be a solution for $\alpha = \beta$ that satisfies the regular constraints. This implies that $h(x_1)\#h(x_2)\#\ldots\#h(x_{n-1}) = h(x_2)\#h(x_3)\#\ldots\#h(x_n)$ and, since $|h(x_i)|_\# = 0$, $1 \leq i \leq n$, $h(x_1) = h(x_2) = \ldots = h(x_n)$ follows. Thus, $h(x_1) \in \bigcap_{i=1}^{n} L(M_i)$. □

Recall that we mentioned in Sect. 3 that word equations without repeated variables can be solved in polynomial time. This also holds for word equations with regular constraints.

Theorem 8. *Solving word equations with regular constraints and without repeated variables can be done in polynomial time.*

Word equations with only one variable can be solved in linear time (see Jeż [7]). If we add regular constraints to equations with only one variable, then the solvability problem is still in P.

Theorem 9. *Solving word equations with regular constraints and with only one variable can be done in polynomial time.*

Word equations with two variables can be solved in polynomial-time (see [2]). We shall see next that for word equations with regular constraints this is no longer the case (assuming P $\neq$ NP). More precisely, solving equations with two variables and regular constraints is NP-hard, even if only one variable is repeated and the equations are variable disjoint. Moreover, we can show that the existence of an algorithm solving word equations with two variables and with regular constraints in time $2^{o(n+m)}$ (where n is the length of the equation and m is the size of the regular constraints) is very unlikely, since it would refute the well-known *exponential-time hypothesis* (ETH, for short).

We conduct a linear reduction from 3-SAT to the problem of solving word equations with regular constraints.[3] Let $C = \{c_1, c_2, \ldots, c_m\}$ be a Boolean formula in conjunctive normal form (CNF) with 3 literals per clause over the variables $\{v_1, v_2, \ldots, v_n\}$. We first transform C into a CNF C' such that C is satisfiable if and only if C' has an assignment that satisfies exactly one literal per clause (in the following, we call such an assignment a *1-in-3 assignment*). For every i, $1 \leq i \leq m$, we replace $c_i = \{y_1, y_2, y_3\}$ by 5 new clauses

$$\{y_1, z_1, z_2\}, \{y_2, z_2, z_3\}, \{z_1, z_3, z_4\}, \{z_2, z_5, z_6\}, \{y_3, z_5\},$$

where z_i, $1 \leq i \leq 6$, are new variables.[4] We note that C' has $5m$ clauses and $n + 6m$ variables. Next, we obtain C'' from C' by adding, for every i, $1 \leq i \leq n$, a new clause $\{v_i, \widehat{v}_i\}$, where $\widehat{v}_i$ is a new variable, and we replace all occurrences of $\overline{v_i}$ (i. e., the variable v_i in negated form) by $\widehat{v}_i$.

The following proposition can be easily verified.

Proposition 10. *There is a satisfying assignment for C if and only if C'' has a 1-in-3 assignment. Furthermore, C'' has no negated variables, C'' has $5m + n$ clauses and $2n + 6m$ variables.*

For the sake of convenience, we set $n' = 2n + 6m$, $m' = 5m + n$, $C'' = \{c'_1, c'_2, \ldots, c'_{m'}\}$ and let $\{v'_1, v'_2, \ldots, v'_{n'}\}$ be the variables of C''. Furthermore, for every i, $1 \leq i \leq n'$, let k_i be the number of occurrences of variable v'_i in C''.

Next, we transform C'' into a word equation with regular constraints as follows. Let $\Sigma = \{v'_1, v'_2, \ldots, v'_{n'}, \#\}$ and let the equation $\alpha = \beta$ be defined by $\alpha = (x_1\,\#)^{n'-1}\,x_1$ and $\beta = x_2$. For the variables x_1 and x_2, we define the following regular constraints over Σ:

$$L_{x_1} = \{w \mid |w| = m', w[i] \in c'_i, 1 \leq i \leq m'\},$$
$$L_{x_2} = \{u_1\#u_2\#\ldots\#u_{n'} \mid u_i \in (\Sigma\backslash\{\#\})^*, |u_i|_{v'_i} \in \{k_i, 0\}, 1 \leq i \leq n'\}.$$

Proposition 11. *There are DFA M_{x_1} and M_{x_2} accepting the languages L_{x_1} and L_{x_2}, respectively, with $5m + n + 2$ and $21m + 5n + 1$ states, respectively.*

By definition, only NFA are required to represent the regular constraints, but our use of DFA here points out that the following hardness result (and the ETH lower bound) also holds for the case that we require the regular constraints to be represented by DFA. So the hardness of the problem does not result from the fact that NFA can be exponentially smaller than DFA.

Lemma 12. *The Boolean formula C'' has a 1-in-3 assignment if and only if $\alpha = \beta$ has a solution that satisfies the regular constraints L_{x_1} and L_{x_2}.*

---

[3] In order to prove NP-hardness, a simpler production would suffice, but we need a linear reduction in order to obtain the ETH lower bound.

[4] Note that this is just the reduction used by Schaefer [15] in order to reduce 3-SAT to 1-IN-3 3-SAT. We recall it here to observe that this reduction is linear.

Proof. We start with the *only if* direction. To this end, let $\pi : \{v'_1, v'_2, \ldots, v'_n\} \to \{0,1\}$ be a 1-in-3 assignment for C'', where, for every i, $1 \leq i \leq m'$, y_i is the unique variable with $y_i \in c'_i$ and $\pi(y_i) = 1$. Let h be a substitution defined by $h(x_1) = y_1 y_2 \ldots y_{m'}$ and $h(x_2) = (h(x_1)\,\#)^{n-1}\, h(x_1)$. Obviously, h is a solution for $\alpha = \beta$, $h(x_1) \in L_{x_1}$ and, since every v'_i has either 0 occurrences in $h(x_1)$ (in case that $\pi(v'_i) = 0$) or k_i occurrences (in case that $\pi(v'_i) = 1$), also $h(x_2) \in L_{x_2}$.

For the *if* direction, let h be a solution for $\alpha = \beta$ that satisfies the regular constraints. Consequently, $h(x_1) = y_1 y_2 \ldots y_{m'}$, where $y_i \in c'_i$, $1 \leq i \leq m'$, and, furthermore, for every i, $1 \leq i \leq n$, $|h(x_2)|_{v'_i} \in \{k_i, 0\}$. This directly implies that $\pi : \{v'_1, v'_2, \ldots, v'_n\} \to \{0,1\}$, defined by $h(v'_i) = 1$ if $|h(x_2)|_{v'_i} = k_i$ and $h(v'_i) = 0$ if $|h(x_2)|_{v'_i} = 0$, is a 1-in-3 assignment for C''. □

The exponential-time hypothesis, mentioned above, roughly states that 3-SAT cannot be solved in subexponential-time. For more informations on the ETH, the reader is referred to Chapter 14 of the textbook [1]. For our application of the ETH, it is sufficient to recall the following result.

Theorem 13 (Impagliazzo et al. [6]). *Unless ETH fails, 3-*SAT *cannot be solved in time $2^{o(n+m)}$, where n is the number of variables and m is the number of clauses.*

The reduction from above implies that a subexponential algorithm for solving word equations with two variables and regular constraints can be easily turned into a subexponential algorithm for 3-SAT; thus, the existence of such an algorithm contradicts ETH.

Theorem 14. *Solving word equations with two variables and with regular constraints is* NP*-hard, even if only one variable is repeated and the equations are variable disjoint. Furthermore, unless ETH fails, such word equations cannot be solved in time $2^{o(n+m)}$ (where n is the length of the equation and m is the size of the regular constraints).*

4.1 Bounded Word Equations

We first note that bounded word equations can be considered as a special case of word equations with regular constraints, since the bound m functions as regular constraints of the form $\{w \in \Sigma^* \mid |w| \leq m\}$ for every variable. However, there is an important difference: the length of a binary encoding of m is logarithmic in the size of an NFA for $\{w \in \Sigma^* \mid |w| \leq m\}$; thus, NP-hardness of a class of bounded word equations does not necessarily carry over to word equations with regular constraints. As usual, we call the solvability problem for a class of bounded word equations NP-*hard in the strong sense*, if the NP-hardness remains if the bound m is given in unary.

Theorem 15. *Solving bounded word equations is* NP*-hard (in the strong sense), even for equations $\alpha = \beta$ satisfying $|\mathsf{var}(\alpha)| = 1$, $\mathsf{var}(\alpha) \cap \mathsf{var}(\beta) = \emptyset$ and β is regular.*

Proof. We reduce from the shortest common superstring problem, i. e., deciding for given $k \in \mathbb{N}$ and strings $v_1, v_2, \ldots, v_n \in \Sigma^*$ whether there is a string u with $|u| \leq k$ that contains each v_i as a factor. Let $v_1, v_2, \ldots, v_n \in \Sigma^*$, $k \in \mathbb{N}$ be an instance of the shortest common superstring problem. Furthermore, let $\#$ be a new symbol, i. e., $\# \notin \Sigma$. We construct a word equation $\alpha = \beta$, where

$$\begin{array}{ccccccccc} \alpha & = & x & \# & x & \# & \ldots & \# & x\,, \\ \beta & = & y_1 v_1 y_1' & \# & y_2 v_2 y_2' & \# & \ldots & \# & y_n v_n y_n'\,. \end{array}$$

Furthermore, we let k be the upper bound on the substitution word lengths.

If there exists a word $w \in \Sigma^*$ with $|w| \leq k$ and, for every i, $1 \leq i \leq n$, $w = u_i v_i u_i'$, then we define a substitution h by $h(x) = w$, $h(y_i) = u_i$ and $h(y_i') = u_i'$, $1 \leq i \leq n$. Obviously, h satisfies the length bound and, for every i, $1 \leq i \leq n$, $h(x) = h(y_i v_i y_i')$; thus, $h(\alpha) = h(\beta)$.

Let h be a solution for $\alpha = \beta$ that satisfies the length bound. We observe that since $h(\beta)$ contains every v_i as a factor, also $h(\alpha) = h(x)\#h(x)\#\ldots\#h(x)$ contains every v_i as a factor and, furthermore, since $|v_i|_\# = 0$, $1 \leq i \leq n$, every v_i is also a factor of $h(x)$. Consequently, $|h(x)| \leq k$ and $h(x)$ contains every v_i, $1 \leq i \leq n$, as a factor.

For the shortest common superstring problem, we can assume that $k \leq \sum_{i=1}^{n} |v_i|$, since otherwise $v_1 v_2 \ldots v_n$ would also be a solution. Consequently, we can assume that k is given in unary, which means that solving bounded word equations of the form mentioned in the statement of the theorem is NP-hard in the strong sense.

Due to the strong NP-hardness in Theorem 15, we can conclude the following.

Corollary 16. *Solving word equations with regular constraints is* NP*-hard, even for equations* $\alpha = \beta$ *satisfying* $|\mathsf{var}(\alpha)| = 1$, $\mathsf{var}(\alpha) \cap \mathsf{var}(\beta) = \emptyset$ *and* β *is regular.*

By using 1 as the bound on the substitution words and by a minor modification of the reduction for Theorem 3, we can obtain a hardness reduction for bounded regular word equations.

Corollary 17. *Solving bounded regular word equations is* NP*-hard.*

4.2 Individual Alphabets

The least restrictive regular constraints are probably constraint languages of the form Γ^* for some $\Gamma \subseteq \Sigma$, i. e., word equations with individual alphabets, which we shall investigate in this section.

We first note that if $|\Sigma| = 1$, then general word equations and word equations with individual alphabets coincide and, furthermore, the solvability problem for word equations can be solved in polynomial-time, if $|\Sigma| = 1$.

Theorem 18. *Solving word equations can be done in polynomial time if* $|\Sigma| = 1$.

However, if $\Sigma = \{\mathtt{a}, \mathtt{b}\}$ and $\{\mathtt{a}\}$ is used as individual alphabet for all variables, then solving word equations becomes NP-hard again, simply because the matching problem is already NP-hard for this case (as can be easily concluded from the reduction of Lemma 5 in [5]).

By using individual alphabets, the reduction for Theorem 3 can be easily transformed to a hardness reduction for the solvability problem of regular equations with individual alphabets.

Corollary 19. *Solving regular word equations with individual alphabets is* NP-*hard.*

5 Conclusions

We conclude this work by summarising our main results and by suggesting some further research directions.

First of all, the polynomial-time decidability of the matching problem for non-cross patterns does not carry over to non-cross equations (which also means that the concept of the scope coincidence degree, briefly mentioned in Sect. 1, will not help, since it is a generalisation of the non-cross concept), while for regular equations, this is still open (see Open Problem 1), which constitutes the most important question left open in this work.

As soon as we allow regular constraints, it is possible to prove hardness results for strongly restricted variants of the solvability problem, often including the regular case. More precisely, for general regular constraints, the solvability problem is PSPACE-complete, even for regular equations (Theorem 7), and NP-hard for variable disjoint equations with only one repeated variable and two variables in total (Theorem 14). Especially this latter result, for which we can also obtain an ETH lower bound, points out a drastic difference in terms of complexity between general word equations and equations with regular constraints: both the tractable cases of equations with only two variables or with only one repeated variable and at least one non-repeated variable on both sides (Theorem 2) become NP-hard if we allow regular constraints.[5] Moreover, the case with only one repeated variable remains intractable, even if the constraints are only bounding the length of the substitution words (Theorem 15). In particular, even if it turns out that, for some k, $k \geq 3$, or even for all constant k, general word equations with at most k variables can be solved in polynomial-time, Theorem 14 severely limits their practical application, since it shows that these polynomial-time algorithms cannot cope with regular constraints (unless $\mathsf{P} = \mathsf{NP}$).

As for regular equations, allowing a system of only two equations (and no further constraints), allowing bounds on the substitution words or allowing individual alphabets is enough to make the solvability problem NP-hard.

Our choice of restrictions for word equations is motivated by polynomial-time solvable cases of the matching problem. In order to obtain tractable classes

[5] For the latter case, note that in the reduction of Theorem 14, we can add a non-repeated variable with regular constraint $\emptyset$ to the left side.

of word equations, it might be worthwhile to strengthen the concept of non-cross and regularity by requiring $\alpha\beta$ to be regular or non-cross, instead of only requiring this for α and β separately. Another possible further restriction would be to require the order of the variables on the left and on the right side to be the same (e. g., $x_1\mathtt{ab}x_2\mathtt{c}x_3 = x_1\mathtt{c}x_3$ is *ordered regular*, while $x_1\mathtt{ab}x_2\mathtt{c}x_3 = x_3\mathtt{c}x_2$ is not). In this regard, it is interesting to note that the patterns produced by the reduction of Theorem 3 are not ordered non-cross (and not ordered regular for the corresponding corollaries), while Theorem 7, the PSPACE-completeness of solving word equations with regular constraints, also holds for ordered regular equations. Additionally requiring $\mathsf{var}(\alpha) = \mathsf{var}(\beta)$ for ordered regular equations would be a further restriction that might be useful.

Acknowledgements. We are indebted to Artur Jeż for valuable discussions. Markus L. Schmid gratefully acknowledges partial support for this research from DFG, that in particular enabled his visit at the University of Kiel.

References

1. Cygan, M., Fomin, F.V., Kowalik, L., Lokshtanov, D., Marx, D., Pilipczuk, M., Pilipczuk, M., Saurabh, S.: Parameterized Algorithms. Springer International Publishing AG, Cham (2015)
2. Dąbrowski, R., Plandowski, W.: Solving two-variable word equations. In: Díaz, J., Karhumäki, J., Lepistö, A., Sannella, D. (eds.) ICALP 2004. LNCS, vol. 3142, pp. 408–419. Springer, Heidelberg (2004)
3. Fernau, H., Manea, F., Mercaş, R., Schmid, M.L.: Pattern matching with variables: fast algorithms and new hardness results. In: Proceedings of 32nd Symposium on Theoretical Aspects of Computer Science, STACS 2015, Leibniz International Proceedings in Informatics (LIPIcs), vol. 30, pp. 302–315 (2015)
4. Fernau, H., Schmid, M.L.: Pattern matching with variables: a multivariate complexity analysis. Inf. Comput. **242**, 287–305 (2015)
5. Fernau, H., Schmid, M.L., Villanger, Y.: On the parameterised complexity of string morphism problems. Theory of Computing Systems (2015). http://dx.doi.org/10.1007/s00224-015-9635-3
6. Impagliazzo, R., Paturi, R., Zane, F.: Which problems have strongly exponential complexity? J. Comput. Syst. Sci. **63**, 512–530 (2001)
7. Jeż, A.: One-variable word equations in linear time. Algorithmica **74**, 1–48 (2016)
8. Jeż, A.: Recompression: a simple and powerful technique for word equations. J. ACM **63**(1), 4:1–4:51 (2016)
9. Kratochvíl, J., Křivánek, M.: On the computational complexity of codes in graphs. In: Chytil, M.P., Koubek, V., Janiga, L. (eds.) MFCS 1988. LNCS, vol. 324, pp. 396–404. Springer, Heidelberg (1988)
10. Lothaire, M.: Algebraic Combinatorics on Words. Cambridge University Press, Cambridge (2002)
11. Makanin, G.: The problem of solvability of equations in a free semigroup. Matematicheskii Sbornik **103**, 147–236 (1977)
12. Plandowski, W.: An efficient algorithm for solving word equations. In: Proceedings of the 38th Annual ACM Symposium on Theory of Computing, STOC 2006, pp. 467–476 (2006)

13. Reidenbach, D., Schmid, M.L.: Patterns with bounded treewidth. Inf. Comput. **239**, 87–99 (2014)
14. Robson, J.M., Diekert, V.: On quadratic word equations. STACS 1999. LNCS, vol. 1563, pp. 217–226. Springer, Heidelberg (1999)
15. Schaefer, T.J.: The complexity of satisfiability problems. In: Proceedings of 10th Annual ACM Symposium on Theory of Computing, STOC 1978, pp. 216–226. ACM (1978)

Unambiguous Büchi Is Weak

Henryk Michalewski and Michał Skrzypczak(✉)

University of Warsaw, Banacha 2, Warsaw, Poland
{h.michalewski,m.skrzypczak}@mimuw.edu.pl

Abstract. A non-deterministic automaton on infinite trees is *unambiguous* if it has at most one accepting run on every tree. For a given unambiguous parity automaton $\mathcal{A}$ of index $(i, 2j)$ we construct an alternating automaton $\textsc{Transformation}(\mathcal{A})$ which accepts the same language, but is simpler in terms of alternating hierarchy of automata. If $\mathcal{A}$ is a Büchi automaton $(i = 0, j = 1)$, then $\textsc{Transformation}(\mathcal{A})$ is a weak alternating automaton. In general, $\textsc{Transformation}(\mathcal{A})$ belongs to the class $\mathrm{Comp}(i + 1, 2j)$, in particular it is simultaneously of alternating index $(i, 2j)$ and of the dual index $(i+1, 2j+1)$. The main theorem of this paper is a correctness proof of the algorithm Transformation. The transformation algorithm is based on a separation algorithm of Arnold and Santocanale (2005) and extends results of Finkel and Simonnet (2009).

Keywords: Infinite trees · Unambiguity · Rabin-Mostowski index

1 Introduction

Determinising a given computation typically leads to an additional cost. Presence of such cost inspires investigation of intermediate models of computations. Here we focus on *unambiguity*, that is the requirement that there are no two distinct accepting computations on the same input. In the case of finite and infinite words a given automaton can be determinised at an exponential cost, but in the case of infinite trees there are automata which cannot be determinised at all. Moreover, there are automata for which one cannot find an equivalent unambiguous automaton [13] (see also [4]). Also, there exist unambiguous automata which cannot be simulated by deterministic ones [9] (see Fig. 1).

Most questions about automata on finite or infinite words are decidable. However, in the case of automata on infinite trees many fundamental decidability problems are open, unless we limit attention to deterministic automata. Then it is decidable whether a given language is recognisable by a deterministic automaton [16], the non-deterministic index problem is decidable [14,15], as well as it is possible to locate the language in the Wadge hierarchy [12]. Moving beyond deterministic automata is a topic of an on-going research [6,7] and the study of unambiguous automata is a part of this effort. Admittedly, problems for

The authors were supported by the Polish National Science Centre grant no. 2014-13/B/ST6/03595.

S. Brlek and C. Reutenauer (Eds.): DLT 2016, LNCS 9840, pp. 319–331, 2016.
DOI: 10.1007/978-3-662-53132-7_26

this class seem to be much harder than for deterministic automata, in particular one can decide if a given automaton is unambiguous, but it is an open problem, whether a given regular language is unambiguous. Additionally, there are no upper bounds on the descriptive complexity (e.g. the parity index) or topological complexity of unambiguous languages among all regular tree languages.

In this work we focus on descriptive complexity and a fortiori also on topological complexity of languages defined by unambigous automata. The most canonical measure of descriptive complexity of regular tree languages is the *parity index*. A parity automaton $\mathcal{A}$ has index (i, j) if the priorities of the states of the automaton belong to the set $\{i, i+1, \ldots, j\}$. In particular, the Büchi acceptance condition corresponds to the index $(1, 2)$. By $\mathrm{Comp}(i, j)$ we denote the class of alternating automata where each strongly-connected component is of index (i, j) or $(i+1, j+1)$, see p. x. It was shown in [1,3] that some languages require big indices: for every pair (i, j) there exists a regular language of infinite trees that is of index (i, j) and cannot be recognised by any alternating nor non-deterministic automaton of a lower index. It means that the non-deterministic and alternating index hierarchies are *strict*.

We will show that the fact that a given automaton is unambiguous allows to effectively find another equivalent automaton with a simpler acceptance condition. More precisely, in Sect. 4 we propose an algorithm TRANSFORMATION with the following properties:

Theorem 1. *For an unambiguous Büchi automaton $\mathcal{A}$,* TRANSFORMATION($\mathcal{A}$) *is a weak alternating automaton recognising the same language. More generally, if $\mathcal{A}$ is an unambiguous automaton of index $(i, 2j)$ then* TRANSFORMATION($\mathcal{A}$) *accepts the same language as $\mathcal{A}$ and belongs to the class* $\mathrm{Comp}(i+1, 2j)$*, in particular it is simultaneously of alternating index $(i, 2j)$ and of the dual index $(i+1, 2j+1)$.*

Additionally, the number of states of TRANSFORMATION($\mathcal{A}$) *is polynomial in the number of states of $\mathcal{A}$.*

This theorem implies in particular that there is no unambiguous Büchi automaton which is strictly of index $(1, 2)$. Since a language accepted by an unambiguous Büchi automaton is also accepted by a weak alternating automaton, topologically such languages must be located at a finite level of the Borel hierarchy. One should note that in the above theorem and in the algorithm TRANSFORMATION, the automaton must be simultaneously unambiguous and of appropriate index. It is still possible for a regular tree language to be both: recognised by some unambiguous automaton and by some other Büchi automaton. An example of such a language is the H-language proposed in [9]: "there exists a branch containing only a's and turning infinitely many times right", see Fig. 1.

1.1 Related Work

There exist two estimates on descriptive complexity of unambiguous languages. Firstly, a result of Hummel [9] shows that unambiguous languages are topologically harder than deterministic ones, see Fig. 1. Secondly, Finkel and Simonnet [8] proved using the Lusin-Souslin Theorem [10, Theorem 15.1], that any language recognised by an unambiguous Büchi automaton must be Borel.

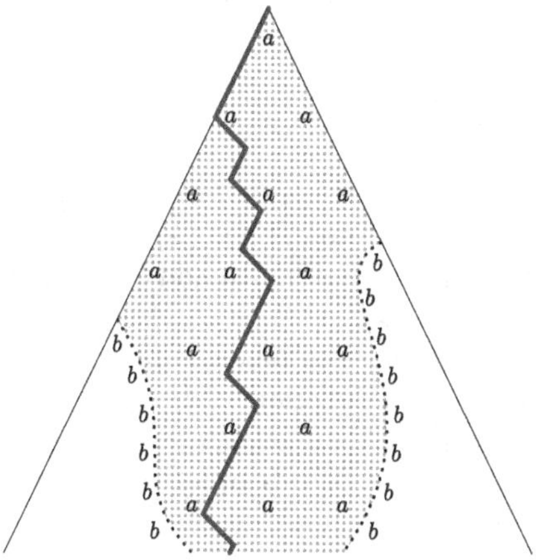

Fig. 1. A tree from the language H—the tree is labelled by letters a and b, the dotted region contains vertices reachable from the root by a-vertices. The blue thick branch is a branch consisting of a-vertices that turns R infinitely many times. (Color figure online)

Our theorem involves not only a set-theoretical argument but also an automata construction encapsulated by the algorithm Transformation. Our result also gives a stronger information about the descriptive complexity, since (1) it is an open problem whether for a given regular Borel language of infinite trees does exist a weak alternating automaton accepting this language, (2) our Transformation algorithm works for arbitrary parities and it is not clear how to generalize the set-theoretical method of Finkel and Simonnet [8] beyond Büchi automata.

The Lusin-Souslin Theorem that is used in [8] says that if $f\colon X \to Y$ is injective and Borel then the image $f[X]$ is Borel in Y. The proof of this theorem is based on the Lusin Separation Theorem [10, Theorem 14.7]. These theorems are set-theoretical in nature and the result in this work can be considered as an automata-theoretic counterpart of the former. As a sub-procedure in the algorithm Transformation we use an algorithm Separation from [2], which itself is an automata-theoretic counterpart of the Lusin Separation Theorem.

To the authors' best knowledge this is the first work where it is shown how to use the fact that a given automaton is unambiguous to derive upper bounds on the parity index of the recognised language. Therefore, this work should be treated as a first step towards descriptive complexity bounds for unambiguous languages, and generally better understanding of this class of automata.

1.2 Outline of the Paper

We first prove Lemma 3 which states that if an automaton is unambiguous then the transitions of the automaton correspond to disjoint languages. In the algorithm Partition we use an algorithm of Arnold and Santocanale and show that these disjoint languages can be separated by Comp$(i+1, 2j)$ languages (Fig. 2).

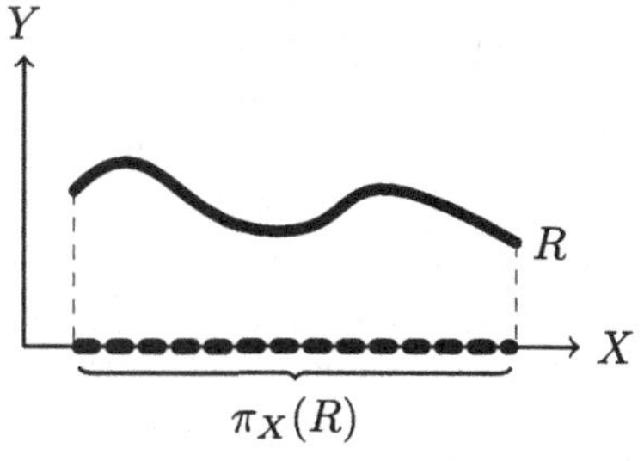

Fig. 2. An illustration of Lusin-Souslin Theorem. A relation $R \subseteq X \times Y$ is Borel and *uniformised*. The theorem implies that $\pi_X(R) \subseteq X$ is Borel as well.

In Sect. 4 we provide a construction of the automaton TRANSFORMATION($\mathcal{A}$) and in Sect. 5.1 we conclude the proof of Theorem 1 by proving correctness of this construction.

2 Basic Notions

In this section we introduce basic notions used in the rest of the paper. A good survey of the relations between deterministic, unambiguous, and non-deterministic automata is [5]. A general background on automata and logic over infinite trees can be found in [18].

Our models are infinite, labelled, full binary trees. The labels come from a non-empty finite set A called *alphabet*. A tree t is a function $t\colon \{\mathtt{L},\mathtt{R}\}^* \to A$. The set of all such trees is Tr_A. Vertices of a tree are denoted $u, v, w \in \{\mathtt{L},\mathtt{R}\}^*$. The prefix-order on vertices is $\preceq$, the minimal element of this order is the *root* $\epsilon \in \{\mathtt{L},\mathtt{R}\}^*$. The label of a tree $t \in Tr_A$ in a vertex $u \in \{\mathtt{L},\mathtt{R}\}^*$ is $t(u) \in A$. $t{\restriction}_u$. stands for the subtree of a tree t rooted in a vertex u. Infinite branches of a tree are denoted as $\alpha, \beta \in \{\mathtt{L},\mathtt{R}\}^\omega$. We extend the prefix order to them, thus $u \prec \alpha$ if u is a prefix of α. For an infinite branch $\alpha \in \{\mathtt{L},\mathtt{R}\}^\omega$ and $k \in \omega$ by $\alpha{\restriction}_k$ we denote the prefix of α of length k (e.g. $\alpha{\restriction}_0 = \epsilon$).

A *non-deterministic tree automaton* $\mathcal{A}$ is a tuple $\langle Q, A, q_0, \Delta, \Omega\rangle$ where: Q is a finite set of *states*; A is an alphabet; $q_\mathrm{I} \in Q$ is an *initial state*; $\Delta \subseteq Q \times A \times Q \times Q$ is a *transition relation*; $\Omega\colon Q \to \mathbb{N}$ is a *priority function*.

If the automaton $\mathcal{A}$ is not known from the context we explicitly put it in the superscript, i.e. $Q^{\mathcal{A}}$ is the set of states of $\mathcal{A}$.

A *run* of an automaton $\mathcal{A}$ on a tree t is a tree $\rho \in Tr_Q$ such that for every vertex u we have $\big(\rho(u), t(u), \rho(u\mathtt{L}), \rho(u\mathtt{R})\big) \in \Delta$. A run ρ is *parity-accepting* if on every branch α of the tree we have

$$\limsup_{n\to\infty} \Omega\big(\rho(\alpha{\restriction}_n)\big) \equiv 0 \mod 2. \tag{$\triangle$}$$

We say that a run ρ *starts* from the state $\rho(\epsilon)$. A run ρ is *accepting* if it is parity-accepting and starts from q_I. The *language recognised* by $\mathcal{A}$ (denoted $\mathcal{L}(\mathcal{A})$) is the set of all trees t such that there is an accepting run ρ of $\mathcal{A}$ on t.

A non-deterministic automaton $\mathcal{A}$ is *unambiguous* if for every tree t there is at most one accepting run of $\mathcal{A}$ on t.

An *alternating tree automaton* $\mathcal{C}$ is a tuple $\langle Q, A, Q_\exists, Q_\forall, q_0, \Delta, \Omega\rangle$ where: Q is a finite set of *states*; A is an alphabet; $Q_\exists \sqcup Q_\forall$ is a partition of Q into sets of positions of the players $\exists$ and $\forall$; $q_\mathrm{I} \in Q$ is an *initial state*; $\Delta \subseteq Q \times A \times \{\epsilon,\mathtt{L},\mathtt{R}\} \times Q$ is a *transition relation*; $\Omega\colon Q \to \mathbb{N}$ is a *priority function*. For technical reasons we assume that for every $q \in Q$ and $a \in A$ there is at least one transition $(q, a, d, q') \in \Delta$ for some $q' \in Q$ and $d \in \{\epsilon,\mathtt{L},\mathtt{R}\}$.

An alternating tree automaton $\mathcal{C}$ induces, for every tree $t \in Tr_A$, a parity game $\mathcal{G}(\mathcal{C}, t)$. The positions of this game are of the form $(u, q) \in \{\mathtt{L},\mathtt{R}\}^* \times Q$. The initial position is (ϵ, q_I). A position (u, q) belongs to the player $\exists$ if $q \in Q_\exists$, otherwise (u, q) belongs to $\forall$. The priority of a position (u, q) is $\Omega(q)$. There is an edge between (u, q) and (ud, q') whenever $(q, t(u), d, q') \in \delta$. An infinite play π in $\mathcal{G}(\mathcal{C}, t)$ is winning for $\exists$ if the highest priority occurring infinitely often on π is even, as in condition $(\triangle)$.

We say that an alternating tree automaton $\mathcal{C}$ *accepts* a tree t if the player $\exists$ has a winning strategy in $\mathcal{G}(\mathcal{C}, t)$. The language of trees accepted by $\mathcal{C}$ is denoted by $\mathcal{L}(\mathcal{C})$. A non-deterministic or alternating automaton $\mathcal{A}$ *has index* (i, j) if the priorities of $\mathcal{A}$ are among $\{i, i+1, \ldots, j\}$. An automaton of index $(1, 2)$ is called a *Büchi automaton*. Every alternating tree automaton can be naturally seen as a graph — the set of nodes is Q and there is an edge (q, q') if $(q, a, d, q') \in \Delta$ for some $a \in A$ and $d \in \{\epsilon, \mathtt{L}, \mathtt{R}\}$. We say that an alternating tree automaton $\mathcal{D}$ is a $\mathrm{Comp}(i, j)$ *automaton* if every strongly-connected component of the graph of $\mathcal{D}$ is of index (i, j) or $(i+1, j+1)$, see [2].

Note that an alternating automaton $\mathcal{C}$ is $\mathrm{Comp}(0, 0)$ if and only if $\mathcal{C}$ is a weak alternating automaton in the meaning of [11]. The following fact gives a connection between these automata and weak MSO (the variant of monadic second-order logic where set quantifiers are restricted to finite sets).

Theorem 2 (Rabin [17], also Kupferman Vardi [11]). *If $\mathcal{C}$ is an alternating* $\mathrm{Comp}(0, 0)$ *automaton then $\mathcal{L}(\mathcal{C})$ is definable in weak* MSO. *Similarly, if $L \subseteq Tr_A$ is definable in weak* MSO *then there exists a* $\mathrm{Comp}(0, 0)$ *automaton recognising L.*

The crucial technical tool in our proof is the SEPARATION algorithm by Arnold and Santocanale [2]. A particular case of this algorithm for $i = j = 1$ is the classical Rabin separation construction (see [17]): if L_1, L_2 are two disjoint languages recognisable by Büchi alternating tree automata then one can effectively construct a weak MSO-definable language L_S that separates them.

Algorithm 1. SEPARATION

Input: Two non-deterministic automata $\mathcal{A}_1$, $\mathcal{A}_2$ of index $(i, 2j)$ such that $\mathcal{L}(\mathcal{A}_1) \cap \mathcal{L}(\mathcal{A}_2) = \emptyset$.

Output: An alternating $\mathrm{Comp}(i{+}1, 2j)$ automaton $\mathcal{S}$ such that

$$\mathcal{L}(\mathcal{A}_1) \subseteq \mathcal{L}(\mathcal{S}) \quad \text{and} \quad \mathcal{L}(\mathcal{A}_2) \cap \mathcal{L}(\mathcal{S}) = \emptyset.$$

---

3 Partition Property

In this section we will prove Lemma 3 stating that if an automaton $\mathcal{A}$ is unambiguous then the transitions of $\mathcal{A}$ need to induce disjoint languages. This will be important in the algorithm PARTITION which for a given unambiguous automaton of index $(i, 2j)$, constructs a family of $\mathrm{Comp}(i+1, 2j)$ automata that split the set of all trees into disjoint sets corresponding to the respective transitions of $\mathcal{A}$. PARTITION will be used in TRANSFORMATION.

Let us fix an unambiguous automaton $\mathcal{A}$ of index $(i, 2j)$. Let Q be the set of states of $\mathcal{A}$ and A be its working alphabet. We say that a transition $\delta = (q, a, q_{\mathrm{L}}, q_{\mathrm{R}})$ of $\mathcal{A}$ *starts* from (q, a); let $\Delta_{q,a}$ be the set of such transitions.

A pair $(q, a) \in Q \times A$ is *productive* if it appears in some accepting run: there exists a tree $t \in Tr_A$ and an accepting run ρ of $\mathcal{A}$ on t such that for some vertex u we have $\rho(u) = q$ and $t(u) = a$. This definition combines two requirements: that there exists an accepting run that leads to the pair (q, a) and that some tree can be parity-accepted starting from (q, a). Note that if (q, a) is productive then there exists at least one transition starting from (q, a). Without changing the language $\mathcal{L}(\mathcal{A})$ we can assume that if a pair is not productive then there is no transition starting from this pair.

For every transition $\delta = (q, a, q_{\mathrm{L}}, q_{\mathrm{R}})$ of $\mathcal{A}$ we define L_δ as the language of trees such that there exists a run ρ of $\mathcal{A}$ on t that is parity-accepting and *uses δ in the root of t* $\rho(\epsilon) = q$, $t(\epsilon) = a$, $\rho(\mathtt{L}) = q_{\mathrm{L}}$, and $\rho(\mathtt{R}) = q_{\mathrm{R}}$. Clearly the language L_δ can be recognised by an unambiguous automaton of index $(i, 2j)$. If (q, a) is not productive then $L_{(q,a,q_{\mathrm{L}},q_{\mathrm{R}})} = \emptyset$. The following lemma is a simple consequence of unambiguity of the given automaton $\mathcal{A}$.

Lemma 3. *If $\delta_1 \neq \delta_2$ are two transitions starting from the same pair (q, a) then the languages L_{δ_1}, L_{δ_2} are disjoint.*

Proof. First, if (q, a) is not productive then by our assumption $L_{\delta_1} = L_{\delta_2} = \emptyset$. Assume contrary that (q, a) is productive and there exists a tree $r \in L_{\delta_1} \cap L_{\delta_2}$ with two respective parity-accepting runs ρ_1, ρ_2. Since (q, a) is productive so there exists a tree t and an accepting run ρ on t such that $\rho(u) = q$ and $t(u) = a$ for some vertex u. Consider the tree $t' = t[u \leftarrow r]$ — the tree obtained from t by substituting r as the subtree under u. Since $\rho(u) = q$ and both ρ_1, ρ_2 start from (q, a), we can construct two accepting runs $\rho[u \leftarrow \rho_1]$ and $\rho[u \leftarrow \rho_2]$ on t'. Since these runs differ on the transition used in u, we obtain a contradiction to the fact that $\mathcal{A}$ is unambiguous. □

The above lemma will be important in the algorithm PARTITION, because it uses the SEPERATION algorithm which in turn requires disjointness of the languages.

The following lemma summarizes properties of the algorithm PARTITION.

Lemma 4. *Assume that $\mathcal{A}$ is an unambiguous automaton of index $(i, 2j)$ and let $(q, a) \in Q \times A$. Take the automata $\left(\mathcal{C}_\delta\right)_{\delta \in \Delta_{q,a}}$ constructed by PARITION($\mathcal{A}$). Then the languages $\mathcal{L}(\mathcal{C}_\delta)$ for $\delta \in \Delta_{q,a}$ are pairwise disjoint and $L_\delta \subseteq \mathcal{L}(\mathcal{C}_\delta)$.*

A proof of this lemma follows directly from the definition of the respective automata, see Fig. 3 for an illustration of this construction.

4 Construction of the Automaton

In this and the following section we will describe the algorithm TRANSFORMATION and prove Theorem 1 which states correctness and properties of this

Algorithm 2. PARTITION

Input: An unambiguous automaton $\mathcal{A}$ of index $(i, 2j)$
Output: for every $\delta \in \Delta$ an automaton $\mathcal{C}_\delta$

foreach $(q, a) \in Q \times A$, *productive* **do**
  foreach $\delta \in \Delta_{q,a}$ **do**
    $\mathcal{E}_\delta \leftarrow$ *non-det.* $(i, 2j)$ *automaton recognising* L_δ
    $\mathcal{F}_\delta \leftarrow$ *non-det.* $(i, 2j)$ *automaton recognising* $\bigcup_{\eta \in \Delta_{q,a}, \eta \neq \delta} L_\eta$
  foreach $\delta \in \Delta_{q,a}$ **do**
    $\mathcal{D}_\delta \leftarrow$ SEPARATION(E_δ, F_δ)
  foreach $\delta \in \Delta_{q,a}$ **do**
    $\mathcal{C}_\delta \leftarrow$ Comp$(i+1, 2j)$ *automaton recognising* $\mathcal{L}(\mathcal{D}_\delta) \setminus \bigcup_{\eta \neq \delta} \mathcal{L}(\mathcal{D}_\eta)$.
  $\mathcal{B}_{q,a} \leftarrow$ Comp$(i+1, 2j)$ *automaton recognising* $Tr_A \setminus \bigcup_{\delta \in \Delta_{q,a}} \mathcal{L}(\mathcal{D}_\delta)$.
foreach $\delta = (q, a, q_L, q_R) \in \Delta_{q,a}$ *with* (q, a) *non-productive* **do**
  $\mathcal{C}_\delta \leftarrow$ Comp$(0, 0)$ *automaton recognising the empty language.*

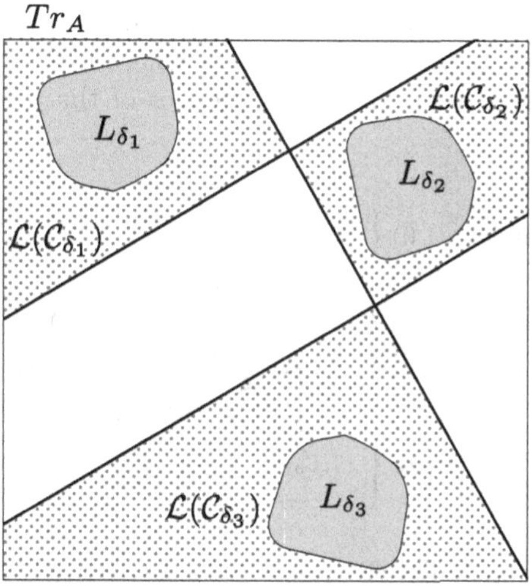

Fig. 3. An illustration of the output of the algorithm PARTITION. The three circles are the languages L_{δ_i} for the transitions starting in a fixed pair (q, a). Each straight line represents the language $\mathcal{L}(\mathcal{D}_{\delta_i})$ that separates the respective language L_{δ_i} from the others. Our construction provides the automata $\mathcal{C}_{\delta_i}$ recognising the dotted regions.

algorithm. Given an automaton $\mathcal{A}$ of index $(i, 2j)$, the algorithm TRANSFORMATION constructs an alternating Comp$(i+1, 2j)$ automaton $\mathcal{R}$ recognising $\mathcal{L}(\mathcal{A})$. It will consist of two sub-automata running in parallel:

1. In the first sub-automaton the role of $\exists$ will be to propose a partial run $\rho\colon \{\mathtt{L}, \mathtt{R}\}^* \rightharpoonup Q$ on a given tree t. She will be forced to propose certain unique run ρ_t that depends only on the tree t, see Definition 6. At any moment $\forall$

can *challenge* the currently proposed transition and check if it agrees with the definition of ρ_t (namely Condition ($\diamond$)).

2. In the second sub-automaton the role of $\forall$ will be to prove that the partial run ρ_t is not parity-accepting. That is, he will find a leaf in ρ_t or an infinite branch of ρ_t that does not satisfy the parity condition. Since the run ρ_t is unique, $\forall$ can declare in advance what will be the odd priority n that is the limes superior (i.e. lim sup) of priorities of ρ_t on the selected branch.

The automaton $\mathcal{R}$ consists of an *initial component* $\mathcal{I}$ and of the union of the automata $\mathcal{C}_\delta$ constructed by the procedure PARTITION.

The idea of the automaton $\mathcal{R}$ is to simulate the following behaviour. Assume that the label of the current vertex is a and the current state is $(q, n) \in Q_{\mathcal{I},\exists}$:

- if $n \neq \star$ and $\Omega^{\mathcal{A}}(q) > n$ then $\forall$ loses, see line 9;
- $\exists$ declares a transition $\delta = (q, a, q_{\mathrm{L}}, q_{\mathrm{R}})$ of $\mathcal{A}$, see line 11;
- $\forall$ can decide to *challenge* this transition, see line 13;
- if $n \neq \star$ then $\forall$ chooses a direction and the game proceeds, see line 15;
- if $n = \star$ then $\forall$ chooses a direction and a new value $n' \in N$, see line 17.

Figure 4 depicts the structure of the automaton $\mathcal{R}$. The initial component $\mathcal{I}$ is split into two parts: $\mathcal{I}_0$ where $n = \star$ and $\mathcal{I}_1$ where $n \neq \star$.

We will now proceed with proving properties of the procedure TRANSFORMATION.

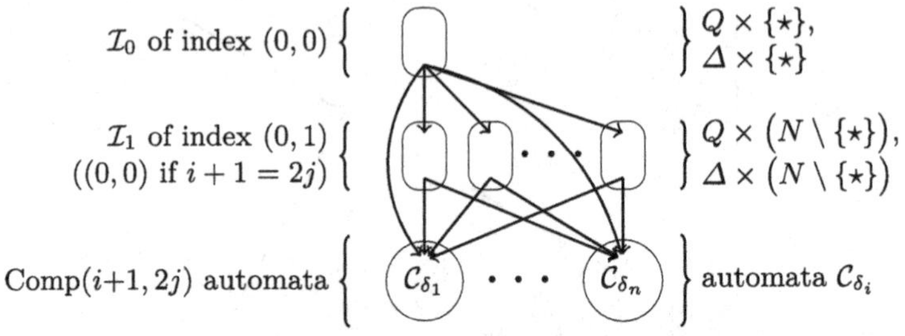

Fig. 4. The structure of the automaton $\mathcal{R}$.

Lemma 5. *If $\mathcal{A}$ is unambiguous and of index $(i, 2j)$ then $\mathcal{R}$ is in* Comp$(i{+}1, 2j)$.

Proof. We first argue that if $i + 1 < 2j$ then $\mathcal{R}$ is a Comp$(i + 1, 2j)$ automaton. Note every strongly-connected component in the graph of $\mathcal{R}$ is either a component of $\mathcal{I}_0$, $\mathcal{I}_1$, or of $\mathcal{C}_\delta$ for $\delta \in \Delta^{\mathcal{A}}$. Recall that all the components $\mathcal{A}_\delta$ are by the construction Comp$(i + 1, 2j)$-automata. By the definition, $\mathcal{I}_0$ and $\mathcal{I}_1$ are Comp$(1, 2)$-automata, so the whole automaton $\mathcal{R}$ is also Comp$(i + 1, 2n)$.

Consider the opposite case: $i + 1 = 2j$. By shifting all the priorities we can assume that $i = j = 1$ (i.e. $\mathcal{A}$ is Büchi). Observe that the only possible odd value n between i and $2j$ is $n = 1$. It means that if $\forall$ declares a value $n \neq \star$ then always $\Omega(q) \geq n$ holds. It means that there are no states in $\mathcal{I}_1$ with priority 1. Therefore, both $\mathcal{I}_0$ and $\mathcal{I}_1$ are Comp$(0, 0)$ automata and $\mathcal{R}$ is a Comp$(0, 0)$ automaton.

Algorithm 3. TRANSFORMATION

Input: An unambiguous automaton $\mathcal{A}$ of index $(i, 2j)$

Output: An automaton $\mathcal{R}$

1 $N \leftarrow \{\star\} \cup \{n \in \{i, \ldots, 2j\} \mid n \text{ is odd}\}$

2 $Q_{\mathcal{I}, \exists} \leftarrow Q^{\mathcal{A}} \times N \sqcup \{\bot, \top\}$

3 $Q_{\mathcal{I}, \forall} \leftarrow \Delta^{\mathcal{A}} \times N$

4 $\Delta_{\mathcal{I}} \leftarrow \{(\bot, a, \epsilon, \bot), (\top, a, \epsilon, \top) \mid a \in A^{\mathcal{A}}\}$

5 $q_{\mathrm{I}}^{\mathcal{R}} \leftarrow (q_{\mathrm{I}}^{\mathcal{A}}, \star)$

6 $(\mathcal{D}_{\delta})_{\delta \in \Delta} \leftarrow \text{PARTITION}(\mathcal{A})$

7 **foreach** $a \in A$, $q \in Q^{\mathcal{A}}$, $n \in N$ **do**

8 $\quad$ **if** $n \neq \star$ and $\Omega^{\mathcal{A}}(q) > n$ **then**

9 $\quad\quad\mid \quad \Delta_{\mathcal{I}} \leftarrow \Delta_{\mathcal{I}} \cup \{((q, n), a, \epsilon, \top)\}$

10 $\quad$ **else**

11 $\quad\quad\Big\lfloor \quad \Delta_{\mathcal{I}} \leftarrow \Delta_{\mathcal{I}} \cup \Big\{((q, n), a, \epsilon, (\delta, n)) \mid \delta \in \Delta_{q, a}^{\mathcal{A}}\Big\}$

12 **foreach** $a \in A$, $\delta = (q, a, q_{\mathrm{L}}, q_{\mathrm{R}}) \in \Delta^{\mathcal{A}}$, $n \in N$ **do**

13 $\quad \Delta_{\mathcal{I}} \leftarrow \Delta_{\mathcal{I}} \cup \Big\{(\delta, a, \epsilon, q_{\mathrm{I}}^{\mathcal{C}_{\delta}})\Big\}$ /* such a transition is a *challenge*   */

14 $\quad$ **if** $n \neq \star$ **then**

15 $\quad\quad\Big\lfloor \quad \Delta_{\mathcal{I}} \leftarrow \Delta_{\mathcal{I}} \cup \{(\delta, a, d, (q_d, n)) \mid d \in \{\mathrm{L}, \mathrm{R}\}\}$

16 $\quad$ **else**

17 $\quad\quad\Big\lfloor \quad \Delta_{\mathcal{I}} \leftarrow \Delta_{\mathcal{I}} \cup \{(\delta, a, d, (q_d, n')) \mid d \in \{\mathrm{L}, \mathrm{R}\}, n' \in N\}$

18 $Q_{\exists}^{\mathcal{R}} \leftarrow Q_{\mathcal{I}, \exists} \sqcup \bigsqcup_{\delta \in \Delta^{\mathcal{A}}} Q_{\exists}^{\mathcal{C}_{\delta}}$

19 $Q_{\forall}^{\mathcal{R}} \leftarrow Q_{\mathcal{I}, \forall} \sqcup \bigsqcup_{\delta \in \Delta^{\mathcal{A}}} Q_{\forall}^{\mathcal{C}_{\delta}}$

20 $\Delta^{\mathcal{R}} \leftarrow \Delta_{\mathcal{I}} \sqcup \bigsqcup_{\delta \in \Delta^{\mathcal{A}}} \Delta^{\mathcal{C}_{\delta}}$

21 **foreach** $q \in Q^{\mathcal{A}}$ **do**

22 $\quad \Omega^{\mathcal{R}}(q, \star) = 0$

23 $\quad$ **foreach** $n \in N \setminus \{\star\}$ **do**

24 $\quad\quad$ **if** $\Omega^{\mathcal{A}}(q) \geq n$ **then**

25 $\quad\quad\quad\big\lfloor \quad \Omega^{\mathcal{R}}(q, n) = 1$

26 $\quad\quad$ **else**

27 $\quad\quad\quad\big\lfloor \quad \Omega^{\mathcal{R}}(q, n) = 0$

28 **foreach** $\delta = (q, a, q_{\mathrm{L}}, q_{\mathrm{R}}) \in Q^{\mathcal{A}}$ **do**

29 $\quad \Omega^{\mathcal{R}}(\delta, \star) = 0$

30 $\quad$ **foreach** $n \in N \setminus \{\star\}$ **do**

31 $\quad\quad$ **if** $\Omega^{\mathcal{A}}(q) \geq n$ **then**

32 $\quad\quad\quad\big\lfloor \quad \Omega^{\mathcal{R}}(\delta, n) = 1$

33 $\quad\quad$ **else**

34 $\quad\quad\quad\big\lfloor \quad \Omega^{\mathcal{R}}(\delta, n) = 0$

5 Correctness of the Construction

In this section we prove that the automaton $\mathcal{R}$ constructed by the algorithm TRANSFORMATION recognises the same language as the given unambiguous automaton $\mathcal{A}$. Let $\mathcal{A}$ be an unambiguous automaton of index $(i, 2j)$.

Definition 6. Let $t \in Tr_A$ be a tree. We define ρ_t as the unique maximal partial run ρ_t of $\mathcal{A}$ on t, i.e. a partial function $\rho_t \colon \{\mathrm{L},\mathrm{R}\}^* \rightharpoonup Q^{\mathcal{A}}$ such that:

- $\rho_t(\epsilon) = q_{\mathrm{I}}^{\mathcal{A}}$;
- if $u \in dom(\rho_t)$ and $t{\restriction}_u \in \mathcal{L}(\mathcal{C}_\delta)$ for some $\delta \in \Delta^{\mathcal{A}}$ then[1]

$$\delta = \big(\rho_t(u), t(u), \rho_t(u\mathrm{L}), \rho_t(u\mathrm{R})\big); \tag{$\diamond$}$$

- if $u \in dom(\rho_t)$ and $t{\restriction}_u \notin \mathcal{L}(\mathcal{C}_\delta)$ for any $\delta \in \Delta^{\mathcal{A}}$ then $u\mathrm{L}, u\mathrm{R} \notin dom(\rho_t)$.

Lemma 7. *$t \in \mathcal{L}(\mathcal{A})$ if and only if ρ_t is total and accepting.*

Proof. If ρ_t is accepting then it is a witness that $t \in \mathcal{L}(\mathcal{A})$. Let ρ be an accepting run of $\mathcal{A}$ on t. We inductively prove that $\rho = \rho_t$. Take a node u of t and define $q = \rho(u)$, $a = t(u)$, $q_\mathrm{L} = \rho_t(u\mathrm{L})$, and $q_\mathrm{R} = \rho_t(u\mathrm{R})$. Observe that ρ is a witness that (q, a) is productive and for $\delta = (q, a, q_\mathrm{L}, q_\mathrm{R})$ we have

$$t \in L_\delta \subseteq \mathcal{L}(\mathcal{C}_\delta).$$

Therefore, $\rho_t(u\mathrm{L}) = \rho(u\mathrm{L})$ and $\rho_t(u\mathrm{R}) = \rho(u\mathrm{R})$. □

5.1 $\mathcal{L}(\mathcal{A} = \mathcal{L}(\mathcal{R})$

Lemma 8. *If $t \in \mathcal{L}(\mathcal{A})$ then $t \in \mathcal{L}(\mathcal{R})$.*

Proof. Assume that $t \in \mathcal{L}(\mathcal{A})$. By Lemma 7 we know that ρ_t is the unique accepting run of $\mathcal{A}$ on t. Consider the following strategy $\sigma_\exists$ for $\exists$ in the initial component $\mathcal{I}$ of the automaton $\mathcal{R}$: always declare δ consistent with ρ_t. Extend it to the winning strategies in $\mathcal{C}_\delta$ whenever they exist. That is, if the current vertex is u and the state of $\mathcal{R}$ is of the form $(q, n) \in \mathcal{I}$ then move to the state (δ, n) for $\delta = (\rho_t(u), t(u), \rho_t(u\mathrm{L}), \rho_t(u\mathrm{R}))$. Whenever the game moves from the initial component $\mathcal{I}$ into one of the automata $\mathcal{C}_\delta$ in a vertex u, fix some winning strategy in $\mathcal{G}(\mathcal{C}_\delta, t{\restriction}_u)$ (if exists) and play according to this strategy; if there is no such strategy, play using any strategy. Take a play consistent with $\sigma_\exists$ in $\mathcal{G}(\mathcal{R}, t)$. There are the following cases:

- $\forall$ loses in a finite time according to the transition from line 9 in the algorithm TRANSFORMATION.
- $\forall$ stays forever in the initial component $\mathcal{I}$ never changing the value of $n = \star$ and loses by the parity criterion.

[1] By Lemma 4 there is at most one such δ.

- In some vertex u of the tree $\forall$ *challenges* the transition δ given by $\exists$ and the game proceeds to the state $q_{\mathrm{I}}^{\mathcal{C}_\delta}$. In that case $t{\restriction}_u \in L_\delta$ by the definition of L_δ (the run $\rho_t{\restriction}_u$ is a witness) and therefore $t{\restriction}_u \in \mathcal{L}(\mathcal{C}_\delta)$. So $\exists$ has a winning strategy in $\mathcal{G}(\mathcal{C}_\delta, t{\restriction}_u)$ and $\exists$ wins the rest of the game.
- $\forall$ declares a value $n \neq \star$ at some point and then never *challenges* $\exists$. In that case the game follows an infinite branch α of t. Since ρ_t is accepting so we know that $k \stackrel{\text{def}}{=} \limsup_{i\to\infty} \Omega^{\mathcal{A}}(\rho_t(\alpha{\restriction}_i))$ is even. If $k > n$ then $\forall$ loses at some point according to the transition from line 9. Otherwise $k < n$ and from some point on all the states of $\mathcal{R}$ visited during the game have priority 0, thus $\forall$ loses by the parity criterion in $\mathcal{I}_1$. □

Lemma 9. *If $t \notin \mathcal{L}(\mathcal{A})$ then $t \notin \mathcal{L}(\mathcal{R})$.*

Proof. We assume that $t \notin \mathcal{L}(\mathcal{A})$ and define a winning strategy for $\forall$ in the game $\mathcal{G}(\mathcal{R}, t)$. Let us fix the run ρ_t as in Definition 6.

Note that either ρ_t is a partial run: there is a vertex u such that $\rho_t(u) = q$ and $(q, t(u))$ is not productive; or ρ_t is a total run. Since $t \notin \mathcal{L}(\mathcal{A})$, ρ_t cannot be a total accepting run. Let α be a finite or infinite branch: either $\alpha \in \{\mathrm{L}, \mathrm{R}\}^*$ and α is a leaf of ρ_t or α is an infinite branch such that $k \stackrel{\text{def}}{=} \limsup_{i\to\infty} \Omega^{\mathcal{A}}(\rho_t(\alpha{\restriction}_i))$ is odd. If α is finite let us put any odd value between i and $2j$ as k. Consider the following strategy for $\forall$:

- $\forall$ keeps $n = \star$ until there are no more states of priority greater than k along α in ρ_t. Then he declares $n' = k$.
- $\forall$ *challenges* a transition δ given by $\exists$ in a vertex u if and only if $t{\restriction}_u \notin \mathcal{C}_\delta$.
- $\forall$ always follows α: in a vertex $u \in \{\mathrm{L}, \mathrm{R}\}^*$ he chooses the direction d in such a way that $ud \preceq \alpha$.

As in the proof of Lemma 8, we extend this strategy to strategies in the components $\mathcal{C}_\delta$ whenever such strategies exist: if the game moves from the component $\mathcal{I}$ into one of the component $\mathcal{C}_\delta$ in a vertex u then $\forall$ uses some winning strategy in the game $\mathcal{G}(\mathcal{C}_\delta, t{\restriction}_u)$ (if it exists); if there is no such strategy, $\forall$ plays using any strategy.

Consider any play π consistent with $\sigma_\forall$. Note that if α is a finite word and the play π reaches the vertex α in a state (δ, n) in $\mathcal{I}$ then by the definition of ρ_t we know that $t{\restriction}_u \notin \mathcal{C}_\delta$ and thus $\forall$ *challenges* this transition and wins in the game $\mathcal{G}(\mathcal{C}_\delta, t{\restriction}_u)$. By the definition of the strategy $\sigma_\forall$, $\forall$ never loses according to the transition from line 9 in the algorithm Transformation — if $\forall$ declared $n \neq \star$ then the play will never reach a state of priority greater than n.

Let us consider the remaining cases. First assume that at some vertex u player $\forall$ *challenged* a transition δ declared by $\exists$. It means that $t{\restriction}_u \notin \mathcal{L}(\mathcal{C}_\delta)$ and $\forall$ has a winning strategy in $\mathcal{G}(\mathcal{C}_\delta, t{\restriction}_u)$ and wins in that case.

The last case is that $\forall$ did not *challenge* any transition declared by $\exists$ and the play followed the branch α. Then, for every $i \in \mathbb{N}$ the game reached the vertex $\alpha{\restriction}_i$ in a state (q, n) satisfying $q = \rho_t(\alpha{\restriction}_i)$. In that case there is some vertex u along α where $\forall$ declared $n = k$. Therefore, infinitely many times $\Omega(q) = n$ in π so $\forall$ wins that play by the parity criterion. □

6 Conclusion

We presented a new algorithm Transformation which for a given unambiguous automaton $\mathcal{A}$ of index $(i, 2j)$ outputs an automaton Transformation($\mathcal{A}$) which accepts the same language and belongs to the class Comp$(i+1, 2j)$. In particular, if $\mathcal{A}$ is an unambiguous Büchi automaton, then Transformation($\mathcal{A}$) is a weak alternating automaton. This can be considered an automata-theoretic counterpart of the Lusin-Souslin Theorem [10, Theorem 15.1].

Further Work. This paper is a part of a broader project intended to understand better the descriptive complexity of unambiguous languages of infinite trees. In our view the crucial question is whether unambiguous automata can reach arbitrarily high levels in the alternating index hierarchy.

Conjecture. There exists a pair (i, j) such that if $\mathcal{A}$ is an unambiguous automaton on infinite trees then the language recognised by $\mathcal{A}$ can be recognised by an alternating automaton of index (i, j).

References

1. Arnold, A.: The μ-calculus alternation-depth hierarchy is strict on binary trees. ITA **33**(4/5), 329–340 (1999)
2. Arnold, A., Santocanale, L.: Ambiguous classes in μ-calculi hierarchies. TCS **333**(1–2), 265–296 (2005)
3. Bradfield, J.: Simplifying the modal mu-calculus alternation hierarchy. In: Morvan, M., Meinel, C., Krob, D. (eds.) STACS 1998. LNCS, vol. 1373, pp. 39–49. Springer, Heidelberg (1998)
4. Carayol, A., Löding, C., Niwiński, D., Walukiewicz, I.: Choice functions and well-orderings over the infinite binary tree. Cent. Eur. J. Math. **8**, 662–682 (2010)
5. Colcombet, T.: Forms of determinism for automata (invited talk). In: STACS, pp. 1–23 (2012)
6. Colcombet, T., Kuperberg, D., Löding, C., Vanden Boom, M.: Deciding the weak definability of Büchi definable tree languages. In: CSL, pp. 215–230 (2013)
7. Facchini, A., Murlak, F., Skrzypczak, M.: Rabin-Mostowski index problem: a step beyond deterministic automata. In: LICS, pp. 499–508 (2013)
8. Finkel, O., Simonnet, P.: On recognizable tree languages beyond the Borel hierarchy. Fundam. Informaticae **95**(2–3), 287–303 (2009)
9. Hummel, S.: Unambiguous tree languages are topologically harder than deterministic ones. In: GandALF, pp. 247–260 (2012)
10. Kechris, A.: Classical Descriptive Set Theory. Springer, New York (1995)
11. Kupferman, O., Vardi, M.Y.: The weakness of self-complementation. In: Meinel, C., Tison, S. (eds.) STACS 1999. LNCS, vol. 1563, pp. 455–466. Springer, Heidelberg (1999)
12. Murlak, F.: The Wadge hierarchy of deterministic tree languages. Log. Methods Comput. Sci. **4**(4), 1–44 (2008)
13. Niwiński, D., Walukiewicz, I.: Ambiguity problem for automata on infinite trees (1996, unpublished)
14. Niwiński, D., Walukiewicz, I.: Relating hierarchies of word and tree automata. In: Morvan, M., Meinel, C., Krob, D. (eds.) STACS 1998. LNCS, vol. 1373, pp. 320–331. Springer, Heidelberg (1998)

15. Niwiński, D., Walukiewicz, I.: A gap property of deterministic tree languages. Theor. Comput. Sci. **1**(303), 215–231 (2003)
16. Niwiński, D., Walukiewicz, I.: Deciding nondeterministic hierarchy of deterministic tree automata. Electr. Notes Theor. Comput. Sci. **123**, 195–208 (2005)
17. Rabin, M.O.: Weakly definable relations and special automata. In: Proceedings of the Symposium on Mathematical Logic and Foundations of Set Theory, pp. 1–23. North-Holland (1970)
18. Thomas, W.: Languages, automata, and logic. In: Rozenberg, G., Salomaa, A. (eds.) Handbook of Formal Languages, pp. 389–455. Springer, Heidelberg (1996)

One-Unknown Word Equations and Three-Unknown Constant-Free Word Equations

Dirk Nowotka[1] and Aleksi Saarela[2(✉)]

[1] Department of Computer Science, Kiel University, 24098 Kiel, Germany
dn@zs.uni-kiel.de

[2] Department of Mathematics and Statistics, University of Turku, 20014 Turku, Finland
amsaar@utu.fi

Abstract. We prove connections between one-unknown word equations and three-unknown constant-free word equations, and use them to prove that the number of equations in an independent system of three-unknown constant-free equations is at most logarithmic with respect to the length of the shortest equation in the system. We also study two well-known conjectures. The first conjecture claims that there is a constant c such that every one-unknown equation has either infinitely many solutions or at most c. The second conjecture claims that there is a constant c such that every independent system of three-unknown constant-free equations with a nonperiodic solution is of size at most c. We prove that the first conjecture implies the second one, possibly for a different constant.

Keywords: Combinatorics on words · Word equations · Independent systems

1 Introduction

One of the most important open problems in combinatorics on words is the following question: For a given n, what is the maximal size of an independent system of constant-free word equations on n unknowns? It is known that every system of word equations is equivalent to a finite subsystem and, consequently, every independent system is finite. This is known as *Ehrenfeucht's compactness property*. It was conjectured by Ehrenfeucht in a language theoretic setting, formulated in terms of word equations by Culik and Karhumäki [3], and proved by Albert and Lawrence [1] and independently by Guba [6]. If $n > 2$, no finite upper bound for the size of independent systems is known. The largest known independent systems have size $\Theta(n^4)$ [10]. Some related results and variations of the problem are discussed in [11].

This work has been supported by the DFG Heisenberg grant 590179 (Dirk Nowotka), the DFG research grant 614256 and the Vilho, Yrjö and Kalle Väisälä Foundation (Aleksi Saarela).

S. Brlek and C. Reutenauer (Eds.): DLT 2016, LNCS 9840, pp. 332–343, 2016.
DOI: 10.1007/978-3-662-53132-7_27

The difference between the best known lower and upper bounds is particularly striking in the case of three unknowns: The largest known independent systems consist of just three equations, but it is not even known whether there exists a constant c such that every independent system has size c or less. When studying independent systems, it is often additionally required that the system has a nonperiodic solution; then the largest known example consists of two equations.

There have been some recent advances regarding this topic. The first nontrivial upper bound was proved by Saarela [14]: The size of an independent system on three unknowns is at most quadratic with respect to the length of the shortest equation in the system. This bound was improved to a linear one by Holub and Žemlička [8]; this is currently the best known result.

Another well-known but less central open problem on word equations is the following question: If a one-unknown word equation with constants has only finitely many solutions, then what is the maximal number of solutions it can have? The answer is at least two, and it has been conjectured that it is exactly two. The best known upper bound, proved by Laine and Plandowski [12], is logarithmic with respect to the number of occurrences of the unknown in the equation. Similar but slightly weaker results were proved in [4,5].

In this article we establish a connection between three-unknown constant-free equations and one-unknown equations with constants. This is done by using an old result by Budkina and Markov [2], or a similar result by Spehner [16]. We use this connection to prove two main results.

The first main result is that the size of an independent system of three-unknown equations is logarithmic with respect to the length of the shortest equation in the system. This result is based on the logarithmic bound for the number of solutions of one-unknown equations.

The second main result is an explicit link between two existing conjectures: If there exists a constant c such that the number of solutions of a one-unknown equation is either infinite or at most c, then there exists a constant c' such that the size of an independent system of three-unknown constant-free equations with a nonperiodic solution is at most c'. Furthermore, if $c = 2$, then we can let $c' = 17$. The number 17 here is very unlikely to be optimal, and we expect that the result could be improved by a more careful analysis.

2 Preliminaries

Let Ξ be an alphabet of unknowns and Σ an alphabet of constants. A *constant-free word equation* is a pair $(u, v) \in \Xi^* \times \Xi^*$, and the solutions of this equation are the morphisms $h : \Xi^* \to \Sigma^*$ such that $h(u) = h(v)$. A *word equation with constants* is a pair $(u, v) \in (\Xi \cup \Sigma)^* \times (\Xi \cup \Sigma)^*$, and the solutions of this equation are the constant-preserving morphisms $h : (\Xi \cup \Sigma)^* \to \Sigma^*$ such that $h(u) = h(v)$. We will state many definitions that work for both types of equations.

A solution h is *periodic* if $h(pq) = h(qp)$ for all words p, q in the domain of h, and *nonperiodic* otherwise.

Usually we assume that the alphabet of constants is $\Sigma = \{a, b\}$. The case of a unary alphabet is not interesting, and if there are more than two constant letters, they can be encoded using a binary alphabet. We are specifically interested in equations with constants on one unknown x, and in constant-free equations on three unknowns x, y, z. We use the notation $[u, v, w]$ for the morphism $h : \{x, y, z\}^* \to \Sigma^*$ defined by $(h(x), h(y), h(z)) = (u, v, w)$, and the notation $[u]$ for the constant-preserving morphism $h : (\{x\} \cup \Sigma)^* \to \Sigma^*$ defined by $h(x) = u$. If U is a set of words, we use the notation $[U] = \{[u] \mid u \in U\}$.

Example 1. The equation (xab, bax) has infinitely many solutions $[(ab)^i]$, where $i \geq 0$. The equation $(xaxbab, abaxbx)$ has exactly two solutions, $[\varepsilon]$ and $[ab]$. The constant-free equation (xyz, zyx) has solutions $[(pq)^i p, (qp)^j q, (pq)^k p]$, where $p, q \in \Sigma^*$ and $i, j, k \geq 0$. It has no other nonperiodic solutions.

A set of equations is a *system of equations.* A system $\{E_1, \ldots, E_n\}$ is often written without the braces as $E_1, \ldots, E_n$. A morphism is a solution of this system if it is a solution of every E_i.

The set of all solutions of an equation E is denoted by $\mathrm{Sol}(E)$. Two equations E_1 and E_2 are *equivalent* if $\mathrm{Sol}(E_1) = \mathrm{Sol}(E_2)$. These notions can naturally be extended to systems of equations.

The set of all equations satisfied by a solution h is denoted by $\mathrm{Eq}(h)$. Two solutions h_1 and h_2 are *equivalent* if $\mathrm{Eq}(h_1) = \mathrm{Eq}(h_2)$.

A system of equations $E_1, \ldots, E_n$ is *independent* if it is not equivalent to any of its proper subsystems. Another equivalent definition would be that $E_1, \ldots, E_n$ is independent if there are solutions $h_1, \ldots, h_n$ such that $h_i \in \mathrm{Sol}(E_j)$ if and only if $i \neq j$. The sequence $(h_1, \ldots, h_n)$ is then called an *independence certificate.* (A system is a set, so the order of the equations is not formally specified, but whenever talking about certificates, it is to be understood that the order of the solutions corresponds to the order in which the equations have been written.)

If an independent system has a nonperiodic solution h, it is called *strictly independent.* If $(h_1, \ldots, h_n)$ is its independence certificate, then $(h_1, \ldots, h_n, h)$ is a *strict independence certificate.*

The above definitions can also be stated for infinite systems. However, by Ehrenfeucht's compactness property, every system of word equations is equivalent to a finite subsystem. We will consider only finite systems in this article.

Example 2. The pair of constant-free equations $(xyz, zyx), (xyyz, zyyx)$ is strictly independent. It has a strict independence certificate $([a, b, abba], [a, b, aba], [a, b, a])$. The system of constant-free equations $(x, \varepsilon), (y, \varepsilon), (z, \varepsilon)$ is independent, but not strictly independent. It has an independence certificate $([a, \varepsilon, \varepsilon], [\varepsilon, a, \varepsilon], [\varepsilon, \varepsilon, a])$.

The *length* of an equation $E = (u, v)$ is $|uv|$ and it is denoted by $|E|$. If h is a morphism, we use the notation $h(E) = (h(u), h(v))$. The equation E is *reduced* if u and v do not have a common nonempty prefix or suffix. We can always replace an equation with an equivalent reduced equation.

3 Main Questions

The following question is one of the biggest open problems on word equations:

Question 3. Let S be a strictly independent system of constant-free equations on three unknowns. How large can S be?

The largest known examples are of size two, and it has been conjectured that these examples are optimal. Even the following weaker conjecture is open:

Conjecture 4. There exists a number c such that every strictly independent system of constant-free equations on three unknowns is of size c or less.

We will refer to this conjecture as SIND-XYZ, or as SIND-XYZ(c) for a specific value of c. Currently, the best known result is the following [8]:

Theorem 5. *Every strictly independent system of constant-free equations on three unknowns is of size $O(n)$, where n is the length of the shortest equation.*

Another well-known open problem is the following:

Question 6. Let E be a one-unknown equation with only finitely many solutions. How many solutions can E have?

The best known examples have two solutions, and it has been conjectured that these examples are optimal. Even the following weaker conjecture is open:

Conjecture 7. There exists a number c such that every one-unknown equation has either infinitely many solutions or at most c.

We will refer to this conjecture as SOL-XAB, or as SOL-XAB(c) for a specific value of c. Currently, the best known result is the following [12]:

Theorem 8. *The solution set of a nontrivial one-unknown equation is either of the form $[(pq)^*p]$, where pq is primitive, or a finite set of size at most $8\log n + O(1)$, where n is the number of occurrences of the unknown.*

As a question between Questions 3 and 6, we can state the following problem and conjecture (we are not aware of any previous research on this problem):

Question 9. Let S be a strictly independent system of one-unknown equations. How large can S be?

Conjecture 10. There exists a number c such that every strictly independent system of one-unknown equations is of size c or less.

We will refer to this conjecture as SIND-XAB, or as SIND-XAB(c) for a specific value of c.

We will prove the following implications between the three conjectures:

$$\text{SOL-XAB} \Rightarrow \text{SIND-XAB} \Leftrightarrow \text{SIND-XYZ},$$

or more specifically,

$$\text{SOL-XAB}(c) \Rightarrow \text{SIND-XAB}(c) \begin{cases} \Leftarrow \text{SIND-XYZ}(c) \\ \Rightarrow \text{SIND-XYZ}(5c+7). \end{cases}$$

Using the same ideas, we will turn Theorem 8 into a result on constant-free equations on three unknowns.

4 One-Unknown Equations with Constants

In this section we prove that Conjectures SIND-XYZ and SOL-XAB imply Conjecture SIND-XAB. The next lemma is from [5].

Lemma 11. *Let E be a one-unknown equation and let pq be primitive. The set $\mathrm{Sol}(E) \cap [(pq)^+p]$ is either $[(pq)^+p]$ or has at most one element.*

Lemma 12. *Let $N \geq 3$ and let $E_1, \ldots, E_N$ be a strictly independent system of one-unknown equations. All of these equations have at least N solutions, and at most one of them has infinitely many solutions. If $N \geq 4$, then none of them has infinitely many solutions.*

Proof. If $(h_1, \ldots, h_{N+1})$ is a strict independence certificate, then E_i has solutions h_j for all $j \neq i$. Thus every equation has at least N solutions.

Let one of the equations, say E_1, have infinitely many solutions. By Theorem 8, $\mathrm{Sol}(E_1) = [(pq)^*p]$ for a primitive word pq.

Let another of the equations, say E_2, have infinitely many solutions, so $\mathrm{Sol}(E_2) = [(p'q')^*p']$ for a primitive word $p'q'$. The equations E_1 and E_2 have at least two common solutions h_3, h_4, so $(pq)^i p = (p'q')^{i'} p'$ and $(pq)^j p = (p'q')^{j'} p'$ for some $i < j$ and $i' < j'$. Then $(pq)^{j-i} = (p'q')^{j'-i'}$. By primitivity, $pq = p'q'$, and then $p = p'$ and $q = q'$, so E_1 and E_2 are equivalent, which is a contradiction. This proves that $E_2, \ldots, E_N$ have only finitely many solutions.

If $N \geq 4$, then $\mathrm{Sol}(E_1, E_2) = \mathrm{Sol}(E_2) \cap [(pq)^*p]$ is finite but contains at least three solutions h_3, h_4, h_5, which contradicts Lemma 11, so none of the equations can have infinitely many solutions in this case. □

Theorem 13. *Every strictly independent system of one-unknown equations is of size at most $8 \log n + O(1)$, where n is the length of the shortest equation. Furthermore, Conjecture SOL-XAB(c) implies Conjecture SIND-XAB(c).*

Proof. Follows from Theorem 8 and Lemma 12. □

Lemma 14. *Let $\Sigma = \{a_1, \ldots, a_k\}$ be the alphabet of constants and*

$$\alpha : (\{x\} \cup \Sigma)^* \to \{x, y, z\}^*, \ \alpha(x) = x, \ \alpha(a_i) = y^i z$$

be a morphism. Let $E_1, \ldots, E_N$ be a strictly independent system of equations on $\{x\}$. The system $\alpha(E_1), \ldots, \alpha(E_N)$ of three-unknown constant-free equations is strictly independent.

Proof. Let

$$\beta : \Sigma^* \to \{a, b\}^*, \ \beta(a_i) = a^i b$$

be a morphism. A constant-preserving morphism $h : (\{x\} \cup \Sigma)^* \to \Sigma^*$ is a solution of E_i if and only if the nonperiodic morphism

$$g_h : \{x, y, z\}^* \to \{a, b\}^*, \ g_h(x) = \beta(h(x)), \ g_h(y) = a, \ g_h(z) = b$$

is a solution of $\alpha(E_i)$ (this follows from the fact that $g_h \circ \alpha = \beta \circ h$ and the injectivity of β). So if $(h_1, \ldots, h_{N+1})$ is a strict independence certificate for $E_1, \ldots, E_N$, then $(g_{h_1}, \ldots, g_{h_{N+1}})$ is a strict independence certificate for $\alpha(E_1), \ldots, \alpha(E_N)$. □

Theorem 15. *Conjecture SIND-XYZ(c) implies Conjecture SIND-XAB(c).*

Proof. Follows from Lemma 14. □

5 Classification of Solutions

We are interested in strictly independent systems and their certificates. Every morphism in a certificate can be replaced by an equivalent morphism, so it would be beneficial for us if there was a simple subclass of morphisms containing a representative of every equivalence class. In the three-unknown case, this kind of a result follows from a characterization of three-generator subsemigroups of a free semigroup by Budkina and Markov [2], or alternatively from a similar result by Spehner [15,16]. A comparison of these two results can be found in [7]. The result we present here in Theorem 16 is a simplified version that is perhaps slightly weaker, but sufficiently strong for our purposes and easier to work with.

We define classes of morphisms $\{x, y, z\}^* \to \{a, b, c\}^*$:

$$\begin{aligned}
&\mathcal{A} = \{[a, b, c]\},\\
&\mathcal{B} = \{[a^i, a^j, a^k] \mid i, j, k \geq 0\},\\
&\mathcal{C}_{xyz}(i, j) = \{[a, a^i b a^j, w] \mid w \in \{a, b\}^* \wedge (i = 0 \vee w \in b\{a, b\}^*)\\
&\qquad\qquad\qquad\qquad\qquad\qquad\qquad\qquad \wedge (j = 0 \vee w \in \{a, b\}^* b)\},\\
&\mathcal{C}_{xyz} = \bigcup_{i,j \geq 0} \mathcal{C}_{xyz}(i, j),\\
&\mathcal{D}_{xyz}(i, j, k, l, m, p, q) = \{[a, a^i b (a^m b)^p a^j, a^k b (a^m b)^q a^l]\},\\
&\mathcal{D}_{xyz} = \bigcup \mathcal{D}_{xyz}(i, j, k, l, m, p, q),
\end{aligned}$$

where the last union is taken over all $i, j, k, l, m \geq 0$ and $p, q \geq 1$ such that $ik = jl = 0$ and $\gcd(p+1, q+1) = 1$. If (X, Y, Z) is a permutation of (x, y, z), then $\mathcal{C}_{XYZ}(i, j)$, $\mathcal{C}_{XYZ}$, $\mathcal{D}_{XYZ}(i, j, k, l, m, p, q)$ and $\mathcal{D}_{XYZ}$ are defined similarly, with the images of the unknowns permuted in a corresponding way. For example, in the case of $\mathcal{C}_{XYZ}(i, j)$, X maps to a, Y to $a^i b a^j$, and Z to w. Then we also define

$$\begin{aligned}
\mathcal{C} &= \mathcal{C}_{xyz} \cup \mathcal{C}_{yzx} \cup \mathcal{C}_{zxy} \cup \mathcal{C}_{zyx} \cup \mathcal{C}_{xzy} \cup \mathcal{C}_{yxz},\\
\mathcal{D} &= \mathcal{D}_{xyz} \cup \mathcal{D}_{yzx} \cup \mathcal{D}_{zxy}.
\end{aligned}$$

For $\mathcal{A}$ and $\mathcal{B}$, we do not need to consider different permutations of the unknowns because the images of the unknowns are symmetric. For $\mathcal{D}$, we need only three of the six permutations, because the images of the latter two unknowns are symmetric.

Theorem 16. *Every morphism $\{x, y, z\}^* \to \{a, b, c\}^*$ is equivalent to a morphism in $\mathcal{A} \cup \mathcal{B} \cup \mathcal{C} \cup \mathcal{D}$.*

Proof. Follows from the characterization of Budkina and Markov [2], or alternatively from the characterization of Spehner [16]. □

By the following lemma, we can concentrate on solutions in classes $\mathcal{C}$ and $\mathcal{D}$.

Lemma 17. *A strictly independent system of $N \geq 2$ constant-free equations on $\{x, y, z\}$ has a strict independence certificate in $(\mathcal{C} \cup \mathcal{D})^{N+1}$.*

Proof. Every solution in a certificate can be replaced by an equivalent solution, so the system has a certificate in $(\mathcal{A} \cup \mathcal{B} \cup \mathcal{C} \cup \mathcal{D})^{N+1}$ by Theorem 16.

The morphism in $\mathcal{A}$ is a solution of only the trivial equations (u, u), and these equations cannot be part of any independent system, so none of the solutions in the certificate can be in $\mathcal{A}$.

It was proved by Harju and Nowotka [7] that if an independent pair of equations has a nonperiodic solution, then both of the equations are balanced, that is, every unknown appears on the left-hand side as often as on the right-hand side. Every morphism in $\mathcal{B}$ is periodic and thus a solution of every balanced equation, so none of the solutions in the certificate can be in $\mathcal{B}$. □

Example 18. The nonperiodic solutions of the equation (xyz, zyx) are of the form $[(pq)^i p, q(pq)^j, (pq)^k p]$. For example, we have the following solutions:

- $[a, b, (ab)^k a] \in \mathcal{C}_{xyz}(0, 0)$ and $[b, a, (ba)^k b] \in \mathcal{C}_{yxz}(0, 0)$ (these are equivalent),
- $[a, b(ab)^j, aba] \in \mathcal{C}_{xzy}(1, 1)$,
- $[a, b(ab)^j, (ab)^k a] \in \mathcal{D}_{xyz}(0, 0, 1, 1, 1, j, k - 1)$ $(j, k - 1 \geq 1, \gcd(j + 1, k) = 1)$,
- $[(ba)^i b, a, (ba)^k b] \in \mathcal{D}_{yzx}(1, 1, 1, 1, 1, k, i)$ $(i, k \geq 1, \gcd(i + 1, k + 1) = 1)$.

6 Class $\mathcal{C}$

In this section we study morphisms in class $\mathcal{C}$. This leads to a natural connection between three-unknown constant-free equations and one-unknown equations with constants.

Lemma 19. *Let E be a nontrivial constant-free equation on $\{x, y, z\}$. There is at most one pair (i, j) such that E has a solution in $\mathcal{C}_{xyz}(i, j)$. For this pair, $i + j \leq |E| - 1$.*

Proof. Let $E = (u, v)$ and $h \in \mathrm{Sol}(E) \cap \mathcal{C}_{xyz}(i, j)$. We can assume that one of the following is true:

1. $v = \varepsilon$.
2. $u = x^k$, $k \geq 1$, and v begins with y.
3. u begins with $x^k y$, $k \geq 1$, and v begins with y.
4. u begins with $x^k z$, $k \geq 1$, and v begins with y.
5. u begins with x and v begins with z.
6. u begins with y and v begins with z.

In all cases, we get either a contradiction or a single possible value for i as follows:

1. $u \neq \varepsilon$, so at least one of $h(x), h(y), h(z)$ is ε. The only possibility is $h(z) = \varepsilon$, and then $i = j = 0$.
2. $h(u) = a^k$ and $h(v)$ contains the letter b, which is a contradiction.
3. $h(u)$ begins with $a^{k+i}b$ and $h(v)$ begins with $a^i b$, which is a contradiction.
4. $h(y)$ must begin with a and thus $h(z)$ must begin with b, so $h(u)$ begins with $a^k b$ and $h(v)$ begins with $a^i b$. Thus $i = k$.
5. $h(z)$ cannot begin with b and thus $h(y)$ must begin with b, so $i = 0$.
6. It is not possible that $h(y)$ would begin with a and $h(z)$ with b, so $h(y)$ must begin with b and $i = 0$.

By looking at the suffixes of u and v, we will similarly see that j is uniquely determined. Moreover, $i + j \leq |E| - 1$. □

Lemma 20. *Let $S = \{E_1, \dots, E_N\}$ be a system of constant-free equations on $\{x, y, z\}$. Let S have a strict independence certificate $(h_1, \dots, h_{N+1}) \in \mathcal{C}_{xyz}^{N+1}$. There is a strictly independent system $E'_1, \dots, E'_N$ of one-unknown equations such that $|E'_n| \leq |E_n|^2$ for all n.*

Proof. The case $N < 2$ is trivial, so let $N \geq 2$. Let i, j be such that $h_{N+1} \in \mathcal{C}_{xyz}(i, j)$. By Lemma 19, $(h_1, \dots, h_N) \in \mathcal{C}_{xyz}(i, j)^N$. Let

$$\alpha : \{x, y, z\}^* \to \{a, b, z\}^*, \ \alpha(x) = a, \ \alpha(y) = a^i b a^j, \ \alpha(z) = z$$

be a morphism and let

$$h'_n : \{a, b, z\}^* \to \{a, b\}^*, \ h'_n(z) = h_n(z)$$

be a constant-preserving morphism. For every n, $h_n = h'_n \circ \alpha$ and $\alpha(E_n)$ is a one-unknown equation with constants. Then $(h'_1, \dots, h'_{N+1})$ is a strict independence certificate of the system $\alpha(E_1), \dots, \alpha(E_N)$. The length of $\alpha(E_n)$ is at most $(i + j + 1)|E_n|$, which is at most $|E_n|^2$ by Lemma 19. □

7 Class $\mathcal{D}$

In this section we study morphisms in class $\mathcal{D}$. This class looks more complicated than class $\mathcal{C}$, but actually there is a lot of structure in the morphisms in $\mathcal{D}$, which allows us to prove stronger results than for $\mathcal{C}$.

Lemma 21. *Let E be a nontrivial constant-free equation on $\{x, y, z\}$. There are i, j, k, l, m, p', q' such that $\mathrm{Sol}(E) \cap \mathcal{D}_{xyz}$ is either $\varnothing$, $\mathcal{D}_{xyz}(i, j, k, l, m, p', q')$, or the union of $\mathcal{D}_{xyz}(i, j, k, l, m, p, q)$ over all $p, q \geq 1$ such that $\gcd(p+1, q+1) = 1$.*

Proof. Let $E = (u, v)$. If $u = \varepsilon$ or $v = \varepsilon$, then $\mathrm{Sol}(E) \cap \mathcal{D}_{xyz} = \varnothing$, so let $u \neq \varepsilon \neq v$. We can assume that E is reduced and write it as

$$(x^{a_0} y_1 x^{a_1} \cdots y_r x^{a_r}, x^{b_0} z_1 x^{b_1} \cdots z_s x^{b_s}),$$

where $y_1, \ldots, y_r, z_1, \ldots, z_s \in \{y, z\}$. We can also assume that $r, s \geq 2$. Let $h \in \mathrm{Sol}(E) \cap \mathcal{D}_{xyz}$ and

$$h(x) = a,\ h(y_t) = a^{i_t} b (a^m b)^{p_t} a^{j_t},\ h(z_t) = a^{k_t} b (a^m b)^{q_t} a^{l_t},$$

$$(i_t, j_t, p_t) = \begin{cases} (i, j, p) & \text{if } y_t = y, \\ (k, l, q) & \text{if } y_t = z, \end{cases} \qquad (k_t, l_t, q_t) = \begin{cases} (i, j, p) & \text{if } z_t = y, \\ (k, l, q) & \text{if } z_t = z. \end{cases}$$

The left-hand side $h(u)$ begins with $a^{a_0 + i_1} b$ and the right-hand side $h(v)$ begins with $a^{b_0 + k_1} b$, so $a_0 + i_1 = b_0 + k_1$. If $y_1 = z_1$, then $i_1 = k_1$, $a_0 = b_0$, and E is not reduced, a contradiction. Thus $y_1 \neq z_1$ and $i_1 k_1 = ik = 0$. From $a_0 + i_1 = b_0 + k_1$, $i_1 k_1 = 0$, $a_0 b_0 = 0$ it then follows that $k_1 = a_0$ and $i_1 = b_0$. Similarly, by looking at the suffixes of $h(u)$ and $h(v)$ we find out that $y_r \neq z_s$, $l_s = a_r$, and $j_r = b_s$. Thus i, j, k, l are uniquely determined by the equation E.

It must be $\{p_1, q_1\} = \{p, q\}$, and $\gcd(p+1, q+1) = 1$, so $p_1 \neq q_1$. If $p_1 < q_1$, then $h(u)$ and $h(v)$ begin with

$$a^{a_0 + i_1} b (a^m b)^{p_1} a^{j_1 + a_1 + i_2} b \text{ and } a^{b_0 + k_1} b (a^m b)^{p_1 + 1},$$

respectively, so $j_1 + a_1 + i_2 = m$. Similarly, if $p_1 > q_1$, then $l_1 + b_1 + k_2 = m$. Thus $m \in \{j_1 + a_1 + i_2, l_1 + b_1 + k_2\}$. If $j_1 + a_1 + i_2 = m \neq l_1 + b_1 + k_2$, then there are $n \neq m$, $A \geq 1$, $B \geq 0$ such that $h(u)$ and $h(v)$ begin with

$$a^{a_0 + i_1} b (a^m b)^{A(p_1+1) + B(q_1+1) - 1} a^n b \text{ and } a^{b_0 + k_1} b (a^m b)^{q_1} a^{l_1 + b_1 + k_2} b,$$

respectively. It must be $A(p_1 + 1) + B(q_1 + 1) = q_1 + 1$. But then $B > 0$ would be a contradiction, and $B = 0$ would contradict $\gcd(p+1, q+1) = 1$. Similarly, $j_1 + a_1 + i_2 \neq m = l_1 + b_1 + k_2$ would lead to a contradiction. Thus it must be $j_1 + a_1 + i_2 = m = l_1 + b_1 + k_2$.

We can write

$$h(u) = a^{c_0} b (a^m b)^{A_1(p+1) + C_1(q+1) - 1} a^{c_1} b \cdots b (a^m b)^{A_R(p+1) + C_R(q+1) - 1} a^{c_R},$$
$$h(v) = a^{d_0} b (a^m b)^{B_1(p+1) + D_1(q+1) - 1} a^{d_1} b \cdots b (a^m b)^{B_S(p+1) + D_S(q+1) - 1} a^{d_S},$$

where $c_1, \ldots, c_{R-1}, d_1, \ldots, d_{S-1} \neq m$. It must be $R = S$, $c_t = d_t$, and

$$A_t(p+1) + C_t(q+1) = B_t(p+1) + D_t(q+1)$$

for all t. Moreover, all values p, q that satisfy these linear relations lead to a solution of the equation. If there are two linearly independent relations, there are no solutions. If there is one nontrivial relation $A(p+1) = C(q+1)$, then there is exactly one solution with $\gcd(p+1, q+1) = 1$. If all relations are trivial, all values of p, q satisfy them. This concludes the proof. □

The next lemma is a special case of Theorem 5.3 in [14]. Here, the *length type* of a solution h is the vector $(|h(x)|, |h(y)|, |h(z)|)$.

Lemma 22. *The length types of nonperiodic solutions of an independent pair of constant-free equations on three unknowns are covered by a finite union of two-dimensional subspaces of* $\mathbb{Q}^3$.

Lemma 23. *Let E_1, E_2, E_3, E_4 be a system of constant-free equations on $\{x, y, z\}$ with a strict independence certificate $(h_1, h_2, h_3, h_4, h_5)$. At most one of the h_i can be in $\mathcal{D}_{xyz}$.*

Proof. Let $h_r, h_s \in \mathcal{D}_{xyz}$, $r \neq s$. Without loss of generality, let $r, s \geq 3$. Then $h_r, h_s \in \mathrm{Sol}(E_1, E_2) \cap \mathcal{D}_{xyz}$, so the third option of Lemma 21 must be true for this set. We will show that the length types of solutions of E_1, E_2 cannot be covered by finitely many two-dimensional spaces, which contradicts Lemma 22.

The length type of $[a, a^i b(a^m b)^p a^j, a^k b(a^m b)^q a^l] \in \mathrm{Sol}(E_1, E_2) \cap \mathcal{D}_{xyz}$ is

$$(1, i + 1 + (m + 1)p + j, k + 1 + (m + 1)q + l).$$

Here i, j, k, l, m are fixed, but p, q can be arbitrary positive integers such that $\gcd(p + 1, q + 1) = 1$. For every p, there are infinitely many possible values of q, giving infinitely many length types on the line

$$L_p = \{(1, i + 1 + (m + 1)p + j, Z) \mid Z \in \mathbb{Q}\}.$$

The only way to cover these with a finite number of two-dimensional spaces is to have one of them be the unique two-dimensional space containing the whole line. This is true for any p, and different values of p give different spaces, so all length types cannot be covered by finitely many two-dimensional spaces. □

8 Main Results

Putting our results together gives the following theorem, which improves the linear bound of Theorem 5 to a logarithmic one.

Theorem 24. *A strictly independent system of constant-free equations on three unknowns has at most $O(\log n)$ equations, where n is the length of the shortest equation.*

Proof. Let the system be $E_1, \ldots, E_N$, where E_1 is the shortest equation. By Lemma 17, it has a strict independence certificate $(h_1, \ldots, h_{N+1}) \in (\mathcal{C} \cup \mathcal{D})^{N+1}$. By Lemma 23, at most three of the h_i can be in $\mathcal{D}$. Let k of the solutions be in $\mathcal{C}_{xyz}$. If h_1 is one of them, we get a system of size $k - 1$, for which we can use Lemma 20, and then Theorem 13 to conclude that $k = O(\log n)$. Otherwise, we can still use the arguments in the proof of Lemma 20 to turn E_1 into a one-unknown equation E_1' with k solutions. Then, by Theorem 13, either $k = O(\log n)$ or E_1' has infinitely many solutions, but the latter leads to a contradiction like in the proof of Lemma 12. Similarly, we can prove that the number of i such that $h_i \in \mathcal{C}_{XYZ}$ is $O(\log n)$ for all permutations (X, Y, Z) of (x, y, z). □

We say that two words *begin in the same way* if they begin with the same letter or are both empty. We say that equations (u_1, v_1) and (u_2, v_2) *begin in the same way* if either u_1 and u_2 begin in the same way and v_1 and v_2 begin in the same way, or u_1 and v_2 begin in the same way and v_1 and u_2 begin in the same way. Equations *ending the same way* is defined analogously.

Lemma 25. *Let $N \geq 3$ and let $E_1, \dots, E_N$ be a strictly independent system of reduced constant-free equations on $\{x, y, z\}$. All of the equations begin and end in the same way.*

Proof. Assume that all of the equations do not begin and end in the same way. Without loss of generality, we can assume that E_1 and E_2 do not begin in the same way and that they are of the form (xu, yv) and (xu', zv'), respectively. By the well-known graph lemma about word equations, every common solution of these two equations is periodic or maps one of the unknowns to the empty word. The equations E_1 and E_2 have two nonequivalent nonperiodic solutions, and these solutions must map x to the empty word. But all nonperiodic solutions mapping x to the empty word are equivalent, which is a contradiction. □

By Theorem 13, Conjecture SIND-XAB could be replaced by Conjecture SOL-XAB in the next theorem. The constants are probably not optimal.

Theorem 26. *Conjecture SIND-XAB(c) implies Conjecture SIND-XYZ(5c + 7). In particular, if SIND-XAB(2) is true, then a strictly independent system of constant-free equations on $\{x, y, z\}$ has at most 17 equations.*

Proof. Let $E_1, \dots, E_N$ be a system of reduced constant-free equations on $\{x, y, z\}$ with a strict independence certificate $(h_1, \dots, h_{N+1})$. For an equation $E_m = (u, v)$, at least one of the unknowns appears both at the beginning of u or v and at the end of u or v. By Lemma 25, this unknown does not depend on m. Without loss of generality, we can assume it is z. By Lemma 17, we can assume that $h_n \in \mathcal{C} \cup \mathcal{D}$ for all n. Because $\mathcal{C}_{xyz}(0,0)$ and $\mathcal{C}_{yxz}(0,0)$ are the same up to swapping a and b, we can assume that $h_n \notin \mathcal{C}_{yxz}(0,0)$ for all n.

By Lemma 20 and the assumption about Conjecture SIND-XAB, at most $c+1$ of the solutions h_n can be in $\mathcal{C}_{xyz}$, and the same is true for the other five permutations of the unknowns. By the assumption about z and the proof of Lemma 19, $\mathrm{Sol}(E_m) \cap \mathcal{C}_{yxz} \subseteq \mathcal{C}_{yxz}(0,0)$ for all m, so $h_n \notin \mathcal{C}_{yxz}$ for all n. Thus at most $5c+5$ of the solutions h_n can be in $\mathcal{C}$.

By Lemma 23, at most one of the solutions h_n can be in $\mathcal{D}_{xyz}$, and the same is true for $\mathcal{D}_{yzx}$ and $\mathcal{D}_{zxy}$. Thus at most three of the solutions h_n can be in $\mathcal{D}$.

This proves that the total number of the solutions h_n, which is $N+1$, cannot be more than $5c+8$. □

9 Conclusion

We can mention several further research goals. Two obvious ones are improving the constants in Theorem 26, ideally so that Conjecture SIND-XAB(c) implies Conjecture SIND-XYZ(c), and proving Conjecture SOL-XAB or Conjecture SIND-XAB (ideally SOL-XAB(2)), and thus also Conjecture SIND-XYZ. Proving similar results for chains of equations instead of independent systems might be possible (see [11] for definitions).

A different topic would be to study the complexity of determining whether a three-unknown constant-free equation has a nonperiodic solution. This decision problem is known to be in NP [13]. Based on the connection to one-unknown equations, a better result could probably be obtained, because one-unknown equations can be solved efficiently, even in linear time, as proved by Jez [9].

Finally, Question 3 could be studied for more than three unknowns. This is of course a big question, and our techniques do not help here, because they are specific to the three-unknown case.

References

1. Albert, M.H., Lawrence, J.: A proof of Ehrenfeucht's conjecture. Theoret. Comput. Sci. **41**(1), 121–123 (1985)
2. Budkina, L.G., Markov, A.A.: F-semigroups with three generators. Mat. Zametki **14**, 267–277 (1973)
3. Culik, K., Karhumäki, J.: Systems of equations over a free monoid and Ehrenfeucht's conjecture. Discrete Math. **43**(2–3), 139–153 (1983)
4. Dąbrowski, R., Plandowski, W.: On word equations in one variable. Algorithmica **60**(4), 819–828 (2011)
5. Eyono Obono, S., Goralcik, P., Maksimenko, M.: Efficient solving of the word equations in one variable. In: Privara, I., Ružička, P., Rovan, B. (eds.) MFCS 1994. LNCS, vol. 841, pp. 336–341. Springer, Heidelberg (1994)
6. Guba, V.S.: Equivalence of infinite systems of equations in free groups and semigroups to finite subsystems. Mat. Zametki **40**(3), 321–324 (1986)
7. Harju, T., Nowotka, D.: On the independence of equations in three variables. Theoret. Comput. Sci. **307**(1), 139–172 (2003)
8. Holub, Š., Žemlička, J.: Algebraic properties of word equations. J. Algebra **434**, 283–301 (2015)
9. Jez, A.: One-variable word equations in linear time. Algorithmica **74**(1), 1–48 (2016)
10. Karhumäki, J., Plandowski, W.: On the defect effect of many identities in free semigroups. In: Paun, G. (ed.) Mathematical Aspects of Natural and Formal Languages, pp. 225–232. World Scientific (1994)
11. Karhumäki, J., Saarela, A.: On maximal chains of systems of word equations. Proc. Steklov Inst. Math. **274**, 116–123 (2011)
12. Laine, M., Plandowski, W.: Word equations with one unknown. Int. J. Found. Comput. Sci. **22**(2), 345–375 (2011)
13. Saarela, A.: On the complexity of Hmelevskii's theorem and satisfiability of three unknown equations. In: Diekert, V., Nowotka, D. (eds.) DLT 2009. LNCS, vol. 5583, pp. 443–453. Springer, Heidelberg (2009)
14. Saarela, A.: Systems of word equations, polynomials and linear algebra: a new approach. Eur. J. Comb. **47**, 1–14 (2015)
15. Spehner, J.C.: Quelques problémes d'extension, de conjugaison et de présentation des sous-monoïdes d'un monoïde libre. Ph.D. thesis, Univ. Paris (1976)
16. Spehner, J.C.: Les systemes entiers d'équations sur un alphabet de 3 variables. In: Semigroups, pp. 342–357 (1986)

Avoidability of Formulas with Two Variables

Pascal Ochem[1] and Matthieu Rosenfeld[2(✉)]

[1] LIRMM, CNRS, University of Montpellier, Montpellier, France
ochem@lirmm.fr

[2] LIP, ENS de Lyon, CNRS, UCBL, Université de Lyon, Lyon, France
matthieu.rosenfeld@ens-lyon.fr

Abstract. In combinatorics on words, a word w over an alphabet Σ is said to avoid a pattern p over an alphabet Δ of variables if there is no factor f of w such that $f = h(p)$ where $h: \Delta^* \rightarrow \Sigma^*$ is a non-erasing morphism. A pattern p is said to be k-avoidable if there exists an infinite word over a k-letter alphabet that avoids p. We consider the patterns such that at most two variables appear at least twice, or equivalently, the formulas with at most two variables. For each such formula, we determine whether it is 2-avoidable.

Keywords: Word · Pattern avoidance

1 Introduction

A *pattern* p is a non-empty finite word over an alphabet $\Delta = \{A, B, C, \ldots\}$ of capital letters called *variables*. An *occurrence* of p in a word w is a non-erasing morphism $h : \Delta^* \rightarrow \Sigma^*$ such that $h(p)$ is a factor of w. The *avoidability index* $\lambda(p)$ of a pattern p is the size of the smallest alphabet Σ such that there exists an infinite word over Σ containing no occurrence of p. Bean, Ehrenfeucht, and McNulty [3] and Zimin [11] characterized unavoidable patterns, i.e., such that $\lambda(p) = \infty$. We say that a pattern p is *t-avoidable* if $\lambda(p) \leqslant t$. For more informations on pattern avoidability, we refer to Chapter 3 of Lothaire's book [6].

A variable that appears only once in a pattern is said to be *isolated*. Following Cassaigne [4], we associate to a pattern p the *formula* f obtained by replacing every isolated variable in p by a dot. The factors between the dots are called *fragments*.

An *occurrence* of f in a word w is a non-erasing morphism $h : \Delta^* \rightarrow \Sigma^*$ such that the h-image of every fragment of f is a factor of w. As for patterns, the avoidability index $\lambda(f)$ of a formula f is the size of the smallest alphabet allowing an infinite word containing no occurrence of p. Clearly, every word avoiding f also avoids p, so $\lambda(p) \leqslant \lambda(f)$. Recall that an infinite word is *recurrent* if every finite factor appears infinitely many times. If there exists an infinite word over Σ avoiding p, then there exists an infinite recurrent word over Σ avoiding p. This recurrent word also avoids f, so that $\lambda(p) = \lambda(f)$. Without loss of generality, a formula is such that no variable is isolated and no fragment is a factor of another fragment.

S. Brlek and C. Reutenauer (Eds.): DLT 2016, LNCS 9840, pp. 344–354, 2016.
DOI: 10.1007/978-3-662-53132-7_28

Cassaigne [4] began and Ochem [7] finished the determination of the avoidability index of every pattern with at most 3 variables. A *doubled* pattern contains every variable at least twice. Thus, a doubled pattern is a formula with exactly one fragment. Every doubled pattern is 3-avoidable [9]. A formula is said to be *binary* if it has at most 2 variables. In this paper, we determine the avoidability index of every binary formula.

We say that a formula f is *divisible* by a formula f' if f does not avoid f', that is, there is a non-erasing morphism such that the image of any fragment of f' by h is a factor of a fragment of f. If f is divisible by f', then every word avoiding f' also avoids f and thus $\lambda(f) \leqslant \lambda(f')$. For example, the fact that $ABA.AABB$ is 2-avoidable implies that $ABAABB$ and $ABAB.BBAA$ are 2-avoidable. Moreover, the reverse f^R of a formula f satisfies $\lambda(f^R) = \lambda(f)$. See Cassaigne [4] and Clark [5] for more information on formulas and divisibility.

First, we check that every avoidable binary formula is 3-avoidable. Since $\lambda(AA) = 3$, every formula containing a square is 3-avoidable. Then, the only square free avoidable binary formula is $ABA.BAB$ with avoidability index 3 [4]. Thus, we have to distinguish between avoidable binary formulas with avoidability index 2 and 3. A binary formula is minimally 2-avoidable if it is 2-avoidable and is not divisible by any other 2-avoidable binary formula. A binary formula f is maximally 2-unavoidable if it is 2-unavoidable and every other binary formula that is divisible by f is 2-avoidable.

Theorem 1. *Up to symmetry, the maximally 2-unavoidable binary formulas are:*

- $AAB.ABA.ABB.BBA.BAB.BAA$
- $AAB.ABBA$
- $AAB.BBAB$
- $AAB.BBAA$
- $AAB.BABB$
- $AAB.BABAA$
- $ABA.ABBA$
- $AABA.BAAB$

Up to symmetry, the minimally 2-avoidable binary formulas are:

- $AA.ABA.ABBA$
- $ABA.AABB$
- $AABA.ABB.BBA$
- $AA.ABA.BABB$
- $AA.ABB.BBAB$
- $AA.ABAB.BB$
- $AA.ABBA.BAB$
- $AAB.ABB.BBAA$
- $AAB.ABBA.BAA$
- $AABB.ABBA$
- $ABAB.BABA$
- $AABA.BABA$
- AAA

- $ABA.BAAB.BAB$
- $AABA.ABAA.BAB$
- $AABA.ABAA.BAAB$
- $ABAAB$

To obtain the 2-unavoidability of the formulas in the first part of Theorem 1, we use a standard backtracking algorithm. In the rest of the paper, we consider the 2-avoidable formulas in the second part of Theorem 1. Figure 1 gives the maximal length and number of binary words avoiding each maximally 2-unavoidable formula.

We show in Sect. 3 that the first three of these formulas are avoided by polynomially many binary words only. The proof uses a technical lemma given in Sect. 2. Then we show in Sect. 4 that the other formulas are avoided by exponentially many binary words.

Formula	Maximal length of a binary word avoiding this formula	Number of binary words avoiding this formula
$AAB.BBAA$	22	1428
$AAB.ABA.ABB.BBA.BAB.BAA$	23	810
$AAB.BBAB$	23	1662
$AABA.BAAB$	26	2124
$AAB.ABBA$	30	1684
$AAB.BABAA$	42	71002
$AAB.BABB$	69	9252
$ABA.ABBA$	90	31572

Fig. 1. The number and maximal length of binary words avoiding the maximally 2-unavoidable formulas.

2 The Useful Lemma

Let us define the following words:

- b_2 is the fixed point of $\mathtt{0} \mapsto \mathtt{01}$, $\mathtt{1} \mapsto \mathtt{10}$.
- b_3 is the fixed point of $\mathtt{0} \mapsto \mathtt{012}$, $\mathtt{1} \mapsto \mathtt{02}$, $\mathtt{2} \mapsto \mathtt{1}$.
- b_4 is the fixed point of $\mathtt{0} \mapsto \mathtt{01}$, $\mathtt{1} \mapsto \mathtt{03}$, $\mathtt{2} \mapsto \mathtt{21}$, $\mathtt{3} \mapsto \mathtt{23}$.
- b_5 is the fixed point of $\mathtt{0} \mapsto \mathtt{01}$, $\mathtt{1} \mapsto \mathtt{23}$, $\mathtt{2} \mapsto \mathtt{4}$, $\mathtt{3} \mapsto \mathtt{21}$, $\mathtt{4} \mapsto \mathtt{0}$.

Let w and w' be infinite (right infinite or bi-infinite) words. We say that w and w' are equivalent if they have the same set of finite factors. We write $w \sim w'$ if w and w' are equivalent. A famous result of Thue [10] can be stated as follows:

Theorem 2 [10]**.** *Every bi-infinite ternary word avoiding* $\mathtt{010}$*,* $\mathtt{212}$*, and squares is equivalent to* b_3*.*

Given an alphabet Σ and forbidden structures S, we say that a finite set W of infinite words over Σ *essentially avoids* S if every word in W avoids S and every bi-infinite words over Σ avoiding S is equivalent to one of the words in S. If W contains only one word w, we denote the set W by w instead of $\{w\}$. Then we can restate Theorem 2: b_3 essentially avoids 010, 212, and squares

The results in the next section involve b_3. We have tried without success to prove them by using Theorem 2. We need the following stronger property of b_3:

Lemma 3 *b_3 essentially avoids 010, 212, XX with $1 \leqslant |X| \leqslant 3$, and 2$YY$ with $|Y| \geqslant 4$.*

Proof We start by checking by computer that b_3 has the same set of factors of length 100 as every bi-infinite ternary word avoiding 010, 212, XX with $1 \leqslant |X| \leqslant 3$, and $2YY$ with $|Y| \geqslant 4$. The set of the forbidden factors of b_3 of length at most 4 is $F = \{00, 11, 22, 010, 212, 0202, 2020, 1021, 1201\}$. To finish the proof, we use Theorem 2 and we suppose for contradiction that w is a bi-infinite ternary word that contains a large square MM and avoids both F and large factors of the form $2YY$.

- Case $M = 0N$. Then w contains $MM = 0N0N$. Since $00 \in F$ and $2YY$ is forbidden, w contains $10N0N$. Since $\{11, 010\} \subset F$, w contains $210N0N$. If $N = P1$, then w contains $210P10P1$, which contains $2YY$ with $Y = 10P$. So $N = P2$ and w contains $210P20P2$. If $P = Q1$, then w contains $210Q120Q12$. Since $\{11, 212\} \subset F$, the factor $Q12$ implies that $Q = R0$ and w contains $210R0120R012$. Moreover, since $\{00, 1201\} \subset F$, the factor $120R$ implies that $R = 2S$ and w contains $2102S01202S012$. Then there is no possible prefix letter for S: 0 gives 2020, 1 gives 1021, and 2 gives 22. This rules out the case $P = Q1$. So $P = Q0$ and w contains $210Q020Q02$. The factor $Q020Q$ implies that $Q = 1R1$, so that w contains $2101R10201R102$. Since $\{11, 010\} \subset F$, the factor $01R$ implies that $R = 2S$, so that w contains $21012S102012S102$. The only possible right extension with respect to F of 102 is 102012. So w contains $21012S102012S102012$, which contains $2YY$ with $Y = S102012$.
- Case $M = 1N$. Then w contains $MM = 1N1N$. In order to avoid 11 and $2YY$, w must contain $01N1N$. If $N = P0$, then w contains $01P01P0$. So w contains the large square $01P01P$ and this case is covered by the previous item. So $N = P2$ and w contains $01P21P2$. Then there is no possible prefix letter for P: 0 gives 010, 1 gives 11, and 2 gives 212.
- Case $M = 2N$. Then w contains $MM = 2N2N$. If $N = P1$, then w contains $2P12P1$. This factor cannot extend to $2P12P12$, since this is $2YY$ with $Y = P12$. So w contains $2P12P10$. Then there is no possible suffix letter for P: 0 gives 010, 1 gives 11, and 2 gives 212. This rules out the case $N = P1$. So $N = P0$ and w contains $2P02P0$. This factor cannot extend to $02P02P0$, since this contains the large square $02P02P$ and this case is covered by the first item. Thus w contains $12P02P0$. If $P = Q1$, then w contains $12Q102Q10$. Since $\{22, 1021\} \subset F$, the factor $102Q$ implies that $Q = 0R$, so that w contains $120R1020R10$. Then there is no possible prefix letter for R: 0 gives 00, 1 gives 1201, and 2 gives 0202. This rules out the case $P = Q1$. So $P = Q2$

and w contains $12Q202Q20$. The factor $Q202$ implies that $Q = R1$ and w contains $12R1202R120$. Since $\{00, 1201\} \subset F$, w contains $12R1202R1202$, which contains $2YY$ with $Y = R1202$.

3 Formulas Avoided by Few Binary Words

The first three 2-avoidable formulas in Theorem 1 are not avoided by exponentially many binary words:

- $\{g_x(b_3), g_y(b_3), g_z(b_3), g_{\overline{z}}(b_3)\}$ essentially avoids $AA.ABA.ABBA$.
- $\{g_x(b_3), g_t(b_3)\}$ essentially avoids $ABA.AABB$.
- $g_x(b_3)$ essentially avoids $AABA.ABB.BBA$.

The words avoiding these formulas are morphic images of b_3 by the morphisms given below. Let $\overline{w}$ denote the word obtained from the (finite or bi-infinite) binary word w by exchanging 0 and 1. Obviously, if w avoids a given formula, then so does $\overline{w}$. A (bi-infinite) binary word w is *self-complementary* if $w \sim \overline{w}$. The words $g_x(b_3)$, $g_y(b_3)$, and $g_t(b_3)$ are self-complementary. Since the frequency of 0 in $g_z(b_3)$ is $\frac{5}{9}$, $g_z(b_3)$ is not self-complementary. Then $g_{\overline{z}}$ is obtained from g_z by exchanging 0 and 1, so that $g_{\overline{z}}(b_3) = \overline{g_z(b_3)}$.

$$\begin{array}{llll} g_x(0) = 01110, & g_y(0) = 0111, & g_z(0) = 0001, & g_t(0) = 01011011010, \\ g_x(1) = 0110, & g_y(1) = 01, & g_z(1) = 001, & g_t(1) = 01011010, \\ g_x(2) = 0. & g_y(2) = 00. & g_z(2) = 11. & g_t(2) = 010. \end{array}$$

To prove the avoidability, we have implemented Cassaigne's algorithm that decides, under mild assumptions, whether a morphic word avoids a formula [4]. For the first two formulas, we have to explain how the long enough binary words split into 4 or 2 distinct incompatible types. A similar phenomenon has been described for $AABB.ABBA$ [8].

First, consider any infinite binary word w avoiding $AA.ABA.ABBA$. A computer check shows by backtracking that w must contain the factor 01110001110. In particular, w contains 00. Thus, w cannot contain both 010 and 0110, since it would produce an occurrence of $AA.ABA.ABBA$. Moreover, a computer check shows by backtracking that w cannot avoid both 010 and 0110. So, w must contain either 010 or 0110 (this is an exclusive or). Similarly, w must contain either 101 or 1001. There are thus at most 4 possibilities for w, depending on which subset of $\{010, 0110, 101, 1001\}$ appears among the factors of w, see Fig. 2a.

Now, consider any infinite binary word w avoiding $ABA.AABB$. Notice that w cannot contain both 010 and 0011. Also, a computer check shows by backtracking that w cannot avoid both 010 and 1100. By symmetry, there are thus at most 2 possibilities for w, depending on which subset of $\{010, 0011, 101, 1100\}$ appears among the factors of w, see Fig. 2b.

Let us first prove that $g_y(b_3)$ essentially avoids $AA.ABA.ABBA$, 0110, and 1001. We check that the set of prolongable binary words of length 100 avoiding $AA.ABA.ABBA$, 0110, and 1001 is exactly the set of factors of length 100 of

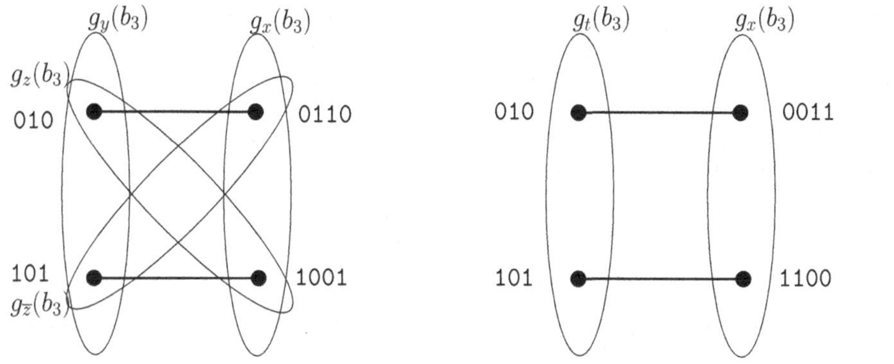

(a) The four bi-infinite binary words avoiding $AA.ABA.ABBA$.

(b) The two bi-infinite binary words avoiding $ABA.AABB$.

Fig. 2. The different possibilities for words avoiding AA.ABA.ABBA or ABA.AABB.

$g_y(b_3)$. Using Cassaigne's notion of circular morphism [4], this is sufficient to prove that every bi-infinite binary word of this type is the g_y-image of some bi-infinite ternary word w_3. It also ensures that w_3 and b_3 have the same set of small factors. Suppose for contradiction that $w_3 \neq b_3$. By Lemma 3, w_3 contains $2YY$. Then w_3 contains $2YYa$ with $a \in \Sigma_3$. Notice that 0 is a prefix of the g_y-image of every letter. So $g_y(w_3)$ contains $g_y(2YYa) = 000U0U0V$ with $U, V \in \Sigma_3^+$, which contains an occurrence of $AA.ABA.ABBA$ with $A = 0$ and $B = 0U$. This shows that $w_3 \sim b_3$, and thus $g_y(w_3) \sim g_y(b_3)$. Thus $g_y(b_3)$ essentially avoids $AA.ABA.ABBA$, 0110, and 1001. The argument is similar for the other types and we only detail the final contradiction:

- Since 1 is a suffix of the g_z-image of every letter, $g_z(2YY) = 11U1U1$ contains an occurrence of $AA.ABA.ABBA$ with $A = 1$ and $B = 1U$.
- Since 010 is a prefix and a suffix of the g_t-image of every letter, $g_z(u2YY) = 010V010010010U010010U010$ contains an occurrence of $ABA.AABB$ with $A = 010$ and $B = 010U010$.
- Since 0 is a prefix and a suffix of the g_x-image of every letter, $g_x(u2YYa) = V000U00U00W$ contains an occurrence of $AABA.AABBA$ with $A = 0$ and $B = 0U0$. Therefore, $g_x(u2YYa)$ contains an occurrence of $AA.ABA.ABBA$, $ABA.AABB$, and $AABA.ABB.BBA$.

4 Formulas Avoided by Exponentially Many Binary Words

The other 2-avoidable formulas in Theorem 1 are avoided by exponentially many binary words. For every such formula f, we give below a uniform morphism g that maps every ternary square free word to a binary word avoiding f. If possible, we simultaneously avoid the reverse formula f^R of f. We also avoid large squares.

Let SQ_t denote the pattern corresponding to squares of period at least t, that is, $SQ_1 = AA$, $SQ_2 = ABAB$, $SQ_3 = ABCABC$, and so on. The morphism g produces words avoiding SQ_t with t as small as possible.

- $AA.ABA.BABB$ is avoided with its reverse by the following 22-uniform morphism which also avoids SQ_6:

  $$\begin{aligned} 0 &\mapsto \texttt{0001101101110011100011} \\ 1 &\mapsto \texttt{0001101101110001100011} \\ 2 &\mapsto \texttt{0001101101100011100111} \end{aligned}$$

  Notice that $\{AA.ABA.BABB, AA.ABA.BBAB, SQ_5\}$ is 2-unavoidable. However, $\{AA.ABA.BABB, SQ_4\}$ is 2-avoidable:

  $$\begin{aligned} 0 &\mapsto \texttt{0001001001100011100100110001001110010010 0111} \\ 1 &\mapsto \texttt{00010010011000100111001001100011100100100111} \\ 2 &\mapsto \texttt{00010010011000100111001001001100011100100111} \end{aligned}$$

- $AA.ABB.BBAB$ is avoided with its reverse, 60-uniform morphism, avoids SQ_{11}:

  $$\begin{aligned} 0 &\mapsto \texttt{000110011100011001110011000111000110011100011100110001110011} \\ 1 &\mapsto \texttt{000110011100011001110001110011000111000110011100110001110011} \\ 2 &\mapsto \texttt{000110011100011001110001100111000111001100011100110001110011} \end{aligned}$$

  Notice that $\{AA.ABB.BBAB, SQ_{10}\}$ is 2-unavoidable.
- $AA.ABAB.BB$ is self-reverse, 11-uniform morphism, avoids SQ_4:

  $$\begin{aligned} 0 &\mapsto \texttt{00100110111} \\ 1 &\mapsto \texttt{00100110001} \\ 2 &\mapsto \texttt{00100011011} \end{aligned}$$

- $AA.ABBA.BAB$ is self-reverse, 30-uniform morphism, avoids SQ_6:

  $$\begin{aligned} 0 &\mapsto \texttt{000110001110011000110011100111} \\ 1 &\mapsto \texttt{000110001100111001100011100111} \\ 2 &\mapsto \texttt{000110001100011001110011100111} \end{aligned}$$

- $AAB.ABB.BBAA$ is self-reverse, 30-uniform morphism, avoids SQ_5:

  $$\begin{aligned} 0 &\mapsto \texttt{000100101110100010110111011101} \\ 1 &\mapsto \texttt{000100101101110100010111011101} \\ 2 &\mapsto \texttt{000100010001011101110111010001} \end{aligned}$$

- $AAB.ABBA.BAA$ is self-reverse, 38-uniform morphism, avoids SQ_5:

  $$\begin{aligned} 0 &\mapsto \texttt{00010001000101110111010001011100011101} \\ 1 &\mapsto \texttt{00010001000101110100011100010111011101} \\ 2 &\mapsto \texttt{00010001000101110001110100010111011101} \end{aligned}$$

- $AABB.ABBA$ is unavoidable with its reverse, 193-uniform morphism, avoids SQ_{16}:

  $0 \mapsto$ 0001000101101110110001011011100010110111011100010110001000101
  1011101100010110111011100010110111011000101101110001011011101110001
  011000100010110111000101101110111000101101110110001011011100010 11
  $1 \mapsto$ 0001000101101110110001011011100010110111011100010110001000101
  1011100010110111011100010110111011000101101110001011011101110001011
  0001000101101110110001011011101110001011011101100010110111000101 1
  $2 \mapsto$ 0001000101101110001011011101110001011000100010110111011000101
  1011101110001011011101100010110111000101101110111000101100010001011
  01110110001011011100010110111011100010110111011000101101110001011

  Previous papers [7,8] have considered a 102-uniform morphism to avoid $AABB.ABBA$ and SQ_{27}. No infinite binary word avoids $AABB.ABBA$ and SQ_{15}.
- $ABAB.BABA$ is self-reverse, 50-uniform morphism, avoids SQ_3, see [7]:

  $0 \mapsto$ 00011001011000111001011001110001011100101100010111
  $1 \mapsto$ 00011001011000101110010110011100010110001110010111
  $2 \mapsto$ 00011001011000101110010110001110010111000101100111

  Notice that a binary word avoiding $ABAB.BABA$ and SQ_3 contains only the squares 00, 11, and 0101 (or 00, 11, and 1010).
- $AABA.BABA$: A case analysis of the small factors shows that a recurrent binary word avoids $AABA.BABA$, $ABAA.ABAB$, and SQ_3 if and only if it contains only the squares 00, 11, and 0101 (or 00, 11, and 1010). We thus obtain the same morphism as for $ABAB.BABA$.
- AAA is self-reverse, 32-uniform morphism, avoids SQ_4:

  $0 \mapsto$ 00101001101101001011001001101011
  $1 \mapsto$ 00101001101100101101001001101011
  $2 \mapsto$ 00100101101001001101101001011011

- $ABA.BAAB.BAB$ is self-reverse, 10-uniform morphism, avoids SQ_3:

  $0 \mapsto$ 0001110101
  $1 \mapsto$ 0001011101
  $2 \mapsto$ 0001010111

- $AABA.ABAA.BAB$ is self-reverse, 57-uniform morphism, avoids SQ_6:

  $0 \mapsto$ 000101011100010110010101100010111001011000101011100101011
  $1 \mapsto$ 000101011100010110010101100010101110010110001011100101011
  $2 \mapsto$ 000101011100010110010101100010101110010101100010111001011

- $AABA.ABAA.BAAB$ is self-reverse, 30-uniform morphism, avoids SQ_3:

  $0 \mapsto$ 000101110001110101000101011101
  $1 \mapsto$ 000101110001110100010101110101
  $2 \mapsto$ 000101110001010111010100011101

– $ABAAB$ is avoided with its reverse, 10-uniform morphism, avoids SQ_3, see [7]:

$$\begin{aligned}0 &\mapsto \mathtt{0001110101}\\ 1 &\mapsto \mathtt{0000111101}\\ 2 &\mapsto \mathtt{0000101111}\end{aligned}$$

For every q-uniform morphism g above, we say that a binary word is an sqf-g-image if it is the g-image of a ternary square free word. Let us show that for every minimally 2-avoidable formula f and corresponding morphism g, every sqf-g-image avoids f.

We start by checking that every morphism is synchronizing, that is, for every letters $a, b, c \in \Sigma_3$, the factor $g(a)$ only appears as a prefix or a suffix in $g(bc)$.

For every morphism g, the sqf-g-images are claimed to avoid SQ_t with $2t < q$. Let us prove that SQ_t is avoided. We first check exhaustively that the sqf-g-images contain no square uu such that $t \leqslant |u| < 2q - 1$. Now suppose for contradiction that an sqf-g-image contains a square uu with $|u| \geqslant 2q - 1$. The condition $|u| \geqslant 2q - 1$ implies that u contains a factor $g(a)$ with $a \in \Sigma_3$. This factor $g(a)$ only appears as the g-image of the letter a because g is synchronizing. Thus the distance between any two factors u in an sqf-g-image is a multiple of q. Since uu is a factor of an sqf-g-image, we have $q \mid |u|$. Also, the center of the square uu cannot lie between the g-images of two consecutive letters, since otherwise there would be a square in the pre-image. The only remaining possibility is that the ternary square free word contains a factor $aXbXc$ with $a, b, c \in \Sigma_3$ and $X \in \Sigma_3^+$ such that $g(aXbXc) = bsYpsYpe$ contains the square $uu = sYpsYp$, where $g(X) = Y$, $g(a) = bs$, $g(b) = ps$, $g(c) = pe$. Then, we also have $a \neq b$ and $b \neq c$ since $aXbXc$ is square free. Then abc is square free and $g(abc) = bspspe$ contains a square with period $|s| + |p| = |g(a)| = q$. This is a contradiction since the sqf-g-images contain no square with period q.

Notice that f is not square free, since the only avoidable square free binary formula is $ABA.BAB$, which is not 2-avoidable. Now, we distinguish two kinds of formula. A formula is *easy* if every appearing variable is contained in at least one square. Every potential occurrence of an easy formula then satisfies $|A| < t$ and $|B| < t$ since SQ_t is avoided. The longest fragment of every easy formula has length 4. So, to check that the sqf-g-images avoids an easy formula, it is sufficient to consider the set of factors of the sqf-g-images with length at most $4(t - 1)$.

A *tough* formula is such that one of the variables is not contained in any square. The tough formulas have been named so that this variable is B. The tough formulas are $ABA.BAAB.BAB$, $ABAAB$, $AABA.ABAA.BAAB$, and $AABA.ABAA.BAB$. As before, every potential occurrence of a tough formula satisfies $|A| < t$ since SQ_t is avoided. Suppose for contradiction that $|B| \geqslant 2q - 1$. By previous discussion, the distance between any two occurrences of B in an sqf-g-image is a multiple of q. The case of $ABA.BAAB.BAB$ can be settled as follows. The factor $BAAB$ implies that $q \mid |BAA|$ and the factor BAB implies that $q \mid |BA|$. This implies that $q \mid |A|$, which contradicts $|A| < t$. For the other formulas, only one fragment contains B twice. This fragment is said to be *important*. Since $|A| < t$, the important fragment is a repetition which is "almost" a square. The important fragment is $\boldsymbol{BAB}$ for $AABA.ABAA.BAB$, $\boldsymbol{BAAB}$ for $AABA.ABAA.BAAB$, and $\boldsymbol{ABAAB}$ for $ABAAB$. Informally, this almost square implies a factor $aXbXc$ in the ternary pre-image, such that $|a| = |c| = 1$ and $1 \leqslant |b| \leqslant 2$. If $|X|$ is small, then $|B|$ is small and we check exhaustively that there exists no small occurrence of f. If $|X|$ is large, there would exist a ternary square free factor $aYbYc$ with $|Y|$ small, such that $g(aYbYc)$ contains the important fragment of an occurrence of f if and only if $g(aXbXc)$ contains the important fragment of a smaller occurrence of f.

5 Concluding Remarks

From our results, every minimally 2-avoidable binary formula, and thus every 2-avoidable binary formula, is avoided by some morphic image of b_3.

What can we forbid so that there exists only few infinite avoiding words? The known examples from the literature [1,2,10] are:

- one pattern and two factors:
  - b_3 essentially avoids AA, 010, and 212.
  - A morphic image of b_5 essentially avoids AA, 010, and 020.
  - A morphic image of b_5 essentially avoids AA, 121, and 212.
  - b_2 essentially avoids $ABABA$, 000, and 111.
- two patterns: b_2 essentially avoids $ABABA$ and AAA.
- one formula over three variables: b_4 and two words from b_4 obtained by letter permutation essentially avoid $AB.AC.BA.BC.CA$.

Now we can extend this list:

- one formula over two variables:
  - $g_x(b_3)$ essentially avoids $AAB.BAA.BBAB$.
  - $\{g_x(b_3), g_t(b_3)\}$ essentially avoids $ABA.AABB$.
  - $\{g_x(b_3), g_y(b_3), g_y(b_3), g_z(b_3), g_{\overline{z}}(b_3)\}$ essentially avoids $AA.ABA.ABBA$.
- one pattern over three variables: $ABACAABB$ (same as $ABA.AABB$).

References

1. Badkobeh, G., Ochem, P.: Characterization of some binary words with few squares. Theoret. Comput. Sci. **588**, 73–80 (2015)
2. Baker, K.A., McNulty, G.F., Taylor, W.: Growth problems for avoidable words. Theoret. Comput. Sci. **69**(3), 319–345 (1989)
3. Bean, D.R., Ehrenfeucht, A., McNulty, G.F.: Avoidable patterns in strings of symbols. Pacific J. Math. **85**, 261–294 (1979)
4. Cassaigne, J.: Motifs évitables et régularité dans les mots. Ph.D. Thesis, Université Paris VI (1994)
5. Clark, R.J.: Avoidable formulas in combinatorics on words. Ph.D. Thesis, University of California, Los Angeles (2001)
6. Lothaire, M.: Algebraic Combinatorics on Words. Cambridge University Press, Cambridge (2002)

7. Ochem, P.: A generator of morphisms for infinite words. RAIRO - Theoret. Inform. Appl. **40**, 427–441 (2006)
8. Ochem, P.: Binary words avoiding the pattern AABBCABBA. RAIRO - Theoret. Inform. Appl. **44**(1), 151–158 (2010)
9. Ochem, P.: Doubled patterns are 3-avoidable. Electron. J. Combinatorics **23**(1) (2016)
10. Thue, A.: Über unendliche Zeichenreihen. 'Norske Vid. Selsk. Skr. I. Mat. Nat. Kl. **7**, 1–22 (1906). Christiania
11. Zimin, A.I.: Blocking sets of terms. Math. USSR Sbornik **47**(2), 353–364 (1984)

Deciding Equivalence of Linear Tree-to-Word Transducers in Polynomial Time

Adrien Boiret[1] and Raphaela Palenta[2(✉)]

[1] CRIStAL, University Lille 1, Avenue Carl Gauss, 59655 Villeneuve d'Ascq Cedex, France
adrien.boiret@inria.fr

[2] Department of Informatics, Technical University of Munich, Boltzmannstr. 3, 85748 Garching, Germany
palenta@in.tum.de

Abstract. We show that the equivalence of linear top-down tree-to-word transducers is decidable in polynomial time. Linear tree-to-word transducers are non-copying but not necessarily order-preserving and can be used to express XML and other document transformations. The result is based on a partial normal form that provides a basic characterization of the languages produced by linear tree-to-word transducers.

Keywords: Tree transducer · Deciding equivalence · Partial normal form

1 Introduction

Tree transformations are widely used in functional programming and document processing. Tree transducers are a general model for transforming structured data like a database in a structured or even unstructured way. Consider the following internal representation of a client database that should be transformed to a table in HTML.

```
{ {
    name: "Alexander"
    surname: "Walker"
    nickname: "Alex"
    title: "Prof."
    salutation: "Mr."
  }
  ...
}
```

⟶

```
<table>
  <tr>  <th> name <\th>
        <th> surname <\th>
        <th> nickname <\th>
        <th> title <\th>
        <th> salutation <\th> <\tr>
  <tr>  <td> Alexander <\td>
        <td> Walker <\td>
        <td> Alex <\td>
        <td> Prof. <\td>
        <td> Mr. <\td> <\tr>
  ...
<\table>
```

A. Boiret—This work was partially supported by a grant from CPER Nord-Pas de Calais/FEDER DATA Advanced data science and technologies 2015–2020.

S. Brlek and C. Reutenauer (Eds.): DLT 2016, LNCS 9840, pp. 355–367, 2016.
DOI: 10.1007/978-3-662-53132-7_29

Deterministic top-down tree transducers can be seen as functional programs that transform trees from the root to the leaves with finite memory. Transformations where the output is not produced in a structured way or where, for example, the output is a string, can be modeled by tree-to-word transducers.

In this paper, we study deterministic linear tree-to-word transducers (LTWs), a subset of deterministic tree-to-word transducers that are non-copying, but not necessarily order-preserving. Processing the subtrees in an arbitrary order is important to avoid reordering of the internal data for different use cases. In the example of the client database the names may be needed in different formats, e.g.

<salutation> <name> <surname>
<surname>, <name>
<title> <surname>
<title> <surname>, <name>

The equivalence of unrestricted tree-to-word transducers was a long standing open problem that was recently shown to be decidable [12]. The algorithm by [12] provides an co-randomized polynomial algorithm for linear transducers. We show that the equivalence of LTWs is decidable in polynomial time and provide a partial normal form.

To decide equivalence of LTWs, we start in Sect. 3 by extending the methods used for sequential (linear and order-preserving) tree-to-word transducers (STWs), discussed in [13]. The equivalence for these transducers is decidable in polynomial time [13]. Moreover a normal form for sequential and linear tree-to-word transducers, computable in exponential time, is known [1,7]. Two equivalent LTWs do not necessarily transform their trees in the same order. However, the differences that can occur are quite specific and characterized in [1]. We show how they can be identified. We use the notion of *earliest* states, inspired by the existing notion of earliest sequential transducers [7]. In this earliest form, two equivalent STWs can transform subtrees in different orders only if they fulfill specific properties pertaining to the periodicity of the words they create. Computing this normal form is exponential in complexity as the number of states may increase exponentially. To avoid this size increase we do not compute these earliest transducers fully, but rather locally. This means we transform two LTWs with different orders to a *partial normal form* in polynomial time (see Sect. 4) where the order of their transformation of the different subtrees are the same. LTWs that transform the subtrees of the input in the same order can be reduced to sequential tree-to-word transducers as the input trees can be reordered according to the order in the transformation.

Due to space constraints some proofs are omitted. The full version of the paper can be found at http://arxiv.org/abs/1606.03758.

Related Work. Different other classes of transducers, such as tree-to-tree transducers [5], macro tree transducers [6] or nested-word-to-word transducers [13] have been studied. Many results for tree-to-tree transducers are known, e.g. deciding equivalence [10], minimization algorithms [10] and Gold-style learning algorithms [8]. In contrast, transformations where the output is not generated in a structured way like a tree are not that well understood. In macro-tree transducers, the decidability of equivalence is a well-known and long-standing question [2]. However, the equivalence of linear size increase macro-tree transducers that are equivalent to MSO definable transducers is decidable [3,4].

2 Preliminaries

Let Σ be a ranked alphabet with $\Sigma^{(n)}$ the symbols of rank n. Trees on Σ ($\mathcal{T}_\Sigma$) are defined inductively: if $f \in \Sigma^{(n)}$, and $t_1, ..., t_n \in \mathcal{T}_\Sigma$, then $f(t_1, ..., t_n) \in \mathcal{T}_\Sigma$ is a tree. Let Δ be an alphabet. An element $w \in \Delta^*$ is a word. For two words u, v we denote the concatenation of these two words by uv. The length of a word w is denoted by $|w|$. We call ε the empty word. We denote a^{-1} the inverse of a symbol a where $aa^{-1} = a^{-1}a = \varepsilon$. The inverse of a word $w = u_1 \ldots u_n$ is $w^{-1} = u_n^{-1} \ldots u_1^{-1}$.

A *context-free grammar* (CFG) is defined as a tuple (Δ, N, S, P), where Δ is the alphabet of G, N is a finite set of *non-terminal symbols*, $S \in N$ is the initial non-terminal of G, P is a finite set of rules of form $A \to w$, where $A \in N$ and $w \in (\Delta \cup N)^*$. A CFG is deterministic if each non-terminal has at most one rule.

We define the language $L_G(A)$ of a non-terminal A recursively: if $A \to u_0A_1u_1...A_nu_n$ is a rule of P, with u_i words of Δ^* and A_i non-terminals of N, and w_i a word of $L_G(A_i)$, then $u_0w_1u_1...w_nu_n$ is a word of $L_G(A)$. We define the context-free language L_G of a context-free grammar G as $L_G(S)$.

A *straight-line program* (SLP) is a deterministic CFG that produces exactly one word. The word produced by an SLP (Δ, N, S, P) is called w_S.

We denote the *longest common prefix of all words* of a language L by $\mathsf{lcp}(L)$. Its *longest common suffix* is $\mathsf{lcs}(L)$.

A word u is said to be *periodic* of period w if w is the smallest word such that $u \in w^*$. A language L is said to be *periodic* of period w if w is the smallest word such that $L \subseteq w^*$.

A language L is *quasi-periodic* on the left (resp. on the right) of handle u and period w if w is the smallest word such that $L \subseteq uw^*$ (resp. if $L \subseteq w^*u$). A language is quasi-periodic if it is quasi-periodic on the right or left. If L is a singleton or empty, it is periodic of period ε. Iff L is periodic, it is quasi-periodic on the left and the right of handle ε. If L is quasi-periodic on the left (resp. right) then $\mathsf{lcp}(L)$ (resp. $\mathsf{lcs}(L)$) is the shortest word of L.

3 Linear Tree-to-Word Transducers

A *linear tree-to-word transducer* (LTW) is a tuple $M = (\Sigma, \Delta, Q, \mathsf{ax}, \delta)$ where

- Σ is a ranked alphabet,
- Δ is an alphabet of output symbols,
- Q is a finite set of states,
- the axiom ax is of the form $u_0q(x)u_1$, where $q \in Q$ and $u_0, u_1 \in \Delta^*$,
- δ is a set of rules of the form $q, f \to u_0q_1(x_{\sigma(1)}) \cdots q_n(x_{\sigma(n)})u_n$ where $q, q_1, \ldots, q_n \in Q$, $f \in \Sigma$ of rank n, $u_0, \ldots, u_n \in \Delta^*$ and σ is a permutation from $\{1, \ldots, n\}$ to $\{1, \ldots, n\}$. There is at most one rule per pair q, f.

The partial function $[\![M]\!]_q$ of a state q on an input tree $f(t_1, \ldots, t_n)$ is defined inductively as

- $u_0[\![M]\!]_{q_1}(t_{\sigma(1)}) \ldots [\![M]\!]_{q_n}(t_{\sigma(n)})u_n$, if $q, f \to u_0 q_1(x_{\sigma(1)}) \ldots q_n(x_{\sigma(n)})u_n \in \delta$
- undefined, if q, f is not defined in δ.

The partial function $[\![M]\!]$ of an LTW M with axiom $u_0 q(x) u_1$ on an input tree t is defined as $[\![M]\!](t) = u_0[\![M]\!]_q(t)u_1$.

Two LTWs M and M' are equivalent if $[\![M]\!] = [\![M']\!]$.

A *sequential tree-to-word transducer* (STW) is an LTW where for each rule of the form $q, f \to u_0 q_1(x_{\sigma(1)})u_1 \ldots q_n(x_{\sigma(n)})u_n$, σ is the identity on $1 \ldots n$.

We define *accessibility* of states as the transitive and reflexive closure of appearance in a rule. This means state q is accessible from itself, and if $q, f \to u_0 q_1(x_{\sigma(1)}) \ldots q_n(x_{\sigma(n)})u_n$, and q is accessible from q', then all states q_i, $1 \leq i \leq n$, are accessible from q'.

We denote by $\mathsf{dom}(M)$ (resp. $\mathsf{dom}(q)$) the domain of an LTW M (resp. a state q), i.e. all trees $t \in \mathcal{T}_\Sigma$ such that $[\![M]\!](t)$ is defined (resp. $[\![M]\!]_q(t)$). We only consider LTWs with non-empty domains and assume w.l.o.g. that no state q in an LTW has an empty domain by eliminating transitions using states with empty domain.

We denote by L_M (resp. L_q) the range of $[\![M]\!]$ (resp. $[\![M]\!]_q$), i.e. the set of all images $[\![M]\!](t)$ (resp. $[\![M]\!]_q(t)$). The languages L_M and L_q for each $q \in Q$ are all context-free languages. We call a state q *(quasi-)periodic* if L_q is (quasi-)periodic.

Note that a word u in a rule of an LTW can be represented by an SLP without changing the semantics of the LTW. Therefore a set of SLPs is added to the transducer and a word on the right-hand side of a rule is represented by an SLPs. The decidability of equivalence of STWs in polynomial time still holds true with the use of SLPs.

The results of this paper require SLP compression to avoid exponential blow-up. SLPs are used to prevent exponential blow-up in [11], where morphism equivalence on context-free languages is decided in polynomial time.

The equivalence problem for sequential tree-to-word transducer can be reduced to the morphism equivalence problem for context-free languages [13]. This reduction relies on the fact that STWs transform their subtrees in the same order. As LTWs do not necessarily transform their subtrees in the same order the result cannot be applied on LTWs in general. However, if two LTWs transform their subtrees in the same order, then the same reduction can be applied. To formalize that two LTWs transform their subtrees in the same order we introduce the notion of state co-reachability. Two states q_1 and q_2 of LTWs M_1, M_2, respectively, are co-reachable if there is an input tree such that the two states are assigned to the same node of the input tree in the translations of M_1, M_2, respectively.

Two LTWs are *same-ordered* if for each pair of co-reachable states q_1, q_2 and for each symbol $f \in \Sigma$, neither q_1 nor q_2 have a rule for f, or if $q_1, f \to u_0 q_1'(x_{\sigma_1(1)}) \ldots q_n'(x_{\sigma_1(n)})u_n$ and $q_2, f \to v_0 q_1''(x_{\sigma_2(1)}) \ldots q_n''(x_{\sigma_2(n)})v_n$ are rules of q_1 and q_2, then $\sigma_1 = \sigma_2$.

If two LTWs are same-ordered the input trees can be reordered according to the order in the transformations. Therefore for each LTW a tree-to-tree transducer is constructed that transforms the input tree according to the transformation in the LTW. Then all permutations σ in the LTWs are replaced by the identity. Thus the LTWs can be handled as STWs and therefore the equivalence is decidable in polynomial time [13].

Theorem 1. *The equivalence of same-ordered* LTW*s is decidable in polynomial time.*

3.1 Linear Earliest Normal Form

In this section we introduce the two key properties that are used to build a normal form for linear tree-to-word transducers, namely the *earliest* and *erase-ordered* properties. The earliest property means that the output is produced as early as possible, i.e. the longest common prefix (resp. suffix) of L_q is produced in the rule in which q occurs, and as left as possible. The erase-ordered property means that all states that produce no output are ordered according to the input tree and pushed to the right in the rules.

An LTW is in *earliest form* if

- each state q is *earliest*, i.e. $\mathsf{lcp}(L_q) = \mathsf{lcs}(L_q) = \varepsilon$,
- and for each rule $q, f \to u_0 q_1(x_{\sigma(1)}) \dots q_n(x_{\sigma(n)}) u_n$, for each $i, 1 \leq i \leq n$, $\mathsf{lcp}(L_{q_i} u_i) = \varepsilon$.

In [1, Lemma 9] it is shown that for each LTW M an equivalent earliest LTW M' can be constructed in exponential time. Intuitively, if $\mathsf{lcp}(L_q) = v \neq \varepsilon$ (resp. $\mathsf{lcs}(L_q) = v \neq \varepsilon$) then q' is constructed with $L_{q'} = v^{-1} L_q$ (resp. $L_{q'} = L_q v^{-1}$) and $q(x)$ is replaced by $vq'(x)$ (resp. $q'(x)v$). If $\mathsf{lcp}(L_q u) = v \neq \varepsilon$ and v is a prefix of $u = vv'$ then we push v through L_q by constructing q' with $L_{q'} = v^{-1} L_q v$ and replace $q(x)u$ by $vq'(x)v'$.

Note that the construction to build the earliest form M' of an LTW M creates a same-ordered M'. Furthermore, if a state q of M and a state q' of M' are co-reachable, then q' is an "earliest" version of q, where some word u was pushed out of the production of q to make it earliest, and some word v was pushed through the production of q to ensure that the rules have the right property: there exists u, $v \in \Delta^*$ such that for all $t \in \mathsf{dom}(q)$, $[\![M']\!]_{q'}(t) = v^{-1} u^{-1} [\![M]\!]_q(t) v$.

Theorem 2. *For each* LTW *an equivalent same-ordered and earliest* LTW *can be constructed in exponential time.*

The exponential time complexity is caused by a potential exponential size increase in the number of states as it is shown in [7, Example 5].

We call a state q that produces only the empty word, i.e. $L_q = \{\varepsilon\}$, an *erasing state*. As erasing states do not change the transformation and can occur at any position in a rule we need to fix their position for a normal form.

An LTW M is *erase-ordered* if for each rule $q, f \to u_0 q_1(x_{\sigma(1)}) \dots q_n(x_{\sigma(n)}) u_n$ in M, if q_i is erasing then for all $j \geq i$, q_j is erasing, $\sigma(i) \leq \sigma(j)$ and $u_j = \varepsilon$.

We test whether $L_q = \{\varepsilon\}$ in polynomial time and then reorder a rule according to the erase-ordered property. If an LTW is earliest it is still earliest after the reordering.

Lemma 3 (extended from [1, Lemma 18]). *For each (earliest) LTW an equivalent (earliest) erase-ordered LTW can be constructed in polynomial time.*

Example 4. Consider the rule $q_0, f \to q_1(x_4)q_2(x_3)q_1(x_2)q_4(x_1)$ where q_2 translates trees of the form $f^n(g), n \geq 0$ to $(abc)^n$, q_4 translates trees of the form $f^n(g), n \geq 0$ to $(abc)^{2n}$, q_1 translates trees of the form $f^n(g), n \geq 0$ to ε. Thus the rule is not erase-ordered. We reorder the rule to the equivalent and erase-ordered rule $q_0, f \to q_2(x_3)q_4(x_1)q_1(x_2)q_1(x_4)$.

If two equivalent LTWs are earliest and erase-ordered, then they are not necessarily same-ordered. For example, the rule $q, f \to q_4(x_1)q_2(x_3)q_1(x_2)q_1(x_4)$ is equivalent to the rule in the above example but the two rules are not same-ordered. However, in earliest and erase-ordered LTWs, we can characterize the differences in the orders of equivalent rules: Just as two words u, v satisfy the equation $uv = vu$ if and only if there is a word w such that $u \in w^*$ and $v \in w^*$, the only way for equivalent earliest and erase-ordered LTWs to not be same-ordered is to switch periodic states.

Theorem 5 [1]. *Let M and M' be two equivalent erase-ordered and earliest LTWs and q, q' be two co-reachable states in M, M', respectively. Let $q, f \to u_0q_1(x_{\sigma_1(1)}) \dots q_n(x_{\sigma_1(n)})u_n$ and $q', f \to v_0q'_1(x_{\sigma_2(1)}) \dots q'_n(x_{\sigma_2(n)})v_n$ be two rules for q, q'. Then*

- *for $k < l$ such that $\sigma_1(k) = \sigma_2(l)$, all q_i, $k \leq i \leq l$, are periodic of the same period and all $u_j = \varepsilon$, $k \leq j < l$,*
- *for k, l such that $\sigma_1(k) = \sigma_2(l)$, $[\![M]\!]_{q_k} = [\![M']\!]_{q'_l}$.*

As the subtrees that are not same-ordered in two equivalent earliest and erase-ordered states are periodic of the same period the order of these can be changed without changing the semantics. Therefore the order of these subtrees can be fixed such that equivalent earliest and erase-ordered LTWs are same-ordered. Then the equivalence is decidable in polynomial time, see Theorem 1. However, building the earliest form of an LTW is in exponential time.

To circumvent this difficulty, we will show that the first part of Theorem 5 still holds even on a *partial normal form*, where only quasi-periodic states are earliest and the longest common prefix of parts of rules $q(x)u$ with $L_q u$ being quasi-periodic is the empty word.

Theorem 6. *Let M and M' be two equivalent erase-ordered LTWs such that*

- *all quasi-periodic states q are earliest, i.e. $\mathsf{lcp}(q) = \mathsf{lcs}(q) = \varepsilon$*
- *for each part $q(x)u$ of a rule where $L_q u$ is quasi-periodic, $\mathsf{lcp}(L_q u) = \varepsilon$*

Let q, q' be two co-reachable states in M, M', respectively and
$q, f \to u_0q_1(x_{\sigma_1(1)}) \dots q_n(x_{\sigma_1(n)})u_n$ and $q', f \to v_0q'_1(x_{\sigma_2(1)}) \dots q'_n(x_{\sigma_2(n)})v_n$ be two rules for q, q'. Then for $k < l$ such that $\sigma_1(k) = \sigma_2(l)$, all q_i, $k \leq i \leq l$, are periodic of the same period and all $u_j = \varepsilon$, $k \leq j < l$.

4 Partial Normal Form

In this section we introduce a partial normal form for LTWs that does not suffer from the exponential blow-up of the earliest form. Inspired by Theorem 6, we wish to solve order differences by switching adjacent periodic states of the same period. Remember that the earliest form of a state q is constructed by removing the longest common prefix (suffix) of L_q to produce this prefix (suffix) earlier. It follows that all non-earliest states from which q can be constructed following the earliest form are quasi-periodic.

We show that building the earliest form of a quasi-periodic state or a part of a rule $q(x)u$ with $L_q u$ being quasi-periodic is in polynomial time. Therefore building the following partial normal form is in polynomial time.

Definition 7. *A linear tree-to-word transducer is in* partial normal form *if*

1. *all quasi-periodic states are earliest,*
2. *it is erase-ordered and*
3. *for each rule* $q, f \to u_0 q_1(x_{\sigma(1)}) \dots q_n(x_{\sigma(n)}) u_n$ *if* $L_{q_i} u_i L_{q_{i+1}}$ *is quasi-periodic then* $q_i(x_{\sigma(i)}) u_i q_{i+1}(x_{\sigma(i+1)})$ *is earliest and* $\sigma(i) < \sigma(i+1)$.

4.1 Eliminating Non-Earliest Quasi-Periodic States

In this part, we show a polynomial time algorithm to build an earliest form of a quasi-periodic state. From which an equivalent LTW can be constructed in polynomial time such that any quasi-periodic state is earliest, i.e. $\mathsf{lcp}(L_q) = \mathsf{lcs}(L_q) = \varepsilon$. Additionally, we show that the presented algorithm can be adjusted to test if a state is quasi-periodic in polynomial time.

As quasi-periodicity on the left and on the right are symmetric properties we only consider quasi-periodic states of the form uw^* (quasi-periodic on the left). The proofs in the case w^*u are symmetric and therefore omitted here. In the end of this section we shortly discuss the introduced algorithms for the symmetric case w^*u.

To build the earliest form of a quasi-periodic state we use the property that each state accessible from a quasi-periodic state is as well quasi-periodic. However, the periods can be shifted as the following example shows.

Example 8. Consider states q, q_1 and q_2 with rules $q, f \to aq_1(x_1)c$, $q_1, f \to aaq_2(x_1)ab$, $q_2, f \to q_2(x_1)abc$, $q_2, g \to abc$. State q accepts trees of the form $f^n(g)$, $n \geq 2$, and produces the language $aaa(abc)^n$, i.e. q is quasi-periodic of period abc. State q_1 accepts trees of the form $f^n(g)$, $n \geq 1$, and produces the language $aa(abc)^n ab$, i.e. q_1 is quasi-periodic of period cab. State q_2 accepts trees of the form $f^n(g)$, $n \geq 0$ and produces the language $(abc)^{n+1}$, i.e. q_2 is (quasi-)periodic of period abc.

We introduce two definitions to measure the shift of periods. We denote by $\rho_n[u]$ the *from right-to-left shifted word of* u *of shift* n, $n \leq |u|$, i.e. $\rho_n[u] =$

$u'^{-1}uu'$ where u' is the prefix of u of size n. If $n \geq |u|$ then $\rho_n[u] = \rho_m[u]$ with $m = n \mod |u|$.

For two quasi-periodic states q_1, q_2 of period $u = u_1u_2$ and $u' = u_2u_1$, respectively, we denote the *shift in their period* by $s(q_1, q_2) = |u_1|$.

The size of the periods of a quasi-periodic state and the states accessible from this state can be computed from the size of the shortest words of the languages produced by these states.

Lemma 9. *If q is quasi-periodic on the left with period w, and q' accessible from q, then q' is quasi-periodic with period ε or a shift of w. Moreover we can calculate the shift $s(q, q')$ in polynomial time.*

We now use these shifts to build, for a state q in M that is quasi-periodic on the left, a transducer M^q equivalent to M where each occurrence of q is replaced by its equivalent earliest form, i.e. a periodic state and the corresponding prefix.

Algorithm 1. *Let q be a state in M that is quasi-periodic on the left. M^q starts with the same states, axiom, and rules as M.*

- *For each state p accessible from q, we add a copy p^e to M^q.*
- *For each rule $p, f \rightarrow u_0q_1(x_{\sigma(1)}) \dots q_n(x_{\sigma(n)})u_n$ in M with p accessible from q, we add a rule $p^e, f \rightarrow u_pq_1^e(x_{\sigma(1)})q_2^e(x_{\sigma(2)}) \dots q_n^e(x_{\sigma(n)})$ with $u_p = \rho_{s(q,p)}\left[\mathsf{lcp}(p)^{-1}u_0\mathsf{lcp}(q_1) \dots \mathsf{lcp}(q_n)u_n\right]$ in M^q.*
- *We delete state q in M^q and replace any occurrence of $q(x)$ in a rule or the axiom of M^q by $\mathsf{lcp}(q)q^e(x)$.*

Note that $\mathsf{lcp}(p)^{-1}u_0\mathsf{lcp}(q_1) \dots \mathsf{lcp}(q_n)u_n$ is equivalent to deleting the prefix of size $|\mathsf{lcp}(p)|$ from the word $u_0\mathsf{lcp}(q_1) \dots \mathsf{lcp}(q_n)u_n$.

Intuitively, to build the earliest form of a state q that is quasi-periodic on the left we need to push all words and all longest common prefixes of states on the right-hand side of a rule of q to the left. Pushing a word to the left through a state needs to shift the language produced by this state. We explain the algorithm in detail on state q from Example 8.

Example 10. Remember that q produces the language $aaa(abc)^n, n \geq 2$ and q_1, q_2 accessible from q produce languages $aa(abc)^nab, n \geq 1$ and $(abc)^{n+1}, n \geq 0$, respectively. Therefore $\mathsf{lcp}(q) = aaaabcabc$, $\mathsf{lcp}(q_1) = aaabcab$ and $\mathsf{lcp}(q_2) = abc$. We start with state q. As there is only one rule for q the longest common prefix of q and the longest common prefix of this rule are the same and therefore eliminated.

$$\begin{aligned} q^e, f &\rightarrow \rho_{s(q,q)}[\mathsf{lcp}(q)^{-1}a\mathsf{lcp}(q_1)c]q_1^e(x_1) \\ &\rightarrow \rho_{s(q,q)}[(aaaabcabc)^{-1}aaaabcabc]q_1^e(x_1) \\ &\rightarrow q_1^e(x_1) \end{aligned}$$

As there is only one rule for q_1 the argumentation is the same and we get $q_1^e, f \rightarrow q_2^e$. For the rule q_2, f we calculate the longest common prefix of the right-hand side $\mathsf{lcp}(q_2)abc = abcabc$ that is larger than the longest common prefix of q_2. Therefore we need to calculate the shift $s(q, q_2) = s(q, q_1) + s(q_1, q_2) = |c| + |ab| = 3$ as q_1 is accessible from q in rule q, f and q_2 is accessible from q_1 in rule q_1, f. This leads to the following rule.

$$\begin{aligned} q_2^e, f &\to \rho_{s(q,q_2)}[\mathsf{lcp}(q_2)^{-1}\mathsf{lcp}(q_2)abc]q_2^e(x_1) \\ &\to \rho_3[(abc)^{-1}abcabc]q_2^e(x_1) \\ &\to abcq_2^e(x_1) \end{aligned}$$

As the longest common prefix of q_2 is the same as the longest common prefix of the right-hand side of rule q_2, g we get $q_2^e, g \to \varepsilon$. The axiom of M^q is $\mathsf{lcp}(q)q^e(x_1) = aaaabcabcq^e(x_1)$.

Lemma 11. *Let M be an* LTW *and q be a state in M that is quasi-periodic on the left. Let M^q be constructed by Algorithm 1 and p^e be a state in M^q accessible from q^e. Then M and M^q are equivalent and p^e is earliest.*

To replace all quasi-periodic states by their equivalent earliest form we need to know which states are quasi-periodic. Algorithm 1 can be modified to test an arbitrary state for quasi-periodicity on the left in polynomial time. The only difference to Algorithm 1 is that we do not know how to compute $\mathsf{lcp}(p)$ in polynomial time and $s(q, p)$ does not exist. We therefore substitute $\mathsf{lcp}(p)$ by some smallest word of L_p and we define a mock-shift $s'(q, p)$ as follows

- $s'(q, q) = 0$ for all q,
- if $q, f \to u_0 q_1(x_{\sigma(1)}) \dots q_n(x_{\sigma(n)}) u_n$, we say $s'(q, q_i) = |u_i w_{q_{i+1}} \dots w_{q_n} u_n|$, where w_q is a shortest word of L_q,
- if $s'(q_1, q_2) = n$ and $s'(q_2, q_3) = m$ then $s'(q_1, q_3) = n + m$.

If several definitions of $s'(q, p)$ exist, we use the smallest. If p is accessible from a quasi-periodic q, then $s'(q, p) = s(q, p)$.

Algorithm 2. *Let $M = (\Sigma, \Delta, Q, \mathsf{ax}, \delta)$ be an* LTW *and q be a state in M. We build an* LTW *T^q as follows.*

- *For each state p accessible from q, we add a copy p^e to T^q.*
- *The axiom is $w_q q^e(x)$ where w_q is a shortest word of L_q.*
- *For each rule $p, f \to u_0 q_1(x_{\sigma(1)}) \dots q_n(x_{\sigma(n)}) u_n$ in M with p accessible from q, we add a rule*

$$p^e, f \to u_p q_1^e(x_{\sigma(1)}) q_2^e(x_{\sigma(2)}) \dots q_n^e(x_{\sigma(n)})$$

  in T^q, where u_p is constructed as follows.
  - *We define $u = u_0 w_1 \dots w_n u_n$, where w_i is a shortest word of L_{q_i}.*
  - *Then we remove from u its prefix of size $|w'|$, where w' is a shortest word of L_p. We obtain a word u'.*
  - *Finally, we set $u_p = \rho_{s'(q,p)}[u']$.*

As the construction of Algorithms 1 and 2 are the same if the state q is quasi-periodic, $[\![M]\!]_q$ and $[\![T^q]\!]$ are equivalent if q is quasi-periodic. Moreover, q is quasi-periodic if $[\![M]\!]_q$ and $[\![T^q]\!]$ are equivalent.

Lemma 12. *Let q be a state of an* LTW *M and T^q be constructed by Algorithm 2. Then M and T^q are same-ordered and q is quasi-periodic on the left if and only if $[\![M]\!]_q = [\![T^q]\!]$ and q^e is periodic.*

As M and T^q are same-ordered we can test the equivalence in polynomial time, cf. Theorem 1. Moreover testing a CFG for periodicity is in polynomial time and therefore testing a state for quasi-periodicity is in polynomial time.

Algorithm 2 can be applied to a part $q(x)u$ of a rule to test $L_q u$ for quasi-periodicity on the left. In this case for each rule $q, f \to u_0 q_1(x_{\sigma(1)}) \cdots q_n(x_{\sigma(n)}) u_n$ a rule $\hat{q}, f \to u_0 q_1(x_{\sigma(1)}) \cdots q_n(x_{\sigma(n)}) u_n u$ is added to M and each occurrence of the part $q(x)u$ in a rule of M is replaced by $\hat{q}(x)$. We then apply the above algorithm to $\hat{q}$ and test $[\![M]\!]_{\hat{q}}$ and $[\![T^{\hat{q}}]\!]$ for equivalence and $\hat{q}^e$ for periodicity.

We introduced algorithms to test states for quasi-periodicity on the left and to build the earliest form for such states. These two algorithms can be adapted for states that are quasi-periodic on the right. There are two main differences. First, as the handle is on the right the shortest word of a language L that is quasi-periodic on the right is $\mathsf{lcs}(L)$. Second, instead of pushing words through a periodic language to the left we need to push words through a periodic language to the right.

Hence, we can test each state q of an LTW M for quasi-periodicity on the left and right. If the state is quasi-periodic we replace q by its earliest form. Algorithms 1 and 2 run in polynomial time if SLPs are used. This is crucial as the shortest word of a CFG can be of exponential size. However, the operations that are needed in the algorithms, namely constructing the shortest word of a CFG and removing the prefix or suffix of a word, are in polynomial time using SLPs, cf. [9].

Theorem 13. *Let M be an* LTW. *Then an equivalent* LTW *M' where all quasi-periodic states are earliest can be constructed in polynomial time.*

4.2 Switching Periodic States

In this part we obtain the partial normal form by ordering periodic states of an erase-ordered transducer where all quasi-periodic states are earliest. Ordering means that if the order of the subtrees in the translation can differ, we choose the one similar to the input, i.e. if $q(x_3)q'(x_1)$ and $q'(x_1)q(x_3)$ are equivalent, we choose the second order. We already showed how we can build a transducer where each quasi-periodic state is earliest and therefore periodic. However, we need to make parts of rules earliest such that periodic states can be switched as the following example shows.

Example 14. Consider the rule $q, h \to q_1(x_2) b q_2(x_1)$ where q_1, q_2 have the rules $q_1, f \to bcabca q_1(x)$, $q_1, g \to \varepsilon$, $q_2, f \to cab q_2(x)$, $q_2, g \to \varepsilon$. States q_1 and q_2 are earliest and periodic but not of the same period as a subword is produced in between. We replace the non-earliest and quasi-periodic part $q_1(x_2)b$ by their earliest form. This leads to $q, h \to b q_1^e(x_2) q_2(x_1)$ with $q_1^e, f \to cabcab q_1^e(x)$, $q_1^e, g \to \varepsilon$. Hence, q_1^e and q_2 are earliest and periodic of the same period and can be switched in the rule.

To build the earliest form of a quasi-periodic part of a rule $q(x)u$ each occurrence of this part is replaced by a state $\hat{q}(x)$ and for each rule $q, f \to u_0 q_1(x_{\sigma(1)}) \dots q_n(x_{\sigma(n)}) u_n$ a rule $\hat{q}, f \to u_0 q_1(x_{\sigma(1)}) \dots q_n(x_{\sigma(n)}) u_n u$ is added. Then we apply Algorithm 1 on $\hat{q}$ to replace $\hat{q}$ and therefore $q(x)u$ by their earliest form. Iteratively this leads to the following theorem.

Theorem 15. *For each* LTW *M where all quasi-periodic states are earliest we can build in polynomial time an equivalent* LTW *M' such that each part $q(x)u$ of a rule in M where $L_q u$ is quasi-periodic is earliest.*

In Theorem 6 we showed that order differences in equivalent erase-ordered LTWs where all quasi-periodic states are earliest and all parts of rules $q(x)u$ are earliest are caused by adjacent periodic states. As these states are periodic of the same period and no words are produced in between these states can be reordered without changing the semantics of the LTWs.

Lemma 16. *Let M be an* LTW *such that*

- *M is erase-ordered,*
- *all quasi-periodic states in M are earliest and*
- *each $q_i(x_{\sigma(i)})u_i$ in a rule of M that is quasi-periodic is earliest.*

Then we can reorder adjacent periodic states $q_i(x_{\sigma(i)})q_{i+1}(x_{\sigma(i+1)})$ of the same period in the rules of M such that $\sigma(i) < \sigma(j)$ in polynomial time. The reordering does not change the transformation of M.

We showed before how to construct a transducer with the preconditions needed in Lemma 16 in polynomial time. Note that replacing a quasi-periodic state by its earliest form can break the erase-ordered property. Thus we need to replace all quasi-periodic states by its earliest form *before* building the erase-ordered form of a transducer. Then Lemma 16 is the last step to obtain the partial normal form for an LTW.

Theorem 17. *For each* LTW *we can construct an equivalent* LTW *that is in partial normal form in polynomial time.*

4.3 Testing Equivalence in Polynomial Time

It remains to show that the equivalence problem of LTWs in partial normal form is decidable in polynomial time. The key idea is that two equivalent LTWs in partial normal form are same-ordered.

Consider two equivalent LTWs M_1, M_2 where all quasi-periodic states and all parts of rules $q(x)u$ with $L_q u$ is quasi-periodic are earliest. In Theorem 6 we showed if the orders σ_1, σ_2 of two co-reachable states q_1, q_2 of M_1, M_2, respectively, for the same input differ then the states causing this order differences are periodic with the same period. The partial normal form solves this order differences such that the transducers are same-ordered.

Lemma 18. *If M and M' are equivalent and in partial normal form then they are same-ordered.*

As the equivalence of same-ordered LTWs is decidable in polynomial time (cf. Theorem 1) we conclude the following.

Corollary 19. *The equivalence problem for* LTW*s in partial normal form is decidable in polynomial time.*

To summarize, the following steps run in polynomial time and transform a LTW M into its partial normal form.

1. Test each state for quasi-periodicity. If it is quasi-periodic replace the state by its earliest form.
2. Build the equivalent erase-ordered transducer.
3. Test each part $q(x)u$ in each rule from right to left for quasi-periodicity on the left. If it is quasi-periodic replace the part by its earliest form.
4. Order adjacent periodic states of the same period according to the input order.

This leads to our main theorem.

Theorem 20. *The equivalence of* LTW*s is decidable in polynomial time.*

Acknowledgement. We would like to thank the reviewers for their very helpful comments.

References

1. Boiret, A.: Normal form on linear tree-to-word transducers. In: Dediu, A.-H., et al. (eds.) Language and Automata Theory and Applications. LNCS, vol. 9618, pp. 439–451. Springer, Heidelberg (2016)
2. Engelfriet, J.: Some open question and recent results on tree transducers and tree languages. In: Book, R.V. (ed.) Formal Language Theory, Perspectives and Open Problems, pp. 241–286. Academic Press, New York (1980)
3. Engelfriet, J., Maneth, S.: Macro tree translations of linear size increase are MSO definable. SIAM J. Comput. **32**(4), 950–1006 (2003)
4. Engelfriet, J., Maneth, S.: The equivalence problem for deterministic MSO tree transducers is decidable. Inf. Process. Lett. **100**(5), 206–212 (2006)
5. Engelfriet, J., Rozenberg, G., Slutzki, G.: Tree transducers, L systems and two-way machines. In: Proceedings of the Tenth Annual ACM Symposium on Theory of Computing, pp. 66–74. ACM (1978)
6. Engelfriet, J., Vogler, H.: Macro tree transducers. J. Comput. Syst. Sci. **31**(1), 71–146 (1985)
7. Laurence, G., Lemay, A., Niehren, J., Staworko, S., Tommasi, M.: Normalization of sequential top-down tree-to-word transducers. In: Dediu, A.-H., Inenaga, S., Martín-Vide, C. (eds.) LATA 2011. LNCS, vol. 6638, pp. 354–365. Springer, Heidelberg (2011)

8. Lemay, A., Maneth, S., Niehren, J.: A learning algorithm for top-down XML transformations. In: Proceedings of the Twenty-Ninth ACM SIGMOD-SIGACT-SIGART Symposium on Principles of Database Systems, pp. 285–296 (2010)
9. Lohrey, M.: The Compressed Word Problem for Groups. Springer, New York (2014)
10. Maneth, S., Seidl, H.: Deciding equivalence of top-down XML transformations in polynomial time. In: Programming Language Technologies for XML, pp. 73–79 (2007)
11. Plandowski, W.: The complexity of the morphism equivalence problem for context-free languages. Ph.D. thesis, Warsaw University (1995)
12. Seidl, H., Maneth, S., Kemper, G.: Equivalence of deterministic top-down tree-to-string transducers is decidable. In: IEEE 56th Annual Symposium on Foundations of Computer Science, pp. 943–962 (2015)
13. Staworko, S., Laurence, G., Lemay, A., Niehren, J.: Equivalence of deterministic nested word to word transducers. In: Charatonik, W., Gębala, M., Kutyłowski, M. (eds.) FCT 2009. LNCS, vol. 5699, pp. 310–322. Springer, Heidelberg (2009)

On Finite and Polynomial Ambiguity of Weighted Tree Automata

Erik Paul(✉)

Institute of Computer Science, Leipzig University, 04109 Leipzig, Germany
epaul@informatik.uni-leipzig.de

Abstract. We consider finite and polynomial ambiguity of weighted tree automata. Concerning finite ambiguity, we show that a finitely ambiguous weighted tree automaton can be decomposed into a sum of unambiguous automata. For polynomial ambiguity, we show how to decompose a polynomially ambiguous weighted tree automaton into simpler polynomially ambiguous automata and then analyze the structure of these simpler automata. We also outline how these results can be used to capture the ambiguity of weighted tree automata with weighted logics.

Keywords: Weighted tree automata · Quantitative tree automata · Finite ambiguity · Polynomial ambiguity · Weighted logics

1 Introduction

Weighted automata, a generalization of non-deterministic finite automata (NFA), have first been investigated by Schützenberger [22]. Since then, a large amount of further research has been conducted on them, cf. [3,9,18,21]. When considering complexity and decidability problems for these automata, the concept of ambiguity plays a large role. For instance, in [13] the equivalence problem for finitely ambiguous automata over the max-plus semiring is shown to be decidable, whereas for general non-deterministic automata over the max-plus semiring this problem is undecidable [17]. The ambiguity of an automaton is a measure for the maximum number of accepting runs on a given input. For example, if the number of accepting paths is bounded by a global constant for every word, we say that the automaton is finitely ambiguous. In the case that the number of accepting paths is bounded polynomially in the word length, we speak of polynomial ambiguity.

In this paper, we investigate these two types of ambiguity for weighted tree automata (WTA), a weighted automata model with trees as input. Our main results are the following:

- A finitely ambiguous WTA can be decomposed into a sum of several unambiguous WTA.

E. Paul—Supported by Deutsche Forschungsgemeinschaft (DFG), Graduiertenkolleg 1763 (QuantLA).

S. Brlek and C. Reutenauer (Eds.): DLT 2016, LNCS 9840, pp. 368–379, 2016.
DOI: 10.1007/978-3-662-53132-7_30

- A polynomially ambiguous WTA can be decomposed into a sum of "simpler" polynomially ambiguous WTA. Here, for each of these simpler automata we can identify a set of transitions such that, intuitively speaking, in every run on any tree each of these transitions occurs at exactly one position of the tree. Furthermore, the possible number of runs on any tree is bounded if we specify the position of each of these transitions. The bound does not depend on the given tree.
- To each of the classes of unambiguous, finitely ambiguous and polynomially ambiguous WTA, we relate a class of sentences from a weighted MSO logic expressively equivalent to it.

Weighted tree automata have been considered by a number of researchers [2,4,19], see [12] for a survey. Likewise, the ambiguity of finite automata has been studied numerous times. For example, [1,23,24] present criteria for ambiguity and algorithms to determine the ambiguity of automata. For weighted automata on words (WA), it has also been shown that expressive power increases with growing degree of ambiguity. It is shown in [15] that the inclusions *deterministic WA* $\subsetneq$ *unambiguous WA* $\subsetneq$ *finitely ambiguous WA* are strict and in [14] it is shown that the inclusion *finitely ambiguous WA* $\subsetneq$ *polynomially ambiguous WA* is strict.

Our first two results give a deeper insight into the structure of WTA and generalize results by Seidl and Weber [24] and Klimann et al. [15] from words to trees. As trees do not have the linear structure of words, however, the corresponding proofs from the word case can not be adapted to the tree case in a trivial way. Both results are new even for WTA over the boolean semiring, i.e. for tree automata without weights.

The initial motivation for our investigations lies with logics and the third result. Weighted logics can be used to describe weighted automata over words and trees, as was shown by Droste, Gastin and Vogler [8,10]. Kreutzer and Riveros [16] later showed that weighted logics can even be used to characterize different degrees of ambiguity of weighted automata over words. With the help of the first two results, we can generalize Kreutzer's and Riveros's result to WTA. For polynomial ambiguity, we even obtain a stronger result, as we are able to capture the polynomial degree of a WTA not only in the boolean semiring, but in any commutative semiring.

2 Weighted Tree Automata

Let $\mathbb{N} = \{0, 1, 2, \ldots\}$. A *ranked alphabet* is a pair (Γ, rk_Γ), often abbreviated by Γ, where Γ is a finite set and $rk_\Gamma \colon \Gamma \to \mathbb{N}$. For every $m \geq 0$ we define $\Gamma^{(m)} = rk_\Gamma^{-1}(m)$ as the set of all symbols of rank m. The rank $rk(\Gamma)$ of Γ is defined as $\max\{rk_\Gamma(a) \mid a \in \Gamma\}$. The set of (*finite, labeled and ordered*) Γ-*trees*, denoted by T_Γ, is the smallest subset T of $(\Gamma \cup \{(,)\} \cup \{,\})^*$ such that if $a \in \Gamma^{(m)}$ with $m \geq 0$ and $s_1, \ldots, s_m \in T$, then $a(s_1, \ldots, s_m) \in T$. In case $m = 0$, we identify $a()$ with a.

We define the set of *positions in a tree* by means of the mapping pos: $T_\Gamma \to \mathcal{P}(\mathbb{N}^*)$ inductively as follows: (i) if $t \in \Gamma^{(0)}$, then $\text{pos}(t) = \{\varepsilon\}$, and (ii) if $t = a(s_1, \ldots, s_m)$ where $a \in \Gamma^{(m)}$, $m \geq 1$ and $s_1, \ldots, s_m \in T_\Gamma$, then $\text{pos}(t) = \{\varepsilon\} \cup \{iv \mid 1 \leq i \leq m,\ v \in \text{pos}(s_i)\}$. Note that pos($t$) is partially ordered by the prefix relation $\leq_p$ and totally ordered with respect to the lexicographic ordering $\leq_l$. We also refer to the elements of pos(t) as *nodes*, to ε as the *root* of t and to prefix-maximal nodes as *leaves.*

Now let $t, s \in T_\Gamma$, $w \in \text{pos}(t)$ and $t = a(s_1, \ldots, s_m)$ for some $a \in \Gamma^{(m)}$ with $m \geq 0$ and $s_1, \ldots, s_m \in T_\Gamma$. The *label of t at w* and the *subtree of t at w*, denoted by $t(w)$ and $t|_w$, respectively, are defined inductively as follows: $t(\varepsilon) = a$ and $t|_\varepsilon = t$, and if $w = iv$ and $1 \leq i \leq m$, then $t(w) = s_i(v)$ and $t|_w = s_i|_v$.

A *commutative semiring* is a tuple $(K, \oplus, \odot, \mathbb{0}, \mathbb{1})$, abbreviated by K, with operations sum $\oplus$ and product $\odot$ and constants $\mathbb{0}$ and $\mathbb{1}$ such that $(K, \oplus, \mathbb{0})$ and $(K, \odot, \mathbb{1})$ are commutative monoids, multiplication distributes over addition, and $k \odot \mathbb{0} = \mathbb{0} \odot k = \mathbb{0}$ for every $k \in K$. In this paper, we only consider commutative semirings. Important examples of semirings are

- the *boolean semiring* $\mathbb{B} = (\{0, 1\}, \vee, \wedge, 0, 1)$ with disjunction $\vee$ and conjunction $\wedge$
- the *semiring of natural numbers* $(\mathbb{N}, +, \cdot, 0, 1)$, abbreviated by $\mathbb{N}$, with the usual addition and multiplication
- the *tropical semiring* $\text{Trop} = (\mathbb{N} \cup \{\infty\}, \min, +, \infty, 0)$ where the sum and the product operations are min and +, respectively, extended to $\mathbb{N} \cup \{\infty\}$ in the usual way.

A *(formal) tree series* is a mapping $S\colon T_\Gamma \to K$. The set of all tree series (over Γ and K) is denoted by $K\langle\langle T_\Gamma \rangle\rangle$. For two tree series $S, T \in K\langle\langle T_\Gamma \rangle\rangle$ and $k \in K$, the sum $S \oplus T$, the *Hadamard product* $S \odot T$, and the product $k \odot S$ are each defined pointwise.

Let $(K, \oplus, \odot, \mathbb{0}, \mathbb{1})$ be a commutative semiring. A *weighted bottom-up finite state tree automaton (short: WTA) over K and Γ* is a tuple $\mathcal{A} = (Q, \Gamma, \mu, \gamma)$ where Q is a finite set (of states), Γ is a ranked alphabet (of input symbols), $\mu\colon \bigcup_{m=0}^{rk(\Gamma)} Q^m \times \Gamma^{(m)} \times Q \to K$ (the weight function) and $\gamma\colon Q \to K$ (the function of final weights). We set $\Delta_\mathcal{A} = \bigcup_{m=0}^{rk(\Gamma)} Q^m \times \Gamma^{(m)} \times Q$. A tuple $(\vec{p}, a, q) \in \Delta_\mathcal{A}$ is called a *transition* and $(\vec{p}, a, q)$ is called *valid* if $\mu(\vec{p}, a, q) \neq \mathbb{0}$. The state q is referred to as the *parent state* of the transition and the states from $\vec{p}$ are referred to as the *child states* of the transition. A state $q \in Q$ is called *final* if $\gamma(q) \neq \mathbb{0}$.

A mapping $r\colon pos(t) \to Q$ is called a *quasi-run of $\mathcal{A}$ on t*. For $t \in T_\Gamma$, a quasi-run r and $w \in \text{pos}(t)$ with $t(w) = a \in \Gamma^{(m)}$, the tuple

$$\mathfrak{t}(r, w) = (r(w1), \ldots, r(wm), a, r(w))$$

is called the *transition with* base point w or *transition at w*. The quasi-run r is called a *(valid) run* if for every $w \in \text{pos}(t)$ the transition $\mathfrak{t}(r, w)$ is valid with respect to $\mathcal{A}$. We call a run r *accepting* if $r(\varepsilon)$ is final. If $r(\varepsilon) = q$ then a run r is also called a *q-run*. By $\mathrm{Run}_{\mathcal{A}}(t)$, $\mathrm{Run}_{\mathcal{A},q}(t)$, $\mathrm{Run}_{\mathcal{A},\mathbb{F}}(t)$ we denote the sets of all runs, all q-runs and all accepting runs of $\mathcal{A}$ on t, respectively.

For $r \in \mathrm{Run}_{\mathcal{A}}(t)$ the *weight of* r is defined by

$$\mathrm{wt}_{\mathcal{A}}(t, r) = \bigodot_{w \in \mathrm{pos}(t)} \mu(\mathtt{t}(r, w)).$$

The *tree series accepted by* $\mathcal{A}$, denoted by $[\![\mathcal{A}]\!] \in K\langle\!\langle T_\Gamma \rangle\!\rangle$, is the tree series defined for every $t \in T_\Gamma$ by $[\![\mathcal{A}]\!](t) = \bigoplus_{r \in \mathrm{Run}_{\mathcal{A},\mathbb{F}}(t)} \mathrm{wt}_{\mathcal{A}}(t, r) \odot \gamma(r(\varepsilon))$ where the sum over the empty set is $\mathbb{0}$ by convention.

An automaton $\mathcal{A}$ is called *trim* if (i) for every $q \in Q$ there exist $t \in T_\Gamma$, $r \in \mathrm{Run}_{\mathcal{A},\mathbb{F}}(t)$ and $w \in \mathrm{pos}(t)$ such that $q = r(w)$ and (ii) for every valid $d \in \Delta_{\mathcal{A}}$ there exist $t \in T_\Gamma$, $r \in \mathrm{Run}_{\mathcal{A},\mathbb{F}}(t)$ and $w \in \mathrm{pos}(t)$ such that $d = \mathtt{t}(r, w)$. The *trim part of* $\mathcal{A}$ is the automaton obtained by removing all states $q \in Q$ which do not satisfy (i) and setting $\mu(d) = \mathbb{0}$ for all valid $d \in \Delta_{\mathcal{A}}$ which do not satisfy (ii). This process obviously has no influence on $[\![\mathcal{A}]\!]$.

An automaton $\mathcal{A}$ is called *deterministic* if for every $m \geq 0$, $a \in \Gamma^{(m)}$ and $\vec{p} \in Q^m$ there exists at most one $q \in Q$ with $\mu(\vec{p}, a, q) \neq \mathbb{0}$. We call $\mathcal{A}$ *(k-)polynomially ambiguous* if $|\mathrm{Run}_{\mathcal{A},\mathbb{F}}(t)| \leq P(|\mathrm{pos}(t)|)$ for some polynomial P (of degree k) and every $t \in T_\Gamma$. If P can be chosen constant, i.e. $P \equiv m$, we call $\mathcal{A}$ *finitely ambiguous* or *m-ambiguous*. If we can put $P \equiv 1$, we call $\mathcal{A}$ *unambiguous*.

Example 1. We consider the alphabet $\Gamma = \{a, b\}$ where $rk_\Gamma(a) = 2$ and $rk_\Gamma(b) = 0$. Over the tropical semiring $(\mathbb{N} \cup \{\infty\}, \min, +, \infty, 0)$ we construct a WTA $\mathcal{A} = (Q, \Gamma, \mu, \gamma)$ with the following idea in mind. Given a tree $t \in T_\Gamma$, there should be exactly one run of $\mathcal{A}$ on t for every leaf b in t, given by mapping all nodes between this leaf and the root to a state q and all other nodes to a filler state p.

We let $Q = \{p, q\}$ and set $\gamma(q) = 0$, $\gamma(p) = \infty$,

$$1 = \mu(p, q, a, q) = \mu(q, p, a, q)$$
$$0 = \mu(p, p, a, p) = \mu(b, p) = \mu(b, q)$$

and all other weights to ∞. It is easy to see that this automaton assigns to every tree the minimum amount of a's we have to visit to reach any leaf b starting from the root. As there is a bijection between the runs of $\mathcal{A}$ on a tree t and the leaves of t, $\mathcal{A}$ is polynomially ambiguous, but not finitely ambiguous.

3 Finite Ambiguity

We come to our first main result, namely that a finitely ambiguous WTA can be written as a sum of unambiguous WTA.

Theorem 2. *Let $\mathcal{A} = (Q, \Gamma, \mu, \gamma)$ be a finitely ambiguous weighted bottom-up finite state tree automaton. Then there exist finitely many unambiguous weighted bottom-up finite state tree automata $\mathcal{A}_1, \ldots, \mathcal{A}_n$ satisfying*

$$[\![\mathcal{A}]\!] = [\![\mathcal{A}_1]\!] \oplus \ldots \oplus [\![\mathcal{A}_n]\!].$$

While the basic idea for the proof is taken from [15, Sect. 4], we have to follow a different line of argumentation due to the non-linear structure of trees. In the first step, we add a deterministic coordinate to our automaton. On the transitions of this new automaton we then define an equivalence relation. Here, two transitions will be equivalent in the following sense. If a run r on a tree t has transition d at some position w, then for every transition d' equivalent to d we can modify r on the subtree at w such that we obtain a new run with transition d' at w. It follows from this that every transition whose equivalence class contains at least two transitions can not occur more than m times in any single run, if $\mathcal{A}$ is m-ambiguous. This contrasts to the word case, where such transitions could occur at most once per run instead of at most m times. For two different runs on the same tree, sorting all transitions occurring in each run first by equivalence class and then lexicographically, shows a difference for at least one equivalence class. This property is the key to the decomposition.

For Γ and m fixed, the number n is exponential in the number of states.

4 Polynomial Ambiguity

We now come to the tree series definable by polynomially ambiguous WTA. Given a polynomially ambiguous WTA $\mathcal{A}$ we define the function $\mathtt{r}_{\mathcal{A}} : \mathbb{N} \to \mathbb{N}$ that counts the maximum number of possible runs for all trees with a limited number of nodes, i.e. $\mathtt{r}_{\mathcal{A}}(n) = \max\{|\mathrm{Run}_{\mathcal{A},\mathbb{F}}(t)| \mid t \in T_\Gamma, |\mathrm{pos}(t)| \leq n\}$. We then define the *degree of polynomial ambiguity of* $\mathcal{A}$ by

$$\begin{aligned}\mathrm{degree}(\mathcal{A}) &= \min\{k \in \mathbb{N} \mid \mathcal{A} \text{ is } k\text{-polynomially ambiguous}\} \\ &= \min\{k \in \mathbb{N} \mid \mathtt{r}_{\mathcal{A}} \in \mathcal{O}(n^k)\}.\end{aligned}$$

This is well defined if $\mathcal{A}$ is polynomially ambiguous.

We will show that the runs of a polynomially ambiguous WTA have a very characteristic structure. Consequently, this structure naturally induces a sort of standard form for polynomially ambiguous WTA. For automata in this standard form it is then much easier to grasp the fundamental principle of polynomial ambiguity for tree automata. A first basic tool we will need for all of this is a form of reachability between states. The second is the degree of a state. For notational purposes we also need a more elaborate concept for runs.

4.1 General Definitions and Observations

For now let $\mathcal{A} = (Q, \Gamma, \mu, \gamma)$ be a polynomially ambiguous WTA. The sets $\mathrm{Run}_{\mathcal{A}}(t; \vec{w}, \vec{q})$ and $\mathrm{Run}_{\mathcal{A}}(t; \vec{w}, \vec{d})$ shall denote the sets of all runs of $\mathcal{A}$ on a tree t such that at the positions $w_1, \ldots, w_n$ we have the states $q_1, \ldots, q_n$ or transitions $d_1, \ldots, d_n$, respectively.

Definition 3. *Let* $t \in T_\Gamma$, $\vec{w} = (w_1, \ldots, w_n) \in \mathrm{pos}(t)^n$, $\vec{q} = (q_1, \ldots, q_n) \in Q^n$ *and* $\vec{d} = (d_1, \ldots, d_n) \in \Delta_{\mathcal{A}}^n$. *Then we let*

$$\mathrm{Run}_{\mathcal{A}}(t; \vec{w}, \vec{q}) = \{r \in \mathrm{Run}_{\mathcal{A}}(t) \mid r(w_i) = q_i \text{ for all } i = 1, \ldots, n\}$$

$$\mathrm{Run}_{\mathcal{A}}(t; \vec{w}, \vec{d}) = \{r \in \mathrm{Run}_{\mathcal{A}}(t) \mid \mathfrak{t}(r, w_i) = d_i \textit{ for all } i = 1, \ldots, n\}.$$

The sets $\mathrm{Run}_{\mathcal{A},\mathbb{F}}(t; \vec{w}, \vec{q})$, $\mathrm{Run}_{\mathcal{A},q}(t; \vec{w}, \vec{q})$, $\mathrm{Run}_{\mathcal{A},\mathbb{F}}(t; \vec{w}, \vec{d})$ and $\mathrm{Run}_{\mathcal{A},q}(t; \vec{w}, \vec{d})$ for $q \in Q$ are defined in a similar manner to these and $\mathrm{Run}_{\mathcal{A},\mathbb{F}}(t)$ and $\mathrm{Run}_{\mathcal{A},q}(t)$.

We define the concept of reachability through a relation $\preccurlyeq$. Intuitively, $q_1 \preccurlyeq q_2$ means that there is a "path" from q_1 down to q_2.

Definition 4. *We define two relations $\preccurlyeq$ and $\approx$ on Q by letting*

$$q_1 \preccurlyeq q_2 \Leftrightarrow \exists t \in T_\Gamma \; \exists w \in \mathrm{pos}(t) : \mathrm{Run}_{\mathcal{A},q_1}(t; w, q_2) \neq \emptyset$$
$$q_1 \approx q_2 \Leftrightarrow q_1 \preccurlyeq q_2 \wedge q_2 \preccurlyeq q_1.$$

The relation $\preccurlyeq$ is reflexive and transitive. Hence, the relation $\approx$ is an equivalence relation inducing equivalence classes $[q]_\approx \in Q_{/\approx}$. One may think of the classes as strongly connected components of states. We set $\mathfrak{C}(q) = [q]_\approx$ and $\mathfrak{Q} = Q_{/\approx}$ and refer to $\mathfrak{C}(q)$ as the *component of* q and to $\mathfrak{Q}$ as the *components of* Q. Then again, $\preccurlyeq$ induces a partial order $\preccurlyeq$ on $\mathfrak{Q}$, defined by $\mathfrak{C}(q_1) \preccurlyeq \mathfrak{C}(q_2) \Leftrightarrow q_1 \preccurlyeq q_2$.

We also need the notion of a *bridge*, similar to the one used in [24]. A bridge is basically a transition which, from a top-down perspective, leaves a component of Q.

Definition 5. *A valid transition $\mathfrak{b} = (p_1, \ldots, p_m, a, q) \in \Delta_{\mathcal{A}}$ is called a* bridge out of $\mathfrak{C}(q)$ *if $\mathfrak{C}(p_i) \neq \mathfrak{C}(q)$ for all $i \in \{1, \ldots, m\}$. Notice that all valid transitions of the form (a, q) with $a \in \Gamma^{(0)}$ and $q \in Q$ are bridges.*

We now define the degree of a state as the degree of the automaton resulting, intuitively, from making this state the only new final state of $\mathcal{A} = (Q, \Gamma, \mu, \gamma)$.

Definition 6. *For every $p \in Q$ we define the WTA $\mathcal{F}_p = (Q, \Gamma, \mu, \gamma_p)$ with $\gamma_p(p) = \mathbb{1}$ and $\gamma_p(q) = \mathbb{0}$ for $q \in Q$, $q \neq p$.*

The intuition is that for $t \in T_\Gamma$ the accepting runs of the automaton $\mathcal{F}_p$ on t are exactly the p-runs of $\mathcal{A}$ on t, i.e. the ones that "begin" with p at the root.

Definition 7. *For a state $p \in Q$ we define* $\mathrm{degree}_{\mathcal{A}}(p) = \mathrm{degree}(\mathcal{F}_p)$ *and we define* $\mathrm{degree}_{\mathcal{A}}(\mathfrak{C}(p)) = \mathrm{degree}_{\mathcal{A}}(p)$. *We will simply write* $\mathrm{degree}(p)$ *and* $\mathrm{degree}(\mathfrak{C}(p))$ *if the automaton $\mathcal{A}$ considered is clear from the context.*

This is well defined, as for $p \approx q$ one can show that $\mathrm{degree}_{\mathcal{A}}(p) = \mathrm{degree}_{\mathcal{A}}(q)$.

It is now easy to show that every valid transition with a parent state q of degree greater than 0 is either (i) a bridge or (ii) exactly one child state belongs to the component of q and all other child states have degree 0. Applying this to a given run r on a tree $t \in T_\Gamma$, we see that states of degree greater than 0 follow branches in the tree. More formally, for $w \in \mathrm{pos}(t)$ with $\mathrm{degree}(t(w)) > 0$ we have $\{v \in \mathrm{pos}(t) \mid w \leq_p v \wedge r(v) \approx r(w)\} = \{v \in \mathrm{pos}(t) \mid w \leq_p v \leq_p w'\}$ for some $w' \in \mathrm{pos}(t)$.

However, for a given component $\mathfrak{c} \in \Omega$ it may still be possible to find a tree t and a run r on t where for two prefix independent positions w and w' we have $r(w) \in \mathfrak{c}$ and $r(w') \in \mathfrak{c}$, or where $r(w) \in \mathfrak{c}$ holds for no position w. For a WTA in *polynomial standard form*, both of these possibilities will be ruled out: for every component $\mathfrak{c}$ it holds that $\{v \in \text{pos}(t) \mid r(v) \in \mathfrak{c}\} = \{v \in \text{pos}(t) \mid w_1 \leq_p v \leq_p w_2\}$ for some $w_1, w_2 \in \text{pos}(t)$, and this set is non-empty.

4.2 Decomposition into a Sum of Standardized Automata

Definition 8. *We call a (polynomially ambiguous) WTA* $\mathcal{A} = (Q, \Gamma, \mu, \gamma)$ standardized *or say it is in* polynomial standard form *if*

(i) *$\mathcal{A}$ is polynomially ambiguous, trim and possesses only one final state $q_f \in Q$ and*

(ii) *for every $p \in Q$ with $\text{degree}_{\mathcal{A}}(p) > 0$ there is exactly one bridge out of $\mathfrak{C}(p)$ and this bridge occurs exactly once in every accepting run r. Formally*

$$\{d \in \Delta_{\mathcal{A}} \mid d \textit{ is a bridge out of } \mathfrak{C}(p)\} = \{\mathfrak{b}(p)\}$$

for some $\mathfrak{b}(p) \in \Delta_{\mathcal{A}}$ and

$$\forall t \in T_\Gamma \; \forall r \in \text{Run}_{\mathcal{A},\mathbb{F}}(t) : |\{w \in \text{pos}(t) \mid \mathfrak{t}(r, w) = \mathfrak{b}(p)\}| = 1.$$

The fundamental concept of standardized WTA is close to the notion of *chain NFAs* as introduced in [24].

Theorem 9. *Let $\mathcal{A} = (Q, \Gamma, \mu, \gamma)$ be a polynomially ambiguous WTA. Then there exist $n \in \mathbb{N}$ and WTA $\mathcal{A}_1, \ldots, \mathcal{A}_n$ in polynomial standard form such that* $\text{degree}(\mathcal{A}_i) \leq \text{degree}(\mathcal{A})$ *for all $i \in \{1, \ldots, n\}$ and*

$$[\![\mathcal{A}]\!] = \bigoplus_{i=1}^{n} [\![\mathcal{A}_i]\!].$$

For a fixed alphabet the number n of automata needed for this is double exponential in the number of states.

Example 10. The WTA from Example 1 is in polynomial standard form. There are two components, $\{p\}$ and $\{q\}$, and the transitions (b, p) and (b, q) are the only bridges. We have $\text{degree}(p) = 0$ and $\text{degree}(q) = 1$.

Proof. (sketch) The theorem is proved in two steps. In the first, we add an entry containing words of bounded length over $\{1, \ldots, rk(\Gamma)\}$ to the states of $\mathcal{A}$. For any such word u and any bridge $(p_1, \ldots, p_m, a, p)$ in $\mathcal{A}$, we will then have a transition $((p_1, u1), \ldots, (p_m, um), a, (p, u))$ in the new automaton $\mathcal{A}'$. For the other transitions, we do the same with the difference that the child states will contain the same word as the parent state.

In the second step, we make copies of $\mathcal{A}'$ and "remove" bridges in the copies appropriately, i.e. we leave at most one bridge out of each component and then trim the automata. The modified copies then fulfill Theorem 9.

4.3 Analysis of the Polynomial Standard Form

Now assume a WTA $\mathcal{A} = (Q, \Gamma, \mu, \gamma)$ in polynomial standard form. We can show that there exist degree($\mathcal{A}$) many bridges in $\mathcal{A}$ such that, given any tree, the number of runs on that tree is bounded by a constant if we fix the position of these bridges. The constant does not depend on the given tree. This property gives a rather intuitive understanding of what polynomial ambiguity means: if our automaton has degree n, then fixing the positions of n predetermined transitions will determine every run up to a bounded number of possibilities.

We consider the set Λ of all bridges that leave components of non-trivial degree, defined as follows.

Definition 11. *Fix $p \in Q$ with $\text{degree}_{\mathcal{A}}(p) > 0$. As there is exactly one bridge $\mathfrak{b} \in \Delta_{\mathcal{A}}$ out of $\mathfrak{C}(p)$ we define $\mathfrak{b}(\mathfrak{C}(p)) = \mathfrak{b}$ and $\mathfrak{b}(p) = \mathfrak{b}$ as this bridge. We set $\Lambda = \{\mathfrak{b}(q) \mid q \in Q,\ \text{degree}_{\mathcal{A}}(q) > 0\}$.*

The degree inherent to an automaton in standard form can now be captured in the following way.

Theorem 12. *Let $p \in Q$ with $l = \text{degree}_{\mathcal{A}}(p) \geq 0$.*

(I) *There exists a set $\mathfrak{N}(p) = \{\mathfrak{b}_1, \ldots, \mathfrak{b}_l\} \subseteq \Lambda$ and a constant $C > 0$ such that for all $t \in T_\Gamma$ and $w_1, \ldots, w_l \in \text{pos}(t)$ we have*

$$|\text{Run}_{\mathcal{A},p}(t; w_1, \ldots, w_l, \mathfrak{b}_1, \ldots, \mathfrak{b}_l)| \leq C.$$

(II) *Furthermore there exists a sequence of trees $(t_n)_{n \in \mathbb{N}}$ in T_Γ and a constant $C' > 0$ such that for all $n \in \mathbb{N}$:*
- *$|\text{pos}(t_n)| \leq C' \cdot n$ and*
- *$|\text{Run}_{\mathcal{A},p}(t_n)| \geq n^l$.*

That is, we can show that if the WTA $\mathcal{F}_p$ (cf. Definition 6) is of degree l, then for all trees the runs of $\mathcal{F}_p$ on those trees are determined up to a constant C by fixing the location of l bridges. Furthermore, the degree of $\mathcal{F}_p$ is not only an upper bound on the amount of runs for a given tree, but also a lower bound. By considering this theorem for the only final state of $\mathcal{A}$, we easily see that it is true for the whole automaton $\mathcal{A}$ as well.

Example 13. In the WTA from Example 1 we have $\mathfrak{N}(p) = \emptyset$ and $\mathfrak{N}(q) = \{(b, q)\}$. By choosing which leaf to "mark" with q, we uniquely determine a run. Therefore, (I) clearly holds for both p and q in this automaton.

For (II) in the case of q we consider the trees $t_0 = b()$ and $t_{n+1} = a(t_n, b)$ for $n \geq 0$. We have $|\text{pos}(t_n)| = 2n + 1$ and one run for every leaf in a tree, i.e. $|\text{Run}_{\mathcal{A},q}(t_n)| = n + 1$.

As a corollary of Theorem 12 we also get that the ambiguity of a WTA $\mathcal{A}$ is either bounded below and above by a fixed polynomial or has a lower exponential bound. While this is a well known result for word automata [24], we could not find a similar result for tree automata in the literature.

Corollary 14. *Let $\mathcal{A} = (Q, \Gamma, \mu, \gamma)$ be a weighted bottom-up finite state tree automaton. Either $\mathcal{A}$ is polynomially ambiguous and $\mathtt{r}_{\mathcal{A}} \in \Theta(n^k)$ for $k =$ degree$(\mathcal{A})$ or there exists a sequence of trees $(t_n)_{n\in\mathbb{N}}$ in T_Γ and a constant $C > 0$ such that for all $n \in \mathbb{N}$ **(i)** $|\mathrm{pos}(t_n)| \leq C \cdot n$ and **(ii)** $|\mathrm{Run}_{\mathcal{A},\mathbb{F}}(t_n)| \geq 2^n$.*

5 Application: Weighted Logics

As stated in the introduction, our investigations were part of the attempt to characterize weighted tree automata with weighted logics. Therefore, we briefly outline how our weighted logic works, which results we obtained with it and what the significance of our investigations is to these results. For further details see [8,10,20].

The standard MSO-logic for trees is given by the following grammar.

$$\varphi ::= \mathrm{label}_a(x) \mid \mathrm{edge}_i(x,y) \mid x \in X \mid \neg\varphi \mid \varphi \vee \varphi \mid \exists x.\varphi \mid \exists X.\varphi$$

where $a \in \Gamma$, x, y are first order variables, $1 \leq i \leq rk(\Gamma)$, and X is a second order variable. The set of free variables of φ is denoted by Free(φ). Let $t \in T_\Gamma$ be a tree and $\mathcal{V}$ be a set of first order and second order variables with Free$(\varphi) \subseteq \mathcal{V}$. A mapping which assigns to every first order variable $x \in \mathcal{V}$ a position $w \in \mathrm{pos}(t)$ and to every second order variable $X \in \mathcal{V}$ a set of positions $I \subseteq \mathrm{pos}(t)$ is called a $(\mathcal{V}, t)$*-assignment*. For a first order variable $x \in \mathcal{V}$ and a position $w \in \mathrm{pos}(t)$ we write $\rho[x \to w]$ to denote the $(\mathcal{V} \cup \{x\}, t)$-assignment given by $\rho[x \to w](x) = w$ and $\rho[x \to w](y) = \rho(y)$ for all variables $y \neq x$. The assignment $\rho[X \to I]$, where X is a second-order variable and $I \subseteq \mathrm{pos}(t)$, is defined analogously. We write $(t, \rho) \models \varphi$ if (t, ρ) satisfies φ using standard MSO-semantics. We then have the generalization of Büchi's and Elgot's fundamental theorems [5,11] to trees, namely that MSO-definable tree languages are exactly the recognizable tree languages [7,25].

On top of the MSO-logic we construct a weighted logic, called wMSO-logic, with the following grammar.

$$\theta ::= \varphi \mid k \mid \theta \oplus \theta \mid \theta \odot \theta \mid \Sigma x.\theta \mid \Sigma X.\theta \mid \Pi x.\theta$$

where $\varphi \in \mathrm{MSO}(\Gamma)$, $k \in K$, x is a first order variable and X is a second order variable. The operators Σx and ΣX are referred to as first order sum quantification and second order sum quantification, respectively, and Πx is referred to as (first order) product quantification. Moreover, the operators Σx, ΣX and Πx also bind the variables x and X, respectively. A wMSO-formula without free variables is also called a *sentence*.

For a formula $\theta \in \mathrm{wMSO}(\Gamma)$, a tree $t \in T_\Gamma$, a set $\mathcal{V}$ of first and second order variables with Free$(\theta) \subseteq \mathcal{V}$ and a $(\mathcal{V}, t)$-assignment ρ we define the value $[\![\theta]\!](t, \rho)$ inductively in the following way.

$$[\![\theta]\!](t,\rho) = \begin{cases} \mathbb{1} & \text{if } (t,\rho) \models \theta \\ \mathbb{0} & \text{otherwise} \end{cases} \quad \text{for } \theta \in \mathrm{MSO}(\Gamma)$$

$$\begin{aligned}
[\![k]\!](t,\rho) &= k \\
[\![\theta_1 \oplus \theta_2]\!](t,\rho) &= [\![\theta_1]\!](t,\rho) \oplus [\![\theta_2]\!](t,\rho) \\
[\![\theta_1 \odot \theta_2]\!](t,\rho) &= [\![\theta_1]\!](t,\rho) \odot [\![\theta_2]\!](t,\rho) \\
[\![\Sigma x.\tau]\!](t,\rho) &= \bigoplus_{w \in \mathrm{pos}(t)} [\![\tau]\!](t,\rho[x \to w]) \\
[\![\Pi x.\tau]\!](t,\rho) &= \bigodot_{w \in \mathrm{pos}(t)} [\![\tau]\!](t,\rho[x \to w]) \\
[\![\Sigma X.\tau]\!](t,\rho) &= \bigoplus_{I \subseteq \mathrm{pos}(t)} [\![\tau]\!](t,\rho[X \to I])
\end{aligned}$$

where $k \in K$ and $\theta_1, \theta_2, \tau \in \mathrm{wMSO}(\Gamma)$.

Example 15. We consider the semiring $(\mathbb{N}, +, \cdot, 0, 1)$ and the alphabet $\Gamma = \{a, b\}$ where $rk_\Gamma(a) = 2$ and $rk_\Gamma(b) = 0$. The following formula outputs for every $t \in T_\Gamma$ the amount of a's taking two b's as child nodes.

$$\Sigma x.\Big(\mathrm{label}_a(x) \wedge \exists y.(\mathrm{edge}_1(x,y) \wedge \mathrm{label}_b(y)) \wedge \exists y.(\mathrm{edge}_2(x,y) \wedge \mathrm{label}_b(y))\Big)$$

In order to characterize different degrees of ambiguity, we use restrictions of above logic. The formulas given by the grammar

$$\theta_b ::= \varphi \mid k \mid \theta_b \oplus \theta_b \mid \theta_b \odot \theta_b$$

with $\varphi \in \mathrm{MSO}(\Gamma)$ and $k \in K$ are called *almost boolean* and define so-called *recognizable step functions* [8,10]. We call a formula *unambiguous* if it is almost boolean, a product quantifier followed by an almost boolean formula or a finite product of such formulas. A formula containing no sum quantifiers, and in which for every subformula $\Pi.\theta$ the formula θ is almost boolean, is called *finitely ambiguous.* This class of formulas is actually the closure of unambiguous formulas under $\oplus$ and $\odot$. Finally, a formula is called *polynomially ambiguous* if it does not contain second order sum quantification and for every subformula $\Pi.\theta$ the formula θ is almost boolean. We have the following theorem.

Theorem 16. *The following classes of automata and sets of sentences are expressively equivalent:*

(a) *unambiguous WTA and unambiguous sentences*
(b) *finitely ambiguous WTA and finitely ambiguous sentences*
(c) *polynomially ambiguous WTA of polynomial degree k and polynomially ambiguous formulas with first order sum quantifier depth k.*

Example 17. The WTA from Example 1, calculating the minimum amount of a's between the root and any leaf, is described by the formula

$$\Sigma x.\Pi y.\left(\mathrm{label}_b(x) \odot ((1 \odot (\mathrm{label}_a(y) \wedge y \leq_p x)) \oplus \neg(\mathrm{label}_a(y) \wedge y \leq_p x))\right).$$

The prefix relation is MSO-definable [6, Sect. 3.3].

A result similar to Theorem 16 has been shown by Kreutzer and Riveros [16] to hold true for weighted automata over words. Most of their proofs can easily be adapted to work for tree automata, but not all. To be precise, we need the results of Sects. 3 and 4 for the translation of automata to logics in (b) and (c). For polynomial ambiguity, we even obtain a stronger result, as we are able to capture polynomial degree not only in the boolean semiring, but in any commutative semiring. For this, we show by induction on the polynomial degree that for a WTA in polynomial standard form, first order sum quantifiers can be used to sum over all possible positions for the bridges identified in Theorem 12 (I). Having specified the positions of all these bridges, we are then essentially in the case of finite ambiguity, and can apply (b). The number of first order sum quantifiers needed to describe the WTA with a wMSO-formula hence equals its polynomial degree.

6 Conclusion

As shown, our results about the structure of weighted tree automata have proven to be useful in the context of weighted logics for trees. Two questions now arise. First, which other problems could be tackled with the newly gained knowledge? Decidability problems for WTA are an obvious candidate here. Second, can we get similar results for other automata models? For example, one might intuitively assume picture automata and graph acceptors to behave in a similar manner, but this is in no way obvious and calls for further investigation.

Acknowledgements. I would like to thank Professor Manfred Droste, Peter Leupold and Vitaly Perevoshchikov for helpful discussions and suggestions.

References

1. Allauzen, C., Mohri, M., Rastogi, A.: General algorithms for testing the ambiguity of finite automata. In: Ito, M., Toyama, M. (eds.) DLT 2008. LNCS, vol. 5257, pp. 108–120. Springer, Heidelberg (2008)
2. Berstel, J., Reutenauer, C.: Recognizable formal power series on trees. Theor. Comput. Sci. **18**(2), 115–148 (1982)
3. Berstel, J., Reutenauer, C.: Rational Series and Their Languages. Monographs in Theoretical Computer Science. An EATCS Series, vol. 12. Springer, Heidelberg (1988)
4. Bozapalidis, S.: Effective construction of the syntactic algebra of a recognizable series on trees. Acta Informatica **28**(4), 351–363 (1991)
5. Büchi, J.: Weak second-order arithmetic and finite automata. Z. Math. Logik und Grundl. Math. **6**, 66–92 (1960)
6. Comon, H., Dauchet, M., Gilleron, R., Löding, C., Jacquemard, F., Lugiez, D., Tison, S., Tommasi, M.: Tree automata techniques and applications (2007). http://www.grappa.univ-lille3.fr/tata
7. Doner, J.: Tree acceptors and some of their applications. J. Comput. Syst. Sci. **4**(5), 406–451 (1970)

8. Droste, M., Gastin, P.: Weighted automata and weighted logics. Theor. Comput. Sci. **380**(1–2), 69–86 (2007)
9. Droste, M., Kuich, W., Vogler, H. (eds.): Handbook of Weighted Automata. Monographs in Theoretical Computer Science. An EATCS Series. Springer, Heidelberg (2009)
10. Droste, M., Vogler, H.: Weighted tree automata and weighted logics. Theor. Comput. Sci. **366**(3), 228–247 (2006)
11. Elgot, C.: Decision problems of finite automata design and related arithmetics. Trans. Am. Math. Soc. **98**, 21–52 (1961)
12. Fülöp, Z., Vogler, H.: Weighted tree automata and tree transducers. In: Droste, M., Kuich, W., Vogler, H. (eds.) Handbook of Weighted Automata. Monographs in Theoretical Computer Science. An EATCS Series, pp. 313–403. Springer, Heidelberg (2009)
13. Hashiguchi, K., Ishiguro, K.: Decidability of the equivalence problem for finitely ambiguous finance automata. Surikaisekikenkyusho Kokyuroku **960**, 23–36 (1996)
14. Kirsten, D.: A Burnside approach to the termination of Mohri's algorithm for polynomially ambiguous min-plus-automata. RAIRO-Inf. Theor. Appl. **42**(3), 553–581 (2008)
15. Klimann, I., Lombardy, S., Mairesse, J., Prieur, C.: Deciding unambiguity and sequentiality from a finitely ambiguous max-plus automaton. Theor. Comput. Sci. **327**(3), 349–373 (2004)
16. Kreutzer, S., Riveros, C.: Quantitative monadic second-order logic. In: 28th Annual ACM/IEEE Symposium on Logic in Computer Science, LICS, pp. 113–122. IEEE Computer Society (2013)
17. Krob, D.: The equality problem for rational series with multiplicities in the tropical semiring is undecidable. Int. J. Algebra Comput. **04**(03), 405–425 (1994)
18. Kuich, W., Salomaa, A.: Semirings, Automata, Languages. EATCS Monographs on Theoretical Computer Science, vol. 5. Springer, Heidelberg (1986)
19. Maletti, A.: Relating tree series transducers and weighted tree automata. Int. J. Found. Comput. Sci. **16**(4), 723–741 (2005)
20. Rahonis, G.: Weighted Muller tree automata and weighted logics. J. Autom. Lang. Comb. **12**(4), 455–483 (2007)
21. Salomaa, A., Soittola, M.: Automata-Theoretic Aspects of Formal Power Series. Texts and Monographs in Computer Science. Springer, New York (1978)
22. Schützenberger, M.: On the definition of a family of automata. Inf. Control **4**(2–3), 245–270 (1961)
23. Seidl, H.: On the finite degree of ambiguity of finite tree automata. Acta Informatica **26**(6), 527–542 (1989)
24. Seidl, H., Weber, A.: On the degree of ambiguity of finite automata. Theor. Comput. Sci. **88**(2), 325–349 (1991)
25. Thatcher, J., Wright, J.: Generalized finite automata theory with an application to a decision problem of second-order logic. Math. Syst. Theor. **2**(1), 57–81 (1968)

An Extremal Series of Eulerian Synchronizing Automata

Marek Szykuła[1(✉)] and Vojtěch Vorel[2(✉)]

[1] Institute of Computer Science, University of Wrocław,
Joliot-Curie 15, Wrocław, Poland
msz@cs.uni.wroc.pl

[2] Faculty of Mathematics and Physics, Charles University,
Malostranské nám. 25, Prague, Czech Republic
vorel@ktiml.mff.cuni.cz

Abstract. We present an infinite series of n-state Eulerian automata whose reset words have length at least $(n^2 - 3)/2$. This improves the current lower bound on the length of shortest reset words in Eulerian automata. We conjecture that $(n^2 - 3)/2$ also forms an upper bound for this class and we experimentally verify it for small automata by an exhaustive computation.

Keywords: Eulerian automaton · Reset threshold · Reset word · Synchronizing automaton

1 Introduction

A complete deterministic finite automaton is *synchronizing* if there exists a word whose action maps all states to a single one. Such words are called *reset words*. Synchronizing automata find applications in various fields such as robotics, coding theory, bioinformatics, and model-based testing. Besides of these, synchronizing automata are of great theoretical interest, mainly because of the famous Černý conjecture [9], which is one of the most long-standing open problems in automata theory. The conjecture states that each synchronizing n-state automaton has a reset word of length at most $(n-1)^2$. The best known general upper bound on this length is $\frac{1}{6}n^3 - \frac{1}{6}n - 1$ for each $n \geq 4$ [21]. Surveys on the field can be found in [16,26].

Major research directions in this field include proving the Černý conjecture for special classes of automata or showing specific upper bounds for them. For example, the Černý conjecture has been positively solved for the classes of monotonic automata [11], circular automata [10], Eulerian automata [15], aperiodic automata [25], one-cluster automata with a prime-length cycle [24],

M. Szykuła—Supported in part by the National Science Centre, Poland under project number 2015/17/B/ST6/01893.

V. Vorel—Research supported by the Czech Science Foundation grant GA14-10799S and the GAUK grant No. 52215.

S. Brlek and C. Reutenauer (Eds.): DLT 2016, LNCS 9840, pp. 380–392, 2016.
DOI: 10.1007/978-3-662-53132-7_31

automata respecting intervals of a directed graph [12] (under an inductive assumption), and automata with a letter of rank at most $\sqrt[3]{6n-6}$ [6]. Moreover, there are many improvements of upper bounds for important special classes, for example, generalized and weakly monotonic automata [2,27], one-cluster automata [4], quasi-Eulerian and quasi-one-cluster automata [5], and decoders of finite prefix codes [2,6,7]. On the other hand, several lower bounds have been established by showing extremal series of automata for particular classes [2,7,9,13]. Still, for many classes the best known upper bound does not match the lower bound.

In this paper we deal with the class of Eulerian automata, which is one of the most remarkable classes due to its properties with regard to synchronization. In particular, the lengths of shortest words extending subsets are at most $n-1$ for each n-state Eulerian automaton [15], whereas they can be quadratic in general [19]. An upper bound $(n-1)(n-2)+1$ on the length of the shortest reset words for Eulerian automata was obtained by Kari [15]. Several generalizations of Eulerian automata were proposed: the class of pseudo-Eulerian automata [23], for which the same bound $(n-1)(n-2)+1$ was obtained, unambiguous Eulerian automata [8] for which the Černý bound $(n-1)^2$ was obtained, and quasi-Eulerian automata [5], for which a quadratic upper bound was obtained. The best lower bound so far was $\frac{1}{2}n^2 - \frac{3}{2}n + 2$, found by Gusev [13]. A series whose shortest reset words seem to have length $\frac{1}{2}n^2 - \frac{5}{2}$ was found by Martyugin (unpublished), but no proof has been established. Further discussion on the bounds for Eulerian automata can be found in the survey [16].

Here we improve the lower bound by introducing an extremal series of Eulerian automata over a quaternary alphabet with the shortest reset words of length $\frac{1}{2}n^2 - \frac{3}{2}$. To prove that, we use a technique of *backward tracing*, which turns out to be very useful in analysis of extremal series of automata in general. We conjecture that the new lower bound is tight for the class of Eulerian automata. Our exhaustive search over small automata did not find any counterexample.

The new series exhibits the extremal property that some of its subsets require extending words of length exactly $n-1$. This matches the upper bound, which was used in [15] to obtain the best known upper bound $(n-1)(n-2)+1$ on the length of shortest reset words. Thus, possible improvements of the upper bound require a more subtle method.

2 Preliminaries

A *deterministic finite automaton* (*DFA*) is a triple $\mathcal{A} = (Q, \Sigma, \delta)$, where Q is a finite non-empty set of *states*, Σ is a finite non-empty *alphabet*, and $\delta \colon Q \times \Sigma \mapsto Q$ is a complete *transition function*. We extend δ to $Q \times \Sigma^*$ and $2^Q \times \Sigma^*$ as usual. When $\mathcal{A}$ is fixed, we write shortly $q \cdot w$ and $S \cdot w$ for $\delta(q, w)$ and $\delta(S, w)$ respectively. The *preimage* of $S \subset Q$ by $w \in \Sigma^*$ is defined as

$$\delta^{-1}(S, w) = \{q \in Q \mid q \cdot w \in S\},$$

which is also denoted by $S \cdot w^{-1}$. If $S = \{q\}$ is a singleton, we write $q \cdot w^{-1}$.

A word $w \in \Sigma^*$ is a *reset word* if $|Q \cdot w| = 1$. Note that in this case $Q \cdot w = \{q_0\}$ and $\{q_0\} \cdot w^{-1} = Q$ for some $q_0 \in Q$. A DFA is called *synchronizing* if it admits a reset word. The *reset threshold* of a synchronizing DFA $\mathcal{A}$ is the length of the shortest reset words and is denoted by $\mathrm{rt}(\mathcal{A})$.

A word w *extends* a subset $S \subset Q$ if $|S \cdot w^{-1}| > |S|$. In this case we say that S is *w-extensible.*

A DFA $\mathcal{A}$ is *Eulerian* if the underlying digraph of $\mathcal{A}$ is strongly connected and the in-degree equals the out-degree for each vertex of the underlying digraph. Equivalently, at every vertex there must be exactly $|\Sigma|$ incoming edges.

We say that a word $w \in \Sigma^*$ is:

- *permutational* if $Q \cdot w = Q$,
- *involutory* if $q \cdot w^2 = q$ for each $q \in Q$,
- *unitary* if $p \cdot w \neq p$ holds for exactly one $p \in Q$.

Note that each involutory word is permutational. Also, w is unitary if and only if its action maps exactly one state to another one and fixes all the other states. For $p, r \in Q$, we write $w = (p \to r)$ if the action of $w \in \Sigma^*$ is defined as $p \cdot w = r$ and $q \cdot w = q$ for each $q \in Q \setminus \{p\}$.

The reversal of a word w is denoted by w^{R}.

Lemma 1. *Let $\mathcal{A} = (Q, \Sigma, \delta)$ be a DFA. Let $w \in \Sigma^*$ contain only involutory letters. Then $S \cdot w^{-1} = S \cdot w^{\mathrm{R}}$ for each $S \subseteq Q$.*

Proof. If $|w| = 0$, the claim is trivial. Inductively, let $w = xv$ for $x \in \Sigma$. We have $S \cdot (xv)^{-1} = (S \cdot v^{-1}) \cdot x^{-1} = (S \cdot v^{\mathrm{R}}) \cdot x^{-1}$ by the inductive assumption, which is equal to $(S \cdot v^{\mathrm{R}}) \cdot x^{-1} \cdot x^2 = (S \cdot v^{\mathrm{R}}) \cdot x$ since x is involutory. □

3 Backward Tracing

There exist several methods of proving reset thresholds of particular series of automata. Here we discuss one of them as a general approach, which we call *backward tracing.*

Definition 2. *Let $\mathcal{A}$ be a synchronizing DFA and let u be a reset word for $\mathcal{A}$ with $Q \cdot u = \{q_0\}$. We say that u is* straight *if*

$$q_0 \cdot (u_{\mathbf{m}} u_{\mathbf{s}})^{-1} \nsubseteq q_0 \cdot (u_{\mathbf{s}})^{-1}$$

for each $u_{\mathbf{p}}, u_{\mathbf{m}}, u_{\mathbf{s}} \in \Sigma^*$ with $u_{\mathbf{p}} u_{\mathbf{m}} u_{\mathbf{s}} = u$.

The following is a simple observation (cf. [17, Theorem 1]):

Proposition 3. *In a synchronizing DFA each shortest reset word is straight.*

The observation above leads to a method of proving reset thresholds of particular DFA series by analyzing subsets that are preimages of a singleton under the action of suffixes of length $i = 1, 2, \ldots, \mathrm{rt}(\mathcal{A})$ of straight reset words. This works well if the number of such subsets is small in every step, i.e., for each i. Note that in general it can grow exponentially.

Interestingly, all known series of most extremal automata, such as the Černý automata having reset threshold $(n-1)^2$ [9], the twelve known *slowly synchronizing series* having only slightly smaller reset thresholds [1,3,19], and DFAs with cycles of two different lengths [14], have the property that the number of possible subsets in each step is bounded by a constant. We call such series *backward tractable*. It is worth mentioning that for such automata we can compute shortest reset words in polynomial time [17].

In this paper, we apply this method to a new series of Eulerian automata, which is backward tractable as well, and whose construction is different from the other known extremal series; in particular, the letters act in many short cycles instead of few large ones.

The new DFAs use only permutational and unitary letters. This property (which also implies an upper bound $2(n-1)^2$ on the reset threshold [22]) allows us to strengthen the restriction on suffixes to be considered within the backward tracing:

Definition 4. *With respect to a fixed DFA $\mathcal{A}$, a reset word $u \in \Sigma^*$ with $Q \cdot u = \{q_0\}$ is* greedy, *if for each suffix v of u it holds that: if some $x \in \Sigma$ extends $q_0 \cdot v^{-1}$, then yv is a suffix of u for some $y \in \Sigma$ that extends $q_0 \cdot v^{-1}$.*

Lemma 5. *If a synchronizing DFA $\mathcal{A}$ has only permutational and unitary letters, then there exists a shortest reset word that is greedy.*

Proof. Let $\Sigma = \Sigma_{\mathrm{p}} \cup \Sigma_{\mathrm{u}}$, where Σ_{p} contains permutational letters and Σ_{u} contains unitary letters.

Suppose for a contradiction that there is no shortest reset word that is greedy. Let u be a shortest reset word of $\mathcal{A}$ with the property that its shortest suffix v violating the greediness is the longest possible. In other words, the shortest suffix yv of u, $y \in \Sigma$, such that some $x \in \Sigma$ extends $q_0 \cdot v^{-1}$ but y doesn't, is the longest possible.

For each suffix zt of u with $z = (p \to q) \in \Sigma_{\mathrm{u}}$, the set $S = q_0 \cdot t^{-1}$ is necessarily z-extensible. Indeed, if $q \in S$ and $p \notin S$, then S is clearly z-extensible. If $q \notin S$ or $p \in S$, then $S \cdot z^{-1} \subseteq S$, which contradicts Proposition 3. Since the inverse actions of the letters from Σ_{p} preserve sizes of subsets, it follows that u contains exactly $|Q| - 1$ occurrences of unitary letters.

Write $u = v'v$ and let $u' = v'xv$. Observe that u' is also a reset word for $\mathcal{A}$: $q_0 \cdot (xv)^{-1}$ is a (possibly proper) superset of $q_0 \cdot v^{-1}$; hence, we still have $q_0 \cdot (v'xv)^{-1} = Q$. Since u' contains $|Q|$ occurrences of letters from Σ_{u}, and letters from Σ_{p} do not decrease the size of a subset, at least one occurrence of $y \in \Sigma_{\mathrm{u}}$ is not applied to an y-extensible subset. Moreover, this occurrence lies within v', because v is the shortest suffix violating the greediness. Let v'' be the word obtained by removing that occurrence of y. We have $u'' = v''xv$, $|u''| = |u|$, and the shortest suffix violating the greediness is longer than v. This yields a contradiction with the choice of u. □

4 The Extremal Series of Eulerian Automata

Fix an arbitrary $m \geq 1$. Let $N = 4m + 1$ and $\mathcal{A}_m = \langle Q, \Sigma, \delta \rangle$, where $Q = \{0, 1, \ldots, N-1\}$, $\Sigma = \{\alpha, \beta, \omega_0, \omega_1\}$. The action of α and β is defined by

$$q \cdot \alpha = (-q - 1) \bmod N,$$
$$q \cdot \beta = (-q + 1) \bmod N,$$

for $q \in Q$, while the action of ω_0, ω_1 is defined by

$$\omega_0 = (1 \rightarrow 0),$$
$$\omega_1 = (0 \rightarrow 1).$$

The automaton $\mathcal{A}_m$ is illustrated in Fig. 1. We are going to prove that

$$\mathrm{rt}(\mathcal{A}_m) = \frac{N^2 - 3}{2}.$$

Throughout the proof we use usual operations and inequalities on integers. Each use of modular arithmetic is described explicitly using the binary operator "mod".

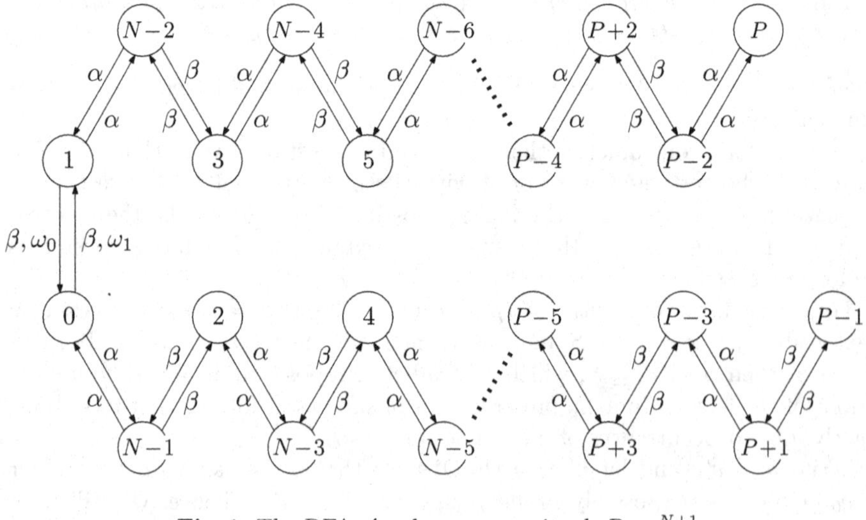

Fig. 1. The DFA $\mathcal{A}_m$, loops are omitted, $P = \frac{N+1}{2}$

We use backward tracing to show that there is a unique optimal way to extend a singleton to Q. Note that ω_0 and ω_1 are unitary, while α and β are

involutory. The following notation will be very useful in the analysis of reset words for $\mathcal{A}_m$:
For $j = 0, \ldots, N$ we set:

$$Q_j = \{q \mid 0 \leq q \leq j-1\}, \qquad R_j = Q_j \cdot \beta,$$
$$Q_j^\diamond = Q_j \setminus \{0\} = \{q \mid 1 \leq q \leq j\}, \quad R_j^\diamond = R_j \setminus \{1\} = Q_j^\diamond \cdot \beta.$$

4.1 Construction of a Reset Word

For an odd $i \geq 1$, we define

$$t_i = \alpha\,(\beta\alpha)^{\frac{i-1}{2}}.$$

Note that:

1. $|t_i| = i$,
2. t_i is a palindrome (i.e., $t_i = t_i^{\mathrm{R}}$).

By Lemma 1, $S \cdot t_i = S \cdot t_i^{\mathrm{R}} = S \cdot t_i^{-1}$ for each $S \subseteq Q$, and we will often interchange t_i^{-1} with t_i. It follows that $qt_i^2 = qt_it_i^{-1} = q$ for every $q \in Q$ and t_i is involutory.

Lemma 6. *Let $q \in Q$. It holds that:*

1. *$q \cdot (\beta\alpha)^h = (q - 2h) \bmod N$ for each $h \geq 0$,*
2. *$q \cdot t_i = (-q - i) \bmod N$ for each $i \geq 1$.*

Proof. The first claim follows trivially from the case of $h = 1$. In this case we have $(q \cdot \beta) \cdot \alpha = (-(-q+1)-1) \bmod N = (q-2) \bmod N$. For the second claim we observe $k \cdot t_i = (k \cdot \alpha) \cdot (\beta\alpha)^{\frac{i-1}{2}}$, which equals $(-q-1-(i-1)) \bmod N = (-q-i) \bmod N$. □

Lemma 7. *Let $2 \leq j \leq N-2$. It holds that:*

1. *$Q_j \cdot t_{N-j} = Q_{j+1}^\diamond$ if j is even,*
2. *$R_j \cdot t_{j-2} = R_{j+1}^\diamond$ if j is odd,*
3. *$Q_{j+1} \cdot t_{N-j} = Q_{j+1}$ if j is even,*
4. *$R_{j+1} \cdot t_{j-2} = R_{j+1}$ if j is odd.*

Proof. For (1) and (2) we use Lemma 6(2) with $i = N-j$ and $i = j-2$ respectively and then substitute $d = j - q$:

$$\begin{aligned} Q_j \cdot t_{N-j} &= \{j - q \mid q \in Q_j\} = \{d \mid 1 \leq d \leq j\} = Q_{j+1}^\diamond, \\ R_j \cdot t_{j-2} &= \{q \cdot \beta t_{j-2} \mid q \in Q_j\} = \{(-(-q+1)-(j-2)) \bmod N \mid q \in Q_j\} \\ &= \{(q-j+1) \bmod N \mid q \in Q_j\} = \{(-d+1) \bmod N \mid 1 \leq d \leq j\} \\ &= \{d \cdot \beta \mid 1 \leq d \leq j\} = R_{j+1}^\diamond. \end{aligned}$$

For (3) and (4) we use (1) and (2) respectively and the fact that t_{N-j} and t_{j-2} are involutory. We have:

$$\begin{aligned} Q_{j+1} \cdot t_{N-j} &= Q_{j+1}^\diamond \cdot t_{N-j} \cup \{0 \cdot t_{N-j}\} = Q_j \cup \{j\} = Q_{j+1}, \\ R_{j+1} \cdot t_{j-2} &= R_{j+1}^\diamond \cdot t_{j-2} \cup \{1 \cdot t_{j-2}\} = R_j \cup \{N-j+1\} \\ &= R_j \cup \{j \cdot \beta\} = R_{j+1}. \end{aligned}$$

□

Let

$$w = v_{N-1}\beta v_{N-2}\beta v_{N-3}\dots\beta v_3\beta v_2,$$

where

$$v_j = \begin{cases} \omega_1 t_{N-j} & \text{if } j \text{ is even,} \\ \omega_0 t_{j-2} & \text{if } j \text{ is odd.} \end{cases}$$

In Lemma 10 below, we show that w extends Q_2 to Q_N according to the following scheme:

$$Q_2 \overset{v_2^{-1}}{\mapsto} Q_3 \overset{\beta^{-1}}{\mapsto} R_3 \overset{v_3^{-1}}{\mapsto} R_4 \overset{\beta^{-1}}{\mapsto} Q_4 \overset{v_4^{-1}}{\mapsto} Q_5 \overset{\beta^{-1}}{\mapsto} R_5 \overset{v_5^{-1}}{\mapsto} R_6 \overset{\beta^{-1}}{\mapsto} Q_6 \mapsto \cdots$$
$$\cdots \mapsto Q_{N-3} \overset{v_{N-3}^{-1}}{\mapsto} Q_{N-2} \overset{\beta^{-1}}{\mapsto} R_{N-2} \overset{v_{N-2}^{-1}}{\mapsto} R_{N-1} \overset{\beta^{-1}}{\mapsto} Q_{N-1} \overset{v_{N-1}^{-1}}{\mapsto} Q_N,$$

and thus the word $w\omega_0$ is a reset word for $\mathcal{A}_m$.

Remark 8. The word w ends with α. The other occurrences of α in w are directly followed by β.

Remark 9. A set $S \subseteq Q$ is:

- ω_0-extensible if and only if $S \cap \{0,1\} = \{0\}$,
- ω_1-extensible if and only if $S \cap \{0,1\} = \{1\}$.

We say that a set $S \subseteq Q$ is *ω-extensible* if it is ω_0-extensible or ω_1-extensible.

Lemma 10. *Let $2 \le j \le N-1$. It holds that:*

1. $Q_2 \cdot (v_j\beta v_{j-1}\dots\beta v_2)^{-1} = Q_{j+1}$ *if j is even,*
2. $Q_2 \cdot (v_j\beta v_{j-1}\dots\beta v_2)^{-1} = R_{j+1}$ *if j is odd,*
3. *$w\omega_0$ is a reset word of $\mathcal{A}_m$.*

Proof. We prove the first two claims by induction. For $j = 2$, using Lemma 7(1) we have:

$$Q_2 \cdot v_2^{-1} = (Q_2 \cdot t_{N-2}^{-1}) \cdot \omega_1^{-1} = Q_3^{\diamond} \cdot \omega_1^{-1} = Q_3.$$

Next, take $j \ge 2$ and suppose that both the claims hold for $j-1$. We use the induction hypothesis and, depending on the parity of j, Lemma 7(1) or Lemma 7(2) respectively. For an even j we have:

$$\begin{aligned} Q_2 \cdot (v_j\beta v_{j-1}\dots\beta v_2)^{-1} &= R_j \cdot (v_j\beta)^{-1} = Q_j \cdot v_j^{-1} = Q_j \cdot (\omega_1 t_{N-j})^{-1} \\ &= Q_{j+1}^{\diamond} \cdot \omega_1^{-1} = Q_{j+1}, \end{aligned}$$

and for an odd j we have:

$$\begin{aligned} Q_2 \cdot (v_j\beta v_{j-1}\dots\beta v_2)^{-1} &= Q_j \cdot (v_j\beta)^{-1} = R_j \cdot v_j^{-1} = R_j \cdot (\omega_0 t_{j-2})^{-1} \\ &= R_{j+1}^{\diamond} \cdot \omega_0^{-1} = R_{j+1}. \end{aligned}$$

The Claim (3) follows from $Q_1 \cdot (w\omega_0)^{-1} = Q_2 \cdot w^{-1} = Q_N$, according to the first claim with $j = N-1$. ☐

It remains to calculate the length of w.

Lemma 11. *The length of w is $\frac{N^2-5}{2}$.*

Proof. The sum of $|v_i|$ with even i is

$$\sum_{i=1}^{\frac{N-1}{2}} (1+N-2i) = \frac{(N-1)(1+N)}{2} - \frac{(N-1)(1+N)}{4} = \frac{1}{4}\left(N^2-1\right),$$

and the sum of $|v_i|$ with odd i is

$$\sum_{i=1}^{\frac{N-3}{2}} 2i = \frac{(N-3)(N-1)}{4} = \frac{1}{4}\left(N^2-4N+3\right).$$

Together with the $N-3$ occurrences of β, we have $|w| = \frac{N^2-5}{2}$. □

Thus, we have that $w\omega_0$ is a reset word for $\mathcal{A}_m$ with length $|w\omega_0| = \frac{N^2-3}{2}$.

4.2 Lower Bound on the Reset Threshold

Finally, let us show that no reset word for $\mathcal{A}_k$ is shorter than $w\omega_0$.

Lemma 12. *If $v \in \Sigma^*$ is greedy and straight reset word with $Q \cdot v = \{q_0\}$, then v does not contain $\omega_0\beta$ nor $\omega_1\beta$ as a factor.*

Proof. Suppose for a contradiction that $v = u''x\beta u'$ for $x \in \{\omega_0, \omega_1\}$. Since v is greedy, $\{q_0\} \cdot (u')^{-1}$ is not ω-extensible, so it contains both 0 and 1 or neither of them. Since β switches these states, $\{q_0\} \cdot (\beta u')^{-1}$ has the same property. Then $\{q_0\} \cdot (\beta u')^{-1} = \{q_0\} \cdot (x\beta u')^{-1}$ and so v is not straight. □

Lemma 13. *Let $2 \le j \le N-1$. It holds that:*

1. $\{0,1\} \cap (Q_j \cdot t_h) = \emptyset$ *for* $1 \le h < N-j$ *if* j *is even,*
2. $\{0,1\} \subseteq (R_j \cdot t_h)$ *for* $1 \le h < j-2$ *if* j *is odd.*

Proof. As t_h is involutory, it is enough to show for $q \in \{0,1\}$ that $q \cdot t_h \notin Q_j$ or $q \cdot t_h \in R_j$ respectively.

As for (1), by Lemma 6(2) we have $q \cdot t_h = N-q-h > j-1$, thus $q \cdot t_h \notin Q_j$. As for (2), denoting $q' = q \cdot t_h$, we have $q' = N-q-h$. Then $q' \cdot \beta = (q+h+1) \bmod N$, and since $q \le 1$ and $h < j-2$, we get $q' \cdot \beta < j$, which implies $q' \cdot \beta \in Q_j$ and $q' \in R_j$.

Lemma 14. *For each suffix xu of w with $x \in \{\omega_0, \omega_1\}$ and $u \in \Sigma^*$ it holds that x extends $Q_2 \cdot u^{-1}$.*

Proof. For every suffix $\omega_1 u$ we have $Q_2 \cdot u^{-1}\omega_1^{-1} = Q_{j+1}^{\diamond} \cdot \omega_1^{-1} = Q_{j+1}$ for some even j, and for every suffix $\omega_0 u$ we have $Q_2 \cdot u^{-1}\omega_0^{-1} = R_{j+1}^{\diamond} \cdot \omega_0^{-1} = R_{j+1}$ for some odd j. □

Lemma 15. *The word $w\omega_0$ is greedy.*

Proof. Let u be the shortest suffix of $w\omega_0$ that violates the greediness, i.e., suppose that $Q_1 \cdot u^{-1}$ is z-extensible for $z \in \{\omega_0, \omega_1\}$, but zu is not a suffix of $w\omega_0$. This simplification works because $Q_1 \cdot u^{-1}$ cannot be both ω_0-extensible and ω_1-extensible. Fix $x \in \Sigma$ such that xu is a suffix of w. Let $u = yu_{\mathbf{s}}$ with $y \in \Sigma$.

If $y \in \{\omega_0, \omega_1\}$ then $Q_1 \cdot (yu_{\mathbf{s}})^{-1}$ is not ω-extensible. If $x \in \{\omega_0, \omega_1\}$, then $Q_1 \cdot (yu_{\mathbf{s}})^{-1}$ is x-extensible due to Lemma 14. Thus, necessarily $x, y \in \{\alpha, \beta\}$.

Assume $y = \beta$. If $Q_1 \cdot (yu_{\mathbf{s}})^{-1}$ is ω-extensible, then $Q_1 \cdot (u_{\mathbf{s}})^{-1}$ is ω-extensible as well due to $0 \cdot \beta = 1$ and $1 \cdot \beta = 0$, implying that $u_{\mathbf{s}}$ is a shorter suffix violating the greediness.

Assume $y = \alpha$. Because w does not contain the factor $\alpha\alpha$, it follows that $x = \beta$. According to (Sect. 4.1), i.e., the definition of w, and the fact that $v_{N-1} = \alpha$, the factor $xy = \beta\alpha$ occurs only within the factors $v_2 \ldots, v_{N-2}$. Thus,

$$yu_{\mathbf{s}} = \alpha\,(\beta\alpha)^i\,\beta\,(v_{j-1}\beta v_{j-2}\ldots\beta v_3\beta v_2)\,\omega_0,$$

where $\alpha\,(\beta\alpha)^i$ is a suffix of v_j. We apply Lemma 10:

1. If j is odd, we get $Q_2 \cdot (v_{j-1}\beta v_{j-2}\ldots\beta v_2)^{-1} = Q_j$, while $v_j = \omega_0 t_{j-2}$ and $i \le \frac{j-3}{2}$. Then

$$Q_1 \cdot (yu_{\mathbf{s}})^{-1} = Q_j \cdot \left(\alpha\,(\beta\alpha)^i\,\beta\right)^{-1} = Q_j \cdot (t_h\beta)^{-1} = R_j \cdot t_h^{-1} = R_j \cdot t_h,$$

   where $h = 2i+1$. We see that $1 \le h \le j-2$. If $h = j-2$, then $\omega_0 t_h = v_j$, so $x = \omega_0$. Otherwise we apply Lemma 13(2) to get $\{0,1\} \subseteq R_j \cdot t_h$, which contradicts that $Q_1 \cdot (yu_{\mathbf{s}})^{-1}$ is ω-extensible.
2. If j is even, we get $Q_2 \cdot (v_{j-1}\beta v_{j-2}\ldots\beta v_2)^{-1} = R_j$, while $v_j = \omega_1 t_{N-j}$ and $i \le \frac{N-j-1}{2}$. Then

$$Q_1 \cdot (yu_{\mathbf{s}})^{-1} = R_j \cdot \left(\alpha\,(\beta\alpha)^i\,\beta\right)^{-1} = R_j \cdot (t_h\beta)^{-1} = Q_j \cdot t_h,$$

   where $h = 2i+1$. We see that $1 \le h \le N-j$. If $h = N-j$, then $\omega_1 t_h = v_j$, so $x = \omega_1$. Otherwise we apply Lemma 13(1) to get $\{0,1\} \cap Q_j \cdot t_h = \emptyset$, which contradicts that $Q_1 \cdot (yu_{\mathbf{s}})^{-1}$ is ω-extensible. □

Lemma 16. *There exists a shortest reset word for $\mathcal{A}_m$ that ends with ω_0 and is greedy.*

Proof. Lemma 5 gives a shortest reset word that is greedy. Clearly, a shortest reset word ends with a non-permutational letter, i.e., ω_0 or ω_1. In the latter case, replacing the ending ω_1 with ω_0 yields a reset word of the same length and preserves greediness. □

Theorem 17. *The word $w\omega_0$ is a shortest reset word for $\mathcal{A}_m$.*

Proof. Using Lemma 5, let $w'\omega_0$ be a greedy shortest reset word of $\mathcal{A}_m$. If $w' = w$, we are done, so let $w' \neq w$ and let $w_{\mathbf{s}}$ be the longest common suffix of w' and w.

If $w_{\mathbf{s}} = w'$, then w' is a proper suffix of w and so it contains at most $N-3$ letters from $\{\omega_0, \omega_1\}$, which contradicts that $w_{\mathbf{s}}\omega_0$ is a reset word. So we can write $w = w_{\mathbf{p}} x w_{\mathbf{s}}$ and $w' = w'_{\mathbf{p}} x' w_{\mathbf{s}}$, where $x, x' \in \Sigma$ and $x' \neq x$. We will show that each of the following cases according to x and x' leads to a contradiction:

1. Suppose that $x \in \{\omega_0, \omega_1\}$. Then Lemma 14 implies that $Q_2 \cdot w_{\mathbf{s}}$ is x-extensible, which contradicts $x' \neq x$ and the greediness of $w'\omega_0$.
2. Suppose that $x' \in \{\omega_0, \omega_1\}$. According to Proposition 3, $w'\omega_0$ is straight, which implies that $Q_2 \cdot w_{\mathbf{s}}$ is x'-extensible, which contradicts $x' \neq x$ and Lemma 15, i.e., the greediness of $w\omega_0$.
3. Suppose that $x = \alpha$ and $x' = \beta$. According to Remark 8, $w_{\mathbf{s}} = \epsilon$ or $w_{\mathbf{s}}$ starts with β. The case of $w_{\mathbf{s}} = \epsilon$ contradicts the straightness of $w'\omega_0$ because each $x' \in \Sigma \setminus \{\alpha\}$ satisfies $Q_2 \cdot (x')^{-1} = Q_2$. The other case implies $\beta\beta$ occurring in w' and thus also contradicts the straightness of $w'\omega_0$.
4. Suppose that $x = \beta$ and $x' = \alpha$. Then $w_{\mathbf{s}} \neq \epsilon$. If $w_{\mathbf{s}}$ starts with α or β, then either w' or w contains the factor $\alpha\alpha$ or $\beta\beta$, which contradicts the straightness of $w'\omega_0$ or the definition of w. Hence, $w_{\mathbf{s}}$ starts with ω_0 or ω_1. Since this starts a factor v_j for some $j \geq 2$, we can write
$$w_{\mathbf{s}} = v_j \beta v_{j-1} \ldots \beta v_3 \beta v_2.$$
We consider the following two subcases:
   (a) Suppose that $w_{\mathbf{s}}$ starts with ω_1. Note that $j \geq 2$ is even and $Q_2 \cdot w_{\mathbf{s}}^{-1} = Q_{j+1}$ by Lemma 10. Let $w_{\mathbf{m}}$ be the longest common suffix of $w'_{\mathbf{p}}x' = w'_{\mathbf{p}}\alpha$ and t_{N-j}. Clearly, $|w_{\mathbf{m}}| \geq 1$. If $w_{\mathbf{m}} = t_{N-j}$, then from Lemma 7(3) we have $Q_{j+1} \cdot t_{N-j} = Q_{j+1}$, which contradicts the straightness of $w'\omega_0$. If $w_{\mathbf{m}} = w'_{\mathbf{p}}x'$, then w' starts with α or β, which contradicts that $w'\omega_0$ is a shortest reset word. It follows that we can write $w' = w'_{\mathbf{pp}} y' w_{\mathbf{m}} w_{\mathbf{s}}$ for $y' \in \Sigma$. Moreover, as w' does not contain the factors $\alpha\alpha$ and $\beta\beta$, we have $y' \neq \alpha$ and $y' \neq \beta$, so $y' \in \{\omega_0, \omega_1\}$. Due to Lemma 12, $w_{\mathbf{m}}$ cannot start with β, and from the construction of t_{N-j} we have $w_{\mathbf{m}} = t_h$ for $h \leq N-j-2$. It holds that $Q_{j+1} \cdot w_{\mathbf{m}}^{-1} = Q_{j+1} \cdot t_h = Q_j \cdot t_h \cup \{j \cdot t_h\}$. Lemma 13(1) provides that $\{0,1\} \cap Q_j \cdot t_h = \emptyset$. Also, $j \cdot t_h = N-j-h \geq 2$. Together, $Q_{j+1} \cdot w_{\mathbf{m}}^{-1} \cap \{0,1\} = \emptyset$, and thus this set is not ω-extensible, which contradicts $y' \in \{\omega_0, \omega_1\}$ and the straightness of $w'\omega_0$.
   (b) Suppose that $w_{\mathbf{s}}$ starts with ω_0. Note that $j \geq 3$ is odd and $Q_2 \cdot w_{\mathbf{s}}^{-1} = R_{j+1}$ by Lemma 10. Let $w_{\mathbf{m}}$ be the longest common suffix of $w'_{\mathbf{p}}x' = w'_{\mathbf{p}}\alpha$ and t_{j-2}. Clearly, $|w_{\mathbf{m}}| \geq 1$. If $w_{\mathbf{m}} = t_{j-2}$, then from Lemma 7(4) we have $R_{j+1} \cdot t_{j-2} = R_{j+1}$, which contradicts the straightness of $w'\omega_0$. If $w_{\mathbf{m}} = w'_{\mathbf{p}}x'$, then w' starts with α or β, which contradicts that $w'\omega_0$ is a shortest reset word. It follows that we can write $w' = w'_{\mathbf{pp}} y' w_{\mathbf{m}} w_{\mathbf{s}}$ for $y' \in \Sigma$. Moreover, as w' does not contain the factors $\alpha\alpha$ and $\beta\beta$, we have $y' \neq \alpha$ and $y' \neq \beta$, so $y' \in \{\omega_0, \omega_1\}$. Due to Lemma 12, $w_{\mathbf{m}}$ cannot start with β, and from construction of t_{j-2} we have $w_{\mathbf{m}} = t_h$ for $h \leq j-4$.

We have $R_{j+1} \cdot w_{\mathbf{m}}^{-1} = R_{j+1} \cdot t_h \supseteq R_j \cdot t_h$. Lemma 13(2) gives $\{0,1\} \subseteq R_j \cdot t_h$. Thus, $\{0,1\} \subseteq R_{j+1} \cdot w_{\mathbf{m}}^{-1}$, and thus this set is not ω-extensible, which contradicts $y' \in \{\omega_0, \omega_1\}$ and the straightness of $w'\omega_0$. □

Theorem 17 implies that $\mathrm{rt}(\mathcal{A}_m) = |w\omega_0| = \frac{N^2-3}{2}$.

4.3 Extending Words

The general upper bound $(n-2)(n-1)+1$ for reset thresholds of synchronizing Eulerian DFAs comes from the fact that any proper and non-empty subset of Q is extended by a word of length at most $n-1$ [15], while in the general case the minimum length of extending words can be quadratic (this was shown recently – see [19]). In view of this, our series shows that this bound is tight for infinitely many n, and so the upper bound for reset thresholds for this class cannot be improved only by reducing this particular bound. The following remark follows from the analysis in the proof of Theorem 17:

Remark 18. The shortest extending word of $\{0,1\}$ in $\mathcal{A}_m$ is $v_2 = \omega_1\alpha(\beta\alpha)^{(N-3)/2}$ of length $N-1$.

5 Experiments

Using the algorithm from [18,20], we have performed an exhaustive search over small synchronizing Eulerian DFAs. We verified the bound $(n^2-3)/2$ for the case of binary DFAs with $n \leq 11$ states, automata with four letters and $n \leq 7$ states, DFAs with eight letters and $n \leq 5$ states, and all DFAs with $n \leq 4$ states.

For $n \in \{3,4,5,7\}$ the bound $(n^2-3)/2$ is reachable. For $n=7$, up to isomorphism, we identified 2 ternary examples and 12 quaternary examples which also meet the bound. It seems that our series $\mathcal{A}_m$ is not unique meeting the bound, as some of the quaternary examples could be generalizable to series with the same reset thresholds. Also, for the binary case we found that for $n \in \{5,7,8,9,11\}$ the bound $(n^2-5)/2$ is met uniquely by DFAs from the Martyugin's series, but it is not reachable for $n \in \{6,10\}$.

Conjecture 19. For $n \geq 3$, $(n^2-3)/2$ is an upper bound for the reset threshold of an n-state Eulerian synchronizing automaton. If $|\Sigma|=2$, then the bound can be improved to $(n^2-5)/2$.

References

1. Ananichev, D., Gusev, V., Volkov, M.: Slowly synchronizing automata and digraphs. In: Hliněný, P., Kučera, A. (eds.) MFCS 2010. LNCS, vol. 6281, pp. 55–65. Springer, Heidelberg (2010)
2. Ananichev, D.S., Volkov, M.V.: Synchronizing generalized monotonic automata. Theor. Comput. Sci. **330**(1), 3–13 (2005)

3. Ananichev, D.S., Volkov, M.V., Gusev, V.V.: Primitive digraphs with large exponents and slowly synchronizing automata. J. Math. Sci. **192**(3), 263–278 (2013)
4. Béal, M.P., Berlinkov, M.V., Perrin, D.: A quadratic upper bound on the size of a synchronizing word in one-cluster automata. Int. J. Found. Comput. Sci. **22**(2), 277–288 (2011)
5. Berlinkov, M.V.: Synchronizing quasi-Eulerian and quasi-one-cluster Automata. Int. J. Found. Comput. Sci. **24**(6), 729–745 (2013)
6. Berlinkov, M., Szykuła, M.: Algebraic synchronization criterion and computing reset words. In: Italiano, G.F., Pighizzini, G., Sannella, D.T. (eds.) MFCS 2015. LNCS, vol. 9234, pp. 103–115. Springer, Heidelberg (2015)
7. Biskup, M.T., Plandowski, W.: Shortest synchronizing strings for Huffman codes. Theor. Comput. Sci. **410**(38–40), 3925–3941 (2009)
8. Carpi, A., D'Alessandro, F.: Strongly transitive automata and the Černý conjecture. Acta Informatica **46**(8), 591–607 (2009)
9. Černý, J.: Poznámka k homogénnym experimentom s konečnými automatmi. Matematicko-fyzikálny Časopis Slovenskej Akadémie Vied **14**(3), 208–216 (1964)
10. Dubuc, L.: Sur les automates circulaires et la conjecture de Černý. Informatique Théorique et Appl. **32**, 21–34 (1998)
11. Eppstein, D.: Reset sequences for monotonic automata. SIAM J. Comput. **19**, 500–510 (1990)
12. Grech, M., Kisielewicz, A.: The Černý conjecture for automata respecting intervals of a directed graph. Discr. Math. Theor. Comput. Sci. **15**(3), 61–72 (2013)
13. Gusev, V.: Lower bounds for the length of reset words in Eulerian automata. Int. J. Found. Comput. Sci. **24**(2), 251–262 (2013)
14. Gusev, V.V., Pribavkina, E.V.: Reset thresholds of automata with two cycle lengths. In: Holzer, M., Kutrib, M. (eds.) CIAA 2014. LNCS, vol. 8587, pp. 200–210. Springer, Heidelberg (2014)
15. Kari, J.: Synchronizing finite automata on Eulerian digraphs. Theor. Comput. Sci. **295**(1–3), 223–232 (2003)
16. Kari, J., Volkov, M.V.: Černý's conjecture and the road coloring problem. In: Handbook of Automata, European Science Foundation (2013)
17. Kisielewicz, A., Kowalski, J., Szykuła, M.: Computing the shortest reset words of synchronizing automata. J. Combin. Optim. **29**(1), 88–124 (2015)
18. Kisielewicz, A., Szykuła, M.: Generating small automata and the Černý conjecture. In: Konstantinidis, S. (ed.) CIAA 2013. LNCS, vol. 7982, pp. 340–348. Springer, Heidelberg (2013)
19. Kisielewicz, A., Szykuła, M.: Synchronizing automata with extremal properties. In: Italiano, G.F., Pighizzini, G., Sannella, D.T. (eds.) MFCS 2015. LNCS, vol. 9234, pp. 331–343. Springer, Heidelberg (2015)
20. Kisielewicz, A., Kowalski, J., Szykula, M.: Experiments with synchronizing automata. In: Han, Y.-S., Salomaa, K. (eds.) CIAA 2016. LNCS, vol. 9705, pp. 176–188. Springer, Heidelberg (2016). doi:10.1007/978-3-319-40946-7_15
21. Pin, J.E.: On two combinatorial problems arising from automata theory. In: Proceedings of the International Colloquium on Graph Theory and Combinatorics. North-Holland Mathematics Studies, vol. 75, pp. 535–548 (1983)
22. Rystsov, I.K.: Estimation of the length of reset words for automata with simple idempotents. Cybern. Syst. Anal. **36**(3), 339–344 (2000)
23. Steinberg, B.: The averaging trick and the Černý conjecture. Int. J. Found. Comput. Sci. **22**(7), 1697–1706 (2011)
24. Steinberg, B.: The Černý conjecture for one-cluster automata with prime length cycle. Theor. Comput. Sci. **412**(39), 5487–5491 (2011)

25. Trahtman, A.N.: The Černý conjecture for aperiodic automata. Discr. Math. Theor. Comput. Sci. **9**(2), 3–10 (2007)
26. Volkov, M.V.: Synchronizing automata and the Černý conjecture. In: Martín-Vide, C., Otto, F., Fernau, H. (eds.) LATA 2008. LNCS, vol. 5196, pp. 11–27. Springer, Heidelberg (2008)
27. Volkov, M.V.: Synchronizing automata preserving a chain of partial orders. Theor. Comput. Sci. **410**(37), 3513–3519 (2009)

Monoid-Based Approach to the Inclusion Problem on Superdeterministic Pushdown Automata

Yuya Uezato[1(✉)] and Yasuhiko Minamide[2]

[1] Department of Computer Science, University of Tsukuba, Tsukuba, Ibaraki, Japan
uezato@logic.cs.tsukuba.ac.jp

[2] Department of Mathematical and Computing Sciences, Tokyo Institute of Technology, Tokyo, Japan

Abstract. We present a new and simple decidability proof for the language inclusion problem between context-free languages and languages accepted by superdeterministic pushdown automata (SDPDAs). The language class of SDPDAs is one of the largest language classes $\mathcal{C}$ for which the inclusion $L_{\text{cfl}} \subseteq L_{\mathcal{C}}$ is decidable for an arbitrary context-free language L_{cfl} and arbitrary language $L_{\mathcal{C}}$ in $\mathcal{C}$. We introduce generalized pushdown automata and reformulate SDPDAs as a subclass of them. This reformulation naturally leads to a monoid that captures SDPDAs. The monoid is key to our simple decidability proof because we translate the inclusion problem on SDPDAs to the corresponding monoid inclusion problem. In addition to the decidability result, we present a new undecidability result regarding the inclusion problem on indexed languages.

1 Introduction

A superdeterministic pushdown automaton (SDPDA) is a deterministic pushdown automaton that is finite delay and satisfies a peculiar condition—for any state p and word w, there is a state q and $z \in \mathbb{Z}$ such that for any configuration $\langle p, \alpha \rangle$, if a computation starting from $\langle p, \alpha \rangle$ ends in $\langle p', \beta \rangle$ after consuming w, then $p' = q$ and $|\alpha| - |\beta| = z$. Greibach and Friedman showed the decidability of *Incl*(**CFL**, **SDPDA**), i.e., the inclusion $L_{\text{cfl}} \subseteq L(M)$ is decidable for an arbitrary context-free language L_{cfl} and arbitrary SDPDA M [6]. They also showed that the language class **SDPDA** includes some important classes, i.e., the class of regular languages (**REG**), Dyck languages (**DYCK**), and generalized parenthesis languages [13]. Moreover, a language class $\mathcal{C}$ for which *Incl*(**CFL**, $\mathcal{C}$) is decidable and $\mathcal{C}$ strictly includes **SDPDA** is not yet known. Our aim is to obtain a simple decidability proof for *Incl*(**CFL**, **SDPDA**) and extend it to a larger class. The original proof [6], however, is elaborated by pumping arguments and it remains unclear why *Incl*(**CFL**, **SDPDA**) is decidable.

We introduce generalized pushdown automata (GPDAs) and a subclass, real-time GPDAs (RGPDAs). Each transition rule of GPDAs is of the form $p \xrightarrow[\sigma]{\alpha\,/\,\beta} q$, which consumes an input σ, pops a sequence of symbols α, and then pushes β

S. Brlek and C. Reutenauer (Eds.): DLT 2016, LNCS 9840, pp. 393–405, 2016.
DOI: 10.1007/978-3-662-53132-7_32

to a stack. Although usual pushdown automata require $|\alpha| = 1$, GPDAs allow $|\alpha| > 1$ and pop multiple symbols in one transition. On the basis of the multiple pop feature, we translate SDPDAs to RGPDAs that satisfy the following property: if we have $p \xrightarrow[\sigma]{\alpha / \beta} q$ and $p \xrightarrow[\sigma]{\alpha' / \beta'} q'$, then $q = q'$, $|\alpha| = |\alpha'|$, and $|\beta| = |\beta'|$. This translation simplifies our decidability proof; thus, RGPDAs with the above condition are adequate normal forms of SDPDAs.

The formalization of RGPDAs instinctively leads to a monoid for RGPDAs and this monoid is the basis of our decidability proof. Our approach generalizes the monoid-based approach for *Incl*(**CFL**, **REG**) and *Incl*(**CFL**, **DYCK**) [11] where the author applied the classical notion of language recognition by monoids to translate inclusion problems to corresponding monoid inclusion problems. For a finite automaton A, there is a finite monoid $\mathbf{M}$, a subset $U \subseteq \mathbf{M}$, and a homomorphism $\mathcal{H} : \Sigma^* \to \mathbf{M}$ that recognize $L(A)$ as $L(A) = \mathcal{H}^{-1}(U)$. This equation translates the inclusion $L(G) \subseteq L(A)$ to the monoid inclusion $\mathcal{H}(L(G)) \subseteq U$ where G is a context-free grammar. Since $\mathbf{M}$ is a finite monoid, we can decide whether $\mathcal{H}(L(G)) \subseteq U$ and this implies the decidability of *Incl*(**CFL**, **REG**). This argument, however, cannot be directly applied to *Incl*(**CFL**, **DYCK**) because a monoid that recognizes a Dyck language is infinite. In [11], to manage this unavoidable infiniteness, the author rephrased an argument given by Berstel and Boasson [2] for the decidability of *Incl*(**CFL**, **DYCK**) in terms of monoids. The present paper generalizes their argument to accommodate *Incl*(**CFL**, **SDPDA**) on the basis of our monoid for RGPDAs. Tsukada and Kobayashi gave a procedure similar to ours in a type-theoretical framework [14]. We compare our approach with their type-theoretical approach in Sect. 6.

Recently, higher-order PDAs [10] have received much attention for their use in higher-order program verification [8]. Hence, it is a natural attempt to extend the decidability of *Incl*(**CFL**, **SDPDA**) to a class of languages accepted by higher-order PDAs. However, unfortunately, we show that such an attempt is even undecidable for *Incl*(**IL**, **DYCK**) where **IL** is the class of indexed languages [1,7] accepted by second-order PDAs [10].

Context-Free Grammar and Normal Form. A *context-free grammar* (CFG) is a 4-tuple $G = (V, \Sigma, P, S)$ where V is a finite set of *variables*, Σ is a finite set of *terminal symbols*, $P \subseteq V \times (V \cup \Sigma)^*$ is a finite set of *production rules*, and $S \in V$ is the *start variable*. We write $X \to \alpha$ instead of $(X, \alpha) \in P$. To denote a one-step derivation, we write $\alpha X \beta \Rightarrow \alpha \xi \beta$ if there is a rule $X \to \xi \in P$ where $\alpha, \beta, \xi \in (V \cup \Sigma)^*$. The words generated by a variable X is $L(X) := \{w \in \Sigma^* : X \Rightarrow^* w\}$ and the language of G, $L(G)$, is defined by $L(G) := L(S)$.

We primarily use the Chomsky normal form (CNF). A CFG is in CNF if all of its production rules are of the form $X \to YZ$, $X \to \sigma$, or $S \to \epsilon$ where $X, Y, Z \in V$ and $\sigma \in \Sigma$. By the standard translation from CFG to CNF [7], we assume that each variable X is *reachable*, i.e., there exists a pair of terminal strings (w_1, w_2) such that $S \Rightarrow^* w_1 X w_2$.

2 Generalized PDA and Superdeterministic PDA

First, we introduce *generalized pushdown automata* (GPDAs). Next, we define *realtime* GPDAs (RGPDAs) and a monoid for GPDAs. Third, we define pushdown automata (PDAs) as a subclass of GPDAs and superdeterministic PDAs (SDPDAs) by following [6]. Finally, after pointing out a problem in the formalization of SDPDAs, we translate SDPDAs into RGPDAs.

Generalized PDA. A GPDA is a 7-tuple $M = (Q, \Sigma, \Gamma, \Delta, q_{\text{init}}, \mathcal{F}, Z)$ where Q is a finite set of *states*, Σ is a finite *input alphabet*, Γ is a finite *stack alphabet*, $\Delta \subseteq (Q \times (\Sigma \cup \{\epsilon\}) \cup \Gamma^+) \times (Q \times \Gamma^*)$ is a finite set of *transition rules*, $q_{\text{init}} \in Q$ is the *initial state*, $\mathcal{F} \subseteq Q \times \Gamma^*$ is a finite set of *final configurations*, and $Z \in \Gamma$ is the *initial stack symbol*. We use Γ^+ to denote $\Gamma^* \setminus \{\epsilon\}$.

A *configuration* $\mathbf{c}$ is a pair $\langle p, \alpha \rangle \in Q \times \Gamma^*$ of a state p and a stack α. We define the set of *transitions* $\mathbb{T} := Q \times \Gamma^* \times \Gamma^* \times Q$ and write $p \xrightarrow{\alpha\,/\,\beta} q$ to denote a transition $(p, \alpha, \beta, q) \in \mathbb{T}$. A transition $\delta = p \xrightarrow{\alpha\,/\,\beta} q$ rewrites a configuration $\mathbf{c}$ to another one $\mathbf{c}'$ as $\mathbf{c} \overset{\delta}{\hookrightarrow} \mathbf{c}'$ if $\mathbf{c} = \langle p, \alpha\xi \rangle$, $\mathbf{c}' = \langle q, \beta\xi \rangle$, and $\xi \in \Gamma^*$. We use Δ as a function from $\Sigma \cup \{\epsilon\}$ to $2^{\mathbb{T}}$ defined by $\Delta(a) := \{p \xrightarrow{\alpha\,/\,\beta} q : ((p, a, \alpha), (q, \beta)) \in \Delta\}$.

We define a *single move* $\mathbf{c} \vdash_{a} \mathbf{c}'$ if $\mathbf{c} \overset{\delta}{\hookrightarrow} \mathbf{c}'$ for some $\delta \in \Delta(a)$ where $a \in \Sigma \cup \{\epsilon\}$ and a *multiple move* $\mathbf{c}_1 \vdash^{*}_{a_1 a_2 \ldots a_n} \mathbf{c}_{n+1}$ if $\mathbf{c}_i \vdash_{a_i} \mathbf{c}_{i+1}$ for all $i \in [1..n]$. The language of M, $L(M)$, is defined as follows:

$$L(M) := \{w \in \Sigma^* : \langle q_{\text{init}}, Z \rangle \vdash^{*}_{w} \langle q_f, \xi \rangle,\ \langle q_f, \xi \rangle \in \mathcal{F}\}.$$

Realtime GPDA and Transition Monoid. We introduce a subclass of GPDAs, *realtime* GPDAs, and define a monoid and homomorphism to recognize the language accepted by a realtime GPDA.

A GPDA $M = (Q, \Sigma, \Gamma, \Delta, q_{\text{init}}, \mathcal{F}, Z)$ is *realtime* (RGPDA) if there are no ϵ-moves; namely, $\Delta \subseteq (Q \times \Sigma \times \Gamma^+) \times (Q \times \Gamma^*)$.

We define a *composition* operator $\odot$ on $\mathbb{T}_\perp (= \mathbb{T} \cup \{\perp\})$ as follows:

$$\delta_1 \odot \delta_2 := \begin{cases} p \xrightarrow{\alpha\,/\,\zeta\xi} r & \text{if } \delta_1 = p \xrightarrow{\alpha\,/\,\beta\xi} q \text{ and } \delta_2 = q \xrightarrow{\beta\,/\,\zeta} r, \\ p \xrightarrow{\alpha\xi\,/\,\zeta} r & \text{if } \delta_1 = p \xrightarrow{\alpha\,/\,\beta} q \text{ and } \delta_2 = q \xrightarrow{\beta\xi\,/\,\zeta} r, \\ \perp & \text{otherwise,} \end{cases} \qquad \begin{array}{l} \perp \odot \_ := \perp, \\ \\ \_ \odot \perp := \perp, \end{array}$$

where the element $\perp$ denotes a composition failure, e.g., $p \xrightarrow{a\,/\,b} q \odot q \xrightarrow{c\,/\,d} r = \perp$ because it means to push b to a stack and then try to pop c but we cannot.

The operator $\odot : \mathbb{T}_\perp \times \mathbb{T}_\perp \to \mathbb{T}_\perp$ is associative; thus, the pair $(\mathbb{T}_\perp, \odot)$ forms a semigroup. This semigroup leads to a monoid $\mathbf{T}_M$ for the RGPDA M: $\mathbf{T}_M := (2^{\mathbb{T}}, \otimes, \mathbf{1} = \{q \xrightarrow{\epsilon\,/\,\epsilon} q : q \in Q\})$ where the multiplication $\otimes$ is defined by extending $\odot$ to the sets of transitions:

$$T_1 \otimes T_2 := \{\delta_1 \odot \delta_2 : \delta_1 \in T_1, \delta_2 \in T_2, \delta_1 \odot \delta_2 \neq \perp\}.$$

Since there are no ϵ-moves in RGPDAs, we can see Δ as a function $\Delta : \Sigma \to 2^{\mathbb{T}}$ and this derives a homomorphism $\widetilde{\Delta} : \Sigma^* \to \mathbf{T}_M$ as follows:

$$\widetilde{\Delta}(\epsilon) := \mathbf{1}, \quad \widetilde{\Delta}(\sigma) := \Delta(\sigma), \quad \widetilde{\Delta}(\sigma_1 \dots \sigma_n) := \Delta(\sigma_1) \otimes \dots \otimes \Delta(\sigma_n).$$

The homomorphism $\widetilde{\Delta}$ naturally interprets moves of M as follows.

Proposition 1.

- *If* $p \xrightarrow{\alpha / \beta} q \in \widetilde{\Delta}(w)$, *then* $\langle p, \alpha\xi\rangle \vdash_{w}^{*} \langle q, \beta\xi\rangle$ *for any* $\xi \in \Gamma^*$.
- *If* $\langle p, \alpha\rangle \vdash_{w}^{*} \langle q, \beta\rangle$, *then there exists* $\xi \in \Gamma^*$ *such that* $\alpha = \alpha'\xi$, $\beta = \beta'\xi$, *and* $p \xrightarrow{\alpha' / \beta'} q \in \widetilde{\Delta}(w)$.

Thus, the homomorphism $\widetilde{\Delta}$ recognizes the language $L(M)$.

Lemma 2. $L(M) = \widetilde{\Delta}^{-1}(\{T : q_{init} \xrightarrow{Z / \xi} q_f \in T,\ \langle q_f, \xi\rangle \in \mathcal{F}\})$.

Note that $\widetilde{\Delta}(w)$ is finite for any $w \in \Sigma^*$ because $\Delta(\sigma)$ is finite for any $\sigma \in \Sigma$. This finiteness is important to obtain a decision procedure for inclusion problems. Although we can show properties similar to Proposition 1 for GPDAs, we require ϵ-closures to build a homomorphism $\mathcal{H}$ and then $\mathcal{H}(w)$ is infinite in general. The existence of ϵ-moves is harmful to give a decision procedure.

PDA and Deterministic PDA. A GPDA $M = (Q, \Sigma, \Gamma, \Delta, q_{\text{init}}, \mathcal{F}, Z)$ is a *pushdown automaton* (PDA) if $\Delta \subseteq (Q \times (\Sigma \cup \{\epsilon\}) \times \Gamma) \times (Q \times \Gamma^*)$. In contrast to GPDAs, PDAs cannot pop multiple symbols in a single move.

A PDA M is *deterministic* (DPDA) if M satisfies the following conditions:

- For each $a \in \Sigma \cup \{\epsilon\}$, if $\mathbf{c}_1 \vdash_{a} \mathbf{c}_2$ and $\mathbf{c}_1 \vdash_{a} \mathbf{c}_3$, then $\mathbf{c}_2 = \mathbf{c}_3$.
- If we have $\mathbf{c}_1 \vdash_{\epsilon} \mathbf{c}_2$, then $\mathbf{c}_1 \not\vdash_{\sigma} \mathbf{c}_3$ for all $\sigma \in \Sigma$ and $\mathbf{c}_3 \in Q \times \Gamma^*$.

A configuration $\mathbf{c}$ is a *reading configuration* if $\mathbf{c} \vdash_{\sigma} \mathbf{c}'$ for some $\sigma \in \Sigma$. To emphasize a move between reading configurations, we write $\mathbf{c} \Vdash_{w}^{*} \mathbf{c}'$ instead of $\mathbf{c} \vdash_{w}^{*} \mathbf{c}'$ where $\mathbf{c}$ and $\mathbf{c}'$ are reading configurations.

Superdeterministic PDA. A DPDA M is *superdeterministic* (SDPDA) [6] if M satisfies the following conditions:

1. M accepts words with the empty stack: if $\langle q_f, \xi\rangle \in \mathcal{F}$, then $\xi = \epsilon$.
2. M is *finite delay*, i.e., any sequence of ϵ-moves is d-bound for some $d \in \mathbb{N}$: there are no configurations $\mathbf{c}$ such that $\mathbf{c} = \mathbf{c}_0 \vdash_{\epsilon} \mathbf{c}_1 \vdash_{\epsilon} \cdots \vdash_{\epsilon} \mathbf{c}_{d-1} \vdash_{\epsilon} \mathbf{c}_d$.
3. Let $\sigma \in \Sigma$ and $\langle p, \alpha\rangle$ and $\langle p, \alpha'\rangle$ be reading configurations with the same state. If $\langle p, \alpha\rangle \Vdash_{\sigma}^{*} \langle q, \beta\rangle$ and $\langle p, \alpha'\rangle \Vdash_{\sigma}^{*} \langle r, \beta'\rangle$, then $q = r$ and $|\alpha| - |\beta| = |\alpha'| - |\beta'|$.

As mentioned above, the presence of ϵ-moves in the formalization of SDPDAs prevents us from directly defining a homomorphism that recognizes $L(M)$. Thus, we remove ϵ-moves from SDPDAs by translating them to RGPDAs.

Theorem 3. *Let* $M = (Q, \Sigma, \Gamma, \Delta, q_{init}, \mathcal{F}, Z)$ *be an* SDPDA. *There exists* RGPDA N *such that (1)* $\$L(M) = L(N)$ *where* $\$ \notin \Sigma$ *and (2) if* $p \xrightarrow{\alpha / \beta} q \in \Delta_N(\sigma)$ *and* $p \xrightarrow{\alpha' / \beta'} r \in \Delta_N(\sigma)$, *then* $q = r$, $|\alpha| = |\alpha'|$, *and* $|\beta| = |\beta'|$.

Proof. We can easily translate M to an SDPDA M' that satisfies the following:

$$M : \langle q_{\text{init}}, Z\rangle \vdash^{*}_{w} \langle q_f, \epsilon\rangle \iff M' : \langle q'_{\text{init}}, \natural\rangle \vdash_{\epsilon} \langle q_{\text{init}}, Z\natural\rangle \vdash^{*}_{w} \langle q_f, \natural\rangle \vdash_{\epsilon} \langle q'_f, \epsilon\rangle$$

where $\langle q_f, \epsilon\rangle \in \mathcal{F}$ and the two states $q'_{\text{init}}, q'_f \notin Q$ are the unique initial and final states of M'. Thus, $L(M) = L(M')$. We assume that M' is d-bound and the special symbol $\natural \notin \Gamma$ is the stack bottom symbol of M'.

We take the ϵ-closure of the initial configuration of M'. If $\langle q'_{\text{init}}, \natural\rangle \vdash^{*}_{\epsilon} \langle q, \epsilon\rangle$, then $L(M') = \emptyset$ or $L(M') = \{\epsilon\}$ and these cases are easy. We assume $\langle q'_{\text{init}}, \natural\rangle \vdash^{*}_{\epsilon} \langle q_\star, \alpha_\star\natural\rangle$ where $\langle q_\star, \alpha_\star\natural\rangle$ is a reading configuration. If $\sigma_1 \dots \sigma_n\sigma_{n+1} \in L(M')$, then we have $\langle q'_{\text{init}}, \natural\rangle \vdash^{*}_{\epsilon} \langle q_\star, \alpha_\star\natural\rangle \Vdash^{*}_{\sigma_1} \langle q_1, \alpha_1\natural\rangle \Vdash^{*}_{\sigma_2} \cdots \Vdash^{*}_{\sigma_n} \langle q_n, \alpha_n\natural\rangle \vdash^{*}_{\sigma_{n+1}} \langle q'_f, \epsilon\rangle$.

We build an RGPDA $N = (Q \cup \{q''_{\text{init}}, q'_f\}, \Sigma \cup \{\$\}, \Gamma \cup \{\natural, \sharp\}, \Delta_N, q''_{\text{init}}, \mathcal{F}_N, \natural)$ as follows. Let $p \in Q$ and $\sigma \in \Sigma$. Since M' is finite delay, we can compute two sets $A = \{(q, \alpha, \beta) : \langle p, \alpha\rangle \Vdash^{*}_{\sigma} \langle q, \beta\rangle, \beta \neq \epsilon\}$ and $B = \{(q'_f, \xi\natural) : \langle p, \xi\natural\rangle \vdash_{\sigma} \mathbf{c} \vdash^{*}_{\epsilon} \langle q'_f, \epsilon\rangle\}$. Since M' is an SDPDA, (when $A \neq \emptyset$) there are $r \in Q$ and $k \in \mathbb{Z}$ such that if $(q, \alpha, \beta) \in A$, then $q = r$ and $|\alpha| - |\beta| = k$. We add transitions to Δ_N as follows:

- Let $(r, \alpha, \beta) \in A$. Add $p \xrightarrow{\alpha\zeta\,/\,\beta\zeta} r \in \Delta_N(\sigma)$ where $\alpha\zeta \in \Gamma^*\natural^*$ and $|\alpha\zeta| = d$.
- Let $(q'_f, \xi\natural) \in B$.

Case $A \neq \emptyset$: Add $p \xrightarrow{\xi\natural\natural^i\,/\,\sharp\natural^j} r \in \Delta_N(\sigma)$ where $i = d - |\xi\natural|$ and $j = (d - k) - 1$.
Case $A = \emptyset$: Add $p \xrightarrow{\xi\natural\natural^i\,/\,\sharp} q'_f \in \Delta_N(\sigma)$ where $i = d - |\xi\natural|$.

This construction ensures that (i) if $p \xrightarrow{\alpha\,/\,\beta} q, p \xrightarrow{\alpha'\,/\,\beta'} q' \in \Delta_N(\sigma)$, then $q = q'$, $|\alpha| = |\alpha'|$, and $|\beta| = |\beta'|$, (ii) $\langle q_\star, \alpha_\star\natural\rangle \Vdash^{*}_{\sigma_1\dots\sigma_n} \langle q_n, \alpha_n\natural\rangle \vdash^{*}_{\sigma_{n+1}} \langle q'_f, \epsilon\rangle$ in M' iff $q_\star \xrightarrow{\alpha_\star\natural^d\,/\,\sharp\natural^c} r \in \Delta_N(\sigma_1 \dots \sigma_n\sigma_{n+1})$ for some $c \in [1..d]$ and $r \in Q$.

We define $\Delta_N(\$) := \{q''_{\text{init}} \xrightarrow{\natural\,/\,\alpha_\star\natural^d} q_\star\}$ so that $\langle q'_{\text{init}}, \natural\rangle \vdash^{*}_{\sigma_1\dots\sigma_{n+1}} \langle q'_f, \epsilon\rangle$ in M' iff $q''_{\text{init}} \xrightarrow{\natural\,/\,\sharp\natural^c} r \in \Delta_N(\sigma_1 \dots \sigma_{n+1})$ for some $c \in [1..d]$ and $r \in Q$. By defining $\mathcal{F}_N := \{\langle p, \sharp\natural^c\rangle : p \in Q \cup \{q'_f\}, c \leq d\}$, we have $\$L(M') = L(N)$. $\square$

We call the following condition of Theorem 3 a *uniformity condition* that is stronger than the third condition of SDPDAs:

$$\text{If } p \xrightarrow{\alpha\,/\,\beta} q,\ p \xrightarrow{\alpha'\,/\,\beta'} r \in \Delta(\sigma), \text{ then } |\alpha| = |\alpha'|,\ |\beta| = |\beta'|, \text{ and } q = r.$$

We denote RGPDAs that satisfy the uniformity condition as RGPDA+**U**.

The normalization through Theorem 3 and the uniformity condition are crucial for the proof of our key lemma, Lemma 9 of Sect. 4. A construction similar to Theorem 3 appears in [14, Theorem 10]. However, they normalized an SDPDA to a corresponding SDPDA that satisfies a property like the uniformity condition; thus, they did not remove ϵ-moves in their proof.

3 Decidability of *Incl*(Cfl, Dyck) Revisited

Before giving a decision procedure for *Incl*(**Cfl**, RGPDA+**U**), we consider the subcase *Incl*(**Cfl**, **Dyck**). We rephrase the decidability proof of *Incl*(**Cfl**,

DYCK) given by Berstel and Boasson [2] as a constraint solving problem on a monoid for the Dyck languages. The argument of this section will be naturally extended to *Incl*(CFL, RGPDA+U) in the next section.

Dyck Language. Let Σ be a finite alphabet $\Sigma = \{\sigma_1, \ldots, \sigma_n\}$. We use $\acute{\sigma}$ and $\grave{\sigma}$ to denote an open and a close parenthesis labelled by σ, respectively, by following [13]. We define the open parentheses $\acute{\Sigma}$ and close parentheses $\grave{\Sigma}$ obtained from Σ by $\acute{\Sigma} := \{\acute{\sigma}_1, \ldots, \acute{\sigma}_n\}$ and $\grave{\Sigma} := \{\grave{\sigma}_1, \ldots, \grave{\sigma}_n\}$. A word $w \in (\acute{\Sigma} \cup \grave{\Sigma})^*$ is a *Dyck word* if w is well-matched. For example, $\acute{a}\grave{a}$ and $\acute{a}\acute{b}\grave{b}\grave{a}$ are Dyck words, but $\grave{a}\acute{a}$ and $\acute{a}\grave{b}$ are not. To formally define this, we build a monoid and homomorphism that recognize the set of Dyck words.

We define a function $\mu : \acute{\Sigma} \cup \grave{\Sigma} \to \mathbb{T}_*$ where $\mathbb{T}_* = \{*\} \times \Sigma^* \times \Sigma^* \times \{*\}$ by interpreting a open parenthesis $\acute{\sigma}$ and close parenthesis $\grave{\sigma}$ as transitions of push σ and pop σ: $\mu(\acute{\sigma}) := * \xrightarrow{\epsilon/\sigma} *$ and $\mu(\grave{\sigma}) := * \xrightarrow{\sigma/\epsilon} *$. For the sake of readability, we write α/β to denote $* \xrightarrow{\alpha/\beta} *$. The triple $\mathbf{D} = (\mathbb{T}_* \cup \{\bot\}, \odot, \mathbf{1} = \epsilon/\epsilon)$ forms a monoid because $\delta \odot \epsilon/\epsilon = \epsilon/\epsilon \odot \delta = \delta$ for any $\delta \in \mathbb{T}_* \cup \bot$. We call the monoid $\mathbf{D}$ *Dyck monoid* and deal the function μ as a homomorphism $\mu : (\acute{\Sigma} \cup \grave{\Sigma})^* \to \mathbf{D}$. For example, $\mu(\epsilon) = \mu(\acute{a}\grave{a}) = \mu(\acute{a}\acute{b}\grave{b}\grave{a}) = \epsilon/\epsilon$, $\mu(\grave{a}\acute{a}) = a/a$, and $\mu(\acute{a}\grave{b}) = \bot$.

A word $w \in (\acute{\Sigma} \cup \grave{\Sigma})^*$ is a Dyck word if $\mu(w) = \epsilon/\epsilon$; thus the Dyck language over Σ is defined as $\mathbf{DYCK}(\Sigma) := \mu^{-1}(\{\epsilon/\epsilon\})$.

Inclusion Problem as Constraint Solving. We fix a CFG $G = (V, \acute{\Sigma} \cup \grave{\Sigma}, P, S)$ and provide a procedure to decide whether $L(G) \subseteq \mathbf{DYCK}(\Sigma)$. For this purpose, we consider the equivalent monoid inclusion $\mu(L(G)) \subseteq \{\epsilon/\epsilon\}$ that is obtained from $\mathbf{DYCK}(\Sigma) = \mu^{-1}(\{\epsilon/\epsilon\})$. We introduce a notation to solve this.

A mapping $\varphi : V \to 2^{\mathbf{D}}$ is a *solution* if it satisfies the following constraint over the Dyck monoid $\mathbf{D}$:

$$\forall X \in V. \begin{cases} \varphi(X) \supseteq \{\mu(w)\} & \text{if } X \to w \in P, \\ \varphi(X) \supseteq \varphi(Y) \odot \varphi(Z) & \text{if } X \to YZ \in P. \end{cases}$$

If φ is a solution, then $\varphi(X) \supseteq \mu(L(X))$ holds for all variable X. Thus, it suffices to search a solution φ such that $\varphi(S) \subseteq \{\epsilon/\epsilon\}$ to solve $\mu(L(G)) \subseteq \{\epsilon/\epsilon\}$.

Proposition 4. *$\mu(L(G)) \subseteq \{\epsilon/\epsilon\}$ if and only if $\exists(\varphi$: solution$)$. $\varphi(S) \subseteq \{\epsilon/\epsilon\}$.*

By this proposition, if we find a solution φ satisfying $\varphi(S) \subseteq \{\epsilon/\epsilon\}$, we can decide whether $L(G) \subseteq \mathbf{DYCK}(\Sigma)$. Unfortunately, however, we cannot find such a solution φ directly in general because the Dyck monoid $\mathbf{D}$ is infinite and thus the set of mappings is infinite.

A Procedure for *Incl*(CFL, DYCK). In order to limit the space in which we explore solutions, we rephrase a result of Berstel and Boasson [2], which shows the decidability of *Incl*(CFL, DYCK), in our framework.

For each variable X, there exists a pair of terminal strings $w_1, w_2 \in (\acute{\Sigma} \cup \grave{\Sigma})^*$ such that $S \Rightarrow^* w_1 X w_2$ because each variable of CFGs is reachable. We define $\mathcal{K}(X) := (w_1, w_2)$ by taking such a pair of terminal strings for each variable. Note that our argument does not depend on the choice of terminal strings. The pair of terminal strings $\mathcal{K}(X)$ serves as an upper-bound for $\mu(L(X))$ as follows.

Lemma 5 [2]. *Let $\mathcal{K}(X) = (w_1, w_2)$ and $w \in L(X)$. If the following holds, then $w_1ww_2 \in L(G) \setminus \textbf{Dyck}(\Sigma)$:*

$$\mu(w) = \bot \quad \textit{or} \quad \mu(w) = \alpha/\beta \textit{ such that } |\alpha| > |w_1| \textit{ or } |\beta| > |w_2|.$$

Proof. We use the property $\mu(w_1w_2w_3) = \mu(\mu(w_1)\mu(w_2)\mu(w_3))$. By this property, if $\mu(w_1) = \bot$, $\mu(w) = \bot$, or $\mu(w_2) = \bot$, then $\mu(w_1ww_2) \neq \epsilon/\epsilon$.

By the definition of the operator $\odot$, if $\alpha_1/\beta_1 \odot \alpha_2/\beta_2 = \alpha_3/\beta_3$, then $|\alpha_3| \geq |\alpha_1|$ and $|\beta_3| \geq |\beta_2|$. Thus, we can assume $\mu(w_1) = \epsilon/\sigma_1 \dots \sigma_n$ and $\mu(w_2) = \sigma_1 \dots \sigma_m/\epsilon$ where $n, m \geq 0$. If not, we have $\mu(w_1ww_2) \neq \epsilon/\epsilon$. Moreover, by the definition of μ, we have $|w_1| \geq n$ and $|w_2| \geq m$.

If $|\alpha| > |w_1| \geq n$, then $\mu(w_1w) = \sigma'_1 \dots \sigma'_{|\alpha|-n}/\beta$ or $\mu(w_1w) = \bot$; thus, $\mu(w_1ww_2) \neq \epsilon/\epsilon$. Similarly, if $|\beta| > |w_2| \geq m$, then $\mu(ww_2) = \alpha/\sigma'_1 \dots \sigma'_{|\beta|-m}$ or $\mu(ww_2) = \bot$; thus, $\mu(w_1ww_2) \neq \epsilon/\epsilon$. These arguments complete the proof. □

Lemma 5 states that the space in which we explore solutions is finitely bounded and this leads to the decidability of *Incl*(**Cfl**, **Dyck**). To formally state this argument, we introduce *bounded* mappings.

A mapping φ is *bounded by* $\mathcal{K}$ if it satisfies the following property:

$$\forall X \in V.\ \forall \alpha/\beta \in \varphi(X).\ |\alpha| \leq |w_1| \ \text{ and } |\beta| \leq |w_2| \text{ where } \mathcal{K}(X) = (w_1, w_2).$$

We can solve the inclusion problem by searching an adequate bounded solution.

Theorem 6.

$$\begin{aligned} \mu(L(G)) \subseteq \{\epsilon/\epsilon\} &\iff \exists(\varphi : \textit{solution}).\ \varphi(S) \subseteq \{\epsilon/\epsilon\} \\ &\iff \exists(\varphi' : \textit{bounded solution}).\ \varphi'(S) \subseteq \{\epsilon/\epsilon\}. \end{aligned}$$

Proof. It suffices to show that if a solution φ satisfies $\varphi(S) \subseteq \{\epsilon/\epsilon\}$, then φ must be bounded. If not, then there exists $w \in L(X)$ such that $|w_1| > |\alpha|$ or $|w_2| > |\beta|$ where $\mu(w) = \alpha/\beta$ and $\mathcal{K}(X) = (w_1, w_2)$. By Lemma 5, $w_1ww_2 \notin \textbf{Dyck}(\Sigma)$ and $L(G) \not\subseteq \textbf{Dyck}(\Sigma)$. However, the presence of φ implies $L(G) \subseteq \textbf{Dyck}(\Sigma)$. □

Proposition 7. *The set of bounded mappings $\{\varphi : \varphi \textit{ is bounded by } \mathcal{K}\}$ is finite.*

Since we can decide whether a given bounded mapping is a solution, Theorem 6 and Proposition 7 imply the following result.

Corollary 8 [2]. *The inclusion problem Incl*(**Cfl**, **Dyck**) *is decidable.*

4 Decidability of *Incl*(Cfl, rgPDA+U)

On the basis of the argument of the previous section, this section provides a procedure to decide whether $L(G) \subseteq L(M)$ where $G = (V, \Sigma, P, S)$ is a CFG and $M = (Q, \Sigma, \Gamma, \Delta, q_{\text{init}}, \mathcal{F}, Z)$ is an rgPDA+U.

A Property Corresponding to Lemma 5. We show a crucial property that corresponds to Lemma 5 and limits a space of mappings in which we explore a solution. To state it formally, we use three constants PUSH, POP, and H where

$$\forall \sigma \in \Sigma. \forall p \xrightarrow{\alpha / \beta} q \in \Delta(\sigma). (\textsc{Push} \geq |\beta| \ \wedge\ \textsc{Pop} \geq |\alpha|); \quad \forall \langle q_f, \xi \rangle \in \mathcal{F}. H \geq |\xi|.$$

PUSH and POP are upper bounds of the numbers of symbols that are pushed onto or popped from a stack in a single move. H is an upper bound of the heights of final configurations.

Lemma 9. *Assume that $S \Rightarrow^* w_1 X w_2$, $q_{init} \xrightarrow{\alpha / \beta} p \in \widetilde{\Delta}(w_1)$, and $w \in L(X)$. If the following holds, then $w_1 w w_2 \in L(G) \setminus L(M)$:*

$$\textit{There is } p \xrightarrow{\alpha' / \beta'} q \in \widetilde{\Delta}(w) \textit{ such that } |\alpha'| > \textsc{Push} \cdot |w_1| \textit{ or } |\beta'| > \textsc{Pop} \cdot |w_2| + H.$$

Proof. If M fails to consume w_1, then it means $w_1 w w_2 \notin L(G)$. Hence, we assume M succeeds on consuming w_1 and have $\langle q_{\text{init}}, Z \rangle \vdash^*_{w_1} \langle p, \xi \rangle$ where $|\xi| = |\beta| \leq \textsc{Push} \cdot |w_1|$ by the uniformity condition of M.

First, we consider the case $|\alpha'| > \textsc{Push} \cdot |w_1| \geq |\xi|$. By the uniformity condition, M have to pop just $|\alpha'|$-symbols from the stack while reading w. However, after consuming w_1, i.e., $\langle q_{\text{init}}, Z \rangle \vdash^*_{w_1} \langle p, \xi \rangle$, we have only $|\xi|$-symbols on its stack. Thus, M cannot pop $|\alpha'|$-symbols and $w_1 w w_2 \notin L(M)$.

Next, we consider the case $|\beta'| > \textsc{Pop} \cdot |w_2| + H$. We assume that M succeeds on consuming $w_1 w$ and $\langle q_{\text{init}}, Z \rangle \vdash^*_{w_1 w} \langle q, \zeta \rangle$ where $|\zeta| = |\beta| - |\alpha'| + |\beta'|$ and $|\beta| \geq |\alpha'|$. If M succeeds on consuming w_2 from $\langle q, \zeta \rangle$, then M pops at most ($\textsc{Pop} \cdot |w_2|$)-symbols: $\langle q_{\text{init}}, Z \rangle \vdash^*_{w_1 w w_2} \langle r, \zeta' \rangle$ where $|\zeta'| \geq |\zeta| - \textsc{Pop} \cdot |w_2|$. Furthermore, we have $\zeta' > H$ because $|\zeta| \geq |\beta'| > \textsc{Pop} \cdot |w_2| + H$, and so $w_1 w w_2 \notin L(M)$. □

This lemma provides upper-bounds for states p and variables X if there exists a pair of terminal strings (w_1, w_2) such that $S \Rightarrow^* w_1 X w_2$ and $q_{\text{init}} \xrightarrow{\alpha / \beta} p \in \widetilde{\Delta}(w_1)$; in other words, we need a witness (w_1, w_2) to use this lemma. However, unfortunately, the following proposition says that we cannot compute such pairs (w_1, w_2) in general; thus, we cannot use this lemma directly for *Incl*(**CFL**, RGPDA+U).

Proposition 10. *Let p be a state and X be a variable. It is unsolvable to decide if there is a pair (w_1, w_2) such that $S \Rightarrow^* w_1 X w_2$ and $q_{init} \xrightarrow{\alpha / \beta} p \in \widetilde{\Delta}(w_1)$.*

To bypass this undecidability, we consider an *underlying automaton* of M where properties similar to Lemma 9 and Proposition 10 hold.

Underlying Automaton. We obtain the underlying automaton of M by forgetting the stack contents from the transition rules of M, i.e., a rule $p \xrightarrow{\alpha / \beta} q \in \Delta(\sigma)$ is translated to $p \xrightarrow{\sigma} q$ in the underlying automaton. The underlying automaton is a pair $\mathcal{A}_M = (Q, E)$ where Q is the set of *states* of M and $E \subseteq Q \times \Sigma \times Q$ has an edge $p \xrightarrow{\sigma} q$ if $p \xrightarrow{\alpha / \beta} q \in \Delta(\sigma)$ for some $\alpha, \beta \in \Gamma^*$. As the usual notation of finite automata, we write $p \xrightarrow{w} q$ if $w = \sigma_1 \sigma_2 \ldots \sigma_n$ and $p = p_0 \xrightarrow{\sigma_1} p_1 \xrightarrow{\sigma_2} p_2 \xrightarrow{\sigma_3} \cdots \xrightarrow{\sigma_n} p_n = q$. By the uniformity condition of M, $\mathcal{A}_M$ is deterministic.

A state q is *quasi-reachable* to a variable X if there exists a pair (w_1, w_2) such that $S \Rightarrow^* w_1 X w_2$ and $q_{\text{init}} \xrightarrow{w_1} q$. The next proposition says that we can determine if a given state q is quasi-reachable to X, and it means that Proposition 10 is solvable in the underlying automaton. A similar construction to consider underlying automata appears in [14, Theorem 8]. It seems that in [14] they also adopted such a construction to avoid the undecidable result of Proposition 10.

Proposition 11. *There is a partial function* $\mathcal{K} : Q \times V \rightharpoonup \Sigma^* \times \Sigma^*$ *such that:*

- *If* $\mathcal{K}(p, X) = (w_1, w_2)$, *then* $S \Rightarrow^* w_1 X w_2$ *and* $q_{init} \xrightarrow{w_1} p$ *in* $\mathcal{A}_M$.
- *If* $\mathcal{K}(p, X)$ *is* undefined, *then* p *is not quasi-reachable to* X.

Proof. For a variable X, the language $L_X = \{w_1 \# w_2 : S \Rightarrow^* w_1 X w_2, w_i \in \Sigma^*\}$ is context-free. For a state p, the language $L_p = \{w \# w' : q_{\text{init}} \xrightarrow{w} p,\ w' \in \Sigma^*\}$ is regular. Hence, the intersection $L_X \cap L_p$ is a context-free language. By the decidability of the emptiness problem of context-free languages, we can decide whether p is quasi-reachable to X and compute a witness (w_1, w_2). □

Furthermore, the property corresponding to Lemma 9 holds.

Lemma 12. *Assume that* $\mathcal{K}(p, X) = (w_1, w_2)$ *and* $w \in L(X)$. *If the following holds, then* $w_1 w w_2 \in L(G) \setminus L(M)$:

There is $p \xrightarrow{\alpha' / \beta'} q \in \widetilde{\Delta}(w)$ *such that* $|\alpha'| > \text{PUSH} \cdot |w_1|$ *or* $|\beta'| > \text{POP} \cdot |w_2| + H$.

A Procedure for ***Incl*(CFL, RGPDA+U).** Although Lemma 12 constrains the pairs of a state p and variable X where p is quasi-reachable to X, this lemma does not constrain non quasi-reachable states. To avoid this problem, we redefine bounded mappings whose codomain is bounded by $\mathcal{K}$.

For a mapping $\psi : V \to 2^{\mathbf{T}_M}$, we define the *restriction* $\psi \upharpoonright \mathcal{K} : V \to 2^{\mathbf{T}_M}$:

$$(\psi \upharpoonright \mathcal{K})(X) := \{T \cap \mathcal{K}_X : T \in \psi(X)\}, \quad \mathcal{K}_X := \{p \xrightarrow{\alpha / \beta} q : \mathcal{K}(p, X) \text{ is defined}\}.$$

A mapping ψ is *bound by* $\mathcal{K}$ if $\psi = \psi \upharpoonright \mathcal{K}$ and ψ satisfies the following:

$$\forall X \in V. \forall T \in \psi(X). \forall p \xrightarrow{\alpha / \beta} q \in T. \left[\begin{array}{c} |\alpha| \leq \text{PUSH} \cdot |w_1| \wedge |\beta| \leq \text{POP} \cdot |w_2| + H \\ \text{where } \mathcal{K}(p, X) = (w_1, w_2). \end{array} \right]$$

Proposition 13. *The set of bounded mappings* $\{\psi : \psi$ *is bound by* $\mathcal{K}\}$ *is finite.*

A mapping ψ is a *solution* if it satisfies the following constraint over the transition monoid $\mathbf{T}_M$ of M:

$$\forall X \in V. \begin{cases} \psi(X) \sqsupseteq \{\widetilde{\Delta}(w)\} & \text{if } X \to w \in P, \\ \psi(X) \sqsupseteq \psi(Y) \otimes \psi(Z) & \text{if } X \to YZ \in P. \end{cases}$$

where $\mathcal{F}_1 \sqsupseteq \mathcal{F}_2$ if $\forall T_2 \in \mathcal{F}_2.\ \exists T_1 \in \mathcal{F}_1.\ T_1 \subseteq T_2$. If ψ is a solution, then $\psi(X) \sqsupseteq \widetilde{\Delta}(L(X))$ for all variable X.

The reason why we redefine solutions is to establish the following property. Indeed, even if $\psi(X) \supseteq \widetilde{\Delta}(L(X))$, then $(\psi \upharpoonright \mathcal{K})(X) \not\supseteq \widetilde{\Delta}(L(X))$ in general.

Proposition 14. *If a mapping ψ is a solution then $\psi \restriction \mathcal{K}$ is also a solution.*

This proposition is crucial to obtain our main theorem.

Theorem 15. $L(G) \subseteq L(M) \iff$
$\exists(\psi : solution).\ \psi(S) \subseteq \{T : q_{init} \xrightarrow{Z\,/\,\xi} q_f \in T, \langle q_f, \xi\rangle \in \mathcal{F}\} \iff$
$\exists(\psi' : bounded\ solution.).\ \psi'(S) \subseteq \{T : q_{init} \xrightarrow{Z\,/\,\xi} q_f \in T, \langle q_f, \xi\rangle \in \mathcal{F}\}$.

As with *Incl*(**Cfl**, **Dyck**), to decide if $L(G) \subseteq L(M)$, it suffices to explore a solution in the finite set of bounded mappings. This implies our main result.

Corollary 16. *The inclusion problem Incl*(**Cfl**, RGPDA+**U**) *is decidable.*

5 Attempt at Generalization or Undecidability

We have given a decidability proof of *Incl*(**Cfl**, **Sdpda**) where each problem is of the form $L(G) \subseteq L(M)$. One possible generalization is to consider the inclusion of the form $L(G) \subseteq L'(M)$ where $L'(M)$ is the language defined by $L'(M) := \{w : \langle q_{\text{init}}, Z\rangle \vdash_w^* \langle q_f, \alpha\rangle, \langle q_f, \epsilon\rangle \in \mathcal{F}_M, \alpha \in \Gamma^*\}$. However, as Friedman and Greibach showed in [5], the above problem becomes undecidable. Indeed, the empty-stack acceptance condition is crucial in the proof of Lemma 9.

The second attempt is to replace the third condition of SDPDAs by the condition: if $\langle p, \alpha\rangle \Vdash_\sigma^* \langle q, \beta\rangle$ and $\langle p, \alpha'\rangle \Vdash_\sigma^* \langle r, \beta'\rangle$, then $q = r$ but maybe $|\alpha| - |\beta| \neq |\alpha'| - |\beta'|$. The relaxed condition enables us to simulate simple machines. A realtime DPDA M is a *simple machine* if M has only one state and accepts words by the empty-stack. Since simple machines are realtime DPDAs that satisfy the above relaxed condition, the undecidability of the inclusion problem on simple machines [4] implies the undecidability of the inclusion problem between CFGs and the relaxed SDPDAs. A similar argument shows that the inclusion problem also becomes undecidable if we adopt the following condition: if $\langle p, \alpha\rangle \Vdash_\sigma^* \langle q, \beta\rangle$ and $\langle p, \alpha'\rangle \Vdash_\sigma^* \langle r, \beta'\rangle$, then $|\alpha| - |\beta| = |\alpha'| - |\beta'|$ but maybe $q \neq r$.

These results illustrate the difficulty of finding a class of languages $\mathcal{B}$ that makes *Incl*(**Cfl**, $\mathcal{B}$) decidable with **Sdpda** $\subsetneq \mathcal{B}$. Now we consider finding a class $\mathcal{A}$ such that **Cfl** $\subsetneq \mathcal{A}$ and *Incl*($\mathcal{A}$, **Sdpda**) is decidable. For this purpose, we consider the class **IL** of indexed languages [1,7]. This class and higher-order indexed languages [10] is known as a natural generalization of **Cfl** and have received attention for higher-order program verification [8]. If we could solve *Incl*(**IL**, **Sdpda**), then this becomes a base for program verification. However, unfortunately, the inclusion problem on **IL** is unsolvable for *Incl*(**IL**, **Dyck**).

An *indexed grammar* is a 5-tuple $G = (V, \Sigma, F, S, P)$ where $V = \{A, B, \ldots\}$ is a finite set of *variables*, $S \in V$ is the *start variable*, $\Sigma = \{\sigma_1, \sigma_2, \ldots\}$ is a finite set of *terminal symbols*, and $F = \{f, g, \ldots\}$ is a finite set of *indices*, and P is a finite set of *production rules* [1,7]. Each rule in P is one of the following: $A \to BC$, $A_f \to B$, $A \to B_f$, $A \to \sigma$ where $A, B, C \in V$, $f \in F$, and $\sigma \in \Sigma$. A *sentential form* ψ is of the form $\psi = \alpha_1(A_1, x_1)\alpha_2(A_2, x_2)\ldots\alpha_n(A_n, x_n)\alpha_{n+1} \in (\Sigma \cup (V \times F^*))^*$ where $\alpha_i \in \Sigma^*$ and

$(A_i, x_i) \in V \times F^*$. Instead of $(A, x) \in V \times F^*$, we write A_x. Each production rule rewrites a sentential form as follows: (1) a rule $A \to BC$ rewrites as $\psi_1 A_x \psi_2 \Rightarrow \psi_1 B_x C_x \psi_2$, (2) a rule $A_f \to B$ rewrites as $\psi_1 A_{fx} \psi_2 \Rightarrow \psi_1 B_x \psi_2$, (3) a rule $A \to B_f$ rewrites as $\psi_1 A_x \psi_2 \Rightarrow \psi_1 B_{fx} \psi_2$, and (4) a rule $A \to \sigma$ rewrites as $\psi_1 A_x \psi_2 \Rightarrow \psi_1 \sigma \psi_2$. The language of an indexed grammar G is defined by $L(G) := \{w \in \Sigma^* : S \Rightarrow^* w\}$. The class of indexed languages **IL** is the languages generated by indexed grammars.

To obtain the undecidability of *Incl*(**IL**, **DYCK**), we use an undecidability result of DT0L-systems. A *DT0L-system* is a tuple $G = (\Sigma, g_1, \dots, g_n, \alpha)$ where Σ is a finite *alphabet*, $g_i : \Sigma^* \to \Sigma^*$ is a *homomorphism* for each $i \in [1..n]$, and the non-empty word $\alpha \in \Sigma^+$ is the *axiom* of G [12]. On a DT0L G, we define the function $\mathcal{F}_G : [1..n]^* \to \Sigma^*$ recursively: $\mathcal{F}_G(\epsilon) := \alpha$ and $\mathcal{F}_G(i_1 \dots i_{n-1} i_n) := g_{i_n}(\mathcal{F}_G(i_1 \dots i_{n-1}))$. The following theorem is the immediate consequence of Theorem II.12.1 and III.7.1 of [12] and is key to showing our undecidability result.

Theorem 17 [12]. *Let $G = (\Sigma, g_1, \dots, g_n, \alpha)$ and $H = (\Sigma', h_1, \dots, h_n, \beta)$ be DT0L-systems with n-homomorphisms. It is unsolvable to decide whether $|\mathcal{F}_G(w)| \geq |\mathcal{F}_H(w)|$ for all $w \in [1..n]^*$.*

Theorem 18. *Let L be an indexed language over $\{\acute{a}, \grave{a}\}$. It is unsolvable to decide whether $L \subseteq$ **DYCK**$(\{a\})$.*

Proof (Sketch). Let G and H be DT0L-systems with n-homomorphisms. On the basis of the construction of [7, Theorem 14.8], we can encode G and H into an indexed grammar I as $L(I) = \{\, \acute{a}^i \grave{a}^j \acute{a}^j \grave{a}^i : w \in [1..n]^*, i = |\mathcal{F}_G(w)|, j = |\mathcal{F}_H(w)| \,\}$.

Since $i \geq j \iff \acute{a}^i\, \grave{a}^j\, \acute{a}^j\, \grave{a}^i \in$ **DYCK**$(\{a\})$, we have the following:

$$L(I) \subseteq \textbf{DYCK}(\{a\}) \iff |\mathcal{F}_G(w)| \geq |\mathcal{F}_H(w)| \text{ for all } w \in [1..n]^*.$$

This property and Theorem 17 imply the undecidability of *Incl*(**IL**, **DYCK**). □

6 Related Work

The idea of using monoids to solve inclusion problems follows previous work by the second author [11], which restated the decidability of the inclusion problems *Incl*(**CFL**, **REG**) and *Incl*(**CFL**, **DYCK**). We have extended the previous approach for *Incl*(**CFL**, **SDPDA**) by introducing RGPDAs and using a transition monoid of RGPDAs to accommodate the technique of Berstel and Boasson. In the paper [3], Bertoni et al. also used a monoid to solve the inclusion problem *Incl*(**CFL**, **DYCK**). Unfortunately, their proof is incorrect because they depended heavily on the incorrect assumption that Dyck monoids are *cancellative*. A monoid M is *cancellative* if $xy = xz$ implies $y = z$ and $yx = zx$ implies $y = z$ for every $x, y, z \in M$. However, the Dyck monoid over $\{a\}$ is not cancellative, i.e., $\mu(\grave{a}^3 \acute{a}^3)\mu(\grave{a}^2 \acute{a}^2) = a^3/a^3 = \mu(\grave{a}^3 \acute{a}^3)\mu(\grave{a}^1 \acute{a}^1)$ but $\mu(\grave{a}^2 \acute{a}^2) \neq \mu(\grave{a}^1 \acute{a}^1)$.

Tsukada and Kobayashi showed the decidability of *Incl*(**CFL**, **SDPDA**) by designing a type system for DPDAs [14]. They translated an inclusion problem

$L(G) \subseteq L(M)$ for a CFG G and an SDPDA M into the type-theoretical problem deciding if G is typable under a type system obtained from M. In their type system, a typed term is of the form $w : \tau$ where a type τ is a set of pairs of configurations $\tau = \{\mathbf{c} \to \mathbf{c}' : \mathbf{c}' \mid\!\frac{*}{w}\, \mathbf{c}\}$. Since each type and each type environment are infinite in general, Tsukada and Kobayashi required extra notations to represent infinite objects in a finite form; this makes their proof elaborated overall. Conversely, as mentioned in Sect. 2, we consider the monoids obtained from transition rules of RGPDAs and thus the set $\widetilde{\Delta}(w)$ is finite for any word w. It is worthy to note that the definition of reading configurations of their paper differs from Greibach and Friedman [6] and us. A configuration $\mathbf{c}$ is a reading configuration if $\mathbf{c}$ is not expandable by ϵ-moves in their definitions; thus, each configuration with the empty stack $\langle q, \epsilon \rangle$ becomes a reading configuration in [14]. This difference is significant and their main theorems [14, Theorems 8 and 10] do not hold in their definition; however, it seems that their proofs and result hold in the original definition of reading configurations considered by Greibach and Friedman.

7 Conclusion and Future Work

We have extended the decidability proof of *Incl*(**CFL**, **DYCK**) given by Berstel and Boasson to *Incl*(**CFL**, **SDPDA**) by introducing RGPDAs and considering their monoid. **SDPDA** strictly includes the class of languages accepted by visibly pushdown automata with empty stack. Recently, the class of languages accepted by Floyd automata (operator precedence languages) [9] have received attention because it includes the class of visibly pushdown languages and enjoys many closure properties. To the best of our knowledge, the relationship between SDPDAs and Floyd automata with empty stack and the decidability of the inclusion problem between CFGs and them have not been studied to date. We would like to tackle these problems by extending the arguments presented in this paper.

Acknowledgement. We are grateful to the anonymous reviewers for their careful reading, pointing out some mistakes, and invaluable suggestions. This work was supported by JSPS KAKENHI Grant Number 15J01843 and 15K00087.

References

1. Aho, A.V.: Indexed grammars–an extension of context-free grammars. J. ACM **15**(4), 647–671 (1968)
2. Berstel, J., Boasson, L.: Formal properties of XML grammars and languages. Acta Informatica **38**(9), 649–671 (2002)
3. Bertoni, A., Choffrut, C., Radicioni, R.: The inclusion problem of context-free languages: some tractable cases. In: Diekert, V., Nowotka, D. (eds.) DLT 2009. LNCS, vol. 5583, pp. 103–112. Springer, Heidelberg (2009)
4. Friedman, E.P.: The inclusion problem for simple languages. TCS **1**, 297–316 (1976)
5. Friedman, E.P., Greibach, S.A.: Superdeterministic DPDAs: the method of accepting does affect decision problems. JCSS **19**(1), 79–117 (1979)

6. Greibach, S.A., Friedman, E.P.: Superdeterministic PDAs: a subcase with a decidable inclusion problem. J. ACM **27**(4), 675–700 (1980)
7. Hopcroft, J., Ullman, J.: Introduction to Automata Theory, Languages, and Computation. Addison-Wesley, Reading (1979)
8. Kobayashi, N.: Types and higher-order recursion schemes for verification of higher-order programs. In: POPL, pp. 416–428. ACM (2009)
9. Lonati, V., Mandrioli, D., Panella, F., Pradella, M.: Operator precedence languages: their automata-theoretic and logic characterization. SIAM J. Comput. **44**(4), 1026–1088 (2015)
10. Maslov, A.N.: Multilevel stack automata. Prob. Inf. Trans. **12**, 38–43 (1976)
11. Minamide, Y.: Verified decision procedures on context-free grammars. In: Schneider, K., Brandt, J. (eds.) TPHOLs 2007. LNCS, vol. 4732, pp. 173–188. Springer, Heidelberg (2007)
12. Salomaa, A., Soittola, M.: Automata-Theoretic Aspects of Formal Power Series. Springer, New York (1978)
13. Takahashi, M.: Generalizations of regular sets and their application to a study of context-free languages. Inf. Control **27**(1), 1–36 (1975)
14. Tsukada, T., Kobayashi, N.: An intersection type system for deterministic pushdown automata. In: Baeten, J.C.M., Ball, T., de Boer, F.S. (eds.) TCS 2012. LNCS, vol. 7604, pp. 357–371. Springer, Heidelberg (2012)

Author Index

Printed in the United States
By Bookmasters